Global Mobile Satellite Communications Theory

Stojče Dimov Ilčev

Global Mobile Satellite Communications Theory

For Maritime, Land and Aeronautical Applications

Second Edition

By
Stojče Dimov Ilčev
(Стойчо Димов Илчев)
Durban University of Technology (DUT)
Durban, South Africa

Springer

Stojče Dimov Ilčev
Durban University of Technology (DUT)
Durban
South Africa

ISBN 978-3-319-39169-4 ISBN 978-3-319-39171-7 (eBook)
DOI 10.1007/978-3-319-39171-7

Library of Congress Control Number: 2016942002

Printed on acid-free paper

This Springer imprint is published by Springer Nature
The registered company is Springer International Publishing AG Switzerland

The original version of the bookfrontmatter was revised: The term 'Volume 1' has been removed from bookfrontmatter. The Erratum to the bookfrontmatter is available at 10.1007/978-3-319-39171-7_8

To the memory of my late father

Prof. Dimo Stoev Ilčev
(Димо Стоев Илчев)

Preface

Global Mobile Satellite Communications (GMSC) structures are specific to mobile satellite communication discipline and technique for maritime, land, and aeronautical applications that enable connections between Mobile Earth Stations (MES). These include ships, land vehicles and aircraft on the one hand and ground telecommunications subscribers on the other, through the medium of space segment (satellite constellation), Land Earth Stations (LES), and Terrestrial Telecommunications Network (TTN) or other landline providers.

This book is very important for modern shipping, land (road and rail) and aeronautical concerns, because GMSC are providing more effective business, trade, and prosperity in the new millennium, in the first place for transport safety and security matters and second, for commercial communications. The most considerable marketing and technical point of this book is great deficiency of suitable manuals on the international book market, which completely describe GMSC fundamentals, space segments, ground segments (MES and LES), and Global Mobile Personal Satellite Communications (GMPSC).

The book discusses hot topics in GMSC techniques and technology, which will be useful for technical staff onboard vessels, land vehicles and aircraft, on offshore constructions, and for those possessing satellite handset phones. This includes sea, land, and air transport sets with many other requirements for more effective trade, which need development, design, utility, implementation, and knowledge of Communication, Navigation and Surveillance (CNS) for safety and commercial applications. Otherwise, GMSC solutions are very important to all modern transportation companies, dispatchers, agencies, brokers, and the successful management, commerce, carriage tracking, and logistics of their fleet.

In general, this book may become the manual for a broad range of readers with different levels of technical education and knowledge, for professional staff involved in GMSC and their technical managers, engineers, professors, students, instructors, and participants in GMDSS courses, consultants and supervisors of MES and for military officers and cadets. This book could find an important place in libraries, universities, and institutions as well.

Mobile satellite systems today have become very important topics for students in many maritime, transportation and aviation universities, faculties in telecommunication and electrical engineering, for all modern transportation companies, GMSC manufacturers, providers, operators, and their management staff. Everyone involved in GMSC systems has to know something about these technology and transmission systems. Thus, in writing this book the author has used the expertise, prospects, literature, and manuals of numerous experts, specialists, institutions, and references mentioned at the end of this handbook as well as information from the Internet.

The author has been a professional expert in maritime radiocommunications since 1969, as a radio officer on board ocean-going cargo ships using Morse MF/HF radiotelegraphy and MF/HF/VHF radiotelephony and later as master mariner and Electronics/GMDSS Operator with Inmarsat Ship Earth Stations (SES). In addition, for over 15 years he has managed the Former-IS Marine Radio Company and newly established CNS Systems for research, service, installation, and engineering of GMSC and Global Navigation Satellite Systems (GNSS) systems and equipment onboard ships and integration with modern IT systems. The author has also used his doctoral dissertation, master's theses, technical manuscripts, papers, and practical experience with radiocommunications, navigation, and GMSC systems.

For basic and principal technical information he has drawn heavily mostly on the following sources:

- "Global Mobile Satellite Communications, For Maritime, Land and Aeronautical Applications", 1st Edition published by Springer in 2005 and "Global Aeronautical CNS", published by AIAA in 2013, both written by S.D. Ilcev.
- "Mobile Satellite Communication Networks", written by R. Sheriff and Y.F. Hu; and "Satellite Communications Systems", written by G. Maral and M. Bousquet. Both books were published by Wiley in 2001 and 1994, respectively.
- "Mobile Satellite Communications—Principles & Trends", written by Madhavendra Richharia and published by Addison-Wesley in 2001.
- "Mobile Antenna Systems Handbook", written by K. Fujimoto and J.R. James; "Mobile Satellite Communications", written by S. Ohmori, H. Wakana and S. Kawase; and "Low Earth Orbital Satellites for Personal Communication Networks", written by A. Jamalipour. All three books were published by Artech House, in 1994, 1998 and 1998, respectively.
- "Satellite Communications: Principles and Applications" and "Electronic Aids to Navigation: Position Fixing". Both books written by L. Tetley and D. Calcutt were published by Edward Arnold, in 1994 and 1991, respectively.
- "An Introduction to Satellite Communications", written by D.I. Dalgleish; and "Satellite Communication Systems" edited by B. Evans. Both books were published by IEE, in 1991 and 1993, respectively.
- "Never Beyond Reach", edited by B. Gallagher and published by Inmarsat, in 1989.

- “Спутниковая связь на море”, written by L. Novik, I. Morozov and V. Solovev; and “Международная спутниковая система морской связи—Инмарсат”, written by V. Zhilin. Both books were published by Sudostroenie, Leningrad, in 1987 and 1988, respectively.
- “Telekomunikacije satelitima”, written by R. Galić, Školska Knjiga, Zagreb, 1983.
- “Radio wave Propagation Information for Predictions for Earth-to-Space Path Communications”, edited by C. Wilson and D. Rogers, ITU, Geneva.

Readers will find that this book has been written using up-to-date systems, techniques, and technology in satellite communications. The material has been systematized in such a way to cover satellite development, systematization, definition of all nomenclature, synonyms of mobile satellite communications systems and services, new kinds of launcher systems and the presentation of all types of satellite orbit constellations and spacecrafts. The newest concepts of transmission models and accesses including IP networking, a complete introduction to mobile antenna systems and propagation, Inmarsat, Cospas-Sarsat, Big LEO, Little LEO, navigation and tracking systems, including the forthcoming augmentation satellite system for Communications, Navigation and Surveillance (CNS) mobile solutions, stratospheric platforms as communications systems, including mobile DVB-RCS.

Furthermore, new concepts and innovations in GMSC, such as Inmarsat BGAN, Fleet and Swift Broadband solutions, Global Xpress for maritime and aeronautical applications, new Iridium LEO mobile applications, innovative maritime O3B MEO solutions, and mobile DVB-RCS GEO users segment are covered. Also, modern VSAT broadcasting applications and integration of GMSC systems with new Personal Videophone Technology and Mobile Videophone over IP (VPoIP) will be discussed. Finally, the historic moment is approaching when we can use MES terminals and say: *“Hallo, can you see me, over”?*

Acknowledgments

Above all, the author of this book would like to express his very special appreciation and gratitude to Prof. Ahmed Cassim Bawa, former Vice Chancellor (VC) and principal of Durban University of Technology (DUT), who gave him huge support in Space Science Research and postgraduate studies. He also expresses his special gratitude to Prof. Sibusiso Moyo, the Director of Research and Postgraduate Studies at DUT, and DUT staff for support and encouragement to establish Research and Postgraduate Studies in Space Science and for moral assistance in completing this book.

The author is Research Professor in Space Science and Supervisor at DUT for research and postgraduate studies. He works on very important multinational project African Satellite Augmentation System (ASAS) for all of Africa and the Middle East including many other proposals in Radio and Satellite Communication, Navigation and Surveillance (CNS), Digital Video Broadcasting-Return Channel via Satellite (DVB-RCS), Global Radio and Satellite Tracking of mobile and living beings, Satellite SCADA (M2M), Stratospheric Platform Systems (SPS,) and Space Solar Power (SSP). He also would like to express his special appreciation to DUT for generous contribution as a sponsor of this book.

The Durban University of Technology takes pride on its commitment to academic excellence.

The over 24,000 students who pass through the doors everyday are testament to a growing ethos of learning, research, and community engagement. DUT is a multi-campus university of technology at the cutting edge of higher education, renowned for technological training and academic prowess. The university is characterized as being research-driven with a focus on strategic and applied research that can be translated into professional practice. Furthermore, research output may be commercialized thus providing a source of income for the institution. In striving to create a new and dynamic ethos, the university builds upon current strengths and celebrates the expertise of its staff. DUT provides Web pages for its research and postgraduate studies in Space Science at: www.dut.ac.za/space_science—Showing full study programs and projects whilst including all research and supervisor staff.

The author is also very grateful to the group of authors for various manuals, brochures, and pamphlets issued by IMO, ICAO, ITU, IATA, ARINC, WMO, ESA, SITA, ETSI, ETRI, Intelsat, Intersputnik, Eutelsat, Roscosmos, Eurocontrol, Inmarsat, Cospas-Sarsat, Iridium, Globalstar, Orbcomm, Sea Launch, and other regulatory bodies and operators.

This book is dedicated to all his friends and students working in shipping companies and industry, and to his present postgraduate students of Space Science at DUT. He also wishes specially to acknowledge the valuable support and understanding from publisher of this book Springer, especially to Ms. Mary E. James, Senior Editor in Applied Sciences and her assistants Ms. Zoe Kennedy and Ms. Rebecca R. Hytowitz.

Finally, he would like to express his heartfelt appreciation and gratitude to his lovely wife, Svetlana Mihailovna Ilčeva, and his family for their help and understanding, while the manuscript was written, especially to his dear children and grandchildren living in Montenegro: son Marijan with his wife Vanja and their children Daria and Martin, daughter Tatjana, with her husband Boško and their children Anja and Stefan, to his stepdaughter Olga from Ukraine, her husband Boris and his stepgranddaughter Bažena, to his sister Prof. Tatjana Ilčeva and niece Ivana in Belgrade, Serbia and to his cousin Valentin Boyadžiev and his family in Sofia, Bulgaria.

Contents

About the Author

Prof. Stojče Dimov Ilčev is Chair of Research and Postgraduate Studies in Space Science and CNS at Durban University of Technology (DUT), Durban, South Africa. He studies both maritime radio engineering and nautical science at Montenegro University in Kotor, maritime electronics and communications at Rijeka University in Croatia, postgraduate satellite engineering at Skopie University in Macedonia and Belgrade University in Serbia. Professor Ilčev holds Bachelor (B.Sc.), Master in Electrical Engineering (M.Sc.) and Doctor of Science (Ph.D.) degrees. He also holds the certificates for Radio operator 1st class (Morse), for GMDSS 1st class Radio Electronic Operator and Maintainer and for Master Mariner without limitations. Since 1969, Ilčev has worked onboard merchant ships, in Satellite Earth Station, at Coast Radio Station, shipping company, Nautical School and at Maritime Faculty. Since 2000, he has worked at IS Marine Radio and CNS Systems companies on research and projects related to modern Communication, Navigation and Surveillance (CNS) for maritime, land, and aeronautical applications. He has written three books on CNS engineering and systems for maritime, land, and aeronautical applications and he has many projects and inventions in this field including DVB-RCS and Stratospheric Platforms.

Acronyms

a	Large semi-major axis of elliptical orbit
A	Apogee
A	Azimuth angle
A/D	Analog-to-Digital signal conversion
AAC	Aeronautical Administrative Communications (Advanced Audio Encoding)
AAC-AOC	Airline Administrative and Airline Operational Control
AAIC	Accounting Authority for Satellite Traffic Invoicing
ABR	Available Bit Rate
ABS	Automatic Beam Switching
AC	Alternating Current
ACARS	Aircraft Communications (Addressing) and Reporting System
ACB	Antenna Control Board
ACC	Aeronautical Control Center
ACE	Antenna Control Equipment
ACK	Acknowledgement
ACM	Adaptive Coding and Modulation
ACO	Aeronautical Communications Organization
ACR	Allowed Cell Rate
ACS	Attitude Control System (Assembly and Command Ship)
ACSE	Antenna Control and Signaling Equipment of LES
ACSSB	Amplitude Companded SSB
ACTS	Advanced Communications Technology Satellite
ACU	Above Cockpit Unit (Antenna Equipment of AES)
ACV	Assembly and Command Vessel
ADE	Above Deck Equipment (Antenna Equipment of SES)
ADM	Adaptive Delta Modulation
ADPCM	Adaptive Differential PCM
ADPS	Aeronautical Data Processing System
ADREP	Accident/Incident Data Reporting

ADS	Automatic Dependent Surveillance
ADSL	Asymmetrical Digital Subscriber Line
ADSS	Automatic Dependent Surveillance System
AEEC	Airline Electronic Engineering Committee
AEIS	Aeronautical Enroute Information Service
AES	Aircraft (Aeronautical) Earth Station
AFC	Automatic Frequency Control
AFIS	Airborne Flight Information Service
AFTAX	Aeronautical Fixed Telecommunication Automatic Exchange
AFTN	Aeronautical Fixed Telecommunication Network
AGC	Automatic Gain Control
AHD	Above Haul Device (Antenna Equipment of VES)
AHNIS	Aeronautical Highlights and Navigation Information Services
AHRS	Attitude and Heading Reference System
AIDC	ATS Interfacility Data Communications
AIES	Aeronautical Information Enroute Service
AIRCOM	Aeronautical Communications
AIS	Aeronautical Information Services
AKM	Apogee Kick Motor
AL	Accuracy Lateral
ALC	Automatic Level Control
ALOHA	A random multiple accesses devised at the University of Hawaii
AM	Amplitude Modulation (Air Ministry)
AMBE	Advanced MultiBand Excitation
AMES	Aeronautical and Maritime Engineering Satellite (Japanese program)
AMSC	Aeronautical MSC
AMSS	Aeronautical Mobile Satellite Service (System)
AMVER	Automated Mutual Assistance Vessel Rescue System
ANS	Aeronautical Navigation System
ANSS	Aeronautical Navigation Satellite System
AOC	Aeronautical (Airline) Operational Control (Attitude and Orbit Control)
AOC	Advanced Operational Capability
AOR	Atlantic Ocean Region (Old system)
AOR-E	Atlantic Ocean Region East (New system)
AOR-W	Atlantic Ocean Region West (New system)
APC	Adaptive Predictive Coding (Aeronautical Passenger Communications)
APL	Airport Pseudolites
Apogee	More distant point of satellite from the Earth
APS	Air Passenger Services

APT	Automatic Picture Transmission
AR	Axial Ratio
ARC	Aeronautical Radio Communications (Aircraft Radio Corporation)
ARINC	Aeronautical Radio Inc.
ARMMD	Automatic Remote Monitoring and Messaging Data (M2M)
ARNSS	Aeronautical Radionavigation Satellite Service
ARQ	Automatic Repeat Request
ARTS	Automated Radar Terminal System
ASAS	African Satellite Augmentation System
ASCII	American Standard Code for Information Exchange
ASI	Aeronautical Safety Information
ASIC	Application Specific Integrated Circuit
ASIU	ATM Satellite Interworking Unit
ASK	Amplitude Shift Keying
ASQF	Application Specific Qualification Facility
ASR	Airport Surveillance Radar
ASTP	Aviation Security Training Packages
Astra	Applications of Space Technology to the Requirements of Aviation
ATAS	Aeronautical Transportation Augmentation System
ATB	Antenna Tracking Block
ATC	Air Traffic Control
ATCAS	ATC Automation System
ATCDPS	ATC Data Processing System
ATFM	Air Traffic Flow Management
ATIS	Automatic Terminal Information Service
ATM	Air Traffic Management (Asynchronous Transfer Mode)
ATS	Air Traffic Service
AUSREP	Australian Report (as an AMVER)
AV	Accuracy Vertical
AVCS	Aerial Vehicle Communications System
AVICOM	Avicom Japan Co. Ltd.
AvSat	Aviation Satellite Program
AVHRR	Advanced Very High Resolution Radiometer
AVSEC	Aviation Security
AWGN	Additive White Gaussian Noise
AWRS	Aeronautical Weather Report Services
AWS	Automatic Weather Station
AWT	Applied Weather Technology
b	Small Semi-major Axis of Elliptical Orbit
b/s	Bits per second (Baud)
BA	British Airways
BACS	Broadband Aeronautical Communications Service

Baud	1 b/s
BBFRAME	Baseband Frame
BBMSS	Broadband MSS
BC	Before Christ
BCD	Broadcasting and Content Distribution
BCH	Bose Chadhuri Hocquenghem Code
BCM	Block Coded Modulation
BCMSS	Broadcast MSS
BCU	Below Cockpit Unit (AES Transceiver Unit)
BDC	Block Down Converters
BDE	Below Deck Equipment (SES Transceiver Equipment)
Beacon	All types of emergency satellite beacons used in COSPAS-SARSAT System
BeiDou	Chinese GNSS System (Compass)
BER	Bit Error Rate
BES	Base Earth Station (Gateway)
BEST	Bandwidth Efficient Satellite Transport
BGAN	Broadband Global Area Network
BHD	Below Haul Device (VES Transceiver Equipment)
BISDN	Broadband Integrated Service Digital Network
BITE	Built-In Test Equipment
BMCS	Broadband Maritime Communication Service
BNC	Bayonet Neill–Concelman
BOB	Blackberry On Board
BOL	Beginning Of Life
BPS	Baseband Processing Signals
BPSK	Binary PSK
BSS	Broadcasting Satellite Service
BSU	Beam Steering Unit
BT	British Telecom
BUC	Block UpConverter
BVS	Bonvoyage System (WX)
c	Axis between center of the Earth and center of ellipse
C/M	Carrier-to-Multipath Ratio
C/N	Carrier-to-Noise Ratio
C/No	Carrier to Noise Power Density Ratio
CA	Channel Amplifier (Conditional Access)
CAASD	Corporation's Center for Advanced Aviation System Development
CAB	Civil Aviation Bureau
CAC	Civil Aviation Community
CADIN	Common Aeronautical Data Interchange Network
CATV	Community Antenna Television
c-band	Centimeter band

C-band	Frequency band on 6/4 GHz (Tx/Rx) for Feeder Link (from 4 to 8 GHz)
CBR	Constant Bit Rate
CCF	Central Control Facility
CCIR	International Radio Consultative Committee
CCIT	International Telecommunications Consultative Committee
CCITT	International Telegraph and Telephone Consultative Committee
CCU	Central Control Unit
CD	Compact Disk
CDDA	Crossed-Drooping Dipole Antenna
CDMA	Code Division Multiple Access
CDR	Critical Design Review
CELP	Code Excited Linear Prediction
CES	Coast Earth Station (Maritime and Land Mobile)
CF	Control Facilities
CIMS	Customer Information Management System
CIR	Committed Information Rate
CIS	Commonwealth of Independent States (Former USSR)
CIS	Communication and Information System
CIU	Cabin Interface Unit
cm	Centimeter waves (SHF)
CM	Coded Modulation (Configuration Module)
CMU	Communications Management Unit
CN	Correspondent Nodes
CNES	Centre National d'Etudes Spatiales (France)
CNS	Communications, Navigation and Surveillance
CNSO	Civil Navigation Satellite Overlay (Inmarsat)
CoA	Care-of Address
COE	Committee on ECDIS
Compass	Chinese GNSS System (BeiDou)
COMSAT	Communications Satellite Corporation
Cospas	Space System for Search of Distress Vessels and Airplanes (In Russian КОСПАС: Космическая Система Поиска Аварийных Судов и Самолетов)
CP	Circular Polarization
CPDF	Central Personnel Data File
CPDLC	Control Pilot Data Link Communications
CPF	Central Processing Facility
CPFSK	Continuous Phase Frequency Shift Keying
CPU	Central Processor Unit
CQD	Come Quick Distress (A UK Precursor to SOS)
CRL	Communications Research Laboratory of Japan
CRS	Coast Radio Station (Maritime)
CRT	Cathode Ray Tube

CSC	Common Signaling Channel
CSCF	Called State Control Function
CSS	Circuit Switched Service
CTU	Cabin Telecommunications Unit (Cabin Telephone Unit)
CWnd	Congestion Window
d	Distance Between Satellite and the Earth's Surface
D&E	Demonstration and Evaluation
D/A	Digital-to-Analog signal conversion
DA	Demand Assignment
DAB	Direct Audio Broadcasting
DACE	Dual Antenna Control Equipment
Dam	Decameter waves (HF)
DAMA	Demand Assigned Multiple Access
DASA	Daimler Chrysler Aerospace AG
dB	Decibel
DBMC	Data Base Main Computer
DBPSK	Differential Binary PSK
DBS	Direct Broadcasting Satellite
DC	Direct Current
DCCP	Distribution and Concentration Communication Processor
DCM	Digital Circuit Multiplication
DCPR	Data Collection Platform Repeaters
DCS	Data Collection System
DCT	Discrete Cosine Transform
DCTE	Data-Circuit Terminating Equipment
DDR	Data Distribution Region
deci mm	Deci millimeter waves (VEHF)
DECT	Digital European Cordless Communication
DeMUX	Demultiplexer
DFSK	Differential Frequency Shift Keying
DGPS	Differential GPS
DHBS	Direct-to-Home Broadcast System
DIP/LNA	Diplexer/Low-Noise Amplifier
DIT	Digital Image Transfer
DLNA/BITE	Diplexer/Low-Noise Amplifier/Built-In Test Equipment
DLP	Data Link Processing
DM	Delta Modulation
dm	Decimeter waves (UHF)
DME	Distance Measuring Equipment
DMG	Distress Message Generator
DND	Department of National Defense (Canada)
DOD	Tepartment of Defense
DOE	Department of Energy
DOP	Dilution of Precision
DPCM	Differential PCM

DPSK	Differential PSK
DRS	Direct Readout Service
DS	Direct Sequence
DSB-AM	Double-Side Band-Amplitude Modulation
DSB-SC	Double-Side Band Suppressed Carrier
DSC	Digital Selective Call
DSCS	Defense Satellite Communications System
DSI	Digital Speech Interpolation
DSNG	Digital Satellite News Gathering
DTAX	Domestic Telecommunication Automatic Exchange
DTE	Data Terminal Equipment
DTH	Direct-to-Home
DTVC	Digital TV Contribution
DVB	Direct Video Broadcasting
DVB-SM	Digital Video Broadcast-Satellite Mobile
DVB-SP	Digital Video Broadcast-Satellite Personal
DVB-S	DVB-Satellite
DVB-T	DVB-Terrestrial
DVB-RCS	Digital Video Broadcasting-Return Channel over Satellite
DVP	Development Verification Platform
DVSI	Digital Voice Systems Incorporated
E	Eccentric Anomaly
EC	European Commission
ECAC	European Civil Aviation Conference
ECDIS	Electronic Chart Display and Information Systems
ECS	European Communications Satellite
EDCT	Estimated Departure Clearance Time (Aviation)
EFB	Electronic Flight Bag
EGC	Enhanced Group Call
EGNOS	European Geostationary Navigation Overlay Service
EHF	Extremely High Frequency from 30 to 300 GHz (mm-band)
EIRP	Effective Isotropic Radiated Power
ELMSS	European Land Mobile Satellite System
ELT	Emergency Locator Transmitter (Aeronautical application)
EM	Electromagnetic
EMCA	European Maritime Core Area
EME	Externally Mounted Equipment
EMS	European Mobile System
EMSS	Experimental Mobile Satellite System (Japanese program)
EOL	End of Life
EP	Electric Power
EPC	Electronic Power Conditioner
EPG	Electronic Program Guide

EPIRB	Emergency Position Indicating Radio Beacon (Maritime application)
ERAST	Environmental Research Aircraft and Sensor Technology
ESA	European Space Agency
ESA/PAA	Electronically Steered Array/Phased Array Antenna
ESC	Engineering Service Channel
ESNP	European Satellite Navigation Programme
ESOC	European Space Operation Centre
ESS	ETC Short Burst Data Subsystem
ESTB	EGNOS System Test Bed
ETA	Estimated Time of Arrival (Shipping and Airways)
ETC	Earth Terminal Controller
ETD	Estimated Time of Departure (Shipping)
FTP	File Transport Protocol
ETS	Engineering Test Satellite of Japan (ETC Transmission Subsystem)
ETSI	European Telecommunications Standard Institute
EU	Electronics Unit
EUMETSAT	Exploration of Meteorological Satellites
Eurocontrol	European Organization for the Safety of Air Navigation
Eutelsat	European Telecommunications Satellite Organization
EVGC	Enhanced Voice Group Call
F1	Fleet One
FAA	Federal Aviation Administration of USA
FANS	Future Air Navigation Systems
Fax	Facsimile
FB	FleetBroadband
FCC	Federal Communications Commission (US)
FDM	Frequency Division Multiplexing
FDMA	Frequency Division Multiple Access
FDP	Flight Data Processing
FEC	Forward Error Correction
FES	Fixed Earth Station
FET	Field Effect Transistor
FFPP	Flexible Flight Profile Planning
FH	Frequency Hopping
FIR	Flight Information Region
FIT	Fixed Interactive Terminals
FleetNET	Inmarsat EGC broadcast of ship-owner data to part of all fleet
FLS	Forward Link Subsystem (System)
FM	Frequency Modulation
FMS	Flight Management System
FNBDT	Future Narrowband Digital Terminal
FoIP	Fax over IP

FQR	Factory Qualification Review
FRec	Fast Recovery
FRet	Fast Retransmit
FRU	Field Replaceable Units
FSC	Fixed Satellite Communications
FSK	Frequency Shift Keying
FSS	Fixed Satellite Service/System (Flight Standards Service)
FSSE	Flight Safety Satellite Equipment
FTP	File Transfer Protocol
G	Universal Gravitational Constant
G/T	Ratio of system gain to system noise Temperature (Figure of Merit)
GaAs	Gallium Arsenide
GaAsFET	Gallium Arsenide FET
GABS	Globalstar Accounting & Billing System
GACCS	Global Aeronautical Corporate and Commercial System
GADSS	Global Aeronautical Distress and Safety System
GAGAN	GPS/GLONASS and GEOS Augmented Navigation
GAN	Global Area Network
GASP	Global Aviation Safety Plan (ICAO)
GASSC	Global Aeronautical Safety Satellite Communications
Gateway	GES (BES)
GBAS	Ground Based Augmentation System
GBO	Globalstar Business Office
Gb/s	Gigabit per second
GCC	Gateway Control Center
DCP	GEO Data Collection Platform
GCS	Ground Control Station
GD	General Dynamics
GDE	Group Delay Equalizers
GDL	Gas Dynamic Laboratory
GDN	Globalstar Data Network
GDSS	Global Determination Satellite System
GEO	Geostationary Earth Orbit
GEOLUT	GEO Local User Terminal
GEOSAR	GEO Search and Rescue
GES	Ground Earth Station/Aeronautical Mobile (Gateway Earth Stations)
GET	Ground Earth Terminals
GFC	Ground Forecasting Center
GFR	Guaranteed Frame Rate
GGS	Gateway Ground Station
GHz	Gigahertz is a radio frequency unit of 1000 MHz
GIC	Ground Integrity Channel
GIO	Geosynchronous Inclined Orbit

GIRD	Group for Investigation of Reactive Movement
GLONASS	Global Navigation Satellite System (In Russian: ГЛОНАСС—Глобальная Навигационнаяа Спутниковая Система)
GLS	GNSS Landing System
GMBSS	Global Maritime Broadcasting Satellite System
GMDSS	Global Maritime Distress and Safety System
GMPSC	Global Mobile Personal Satellite Communications
GMR	Geostationary Mobile Radio (of Himawary Japanese Series)
GMS	Ground Monitoring Stations (Gateway Management System)
GMS	Geosynchronous Meteorological Satellite
GMSC	Global Mobile Satellite Communications
GMSS	Global Mobile Satellite Systems
GMT	Global Meteorological Technologies
GNC	Ground Notification Center
GNSS	Global Navigation Satellite System (ICAO Definition)
GNSS-P	GNSS Panel
GOES	Geostationary (Global) Operational Environmental Satellite
GOMS	Geostationary Operational Meteorological Satellite
GPRS	General Packet Radio Service
GPS	Global Positioning System
GR	Guaranteed Rate
GRS	Geostationary Ranging Station
GRS	Ground Radio Station (Aeronautical)
GSAS	Global Satellite Augmentation System
GSSII	Global Sea State Information via Internet
GSM	Global Service for Mobile communications or Group Special Mobile
GSPS	Global Stratospheric Platform Systems
GTE	Ground Telecommunication Equipment
GTO	Geostationary Transfer Orbit
GTS	Global Telemaque Service
GUS	Ground Uplink System
GW	Gateway
GX	Global Xpress
h	Altitude of Satellite Above the Earth's Surface
HAP	High-Altitude Platform
HAIPE	High Assurance Internet Protocol Encryptor
HDTV	High-Definition TV
HEC	Hybrid Error Correction (Header Error Check}
HEMT	High-Electron Mobility Transistor
HEO	Highly (Highinclined) Elliptical Orbit

HF	High Frequency from 3 to 30 MHz (Dam-band)
HGA	High-Gain Antenna
HLP	Horizontal Linear Polarization
Hm	Hectometrer waves (MF)
HoA	Home Address
HPA	High-Power Amplifier
HPA/DLNA	High-Power Amplifier and Diplexer–Low-Noise Amplifier
HPBW	Half Power Beamwidth
HSD	High Speed Data over 9600 b/s, max 56/64 Kb/s
HSO	Hybrid Satellite Orbits
HTTP	Hyper-Text Transfer Protocol
HVAC	Heating, Ventilation and AirConditioning
Hz	Hertz is basic radio frequency unit (1 Cycle/sec)
i	Inclination Angle
IA	Initial Approach
IAMSAR	International Aeronautical and Maritime SAR Manual
IATA	International Air Transport Association
IAWVG	Instrumental Approach with Vertical Guidance
IBF	Input Bandpass Filter
ICAO	International Civil Aviation Organization
ICF	Integrated Communications Facility
ICM	Integral Core Module (Integrated CommBox Modem)
ICO	Intermediate Circular Orbits
IDEC	Integrated Development and Evaluation Center
IDP	Input Data Processing
IDU	InDoor Unit
IEC	International Electrotechnical Commission
IESS	Intelsat Earth Station Standards
IETF	Internet Engineering Task Force
IF	Intermediate Frequency
IFE	In-Flight Entertainment
IFR	Instrument Flight Rule
IFRB	International Frequency Registration Board
IGA	Intermediate-Gain Antenna
IGN	Inmarsat Ground Network
IGP	Ionospheric Grid Points
IH	Inmarsat Hemisphere
IHO	International Hydrographic Organization
IKE	Internet Key Exchange
ILS	Instrument Landing System
IM	Intermodulation
IMAP	Internet Message Access Protocol
IMBE	Improved MultiBand Excitation
IMCO	Intergovernmental Maritime Consultative Organization (Former IMO)

IME	Internally Mounted Equipment
IMO	International Maritime Organization (New name)
IMOSAR	IMO Search and Rescue Manual
INMARSAT	INternational MARitime SATellite
INR	Interference and Noise Ratio
INS	Inertial Navigation System
INSAT	Indian National Satellite System
Intelsat	International Telecommunications Satellite organization
Intersputnik	East European satellite communications organization formed by Russia
IODFT	Inverse Offset Discrete Fourier Transform
IOL	Inter Orbital Link
IOR	Indian Ocean Region
IP	Internet Protocol
IPgen	IP generation
IPSec	IP Security Protocol
IPTV	Internet Protocol TV
IPv4	IP version 4
IPv6	IP version 6
IRD	Integrated Receiver Decoder
ISAS	Inmarsat Satellite Augmentation System
ISDN	Integrated Services Digital Network
ISL	Inter-Satellite Link
ISP	Internet Service Provider
ISPS	International Ship and Port Security
ISTS	Intelligent Satellite Transport System
ISR	Intelligence, Surveillance and Reconnaissance
ISU	Indoor Satellite Unit
ITS	Intelligent Transportation System
ITU	International Telecommunications Union
ITU-R	ITU Radio
IWG	Interoperability Working Group
JAL	Japan Air Lines
JAMSTEC	Japan Marine Science and Technology Centre
JAPREP	Japan Report (like AMVER)
JMA	Japan Meteorological Agency
JTIDS	Joint Tactical Information Distribution System
JX	Jet ConneX (Aeronautical GX)
K	Boltzmann Constant
Ka-band	Frequency band between 24 and 40 GHz
Kb/s	Kilobit per second
K-band	Frequency band between 18 and 24 GHz
KDD	Kokusai Denshin Denwa (Japanese Telecom)
kHz	Kilohertz is a eadio frequency unit of 1000 Hz
km	Kilometer

Km	Kilometer waves (LF)
KPA	Klystron Power Amplifiers
KRx	Ku-band Receiver
Ku-band	Frequency band between 12 and 18 GHz
kW	Kilowatt
LAA	Local Augmentation Area
LAAS	Local Area Augmentation System
LAN	Local Area Network
L-band	Frequency band on 1,6/1,5 GHz (Tx/Rx) for Service Link (from 1 to 2 GHz)
LCC	Launch Control Center
LD-CELP	Low Delay Codebook Excited Linear Prediction
LDM	Linear Delta Modulation
LDPC	Low-Density Parity Check
LEO	Low Earth Orbit
LEOLUT	LEO Local User Terminal
LEOSAR	LEO Search and Rescue
LES	Land Earth Station (All Mobile Applications)
LESO	Land Earth Station Operators
LF	Low Frequency from 30 to 300 kHz (Km-band)
LGA	Low-Gain Antenna
LGF	LAAS Ground Facility
LHCP	Left Hand Circular Polarization
LLM	L-band Land Mobile
LMDS	Local Multipoint Distribution Service
LMGT	Lockheed Martin Global Telecommunications
LMSC	Land MSC
LMSS	Land Mobile Satellite Service (System)
LNA	Low-Noise Amplifier
LNSS	Land Navigation Satellite System
LO	Local Oscillator
LOA	Length Over All
LOS	Line-of-Sight
LP	Launch Platform
LPC	Linear Predictive Coding
LPF	Low-Power Filter
LRIT	Long Range Identification and Tracking
LRNSS	Land Radionavigation Satellite Service
LRPT	Low Rate Picture Transmission
LRU	Line Replaceable Unit
LSD	Low Speed Data up to 2400 b/s
LTAS	Land Transportation Augmentation System
LUT	Local User Terminal (COSPAS-SARSAT receiving GES)
M	Mass of the Earth Body
M	Point of the Observer or Mobile

M2M	Mashine-to-Mashine (SCADA)
MA	Multiple Access
MAA	Microstrip Array Antenna
MAP	Ministry of Aircraft Production (Maximum a Posteriori)
Marecs	Maritime European Communication Satellites
Marisat	Maritime Satellite (system established 1976 by USA)
Marots	Maritime Orbital Test Satellite
MAYDAY	Distress Signal in Mobile Radiotelephony
Mb/s	Megabit per second
MBSS	Mobile Broadcast Satellite Service
m-band	Meter band
MBCM	Multiple BCM
MCC	Mission Control Center (Master Control Center)
MCDU	Multifunction Control and Display Unit
MCPC	Multiple Channels per Carrier
MCR	Minimum Cell Rate
MCS	Master Control Station
MCU	Microcontroller Unit
MDC	Message Delivery Center
MDCRS	Meteorological Data Collection and Reporting System
MDCT	Modified Discrete Cosine Transform
MDSS	Medium Data Rate Satellite System
MELPC	Multipulse Excited LPC
MEM	MicroElectroMechanical
MEO	Medium Earth Orbit
MERSAR	Merchant Ship Search and Rescue Manual
MES	Mobile Earth Station
METAR	Meteorological Aviation Reports (MÉTéorologique Aviation Régulière)
MF	Medium Frequency from 300 to 3000 kHz (Hm-band)
MGA	Medium Gain Antenna
MHz	Megahertz is a radio frequency unit of 1000 kHz
MIFR	Master International Frequency Register
MIPv6	Mobile IP version 6
MIRP	Manipulated Information Rate Processor
MIT	Mobile Interactive Terminals
MLQ	Maximum Likelihood Quantization
MLMSS	Meteorological MSS
mm	Millimeter waves (EHF)
MM	Motion Media
MMMSC	Multimedia MSC
MMSC	Maritime Mobile Satellite Communications
MMSS	Maritime Mobile Satellite Service (System)
MMU	Mission Management Unit
MMW	Millimeter Wavelength

MN	Mobile Node
MNSS	Maritime Navigation Satellite System
MOBSAT	Mobile Satellite Group
MORFLOT	Ministry of Merchant Marine (former USSR)
Morya	Russian (the former USSR) program for MMSS
MOC	Message Origination Center
MOU	Memorandum of Understanding (COSPAS-SARSAT)
MPA	Microstrip Patch Antenna (Medium Power Amplifier)
MPDS	Mobile Packet Data Service
MPEG	Moving Picture Expert Group
MPEG-2	Moving Picture Expert Group 2
MPLS	Multiprotocol Label Switching
MRC	Mobile Radio Communications
MRDU	Multichannel Receiver Decoder Unit
MRNSS	Maritime Radionavigation Satellite Service
MRS	Monitoring and Ranging Station
ms	Millisecond
MSA	Mobile Satellite Antenna
MSAS	MTSAT Satellite-based Augmentation System
MSAT	Mobile Satellite System
MSB	Mobile Satellite Broadcasting (Most Significant Bit)
MSBS	Mobile Satellite Broadband Service
MSC	Mobile Satellite Communications
MSG	Meteosat Second Generation
MSI	Maritime Safety Information (Mobile Satellite Internet)
MSK	Minimum Shift Keying
MSL	Mean Sea Level (Multi-Satellite Link)
MSMC	Message Switching Main Computer
MSMS	Mobile Satellite Meteorological Service
MSS	Mobile Satellite Service (System)
MSU	Magnetic Sensor Unit
MSUA	Mobile Satellite Users Association
MTAS	Maritime Transportation Augmentation System
MTB	Mediterranean Test Bed
MTBF	Mean Time Before Failures
MTC	Message Termination Center (Maritime Traffic Control)
MTCM	Multiple TCM
MToIP	Mobile Transmissions over IP
MTSAT	Multifunctional Transport Satellite
MTU	Mobile Terminal Unit
MUSAT	Canadian program for LMSS
m/v	Motor vessel
MUT	Mobile User Terminal
MUX	Multiplexer
MWIF	Mobile Wireless Internet Forum

NAS	National Air (Airspace) System
NASA	National Aeronautical and Space Administration (formed by the USA)
NASDA	National Space Development Agency (formed by Japan)
NAVAREA	IMO Global Navigational Areas
NAVTEX	Narrowband direct-printing telegraphy system
NCC	Network Control Center
NCF	Network Control Functions
NCS	Network Coordination (Control) Station
NDWS	Noble Denton Weather Services
NES	Navigation Earth Station
Ni-Cd	Nickel-Cadmium
Ni-H	Nickel-Hydrogen
NLES	Navigation LES
NMA	Norwegian Mapping Authority
NMEA	National Marine Electronics Association
NMF	Network Management Functions
NMS	Network Management Station
NOAA	National Oceanic and Atmospheric Administration (USA)
NOC	Network Operations Center
Non-GEO	Non-GEO satellite configuration
NOTAM	Notice to Airman
NPA	Non-Precision Approach
NPM	Non Penetrating Mount
NSTB	National Satellite Test Bed
NVNG MSS	Non-Voice Non-GEO MSS
NWS	National Weather Service
NX	Navigation Report (Warning)
O3b	Other 3 Billion (MEO Maritime MSC Sydstem)
OAM	Operation and Maintenance
OBF	Output Bandpass Filter
OCC	Operation Control Center
OCD	Oceanic Clearance Delivery
ODE	OutDoor Equipment
ODL	Oceanic Data Link
ODP	Oceanic Air Traffic Control Data Processing
ODU	OutDoor Unit
OMC-G	Operational Support Network-Gateway
OMC-R	Operational Support Network-Radio
OMUX	Output Multiplexer
OMT	Orthogonal Mode Transducer
O-QPSK	Offset Quadrature Phase Shift Keying
ORR	Operational Readiness Review
OSN	Operational Support Network
OTG	Ocean Traffic Generator

OTS	Orbital Test Satellite
p	Focal parameter
P	Sub-satellite Point
PA	Precision Approach (Power Amplifier)
PABX	Private Automatic Branch Exchange
PACF	Performance Assessment and System Checkout Facility
PAM	Pulse Amplitude Modulation
PAMA	Pre-Assigned Multiple Access
PANPAN	Urgency signal in mobile radiotelephony
P-band	Frequency band between 200 and 400 MHz
PC	Personal Computer
PCM	Pulse Code Modulation
PCR	Program Clock Reference (Peak Cell Rate)
PCS	Personal Communications Service
PCU	Pedestal Control Unit
PDC	Personal Digital Cellular system designed in Japan for operation in either the 800 MHz and 1,50 GHz band
PDH	Plesiochronous Digital Hierarchy
PDL/CMT	Portable Data Loader/Configuration Management Terminal
PDN	Public Data Network
PDS	Processed Data Stream
PE	Propulsion Engine
PEO	Polar Earth Orbit
Perigee	More nearest point of satellite from the Earth
PES	Portable (Personal) Earth Station
PFD	Power Flux Density
PIB	Pre-flight Information Bulletin
PID	Pager Identity (Packet Identifier)
PIM	Passive Intermodulation
PIREP	Pilot Weather Reports
PL	Physical Layer
PLB	Personal Locator Beacon (Land or Personal Applications)
PLCP	Physical Layer Convergence Protocol
PLFRAME	Physical Layer Frame
PM	Phase Modulation
PMSS	Personal Mobile Satellite Service (System)
PN	Pseudo (Pseudorandom) Noise
PNA	Public Network Access
PoE	Power over Ethernet
POR	Pacific Ocean Region
PPP	Point-to-Point Protocol
PPS	Precise Positioning Service
PRMA	Packed Reserved Multiple Access
PS	Program Streams
PSDN	Public Switched Data Network

PSK	Phase Shift Keying
PSPDN	Public Switched Private Data Network
PSTN	Public Switched Telephone Network
PTT	Post, Telegraph and Telephone
PVT	Position, Velocity and Time
QAM	Quadrature Amplitude Modulation
QFH	Quadrifilar Helix
QHA	Quadrifilar Helix Antenna
QoS	Quality of Service
QPSK	Quadrature PSK
R	Equatorial Radius of the Earth
RA	Random Access
R&D	Research and Development
R&R	Rural and Remote
RAC	Rural Area Coverage
RADAT	Radiosonde Additional Data
RAIM	Receiver Autonomous Integrity Monitoring
RAM	Random Access Memory
RC	Requesting Channel
RCC	Rescue Coordination Center
RCS	Recursive Systematic Convolution
RDDI	Radio Direction Distance Information
RDI	Radio Direction Information
RDMA	Random (Packet) Division Multiple Access
RDP	Radar Data Processing
RDSS	Radio Determination Satellite Service
RES	Radio Earth Stations
RF	Radio Frequency
RFC	Request for Comments (Regenerative Fuel Cells)
RFE	Radio Frequency Equipment
RFG	RF Generator
RFP	Request for Proposal
RFT	Radio Frequency Terminal
RFU	RF Unit
RHCP	Right-Hand Circular Polarization
RIMS	Ranging and Integrity Monitoring Stations
RLS	Return Link Subsystem
RNII	Russian Institute/Реактивный научно-исследовательский институт (РНИИ)
RNSS	Radionavigation Satellite Service
ROM	Read Only Memory
RPY	Roll, Pitch, Yaw
RQHA	Resonant QHA
RR	Radio Regulations
RS	Reed–Solomon

RSAS	Regional Satellite Augmentation System
RSC	Rescue SubCenter
RSS	Reference and Synchronization Subsystem
RSM	Reduction of Separation Minima
RTS	Remote Troubleshooting System
RTT	Round-Trip Time
RUDICS	Router-Based Unrestricted Digital Internetworking Connectivity Solutions
RVR	Runway Visual Range
RW	Radiowaves
RWnd	Receiver Advertised Window
Rx	Receiver
s	Seconds
S	Satellite
S/N	Signal-to-Noise ratio (Signal and Noise)
s/s	Steam Ship
SA	Service Area
SAC	Special Access Code
SAC	Suburban Area Coverage
SACK	Selective Acknowledgment
SafetyNET	Inmarsat EGC Based System for MSI Transmission
SAN	Satellite Access Node
SAO	Surface Aviation Observation
SAR	Search and Rescue
SARP	SAR Receiver-Processor and Memory (Standard and Recommended Practices)
SARR	SAR Repeater
Sarsat	Search and rescue satellite aided tracking
SART	SAR Radar Transponder
SAS	Safety and Security
SAT	Satellite Asset Tracking
SATAFIS	Satellite Aerodrome Flight Information Service
SATC	Satellite Air Traffic Control
SAU	Satellite Antenna Unit
SAW	Surface Acoustic Wave
SB	SwiftBroadband
S-band	Frequency band between 2 and 4 GHz
SBAS	Satellite-Based Augmentation System
SBD	Short Burst Data
SBF	Short Backfire
SBIR	Small Business Innovative Research
SCADA	Supervisory Control And Data Acquisition (M2M)
SC-AM	Suppressed Carrier-AM
SCC	Satellite Control Center
S-CDMA	Synchronous-CDMA

SCDSB-AM	Suppressed Carrier Double-Side Band AM
SCORE	Signal Communicating by Orbiting Relay Equipment
SCP	Stratospheric Communications Platform
SCPC	Signal Channel Per Carrier
SCS	Satellite Control Station/System (System Control Segment)
SCS	Scotty Computer System
SCU	Satellite (Subscriber) Communication Unit
SDCM	System of Differential Correction and Monitoring
SDH	Synchronous Digital Hierarchy
SDK	Software Developers Kit
SDL	Satellite Data Link
SDM	System Definition Manual (of Inmarsat)
SDMA	Space Division Multiple Access
SDTV	Standard-Definition TV
SDU	Satellite Data Unit
SECURITE	Safety signal in mobile radiotelephony
SEHA	Safety Enhancement in High Altitudes
SELA	Safety Enhancement in Low Altitudes
SES	Ship Earth Station
SFIR	Spatial Filtering for Interference Reduction
SHF	Super High Frequency between 3 and 30 GHz (cm-band)
SI	Service/system Information
SID	Standard Interface Description
SILEX	Semiconductor Laser Inter-satellite Link Experiment
SIM	Subscriber Identity Module
SINR	Signal to Interference and Noise Ratio
SIR	Signal-to-Interference Ratio
SIS	Signal in Space
SITA	Societé Internationale de Télécomommunications Aeronautique
SMAP	Spot Management and Planning
SMATV	Satellite Master Antenna Television
SME	Small to Medium-size Enterprise
SMGC	Surface Movement Guidance and Control
SMS	Short Message Service
SMTP	Simple Mail Transfer Protocol
SNAS	Satellite Navigation Augmentation System
SoHo	Small office Home office
SOIT	Satellite Operational Implementation Team
SOLAS	Safety Of Life At Sea Convention
SONET	Synchronous Optical Network
SOS	Distress Signal in Mobile Radiotelegraphy
SOTM	Satcom On The Move
SP	Structure Platform (Sub-satellite Point)
SPECI	Selected Special Weather Report

SPOC	SAR Point Of Contact
SPS	Stratospheric Platform Systems (Standard Positioning Service)
SRS	Ship Radio Station (MF/HF/VHF Conventional Mobile Radio)
SRU	Satcom Reference Unit
SS	Switching Satellite
SSA	Solid-State Amplifier
SSAS	Ship Security Alert Systems
SSB	Single-Side Band
SSB-AM	Single-Side Band-Amplitude Modulation
SSB-SC	SSB Suppressed Carrier
SSI	Sky Station International Inc
SSMA	Spread Spectrum Multiple Access
SSPA	Solid-State Power Amplifier
SSS	Switch Subsystem
SSThres	Slow Start Threshold
SSTV	Slow Scan Television
STE	Secure Telephone/terminal Equipment
STM	Synchronous Transfer Mode (Ship Traffic Management)
STS	Stratospheric Telecommunications Service (Space Transportation System)
STU	Secure Telephone Unit
SUDS	Secondary Users Data Station
SUT	SatLink User Terminal
SV	Space Vehicle
SWAT	SBAS Worldwide Availability Tool
Swift64	Aeronautical HSD solution
TAF	Terminal Area Forecasts
TAO	Telecommunications Advancement Organization of Japan
TAV	Trans Atmospheric Vehicle
Tb/s	Terabit per second
TBUE	Test Bed User Equipment
TC	Telecommand
TC	Thermal Control
TCC	Traffic Control Center
TCE	Traffic Channel Equipment
TCM	Trellis Coded Modulation (Traffic Control and Management)
TCP	Transmission Control Protocol
TCS	Terrestrial Communications Subsystem
TDM	Time Division Multiplexing
TDMA	Time Division Multiple Access
TDMoIP	TDM over IP
TDRSS	Tracking and Data Relay Satellite System

TEC	Total Electron Content
Tel	Telephone
TES	Transportable Earth Station
Tlg	Telegraphy
Tlx	Telex
TM	Telemetry
TMI	Telesat Mobile Inc.
TMS	Test Master Station
TNC	Threaded Neill–Concelman
TRAD	Terminal Radar Alphanumerical Display
TS	Transport Streams
TT&C	Telemetry, Tracking, and Command (Telecommand, Telemetry, and Control)
TTAC	Telemetry Tracking and Control
TTN	Terrestrial Telecommunications Network
TTT	Safety Signal in Mobile Radiotelegraphy
Tty	Teletypewriter (Teleprinter)
TV	TeleVision
TVRO	TV Receive Only
TWIP	Terminal Weather Information for Pilots
TWT	Traveling Wave Tube
TWTA	Traveling Wave Tube Amplifier
Tx	Transmitter
UAC	Urban Area Coverage
UAS	Unmanned Aircraft Systems
UAV	Unmanned Aerial Vehicles
UBR	Undefined (Unspecified) Bit Rate
UDI	Unrestricted Digital Information
UDP	User Datagram Protocol
UHF	Ultra High Frequency between 300 MHz and 3 GHz (dm-band)
UMTS	Universal Mobile Telecommunications Service
UN	United Nations
UPS	Uninterrupted Power Supply
USB	Unified S-band
USB	Universal Serial Bus
USIM	Universal Subscriber Identity Module
USNS	US Navy Ship
USO	Ultra-Stable Oscillator
USS	US Ship
UST	User Service Terminal
UT	User Terminal
VARB	Van Allen Radiation Belts
VAS	Value-Added Service
VBR	Variable Bit Rate

VC	VideoConferencing
VC	Virtual Channels
VCoIP	VideoConference over IP
VCI	Virtual Channel Identifier
VDB	VHF Data Broadcast
VDL	VHF Digital Link
VDU	Video Display Unit
VDV	Voice, Data, and Video
VDVoIP	Voice, Data, and Video over IP
VEHF	Very Extremely High Frequency from 300 GHz to 3 THz (deci mm-band)
VES	Vehicle Earth Station
VFR	Visual Flight Rules
Version-2	Two GAN units linked together via bonding cable double throughput to 128 Kb/s
VGA	Variable Gain Amplifier
VHF	Very High Frequency between 30 and 300 MHz (m-band)
VLP	Vertical Linear Polarization
VoIP	Voice over IP
Volna	Russian (Ex-USSR) program for MMSS and AMSS
VOR	VHF Omnidirectional Ranging
VPI	Virtual Path Identifier
VPN	Virtual Private Networks
VPoIP	VideoPhone over IP
VQ	Vector Quantization
VS/VD	Virtual Source/Virtual Destination
VSAT	Very Small Aperture Terminal
VSWR	Voltage Standing Wave Ratio
VTS	Vessels Traffic Services
WAA	Wide Augmentation Area
WAAS	Wide Area Augmentation System
WAD	Wide Area Differential
WADGNSS	Wide Area Differential GNSS
WAN	Wide Area Network
WANAV	Wide Area Navigation (RNAV)
WAP	Wireless Access Point
WARC	World Administrative Radio Conference
WARC-ST	WARC-Satellite
W-CDMA	Wireless-CDMA
WDCP	Weather Data Collection Platforms
WEFAX	WX Fax
WEMS	Web and Email Messaging Server
WiFi	Wireless Fidelity
WIPP	WAAS Integrity Performance Panel
WLAN	Wireless Local Area Networking

WMO	World Meteorological Organization
WMS	Wide Area Master Stations
WQHA	Wire Quadrifilar Helix Antenna
WRC	World Radio (Radiocommunications) Conference
WRS	Wide Area Reference Stations
WWW	World Wide Web (Internet)
WX	Weather Report (Warning)
X25/X70/X400	Standards for messaging and data systems specified by the CCITT
X-band	Frequency band between 8 and 12 GHz
XPD	Cross-Polar Discrimination
XPI	Cross-Polar Isolation
XSA	Cross-Slot Array Antennas
XXX	Urgency Signal in Maritime Mobile Radiotelegraphy
δ	Angle of Antenna Radiation
ε	Elevation Angle (E)
ϕ	Phase Angle
Θ	True Anomaly
Π	Perigee
ν	Angular Speed of the Earth's Rotation
λ	Geographical Longitude
φ	Geographical Latitude
Ψ	Central or Sub-satellite Angle
ω	Frequency Angle
ω	Argument of the Perigee
Ω	Right Ascension of an Ascending Node Angle

Introduction

The two volumes of this book were written in order to form a bridge between potential readers and current GMSC trends, system concepts, and network architecture by using a very simple style with easily comprehensible technical information, characteristics, graphicons, figures, illustrations, and mathematic equations.

The special part of GMSC for maritime, land, and aeronautical CNS applications is new techniques for modern transportation concerns and their fleets at sea, on land, and in the air for the enhancement of commercial and distress communications and tracking solutions. The modern GMSC infrastructures are greatly important for all transportation companies, their successful commerce, carriage and management of vessels, land vehicles and aircraft. Finally, modern, innovative techniques and technology in GMSC are needed for newly developed mobile broadband and multimedia communications and for IT, CNS, and Global Navigation Satellite Systems (GNSS).

The Volume 1 of this book, nominated as "Theory," consists of seven chapters on the following particular subjects:

Chapter 1: *Introduction* gives a short background to the development of Radio and MSC, overview, concept, and applications of GMSC for maritime, land, and aeronautical applications, including international organizations, operators, and new GMPSC systems.

Chapter 2: *Space Segment* discusses the fundamentals of space platforms and orbital parameters, laws of satellite motions, new types of launching systems, satellite orbits and geometric relations, spacecraft configuration, payload structure, and satellite link design.

Chapter 3: *Transmission Techniques* introduces an essential basic knowledge of modulation and demodulation, coding and error corrections, multiple access, mobile Internet and broadband protocols, and forthcoming mobile DVB-CNS solutions.

Chapter 4: *Mobile Satellite Antennas Systems* explains electromechanical characteristics and basic relations and classifications of antennas, considerable information about mobile antennas for ships, sea platforms, land vehicles and aircraft.

Chapter 5: *Propagation and Interference Consideration* comprises all the particulars about propagation loss effects, path depolarization causes, trans-ionospheric contribution, and propagation effects that are specific and important for GMSC maritime, land, and aeronautical environments and requirements.

Chapter 6: *Ground Segment* in general explains the ground infrastructures of both Land Earth Stations (LES) and Antenna Systems, and in particular presents the definition of GMSC services, components and control systems, radio frequency and communication equipment and general infrastructure of Earth station.

Chapter 7: *Users Segment* provides an overview of GMSC applications, Inmarsat GEO users segment, maritime Ship Earth Stations (SES), FleetBroadband, SES terminal equipment and installation, shipborne antenna mounting and tracking system, land Mobile Earth Station (MES), Broadband Global Area Network (BGAN) Transportable Earth Stations (TES), BGAN Road and Railway Vehicle Earth Stations (VES), Personal Earth Stations (PES), Aircraft Earth Stations (AES), SwiftBroadband, AES Terminal Equipment and Installation, Global Xpress Solutions, Iridium LEO users segment, Maritime O3B MEO users segment and Mobile DVB-RCS GEO users segment.

Chapter 1
Introduction

The safety of navigation through all past ages has been a primary preoccupation for all seamen and shipping owners. Distress and disasters at sea caused by the blind forces of Mother Nature or by human factors have occurred during the course of many centuries on ships and in the life of seafarers. For many centuries, seafarers sailed without incoming information about trip, navigation, and weather conditions at sea. At that time, only audio and visual transfers of information from point to point were used.

On the other hand, the safety of flying in history has been a primary preoccupation for all airmen and airways owners. Disasters of aircraft caused by different or somehow by human factors have occurred during the course of the last century in the life of avionics. At the beginning, avionics flied without incoming information about some air environment and weather conditions. At that time, only audio and visual transfers of information from point to point were used.

However, no earlier than the end of the nineteenth century were developed new disciplines, such as the transmission of news and information via wire initially, then by radio (wireless), and latterly via modern satellite and stratospheric platform communications systems.

The facts about road vehicles and trains are well determined and clear because these transport mediums have more reliable environments and routes than ships and aircraft. After a disaster of train, truck, or bus, it is much easier to find out their positions and to provide alert, search, and assistance.

With the exception of safety demands, an important question is the utilization and development of new mobile radio and satellite communications and navigation systems for commercial and social utilization at sea, on land and in the air.

S.D. Ilčev, *Global Mobile Satellite Communications Theory*,
DOI 10.1007/978-3-319-39171-7_1

1.1 Abstract

Communication satellites provide the bridges for a number of new, specialized markets in commercial and private telecommunications and make closer ties between nations. Thus, in the course of more than 50 years, satellite communications have obtained global links in the public and private terrestrial telecommunication network (TTN). In the next stage, military satellite systems for communication and navigation came to serve navy, ground, and air forces worldwide, and because of economic reasons, soon after they also provided commercial Mobile Satellite Communications (MSC) and Global Navigation Satellite System (GNSS).

For about 40 years, MSC was used, particularly because ocean-going vessels have become dependent for their commercial and safety communications on Mobile Satellite Systems (MSS). Although aircraft and land vehicles started before ships, due to many unsuccessful experiments and projects they have had to follow the evident lead of Inmarsat maritime MSC engineering. The modified ship's mobile Earth stations (MES) are today installed on land (road or railway) vehicles and aircraft for all civil and military applications, including remote or rural locations and industrial onshore and offshore installations.

The GPS, DGPS, GLONASS, and other new satellite navigation and determination systems provide precise positioning data for vessels and aircraft and also serve land navigation and fleet vehicle management. Because of the need for new, enhanced services, these systems will be augmented with satellite communications and ground surveillance facilities. At the end of this race, a new mobile satellite revolution is coming, whereby anyone can carry a personal handheld telephone using simultaneously satellite or cellular/dual systems at sea, in the car, in the air, on the street, in rural areas, and in the desert, that is to say everywhere and in all positions.

These integrated systems will be soon implemented with modern Global Mobile Satellite Communications (GMSC), Global Satellite Augmentation System (GSAS), Digital Video Broadcasting Return Channel via Satellite (DVB-RCS), Stratospheric Platform Systems (SPS), and very modern transportation discipline known as communication, navigation and surveillance (CNS) designed to provide enhanced radio and satellite traffic control and management (TCM) including safety and security (SAS).

1.2 History of Radio

The word **communications** is derived from the Latin phrase "**communication**," which stands for the social process of information exchange and covers the human need for direct contact and mutual understanding. The word "**telecommunications**" means to convey and exchange information at a distance (**tele**) by the medium of electrical signals.

In general, telecommunications are the conveyance of intelligence in some form of signal, sign, sound, or electronic means from one point to a distant second point. In ancient times, that intelligence was communicated with the aid of audible callings, fire and visible vapor or smokes, and image signals. We have come a long way since the first human audio and visual communications, in case you had forgotten, used during many millennia. In the meantime, primitive kinds of communications between individuals or groups of people were invented. Hence, as impressive as this achievement was, the development of more reliable communications and so, wire and radio, had to wait a couple of centuries more.

The invention of the telegraph in 1844 and the telephone in 1876 harnessed the forces of electricity to allow the voice to be heard beyond shouting distance for the first time. The British physicist M. Faraday and the Russian academic E. H. Lenz made experiments with electric and magnetic phenomena and formulized a theory of electromagnetic (EM) induction at the same time. The British physicist James C. Maxwell published in 1873 his classical theory of electromagnetic radiation, proving mathematically that electromagnetic waves travel through space with a speed precisely equal to that of light.

The famous German physicist Heinrich Rudolf Hertz during 1886 experimentally proved Maxwell's theoretical equations. Thus, during Hertz's studies in 1879, his mentor, physicist Hermann von Helmholtz, suggested that Hertz's doctoral dissertation has been on testing Maxwell's theory of electromagnetism, published in 1865, which predicted the existence of electromagnetic waves moving at the speed of light and predicted that light itself was just such a wave.

During his research and studies, Hertz demonstrated that high-frequency (HF) oscillations produce a resonant effect at a very small distance away from the source and that this phenomenon was the result of electromagnetic waves. Practically, Hertz used the damped oscillating currents in a dipole antenna, triggered by a high-voltage electrical capacitive spark discharge, as his transmitter (Tx) source of radio waves (RW).

His detector in some experiments was another dipole antenna connected to a narrow spark gap as receiver (Rx). A small spark in this gap signified detection of the radio waves. When he added cylindrical reflectors behind his dipole antennas, Hertz could detect radio waves about 20 m from the Tx in his laboratory. In Fig. 1.1 (Left) is **illustrated** the electronic diagram of spark generator as a main component of his radio Tx (Ruhmkorff coil with linear vibrator). In fact, the wave character of the spark generator discharge is creating short-lived electromagnetic waves. The waves thus produced were received by a resonator located a short distance away from the generator's aerial, whose radio Rx is illustrated in Fig. 1.1 (Right). At the moment when the resonator picked up the waves, hardly perceptible sparks were produced in the resonator gap that could be detected using a magnifying glass. Therefore, it is after Hertz that the new discipline of radio technology is sourced and after whom the frequency and its measuring unit hertz (Hz) are named.

An English physician, Sir Oliver J. Lodge using the ideas of others, realized that the EM resonator was very insensitive, and because of this phenomenon, he invented a "coherer." A much better coherer was built and devised by a Parisian

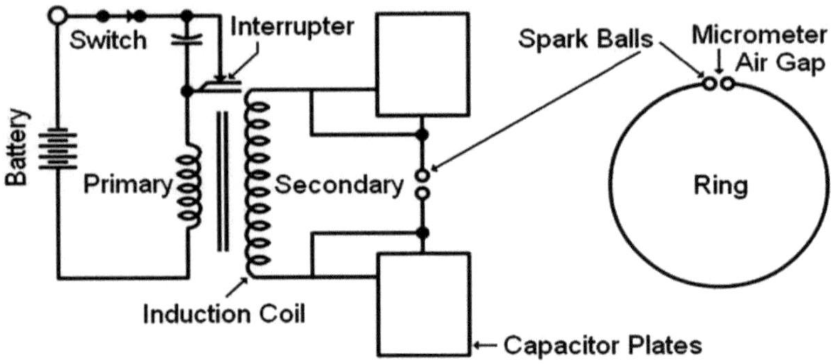

Fig. 1.1 Hertz's spark gaps Tx and Rx—Courtesy of Brochure: by Aatman

professor, Edouard Branly, in 1890. He put metal filings (shut in a glass tubule) between two electrodes, and so a great number of fine contacts were created. This coherer suffered from one disadvantage, and it needed to be "shaken before use." Owing to imperceptible electric discharges, it always got "baked" and blocked.

In early 1889, the Russian professor of physics Aleksandar Stepanovich Popov conducted experiments along the lines of Hertz's research and successfully realized the first practical experiments with EM waves for the transmission of radio signals. Soon after, Professor Popov attended a meeting of the Russian Physical-Chemical Society at which the St. Petersburg Professor N. G. Yegorov reproduced Hertz's experiments, but in a manner that Popov felt to be insufficiently graphic. It took him only a few months to build a more compact and effective device to demonstrate Hertz's experiments, which he then gradually improved on, so that by 1894 he had constructed a working transmitter that generated electromagnetic waves based on Hertz's vibrator and using a Ruhmkorff coil.

This was the first transmission and reception system for remittent electronic waves suitable for the reliable communication of information in the history of telecommunications. The method of Popov invention became the prototype for the subsequent first generation of radio communication systems. At that time, pioneer of radio equipment Popov built his first radio receiver (Rx) with improved version of coherer designed by both Sir Lodge and Branly, so using new inventions, he practically realized the first experiments ever in the world with EM waves for the transmission of radio signals.

As Hertz's early experiments, hardly had Lodge's paper been published in the journal Electrician in July 1894 than his experiments with the coherer were repeated by physicists around the world, but only Popov was able to obtain a practical result. In 1895, he created a system in which the device was mounted on a wooden base and placed in a metal cage to screen it from electrical interference. The aerial used was either a vertical length of wire or an asymmetrical vibrator with a parabolic mirror. The circuit proposed by Popov was also used for thunderstorm detectors and telegraph transmission.

In 1895, Popov upgraded the coherer's sensitivity and invented a special mechanism to automatically reset the device. In fact, he improved Branly's radio Rx by the insertion of choke coils on each side of the relay to protect the coherer and also by replacing the spark gap with a vertical antenna insulated at its upper end and connected to the ground through the coherer. He then mounted a small bell in serial connection with the coherer's relay anchor, whose ringing effected its automatic destabilization and successive unblocked function of the receiver system. On May 7, 1895, he demonstrated his new apparatus to the members of the Russian Physical-Chemical Society: a lightning conductor as an antenna, a metal filings coherer, and detector element with telegraph relay and a bell. The relay was used to activate the bell, which announces the occurrence of transmitting signals and in this way it serves as a decoherer (tapper) to prepare the receiver to detect the next signals. This was the first telegraph station in the world, which could work without any wires.

Popov succeeded in making a reliable generator of EM waves, when the detecting systems in common use still were not at all satisfactory. So his system was extended to function as a wireless telegraph with a Morse telegraph key attached to the transmitter (Ruhmkorff coil and blocking capacitor), shown in Fig. 1.2 (Left). In May of the same year, instead of a bell he contrived to use a clock mechanism to realize direct, fast destabilization of metal filings in the coherer upon receipt of the signals. In the same year, he succeeded in making a more reliable generator of EM waves, when the receiving (detecting) systems in common use still were not at all satisfactory. Thus, using the inventions of his predecessors and on the basis of his proper experiments, Popov elaborated the construction of the world's first radio receiver (coherer, amplifier, electric bell, aerial) with a wire-shaped antenna system in the air attached to a balloon, whose sample is shown in Fig. 1.1 (Right).

In December 1895, he officially announced the success of a regular radio connection and on March 21, 1896, at the St. Petersburg University, and he demonstrated it in the public. Finally, on May 24, 1896, Popov installed a pencil instead of the bell and sent the first wireless message in the world at the distance of 250 m

Fig. 1.2 Popov's spark gaps Tx and Rx—Courtesy of Paper: by Borisova

between two buildings, conveying the name “GENRICH GERZ” (the name of Hertz in Russian) by Morse code using his homemade transmitter and receiver.

In March 1897, Popov equipped a coastal radio station at Kronstadt and the Russian Navy cruiser Africa with his wireless apparatus, and in summer 1897, Popov started to make experiments at sea, using radios on board ships before the entire world. In 1898, he succeeded in relaying information at a distance of 9 km and in 1899, at a distance of 45 km between the island of Gogland and the city of Kotka in Finland. With all his inventions, Popov made advances on the discoveries of Hertz and Branly and created the groundwork for the development of maritime radio.

In 1895, a few months later following the work of Popov, a young Italian experimenter, Guglielmo Marconi, started to use radio and was the first to put EM theory into business application. A detailed report on the Marconi’s experiments was presented by chief of UK engineer telegraph agency V. Pris (1834–1913), who refused to assist him in the works in England (Forum—Yakimenko S). This report shows that the transmitter of G. Marconi was designed by his Italian teacher A. Rigi and that he used the receiver of Professor Popov. Apparently, V. Pris in his report about the invention of G. Marconi was forced to point out that it has been said previously: “G. Marconi did not do anything new.” In conclusion, it was stated that unlike Popov, Marconi was not inventor of radio, but good businessman able to compose something already done, and was able to turn his research into a financial and manufacturing empire.

By the next year, he had sent Morse code messages at a distance of 2 miles. Moving to England to obtain patents on his equipment, he demonstrated radio reception over 8 miles (see photograph of Marconi with his first radio Tx and Rx equipment in Fig. 1.3). In 1897, he exhibited the use of radio between ship and

Fig. 1.3 Marconi and his radio—Courtesy of Paper: by Bertrand

shore and according to Western literature practically started the use of maritime radio. In the same year, Marconi succeeded in getting his wireless telegraphy transmissions officially patented for the first time in the world. The owners of the Dublin Daily Express in Ireland invited Marconi to conduct wireless reports of the Kingston Regatta of July 1898 from the steamer Flying Huntress, the first ship equipped with a commercial wireless system. Using an antenna hung from a kite to increase the effective height of his masts and a LF of 313 kHz at 10 kW of power, on 12 December 1901 Marconi crackled out the first wireless message to span an ocean in the form of Morse code; three dots forming the letter "s". This telegraph signal was sent from Newfoundland in Canada and received in Cornwall on the west coast of England.

In 1897, Marconi registered the company Wireless Telegraph and Signal Company, which then in 1900 was renamed Marconi's Wireless Telegraph Company, Ltd (MWCT). This company started to design and manufacture radio telegraphs transmitting and receiving equipment including antenna, other accessories, and devices for transmitting distress messages for shipping industries and later for commercial ship to shore and vice versa direction at the turn of the twentieth century. In Fig. 1.4 (Left) is shown Marconi of America QMS 1/2 KW Quenched Spark Gap Ship Transmitter in period of 1919, in Fig. 1.4 (Middle Above) is illustrated Marconi America Type-I Antenna Switch, in Fig. 1.4 (Middle Below) is shown Marconi Wireless Keys from 1912, and Fig. 1.4 (Right) **is illustrated** Marconi's Wireless Marine Rx-Tuner Type 226A from 1916.

In 1900, R. Fessenden made the first transmission of voice via radio in the USA, Fleming in 1904 discovered the diode valve, while their countryman and pioneer Lee de Forest developed and used a triode valve, which made it possible to use radio not only for radiotelegraphy but also for voice communications. As early as 1907, he installed a triode valve mobile radio on a ferryboat operating on the Hudson River near New York City.

Fig. 1.4 Marconi's ship radio equipment—Courtesy of Book: by Jenkins

1.2.1 Development of Mobile Radiocommunications

The very impressive development of mobile radio for maritime use at first and later on for aero applications initiated mobile distress and safety radio. Once the principles of radio were understood, mobile radio has been a matter of the steady development of technology to extend communications accessibility, coverage, and reliability by reducing the size, cost, and power consumption of equipment and improve efficiency. With further innovations, an age-old barrier between ships and shore was eliminated and possibility to communicate with mobile radios independent of space and time was created. These early radio devices were primitive by today's standards, incorporating spark transmitters, which blasted their signals across almost the entire radio spectrum. It is supposed that the first vessel to have a ship radio station (SRS) was the American liner St. Paul, equipped in 1899. The next one, early in the following year, was the German vessel s/s Kaiser Wilhelm der Grosse.

Thereafter, mobile radio spread rapidly throughout the shipping and safety business. By 1899, however, A. S. Popov had been the first in the world who successfully carried out demonstrations of mobile wireless telegraphy communications at a distance of 20 miles between warships of the Black Sea fleet. The first recorded use of radio for saving life at sea occurred early in March 1899. The lightship on the Goodwin Sands near Dover on the south coast of England was fitted with one of the first seaborne Marconi SRS and used it to report to the coastal radio station (CRS) that the German steamer Elbe had run aground.

The first distress signal was CQD (come quick distress), used from 1904 only on British ships equipped with Marconi devices. After the collision of two passenger ships, the British s/s Republic and the Italian s/s Florida, running in thick fog in the early hours of January 23, 1909, the radio officer on board the s/s Republic sent for the first time in history the distress signal: "CQD MKC (call sign of s/s Republic), CQD MKC, CQD MKC" and text: "Republic rammed by unknown steamship, 26 miles southwest of Nantucket, badly in need of assistance."

After the catastrophe of the White Star liner s/s Titanic in the early hours of 15 April 1912, the UK proposal for a distress signal was the already established CQD, the USA proposed the NC of International signal codex and Germany preferred SOE. In Fig. 1.5 (Left) is shown Marconi's ship radio station used onboard s/s Titanic. Two men as radio officers employed by the Marconi Company worked in this small windowless room near the bridge. Their main job was to receive and send messages by radio waves using Morse code. Little did they know they would soon be tapping out one of the first SOS signals from a ship in distress and that will happen disaster of their ship Titanic.

However, after the disaster of Titanic, new radio regulation and recommendations required more effective communication equipment and unique distress signals and rescue procedure, so in Fig. 1.5 (Right) is illustrated the prototype of more reliable ship's radio stations used since 1940.

Fig. 1.5 Marconi's ship radio station—Courtesy of Paper: by Richardson/ROA

By 1912, there were 327 established CRS and 1924 SRS available for public, commercial, and safety use. The use of radio at sea became very attractive and indispensable, creating an immediate need for enforcement of the rules and regulations under international radio coordination. Because of this, the first Preliminary Radio Conference was held in Berlin in 1903, where some of the basic principles for the use of radiotelegraphy at sea were established. In the subsequent Berlin Conference of 1906, two radio frequencies, 500 and 1000 kHz, were earmarked for correspondence. This conference also established a standard for international distress signals in radiotelegraphy, SOS.

The radiotelephony distress signal MAYDAY was adopted in 1927 at Hanover. The name of this signal derives from the French phrase "M'aidez," which means "Help me." The first international SOLAS (Safety of Life at Sea) Convention was adopted during 1914 in London, partly as a result of the Titanic disaster. It stipulated Morse telegraphy radio on 500 kHz and auxiliary battery-operated backup radio transceiver unit on 500 kHz as well. In addition, ships carrying more than 50 passengers were required to carry radio devices with a range of at least 100 Nm, and larger ships had to maintain continuous radio watch with a minimum of 3 radio officers.

At the conference in Washington in 1927, the regulations were established as a supplement to bring into force three safety calls in radiotelegraphy for distress, urgency, and security, SOS, XXX, and TTT, respectively. These three signals were obligatory only on 500 kHz, with a silence period of 3 min after every 15th and 45th minute. At that time was also introduced the use of radiotelephony at sea, and soon after that, the first radiotelephone communications between s/s America and coastal radio station Deal Beach, New Jersey, in the USA was realized.

Then, at the Radio Conference in Madrid, radio stations' call signs and frequencies were determined, and the International Telecommunications Union (ITU) was established and Radio Regulations (RR) adopted. At the conference in Atlantic City in 1947, a supplementary ITU RR was adopted and a new radiotelephony distress frequency on 2182 kHz accepted, instead of the old one on 1650 kHz, with silence periods of 3 min after every 00th and 30th minute. Three telephone safety calls were previously used for distress, urgency, and security such as MAYDAY, PANPAN, and SECURITE, respectively, on 2182 kHz and, more recently, on 156.8 MHz (16 VHF channel).

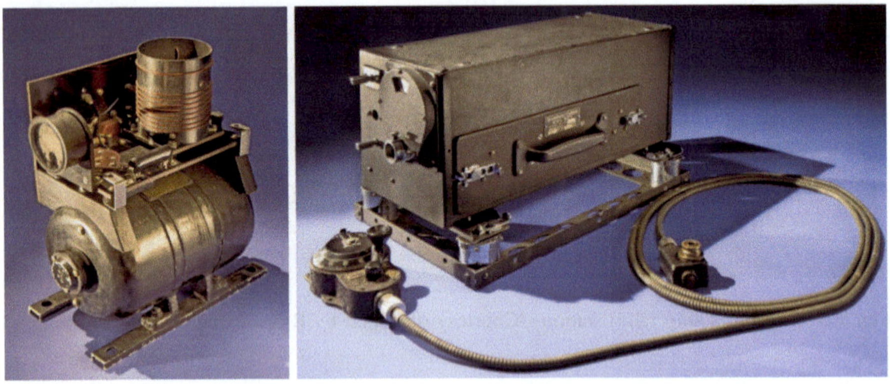

Fig. 1.6 First airborne radio Tx and Rx—Courtesy of Brochure: by Smithsonian

Finally, the new era of transistors commenced, and later on, the period of revolutionary integrated circuits started after 1957. In the meantime, was the change to frequency instead amplitude modulation with a new ARQ system for use in maritime radio telex services.

On the other hand, airmen started to follow success in the use of radio at sea, so in 1928 was developed first aircraft radio Tx, shown in Fig. 1.6 (Left). This was the first lightweight radio transmitter built for use on aircraft. It featured a loop antenna, which could be turned to find the signal direction. It supplanted traditional visual dead reckoning navigation methods. Designed and built by Hugo Leuteritz of Pan American, the transmitter enabled aircraft to navigate accurately along Pan Am's first route, between Key West, Florida, and Havana, Cuba, in 1928. Leuteritz designed other electronic devices as well that allowed Pan American aircraft to navigate safely throughout the Caribbean and across the Pacific and Atlantic. The first aircraft radio Rx was designed by the Aircraft Radio Corporation in 1929, whose ARC Model D is shown in Fig. 1.6 (Right).

The Bendix Radio manufacturing company of airborne radio equipment enjoyed a very close association with the US Navy as early as the mid-1930s. At the beginning, this company designed Bendix TA-12C HF AM Aircraft Transmitter four channels HF Tx first produced in the 1940s and fitted to USAAF B-17 Flying Fortress and RAF Coastal Command B-24 Liberator aircraft in WWII.

The Bendix TA-12C Tx, as shown in Fig. 1.7 (Left), operates in four wavebands from 300 to 600 Kc/s, 3 to 4.8 Mc/s, 4.8 to 7.68 Kc/s, and 7.68 Kc/s to 12 Mc/s. Each band has a separate oscillator and tuning arrangement so that four channels can be preset. Channel selection is made using a motorized changeover, switched from remote control unit MT-51B fitted at the wireless operator's and pilot's positions.

On order from the British Air Ministry (AM), Marconi Wireless Telegraph Company in 1939 started the development of a transmitter/receiver system, the T1154/R1155, based on the AD77/AD6872 that was used in passenger aircraft. By that time, Marconi Company was the leading manufacturer of aircraft radio in

Fig. 1.7 Aircraft radio equipment of Bendix & Marconi—Courtesy of Brochures: by MOYMK/Jenkins

England and perhaps in the world. Thus, the system was initially intended for use in heavier military aircraft, so in the beginning of 1940, the Ministry of Aircraft Production (MAP) accepts the design and the first production set is installed and tested in June of the same year, whose Rx type R1155 is shown in Fig. 1.7 (Right). The R1155 is a single super with one HF stage and oscillator/mixer, a total of three tuned circuits, and two IF stages. With detector, BFO, and AF stages, there are a total of 6 valves plus a magic eye as tuning indicator. There are facilities for using this Rx as direction finding and homing, and the D/F part has 3 valves. Although rather simple, the receiver has a remarkably good performance for short- and medium-wave world radio listening.

1.2.2 Evolution of Satellite Communications

The first known annotation about devices resembling rockets is said to have been used by Archytas of Tarentum, who invented in 426 B.C. a steam-driven reaction jet rocket engine that flew a wooden pigeon around his room. Apparently, the first rockets owe their origin to the invention of gunpowder in China around the tenth century. Devices similar to rockets were also used in China during the year 1232, during the siege of Beijing (according to another source: town of Kai-fung-fu) by the Mongols; the city's defenders fired missiles. It is believed that around the thirteenth century, knowledge of rocketry reached Italy and France.

All through the thirteenth to the fifteenth centuries, there were reports of many rocket experiments. In England, a monk named Roger Bacon worked on improved forms of gunpowder that greatly increased the range of rockets. In France, Jean Froissart found that more accurate flights could be achieved by launching rockets through tubes. Froissart's idea was the forerunner of the modern bazooka. Joanes de Fontana of Italy designed a surface-running rocket-powered torpedo for setting enemy ships on fire.

In the meantime, human space travel had to wait almost a millennium, until Sir Isaac Newton's time, when we understood gravity and how a projectile launched at the right speed could go into Earth orbit. By the sixteenth century, rockets fell into a time of disuse as weapons of war, though they were still used for fireworks displays, and a German fireworks maker, Johann Schmidlap, invented the "step rocket," a multistage vehicle for lifting fireworks to higher altitudes. The use of rockets near the Russian (today Ukrainian) city of Belgorod is recorded in 1516, and the first appearance of rockets in the Russian city of Usury dates around 1675. Following the development and use of military missiles in Europe, the "Rocket Enterprise" was founded in Moscow around 1680 to provide the similar experiments. A signaling rocket developed in Russia in 1717 could reportedly reach an altitude of several hundred meters.

In 1815, Russian artillery engineer Alexander Zasyadko started the development of 3 size types of battlefield missiles for the Russian army with range up to 2700 m. In 1849, another artillery engineer Konstantin Konstantinov developed rockets reaching a range of 4–5 km. Rocket experimenters in Germany and Russia began working with rockets with a mass of more than 45 kg. Some of these rockets were so powerful that their escaping exhaust flames bored deep holes in the ground even before liftoff.

Finally, the twentieth century came with its great progress and the historical age of space communications began to unfold. Russian scientist Konstantin Tsiolkovskiy (1857–1935) published a scientific book on virtually every aspect of space rocketing. He propounded the theoretical basis of liquid-propelled rockets, put forward ideas for multistage launchers and manned space vehicles, space walks by astronauts and a large platform system that could be assembled in space for normal human habitation. Although the Russian scientific elite was primarily concentrated in centers such as Moscow and St. Petersburg, far from two capitals, a modest physics teacher conceived some of the most remarkable ideas about the future of the human race. Almost unknown to its contemporaries, Tsiolkovskiy's work decades later was universally accepted as a theoretical foundation of modern astronautics.

In 1917, the Bolshevik revolution spurred profound changes in Russia. The new leaders of Soviet Russia conducted a ruthless policy of turning an agrarian peasant society into an industrial power. At the same time, the Soviet government spared no effort, in equipping the Red Army with new weapons. In line with this strategy, on March 1, 1921, the Soviet authorities created Gas Dynamic Laboratory (GDL) for research in rocketry, to be led by Nikolai Tikhomirov. Tikhomirov had started studying problems of solid- and liquid-fueled rockets as early as 1894, and in 1915, he patented "self-propelled aerial and water-surface mines." Soviet researchers at GDL worked tirelessly on perfecting military missiles and developing new types of solid rocket fuel, which would allow the new weapons to compete with artillery.

In early 1924, two new Soviet (Russian) rocket programs were founded. They were the Central Bureau for the study of the problems of rockets (TsBIRP) and the All-Union Society for the Study of Interplanetary Flight (OIMS). The main leader of these early Russian rocket efforts was Fridrikh Tsander, who developed liquid-propelled rockets in the 1920s and 1930s.

A little later, the American Robert H. Goddard launched in 1926 the first liquid-propelled engine rocket. The rocket 1.22 m high named "Nell," the world's first liquid-fueled rocket got an altitude of 12 m, a speed of 60 miles per hour, and a distance of 46 m. The German society for space travel or VfR was founded in June of 1927, and it soon had around five hundred members and the first its journal "The Rocket."

At the beginning of the 1930s, the Soviet government sanctioned the creation of several research groups, which united rocket enthusiasts in organizations known as GIRD (Group for Investigation of Reactive Movement) established in Moscow, Leningrad, and other Soviet cities. In Moscow, thanks to the efforts of Sergei Korolev and Fridrikh Tsander, the government-sponsored society for the Advancement of Defense, Aviation and Chemical Technology, Osoaviakhim, agreed to fund GIRD. However, after Stalin took over the Soviet country, these rocket programs became Len-GRID and Mos-GRID, or "Group for the study of reactive motion" based in Leningrad and Moscow, respectively. These two centers, in turn, soon became the State Reaction Scientific Research Institute.

In January 1933, Tsander began development of the GIRD-X missile. It was originally to use a metallic propellant, but after various metals had been tested without success, it was powered by the Project 10 engine, which was first bench-tested at the beginning of March 1933. This design involves burning of liquid oxygen and gasoline and was one of the first engines to be regeneratively cooled by the liquid oxygen, which flowed around the inner wall of the combustion chamber before entering it. However, problems with burn-through during testing prompted a switch from gasoline to less energetic alcohol.

Tsander died unexpectedly from an illness on March 28, 1933, and his engineer, Leonid Konstantinpvich Korneev, became the new leader of his research Brigade. According to the proposal of Stalin, the Revolutionary Military Board established on September 21, 1933, a brand new institution RNII—Реактивный научно-исследовательский институт (РНИИ). The activities of the new institute began on October 31 by merging of GDL and GIRD. In the beginning, the works on a rocket glider were not a part of RNII activities, also the development of rocket engines using a liquid propellant was also not in the focus of activities, and in fact, the main activities were focused on military rockets, using solid fuel.

The first official Soviet rocket launch was the GIRD-9, on August 17, 1933, which reached the modest altitude of 400 m. Then, on November 25, 1933, GIRD-X (ГИРД-Х) became the first Soviet liquid-propelled rocket launched, shown in Fig. 1.8 (Left). The final missile was 2.2 m long by 140 mm in diameter and had a mass of 30 kg, and it was anticipated that it could carry a 2-kg payload to an altitude of 5.5 km. On the other hand, Soviet rocket scientists brought closest to the development of an unmanned cruise missile; thus, on June 16, 1936, Korolev presented project of the "Object 218" powered by liquid-fueled rocket engine and intended for stratospheric flights. On January 29, 1939, a small winged rocket took to the sky at a test range near Moscow code named as Vehicle 212, shown in

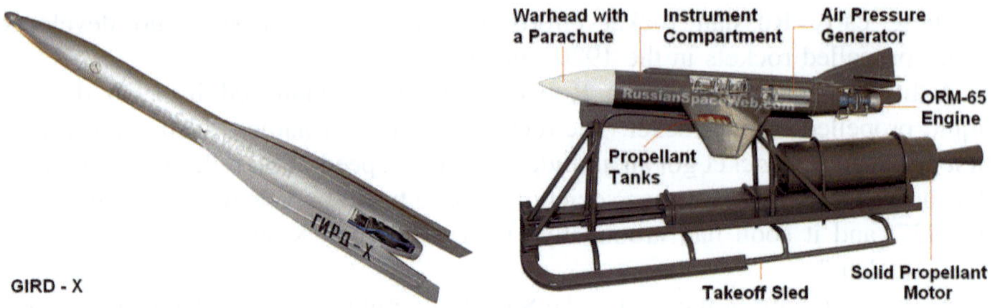

Fig. 1.8 Russian early rocket and winged rocket—Courtesy of Brochures: by Zak

Fig. 1.8 (Right). Vehicle 212 sported many elements of its future successors, such as a three-axis gyroscopic autopilot.

The development of this rocket was also a backdrop for crucial debates among Soviet rocket pioneers, and the missile's arrival on the launchpad coincided with the most horrific events in the Soviet history and in the life of Korolev, the founder of the Soviet space program. He was the main designer responsible for the rocket and who would later go on to mastermind all the early Soviet space triumphs, including wing rockets RP-318, RP-318-1, SK-9, Project 05 used the ORM-50 engine developed and rocket R-7, later Soyuz rocket, Sputnik, Yuri Gagarin's Vostok, and many more. His team attempted to modify Vehicle 212 into an aircraft-launched missile, capable of hitting "primarily" aerial targets as well as those on the ground.

As Tsiolkovskiy was the first to discover that the reaction principle could reach space, it would make sense that the Russians would provide very significant developments in rockets. He published a scientific book on virtually every aspect of space rocketing. He propounded the theoretical basis of liquid-propelled rockets, put forward ideas for multistage launchers and manned space vehicles, space walks by astronauts and a large platform system that could be assembled in space for normal human habitation.

Between the two world wars, Russian and former USSR scientists and constructors used the great experience of Tsiolkovskiy to design many models of rockets and in 1939 to build the first reactive weapon rockets ever, the so-called Katyusha, which Soviet Red Army used against German troops at the beginning of the Great Patriotic War (Second World War).

Thus, toward the end of the Second World War, many military constructors in Germany started with experiments to use their series V1, similar to Soviet wing rockets, and V2, similar to already developed Russian rockets to attack targets in England. Among the people working at these projects was Wernher von Braun as a main constructor, who after Second World War was to work for the US Space Program.

In 1947, Sergey Korolev created one of the most innovative management mechanisms in the early Soviet missile program known as the Council of Chief Designers shown in Fig. 1.9. This photograph, a still from a rare film from the postwar years, shows the members of the council at a meeting: from the left, Boris

Fig. 1.9 Council of Soviet Chief Designers—Courtesy of Manual: by NATO

Chertok, Vladimir Barmin, Mikhail Ryazanskiy, Korolev, Viktor Kuznetsov, Valentin Glushko, and Nikolay Pilyugin (standing), all Russian brains who put for the first time in the history of human being a ballistic rocket in the space.

After that, in October 1945, the British radar expert and writer of science-fictional books Arthur C. Clarke proposed that only three communications satellites in geostationary Earth orbit (GEO) could provide global coverage for TV broadcasting.

The work on rocket techniques in Russia and the former USSR was much extended after the Great Patriotic War, thanks to enthusiastic and productive work of the Russian great rocket constructor and spacecraft designer Sergei Korolev. In 1951, Soviet Union realized the first launch of the "geophysical" rocket carrying live animals onboard.

The satellite era began when the Soviet Union shocked the globe with the launch of the first in the world artificial satellite, Sputnik 1, on October 4, 1957, shown in Fig. 1.10 (Left). Sputnik contained two radio transmitters, which sent back the "beep–beep–beep" signals heard round the world, and in such a way, it provided the first satellite communication link with the control center on the ground. Then a month later only, on November 3, 1957, an even larger and heavier satellite, Sputnik 2, carried the dog Laika into orbit. This launch marked the beginning of the use of artificial satellites to extend and enhance the horizon for communications, navigation, weather monitoring, observation, and remote sensing.

This event signified the announcement of the space race and the development of satellite communications and navigation. That was followed on January 31, 1958, by the launch of US satellite Explorer 1, presented in Fig. 1.10 (Right). Explorer contained a cosmic ray detector, radio transmitter, and temperature and micrometeoroid sensors. Therefore, the development of launchers and satellite systems for future cosmic explorations started, and the space race began.

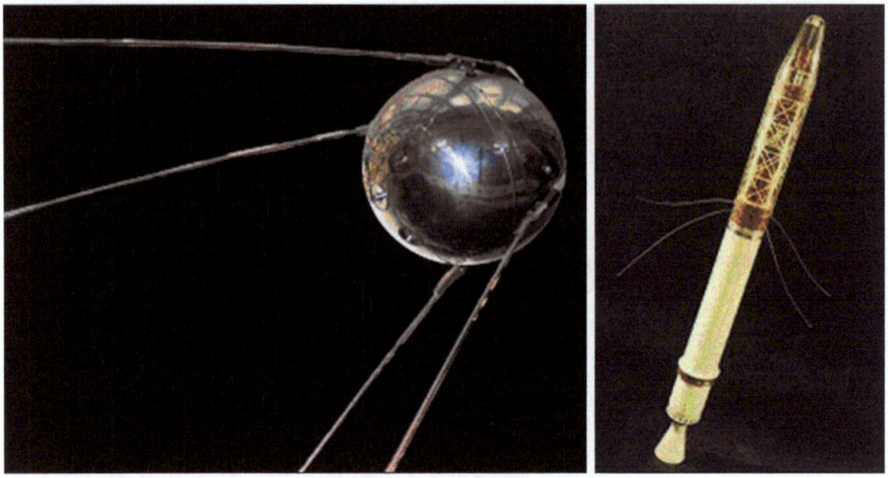

Fig. 1.10 Sputnik-1 and Explorer-1 Satellites—Courtesy of Book: by Ilcev

On January 1–2, 1959, is launched Luna 1 as the first artificial spacecraft to escape Earth orbit. On September 12, 1959, is launched Luna 2 the first man-made object to impact the Moon, and on 3 October of the same year Luna 3 photographs far side of the Moon. On August 19, 1960, two dogs, Belka and Strelka, landed onboard the prototype of the spacecraft Sputnik 5, becoming first animals returning from orbit.

The most significant progress in space technology and engineering was on April 12, 1961, when Yuri Gagarin, officer of the former Soviet Union Air Forces, lifted off aboard the Vostok-1 spaceship from Baikonur Cosmodrome and made the first historical manned orbital flight in space (Fig. 1.11).

In Fig. 1.6 (Left) is illustrated first man flying in the Cosmos Soviet pilot Gagarin and in the same figure (Right) is the picture of the chief designer of Soviet missile and space systems and the founder of modern cosmonautics Sergey Korolev. The twentieth century came with its great progress, and the historical age of space rocketry and satellite communications began to unfold.

1.2.3 Experiments with Active Communications Satellites

After the launch of Sputnik-1, a sustained effort by the USA to catch up with the USSR was started. This was reflected in the first active communications satellite named SCORE launched on December 18, 1958, by the US Air Force. The second satellite, Courier, was launched on October 4, 1960, in high-inclined elliptical orbit (HEO) with its perigee at about 900 km and its apogee at about 1350 km using solar cells and a frequency of 2 GHz.

Fig. 1.11 Gagarin and Korolev—Courtesy of Book: by Ilcev

The maximum emission length was between 10 and 15 min for every successive passage. The third such satellite was Telstar-1 designed by Bell Telephone Laboratories experts and launched by NASA on the July 10, 1962, in HEO configuration with its perigee at about 100 km and apogee at about 6000 km (see Fig. 1.12 (Left)). The plane of the orbit was inclined at about 45° to the equator, and the duration of the orbit was about 2.5 h. Because of the rotation of the Earth, the track of the satellite as seen from the Earth stations appeared to be different on every successive orbit. Thus, over the next two years, Telstar-1 was joined by Relay-1, Telstar-2, and Relay-2. All of these satellites had the same problem and were visible to widely separated Ground Earth Station (GES) for only a few short daily periods, so numbers of GES were needed to provide full-time service.

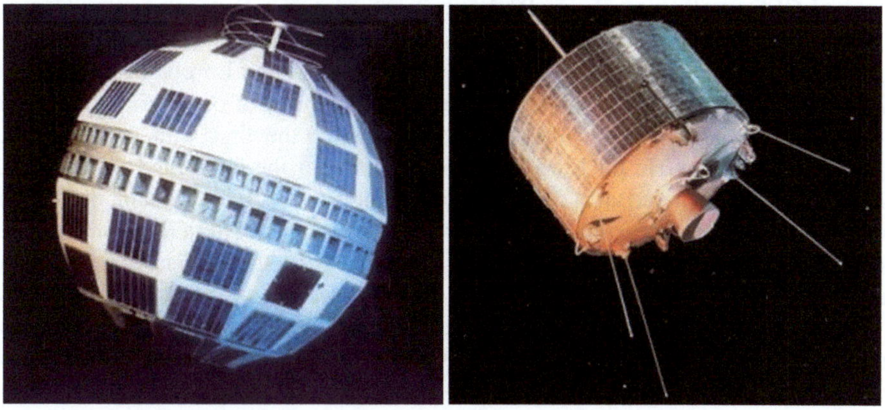

Fig. 1.12 Telstar-1 and Intelsat-1—Courtesy of Brochures: by Launius/Gunter

On the other hand, GEO satellites can be seen 24 h a day from approximately 40 % of the Earth's surface, providing direct and continuous links between large numbers of widely separated locations. The world's first GEO satellite Syncom-1 was launched by NASA on February 14, 1963, which presented a prerequisite for the development of MSC systems. This satellite failed during launch, but Syncom-2 and -3 were successfully placed in orbit on July 26, 1963, and July 19, 1964, respectively.

Both satellites used the military band of 7.360 GHz for the uplink and 1.815 GHz for the downlink. Using frequency modulation (FM) or phase shift keying (PSK) mode, the transponder could support two carriers at a time for full-duplex operation. Syncom-2 was used for direct TV transmission of the Tokyo Olympic Games in August 1964. These spacecraft continued successfully in service until some time after 1965 and they marked the end of the experimental period.

Technically, all these satellites were being used primarily for fixed satellite service (FSS) experimental communications, which were used only to relay signals from fixed Earth stations (FES) at several locations around the world. Hence, one FES was actually located aboard large transport vessel, the USNS Kingsport, home-ported in Honolulu, Hawaii. The ship had been modified by the US Navy to carry a 9.1 m parabolic antenna for tracking the Syncom satellites. The antenna dish was protected, like present mobile antennas, from the marine environment by an inflatable Dacron radome, requiring access to the 3-axis antenna through an air lock within the ship. Otherwise, the Kingsport ship terminal was the world's first true MES and could be considered the first Ship Earth Station (SES). The ITU authorized special frequencies for Syncom communication experiments at around 1.8 GHz for the downlink (space to Earth) and around 7.3 GHz for the uplink (Earth to space). This project and trial was an unqualified success proving only practically the GEO system for satellite communications, but because of the large size of the Kingsport SES antenna, some experts in the 1960s concluded that MSC at sea would never really be practical. However, it was clear that the potential to provide a high-quality line-of-sight path from a ship to the land and vice versa via the satellite transponder existed at this time.

Intelsat was founded in August 1964 as a global FSS operator. The first commercial GEO satellite was Early Bird (renamed as Intelsat-1) developed by Comsat for Intelsat (see Fig. 1.12 (Right)). It was launched on April 6, 1965, and remained active until 1969. Routing operations between the USA and Europe began on June 28, 1965, a date that should be recognized as the birthday of commercial FSS. The satellite had 2 × 25 MHz width transponder bands, the first with 2 Rx uplinks (centered at 6.301 GHz for Europe and 6.390 GHz for the USA) and the second 2 Tx downlinks (centered at 4.081 GHz for Europe and 4.161 GHz for the USA), with maximum transmission power of 10 W for each Tx. This GEO system used several GES located within the USA and Europe, and so, the modern era of satellite communications had begun.

In the meantime, considerable progress in satellite communications had been made by the former USSR, the first of which the Molniya-1 (Lightning) satellite was launched at the same time as Intelsat-1 on April 25, 1965. These satellites were put into an HEO, very different to those used by the early experiments, and were used for voice, fax, and video transmission from central FES near Moscow to a large number of relatively small receive-only stations.

In other words, that time became the era of development of the international and regional FSS with the launch of many communications spacecraft in the USSR, USA, UK, France, Italy, China, Japan, Canada, and other countries. At first, all satellites were put in GEO, but later HEO and polar Earth orbits (PEO) were proposed, because such orbits would be particularly suitable for use with MES at high latitudes. The next step was the development of MSC for maritime and later for land and aeronautical applications. The last step has to be the development of the non-GEO systems of Little and Big low Earth orbits (LEO), HEO, and other GEO constellations for new MSS for personal and other applications.

1.2.4 Early Progress in Mobile Satellite Communications and Navigation

The first successful experiments were carried out in aeronautical MSC. The Pan Am airlines and NASA program in 1964 succeeded in achieving aeronautical satellite links using the Syncom-3 GEO spacecraft. The frequencies used for experiments were the VHF band (117.9–136 MHz), which had been allocated for aeronautical MSC (AMSC).

The first satellite navigation system, called Transit, was developed by the US Navy and became operational in 1964. The great majority of the satellite navigation receivers has worked with this system since 1967 and has already attracted about 100,000 mobile and fixed users worldwide. The former USSR equivalent of the Transit was the Cicada system developed almost at the same time.

Following the first AMSC experiments, the Radiocommunications Subcommittee of the Intergovernmental Maritime Consultative Organization (IMCO), as early as 1966, discussed the applicability of a maritime MSC (MMSC) to improve ship communications. This led to further discussions at the 1967 ITU WARC for the where it was recommended that detailed plan and study be undertaken of the operational requirements and technical aspects of Maritime Mobile Satellite System (MMSS) by the IMCO and CCIR administrations.

A little bit later, the International Civil Aviation Organization (ICAO) performed a similar role to that of IMCO (described earlier) by the fostering of interest in Aeronautical Mobile Satellite System (AMSS) for air traffic control (ATC) purposes. The majority of early work was carried out by the Applications of Space Technology to the Requirement of Aviation (Astra) technical panel. In the proper manner, this panel considered the operational requirements for and the design of suitable systems and much time was spent considering the choice of frequency band. At the 1971 WARC, 2 × 14 MHz of spectrum, contiguous with the MMSC spectrum, was allocated at L-band for safety use.

Hence, the work of the Astra panel led to the definition of the Aerosat project, which aimed to provide an independent and near-global AMSC, navigation and surveillance system for ATC, and airline operational control (AOC) purposes. The Aerosat project unfortunately failed because, whereas both the ICAO authority and world airlines of the International Air Transport Association (IATA) agreed on the operational benefits to be provided by such a system, there was total disagreement concerning the scale, the form, and potential cost to the airlines. Finally, around 1969, the project failed for economic reasons.

The first experiment with land MSC (LMSC) started in 1970 with the MUSAT regional satellite program in Canada for the North American continent. However, in the meantime, it appeared that the costs would be too high for individual countries and that some sort of international cooperation was necessary to make MSS globally available.

In 1971, the ICAO recommended an international program of research, development, and system evaluation. Before all, the L-band was allocated for distress and safety satellite communications and 2 × 4 MHz of frequency spectrum for MMSS and AMSS needs, by the WARC held in 1971. According to the recommendations, Canada, FAA of the USA, and ESA signed a memorandum of understanding in 1974 to develop the Aerosat system, which would be operated in the VHF and L-bands. Although Aerosat was scheduled to be launched in 1979, the program was canceled in 1982 because of financial problems.

The first GEP global MSC system was begun with the launch of the three Marisat satellites in 1976 by Comsat General. Marisat spacecraft had a hybrid payload: one transponder for US Navy ship operating on an UHF-band and another one for commercial merchant fleets utilizing newly allocated MMSC-bands. The first official mobile satellite telephone call in the world was established between vessel oil platform "Deep Sea Explorer" operating close to the coast of Madagascar and the Phillips Petroleum Company in Bartlesville, Oklahoma, USA, on July 9, 1976, using AOR CES and GEO of the Marisat system.

The IMCO convened an international conference in 1973 to consider the establishment of an international organization to operate the MMSC system. The International Conference met in London two years later to set up the structure of the International Maritime Satellite (Inmarsat) organization. The Inmarsat Convention and operating agreements were finalized in 1976 and opened for signature by states wishing to participate. On July 16, 1979, these agreements entered into force and were signed by 29 countries. The Inmarsat officially went into operation on February 1, 1982, with worldwide maritime services in the Pacific, Atlantic, and Indian Ocean regions, using only Inmarsat-A SES at first. Moreover, the Marecs-1 B2A satellite was developed by nine European states in 1984 and launched for the experimental MCS system Prodat, serving all the mobile applications.

In 1985, the COSPAS–SARSAT satellite search and rescue (SAR) mobile system was declared operational. Three years later, the International COSPAS–SARSAT Program Agreement was signed by Canada, France, USA, and the former USSR. In 1992, the Global Maritime Distress and Safety System (GMDSS), developed by the International Maritime Organization (IMO), before IMCO, began its operational phase. Hence, in February 1999, the GMDSS became fully operational as an integration of Radio MF/HF/VHF (DSC), Inmarsat and COSPAS–SARSAT LEOSAR, GEOSAR, and new developed MEOSAR systems.

The Transit system was switched off in 1996–2000 after more than 30 years of reliable service. By then, the US Defense Department was fully converted to the new Global Positioning System (GPS). However, the GPS could not have the market to itself, and the ex-Soviet Union developed a similar system called Global Navigation Satellite System (GLONASS) in 1988 and ceased the previous Cicada system. While both, the Transit or Cicada system provides intermittent two-dimensional (latitude and longitude when altitude is known) position fixes every 90 min on average and was best suited to marine navigation, the GPS or GLONASS system provides continuous position and speed in all three dimensions, equally effective for navigation and tracking at sea, on land and in the air. In the meantime, China started development of own satellite navigation system known as Compass (BeiDou). It consists of two separate satellite constellations that have been operating since 2000 and a full-scale global system that is currently under construction.

While both, the Transit or Cicada system provides intermittent two-dimensional (latitude and longitude when altitude is known) position fixes every 90 min on average and was best suited to marine and aeronautical navigation, the GPS or GLONASS system provides continuous position and speed in all three dimensions, equally effective for navigation and tracking at sea, on land and in the air.

The US Federal Communications Commission (FCC) is reasonably encouraging toward private development of the Radio Determination Satellite System (RDSS), which would combine positioning fixing with short messaging. Thus, in 1985, Inmarsat developed the standard-C system and later examined the feasibility of adding navigational capability by integrating GPS or GLONASS receiver in the mobile unit.

However, the ESA satellite navigation concept, called Navsat, dates back to the 1980s. Moreover, the proposed new project has received relatively little attention and even less financial support from investors. Since 1988, the US-based Company Qualcomm for satellite communications and determination has established the OmniTRACS service for mobile messaging and tracking. Soon after, Eutelsat promoted a very similar system named EuroTRACS integrated with GPS and the Emsat communications system.

At the beginning of this millennium, two-satellite augmentation systems were developed for aeronautical communication, navigation, and surveillance (CNS): the American WAAS, European EGNOS, Japanese MTSAT, Russian SDCM, Chinese SNAS, and Indian GAGAN. The next similar project is ASAS network for Africa and Middle East regions, as first real venture into global satellite navigation systems for Southern Hemisphere.

Hence, it will augment the two military satellite navigation systems now operating, the US GPS and the Russian GLONASS, and make them suitable for safety critical applications such as flying aircraft or navigating ships through narrow channels and port approaches. This system is a joint project of the ESA, the European Commission (EC), and Eurocontrol (the European Organization for the Safety of Air Navigation) and will become fully operational for commercial usage in 2004. The European Union contribution is the Global Navigation Satellite

System (GNSS) as a precursor to a new system known as Galileo. This full GNSS, under development in Europe, is a joint initiative of the EC and the ESA in order to reduce dependency on the GPS service. The target of new Galileo project is to start with operations by 2005 and still is not clear when it will be fully operational.

In the meantime, several very interesting projects are developed in Europe, Japan, and the USA for new mobile and fixed multimedia stratospheric platform communications systems powered by fuel or the Sun's energy and manned or unmanned aircraft or airships equipped with transponders and antenna systems at an altitude of about 20–25 km.

1.3 Development of Global Mobile Satellite Systems (GMSS)

Once the principles of radio systems were understood, mobile radiocommunications have been a matter of steadily developing and perfecting the radio technologies and innovations, extending accessibility and the possibility of radio networks, enhancing range, extending coverage and reliability, reducing the size, cost, and power consumption of radio devices, and improving efficiency.

However, with further MSC innovations, an age-long barrier was eliminated between vessels and shore, vehicles (road and rails) and dispatch centers, and aircraft and airports, and thus, facilities were created to provide mobile offices in ships, land vehicles, and aircraft and to communicate with land Earth stations (LES) independently of space, place, and time.

So, the world is going to reduce communications barriers and move people across borders for business, social, safety, economic, technical, and prosperity purposes. Therefore, the new mobile satellite industry must ensure that mobile communications and navigation services will be responsive to these extraordinary changes and globalization trends.

The MSC systems and technology also offer other benefits and perspectives. In many developing countries, telephone density is still at a low level in urban and non-urban areas, because the cost of upgrading such facilities through wireless or TTN means is prohibitive for much of the world areas. Remote, rural, and mobile service sectors in many regions are outside the reach of communications facilities, so the new MSS technology, with its instant ubiquitous coverage, may provide cost-effective solutions for developing countries.

1.3.1 Definition of Global Mobile Satellite Communications (GMSC)

The GMSC are GEO or non-GEO satellite systems, which refers to all communications solutions that provide global MSC service directly to end users from a

Table 1.1 GMSC position within the telecommunication structure

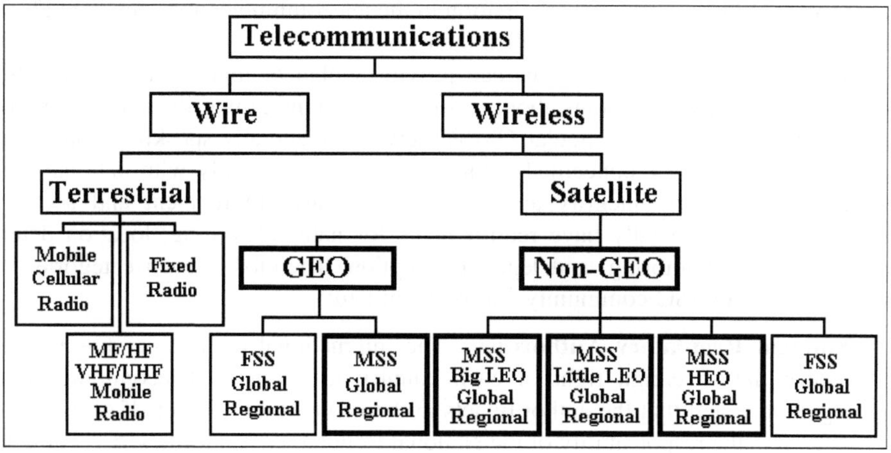

satellite segment, ground satellite network and TTN landline and/or radio infrastructures. The term "GMSC" means not only global coverage but also includes local or regional MSS solutions as an integral part of the worldwide telecommunications village. Namely, some of the regional or local MSS can be afterward integrated to establish a global MSC network. Table 1.1 gives an overview of telecommunication systems showing the respective satellite fit.

The GMSC solution is a modern mobile communication structure system, which began providing communication links to vessels initially in the 1970s and later to aircraft and all kinds of land vehicles. It must be noted that GMSC providing global and regional coverage represents a new technology era in which wire terrestrial and wireless cellular voice, image, video, and data systems are combined with MSC applications to provide services available anytime and anywhere. Additionally, new satellite technologies, such as current Global Mobile Personal Satellite Communication (GMPSC) system and new very small aperture terminals (VSAT), have also allowed global personal and commercial mobility for both satellite communication and determination.

On the other hand, some of the new GEO or non-GEO GMPSC systems have entered the field of MSC solutions, which for the past 20 years has been occupied predominately by intergovernmental satellite organizations. In recent years, a growing number of private entities have been prepared to develop satellite technology, such as Iridium, Globalstar, and Orbcomm. Recently, the modern DVB-RCS solutions are implemented more than a decade ago and are serving fixed and mobile applications.

At the same time, satellite technology and applications continue to advance; so satellite mobile terminals have become smaller, better, and cheaper. Some GMSC systems now being developed are the initiative of the private sector or consortiums.

This implies that there should be changes in policy, particularly in countries that do not foresee sufficient private participation in the telecommunication sectors to allow these systems to thrive and to realize their potential.

As mentioned, GMSC systems can provide global or regional coverage. This capability has raised questions about national sovereignty, integrity, and security of a country covered by a particular GMSC network. Generally speaking, communication networks in the concerned country must always comply with national regulations that govern integrity and assistance to law enforcement and security agencies. These typically have requirements for national routing, location determination, call monitoring, and legal interception. Therefore, seven categories of players in the GMSC community can be identified:

1. **National Regulatory Authorities**—The international community has to recognize the sovereign right of each country to adopt its telecommunication regulations and that the authority acts in the name of and on behalf of a certain state. It is the responsibility of the National Regulatory Authorities, according to their national laws, regulations, and policies, to grant the appropriate authorization to allow GMSC services in a country.
2. **GMSC System Operators**—These are the owners or operators of the space segment, who have assumed all the financial, technical, and commercial risks of developing a GMSC system applications and seek the harmonization of procedures governing the provision of GMSC services to avoid a proliferation of administrative impediments liable to constrain the development of the market.
3. **GMSC Gateway/LES Operators**—Gateways are land Earth stations (LES) or GES links between the space segment and TTN, from which, as main sources, GMSC terminal traffic is drawn. The LES in some cases, depending on the business structure of the GMSC system, can be considered as a part of the space segment and can be managed by the GMSC satellite network operator.
4. **PSTN Operators**—The traditional PSTN operators provide most telecommunication services and networks, both wired and wireless, in a certain country. Furthermore, they are indispensable business partners and responsible for interconnection with terrestrial landline telecommunication networks.
5. **Local and/or Regional Service Providers**—Local service providers are responsible for the local or regional provision of GMSC services, distributing GMSC terminals and billing GMSC customers. Otherwise, GMSC system or gateway operators could also be local or regional service providers.
6. **GMSC Terminal Manufacturers**—These are companies that manufacture MSC and semi-fixed terminals for mobiles, including GEO and non-GEO satellite networks.
7. **GMSC Terminal Users**—These are the customers whom all the other players are called upon to serve. They should receive good-quality service at the best possible price, within the strict confines of the laws and regulations of the host countries.

1.3.2 Definition of Global Navigation Satellite Systems (GNSS)

The first GNSS generation offers augmentation systems in addition to the basic GPS and GLONASS constellations in order to achieve the level of performance suitable for enhanced maritime routing applications worldwide, especially in narrow passages, coastal navigation, and approaching ports, for augmented civil aviation applications in oceanic and approaching flight, for land vehicles, and also for ground positioning and surveying.

The GPS as a first generation of GNSS network is a satellite all-weather, full-jam-resistant, continuous operation radio navigation system, which utilizes precise range measurements from the GPS satellites to determine exact position and time anywhere in the world. This system provides military, civilian and commercial maritime, land, aeronautical, and ground users with highly accurate worldwide three-dimensional, common-grid, position, and location data, as well as velocity and precision timing to accuracies that have not previously been easily attainable. The GPS service is based on the concept of triangulation from known points similar to the technique of "resection" used with a map and compass, except that it is done with radio signals transmitted by satellites. The GPS receiver must determine when a signal is sent from selected GPS satellites and the time it is received. Nothing except a GPS receiver is needed to use the system, which does not transmit any signals; therefore, they are not electronically detectable. Because they only receive RF satellite signals, there is no limit to the number of simultaneous GPS users. The GPS satellite constellation is also providing facilities for new augmentation system known as Global Satellite Augmentation System (GSAS). Today are developed three Regional Satellite Augmentation Systems (RSAS) such as the US Wide Area Augmentation System (WAAS), the Japanese MTSAT Satellite-based Augmentation System (MSAS), and the European Geostationary Navigation Overlay Service (EGNOS). Recently, were projected or are in development stage other four RSAS such as the Russian System of Differential Correction and Monitoring (SDCM), the Chinese Satellite Navigation Augmentation System (SNAS), Indian GPS/GLONASS and GEOS Augmented Navigation (GAGAN), and African Satellite Augmentation System (ASAS) designed by the author.

The Russian Federation (former USSR) provides the GLONASS service from space for accurate determination of position, velocity, and time for mobile or fixed users worldwide and in all weather conditions anywhere. Therefore, three-dimensional position and velocity determinations are based upon the measurement of transit time and Doppler shift of RF signals transmitted by GLONASS satellites. The GLONASS satellite constellation as a first generation of GNSS will also provide signals for all RSAS infrastructures.

At this point, the GPS and GLONASS are parts of the first generation of GNSS known as GNSS-1, and there are in development phase other two GNSS of second generation known as GNSS-2 systems, such as the European Galileo and Chinese Compass.

The GNSS consists of many players with similar GMSC systems and three major segments:

(1) The space segment has 24 satellites (21 functioning satellites and 3 on-orbit spares) and is controlled by a proprietary satellite operator or service provider.
(2) The control segment is operated by master control and monitor stations.
(3) The user segment is represented by the military and civilian authorities for maritime, land, and aeronautical users located worldwide. This segment offers standard positioning service (SPS) and precise positioning service (PPS), available to all users around the world. Access to the SPS does not require approval by a certain service provider, but PPS is only available to authorized users via the service provider administration.

1.3.3 Network Architecture of GMSC

The increased availability of MSS and GNSS solutions means that many mobiles and individuals will have radio connections and determinations at their disposal whenever and wherever they are traveling, including worldwide shipping, long-distance road and railway haulage, transcontinental flights, and universal personal handheld terminals. The new MSC technology is very attractive to mariners, drivers, and aviators alike. As discussed, the MSS began in the 1970s for vessels and ocean rigs; land vehicles have been served at a modest level since the late 1980s, but a stage of rapid development began in the mid-1990s; service to aircraft has started to grow faster than that for land vehicles since 1992. The MSC system consists of space and ground segments, represented by communications satellites, and the ground segment comprises different types of MES and networks.

1.3.3.1 Space Segment and Configuration of MCS Links

The space segment provides the connection between the subscribers onshore and mobile users via LES or gateways. It consists of one or more operational or spare spacecraft in a corresponding constellation. The satellite constellation can be formed by a particular type of orbit, such as GEO, non-GEO (LEO, PEO, and HEO), or combinations of these orbits. The satellites can be independent or connected with each other through intersatellite link (ISL) or interorbit link (IOL). The space segment can be shared among different radio networks in different areas in both time and space. There are also special constellations of multipurpose satellites, whose platform can serve more than two payloads, such as a combination of communication, meteorological, and navigational payloads.

An MSC link is an RF connection between an LES or gateway stations and MES via GEO or non-GEO satellites. The part of the MSC link between LES and satellite is called feeder link, while the link between the MES and the satellite is called the service link. Both feeder and service links consist of an uplink from the ground toward a satellite and a downlink in the opposite direction, and both fall under the category of MSS. The feeder link of the LES is categorized as a fixed utilization in MSS and not at all as a part of FSS as is stated in the MSS Handbook published by ITU, page no. 78. Therefore, the location of LES in MSS is fixed and is not a part of FSS, although some LES in MSS can also provide FSS.

1.3.3.2 Ground Segment and Networks

The ground segment consists of two major network elements: user mobile or portable terminals and ground support stations. The user network comprises four main categories whose characteristics are highly related to its applications and operational environments:

1. **Mobile Earth Stations (MES)**—The MES group of terminals are designed for group usage and installation on board collective transport systems such as Ship Earth Stations (SES) mounted on ships and other floating objects at sea, on lakes, and on rivers; Vehicle Earth Stations (VES) mounted on road or railway vehicles; and Aircraft Earth Stations (AES) mounted on airplanes, helicopters, and other flying objects. Otherwise, the MES terminals are composed of in-mobile and out-of-mobile and outdoor equipment such as above deck equipment (ADE) and below deck equipment (BDE) for ships; above haul device (AHD) and below haul device (BHD) for land vehicles; and above cockpit unit (ACU) and below cockpit unit (BCU) for aircraft.
2. **Personal Earth Stations (PES)**—The PES often refer to handheld or palmtop devices for personal utility and carriage in hand, pocket, or bag. The transceiver and antenna are integrated into one unit together with an additional antenna for cellular roaming. Other PES categories include those situated on board a mobile platform, such as small ships, cars, and airplanes, or they can be transformed as a public payphones in rural areas, ships, oil rigs, and airplanes or can be installed as a fixed office unit with outdoor antenna.
3. **Transportable Earth Stations (TES)**—The TES terminals are typically similar in dimensions to those of a briefcase or laptop computer. As the name implies, these terminals can be transported from one remote or rural site to another; however, operation while mobile will not normally be supported. Every TES contains the transceiver modem and antenna units and can also serve as indoor equipment.
4. **Fixed Earth Stations (FES)**—The FES terminals are similar to public urban payphones or fixed office units in both urban and suburban rural areas. These units with antenna can be mounted in some rural areas for public service, while office units can be mounted like remote household or business sites.

The ground network consists of six main network elements, which support, maintain, and control the space segment and user network as follows:

1. **Land Earth Stations (LES)**—The LES infrastructure is actually a gateway station for MSC service and provides an interface to the satellite access network and existing TTN, such as PSTN/PLMN/ISDN via local exchanges. A single LES can be associated with a particular spot beam or global beam or can access the FSS network. Similarly, LES could provide access to more than one spot or global beam in cases where the coverage of beams overlaps. The main structure of LES can be an external antenna system and the internal sets of RF/IF components, traffic channel equipment (TCE), and gateway subsystems. The synonym LES is usually used for Inmarsat MSS in general, for all types of gateways. The Coast Earth Stations (CES) are for Inmarsat maritime and land applications, while Ground Earth Stations (GES) are for aeronautical applications only.
2. **Network Control Centers (NCC)**—The NCC, also known as the Network Management Station (NMS), is connected to the Customer Information Management System (CIMS) to coordinate access to the satellite resource and perform the logical functions associated with network management (NMF) and control function (NCF) facilities. In fact, the NMF can include the following: development of call traffic profiles; congestion control; system resource management and network synchronization solutions; provision of support in user terminal commissioning; operation and maintenance (OAM) functions; and the management of interstation signaling links.
   The NCF service includes common channel signaling functions, definition of gateway configurations, and gateway selection for any mobile station origination. The CIMS is responsible for maintaining gateway configuration data, performing system billing and accounting, and processing call detail records. Thus, the third element of the entire GMSC network can be the control segment, consisting of NCC, SCC, and CIMS.
3. **Satellite Control Centers (SCC)**—The SCC monitors the current performance of a certain space segment and controls the satellite's position in orbit. The spacecraft control functions include the following elements: generation and distribution of satellite ephemera; generation and transmission of commands for inclined orbit operations, payload, and bus; reception and processing of telemetry; transmission of beam-pointing commands; and performance of range calibration. The call control function is associated with SCC, which includes the provision of real-time switching for mobile-to-mobile calls.
4. **Network Coordination Stations (NCS)**—One NCS serves one ocean region to monitor and control communication traffic within four Inmarsat ocean regions: AORW, AORE, IOR, and POR. Each NCS communicates with LES in its own ocean region using special interstation signaling links, with other NCS sites and with NCC located in Inmarsat headquarters, making possible the transfer of information throughout the system. The NCS is involved in setting up calls to and from MES and in assessing the channel to which both the MES and LES are to tune for the call. For these functions to be performed, each MES Rx must initially be synchronized to the NCS common channel and logged into the NCS for its ocean region.

5. **Local User Terminal (LUT)**—The LUT is a special ground Rx station that receives alert data from COSPAS–SARSAT satellites of old LEOSAT and GEOSAT including new MEOSAT constellations, derives the position of the satellite beacons (EPIRB, PLB, and ELT), retrieves/checks coded information, and forwards the resultant information to the convenient MCC.
6. **Mission Control Center (MCC)**—The MCC is a ground main infrastructure of the COSPAS–SARSAT system which receives alert data from its LUT ground receiving stations and distributes that information to affiliate SAR points of contact or forwards it to other MCC, or vice versa.
7. **Rescue Coordination Centers (RCC)**—The RCC operates a system responsible for promoting the efficient organization of SAR service under the LEOLUT, GEOLUT, and MEOLUT COSPAS–SARSAT system and for coordinating the conduct of distress and SAR operations with other MES within a certain on-scene region. All MES have to be equipped with adequate satellite beacons on board ships (EPIRB), vehicles (PLB), and aircraft (ELT).
8. **Terrestrial Telecommunications Network (TTN)**—The TTN element is a local ground exchange service, which provides an interface between subscribers ashore and gateways, i.e., mobile users.

1.4 GMSC Applications

The present GMSC systems are in use for maritime, land, and aeronautical applications. Recently, several personal mobile multipurpose applications using GEO and non-GEO satellites have been developed and introduced. The lately developed regional networks using stratospheric aircraft and airship platforms will be introduced as very low orbital satellite systems.

1.4.1 Maritime Mobile Satellite Communications (MMSC)

The commercial MMSC systems are designated for very large and medium oceangoing vessels, passenger cruisers, small coastal and river ships, fishing boats, pleasure yachts, and rescue boats. These systems are also available for navy vessels, offshore rigs, and platforms, including any kind of off/onshore infrastructures.

The MMSC system is a successor to the conventional maritime radiocommunications system, which for almost a century was very successful on the commercial and distress scene at sea. In fact, the biggest MSS operator for MMSC is Inmarsat, while other global, regional, or local GEO or non-GEO systems providing MMSC are Iridium, Globalstar, Optus, Emsat, Thuraya, MSAT, AMSC, N-Star, Orbcomm, Leo One, and others who have introduced their own MMSS.

1.4.1.1 Maritime Transportation Augmentation System (MTAS)

The development of the MTAS was to identify the possible applications for enhancement of global DCS MF/HF/VHF radio and satellite communication, navigation, surveillance, and safety systems including transport security and control of vessels and freight at sea, on lakes, and on rivers, and the security of passengers on board cruisers and hovercrafts. These enhancements include many applications for the better management and operation of vessels and they are needed more than ever because of world merchant fleet expansion. Just the top 20 world ship registers have about 40,000 units under their national flags. Above all, the biggest problem today is that merchant ships and their crews are targets of the types of crime traditionally associated with the maritime industries, such as piracy, robbery, and recently, a target for terrorist attacks. Thus, IMO and flag states will have a vital role in developing International Ship and Port Security (ISPS). The best way to implement ISPS is to design a port control system by special code augmentation satellite tracking, monitoring, and surveillance of all vehicle circulation in and out of the port area. The establishment of MTAS will meet most of these requirements and will complement the services already provided by marine radio beacons.

1.4.1.2 Service for MMSC Users

The first-class two-way MMSC will be essential for mariners to contact and constantly exchange information between vessels, owners, agents, shippers, port authorities, families, and friends, to deal with emergencies, distress, and rescue situations at sea. Navy ships can use these facilities for fleet defense, tactical, emergency, and information purposes. Therefore, shippers will be nearer to their fleet units, using not only commercial MSC but also reliable distress and intership communications, and will also have important 24-hour maritime safety information (MSI), such as weather (WX) and navigation warning (NX).

1.4.2 Land Mobile Satellite Communications (LMSC)

The development of LMSC application for vehicles first started with the unsuccessful Canadian MUSAT regional program in 1970. The earliest experiment was carried out in the USA, where it was realized that the cellular system could not economically provide coverage of vast rural areas, as could MSC. After the cellular system was allocated the spectrum of 810–960 MHz, the ITU WARC-97 allocated sections of this band to the LMSS in regions 2 and 3 only. This allocation stimulated much research in North America with the initiation of Canadian MSAT and NASA/JPL MSAT-X projects. The MSAT LMSS would be operated by TMI (Telesat Mobile Inc.), a joint venture between Telesat, Canadian Pacific, and C Itoh.

In the meantime, the US-based Marisat network was developed and the next ESA Prosat program, which started in 1982 with the initial phase, involved a number of propagation experiments via the Marecs AOR spacecraft. Furthermore, the American Mobile Satellite Consortium (AMSC) in the USA and TMI in Canada, were both inaugurated in 1988 and started to collaborate on a low-bit-rate messages of MSC system in 1991, in the same way as Inmarsat-C, by leasing Marisat and Inmarsat transponders. The first AMSC and MSAT spacecraft were launched in 1995 and 1996, respectively. Both establish the interim use of Inmarsat-C for LMSS.

However, this is perhaps one of the major reasons why Inmarsat has decided to reconfigure AOR into two regions, AORE and AORW, in order to improve coverage of the North American land mass.

In Europe, a less ambitious scheme has been initiated such as the unsuccessful Swedish Trucksat project. The geographic nature of Europe is very attractive for LMSS because of the predominant high-mountain localities and many small countries, which are not at all convenient for cellular roaming. Therefore, the ESA has been considering, together with Eutelsat, the possibility of including an L-band transponder on the Eutelsat II spacecraft to provide a space segment for a new MSC solution known as the European Land Mobile Satellite System (ELMSS). The discussions about this project were unsuccessful, and ESA is now considering the use of Italsat as a host satellite.

In addition, Australia has become a new domestic MSS provider via the Australian MSC operator Optus (former Aussat), who decided to include an L-band transponder on their second-generation B1-series spacecraft and to allow the provision of LMSS service in rural areas. They introduced the new MSS Optus in 1992, compatible with both AMSC and MSAT systems, launched for the commercial Australian market in 1994 and provided the AMSC service as well. Similarly, Japan commenced with MSS implementation within the large Experimental Mobile Satellite System (EMSS) program funded by the Japanese Ministry of Post and Telecommunications. In effect, this project has included the launch of the ETS-V satellite, the development, experiments, and trial of various speech and data-only LMSC terminals.

Inmarsat also started in 1988 with LMSC service, proposing its second standard Inmarsat-C two-way data-only system. This standard was initially developed for ships but was later adapted for land and air services. For such reasons, several trials were carried out by ESA throughout Europe to establish its performances. After some time, the MOBSAT Group carried out comparative trials of Inmarsat-C and the ESA Prodat and concluded in early 1989 that while Prodat provides a higher throughput under realistic operating conditions, the Inmarsat-C was preferable due to its more flexible network configuration. Soon after, Inmarsat proposed a low-cost MSC telephony system, known as Inmarsat-M, compatible with the service proposed by AMSC, MSAT, and Optus.

The next standard suitable for LMSC network is Inmarsat Mini-M, designed to exploit the spot beam power of the old generation of Inmarsat-3 and new Inmarsat-4 and -5 satellites, as the smallest, lightest, and most cost-effective MSC unit ever made.

All systems discussed above have been developed to use L-band frequency spectrum via the ITU WARK MOB-97. The new initiative, Qualcomm OmniTRACS system has been developed to use the secondary LMSC location at Ku-band (11/14 GH). This US-based project started in 1986, and trials were carried out two years later. The OmniTRACS system has also been trialed in Europe, using Eutelsat I-F1 spacecraft during the summer of 1989. However, soon after, Eutelsat announced the intention to launch their own service known as EutelTRACS, based on the OmniTRACS CDMA system.

1.4.2.1 Land Transportation Augmentation System (LTAS)

The LTAS network has been set up to identify the possible applications for global radio and satellite communication system, navigation, tracking, determination, safety systems, transport security and control of all vehicles and freight on roads and railways, including the security of passengers in buses and trains.

Namely, these enhancements will comprehend the local and regional road and railway transportation, including a drastic density of vehicles on the roads and railways, which in the future will need some regulation and control using augmentation satellite communication, and positioning and tracking solutions. Therefore, this potential benefit will assist vehicle tracking and control to cope with increased traffic and to improve the safety and control of track lines and signaling.

1.4.2.2 Service for LMSC Users

On a worldwide scale, millions of medium to large trucks, buses, cars, trains, and other types of land vehicles lack any form of in-cab communications with their dispatch bases, owners, agents, families, and friends, or the capability to deal with emergencies such as damage to cargo, engine breakdowns, collisions, and rescue situations on the roads. Now, transport companies can locate their vehicle fleet and stay in touch with them, no matter where and when they roam. In the same spirit, bus and railway companies can always be in contact with their rolling stock and coach and train personnel and passengers will have the possibility to make private phone calls. The best LMSC solutions on board road and railway vehicles will be the utilization of MSC service by two-way low-bit-rate data and message facilities in combination with voice-only services such as Inmarsat-C and Inmarsat Mini-M standards, respectively. Besides, there are similar solutions such as Iridium, Globalstar, Optus, AMSC, MSAT, OmniTRACS, and other systems also suited for Army ground forces.

1.4.3 *Aeronautical Mobile Satellite Communications (AMSC)*

Commercial AMSS is being implemented for aircraft and helicopters, though the different service varies widely depending upon the group or aircraft concerned. The general aviation communities with large aircraft flying worldwide are potential users of AMSS, while light airplanes flying short routes are not now considered to be the potential users of MSS for economic reasons. The Army ground forces are also potential users of AMSC terminals.

1.4.3.1 Development of AMSC

As discussed, the MUSAT system was the first LMSS in the VHF 200–400 MHz band, developed by Canadian specialists and started in 1970, also providing AMSC experiments, mainly for military applications, using the ATS-6 spacecraft. This program was changed to the new MSAT for all three applications in 1980 because the 800 MHz band was allocated to MSS by the WARC-79. The next program was designed by Japanese experts to develop an AMSC and MMSC and started with exploitation in 1975.

The ESA started with the Prosat system in 1982 to contribute to the Inmarsat system. The ICAO embarked on a study into the Future Air Navigation System (FANS) and new concept of ADSS using the MSC system in 1985. The ADSS was developed and finally accepted as the future prime means of aircraft surveillance and control over oceanic and remote regions.

The Oasis study carried out and completed a research project in 1982 by Stanford Research Laboratories, which confirmed the benefits of MSS in terms of improving ATC in ocean areas and recommended that "satellites of opportunity" be considered as a way of easing the financial burden of introducing such a system.

Thus, the Inmarsat organization noted this recommendation and commissioned the Oasis study to investigate the applicability of the Inmarsat system to the AMSS. This study was started by Racal Decca and confirmed that a 200 b/s data service could be provided to aircraft using 0 dBi antennas and the existing Inmarsat space constellation, which was considered for ATC and AOC, including the provision of data service to passengers. The Oasis study was practically validated by Racal at the 1984 Farnborough International Air Show using aircraft-to-ground data links via Inmarsat AOR satellite and British BT Goonhilly GES.

Following the successful events at this air show, the Skyphone consortium was formed by Racal, BT, and British Airways (BA) to design and develop a system capable of supporting preoperational trials of a public telephony service for passengers on a few BA long-haul airplanes.

A similar experiment was performed by Comsat Labs, in conjunction with Mitre, Collins, and Ball Aerospace, in 1985. Simultaneously, ESA was working on the Prosat program that at first included propagation measurements of the aeronautical channels. Furthermore, in the second phase, the ESA program will be renamed Prodat in consideration of the design and prolonged demonstration of the two-way low-rate data transmission system for all GMSC applications.

The USA-based program, developed by the ARINC organization, commenced in 1985. This project proposed the AvSat system to provide integrated voice and data AMSS between air and ground services throughout the world. AvSat developed a system in 1986 based on six satellites with the advanced TDMA scheme. However, this program collapsed because of the excessive costs of deploying a six-satellite constellation, leaving Inmarsat as the only provider of L-band links, so in 1986, Inmarsat confirmed its intention to set up a versatile AMSS and invitations were issued to tender for the development of the avionics.

The world's first aeronautical telephone call from a commercial jetliner over transoceanic flight routes was successfully carried out from a Japan Air Lines (JAL) aircraft Boeing 747 in October 1987 using the Inmarsat space and ground satellite segments. Otherwise, this experiment was jointly carried out by KDD, JAL, and CRL of the Ministry of Post and Telecommunications in Japan. It was based on ETS-V/EMSS research and development program of CRL, which was reorganized from AMES in 1984.

The other significant AMSS voice trials were conducted by the Inmarsat, British Telecom (BT), BA, and Racal Decca companies and conducted a Skyphone service in May 1988, using a fully avionics-compliant package located in the bay of BA Boeing 747 aircraft, not in the passenger cabin.

The AMSS L-band frequencies, which had been exclusively allocated for ATC and ACC, were opened to APC at WARC-87. With these trends, L-band became the most suitable spectrum for integrated GMSC, which provides worldwide service to ships, land vehicles, and aircraft. Since 1990 Inmarsat has carried out a program of AMSC tests, demonstrations and trials later started to lead an AMSC service. Accordingly, large airways companies worldwide can now use Inmarsat for reliable corporate, social, and safety AMSC service.

The Inmarsat system is not ideal for AMSC, in particular because some scheduled flights are using lines far over the North Pole, beyond the coverage of GEO. Meanwhile, this problem can be solved similarly to MMSC, by using integrated Radio and Satellite Mobile Communications under GMDSS and establishing a Global Aeronautical Distress and Safety System (GADSS). On the other hand, by using a combination of GEO and PEO or HEO constellations, the coverage problem for both polar zones will be solved.

1.4.3.2 Present Status of Aeronautical Communications

Business or corporate airlines have for several decades used HF communication for long-range voice and Tlx communications during intercontinental flights. For short distances, nearby airports during approach or departures, all aircraft have used the well-known VHF/UHF radio. Data communications are since recently also in use, primarily for flight plan and worldwide weather (WX) reporting in a form similar to the VHF/UHF Airborne Flight Information Service (AFIS) system. Apart from data service for the aircraft cockpit and cabin crew, cabin voice solutions and passenger telephony have also been developed. On the other hand, some airlines servicing major transcontinental routes have provided free telephone services to passengers as marketing ploys, whereas others want the service to generate additional revenue.

The airline data service requirements are divided into two main areas: AAC-AOC (Airline Administrative and Airline Operational Control) and ATC services. Efficient operation of modern aircraft requires that automatic engine airframe health monitoring and control is available wherever the aircraft is flying. Worldwide access to remote databases while in flight is required for optimum flight profile planning. Therefore, the ARINC (Aeronautical Radio Inc.) program named ARINC Communication and Reporting System (ACARS) and the SITA (Societé Internationale de Télécomommunications Aeronautique) system known as AIRCOM still provide these facilities via VHF radio, although they have started to use AMSC to make these services available worldwide and more effectively.

The success of Inmarsat MMSC and LMSC systems together with the rapid growth of VHF air-to-ground radiocommunications has encouraged the development of the AMSC system. Avionics companies have the best solution to organize MSC links and within their own fleets and also to enable reliable voice, fax, video, image, data, and Internet service for their passengers, crew members, and corporate purposes. At present, Inmarsat is the only global operator providing the AMSC service, while other GEO and non-GEO operators also provide AMSC, such as Globalstar, Iridium, Optus, AMSC, and MSAT.

As with Inmarsat's MMSC system, the AMSC service is provided to users via the Inmarsat signatories, who are predominantly national PTT or telecommunications organizations. In the meantime, several competing service provider agreements have been announced with US-based Arinc, Japanese Avicom, Satellite Aircom Consortium of France Telecom, SITA, Teleglobe Canada and Telstra Australia, Skyphone Consortium of `BT, Singapore's SingTel and Telenor and the Multinational Skyways Alliance led by Comsat.

1.4.3.3 Aeronautical Transportation Augmentation System (ATAS)

The ATAS has been set up to identify the possible applications for global radio and satellite communication, navigation, surveillance, and safety systems including security and control of aircraft, freight and passengers, and SAR service in accordance with ICAO regulations and recommendations. The world's commercial airline fleet is expected to double in the next 20 years, which will result in crowded routes, leading to fuel wastage and delays, which could cost millions of dollars annually. In this sense, the new augmentation system for aeronautical satellite CNS/ATM is designed to assist navigation both en route and during landing and in airports. In fact, the potential benefits will assist ATC to cope with increased air traffic as well as improving safety and reducing the infrastructure needed on the ground.

When planning aircraft routes and landing schedules at busy airports, it is essential to ensure that aircraft are always a safe distance from each other. The trouble is that it is not always possible to know where the planes are. Thus, it is necessary to leave a very large safety margin, but when it is possible to know precisely where the planes are, it will be easy to reduce the margins safely and increase the numbers of planes in each corridor. At any rate, the new WAAS, EGNOS, MTSAT, and forthcoming Galileo GNSS will provide a guaranteed service with sufficient accuracy to allow airlines and pilots to know their current position and safety margins reliably and precisely enough to make substantial efficiency savings.

The GNSS can also help pilots to land planes safely, especially in poor weather and dense fog, in which only with DGPS or the augmented satellite CNS system is reliable. Unfortunately, small airports are unlikely to invest in this system, but they can use local augmented system, or when Galileo becomes operational, the need for a differential antenna will reduce costs. Galileo's guaranteed service and use of dual frequencies will increase accuracy and reliability to such an extent that planes will be able to use its navigational signals for guidance with their onboard technology alone.

1.4.3.4 Service for AMSC Users

New aeronautical services will provide mile-high opportunities to bring AMSC to general aviation planes, commercial airlines, helicopters, their crew members, and passengers. They can use air-to-ground AMSC facilities to optimize commercial and safety communications, fuel, maintenance, and revenue demands. All the communication features found in the office are made available in the airborne environment, with direct dialing and continuous global coverage. The AMSC service allows passengers, pilots, and crew to communicate critical business decisions or social, private, and confidential cockpit and seat phone calls.

The airborne equipment enables direct-dial phone calls, data and messaging services, HSD, Internet, e-mail, e-commerce, videoconferencing, databases, and aeronautical information services (AIS), such as news highlights, WX reports, and stock market information. The GMSC system benefits long-, medium-, and short-haul airline customers by facilitating accurate positional information for better aircraft utilization, more productivity, and good air traffic management and they are very important for increasing airline revenues. In this way, except for crew and passenger service, Aircraft Operation Management, ATC, and HSD/ISDN services through the Swift64 Inmarsat system integrated with GNSS solutions will offer two very important services:

1. **Aircraft Security Control**—All planes can be provided with the service of aeronautical safety information (ASI) via satellites, such as turbulence warnings, weather (WX), and ace reports to assure safe flights. Antihijacking actions are important for the safety of lives and property on board aircraft, as hijacking is

usually violent, causing injury or death among crew and passengers or the loss of valuable cargo. All airports have security control of passengers on all gates, but it will also be important to find out how the AMSC system can provide automatic surveillance and monitoring of aircraft and cargo before departure.

2. **Remote Troubleshooting System (RTS)**—For almost one year, Dassault Aviation has been using the Inmarsat-ISDN solution for a new remote analysis and technical assistance system called Telemaque. It enables the technical support specialists at Dassault Aviation to carry out all diagnoses or repair operations in real time with customer sites located anywhere in the world. Using an audio and video system connected to an Inmarsat-ISDN transmitting (Tx) antenna, the aircraft sends images in real time to the technical support team at Dassault, enabling them to identify the fault precisely. Besides, secure software for application sharing and file transfer enables the two teams to connect up and work on the same documents, such as databases and plans of the aircraft located at the Dassault office. Dassault Aviation started with Global Telemaque Service (GTS) to all its customers in 2001, which offered video camera, a digital camera, an Inmarsat antenna, a PC, an earpiece, and remote maintenance software.

1.4.4 Global Mobile Personal Satellite Communications (GMPSC)

At the beginning of the 1990s, several private US firms proposed new concepts for MSC known as GMPSC, using a group of LEO or MEO spacecraft. Typical systems were Iridium, Globalstar, and Odyssey. In 1991, Inmarsat also proposed the new Project 21, named Inmarsat-P, which would provide GMSC for personal applications using non-GEO systems. After feasibility studies, Project 21 is now going to use the ICO, which is the same as MEO and belongs to ICO Global Communications. With regard to this, in 1992, WARC-92 responded to these activities and allocated the L-band (1626.5–631.5 MHz) and S-band (2483.5–2500 MHz) for MMS using LEO or MEO satellites.

Those people living and working far from the reach of cellular wireless networks, for example, roaming construction engineers, exploration workers, medical staff, journalists, rangers, farmers, fishing boat or yacht crew members, and small planes or helicopters, can also have access to the GMSC network. Namely, every customer can choose to use satellite and/or a convenient cellular network when they are in urban roaming areas, because mobile phones will be dual-mode and triple-mode handsets. Therefore, it is possible to produce dual-mode satellite/GSM handheld phones (the waterproof boat kit solution is optional) and triple-mode satellite/cellular models for D-AMPS/AMPS or for CDMA/AMPS systems.

The GMPSC handsets are lightweight, easy to carry, simple to use, and, similar to current mobile phones, are able to provide a GMSC service with both satellite and cellular network accesses. These phones can offer voice, fax, data, SMS, and voice mail. The dimensions of unit are about 54 × 26 × 145 mm and about 240 g in weight. The graphic display is with fixed icons, and keypad consists of 6 menu-driven soft keys. This unit can support European Community, East European, Chinese, English, and other languages. The delivery features include voice, data, and fax messaging in both satellite and cellular modes. The handset has manual or automatic satellite/GSM mode selection, using a small extending antenna.

The special, rugged GMPSC equipment with external antenna can be fitted on board ships, land vehicles, or aircraft similar to the shipborne, vehicleborne, and airborne Inmarsat solutions. Otherwise, similar equipment is designed for use in rural and remote (R&R) environments as well. Three billion people living in remote villages and households have no access to a cellular or landline network using semi-fixed installations, but providing this telephone service to remote rural areas is often very slow, complex, and quite an expensive project. The GMPSC system can extend the existing telephone utilization to any location in the world, with more products and solutions and with much cheaper equipment and usage charges, providing the following services:

1. **Rural Villages and Remote Communities**—The GMPSC model of rural payphone will offer the ability to communicate with any location in the world using voice facilities and emergency numbers for access to medical, fire-fighting, or security services.
   The terminal will be similar to a typical city payphone operated by phone cards, ruggedly constructed to withstand vandalism, simple to use, and able to indicate the amount of money remaining on the card and will be powered using mains network supply or solar panels and batteries. Additional services include remote fault diagnosis and local support to ensure rapid repair and high availability. This payphone can be used in remote suburban and village environments, in desert and forest areas, on cruise ships, onboard aircraft, sea platforms, and oil rigs. They are easy and cost-effective to install and operate, using prepaid chip cards. Local authorities can use payphones for emergency calls, using a special access card.
2. **Phones for Remote Households**—The GMPSC model of an indoor phone offers every remote household connection to TTN via a standard phone set connected to an interface box and external outdoor antenna mounted on the roof. This box can be linked to standard tel/fax or ISDN/Internet lines, using mains power, 12 V battery, or solar cell energy.
3. **Remote Business Sites**—The GMPSC system is giving an advantage to every business based in suburban and R&R areas, providing the developing and developed world the following services:

(a) Basic GMPSC remote service is suitable for companies requiring voice, fax, LSD, and SMS facilities via a compact system, which includes a small interface box and a compact rod antenna. The system will utilize existing TTN devices easily attached to a company's tel, fax, or PC and is ideal for remote post offices, credit card companies, administrative authorities, small private companies, and companies requiring telemetry and SCADA.

(b) Advanced remote service is specified for companies requiring full telecommunication services such as voice, fax, HSD, and efficient data transmissions and SMS facilities, using a GMPSC interface box and an A4-sized antenna. The interface box can easily be fixed to tel, PABX, fax, or PC, while outdoor antenna can be mounted onto an appropriate on-site building. This GMPSC service is ideal for geological, mining, agriculture, manufacturing, construction, energy and gas exploration companies, local government administration, and other organizations requiring high volumes of computer data service.

(c) Portable GMPSC remote service has been determined for the early stage of exploration and survey work, or for journalists using a portable laptop-sized terminal, which can easily be connected to a telephone handset or PC configuration. This service is ideal for the immediate transmission of data from a remote site to a headquarters or research location, for transmitting urgent copies of visual images or photographs, and for any customer requiring advanced telecommunications with easy installation in remote locations.

4. **SCADA (Supervisory Control and Data Acquisition)**—The Inmarsat-D+, -C, GAN, OmniTRACS or Orbcomm and other SCADA (M2M) equipment or systems may be used in the point-to-multipoint broadcast of Automatic Remote Monitoring and Messaging Data (ARMMD), whose installations are fitted with automatic sensors that regularly report back to a control center via satellite. With its greatly reduced power consumption and by being remotely controlled, this device is a very effective way of remotely collecting basic environmental and industrial data via messages of up to 2000 bits, together with one of four alert signals, which may be sent as a single message. Therefore, the setting up of these M2M systems and devices are all remote industrial and manufacturing installations, oil and gas pipeline pumping stations, offshore platforms, meteorological observatories, water treatment and energy stations, maritime lighthouses and buoys signaling and other remote systems to run unattended fairly easy. Keeping this equipment working efficiently, gathering management data and guarding against remote catastrophic failure, all without overspending on expensive technical labor and travel, is a tougher challenge.

5. **Government Services**—The GMPSC system network will also provide possibilities to organizations involved in fast-moving emergencies and interventions such as disaster relief agencies and SAR organizations, military and international peace-keeping forces, police squads, medical teams, and civil authorities. This service offers assured communications by PC data, voice, fax, SMS, and encrypted messages. For personal use, it will offer dual-mode-function portable handsets for both satellite and cellular connections. On the other hand, for group use, the system will offer mobile handset terminals for fitting to ships, vehicles, aircraft, and temporary accommodation at camps or bases. Therefore, Inmarsat, Iridium, Globalstar, and other current GMSC systems have developed both common and special military mobiles and portable satellite tactical and defense communications equipment.

1.5 International Coordination Organizations and Regulatory Procedures

International coordination in MSC has been carried out by the International Coordination Organizations, which include ITU, IMO, ICAO, IHO, WMO, and MSUA.

1.5.1 International Telecommunications Union (ITU) and Radio Regulations

The ITU organization of the UN with all member governments has carried out the entire international coordination and regulation of mobile radio and satellite communications by the ITU RR. The ITU was inaugurated in 1932 and reorganized in 1992, with head office, all committees, and departments located in Geneva, Switzerland. Numerous provisions of the telecommunication services applicable or useful to all stations have been defined and introduced by the general RR articles and manual on mobile radio service and in a special manual for use by the Maritime Mobile and Maritime Mobile Satellite Services.

For instance, the ITU also publishes many additional lists of recommendations concerning RR, systems, radiocommunications, and other technical publications. The administrative structures established by the ITU Convention comprise a Secretariat headed by the Secretary General, an Administrative Council, Registration Board for RF, and Consultative Committees for radio and telecommunications.

The entire terminology, definition of radio and satellite services, technical standards, and frequency allocations are defined in the RR and drafted by the World Administrative Radio Conference (WARC) of the ITU. In effect, the WARC was reorganized in 1992 as a World Radio Conference (WRC) of the ITU. The basic concept of the present RR relevant to satellite communications was proposed by the WARC-ST (WARC satellite) in 1971 and by all other conferences held during later years. The WARC-92 covers the spectrum needs of numerous telecommunications services, from HF/VHF broadcasting to personal satellite communications networks.

Thus, the requirements for MSC systems below 1 GHz, above 1 GHz, and in the 20/30 GHz bands have dominated the agenda of the conference. After Spain, the USA heads off to the WRC-93 in November armed with a clear mandate to seek more spectrums for MSS. Thus, the USA position for the conference, to be held from 15 to 19 November in Geneva, is to establish MSS on the agenda for the 1995 and 1997 WRC.

The task of the International Radio Consultative Committee (CCIR) is to form study groups to consider and report on the operational and technical issues relating to the use of radio and satellite communications. Hence, the International Telecommunications Consultative Committee (CCIT) offers the same telecommunication services.

The study groups produce new recommendations on all aspects of radiocommunications, which are considered by the Plenary Assembly of the CCIR, and if accepted become incorporated into the RR and published separately. Besides, another subgroup of the ITU is the International Frequency Registration Board (IFRB), which considers operating frequencies, transmitter sites, and the location of satellites in orbit. Consequently, satellite communication systems have been internationally authorized by the IFRB and the present ITU-R (ITU-Radio).

The administration of a country that intends to establish satellite communication systems, operated in GEO, HEO, or LEO configuration, must send to the ITU-R, not earlier than five years before the date of introducing the system, information on each satellite network of the planned service, information such as the frequency bands to be used, the modulation type of signals, and the radiation characteristics of antennas, satellite constellations, and Earth stations. The ITU-R publishes the information in a weekly circular distributed to all relevant administrators around the world. If any administration is of the opinion that interference will be caused to its existing or planned space radiocommunication services, it can, within four months after the publication of the relevant weekly circular, send its comments to the administration concerned. If no such comments are received from anyone within the period mentioned above, it might safely be assumed that no administration has any objections to the planned satellite network, for the system in question, on which details have been published. Accordingly, an organization has been set up to receive comments about registration endeavors to resolve any difficulties that may arise and therefore provides any additional information necessary. Soon after completing all coordinations, the planned satellite system will be internationally authorized by recording it in the Master International Frequency Register (MIFR) of frequency assignments.

1.5.2 International Maritime Organization (IMO) and Regulations

The service and regulations reaching development, studies, and agreements of the Maritime Safety Committee regarding distress and safety at sea are provided by the IMO. It consists of an assembly, a council, and 4 main committees: maritime safety; marine environment protection; and legal and technical cooperation. There is also the Facilitation Committee and a number of subcommittees, which support the work of the main technical committees. Since its establishment in 1959, IMO and all its member governments have striven to enhance the International Convention for the Safety of Life at Sea (SOLAS—1974). In 1972, IMO, with the assistance of CCIR, commenced a study of satellite communications systems, which resulted in the establishment, in 1979, of the Inmarsat Organization. With the continuing support of CCIR, ITU, WMO, IHO, COSPAS–SARSAT and Inmarsat, IMO has developed the GMDSS system, which entered very late into force in early 1999, following almost 30 years of careful preparation.

Shipping is perhaps the most international of all the world's great industries and one of the most dangerous. It has always been recognized that the best way of improving safety at sea is by developing international regulations that are followed by all shipping nations and from the mid-nineteenth century onward, a number of such treaties were adopted. In 1948, an international conference in Geneva adopted an Intergovernmental Convention, formally establishing a new international organization named IMCO (the original name of this organization was changed in 1982 to IMO). The IMO Convention entered into force in 1958, and the new organization met for the first time the following year.

The purposes of the organization are to provide the machinery for cooperation among governments in the field of governmental regulation and practices relating to technical matters of all kinds affecting shipping engaged in international trade, to encourage and facilitate the general adoption of the highest practicable standards and rules in matters concerning maritime safety; efficiency of navigation; and prevention and control of marine pollution from ships.

The first IMO task was to adopt a new version of the SOLAS International Convention, the most important of all treaties dealing with maritime safety. This was achieved in 1960, and the IMO then turned its attention to such matters. In 1974, the International Convention for the SOLAS adopted the SOLAS 74 Convention based on the requirement that certain classes of ships, when at sea, keep continuous radio watch on the international distress frequencies assigned in accordance with the ITU RR and carry corresponding radio equipment capable of transmitting over a minimum specified range. Shipping, like all of modern life, has seen many technological innovations and changes. Some of these have presented challenges for the organization, and others have presented opportunities. The enormous strides and efforts made in new radiocommunications technology, for example, have made it possible for the IMO to introduce major improvements to the maritime distress and safety system. In this sense, in the 1970s, a global SAR system was initiated worldwide.

The 1970s also saw the establishment of the Inmarsat at London, which has greatly improved the provision of commercial and distress radio and other messages to ships. A further advance was made in 1992, when the GMDSS began to be phased in. In February 1999, the GMDSS became fully operational, so that now any ship in distress anywhere in the world can be virtually guaranteed assistance because even if the crew does not have time, the message will be sent automatically.

Other measures introduced by the IMO have concerned the safety of container units, bulk cargoes, gas tankers, and other ship types. Special attention has been paid to crew standards and adoption of a special convention on standards of training, certification, and sea watch keeping. The challenge also facing the IMO and its 162 member states is how to maintain this success at a time when shipping is changing more rapidly than ever before. The revised technical annex of the SAR Convention clarifies the responsibility of all governments and puts greater emphasis on the regional or local approach and coordination between maritime and aeronautical SAR operations. In fact, both parties are also required to ensure the closest practicable coordination between mutual services. Each RCC and Rescue Sub-Center

(RSC) should have up-to-date information on the SAR facilities and communications in the area and should have detailed plans for the conduct of SAR operations.

In addition, concurrently with the revision of the SAR Convention, the IMO and ICAO technical experts jointly developed the International Aeronautical and Maritime Search and Rescue Manual (IAMSAR), published in three separate volumes covering Organization and Management, Mission Coordination, and Mobile Facilities. Therefore, the IAMSAR Manual revises and replaces both the IMO Merchant Ship Search and Rescue Manual (MERSAR), first published in 1971, and the IMO Search and Rescue Manual (IMOSAR), first published in 1978. The MERSAR Manual was the first step toward developing the 1979 SAR Convention and it provided guidance for those who, during emergencies at sea, may require assistance from others or who may be able to provide assistance themselves. In particular, it was designed to aid the master of any vessel who might be called upon to conduct SAR operations at sea for persons in distress.

The second IMOSAR Manual was adopted in 1978. It was designed to help governments to implement the SAR Convention and provide guidelines rather than requirements for a common maritime search and rescue policy, encouraging all coastal states to develop their organizations on similar lines and enabling adjacent states to cooperate and obtain mutual assistance. It was also updated in 1992, with the amendments entering into force in 1993. This manual was aligned as closely as possible with the ICAO Search and Rescue Manual to ensure a common policy and to facilitate consultation of the two manuals for administrative or operational reasons. In such a way, MERSAR was also aligned, where appropriate, with IMOSAR.

1.5.3 International Civil Aviation Organization (ICAO) and Regulations

The consequence of the studies initiated by the USA and subsequent consultations between the major allies was that the US government extended an invitation to 55 member states to attend, in November 1944, an International Civil Aviation Conference in Chicago. In fact, 54 states attended this conference, at the end of which a Convention on International Civil Aviation was signed by 32 states, setting up the permanent ICAO policy and task as a means to secure international cooperation and the highest possible degree of uniformity in regulations and standards, procedures, and organization regarding civil aviation matters.

One of the primary objectives of the ICAO with its corresponding expert committees and departments is to promote the safety and security of civil aviation worldwide. With 188 contracting states and its active involvement in global aviation safety issues, the ICAO is well positioned to assume a coordinating role for safety initiatives for the reduction of aviation accidents. Recognizing this, in 1997, the Air Navigation Commission proposed an ICAO Global Aviation Safety Plan (GASP) to the ICAO Council. The GASP serves to focus the safety-related activities within ICAO on those safety initiatives, either planned or in progress, that offer the best safety dividend in terms of reducing accident numbers and rates worldwide. For that reason is necessary to develop new GADSS similar to GMDSS.

The main objective of ICAO's technical aviation security program is to assure the safety of passengers, aircraft crew, ground personnel, and other members of the general public by first attempting to deny offenders access to aircraft. In accordance with the Chicago Convention (Annex 17), the ICAO Council adopts Standards and Recommended Practices (SARP) for safeguarding of international civil aviation security. The Security Manual for Safeguarding Civil Aviation Against Acts of Unlawful Interference contains guidance material on the interpretation and implementation of the SARP of Annex 17. This Manual and Annex 17 are kept under constant review and amended in light of the new trends and recent technical developments which have a bearing on the effectiveness of preventive measures against acts of unlawful interference. In order to assist any state to implement the aviation security standards contained in Annex 17, the ICAO has developed the Training Program for Aviation Security, currently comprising seven important Aviation Security Training Packages (ASTP). Through its global network of ten Aviation Security Training Centers, ICAO promotes regional cooperation with regard to AVSEC training, the main objectives being to improve the quality of training of aviation security personnel worldwide and to assist technically and financially every active member state with their own national AVSEC training programs. The major objectives of the ICAO GASP are to:

(1) Reduce the number of accidents and fatalities irrespective of the volume of air traffic, achieving a significant decrease in worldwide accident rates, placing emphasis on regions;
(2) Urge all contracting state members worldwide to provide the needed support of the ICAO GASP;
(3) Endorse the concept of concentrating the security and safety-related activities of ICAO on those initiatives, planned or currently under way, that offer the best safety and safety dividend in terms of reducing the accident rate;
(4) Instruct the council and secretary general to participate in efforts by states to improve existing safety/security database systems and the exchange of safety-related information and to participate in activities aimed at the development of a comprehensive data analysis and information dissemination network, taking into account the need to adequately protect privileged information and its sources;
(5) Encourage the free communication of security and safety-related information among users of the aviation system, including the reporting of accident and incident data by states to the ICAO accident/incident data reporting (ADREP) system.
(6) Commence a study of MSC systems to enhance air traffic control, to improve the flow of management data, meteorological, security, and safety information and to participate in the development of the Global Satellite Augmentation System for Aeronautical CNS.

The ICAO and IMO have jointly collaborated to study, research, and develop an important IAMSAR Manual. With the revision of the SAR Convention and previous manuals, they published the IAMSAR Manual, which covers all particulars about SAR programs and procedures for maritime and aeronautical SAR applications. This manual was aligned as closely as possible with the previous ICAO Aeronautical Search and Rescue Manual.

1.5.4 International Hydrographic Organization (IHO)

The IHO is an international intergovernmental consultative and technical organization that was established in 1921 to support safety in navigation and the protection of the marine environment. The main IHO activity is standardization of nautical charts and documents, bathymetry and ocean mapping and related publications, and technical assistance and training. The special activity of IHO related to MSS is radio navigational warnings implemented by the GMDSS as an integrated radio and satellite communication system of Inmarsat and COSPAS–SARSAT systems. The GMDSS system has improved the dissemination of maritime safety information (MSI), taking advantage of modern communications technology.

The far offshore navigation warnings are broadcasted via Inmarsat-C Enhanced Group Call (EGC) SafetyNET Service, while polar area warnings are covered by DCS HF radio services and coastal area warnings are transmitted via DCS MF/VHF radio and NAVTEX on a single-frequency, time-shared, and automatic broadcast. The NAVTEX and Digital Selective Call (DSC) VHF and HF radio services are already in operation in most parts of the world in the framework of the GMDSS mission.

The advent of digital data transmission, computers, and video display systems is having a considerable impact on hydrographic and navigation technology. This has made possible the development of ECDIS (Electronic Chart Display and Information Systems), which has become a major focus of activity. Several working groups have been established by the IHO to coordinate these new developments and ensure the standardization of systems and specifications. In 1992, the IHO adopted standards for the formatting of ECDIS data so that it can be readily exchanged between hydrographic offices. At this point, the Committee on ECDIS (COE) examines both the database requirements and the standardization and overall parameters of such systems.

1.5.5 World Meteorological Organization (WMO)

The WMO coordinates all global scientific activity to allow prompt and accurate weather information, WX, and tropical storm forecasting and other services for public, private, and commercial use, including the international shipping and airline industries, sent by GMSC media. More exactly, the major part of WMO's activities contributes to the safety of life and property, the socioeconomic development of nations, and the protection of the entire Earth environment, including the Global Ozone Observing System.

The World Meteorological Convention, by which body the WMO was created and adopted at the 12th Conference of Directors of the International Meteorological Organization, met in the US capital Washington in 1947. Although the Convention itself came into force in 1950, the WMO commenced operations as its successor in 1951 and, later that year, was established as an UN's specialized agency by agreement between the UN and the WMO.

The purposes of the WMO are to facilitate international cooperation in the establishment of networks of stations for making meteorological, hydrological, and other observations and to promote the rapid exchange of meteorological information by means of meteorological radio or satellite communications, the standardization of meteorological observations, and the uniform publication of observations and statistics. The WMO also furthers the specific application of meteorology to maritime, aviation, agriculture and other human activities, climatology, atmospheric sciences, hydrology, and instruments with important methods of observation. Each of them meets every four years, in an effort to promote operational hydrology and to encourage research and training in meteorology.

The main WMO task is to organize, control, collect, and offer up-to-the-minute worldwide meteorological observations and weather information through many member-operated observation systems and telecommunication links with four PEO and five GEO satellites, about 10,000 land observation stations, 7000 ship stations, and 300 moored and drifting buoys carrying automatic weather stations. Namely, each day, high-speed links transmit over 15 million data characters and 2000 weather charts through 3 world, 35 regional, and 183 national meteorological centers cooperating with each other in preparing weather analyses and forecasts in an elaborately engineered fashion. Thus, transoceanic ships and airplanes, research scientists on sea/air pollution or global climate change, the media, and the general public are given a constant supply of timely data, which are very important for safe navigation and flight. Moreover, it is through the WMO that the complex agreements on standards, codes, measurements, and communications are internationally established.

Data from all over the world are needed to provide weather (WX) forecasts and warnings. In other words, an aircraft does not take off, nor does a ship leave a port, without a WX received via agent or by means of radio and/or satellite communication systems for safety utilization. Combining facilities and services provided by all members, the program's primary purpose of the system is to make available meteorological and related geophysical and environmental information enabling countries to maintain efficient meteorological services. Facilities in regions outside any national territory (outer space, wide ocean areas, and Antarctica) are maintained by members on a voluntary basis. The World Weather Watch System comprises the Global Observing System, Global Data Processing System, the Global Telecommunication System, and Data Management and System Support Activities.

The Hydrology and Water Resources Program promotes worldwide cooperation in the evaluation of water resources and the development of hydrological networks and services, including data collection and processing, hydrological forecasting, and weather warnings.

1.5.6 Mobile Satellite Users Association (MSUA)

The MSUA was established in 1992 as a non-profit association to promote the interests of users of MSS worldwide. It fosters effective communication among MSS users, suppliers of equipment and services, operators of the satellite systems, and the various governmental entities that may affect the future of the entire industry. Membership is not limited to American entities; it is open to organizations worldwide engaged in any of these activities. The MSUA has created a Web site that is the premier source of MSS information. Namely, the association maintains a Web site on the World Wide Web—(www.msua.org). The Web site will publicize MSUA activities and provide, to members only, information and analyses that enhance their business opportunities. It is possible to find a description of the association, a great list of links (service providers, manufacturers, consultants, etc.), as well as links to all the major operational systems such as Inmarsat, Iridium, Globalstar, Leo One, Orbcomm, and many others. Three classes of membership are open to applicants: corporate, associate, and personal. In pursuing its mission, the MSUA media informs its membership on issues that affect them and conveys all members' views on such issues to appropriate authorities, and specifically, the association reviews, reports, and analyzes the activities of all members and MSS providers and reports on those executive and legislative activities of governments that affect MSS and navigation services, etc.

1.6 Satellite Communications Organizations and Operators

Government and intergovernmental satellite communications organizations and operators are divided into international, offering almost global service, and regional, usually covering few countries or entire continents and domestic, for local service only.

The GMSC system operators are the only entity responsible for the operation of the GMSC space and ground network configurations providing global, regional, or domestic coverage. However, the GMSC service providers are any entity commissioned by a GMSC system operator to provide GMSC services to the public within a region or country which may require an authorization to do so under the applicable legislation of the country concerned.

1.6.1 International Satellite Communications Organizations

International satellite communications organizations and operators are Intelsat, Inmarsat, Intersputnik, Eutelsat, ESA, and other global, multinational, or intergovernmental operators serving outside domestic and regional boundaries.

1.6.1.1 Intelsat

The FSS were the first to develop, and there was rapid recognition that these new global possibilities necessitated the creation of some kind of international organization. Thus, this led to the creation of the Intelsat international organization based in Washington. On August 20, 1964, 11 countries signed a charter agreement creating the Intelsat, the first open worldwide satellite communications network, and appointed the Comsat Corporation as its first manager. However, Comsat placed a contract for a GEO and the Intelsat fixed system started to offer transatlantic satellite services in 1965 after the successful deployment of Intelsat I (formerly Early Bird), the world's first GEO spacecraft.

Today, more than a fifth generation of Intelsat birds provides international and domestic satellite communication service on behalf of over 112 member nations. Since 1973, Intelsat has operated with an organizational structure that has only four tiers: the Assembly of Parties, the Meeting of Signatories, the Board of Governors, and the Executive Organ.

During the 1990s, users were served by the following satellite constellations: Intelsat K and the Intelsat V to VIII series of spacecraft. The first Intelsat VIII was launched in 1996 on board an Ariane 4 booster. Moreover, the first satellite in the Intelsat IX series, Intelsat 901, was scheduled to be launched in 2000 to the 60°E GEO location. The next series of Intelsat GEO spacecraft 902, 903, and 904 were scheduled in the period between 2000 and 2002.

The Intelsat provided as well as MSS using payloads carried by its GEO spacecraft Intelsat V MCS series: A (F5), B (G6), C (F7) and D (F8). The Inmarsat organization leased capacity on Intelsat V MCS flights F5 through F8 for MMSS, shown in Fig. 1.13. These four satellites used portions of the L-band (from 0.5 to 1.7 GHz) and C-band assigned for such purposes by the ITU. The Intelsat V MCS A (F5) was launched on September 28, 1982, over IOR in orbital location at 63°E; the Intelsat V MCS B (F6) was launched on May 19, 1983, over AOR in orbital location at 18.5°W. The Intelsat V MCS C (F7) was launched on October 19, 1983, as a spare spacecraft over IOR in orbital location at 66°E, and the Intelsat MCS V D (F8) spacecraft was launched on March 3, 1984, over POR in orbital location at 53°E longitude and at the end of 1985 was relocated to the assignment of 179/180°. Thus, MCS uses radio frequencies in the top C-band transponder and an additional L-band spectrum to link CES with ships as part of the Inmarsat network. Because of the additional power requirements by the MCS packages on Intelsat V F5–F8 satellites, the Ku-band operations on these four spacecraft must be limited or

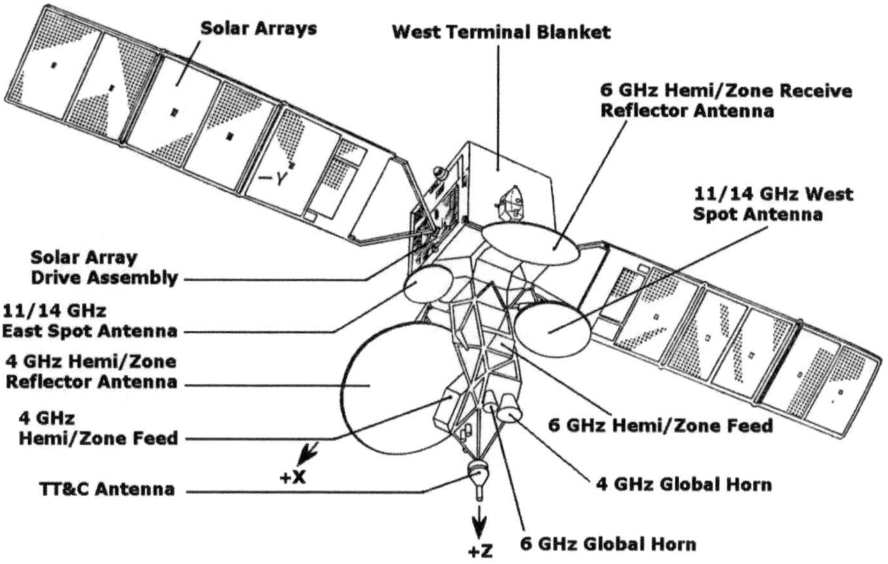

Fig. 1.13 Intelsat V MCS spacecraft—Courtesy of Book: by R. Galic

curtailed when they are activated for Inmarsat use. However, when the L-band signal amplifier is operating in the high-power mode, the 11 GHz Ku-band capacity of these spacecraft is switched off. These satellites after many years of service are deployed and not used by the Inmarsat system any more.

1.6.1.2 Inmarsat

The Inmarsat organization is the only international and nonmilitary MSC provider in the world. The Inmarsat MSC network provides tel; Tlx; fax; slow-, medium-, and high-speed data transfer; image and video; videoconferencing; videophone; new mobile MPDS and ISDN; LAN and IP service; and in the framework of the GMDSS system, enables distress and safety service for maritime, land and aeronautical MES, including government, rural, military and offshore infrastructures.

In the early 1970s, the IMO, then known as the IMCO, began to consider the possibility of using MSS to improve maritime communications, not least for safety purposes. Toward the end of 1973, IMCO convened a conference to decide on the principle of establishing an international maritime safety system and to conclude the necessary agreements. The work of this conference culminated in September 1976 in the adoption of what became the Inmarsat Convention and its complementary operating agreement, requiring always acts exclusively for peaceful purposes. Inmarsat was established on July 16, 1979, by the major maritime nations to finance this project, which is to investigate using satellites to form links with vessels and offshore oil rigs at sea. Inmarsat also owes much of its success to the foresight and commitment of the IMO, which played a crucial role in its creation.

On February 1, 1982, Inmarsat officially took control of satellites previously operated by three Marisat spacecraft (series F1, F2, and F3), a joint venture among nations begun in the early 1970s through the efforts of the Comsat General. Inmarsat has since expanded its space segment by leasing additional capacity from Intelsat and ESA, Intelsat V MCS series of four spacecraft, and two Marecs spacecraft (B and B2A), respectively. However, today an additional Inmarsat second, third, and forthcoming fourth generation of GEO, Inmarsat-2, Inmarsat-3, and Inmarsat-4 spacecraft, respectively, are being exploited.

In many ways, the Inmarsat organization is patterned after the Intelsat system and almost all countries comprise the Inmarsat Assembly as members, each country casting one vote. The assembly meets once every two years to formulate general policy, long-term strategy, and objectives. Besides, each government selects a representative signatory to the Inmarsat Operating Agreement from the public or private sector. The assembly has also primarily to establish financial, technical, and operational standards. The Inmarsat Council, similar to a corporate board of directors, meets three times a year to implement all policy decisions of the assembly. The Director General of Inmarsat oversees the day-to-day management of the organization, with an executive staff headquartered in London.

Inmarsat is financed in two ways: Signatories must pay fees based on their use of the network, or can purchase investment shares in Inmarsat that reduce the user fees in proportion to the investment, and signatories earn a return of 14 % per annum on their investment in Inmarsat. During 1987, the largest numbers of investment shares were held by the USA, UK, Norway, and Japan.

1.6.1.3 Intersputnik

The Russian Federation, that is to say, the former the Soviet Union, is not a member of Intelsat. Instead, in 1971, it created a similar multinational organization named Intersputnik, which provides FSS for its 14 member states and a number of other associated countries. This system uses the various families of former Soviet communications satellites, such as Molniya, Raduga, and Gorizont, using GEO, HEO, and PEO satellite coverage.

The Russian satellite system conceivable can beam radio broadcasting, TV programs, voice, and data traffic to almost any location on Earth, and only Intelsat can supply more global FSS links than Intersputnik. This organization is an open international intergovernmental that any sovereign state can join. The fundamental structure of Intersputnik was determined by the Cooperative Agreement on the Establishment of the Organization. A representative for each member nation serves on the Board of Members, which is the main governing body of Intersputnik organization. Besides, this board selects a director general to chair the Intersputnik Directorate based in Moscow.

The organization also provides three MSS for all three applications, using payloads carried by its GEO spacecraft Gorizont (Horizon), Raduga (Rainbow), and Morya (Seamen):

1. **The Volna (Wave) Network**—The Volna MSC system served to connect maritime and aeronautical MES terminals via space segment constellation to LES and ground-based telecommunication facilities for former USSR ships and aircraft. This MSS consists of communications payloads carried by the spacecraft Gorizont and Raduga. The Volna Network provided radio and TV service for mobile stations on UHF frequency bands between 335 and 399 and 240 and 328 MHz. On the other hand, the Volna Network provided service uplink/downlink on L-band between 1636 and 1644 and 1535 and 1542 MHz for MMSS and also on L-band 1645–1660/1543/158 MHz for AMSS applications, while feeder link used 6/3 GHz uplink/downlink bands for both MSS applications.
2. **The Morya Network**—In 1989, the former Soviet Union expanded its MMSS with the Morya MSC Network, using existing Soviet satellites series Morya for carrying the MSC payload. Namely, the Morya Network provided MMSS on two 2.5 MHz wide frequency uplink/downlink bands centered on 1637.25/1535.75 MHz (service link) and 6084.0/3758.3 MHz (feeder link).
3. **The Gals Network**—Piggybacked on the former Soviet Union's Raduga spacecraft, named Gals (Tack), a special telecommunications payload serves as satellite links for the former USSR (today Russia) military forces. Thus, this network will operate using the X-band spectrum for uplinking (7.9–8.4 GHz) and downlinking (7.25–7.75 GHz), for defense maritime, land, and aeronautical applications.

1.6.1.4 ESA

The idea of creating an independent space organization in Europe goes back to the early 1960s when six European countries, Belgium, France, Germany, Italy, the Netherlands, and the UK, are associated with Australia, to develop and build a heavy launcher called Europa. In 1975, a convention was endorsed at the political intergovernmental level to set up the European Space Agency (ESA). Finally, the convention entered into force on October 31, 1980. Since then, the founding ESA members have been joined by 4 new members from Europe, while some other European countries have expressed their interest to join ESA. In addition, the cooperation agreements have also been signed to allow Canada to participate in certain ESA programs and to sit on the ESA Council.

This makes the dreams of Europe's space scientists come true, by creating and operating new scientific spacecraft for the ESA. The project management interacts with the aerospace and instrument industry in the 15 member states and oversees the construction, launch, and operation of the spacecraft. There is in close liaison with ESA's European Space Operations Center (ESOC) in Darmstadt, Germany, with Arianespace and other cooperative agencies providing launchers and with the GES needed for communication with the spacecraft.

The ESA Research and Scientific Support Department was restructured in 2000/2001 into divisions. In such a way, each division is the home of the project scientists for the study, project development, and operation of the missions and serves the space scientists of Europe, under the supervision of the Science Program Committee, on which all member states are represented. Any scientist or group within the member states can propose a space science mission. Before a final selection, some rival proposals are selected for detailed study of their scientific, technological, and budgetary implications. Under a science communications initiative, ESA is intensifying its efforts to keep the press, public, and schoolteachers well informed about the science program and the progress of its various missions.

Apart from construction, launch, and operation of the spacecraft, ESA has developed Artemis MSS, derived from former Prodat MSS for all three applications. At present, ESA members are developing GSNS named Galileo and satellite augmentation systems EGNOS for maritime, land, and aeronautical satellite CNS.

1.6.1.5 Eutelsat

The European Telecommunications Satellite (Eutelsat) Organization was provisionally founded in 1977 by representatives of 17 members of the European Conference of PTT and telecommunication administrations, with headquarters based in Paris. Its major mandate was to establish and run the European satellite communications system as a regional operator, but because it is currently offering global FSS service on an international basis, it can be classified into the group of global and international satellite organizations.

The constitution and financing of Eutelsat are modeled on those of Intelsat. Although it did not formally come into existence until 1984, Eutelsat started work in 1977 with ESA on the exploitation of the Orbital Test Satellite (OTS) experimental communication birds and on the design of the operational European Communications Satellite (ECS) series. Finally, the latter were in due course taken over and operated by Eutelsat, which is now procuring its second generation of satellites.

At this point, Eutelsat passed from an interim organizational structure to definitive operational status on September 1, 1985, and all shares were divided at that time among about 25 member nations. During the early stages of planning, Eutelsat designers thought their spacecraft primarily would carry voice and high-speed data service. Although Eutelsat was set up to handle long-distance traffic, a significant part of its revenue now comes from the relaying of satellite TV programs for distribution through the terrestrial telecommunications network or into the cable system. Throughout its 25 years of experience, Eutelsat has placed innovation at the center of its development. The company has distinguished itself, notably by being the first in Europe to distribute satellite TV and pioneering the use of the Digital Video Broadcasting (DVB) standard for the transmission of digital TV channels and data.

Eutelsat is one of the world's leading operators of multipurpose satellite infrastructures. It provides capacity on 23 satellites that offer a broad portfolio of services, which include direct TV and radio broadcasting for the consumer public, professional video broadcasts, corporate networks, Internet services, and MSC systems. Therefore, Eutelsat is providing regional MSS for all three applications with the current Emsat MSC system for maritime and land applications and the EutelTRACS system for mobile tracking and messaging. The latter system is developed by technical cooperation with Qualcomm, a US-based company, to establish communication network and equipment infrastructures.

1.6.2 *Former International MSS Operators*

Former international and regional MSS organizations and operators were Marisat, developed by the US-based Comsat Company Marecs, which was formed by European nations, and Prodat was a project of ESA.

1.6.2.1 Marisat MSS

The world's first maritime MSC system as a new application of the GEO system was unveiled in 1976 with only three satellites and ocean networks that are providing MMSC services in the Atlantic, Pacific, and Indian Oceans. The Hughes Aircraft Company, known today as Boeing Satellite Systems Inc, under contract to Comsat General Corp, built three multifrequency communications spacecraft called Marisat (Maritime Satellite), for the space segment of the world's first MMSC operator. The Comsat General was developed a Marisat system for MSS at first for only maritime applications. In 1971, frequency bands around 1.6 GHz were allocated for satellite communications connections with ships and aircraft. The Marisat satellites were designed initially for US Navy vessels and they had a UHF transponder on board in a band from 240 to 400 MHz. Because there was sufficient margin for additional payload, L- and C-band transponders were installed on the Marisat satellite to provide commercial MSC traffic for maritime applications.

These satellites had a dual role at that time to provide space segment facilities that were leased to the US Navy for military communications with naval ships, and it also enabled the use of transponders for Comsat General itself to operate MSS for traffic with merchant ships, virtually worldwide. All three Marisat satellites were launched with the same type of USA rocket "McDonnell Douglas 2914 Delta" during 1976, on 19 February, 9 June, and 14 October, for the needs of the company Comsat General. Marisat F1, F2, and F3 satellites were placed in GEO planes at 15° W, 72.5°E, and 176.5°E longitude, respectively.

All satellites have been leased from Comsat, in effect; Marisat F1 spacecraft served as an in-orbit spare for the Marecs A spacecraft in the Marisat AOR region at a position 15°W. Then, this satellite was leased as a spare in Inmarsat AORE region and removed to 106°W.

Marisat F2 spacecraft served as a spare in IOR region at a position 73°E and lately for the Inmarsat space segment. The Marisat F3 spacecraft served at a position 176.5°E in the POR region and afterward reprogrammed as a spare for the Marecs B2 satellite. This satellite was finally relocated as in-orbit spare to 182.5°E. The service at that time was welcomed by merchant shipping, and by 1982, around one thousand vessels were equipped to use the Marisat system. All three Marisat satellites also served as an emergency backup, one in each of the three ocean regions: AOR, IOR, and POR. Accordingly, after many years, these satellites are no longer in exploitation either by the Comsat or by the Inmarsat systems.

The Navy fleet used P-band frequencies (Tx = 248–260 MHz/Rx = 1300–312 MHz) for the MMSS link, while the merchant-shipping element of the Marisat payload used the newly allocated L-band frequencies (Tx = 1537–1541 MHz/Rx = 1638.5–1642.5 MHz) for its MMSS links and C-band (Tx = 6174.5–6424 MHz/Rx = 3945.5–4199 MHz) for feeder links. Fixed CES for mobile service were located at Santa Paola for POR, at Southbury for AOR, and at Fucino and Yamaguchi for IOR (see Fig. 1.14). The system provided access to the satellites, linking ships at sea through the PSTN with the TTN subscribers ashore for tel, Tlx, fax, data, and HSD transmissions. The Marisat system was controlled by network control center (NCC) located at Washington. Satellite Tracking, Telemetry and Command (TT&C) are also conducted over C-band frequencies.

However, the governments of many other countries were not quite content for control of MSC with their ships to rest with a foreign commercial corporation. Owing to this problem, in 1976, under the aegis of IMCO, an agreement was drawn up for the establishment of an Inmarsat organization, initially for maritime service only.

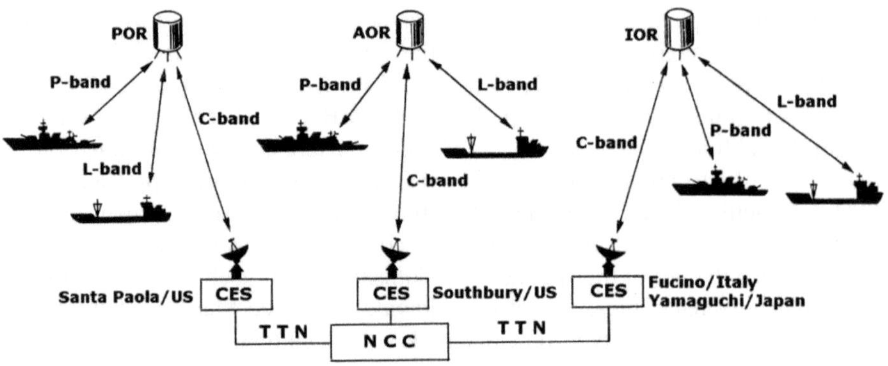

Fig. 1.14 Marisat space and ground segments—Courtesy of Book: by Novik

1.6.2.2 Marecs MSS

The ESA organization's Maritime European Communications Satellite (ECS) or Marecs project covered the study, development, launch, and in-orbit operations of communication spacecraft to be integrated in a global MMSC. Development began in 1973 with funding from Belgium, France, Italy, UK, Spain, and Germany and then later joined by the Netherlands, Norway, and Sweden. In effect, the program started at first as the experimental Maritime Orbital Test Satellite (Marots) but was subsequently changed to an operational system resulting in a name change, satellite redesign, and delayed development.

The Marecs satellite was part of the GMSC system configured to provide high–quality, full-duplex, reliable, real-time voice, fax, Tlx, and data services between SES and CES with automatic connection to the TTN. The Marecs satellites operated by ESA were members of Inmarsat first-generation MMSC network. The Marecs-1 B2A spacecraft was successfully launched via an Ariane rocket from Kourou, French Guiana, on November 9, 1984, at AOR region location 26°W. It was later relocated to the assignment of AORW at 55.5°W as an operational satellite of the Inmarsat network. The next Marecs A satellite was successfully launched via an Ariane rocket from Kourou, French Guiana, on December 20, 1981, on a POR assignment at a position of 177.5°E and was later relocated to POR at 178.5°E, as a spare satellite leased by Inmarsat. The Marecs satellite consists of two modules: a service module, which is a derivative of the ECS bus and a payload module. The spacecraft has a design life of 7 years, a 3-axis altitude control, and TT&C that uses VHF spectrum during transfer orbit and C-band through the communications subsystems on station. The payload is capable of operating without continuous ground control, and it consists of a C- to L-band forward transponder and an L- to C-band return transponder, incorporating SAR channels.

The Marecs spacecraft was based on the British Aerospace ECS 3-axis stabilized platform, with two sun-tracking solar arrays providing 955 W (BOL) with 2 NiCd batteries for eclipse power supply. Payload had three repeaters: shore to ship with 5 MHz bandwidth, ship to shore with 6 MHz bandwidth, and shore to shore with 0.5 MHz bandwidth. These provide 35 two-way voice channels plus search and rescue capabilities. Hence, one 2-m-diameter L-band antenna and 2 horns (one transmit, one receive) for 4/6 GHz channels provide almost 1/3 of the Earth coverage.

1.6.2.3 Prodat MSS

The ESA organization promoted a special Prosat program, which included propagation measurements of the aeronautical channels. The second phase of the Prosat MSS project primarily consisted of the design, development, and prolonged demonstration of a two-way low-rate data system. The latter service is referred to as Promar, while the low-cost digital data-only MSC service included in the configuration is known as Prodat, to be available for maritime, land, and aeronautical mobile operations, on 1.6 GHz band to the satellite, which in turn relays the signals to ground using a 6 GHz carrier.

The Prodat program conducted successful field trials with low data terminals using the Marecs satellite. In parallel, two L-band MSC payloads are being procured by ESA to promote European MSS: the European Mobile System (EMS) payload on the Italsat 1-F2 satellite and the L-band land mobile (LLM) payload on the Artemis satellite. The EMS allows voice and data communications with a capacity of 300 channels. The available capacities are partially being used to demonstrate and evaluate the emerging MSC system. The European MSS operational phase started with the EMS payload in orbit during 1996 and is continuing with the LLM payload, which was planned to be launched in 1997.

The Prodat MSS was operated through ESA ground stations and Inmarsat communications satellites and provided a low-speed, low-power, and narrow-bandwidth service between the terrestrial public data network (PDN) and a population of maritime, land, and aeronautical mobile terminals.

1.7 Frequency Designations and Classification of Services

The assignment of a radio frequency, band, or channels is performed by an authorized administration for radiocommunications via platform, satellite, or space stations to use a radio frequency (RF) spectrum or frequency channels under specified conditions. In this sense, the allotment of a RF or frequency channel comprises the entry of a designed frequency channel in an agreed plan, adopted by a competent conference for use by one or more administrations for a terrestrial or space radiocommunications service in one or more identified countries or geographic areas and under specified conditions.

The allocation of a RF band makes possible its entry in the table of frequency allocations of a given RF band for the purpose of its use by one or more terrestrial or space radiocommunications service or the radio astronomy service under designated and specified conditions.

At this point, in satellite communication fields, the frequency bands are often denoted with alphabetical symbols such as L- to Ka-bands. Frequency band numbers and names are defined by the ITU RR, ITU Tables of Frequency Allocations in general or for a particular band and the mentioned alphabetic symbols by the IEEE Standard Radar Definitions.

The radio spectrum shall be subdivided into nine RF bands and designated by progressive whole numbers, in accordance with Table 1.2. At any rate, in satellite communication fields, the radio frequency bands are often denoted with alphabetical symbols such as L- to Ka-bands.

In a more general sense, frequency designations for MSS are used by a number of different administrations for their national or international MSS networks and can be systematized into two main categories:

Table 1.2 Frequency band designation

Band no.	Abbreviation	Band name	Name	Symbol	Frequency
4	VLF	Very low frequency	Myria m		3–30 kHz
5	LF	Low frequency	Km		30–300 kHz
6	MF	Medium frequency	Hm		300–3000 kHz
7	HF	High frequency	Dam		3–30 MHz
8	VHF	Very high frequency	m		30–300 MHz
9	UHF	Ultrahigh frequency	dm		300–3000 MHz
				L-band	1–2 GHz
				S-band	2–4 GHz
10	SHF	Super high frequency	cm		3–30 GHz
				C-band	4–8 GHz
				X-band	8–12 GHz
				Ku-band	12–18 GHz
				K-band	18–27 GHz
				Ka-band	27–40 GHz
11	EHF	Extremely high frequency	mm		30–300 GHz
12	VEHF	Very extremely high frequency	deci mm		300–3000 GHz

1. **Frequency Allocations for Service Links**—The frequency allocation for service links in current commercial use is in L-band at 1.5 GHz (downlinks) and at 1.6 GHz (uplinks). In addition to the L-band designations, there are other MSS frequency band allocations for service links between MES and spacecraft, shown in Table 1.3. New frequency bands below 3 GHz were allocated for MSS at WARC-92 and WRC-95. The new allocations below band of 1 GHz are very narrowly designated for only LEO configurations, while GEO satellites are excluded from most of the terms. Very soon, it will be clear which of the new bands above 1 GHz should be used for MSS via GEO satellites and the rest will be available for Big LEO systems and satellites in other non-GEO constellations. Moreover, there are also LMSS frequency allocations for land vehicles at 14.0–14.5 GHz.
2. **Frequency Allocations for Feeder Links**—The feeder links between LES and satellites are illustrated for most MSS in Table 1.4. In such a manner, all radio frequency allocations can be designated for global and regional satellite coverage. The subcategories of frequency allocation are divided into three regions and among three mobile applications. All frequencies used between satellite and MES are part of MSS and can be classified into three main allocations: MMSS for ships, LMSS for land vehicles, and AMSS for aircraft. Some allocations differentiate among the MMSS, the LMSS, and the AMSS scheme and are somewhat

Table 1.3 Frequency spectrum for MSS service links

Uplink (MHz)—Tx	Downlink (MHz)—Rx	Comments
Earth (MES)–spacecraft	Spacecraft–Earth (MES)	
128 and 240 except		Distress and safety band allocated for former SAMSARS and DRCS system, respectively
406.025	1544.5	Distress/safety bands allocated for COSPAS–SARSAT system
1645.5–1646.5 1610–1621.35 1626.5–1660.5 except 1645.5–1646	2483.5–2500 (Globalstar) 1525-1559 except 1544–1545	Principal current frequency allocations for MSS at L-band using by MMSS, MLSS, and AMSS applications
148–149.9 148–150.05 312–315 399.9–400.05 & 406–406.1 454–460	400.505–400.645 (Faisat) 137–138 (Leo One) 387–390 400.15–401 400.505–400.645	New MSS frequency allocations made in WARC-92, WRC-95, and WRC-97 for GEO and non-GEO MSS and RDSS applications; last allocation will take effect from January 2005
1610–1626.5 2660.2–2690 1980–2010	2483.5–2500 (Faisat) 2500–2535 (N-Star) 2170–2200	Satellite component of IMT-2000 worldwide
1215–1260	17700–20200	Frequency allocations for GNSS applications

complex in detail. Thus, the regional frequency allocations for corresponding countries are systemized into regions 1, 2, and 3, respectively. In addition, there are also allocations for other services within these frequency bands in some countries, but most of these shared allocations are of secondary status.

Finally, new Ka-band allocations are being studied by NASA for commercial MSS and may present a feasible opportunity. The ITU has defined many communication services within its RR that can be carried out by satellite systems and has developed many rules for worldwide RR of the services in order to maximize the peaceful use of outer space. In this instance, depending on the specific purpose and services, satellite communications can be classified into several services, as explained later. In effect, many of these services have been provided under the auspices of special satellite operational organizations set up to develop, operate, and market the service.

As already mentioned, classification of an MSC service can be achieved by considering the role played by the geographic extent of network coverage on international, regional, and local bases. All these organizations set up to handle

Table 1.4 Frequency spectrum for MSS feeder links

Uplink (GHz)—Tx	Downlink (GHz)—Rx	Comments
Earth–spacecraft	Spacecraft–Earth	
0.148–0.14855	0.1370725–0.1379275	E-Sat
0.148–0.15005	0.400150–0.401000	Leo One
0.150–0.15005	0.400505–.400645	Faisat (WARC-92)
6.345–6.425	4.120–4.200	N-Sat
6.425–6.725	3.400–3.629	Inmarsat, Thuraya
5.091–5.250 15.45–15.65	6.875–7.075 5.091–5.150 temporary 5.150–5.216 from 2010 15.40–15.70	New MSS frequency allocations made in WRC-95 for GEO and non-GEO MSS and RDSS applications (ICO, Globalstar)
13–13.25	10.75–10.95	MSAT
14.236–14.250	12.736–12.750	Emsat
14 & 30	11 & 20	MTSAT
29.1–29.3	19.4–19.6	Iridium

these classes of satellite systems having a similar structure, only specific service spans national boundaries. Otherwise, the main classification of the satellite service in connection with types of users has been realized on fixed and mobile units.

1.7.1 Fixed Satellite Service (FSS)

The FSC system enables a radiocommunications link between two or more FES or GES terminals at given positions, when one or more satellites are used, illustrated in Fig. 1.15. In this sense, the given position may be a specified position or any fixed point within a particular area. In some cases, this particular service includes satellite-to-satellite links, which may also be operated in certain intersatellite services. Moreover, the FSS solutions may also include feeder links for other space communication services, including MSS, or MSS GES can provide service for FES as well.

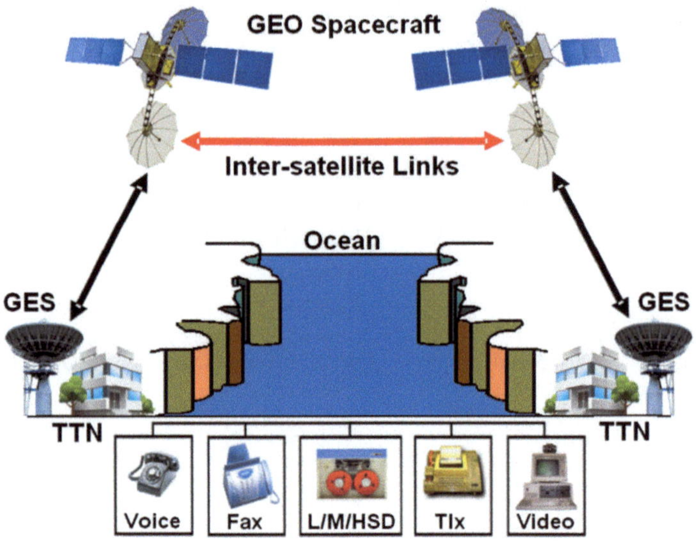

Fig. 1.15 Fixed satellite service—Courtesy of Book: by Gallagher

The FSS signals are relayed between many FES, which are relatively large, complex, and expensive systems. The FES terminals are connected to the conventional TTN, and the service is intended for long-distance voice, video, and data communications. According to the WARC-85/88 principle plan, the FSS shares frequency bands with terrestrial networks in the 4/6, 12/14 and 20/30 GHz, which guarantees every country has equal access to the GEO space constellations.

A typical example of FSS is the Intelsat, one of the pioneers in satellite communications. The first generation of Intelsat system operated in the C-band (4/6 GHz). At present, many global systems, such as Intersputnik, Telesat, Eutelsat, PanAmSat, and others, and regional systems, such as Optus in Australia and JCSAT in Japan, operate in the Ku-band (30/14 GHz), which provide coverage throughout most of Europe and Japan, respectively. This service may include satellite voice, fax, voice, data and video (VDV), low/medium- and high-speed data, Tlx, and video only via VSAT of DVB-RCS networks.

1.7.1.1 Satellite Voice Network

Voice service is inherently interactive communication in nature providing global telephone infrastructures, which represents two-way wire or wireless communications on distance. Voice channels are useful for relatively low-rate data applications such as fax, e-mail, and low-speed Internet access on less than 64 Kb/s. The digital standard for PSTN access at the subscriber level is ISDN, which provides 144 Kb/s of active data subdivided into two 64 Kb/s circuit-switched bearer channels plus one data channel (2B+D). The FSS voice network uses GEO satellites and a bandwidth per channel from 8 to 64 Kb/s with FDMA and TDMA transmissions. The system enables the international telephone trunks to extend the global coverage and thin route and to improve rural services in developing regions.

1.7.1.2 VSAT Network

The VSAT (very small aperture terminals) devices are quite similar to Inmarsat MES. This equipment is Small Earth Stations capable of receiving only from and transmitting and receiving (transceivers) to or from spacecraft. Therefore, VSAT devices are classified as a communications media with either one-way or two-way facilities. In the broadest sense, the term one-way VSAT includes the data terminals designed for the reception only of DSB transmission and conventional TVRO, using PAL and similar TV system. This device can also receive data using modulated subcarriers. Two-way VSAT can transmit and receive signals at rates at approximately 64 Kb/s. The low directivity of VSAT antennas limits the power and hence the boresight EIRP, which may be transmitted. The transceivers are usually completely solid state and can be highly integrated.

This equipment represents an important addition to the telecommunications world, because they can provide a service directly to their customers at virtually any geographic location covered by suitable satellite beam. They do not require any support from a local TTN and can even be run from portable or alternative power supply. This system is useful for private data network within countries and regions to promote business needs. Examples of such networks are Equatorial, Intelnet, Intelsat, and other systems. The VSAT system main applications are serving for data and document distributions, rural communities, and business utilizations and for disaster area communications. Data distribution can be in two-way between central HUB stations for archive and data processing and all VSAT users. Thus, documents distribution by VSAT can be only one-way satellite transmission from Hub library stations to all users. The users can be in touch with HUB via TTN.

Rural application of VSAT devices is very important for improving capability to transmit and receive much information from rural areas to central locations and vice versa. Rural communities would also like to provide speech service via VSAT systems, because of very limited telephone line facilities, if they are available at all. Voice VSAT system operating at 4.8 Kb/s is already in use. High-quality speech can be carried on satellite channels with a data rate of 9.6 Kb/s using modern voice encodes and modems capable of operating at low C/N values, something like Inmarsat aeronautical voice system. A voice circuits set up between two VSAT devices include a double hop and accordingly a delay of about 0.5 s. A HUB located in a city can provide both voice and data services to many villages.

Thus, disaster areas and in generally for emergency situation needs deployment of VSAT devices for alert and security communications, usually regarding the terrain, locations and safety of life. The reason is similar that many ships, vehicles, and aircraft now carry satellite distress beacons.

Business VSAT applications are of an essential interest because of their large potential to provide persons and companies with the competitive edge. In general, business wants to establish private networks to link their all locations and move their information in a safe manner and for the lowest possible cost. The service can be numerous like airlines and bus reservations, car rental conference facilities, insurance, and newsgathering. Although the term VSAT equipment is generally used in connection with very small and fixed location terminals for business use, there are comparable developments in related fields that should be considered. For example, TVRO of very low cost, which can receive MAC signals from high-power satellites, could offer VSAT type service in higher data rates and less cost.

The VSAT technology brings all of the features and benefits of unit or bidirectional FSS down to an extremely economical and usable for business data transmission. The system can provide also very efficient bypass with TTN for voice, data, and video services using sophisticated digital technology and advanced communication network protocols.

The VSAT system enables the use of one or more 56 Kb/s data channels, each of which can be subdivided or applied directly. Voice communication is also possible using 16 or 32 Kb/s, depending on the compression algorithm. The VSAT network uses GEO configuration, a channel bandwidth from 64 to 512 Kb/s with FDMA and TDMA transmissions.

In Fig. 1.16 (Left) is shown GT&T Faraway SES100 VSAT cost-effective, state-of-the-art satellite solution for companies having high communication flow to their clients, agents, or branch offices located abroad or for remote and rural areas where communication services are unavailable, unreliable, or too expensive. This equipment is using V-sat communication network with African C-band and European, African, and Middle East Ku-band coverage. The system is providing on-demand high-quality voice up to 8 telephone lines, group III fax, data transfer compatible via external Hayes-compatible modem with rate from 4.8 up to 64 or 128 Kb/s, IP/X25/X400, Internet (e-mail and Web sites), videoconference, and broadcasting (TV and Radio) services. The main unit with high-capacity chassis allows several telephones (up to 8 interfaces), fax, and PCs to be connected simultaneously via parabolic antenna. The antenna is available with a diameter of 65-, 98-, 120-, 180-, and 240-cm dishes depending on position in coverage area and distance from subsatellite point. This configuration employs a full-mesh, DAMA, and PAMA network architecture that maximizes the use of available space and ground-based resources.

Mobile satellite users can also be included in this category of satellite communications. Most probably in the near future, VSAT system will offer full MSS similar to Inmarsat-C or Mini-M systems. With an additional mobile antenna, VSAT can offer MSS for maritime, land, and aeronautical applications.

Another very last VSAT model is GT&T IPsky2 of V-Sat two-way transceiver system together with Internet modem, and router launched in January 2002 can be easily adapted in mobile unit with slight transformation of antenna system only (see Fig. 1.16 (Right)).

Fig. 1.16 VSAT for multipurpose applications—Courtesy of Brochure: by GT&T

This equipment is a low-cost and highly compact Internet dedicated DVB-RCS-MPEG2 solution that also offers prepaid VoIP and fax by satellite. In reality, customer will have an ultra-fast asymmetric Internet interface that is directly connected to the Internet backbone of 155 and 34 Mb fiber using an E-Mai, FTP, TCP, and Web access service available on 24-hours basis. The maximum available download (outbound) speed of service via GEO is 2 Mb/s including very high level of transmission and reception coding for high speed and reliability. Return path or inbound speed is 33, 76.8, or 153.6 Kb/s depending on the antenna size of 75, 96, or 1.2 m, respectively. Several PCs (from 5 to 13 or more if necessary) and Ethernet LAN can be connected to the IPsky2 modem/router through a proxy server, public IP network, private FTP, VPN, etc.

1.7.2 Mobile Satellite Service (MSS)

The MSS consists of three types of GMSC services: maritime, land, and aeronautical and may include the space segment, Mobile and Land Earth Stations (MES and LES) with all applications (SES, VES, AES, TES, PES), coordination and control stations and centers (NCC, SCC, NCS, RCC, LUT), and TTN interface with subscribers (see Fig. 1.17).

The MSS enables satellite linkage between MES and one or more space stations or between space stations used for this service, or between two or more MES by means of one or more space stations. This service may also include service and feeder links necessary for its operations, and in such a way, MES can be connected via service link to the satellite and from the satellite via feeder link to the LES.

The LES in MSS network can be Coast Earth Station (CES) for both maritime and land mobile applications and Ground Earth Station (GES) for aeronautical mobile application only. In a more general sense, this modern classification can be a

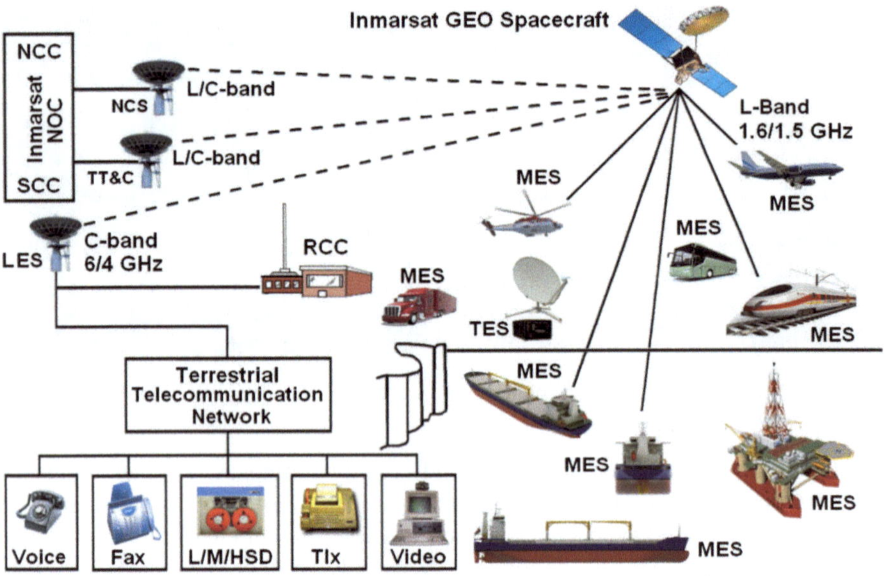

Fig. 1.17 Global Mobile Satellite Communication Systems—Courtesy of Book: by Ilcev

provisional proposal but for the moment gives the reasonable ideas for future establishment some practical universal nomenclature of GMSC's terms.

The LES or GES are special Earth stations in MSC located at a specified fixed point or within a specified area on land, to provide a feeder link for MSS. The Base Earth Station (BES) or gateway is an Earth station in mobile or fixed satellite service located at a specified fixed point, or within a specified area on land, to provide a feeder link for the MSS, especially for non-GEO systems.

Satellite radio beacon for indicating distress emergency position is a special mobile or personal Earth station in the MSS, the emissions of which are intended to facilitate SAR operation for distress in maritime, land, and aeronautical applications, such as EPIRB, PLB, and ELT, respectively.

The service link is a connection between MES and satellite, while the feeder link means duplex connection between CES or GES in the given location and spacecraft of the various MSC services operate in frequency bands allocated to the FSS. Thus, the given location of MSS may be at a specified fixed point or at any point within specified areas. Therefore, a satellite link comprises one uplink and one downlink performing radio linkage between a transmitting and a receiving Earth station.

1.7.2.1 Maritime Mobile Satellite Service (MMSS)

The MMSS is a service in which MES is located on board merchant or military ships, other floating objects, rigs or offshore constructions, hovercrafts, and/or survival craft stations providing commercial, determination, logistic, tactical, defense, and safety communications. In addition, the special maritime emergency position-indicating radio beacon (EPIRB) terminal, either portable or fixed stations on board oceangoing ships, may also participate in this service. The EPIRB is a special Earth station in the MMSS, the emission of which is intended to facilitate urgent SAR operation for vessels in distress for maritime applications. The EPIRB unit can be also used as personal devices of passengers or crew members.

The MMSS service enables mobile satellite links between CES and Ship Earth Station (SES), between two or more SES and/or between associated ships and other satellite communications stations in all positions at sea or in ports. The SES is a mobile Earth station in the MMSS capable of surface movement at sea within the geographic limits of a country or continent.

In distinction from conventional maritime communications, a ship fitted with SES in or near a port may operate with CES or other SES in cases of distress and commercial operations. The CES is a maritime Earth station located at a specified fixed point on the coast to provide a feeder link for MMSS. The SES is a maritime Earth station fixed on board ships or other floating objects, which can provide communications links with subscribers onshore via CES and communications spacecraft.

The ship on scene radiocommunications and alert service performs a distress and safety service in the MMSS between one or more SES and CES, or between two or more nearby SES, or between SES and RCC, or between portable or floating EPIRB and LUT stations in which alert messages are useful to those concerned with the movement and position of ships and of ships in distress.

1.7.2.2 Land Mobile Satellite Service (LMSS)

The LMSS is a service in which MES is located on different types of cars, trucks, buses, trains, and other civil or military vehicles, providing logistics and business communications, or it can be a transportable Earth station (TES). Besides, the land emergency personal locator beacon (PLB) terminal may also participate in this service as a special Earth station in the LMSS, the emissions of which are intended to facilitate SAR operation of vehicles and/or personal distress and emergency on the ground for land applications. This unit is very suitable for military applications, for use in polar expeditions and for desert, remote, and rural environments.

The LMSS enables mobile satellite links between CES and Vehicle Earth Station (VES), between two or more VES, and/or between associated MSC stations. The VES is a mobile Earth station in the LMSS capable of surface movement on dry roads or railways on land within the geographic limits of a certain country or continent. The CES is a maritime Earth station also used for LMSS located at a specified fixed point on the coast to provide a feeder link for MMSS. The VES is a land mobile Earth station fixed on board road or rail or land vehicles, providing communications links with terrestrial subscribers via CES and communications spacecraft.

The land vehicle or persons on scene radiocommunications and alert service perform a distress and safety service in the LMSS between one or more VES and CES, between two or more nearby MES, between VES and RCC, or between PLB and LUT stations in which messages are useful to those concerned with the position of vehicles or persons in distress.

1.7.2.3 Aeronautical Mobile Satellite Service (AMSS)

The AMSS is a service in which MES is located on different types of airplanes, helicopters, and other civil or military aircraft providing logistics, flight regulations, air traffic control, safety, business, and private communications, primarily along national or international civil air routes. This service is also intended for all kinds of satellite communications, including those relating to flight margin coordination in corridors and air traffic control, primarily outside national and international civil aeronautical routes. The special emergency locator transmitter (ELT) terminal either portable or fixed on board aircraft stations may also participate in this service as a special Earth station in the AMSS, the emission of which is intended to facilitate the SAR operations for aircraft in distress and emergencies at sea, on the ground, and for aeronautical applications.

The AMSS enables mobile satellite links between GES and Aircraft Earth Station (AES), between two or more AES, and/or between associated airplanes/aircraft and other satellite communications stations. The AES is a mobile Earth station in the AMSS capable of flying and maneuvering in the air over the geographic limits of a country or continent. The GES is an aeronautical Earth station located at a specified fixed point on land to provide a feeder link for AMSS. The AES is an aeronautical Earth station fixed on board aircraft, which can provide communications links with subscribers on land via GES and spacecraft.

The aircraft on scene radiocommunications and alert service performs a distress and safety service in the AMSS between one or more AES and GES, between two or more nearby MES, between AES and RCC, or between ELT and LUT stations in which messages are useful to those concerned with the movements, flight locations, and/or positions of aircraft on land, in distress.

1.7.3 Personal Mobile Satellite Service (PMSS)

The PMSS is a service in which MES is handled by individuals and serving everyone in wherever position possessing handset satellite phones. The system can serve for both their GMSC and Global Mobile Personal Satellite Communications (GMPSC). The land mobile emergency PLB terminal may also participate in this service or actually may need some coordination with regard to the COSPAS–SARSAT system. This service enables mobile satellite links between BES (gateways) and personal Earth station (PES), between two or more PES, and/or between other satellite communication stations using the same satellite providers or frequency spectrum. The PES is a mobile Earth station in the PMSS carried handheld by individuals for personal and/or business facilities or capable of surface movement at sea, on land, and in the air for professional utilization. As is evident, this new system can also provide three types of GMSC services: maritime, land, and aeronautical.

The BES or gateway is an Earth station located at a specified fixed point to provide a feeder link for PMSS. The PES is a personal Earth station or handheld terminal carried by individuals or fixed on board ships, vehicles, or aircraft, which can provide two-way communications links with subscribers anywhere on Earth via satellites and gateways.

The PES or persons on scene radiocommunications and alert service perform a distress and safety service in the PMSS between one or more PES and BES, between two or more nearby MES or PES, between PES and RCC, or between PLB and LUT stations in which messages are useful to those concerned the position of mobiles or persons in distress.

1.7.4 Radio Navigation Satellite Service (RNSS)

The RNSS can be used for the purposes of safety navigation and secure sailing or flight, including obstruction warnings. The RNSS is a one-directional system, in which only mobile stations can know its own position, and others cannot. This service may include the following:

1. **Maritime Radio Navigation Satellite Service (MRNSS)**—This is a special service in which Earth station or navigation equipment is located on board ships.
2. **Land Radio Navigation Satellite Service (LRNSS)**—This is a special service in which Earth stations or satellite air navigation equipment is located on board land vehicle.
3. **Aeronautical Radio Navigation Satellite Service (ARNSS)**—This is a special service in which Earth stations or satellite navigation equipment is located on board aircraft.

1.7.5 Radio Determination Satellite Service (RDSS)

The RDSS is a special service for the purpose of radio determination involving the use of one or more space stations and may include feeder links necessary for its own operations.

This service provides the satellite radio determination of the position, velocity, and/or other characteristics for each moving object and all mobile applications, to

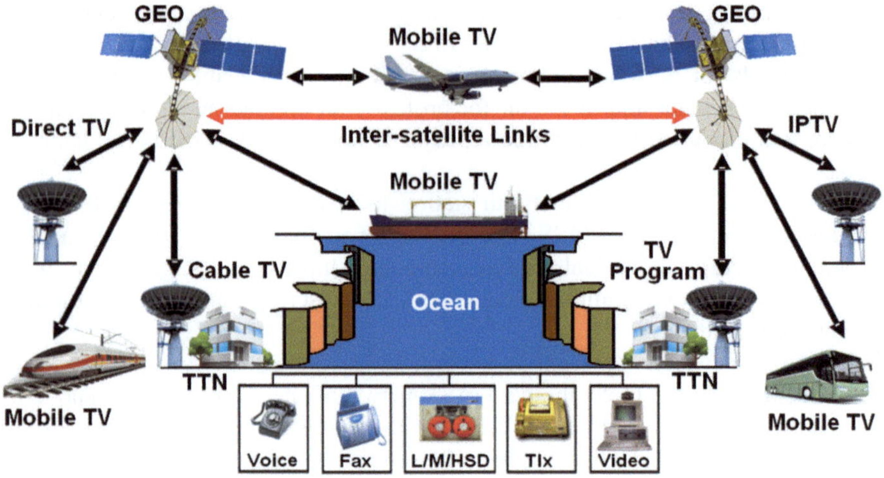

Fig. 1.18 Fixed and mobile satellite broadcasting—Courtesy of Manual: by Ilcev

obtain information relating to these parameters by means of the propagation properties of radio waves.

A typical RDSS is Iridium, which provides two-way voice communications and positioning services in a self-consistent manner. Using this system, a mobile station has the possibility to determinate its own position, when other user stations can also know that position, for example, handheld, fixed and other Earth stations or a station installed in an office or onboard vehicle. In distinction from a radio navigation one-directional system, the RDSS system is a bidirectional system (Fig. 1.18).

1.7.6 Mobile Broadcast Satellite Service (MBSS)

The MBSS is one-way or two-ways satellite solution in which signals transmitted by Earth stations or retransmitted by space stations are intended for direct reception by the general public via satellite receivers and suitable antennas. Direct Broadcasting Satellite (DBS) are one-way signals transmitted from fixed positions on the ground to the satellite and then back to all MES terminals, community distribution by cable network, or direct individual receivers. This service requires the generation of high RF power from the satellite to permit reception by small antennas or onboard mobiles or on ground, shown in Figure 1.18.

Developments in digital audio/video broadcasting are, of course, not solely limited to the terrestrial domain of the home market. Digital services offered by satellite are now making significant inroads into the mobile entertainment market, which especially are maritime and aeronautical. New digital broadcasting service of

CD-quality music programs and DVD-quality video transmissions via satellites to many customers in mobiles has recently been attracting the attention of many people in transportation industries.

The term "direct reception" in the MBSS shall encompass both individual fixed or mobile reception and community reception. Individual reception is any reception of the emissions from space stations in the MBSS by simple domestic installations at home or on board mobiles and in particular those processing small antenna, intended for private or common use on board mobiles. Community reception is any reception of the emissions from a space station in the MBSS by professional receiving equipment, which in some cases may be complex and can have antenna larger than those used for individual reception. This service is intended to be used by a group of the general public at one fixed or mobile location or through a distribution system covering a limited area.

The present MBSS operates at 12 GHz band and is designed for community reception, equipped with fixed satellite terminals and large antennas. Besides, if a satellite has enough power to transmit signals to be received by small antennas suitable for individual reception equipped with satellite very small aperture terminals (VSAT), or very small aperture antennas, the system is called a DBS system.

Although the present system is designed for fixed service and terminals, new systems are also serving all three mobile applications equipped with auto tracking antennas. Some transportation means such as large oceangoing ships, cruisers, airplanes flying on intercontinental airlines, trains, and buses traveling on international or interregional routes can install equipment to receive TV programs from direct video broadcasting (DVB) and direct audio broadcasting (DAB) stations. These mobile stations can be equipped with special tracking receive-only antenna systems as a separate unit or in combination with two-way communications satellite antennas. Therefore, this service may include audio (sound), video (television), and data transmission.

1. **Audio Broadcasting**—Audio broadcasting is a special radio emission via satellite for MSBS, in which signals transmitted by Earth stations or retransmitted by space stations are intended for direct reception of audio signals by the general public or mobile units via satellite receivers and corresponding antennas. In Europe, USA, and Japan, the satellite DAB systems in the frequency L- and S-bands have been investigated to allow the direct transmission of high-quality programs, comparable to CD, to be developed. Usually, road vehicles are equipped with such equipment.
2. **Video Broadcasting**—Video broadcasting is a special TV emission via satellite for MSBS, in which signals transmitted by Earth stations or retransmitted by space stations are intended for direct reception of DVB video and audio signals by the general public or mobile units via satellite TV receivers and corresponding antennas. All mobile units at sea, on land, and in the air can be equipped with suitable video and tracking antenna equipment. The Inmarsat system also provides audio and videoconferencing via HSD MPDS or mobile ISDN interface using Fleet F77 service for ships and Swift64 for aircraft.

3. **Data Broadcasting**—Data broadcasting, using a different speed rate, is a special data transmission via satellite for MBSS, in which signals transmitted by Earth stations or retransmitted by space stations are intended for direct reception of data signals by the general public or MES terminal via special HSD satellite data receivers and corresponding omnidirectional or tracking antennas.

1.7.7 Mobile Satellite Broadband Service (MSBS)

The MSBS is a multimedia one- or two-way radiocommunication service in which signals transmitted by mobile or fixed Earth stations or retransmitted by space stations are intended for direct reception by the general public or mobile units through satellite receivers and corresponding antennas. This new system, similar to the optical fiber terrestrial network, will provide various advanced interactive multimedia services operating at higher bit rates, such as voice (audio), video, different speeds of data, teletext, videoconferencing, distance learning, high-resolution graphics, HiFi audio, high-definition TV, color fax and imaging, mobile Internet, and PC communications. Such advanced networks will soon cover urban, fixed, and mobile environments and most populations living in rural/remote areas, which cannot be covered by landline and optical fiber or wireless TTN.

These signals are transmitted from LES to the satellite and then back to all mobile or fixed applications via optical fiber cable TTN. This service requires the generation of high RF power from the satellite to permit reception by receivers and small antennas on the ground. The MSBS has been developed by broadband MSC operators and will start to offer a multimedia service for personal and all three mobile applications, which allows any person or mobile to communicate anytime and anyplace.

Thus, it will provide two categories of high-speed wireless access communications. The first will be serviced both outdoors and indoors, which can enable a high-speed rate up to 30 Mb/s, and the second will provide ultrahigh-speed indoors only, which can transmit high-speed signals up to 600 Mb/s. The second system cannot provide wide coverage areas or services in mobile environments, so the main application is limited to a "hot spot" of indoor premises.

The next example of MSBS is DVB-RCS for fixed and mobile solutions. This system ties terrestrial networks (broadband, broadcasting, Internet, UMTS/GPRS cellular, private, and public) via satellite HUB (LES) with antenna and C-, Ku-, or Ka-band satellites, with satellite router terminals for remote Internet, VoIP, videoconference, all e-services, interactive TV/radio, broadband LAN/WAN, multicasting, intranet/VPN, etc.

The newly designed Intelligent Satellite Transport System (ISTS) will be also part of the MBSS infrastructure, which comprises an advanced information and telecommunications network for users, roads, and vehicles. The ISTS is expected to greatly contribute to solving problems such as traffic accidents and congestion. Not only it solves such problems, but it ISTS will also provide multimedia services for vehicle drivers and passengers. Otherwise, the ISTS consists of several development areas, including technical advances in MSC and navigation systems, control and tracking system, electronic toll collection system, and assistance for safe driving, which is appreciated as one of the most promising mobile satellite multimedia businesses.

Digital networks used to operate 64 Kb/s channels with the switching (transmission) are based on the so-called synchronous transfer mode (STM). This mode was developed mainly for transmission; thus, switching functions are difficult to handle at different bit rates. At this point, in the interest of more flexible broadband switching, therefore, a new asynchronous transfer mode (ATM) has been developed that can handle traffic relating to services that require widely differing bit rates. In ATM, basically the information is put in fixed-length cells that are switched and transported to the broadband network and at the point of destination reconstituted in its original synchronous form. Typical new services requiring broadband switching equipment operating at higher speeds (bit rates) include desktop publishing, multimedia service, videoconferencing, color fax, and HiFi music/HDTV.

1.8 Mobile Satellite Meteorological Service (MSMS)

The MSMS is a special infrastructure for obtaining meteorological data transmission via satellite, in which data signals transmitted by Earth stations or retransmitted by space stations are intended for direct reception by the meteorological centers. In fact, this service is used for one-way transmission of meteorological data only from meteorological centers through the MSC system, including hydrological, observations, and exploration particular to ships, oil rigs, and aircraft mobile stations. In such a way, this meteorological information, bulletins, and weather warnings are very useful for the safe navigation and flight of ships and aircraft, respectively.

The current need for quick and easy access to marine and aeronautical WX data and forecast information has become increasingly important. The growing demand of new sources of energy has led the offshore oil exploration into more remote and hostile seas. Environmental constraints have narrowed the maneuvering margins of many at-sea operations, so there is a need for an operational WX/forecast system, which can quickly disseminate maritime and aeronautical WX to the users.

The current need for quick and easy access to marine and aeronautical WX data and forecast information has become increasingly important. The growing demand of new sources of energy has led the offshore oil exploration into more remote and hostile seas. Environmental constraints have narrowed the maneuvering margins of many at-sea operations, so there is a need for an operational WX/forecast system, which can quickly disseminate maritime and aeronautical WX to the users.

Much of this information is transmitted from WX satellites via direct readout to receiving ground stations where it can be displayed, analyzed, and prepared for customers. These direct readout service (DRS) was developed and operated by

US NOAA, Russian (CIS), the ESA, Japan, China, India, etc. The most popular of these WX services are WEFAX transmitted by the US Geostationary Operational Environmental Satellites (GOES), and the Automatic Picture Transmission (APT) from PEO NOAA satellites.

The GOES-8, -9, and -10 spacecraft series are carrying as well payloads with GEOSAR transponders of GMSC COSPAS–SARSAT system. Therefore, remotely sensed meteorological data are transmitted directly from GEO or PEO satellites in "real time" to many ground forecasting center (GFC) and GES within signal range of the WX satellite. The WX satellite images are designed with a format so that they could be received, processed, and reproduced by relatively expensive GES equipment and retransmitted free of charge to anyone with the appropriate satellite receiving and display equipment.

The former USSR has been attempting for decades to develop an effective GEO program for WX applications. Namely, due in part to the very high latitude of their launch sites, it has been difficult for Russia to design, build, and launch a dependable GEO WX satellite system. After all, Russia had launched in 1994 a Geostationary Operational Meteorological Satellite (GOMS) called Elektro-1 that is placed in GEO at 76°E. During 1996 and 1997, the GOMS WEFAX service has been operating erratically, and there have been problems with the imaging sensors. An Elektro-2 satellite was scheduled for launch in 2000 to improve GOMAS WEFAX service. This satellite is also projected to be part of COSPAS–SARSAT system, because is currying as well as GEOSAR transponder.

The European agency ESA organization also operates a series of GEO WX satellites called meteorological satellite (Meteosat), which provide low-resolution DRS similar to GOES WEFAX and called secondary users data station (SUDS). Thus, the Meteosat-6 was replaced by Meteosat-7 in 1988, and new series of MSG satellites were scheduled to be launched in 2001 (Fig. 1.19).

Japan has launched a series of geostationary meteorological satellites (GMS). The last one GMS-5 was launched in 1994 at a position of 140°E, which transmits both a primary data stream and WEFAX transmission on 1691 MHz. A new,

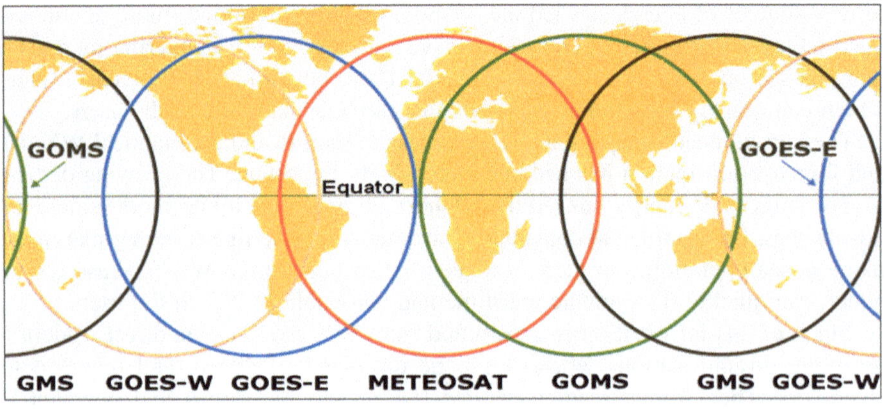

Fig. 1.19 Global GEO meteorological satellite coverage—Courtesy of Book: by Ilcev

advanced series of GEO MTSAT satellites for WX and aeronautical augmentation services will start with operation in 2003 and replace GMS-5 spacecraft, which is described in the same book of Volume 2.

The Indian INSAT-2A, -2B, and -3 multipurpose satellites have several payloads; among the rest, they carry both WX and GEOSAR (COSPAS–SARSAT) transponder. The Chinese GEO WEFAX program began with the launch of the FY-2 (Feng Yun) satellite in 1997.

The basic footprint for the current operational GEO satellites transmitting WEFAX data is presented in Figure 1.19. There are many other MLMSC obtaining WX for mariners like AWT and others, and also for avionics such as METAR.

1.8.1 WEFAX System

The WEFAX (WX Fax) satellite images are designed with the format for the DRS provided by the US GOES system. Otherwise, similar services are transmitted from the European Meteosat, Russian GOMS, Japanese GMS and satellites from other countries. The WEFAX system retransmits data consists of processed images produced by the primary imager on the GOES above mentioned GEO as well as other meteorological data and images relayed from PEO satellites. This System was first incorporated into the GOES satellites in 1975. The format of the WEFAX signal was designed to be received and reproduced by low-cost GES. This satellite delivered WEFAX signal should not be confused with the HF WX radio fax transmission from coastal radio station (CRS), which some of service is still using. Namely, the WEFAX system is a line-of-sight satellite transmission, with different contents, and it uses specific receiving equipment. The GOES WEFAX data are transmitted as an analog signal at 1691 MHz and 240 lines per minute. A large amount and variety of data can than be obtained.

In the current WEFAX schedule, over 100 images can be received in a 24 h period. These consist of scheduled data transmissions of quadrants of the full Earth disk and equatorial regions in visible and infrared spectra, composite images from PEO, WX and ice charts, and operational messages. The WEFAX image is initially formatted by the special ground control station (GCS), using high-resolution GOES image data polar orbiter mosaics and weather charts, and rebroadcasted through the GOES satellite back to DRS of GES. In fact, it is a delayed DRS product or near real time. The basic structure of the GOES network of Tx (transmitted) from weather data collection platforms (WDCP) and Rx (received) weather information by mobiles is shown in Fig. 1.20.

In order to receive the WEFAX transmissions, most stations use a parabolic or dish antenna with stabilized and tracking platform for large ship MES. Thus, continuous tracking of the antenna is not required for a fixed GES. Namely, once properly aligned to the satellite downlink signal, the antenna is locked into permanent position and rarely requires any further adjustment. Except the antenna, a basic WEFAX RDS station is typically comprised of the following electronic components: preamplifier,

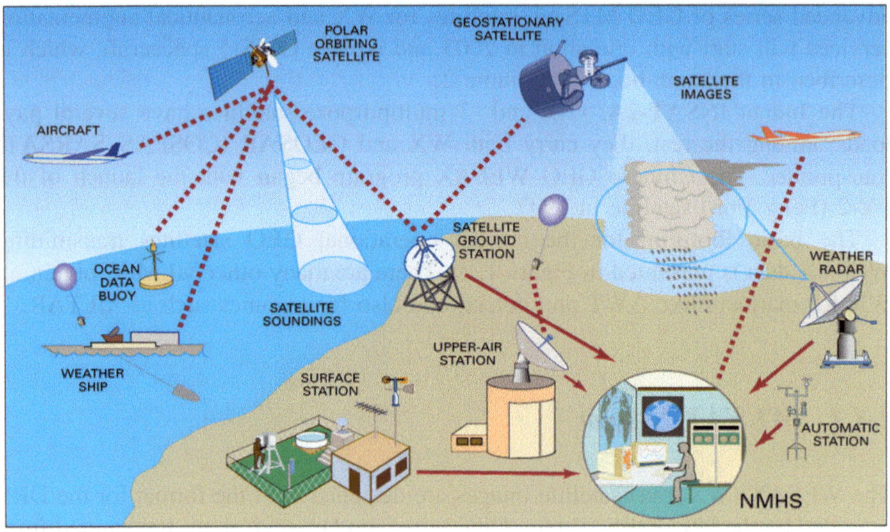

Fig. 1.20 WDCP meteorological satellite system—Courtesy of Paper: by Branski

receiver, demodulator card to decode the satellite signals, PC to display the satellite imagery, a storage system (hard disk, CD writer, or ZIP drive) to memorize and archive the satellite imagery, and computer software applications to manipulate the imagery (image processing and enhancement). The basic WEFAX receiving configuration system can be purchased for about 800/1500 US$ without PC configuration. The commercial DRS now available has been designed for a variety of computers such as IBM-compatible, Apple Macintosh, and even some UNIX systems. While most of the details in the preceding article have emphasized the US NOAA GOES satellites and their WEFAX system, it is important to keep in mind that the WEFAX service is global in nature. The WEFAX transmission of the Russian, Japanese, and European GEO is nearly identical in their technical characteristics. The small differences between these systems are normally easily resolved in the computer software that comes with a commercial purchased receiving system, which will be usable even if moved to different locations.

1.8.2 *Automatic Picture Transmission (APT)*

The APT is using DRS from PEO satellites to provide WX imagery for world customers. The first APT was pioneered on TIROS-VIII satellite launched in December 1963. Today, US TIROS PEO satellites continue to transmit images of the Earth by APT. These have been joined by Russian Meteor and the Chinese Feng Yun spacecraft, providing similar transmissions. Because of that, an international transmission standard has been agreed upon, so a GES capable of receiving data from the US PEO can receive images from satellites of other countries as well. On the latest TIROS-N (ATN) series of PEO birds, the APT images are produced by a special instrument called the advanced very high resolution radiometer (AVHRR). This instrument is designed to detect five channels of energy reflection from the Earth ranging from the visible spectrum and the near-infrared and infrared spectra. The analog APT signal is derived from the original five-channel digital data and multiplexed so that only two of the original channels appear in the APT format.

These two images are selected from GCS, and during daylight passes, they usually consist of the visual channel and one of the infrared channels. At night, two infrared images are usually found in the APT. The APT imagery consists of two pictures, side by side, representing the same view of the Earth in two different special bands. The Russian Meteor is transmitting daylight visible pictures only.

The APT signal is transmitted continuously from the PEO satellites to the GES terminals. The result is image strip as long as the data transmission is received at the GES and as wide as the scanning instrument is designed to operate at a particular altitude. Radio reception of the APT signals is limited to line of sight from GES and can only be received when the PEO satellite is above the horizon, namely when it is in view with GES. This is determined by both the altitude of the satellite and its particular path during the orbit across the GES reception range. At present, the US, Russian, and Chinese PEO operate at an altitude between 810 and 1200 km. In such a manner, at these altitudes, the maximum time of signal reception during an overhead pass is about 16 min. During this time, a GES terminal can receive a picture strip equivalent to about 5800 km along the satellite path under the best reception conditions.

In order to obtain APT video data using direct reception, accurate information concerning locations, movements, and times that the satellite can be received must be available. This is necessary because signal reception is possible only while the satellites are above that GES horizon. Although all PEO satellites have basic orbital characteristics in common, each spacecraft is unique in its orbital parameters and needs to be tracked individually. The data necessary to locate and track the WX satellites are generally not difficult to obtain. At any rate, the generation of future orbits of a given satellite can be easily calculated, and if a directional antenna is used, determining the azimuth and elevation of the satellite as it passes over the GES is not difficult after the basic orbital patterns are understood.

The last generation of US ATN satellites represents the current PEO spacecraft available for receiving DRS data. The basic operational concept of this series is only to maintain two satellites in a polar orbit at all time. One will maintain an orbit so that it will pass over the GES, traveling from north to south during the morning, having a southbound Equator crossing at about 0830 local solar time, while the next satellite will pass from south to north during the afternoon, having a northbound Equator crossing at about 1430 local solar time. In fact, each of these satellites will also pass over the GES circa 12 h later traveling in the opposite direction. During winter 1997/98 of North Hemisphere, the USA is operating NOAA-12 as a morning spacecraft and NOAA-14 as the afternoon spacecraft.

With a two-satellite constellation such as the NOAA-12 and NOAA-14, one of which will pass over the observing/receiving station about every six hours. Most stations can receive two consecutive passes (about 100 min apart) from each satellite, day and night. Being able to receive imagery about every six hours is more than adequate to track the development and movement of large weather system within an 800 to 1000 km radius of the station.

Technological advances in microelectronics and computer software applications over the past two decades have made it rather simple to assemble and use a basic DRS GES suitable for APT. However, a basic DRS station consists of the same components except is additional a method to predict when the PEO satellite has to be in view of the GES. This is because PEO satellites are not stationary like GEO over the adequate positions on the Earth.

The APT direct transmission service from PEO satellites is more suitable for smaller ships. The antenna is small and omnidirectional and therefore does not need to be aimed at the satellite direction. Several different types of antennas may be used for APT reception of slow-speed data. One can be directional and requires tracking of the moving satellite, and the second type is omnidirectional. It has the advantage of being less expensive and not requiring tracking system, but will give a slightly reduced reception range. A turnstile reflector type of omnidirectional antenna is one of the simplest and least expensive antennas to use for APT.

The next quadrifilar helix (QFH) antenna is a special type of omnidirectional antenna that provides a much better radiation pattern compared to previous and does not suffer from the loss of signal strength exhibited in simple turnstile antenna.

The APT signals from PEO satellites are transmitted on radio frequency between 137 and 138 MHz FM. Two US TIROS PEO satellites transmit APT on 137.5 and 137.62 MHz, and Russian Meteor has been using 137.85 MHz. Beginning in the early 2003, the European Organization for the Exploration of Meteorological Satellites (EUMETSAT) will launch their first PEO meteorological satellite Metop-1. It will carry a suite of advanced sensors and DRS transmission system. Otherwise, of special note to the DRS user community is that the current APT service will be replaced by low-rate picture transmission (LRPT) by Metop-1 spacecraft. The LRPT service will be digital rather than analog, requiring same adequate modification to present installed receiving stations.

1.8.3 Applied Weather Technology (AWT)

The AWT focus is on providing high-quality maritime global weather routing data service through Inmarsat-A, -B, and -M SES equipment. This includes providing the shipmaster with initial routing recommendation, continuous en route weather forecast advisories (WX), and postvoyage analyses. The AWT is PC-aided weather routing system and is committed to research and development in worldwide weather information system. Thus, a number of fully operational software packages are under continuous development in order to provide up-to-date technology solution to maritime and other clients:

1. **Voyage Simulation Engine**—Is PC-based computer software for dead reckoning vessel position commensurate with prospective WX and adjusted by true position reported by vessel. An integrated database engine enables instant access to ship's actual and potential performance en route using MSS.
2. **Bonvoyage System (BVS)**—Is actually network compatible shore-based company fleet management and monitoring system or can be onboard ship graphical marine WX data briefing system. Color-enhanced WX maps on PC screen overlaying with vessels routing information not only provides up-to-the-minute fleet status at sea or in port, but also allows visual recognition of en route WX conditions. On the other hand, the master of vessel is routinely facing real-time WX report, not only bearing the paramount burden of safety of his ship, cargo, and crew, but also struggling for the most cost-effective management and economical operations and voyages.

For example, the forecast of an imminent storm system ahead of vessel track is in reality a very difficult task and huge responsibility of ship captain. However, it is impossible for the traditional text-based WX routing to explain fully the detailed shape of the storm system, dangerous wave-generating area, and detail grid information and to find out the best solution for safe and economical ship's routes. To explore the BVS is necessary to supply a modern PC configuration running Windows, minimum 8 Mb RAM and 10 Mb free disk space, Super VGA video card with at least 1 Mb RAM, and popular asynchronous 9600 baud modem or faster would be recommended. The BVS can be easily tailored for pleasure boat and yacht operators.

The AWT system is actively conducting research and development in the area of ocean wave modeling, oceanographic data compression, optimum ship routing, high-speed shipboard data transmission, and so on. These future developments, when available, will become an integral part of the overall weather information service.

1.8.4 Global Meteorological Technologies (GMT)

The GMT system has been in operation for approximately 8 years as a manufacturer and service provider of software supplying WX data. Such data are provided to ships or land-based operations through the Internet, Inmarsat, HF/VHF radio, or PSTRN/PSDN networks. This system utilizes the backup network used by NOAA and Environment Canada for their WX service communication. The GMT WeatherWise software is a more economical and superior alternative to Weatherfax as a means of receiving WX charts. Thus, charts are as near to real-time WX conditions as can be technically be achieved and are available within the hour of their measurement. Charts are presented in full color using NOAA and WMO standard formats and codes and are available 24 h per day. Users have to supply corresponding PC configuration and peripherals. At all events, the faster the PC is, the sooner the analyses can be calculated. Otherwise, all action of the mentioned software can be activated by a mouse, trackball, joystick, or touch screen-equipped PC. A keyboard is not required to use GMT WeatherWise, other than to reconfigure active map and WX databases.

Data WeatherWise services are broadcast via Inmarsat-C or VHF radio and are available on GMT bulletin board via HF radio and Internet network. However, any GMDSS type-approved Inmarsat-C SES can be used to receive Inmarsat FleetNET broadcast. A Hayes-compatible 2400-baud, or higher, modem is required to log onto GMT service via Internet. Thus, the GMT WeatherWise is available in six modular regions: the North Pacific, North Atlantic, Eurasia, Far East, Good Hope, and Horn.

1.8.5 Maritime Noble Denton Weather Services (NDWS)

The NDWS system is private initiative to provide special marine WX forecasting services to the international offshore and marine industry. Global meteorological data are gathered and processed from NDWS center in London within 24 h to enable the production of site- or route-specific forecasts for any location worldwide. The WX forecasts are updated once or twice daily by fax, Tlx, or e-mail. Weather routing for ocean towages can also be provided by Noble Denton in-house master mariners.

1.8.6 Global Sea State Information via Internet (GSSII)

The ERS satellites operated by the ESA measure significant sea wave heights and period, 10 m wind speeds, and direction that benefits the satellites 24 h. These data are continuously received by UK Company Satellite Observing Systems where they are routinely processed and corrected before necessary calibration factors are applied. Results in the form of image maps and text summaries are generally available within 2 to 3 h of data acquisition for direct delivery over the Internet or retrieval on a daily subscription service via the World Wide Web (WWW). However, a sea state alarm system also operates to give immediate WX warning by e-mail messages of regional conditions in any part of the world in excess of 10 m significant wave height or 30 m/sec wind speed or 6 m minimum swell.

1.8.7 Aeronautical Weather Applications

The most popular WX service for aircraft is provided by SITA and ARINC meteorological systems for aircraft in flight or in the airport as follows:

1. **SITA Aircom Weather System**—This WX system offers several alternatives via radio, Inmarsat Aero terminals, Internet, and on-site or hosted solutions such as graphical WX charts created with country boundaries, airports, and various other aeronautical features, thus providing maximum information within one viewable image; graphical representation of WX data, mapped over airline routes, facilitates alternative route selection based on wind patterns and other critical WX information; each instance of wind within user's environment can be programmed to receive a unique set of charts; Aircom surface WX with all meteorological parameters and NOTAM information important for safety flight; and different WX charts adapted to pilot needs such as surface WX, visibility, satellite imaginary, radar imaginary, and lighting. The Aircom WX chart service includes, but is not limited, the following types of weather information: significant weather, upper air weather (wind speed and direction), temperatures, icing, turbulence, and precipitation.
2. **ARINC Value-Added Service (VAS)**—Except other corporate and safety contribution, the VAS system provides similar media as SITA to transfer all meteorological parameters and information to civil aircraft including meteorological aviation reports (METAR) and terminal area forecast (TAF). The ARINC network organize as well as meteorological data collection and reporting system (MDCRS), which collects information, organizes and disseminates real-time automated position and WX reports from participating airlines, and forwards them to the National Weather Service (NWS) for input to their forecast models. Weather products include radar precipitation images, lighting, temperatures, icing, and turbulence and accurate forecast of wind aloft is used to define areas of severe WX and contributes to flight planning efficiencies and aviation safety. Otherwise, this service is similar to GOES meteorological satellite system for WX data collection from aircraft, ships, or other mobile or stationary platforms and transfers them via satellite to NOAA/NESDIS stations, as is shown in Fig. 1.15. On the other hand, Terminal Weather Information for Pilots (TWIP) provides valuable situational awareness of weather conditions within 15 Nm of the airport. On the other hand, the TWIP service collects information from airport sensors and transmits severe WX warnings such as wind shear microbursts, gust fronts, heavy-to-moderate precipitation to aircraft, and ground operations computers.
3. **Aeronautical Weather Reports**—There are two types of weather reports known as a METAR and the aviation Selected Special Weather Report (SPECI). The METAR is observed hourly between 45 min after the hour till the hour and transmitted between 50 min after the hour till the hour. It will be encoded as a METAR even if it meets SPECI criteria.

1.8.7.1 Aviation Routine Weather Report (METAR)

The METAR format was introduced January 1, 1968, internationally and has been modified a number of times since. However, North American countries continued to use a surface aviation observation (SAO) for current weather conditions until June 1, 1996, when this report was replaced with an approved variant of the METAR agreed upon in a 1989 Geneva agreement. The WMO publication No. 782 "Aerodrome Reports and Forecasts" contains the base METAR code as adopted by the WMO member countries.

The name METAR is commonly believed to have its origins in the French phrase: "message d'observation météorologique pour l'aviation regulars" or in English: "Aviation Routine Weather Observation Message" or simply "report" and would therefore be a contraction of MÉTéorologique Aviation Régulière (METAR).

In general, METAR reports typically come from airports or permanent weather observation stations. Reports are typically generated once an hour; if conditions change significantly, however, they can be updated in special reports called SPECI. Some reports are encoded by automated airport weather stations located at airports, military bases, and other sites. Some locations still use augmented observations, which are recorded by digital sensors, encoded via software, and then reviewed by certified weather observers or forecasters prior to being transmitted. However, observations may also be taken by trained observers or forecasters who manually observe and encode their observations prior to transmission.

A typical METAR report contains data for the temperature, dew point, wind speed and direction, precipitation, cloud cover and heights, visibility, and barometric pressure. A METAR report may also contain information on precipitation amounts, lightning, and other information that would be of interest to pilots or meteorologists such as a pilot report or PIREP, color states, and runway visual range (RVR).

Although the METAR code is adopted worldwide, each country is allowed to make modifications or exceptions to the code for use in their particular country, e.g., the USA will continue to use statute miles for visibility, feet for RVR values, knots for wind speed, and inches of mercury for altimetry. However, temperature and dew point will be reported in degrees Celsius.

The USA will continue reporting prevailing visibility rather than lowest sector visibility. Most of the current US observing procedures and policies will continue after the METAR conversion date, with the information disseminated in the METAR code and format. The elements in the body of a METAR report are separated with a space. The only exceptions are RVR, temperature, and dew point, which are separated with a solidus (/). When an element does not occur, or cannot be observed, the preceding space and that element are omitted from that particular report.

A METAR report contains the following elements: (1) type of report, (2) ICAO station identifier, (3) date and time of report, (4) modifier (as required), (5) wind, (6) visibility, (7) runway visual range (RVR) (as required), (8) weather phenomena,

(9) sky condition, (10) temperature and dew point group, (11) altimeter, and (12) remarks (RMK) (as required).

1.8.7.2 Aeronautical Weather Forecast

1. **Pilot Weather Reports (PIREP)**—No observation is more timely than the one made from the flight deck. In fact, aircraft in flight is the only means of observing icing and turbulence. Other pilots welcome PIREP as well as do the briefers and forecasters. A PIREP always helps someone and becomes part of aviation weather. Pilots should report any observation that may be of concern to other pilots. Also, if conditions were forecasted but were not encountered, a pilot should also provide a PIREP. This will help the WX to verify forecast products and create accurate products for the aviation community. Pilots should help themselves, the aviation public, and the aviation weather forecasters by providing PIREP. Required elements for all PIREP are type of report, location, time, flight level, aircraft type, and at least one weather element encountered. All altitude references are mean sea level (MSL) unless otherwise noted, distance in nautical miles, and time in Universal Coordinated Time (UTC). A PIREP is transmitted in a prescribed format as follows: type of report (UUA/UA), location (OV), time (TM), altitude/flight level (FL), aircraft type (TP), sky cover (SK), flight visibility and weather (WX), temperature (TA), wind (WV), turbulence (TB), icing (IC), and remarks (RM).
2. **Radar Weather Report**—General areas of precipitation, including rain, snow, fog, clouds, and thunderstorm, can be observed by radar. Otherwise, on the radar screen cannot be visible anything during very bad weather conditions and in situation when deep clouds are spreading to the Earth surface.
3. **Satellite Weather Pictures**—Prior to weather satellites, WX observations were made only at distinct points within the atmosphere and supplemented by PIREP observation.
4. **Radiosonde Additional Data (RADAT)**—These data are obtained from the radiosonde observations that are conducted twice a day at 00 and 12Z, such as freezing level and the relative humidity associated with the freezing level.
5. **Aviation Weather Forecasts**—Good flight planning involves considering the following available weather information and weather forecasts: (1) Terminal Aviation Forecast (TAF), (2) Aviation Area Forecast (FA), (3) Inflight Aviation Weather Advisories, (4) Alaska, Gulf of Mexico, and International Area Forecasts (FA), (5) Transcribed Weather Broadcasts (TWEB) Text Products, (6) Winds and Temperatures Aloft Forecast (FD), and (7) Centres Weather Service Unit (CWSU), and including (a) Hurricane Advisory (WH), (b) Convective Outlook (AC), and (c) Severe Weather Watch Bulletins (WW) and Alert Messages (AWW).

Table 1.5 SPECI Criteria

Report element	Criteria
Wind	Wind direction changes by 45°; or more in less than 15 min and the wind speed is 10 knots or more throughout the windshift
Visibility	Surface visibility as reported in the body of the report decreases to less than, or if below, increases to equal or exceeds: 3, 2, or 1 mile or the lowest standard instrument approach procedure minimum as published in the National Ocean Service U-S Instrument Procedures. If none is published use 1/2 mile
RVR	Changes to above or below 2400 feet
Tornado, Funnel Cloud, Waterspout	When observed or when disappears from sight (ends)
Thunderstorm	Begins or ends
Precipitation	When freezing precipitation or ice pellets begin, end. or change intensity or hall begins or ends
Squalls	When they occur
Ceilings	The ceiling forms or dissipates below, decreases to less than, or if below, increases to equal or exceeds: 3000, 1500, 1000, or 500 feet or the lowest standard instrument approach procedure minimum as published in the National Ocean Service U-S Instrument Procedures. If none is published use 200 feet
Sky condition	A layer of clouds or obscuring phenomenon aloft that forms below 1000 feet
Volcanic eruption	When an eruption is first noted
Aircraft Mishap	Upon notification of an aircraft mishap, unless there has been an intervening observation
Miscellaneous	Any other meteorological situation designated by the agency, or which, in the opinion of the observer, is critical

1.8.7.3 SPECI Criteria

The SPECI is a non-routine aviation weather report taken when any of the SPECI criteria have been observed. The SPECI criteria of all reported elements are listed in Table 1.5.

1.9 GEO Data Collection Platform (DCP)

Climatologic and synoptic meteorological surface data collected via DCP are used to create weather prediction models that describe the atmosphere and its changes over time. Weather data need to be collected and build a functional and exhaustive meteorological database for accurate models that requires high-quality observations of a variety of relevant measurements, whose process is shown in Fig. 1.21, AWS system (Left) and its components (Right). Automatic weather station (AWS) for

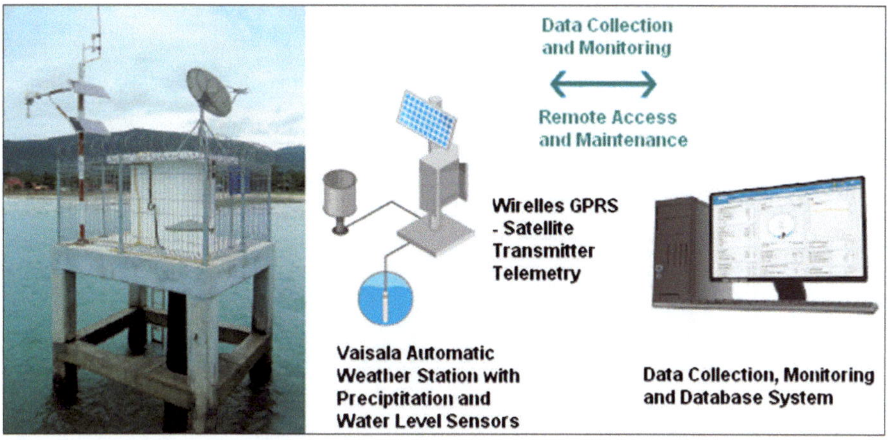

Fig. 1.21 Automatic weather station and components—Courtesy of Brochure: Vaisala

DCP can be installed in offshore or onshore locations and onboard ships, whose main task is to collect local meteorological data and send via GEO satellite to some direct readout station. Each DCP station is specialized for the following tasks:

1. Hydrological measurement data are essential for the prediction and solution of flood, water pollution, drought, and erosion problems. Some areas have plenty of water, while other areas, afflicted by drought, go wanting.
2. Coastal weather stations produce meteorological data for sea weather forecasts and to warn ships and offshore operators about severe weather conditions. The same data are also used in regional and global computer models to help predict atmospheric changes and monitor ocean climate and the state of the oceans.
3. Agrometeorological automatic weather station (AWS) systems measure in-site weather conditions in parallel with local forecasts. The objective is to produce weather data that enable optimal timing and control of all field operations. The data are gathered, stored, viewed, and analyzed to enable more profitable decisions with less risk in farm management.
4. Weather can create significant disruptions in urban areas. Heavy rains can cause severe flooding, snow and freezing rain can disrupt transportation systems, and major storms with accompanying lightning, hail, and high winds can cause power failures.

The main variables of hydrological measurements in AWS are precipitation; water level (rivers, lakes, reservoirs, wells); water temperature; snow depth; water flow; evaporation; soil moisture; and ambient water quality. The sensors used in Vaisala AWS fulfill WMO recommendation for accuracy.

Chapter 2
Space Segment

This chapter describes orbital mechanics and their significance with regard to satellite use for Mobile Satellite Communications (MSC). The space segment of an artificial satellite system is one of its three major operational components, the others being the ground and user segments. It comprises the different satellite constellations and the satellite uplink and downlink. Namely, here are introduced the fundamental laws governing satellite orbits and the principal parameters that describe the motion of the Earth's artificial satellites. The types of satellite orbits are also classified, presented, and compared from the MSC system viewpoint in terms of coverage and link performances. The satellite bus and payloads for communications, broadcasting, and navigation (GNSS) are discussed. The chapter concludes with a brief overview of satellite launch vehicles and orbit insertion. Types of satellite orbits and perturbations are also classified and compared from the communication and navigation systems viewpoint in terms of coverage and link performances.

During the last four decades, military and commercial MSC networks have utilized GEO extensively to the point where orbital portions have become quite crowded and coordination between satellites is becoming constrained and could never solve the problem of polar coverage. On the other hand, non-GEO MSC solutions have recently grown in importance because of their orbit characteristics and coverage capabilities in high latitudes and polar regions.

2.1 Platforms and Orbital Mechanics

The platform is an artificial object located in orbit around the Earth at a minimum altitude of about 20 km in the stratosphere and a maximum distance of about 36,000 km in the Space. The artificial platforms can have a different shape and designation but usually they have the form of aircraft, airship, or spacecraft. In addition, there are special space stations and space ships, which are serving on more

S.D. Ilčev, *Global Mobile Satellite Communications Theory*,
DOI 10.1007/978-3-319-39171-7_2

distant locations from the Earth's surface for scientific exploration and research and for cosmic expeditions.

Orbital mechanics is a specific discipline describing planetary and satellite motion in the solar system, which can solve the problems of calculating and determining the position, speed, path, perturbation, and other orbital parameters of planets and satellites. In fact, a space platform is defined as an unattended object revolving about a larger one. Although it was used to denote a planet's Moon, since 1957 it also means a man-made object put into orbit around a large body (planet), when the former USSR launched its first spacecraft Sputnik-1. Accordingly, man-made satellites are sometimes called artificial satellites.

Orbital mechanics support a communications satellites project in the phases of orbital design and operations. The orbital design is based on a generic survey of orbits and at an early stage in the MSC project is tasked to identify the most suitable orbit for the objective MSC service. The orbital operation is based on rather short-term knowledge of the orbital motion of the satellite and starts with TT&C maintenances after the satellite is located in orbit. In effect, only a few types of satellite orbits are well suited for MSC and navigation systems.

2.1.1 Space Environment

The satellite service begins when a spacecraft is located as a space platform in the desired orbital position in a space environment around the Earth. This space environment is a very specific part of the Universe, where many factors and determined elements affect the planet and satellite motions. The Earth is surrounded by a thick layer of many different gasses and parts known as the atmosphere, whose density decreases as the altitude increases. Hence, there is no air and the atmosphere disappears at about 180 km above the Earth, where the Cosmos begins. The endless environment in space is not very friendly and is extremely destructive, mainly because there is no atmosphere, the cosmic radiation is very powerful, the vacuum creates very high pressure on spacecraft or other bodies, and there is the negative influence of very low temperatures.

The Earth's gravity keeps everything on its surface. All the heavenly bodies such as the Sun, Moon, planets, and stars have gravity and reciprocal reactions. Any object flying in the atmosphere continues to travel until it meets forces due to the Earth's gravity or until it has enough speed to surpass gravity and to hover in the stratosphere. However, to send an object into space, it first has to overcome gravity and then travel at least at a particular minimum speed to stay in space. In this case, an object traveling at about 5 miles/s can circle around the Earth and become an artificial spacecraft.

An enormous amount of energy is necessary to put a satellite into orbit and this is realized by using a powerful rockets or launchers, which are defined as an apparatus consisting of a case containing a propellant (fuel) and reagents by the combustion of which it is projected into the space. As the payload is carried on the top, the rocket is usually separated and drops each stage after burnout and brings a payload up to the required velocity and leaves it in orbit. A rocket is also known as a booster, as a rocket starts with a low velocity and attains some required height, where air drag decreases and it attains a higher velocity.

2.1.2 Laws of Satellite Motion

A satellite is an artificial object located by rocket in space orbit following the same laws in its motion as the planets rotating around the Sun. Thus, Johannes Kepler, a German mathematician, has contributed a great deal to the field of astronomy and astrology. The Laws of Planetary Motion formulated by Kepler proves that the orbits of the planets are ellipses and not circles, as believed by many. The ellipse is a geometrical shape that has two foci, such that the sum of the distance from the focus to any point on the surface of the ellipse is constant. The orbits of planets have small eccentricities (flattening of ellipse), and so, they appear as circles. Based on the properties of ellipses, Johannes Kepler devised three laws that explain the motion of planets around the Sun.

A satellite is an artificial object launched and located by rocket in orbit follows the same laws in its motion as the planets rotating around the Sun. Thus, three important laws for planetary motion were derived by Johannes Kepler, as follows:

1. **First Law**—The first law is also known as The Law of Orbits. As stated, an ellipse has two foci. While studying the motion of planets around the Sun, Kepler explained that the path followed was elliptical, with the Sun as one of the two foci. In simple terms, the law is stated as: The orbit of each planet follows an elliptical path or all planets move in elliptical orbits, with the Sun at one focus, shown in Fig. 2.1a. This indicates that the Sun is one focus, while the other focus is known as the vacant or empty focus.

As shown in the diagram, the Sun and the empty focus lie on the major axis of the ellipse, and the planet lies on the surface of the ellipse. As the planet is continuously moving around the Sun, and as the Sun is not at the center of the ellipse, the Planet–Sun distance will always keep on changing. The Law of Orbits proofs that planet motion lies in the plane around the Sun (1602).

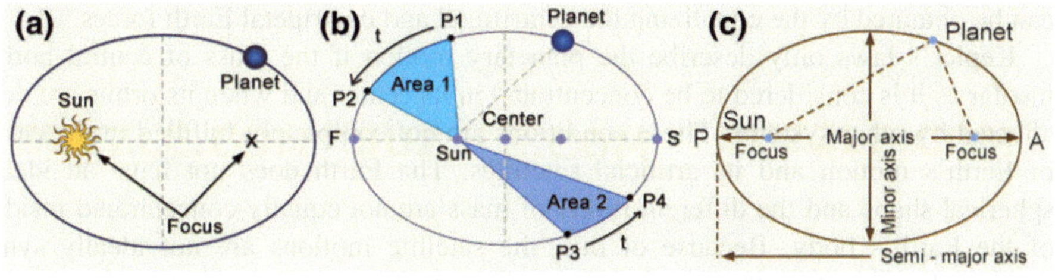

Fig. 2.1 Kepler's laws of satellite motion—Courtesy of Manual: by Ilcev

2. **Second Law**—The second law is also known as The Law of Equal Areas, shown in Fig. 2.1b. As the Sun is one of the foci, it is clear that the Planet–Sun distance will be changing. But, the planet covers up for the increase in the distance by moving faster when it is closer to the Sun. This indicates that planets do not move at a uniform speed. This law states that: The line from the Sun to orbital planet or radius vector (r) sweeps out equal areas in equal intervals of time (t) as the planet travels around the ellipse. The point at which the speed of the planet is fastest is known as Perihelion or Perigee indicated with F (Fastest motion), while the distance with slowest speed is known as Aphelion (Apogee) indicated by S (Slowest motion). The distance measured from the Perihelion to the position of the Sun is known as Perihelion distance, while the distance from the Sun to the Aphelion is known as the Aphelion distance. The law says that while moving in an elliptical path, the planet moves faster when it is closer to the Sun. This way, the radius sweeps equal areas in equal amount of time. If the planet is observed at successive times (P1, P2, P3, P4), it draws the radius vector during the first second observations, showing that the two radius vectors having the same area. So, the area swept during the time (t) by the planet to move from P1 to P2 is the same as the area swept while moving from P3 to P4. This is the Law of Equal Areas (1605).

3. **Third Law**—The third law of planetary motion in ellipse with Perigee (P) and Apogee (A) is alternatively known as The Law of Periods and Harmonic Law, see Fig. 2.1c. This law relates the time required by a planet to make a complete trip around the Sun to its mean distance from the Sun. It can be simply stated as: The square of the planet orbital period is directly proportional to the cube of the semi-major axis of its orbit. The square of the planet's orbital period around the Sun (T) is proportional to the cube of the semi-major axis (a = distance from the Sun) of the ellipse for all planets in the Solar system (1618).

Kepler's laws only describe the planetary motion if the mass of central body insofar as it is considered to be concentrated in its center and when its orbits are not affected by other systems. However, these conditions are not completely fulfilled in the case of Earth's motion and its artificial satellites. Namely, the Earth does not have an ideal spherical shape and the different layers of mass are not equally concentrated inside of the Earth's body. Because of this, the satellite motions are not ideally synchronized and stable, and the motions are namely slower or faster at particular orbital sectors, which present certain exceptions to the rule of Kepler's laws.

Furthermore, in distinction from natural satellites, whose orbits are almost elliptical, the artificial satellites can also have circular orbits, for which the basic relation can be obtained by the equalizing the centrifugal and centripetal Earth forces.

Kepler's laws only describe the planetary motion if the mass of central body insofar as it is considered to be concentrated in its center and when its orbits are not affected by other systems. These conditions are not completely fulfilled in the case of Earth's motion and its artificial satellites. The Earth does not have an ideal spherical shape and the different layers of mass are not equally concentrated inside of the Earth's body. Because of this, the satellite motions are not ideally synchronized and stable; thus, the motions are namely slower or faster at particular orbital sectors, which present certain exceptions to the rule of Kepler's laws.

Kepler's laws were based on observational records and only described the planetary motion without attempting an additional theoretical or mathematical explanation of why the motion takes place in that manner.

In 1687, the English physicist British Sir Isaac Newton published his breakthrough work "Principia Mathematica" with own syntheses, known as the Three Laws of Motion, such as follows:

1. **Law I**—Every body continues in its state of rest or uniform motion in a straight line, unless it is compelled to change that state by forces impressed on it.
2. **Law II**—The change of momentum per unit time of a body is proportional to the force impressed on it and is in the same direction as that force.
3. **Law III**—To every action, there is always an equal and opposite reaction.

On the basis of Law II, Newton also formulated the Law of Universal Gravitation, which states that any two bodies attract one another with a force proportional to the products of their masses and inversely proportional to the square of the distance between them. This law may be expressed mathematically for a circular orbit with the relations:

$$F = m(2\pi/t)^2(R+h) = G\left[M \cdot m/(R+h)^2\right] \quad (2.1)$$

where parameter m = mass of the satellite body; t = time of satellite orbit; R = equatorial radius of the Earth (6.37816×10^6 m); h = altitude of satellite above the Earth's surface; G = Universal gravitational constant (6.67×10^{-11} N m^2/kg^{-2}); M = Mass of the Earth body (5.976032×10^{24} kg) and finally, F = force of mass (m) due to mass (M). Force of mass can be also presented by the following relation:

$$F = ma = dv/dt \quad (2.2)$$

where a = acceleration and v = velocity of satellite orbit.

The force of attraction between two distant point masses m_1 and m_2 separated by a distance r is giving the following relation:

$$F = Gm_1m_2/r^2 \quad (2.3)$$

where G = Newtonian (or universal) gravitation constant.

Consider the simple circular orbit and assuming that the Earth is a sphere, it is possible that can be treated a point mass. The centripetal force F_c required to keep the satellite in a circular orbit = mv^2/r, where v = orbital velocity of the satellite.

The force of gravity that supplies this centripetal force is GMm/r^2, where M = mass of the Earth, and m is the mass of the satellite. Equating the two forces gives relation:

$$F_c = mv^2/r = GMm/r^2 \tag{2.4}$$

Division by m eliminates the mass of the satellite from the equation, which means that the orbit of a satellite is independent of its mass. Thus, the period of the satellite is the orbit circumference divided by the velocity: $T = 2\pi r/v$. Substituting in Eq. 3.3 gives the following relation:

$$T^2 = (4\pi^2/GM)r^3 \tag{2.5}$$

The first-generation NOAA meteorological satellites orbit at approximately 850 km above the Earth's surface. Since the equatorial radius of the Earth is about 6378 km, the orbit radius is about 7228 km. Substituting in Eq. 3.4 shows that the NOAA satellites have a period of about 102 min.

However, radius required for a satellite in GEO has the same angular velocity as the Earth, so the angular velocity mean motion constant of a satellite shows the following:

$$\xi = 2\pi/T \tag{2.6}$$

Substituting Eq. 2.6 in Eq. 2.5 is giving the following formula:

$$r^3 = GM/\xi^2 \tag{2.7}$$

Inserting the angular velocity of the Earth, the required radius for a GEO is 42,164 km or about 35,786 km above the Earth's surface.

2.1.2.1 Geometry of Elliptical Orbit

The satellite in circular orbit undergoes its revolution at a fixed altitude and with fixed velocity, while a satellite in an elliptical orbit can drastically vary its altitude and velocity during one revolution. The elliptical orbit is also subjected to Kepler's Three Laws of satellite motion.

Therefore, the characteristics of elliptical orbit can be determined from elements of the ellipse of the satellite plane with the perigee (Π) and apogee (A) and its position in relation to the Earth, see Fig. 2.2 (Left). The parameters of elliptical orbit are presented as follows:

$$\begin{gathered} e = c/a = \sqrt{\left[1-(b/a)^2\right]} \quad \text{or} \quad e = \left(\sqrt{a^2-b^2}/a\right) \quad p = a\left(1-e^2\right) \quad \text{or} \quad p = b^2/a \\ c = \sqrt{\left(a^2-b^2\right)} \quad a = p/1-e^2 \quad b = a\sqrt{\left(1-e^2\right)} \end{gathered} \tag{2.8}$$

where e = eccentricity, which determines the type of conical section; a = large semi-major axis of elliptical orbit; b = small semi-major axis of elliptical orbit;

c = axis between center of the Earth and center of ellipse, and p = focal parameter. The equation of ellipse derived from polar coordinates can be presented with the resulting trajectory equation as follows:

$$r = p/1 + e\cos\theta[m] \tag{2.9}$$

where r = distance of the satellites from the center of the Earth (r = R + h) or radius of path; Θ = true anomaly and E = eccentric anomaly. In this case, the position of the satellite will be determined by the angle called "the true anomaly," which can be counted positively in the direction of movement of the satellite from 0° to 360°, between the direction of the perigee and the direction of the satellite (S).

The position of the satellite can also be defined by eccentric anomaly, which is the argument of the image in the mapping, which transforms the elliptical trajectory into its principal circle, an angle counted positively in the direction of movement of the satellite from 0 to 360°, between the direction of the perigee and the direction of the satellite. The relations for both mentioned anomalies are given by the following equations:

$$\cos\theta = \cos E - e/1 - e\cos E \quad \cos E = \cos\theta + e/1 + e\cos\theta \tag{2.10}$$

The total mechanical energy of a satellite in elliptical orbit is constant, although there is an interchange between the potential and the kinetic energies. As a result, a satellite slows down when it moves up and gains speed as it loses height. Thus, considering the termed gravitation parameter μ = GM (Kepler's Constant $\mu = 3.99 \times 10^5$ km^3/s^2), the velocity of a satellite in an elliptical orbit may be obtained from the following relation:

$$v = \surd[GM(2/r) - (1/a) = \surd\mu(2/r) - (1/a)] \tag{2.11}$$

Applying Kepler's Third Law, the sidereal time of one revolution of the satellite in elliptical orbit is as follows:

$$\begin{aligned} t &= 2\pi\surd\left(a^3/GM\right) = 2\pi\surd\left(a^3/\mu\right) \\ t &= 3.147099647\surd\left(26{,}628.16 \cdot 10^3\right)^3 \cdot 10^{-7} = 43{,}243.64\,[s] \end{aligned} \tag{2.12}$$

Therefore, the last equation is the calculated period of sidereal day for the elliptical orbit of Russian-based satellite Molniya with apogee = 40,000 km, perigee = 500 km, revolution time = 719 min and a = 0.5 (40,000 + 500 + 2 × 6378.16) = 26628.15 km.

2.1.2.2 Geometry of Circular Orbit

The circular orbit is a special case of elliptical orbit, which is formed from the relations a = b = r and e = 0, see Fig. 2.2 (Right). According to Kepler's Third

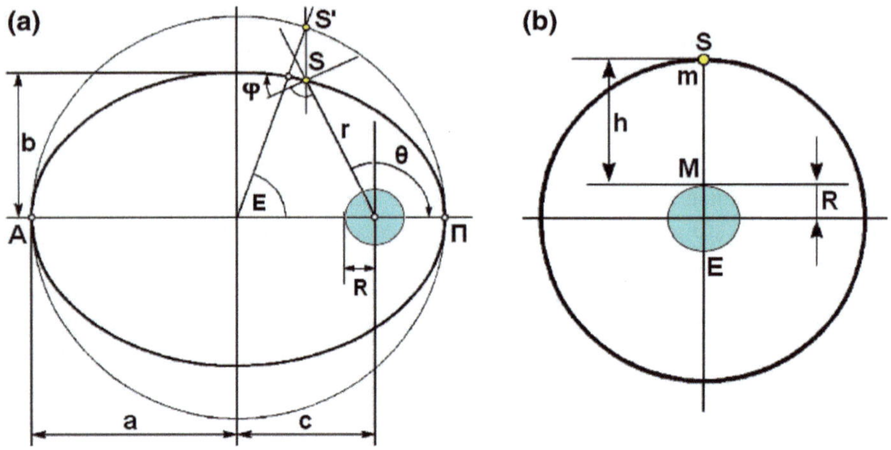

Fig. 2.2 Elliptical and circular satellite orbits—Courtesy of Book: by Galic

Law, the solar time (τ) in relation to the right ascension of an ascending node angle (Ω), the sidereal time (t) with the consideration that μ = GM, and satellite altitude (h), for a satellite in circular orbit, will have the following relations:

$$\begin{aligned} \tau &= t/(1-\Omega t/2\pi) \\ t &= 2\pi\sqrt{(r^3/\mu)} = 3.147099647\sqrt{\left(r^3 \cdot 10^{-7}\right)} \quad [s] \\ h &= \left[{}^3\sqrt{(\mu t^2/4\pi^2)}\right] - R = 2.1613562 \cdot 10^4\left({}^3\sqrt{t^2}\right) - 6.37816 \cdot 10^6[m] \end{aligned} \tag{2.13}$$

The time is measured with reference to the Sun by solar and sidereal day. Thus, a solar day is defined as the time between the successive passages of the Sun over a local meridian. In fact, a solar day is a little bit longer than a sidereal day, because the Earth revolves by more than 360° for successive passages of the Sun over a point 0.986° further. On the other hand, a sidereal day is the time required for the Earth to rotate one circle of 360° around its axis: t_E = 23 h 56 min 4.09 s. Therefore, a geostationary satellite must have an orbital period of one sidereal day in order to appear stationary to an observer on Earth. During rotation, the duration of sidereal day t = 85,164,091 (s) and is considered in such a way for synchronous orbit that h = 35786.04 × 10^3 (m). The speed is conversely proportional to the radius of the path (R + h) and for the satellite in circular orbit, it can be calculated from the following relation:

$$v = \sqrt{(MG/R+h)} = \sqrt{(\mu/r)} = 1.996502 \cdot 10^{-7}/\sqrt{r} = 631.65\sqrt{r}[m/s] \tag{2.14}$$

From Eq. (2.8) and using the duration of sidereal day (t_E) gives the relation for the radius of synchronous or geostationary orbits:

Table 2.1 The values of times different than the synchronous time of orbit

Parameter	Values of time					Unit
t	86164.00	43082.05	21541.23	10770.61	6052.00	s
h	35786.00	20183.62	10354.71	4162.89	800.00	km
(R + h)	42164.00	26561.78	16732.87	10541.05	7178.00	km
v	3075.00	3873.83	4880.72	5584.12	7450.00	km/s^{-1}

$$r = {}^{3}\sqrt{\left[(\mu t)/2\pi)^{2}\right]} \quad (2.15)$$

The satellite trajectory can have any angle of orbital planes in relation to the equatorial plane: in the range from PEO up to GEO plane. Namely, if the satellite is rotating in the same direction of Earth's motion, where (t_E) is the period of the Earth's orbit, the apparent orbiting time (t_a) is calculated by the following relation:

$$t_a = t_E \cdot t/t_E - t \quad (2.16)$$

This means, inasmuch as $t = t_E$ the satellite is geostationary ($t_a = \infty$ or $\tau = 0$). In Table 2.1, several values for times different from synchronous orbital time are presented.

According to Table 2.1 and Eq. (2.9), it is evident that a satellite does not depend so much on its mass but decreases with higher altitude. In addition, satellites in circular orbits with altitudes of a 1700, 10,400 and 36,000 km will have t/ τ values 2/2,18, 6/8, and 24/zero, respectively. In this case, it is evident that only a satellite constellation at altitudes of about 36,000 km can be synchronous or geostationary.

2.1.3 Horizon and Geographic Satellite Coordinates

The horizon system is a type of orbital coordinate parameters that can be used to locate the position of objects in the space. In a satellite, orbits usually use local geographic coordinates, which rotate with the Earth. The horizon and geographical coordinates are very important to find out many satellite parameters and equations for better understanding the problems of orbital plane, satellite distance, visibility of the satellite, coverage areas, etc. The coverage areas of a satellite are illustrated in Fig. 2.3a with the following geometrical parameters: actual altitude (h), radius of Earth (R), angle of elevation (ε), angle of azimuth (A), distance between satellite and the Earth's surface (d), and central angle (Ψ) or subsatellite angle, which is similar to the angle of antenna radiation (δ).

The geographical and horizon coordinates of a satellite are presented in Fig. 2.3b with the following, not yet mentioned, main parameters: angular speed of the Earth's rotation (ν), argument of the perigee (ω), moment of satellite pass across any point on the orbit (t_o), which can be perigee (Π), projection of the perigee point

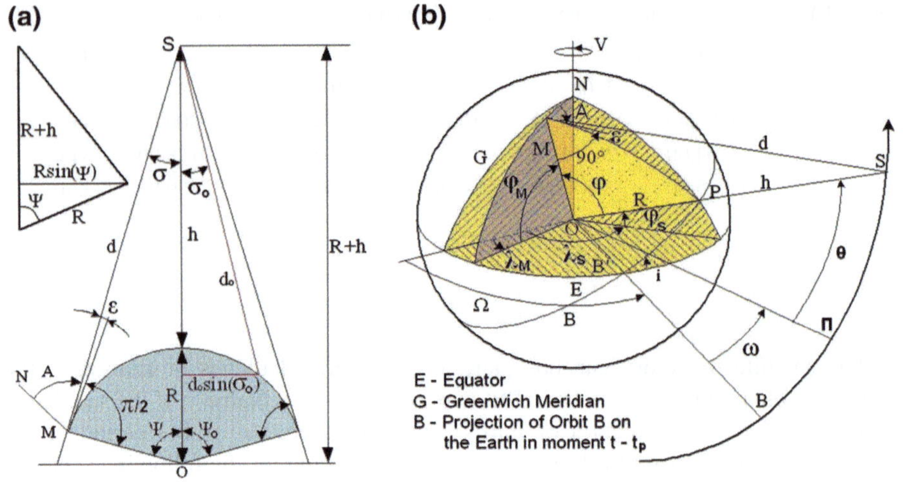

Fig. 2.3 Geometric projection of satellite orbits—Courtesy of Book: by Zhilin

on the Earth's surface (Π'), spherical triangle (B'ΓP), satellite (S), the Point of the Observer or Mobile (M), latitudes of observer and satellite (φ_M and φ_S), longitudes of observer and satellite (λ_M and λ_S), inclination angle (i) of the orbital plane measured between the equatorial and orbital plane, and the right ascension of an ascending node angle in the moment of t_o (Ω_o).

Otherwise, the right ascension of an ascending node angle (Ω) is the angle in the equatorial plane measured counterclockwise from the direction of the vernal equinox to that of the ascending node, while the argument of the perigee (ω) is the angle between the direction of the ascending node and the direction of the perigee.

2.1.3.1 Satellite Distance and Coverage Area

The area coverage or angle of view for each type of satellite depends on orbital parameters, its position in relation to the Land Earth Station (LES), and geographic coordinates. This relation is very simple in the case where the subsatellite point is in the center of coverage, while all other samples are more complicated. Thus, the angle of a GEO satellite inside its range has the following regular reciprocal relation:

$$\delta + \varepsilon + \Psi = 90^\circ \tag{2.17}$$

The circular sector radius can be determined by the following relation:

$$R_s = R \sin \Psi \tag{2.18}$$

When the altitude of orbit h is the distance between satellite and subsatellite point (SP), the relation for the altitude of the circular sector can be written as:

$$h_s = R(1 - \cos \Psi) \tag{2.19}$$

From a satellite communications point of view, there are three key parameters associated with an orbiting satellite, such as (1) Coverage area or the portion of the Earth's surface that can receive the satellite's transmissions with an elevation angle larger than a prescribed minimum angle; (2) The slant range (actual LOS distance from a fixed point on the Earth to the satellite), and (3) The length of time a satellite is visible with a prescribed elevation angle.

Elevation angle is an important parameter, since communications can be significantly impaired if the satellite has to be viewed at a low-elevation angle, that is, an angle too close to the horizon line. In this case, a satellite close to synchronous orbit covers about 40 % of the Earth's surface. Thus, from the diagram in Fig. 2.3a a covered area expressed with central angle (2δ or 2Ψ) or with arc ($MP \approx R\Psi$) as a part of Earth's surface can be derived with the following relation:

$$C = \pi\left(R_s^2 + h_s^2\right) = 2\pi R^2(1 - \cos \Psi) \tag{2.20}$$

Since the Earth's total surface area is $4\pi R^2$, it is easy to rewrite C as a fraction of the Earth's total surface:

$$C/4\pi R^2 = 0,5(1 - \cos \Psi) \tag{2.21}$$

The slant range between a point on Earth and a satellite at altitude (h) and elevation angle can be defined in this way:

$$z = \left[(R \sin \varepsilon)^2 + 2Rh + h^2\right]^{1/2} - R \sin \varepsilon \tag{2.22}$$

This determines the direct propagation length between LES, (h) and (ε) and will also find the total propagation power loss from LES to satellite. In addition, (z) establishes the propagation time (time delay) over the path, which will take an electromagnetic field as:

$$t_d = (3.33)z \quad [\mu \text{ sec}] \tag{2.23}$$

To propagate over a path of length (z) km, it takes about 100 ms to transmit to GEO. If the location of the satellite is uncertain ± 40 km, a time delay of about ±133 μsec is always present in the Earth-to-satellite propagation path. When the satellite is in orbit at altitude (h), it will pass over a point on Earth with an elevation angle (ε) for a time period:

$$t_p = (2\Psi/360)(t/1 \pm (t/t_E) \tag{2.24}$$

The quotations for right ascension of the ascending node angle (Ω) and argument of the perigee (ω) are as follows:

$$\Omega = 9{,}95(R/r)^{3.5}\cos i \quad \text{or} \quad \Omega = \Omega_o + \nu(t - t_o)$$
$$\omega = 4{,}97(R/a)^{3.5}\left[5\cos^2 i - 1/\left(1 - e^2\right)^2\right] \tag{2.25}$$

The limit of the coverage area is defined by the elevation angle from LES above the horizon with angle of view ε = 0°. In this case, the satellite is visible and its maximal central angle for GEO will be as follows:

$$\begin{aligned}
\Psi &= \text{arc}\cos(R\cos\varepsilon/r) - \varepsilon \quad \text{or} \\
\Psi &= \pi/2 - \text{arc}\ \sin(R/r) = \text{arc}\cos(R/r) - \varepsilon = \text{arc}\cos k - \varepsilon \\
\Psi &= \text{arc}\cos 6{,}376.16/42{,}164.20 = \text{arc}\cos 0.15126956 = 81^\circ 17' 58.18'' \\
C_{max} &= 255.61 \cdot 10^6(1 - 0.15126956) = 216.94 \cdot 10^6\left(\text{km}^2\right)
\end{aligned} \tag{2.26}$$

Therefore, all MES and LES with a position above Ψ = 81° will be not covered by GEO satellites. Since the Earth's square area is 510,100,933.5 km^2 and the extent of the equator is 40,076.6 km, only with three GEO mutually moved apart in the orbit by 120° it is possible to cover a great area of the Earth's surface, see Fig. 2.4a, which shows AOR (Atlantic), IOR (Indian), and POR (Pacific) satellite coverages. The zero angles of elevation have to be avoided, even to get maximum coverage, because this increases the noise temperature of the receiving antenna. Owing to this problem, an equation for the central angle with minimum angle of view between 5° and 30° will be calculated with:

$$\Psi_s = \text{arc}\ \cos(k\ \cos\ \varepsilon) - \varepsilon \tag{2.27}$$

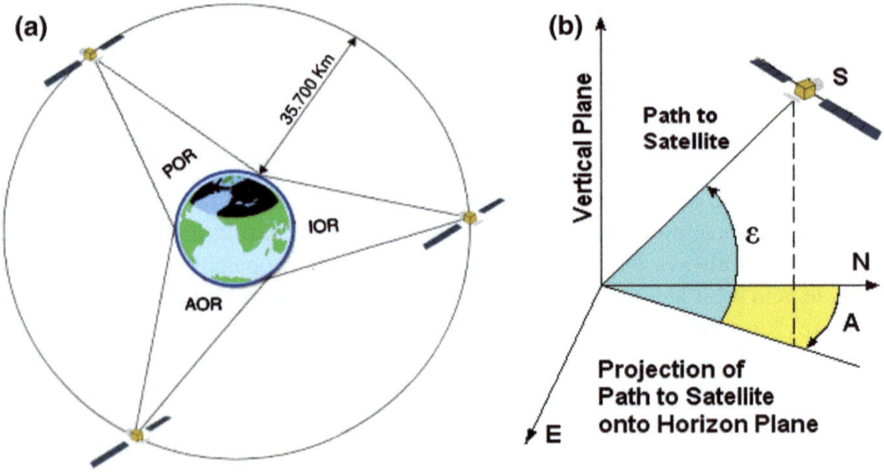

Fig. 2.4 GEO coverage and look angle parameters—Courtesy of Book: by Pratt

The arch length or the maximum distant point in the area of coverage can be determined in the following way:

$$l = 2\pi R(2\Psi/360 = 222.64\Psi \text{ [km]} \tag{2.28}$$

The real altitude of satellite over subsatellite point is as follows:

$$h = r - R = 42,162 - 6,378 = 35,784 \text{[km]} \tag{2.29}$$

The view angle under which a GEO satellite can see LES/MES is called the "subsatellite angle." More distant points in the coverage area for GEO satellites are limited around $\varphi = 70°$ of North and South geographical latitudes and around $\lambda = 70°$ of East and West geographical longitudes, viewed from the subsatellite point. Theoretically, all Earth stations around these positions are able to see satellites by a minimum angle of elevation of $\varepsilon = 5°$. Such access is very easy to calculate, using simple trigonometry relations:

$$\delta_{\varepsilon=0} = \text{arc sin } k \approx 9^{\text{o}} \tag{2.30}$$

At any rate, the angle (Ψ) is in correlation with angle (δ), which can determine the aperture radiation beam. For example, the aperture radiation beam of satellite antenna for global coverage has a radiation beam of $2\delta = 17.3°$. According to Fig. 2.3a, it will be easy to find out relations for GEO satellites as follows:

$$\text{tg}\,\delta = k \sin\Psi/1 - k\cos\Psi = 0.15126956 \sin\Psi/1 - \cos\Psi/1 - 0.15126956\cos\Psi) \tag{2.31}$$

$$\delta_s = 90° - \Psi_s = 8°42' \cdot 1.82''$$

Differently to say, the width of the beam aperture ($2\delta_s$) is providing the maximum possible coverage for synchronous circular orbit. The distance of LES and MES with regard to the satellite can be calculated using Fig. 2.3a and Eqs. (2.13) and (2.22) by:

$$d = R \sin\Psi / \sin\delta = r \sin / \cos\varepsilon \tag{2.32}$$

The parameter (d) is quite important for transmitter power regulation of LES, which can be calculated by the following equation:

$$\begin{gathered} d = \surd\left[(R + r)^2 - 2Rr\cos\Psi\right] \quad \text{or} \\ d = h\surd\left[1 + 2(1/k)(R/h)^2(1 - \cos\varphi\cos\Delta\lambda)\right] \text{or} \\ d = r[1 - (R\cos\varepsilon/r)^2]^{1/2} - R\sin\varepsilon \end{gathered} \tag{2.33}$$

Accordingly, when the position of any MES is near the equator in subsatellite point (P) or right under the GEO satellite, its distance is equal to the satellite altitude and takes out value for d = H of 35,786 km. Thus, every MES will have a further position from (P) when the central angle exceeds $\Psi = 81°$, when $d_{max} = 41,643$ km.

2.1.3.2 Satellite Look Angles (Elevation and Azimuth)

The horizon coordinates are considered to determine satellite position in correlation with an Earth observer, LES, and MES terminals. These specific and important horizon coordinates are angles of satellite elevation and azimuth, illustrated in Figs. 2.3a, b and 2.4b, respectively.

The satellite elevation (ε) is the angle composed upward from the horizon to the vertical satellite direction on the vertical plane at the observer point. From point (M) shown in Fig. 2.3a, the look angle of ε value can be calculated by the following relation:

$$\mathrm{tg}\,\varepsilon = \cos\Psi - k/\sin\Psi \tag{2.34}$$

Figure 2.5a illustrates the Mercator chart of the First-Generation Inmarsat space segment, using three-ocean coverage areas with the projection of elevation angles and with one example of a plotted position of a hypothetical ship (may also be aircraft or any mobile). Thus, it can be concluded that Mobile Earth Station (MES) at designated position ($\varepsilon = 25°$ for IOR and $\varepsilon = 16°$ for AOR) has the possibility to use either GEO satellites over IOR or AOR to communicate with any LES inside the coverage areas of both satellites.

The satellite azimuth (A) is the angle measured eastward from the geographical North line to the projection of the satellite path on the horizontal plane at the observer point. This angle varies between 0 and 360° as a function of the relative positions of the satellite and the point is considered. The azimuth value of the satellite and subsatellite point looking from the point (M) or the hypothetical position of MES can be calculated as follows:

$$\mathrm{tg}\,A' = \mathrm{tg}\,\Delta\lambda_M - k/\sin\Psi \tag{2.35}$$

Otherwise, the azimuth value, looking from subsatellite point (P), can be calculated as:

$$\mathrm{tg}\,A = \sin\Delta\lambda/\mathrm{tg}\,\varphi \text{ or } \sin A = \cos\varphi\sin\Delta\lambda\,\mathrm{cosec}\Psi \tag{2.36}$$

Figure 2.5b illustrates the Mercator chart of First-Generation Inmarsat 3-satellite or ocean coverage areas with the projection of azimuth angles, with one example for the plotted position of a hypothetical ship ($\varepsilon = 47°$ for IOR and $\varepsilon = 303°$ for AOR).

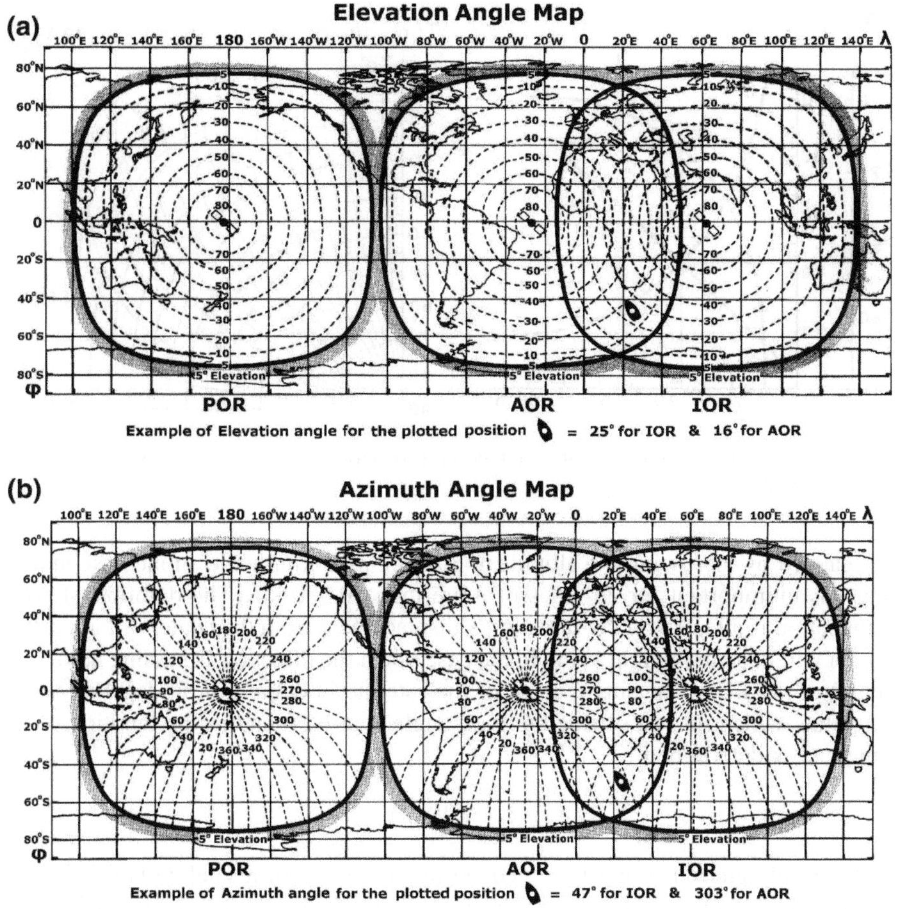

Fig. 2.5 Elevation and Azimuth angle maps—Courtesy of Manual: by EB Communications

Any mobile inside of both satellites' coverage can establish a radio link to the subscribers onshore via any LES.

However, parameter (A') is the angle between the meridian plane of point (M) and the plane of a big circle crossing this point and subsatellite point (P), while the parameter (A) is the angle between a big circle and the meridian plane of point (P). Thus, the elevation and azimuth are, respectively, vertical or horizontal look angles, or angles of view, in which range the satellite can be seen.

Figure 2.6a presents a correlation of the look angle for three basic parameters (δ, Ψ, d) in relation to the altitude of the satellite. Inasmuch as the altitude of the satellite is increasing as the values of central angle (Ψ), distance between satellite and the Earth's surface (d) and duration of communication (t_c) or time length of signals are increasing, while the value of subsatellite angle (δ) is indirectly proportional. An important increase of look angle and duration of communication can

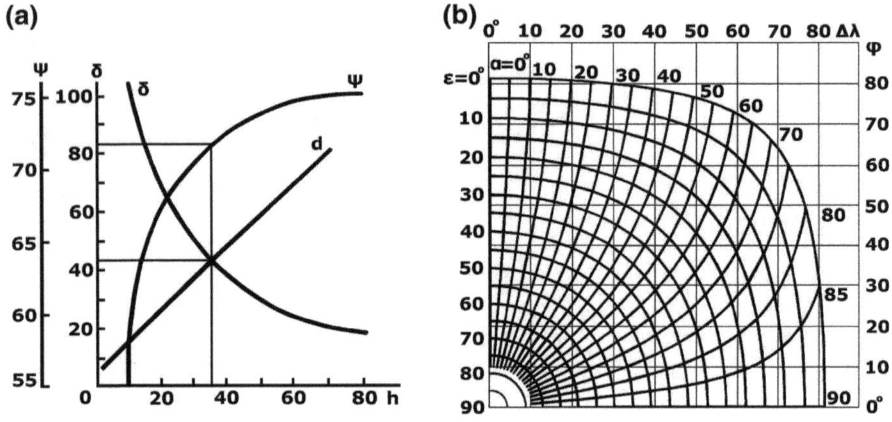

Fig. 2.6 Look angle parameters and graphic of geometric coordinates for GEO—Courtesy of Book: by Zhilin

be realized by increasing the altitude to 30 or 35,000 km, while an increase in look angle is unimportant for altitudes of more than 50,000 km. The duration of communication is affected by the direction's displacement from the center of look angle, which will have maximum value in the case when the direction is passing across the zenith of the LES. The single angle of the satellite in circular orbit depends on the t/2 value, which in area of satellite look angle, can be found in the duration of the time and is determined as:

$$t_c = \Psi t/\pi \qquad (2.37)$$

Practical determination of the geometric parameters of a satellite is possible by using many kinds of plans, graphs, and tables. It is possible to use tables for positions of MES (φ, λ), by the aid of which longitudinal differences can be determined between MES and satellite for four feasible ship's positions: N/W, S/W, N/E, and S/E in relation to GEO.

One of the most important practical pieces of information about a communications satellite is whether it can be seen from a particular location on the Earth's surface. In Fig. 2.6b, a graphic design is shown which can approximately determine limited zones of satellite visibility from the Earth (MES) by using elevation and azimuth angles under the condition that δ = 0. This graphic contains two groups of crossing curves, which are used to compare (φ) and (Δλ,) coordinates of mobile positions. Thus, the first group of parallel concentric curves shows the geometric positions where elevation has the constant value (ε = 0), while the second group of fan-shaped curves starting from the center shows the geometric positions where the difference in azimuth has the constant value (a = 0). This diagram can be used in accordance with Fig. 2.3b in the following order:

Table 2.2 The form for calculation of Azimuth values

The GEO direction in relation to MES	Calculating of Azimuth angles
Course of MES toward S & W	A = a
Course of MES toward N & W	A = 180° − a
Course of MES toward N & E	A = 180° + a
Course of MES toward S & E	A = 360° − a

(1) First, it is necessary to note the longitude values of satellite (λ_S) and mobile (λ_M) and the latitude of the mobile (φ_M), then calculate the difference in longitude ($\Delta\lambda$), and plot the point into the graphic with both coordinates (φ_M & $\Delta\lambda$).
(2) The value of elevation angle (ε) can then be determined by a plotted point from the group of parallel concentric curves.
(3) The difference value of azimuth (a) can be determined by a plotted point from the group of fan-shaped curves starting from the center.
(4) Finally, depending on the mobile position, the value of azimuth (A) can be determined on the basis of the relations presented in Table 2.2.

Inasmuch as the position of Ship Earth Station (SES) or any MES is of significant or greater height above sea level (if the bridge or ship's antenna is in a very high position) or according to the flight altitude of Aircraft Earth Station (AES), the elevation angle will be compensated by the following parameter:

$$x = \arccos(1 - H/R) \tag{2.38}$$

where H = height above sea level of observer or MES. Let us say, if the position of LES is a height of H = 1000 m above sea level, the value of $x \approx 1°$. This example can be used for the determination of AES compensation parameters, depending on actual aircraft altitude. In such a way, the estimated value of elevation angle has to be subtracted for the value of the compensation parameter (x).

2.1.3.3 Satellite Track and Geometry (Longitude and Latitude)

The satellite track on the Earth's surface and the presentation of a satellite's position in correlation with the MES results from a spherical coordinate system, whose center is the middle of Earth, is illustrated in Fig. 2.3b. In this way, the satellite position in any time can be decided by the geographic coordinates, subsatellite point, and range of radius. Thus, the subsatellite point is a determined position on the Earth's surface; above, it is the satellite at its zenith.

The longitude and latitude are geographic coordinates of the subsatellite point, which can be calculated from the spherical triangle (B'ΓP), using the following relations:

$$\begin{aligned} \sin\varphi &= \sin(\theta+\omega)\sin i \\ tg(\lambda_S-\Omega) &= tg(\theta+\omega)\cos i \end{aligned} \tag{2.39}$$

With the presented equation in previous relation, it is possible to calculate the satellite path or trajectory of subsatellite points on the Earth's surface. The GEO track breaks out at the point of coordinates $\varphi = 0$ and λ = const.

Furthermore, considering geographic latitude (φ_M) and longitude (λ_M) of the point (M) on the Earth's surface presented in Fig. 2.3b, what can be the position of the MES, taking into consideration the arc (MP) of the angle illustrated in Fig. 2.3a, the central angle can be calculated by the following relations:

$$\begin{aligned} \cos\Psi &= \cos\varphi_S\cos\Delta\lambda\cos\varphi_M + \sin\varphi_S\sin\varphi_M \text{ or} \\ \cos\Psi &= \cos \text{arc MP} = \cos\varphi_M\cos\Delta\lambda \end{aligned} \tag{2.40}$$

The transition calculations from geographic to spherical coordinates and vice versa can be computed with the following equations:

$$\begin{aligned} \cos\Psi &= \cos\varphi\cos\Delta\lambda \text{ and } tg\,A = \sin\Delta/\lambda tg\,\varphi, \text{ respectively} \\ \sin\varphi &= \sin\Psi\cos A \text{ and } tg\,\Delta\lambda = tg\,\Psi\sin A \end{aligned} \tag{2.41}$$

These relations are useful for any point or area of coverage on the Earth's surface, then for a center of the area if it exists, as well as for spot beam and global area coverage for MSC systems. The optimum number of GEO satellites for global coverage can be determined by:

$$n = 180^\circ/\Psi \tag{2.42}$$

For instance, if $\delta = 0$ and $\Psi = 81^\circ$, it will be necessary to put into orbit only 3 GEO, and to get a global coverage from 75° N to 75° S geographic latitude. Hence, in a similar way the number of satellites can be calculated for other types of satellite orbits.

The trajectory of radio waves on a link between an MES and a satellite at distance (d) and the velocity of light ($c = 3 \times 10^8$ m/s) requires a propagation time equal to:

$$T = d/c\,(s) \tag{2.43}$$

The phenomenon of apparent change in frequency of signal waves at the receiver when the signal source moves with respect to the receivers (Earth) was explained and quantified by Johann Doppler (1803–53). The frequency of the satellite transmission received on the ground increases as the satellite is approaching the ground observer and reduces as the satellite is moving away. This change in frequency is called Doppler effect or shift, which occurs on both the uplink and the downlink. This effect is quite pronounced for LEO and compensating for it requires frequency tracking in a narrowband receiver, while its effects are negligible for GEO satellites. The Doppler shift at a transmitting frequency (f) and radial velocity (v_r) between the ground observer and the transmitter can be calculated by the following relation:

$$\Delta f_D = f\, v_r/c \text{ where } v_r = dR/dt \tag{2.44}$$

For an elliptical orbit, assuming that R = r, the radial velocity is given by:

$$v_r = dr/dt = (dr/\theta)(d\theta/dt) \tag{2.45}$$

The sign of the Doppler shift is positive when the satellite is approaching the observer and vice versa. Doppler effect can also be used to estimate the position of an observer provided that the orbital parameters of the satellite are precisely known. This is very important for the development of Doppler satellite tracking and determination systems.

2.2 Spacecraft Launching and Station-Keeping Techniques

The launch of the satellite and the controlling support services are a very critical point in the creation of space communications and the most expensive phase of the total system cost. At the same time, the need to make a satellite body capable of surviving the stresses of the launch stages is a major element in their design phase. Satellites are also designed to be compatible with more than one model of launch vehicle and launching type.

In a more determined sense, there are multistage expendable and, manned or unmanned, reusable launchers. Owing to location and type of site, there are land-based and sea-based launch systems. Additional rocket motors, such as perigee and apogee kick propulsion systems, may also be required.

The process of launching a satellite is based mostly on launching into an equatorial circular orbit and after in GEO, but similar processes or phases are used for all types of orbits. The processes involved in the launching technique depend on the type of satellite launcher, the geographical position of the launching site, and constraints associated with the payload. In order to successfully put the satellite into the transfer and drift orbit, the launcher must operate with great precision with regard to the magnitude and orientation of the velocity vector. On the other hand, launching operations necessitate either TT&T facilities at the launching base or at the stations distributed along the trajectory.

2.2.1 *Satellite Installation and Launching Operations*

Satellites are usually designed to be compatible with more than one prototype of launchers. Launching, putting, and controlling satellites into orbit are very expensive operations, so the expenses of launcher and support services can exceed the cost of the satellites themselves. The basic principle of any launch vehicle is that the rocket is propelled by reaction to the momentum of hot gas ejected through exhaust nozzles.

Thus, for a spacecraft to achieve synchronous orbit, it must be accelerated to a velocity of 3070 m/s in a zero-inclination orbit and raised a distance of 42,242 km from the center of the Earth. Most rocket engines use the oxygen in the atmosphere to burn their fuel but solid or liquid propellant for a launcher in space must comprise both a fuel and an oxygen agent. There are two techniques for launching a satellite, namely by direct ascent and by Hohmann transfer ellipse.

2.2.1.1 Direct Ascent Launching

A satellite may be launched into a circular orbit by using the direct ascent method, shown in Fig. 2.7a. The thrust of the launch vehicle is used to place the satellite in a trajectory, the turning point of which is marginally above the altitude of the desired orbit. The initial sequence of the ascent trajectory is the boost phase, which is powered by the various stages of the launch vehicle. This is followed by a coasting phase along the ballistic trajectory, the spacecraft at this point consisting of the last launcher stage and the satellite. As the velocity required to sustain an orbit will not have been attained at this point, the spacecraft falls back from the highest point of the ballistic trajectory.

When the satellite and final stage have fallen to the desired injection altitude, having in the meantime converted some of their potential energy into kinetic energy, the final stage of the launcher, called the Apogee Kick Motor (AKM), is

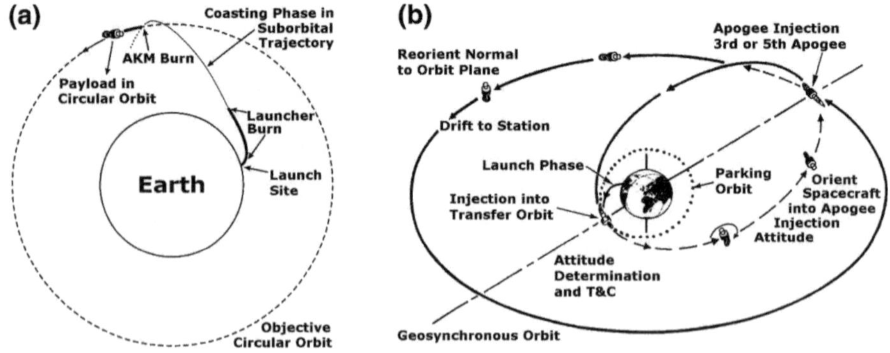

Fig. 2.7 Satellite installation in circular and synchronous orbit—Courtesy of Book: by Pascall

activated to provide the necessary velocity increase for injection into the chosen circular orbit. In effect, the AKM is often incorporated into the satellite itself, where other thrusters are also installed for adjusting the orbit or the altitude of the satellite throughout its operating lifetime in space. The typical launch vehicles for direct ascent satellite launching are US-based Titan IV, Russian-based Proton, and Ukrainian-based Zenit.

2.2.1.2 Indirect Ascent Launching

A satellite may be launched into an elliptical or synchronous orbit by using the successive or indirect ascent sequences, known as the Hohmann transfer ellipse method, illustrated in Fig. 2.7b. The Hohmann transfer ellipse method enables a satellite to be placed in an orbit at the desired altitude using the trajectory that requires the least energy. At the first sequence, the launch vehicle propels the satellite into a low parking orbit by the direct ascent method. The satellite is then injected into an elliptical transfer orbit, the apogee of which is the altitude of the desired circular synchronous orbit. At the apogee, additional thrust is applied by an AKM to provide the velocity increment necessary for the attainment of the required synchronous orbit. In practice, it is usual for the direct ascent method to be used to inject a satellite into a LEO and for the Hohmann transfer ellipse to be used for higher types of satellite orbits.

2.2.2 Satellite Launchers and Launching Systems

Two major types of launch vehicles can be used to put a satellite into LEO, HEO, and GEO constellation: Expendable and Reusable Vehicles. There are also two principal locations or site-based types of launching centers: Land-based and Sea-based launch systems.

2.2.2.1 Expendable Launching Vehicles

The great majority of communications satellites have been launched by expendable vehicles and this is likely to continue to be the case for many years to come. There are two types of these vehicles: expendable three-stage vehicles and expendable direct-injection vehicles.

1. **Expendable Three-Stage Vehicles**—Typical series of three-stage vehicles are Delta and Atlas (USA), Ariane (Europe), Long March (China), and H-II (Japan). In addition, a new generation of launchers has already been developed with two stages such as Delta III and Ariane 5. Both stages are propellant systems using cryogenic liquid fuel, while the first stage is assisted by nine strap-on solid-fuel motors.

The first and second stages of three-stage expandable launch vehicles are usually designed to lift it clear of the Earth's atmosphere, to accelerate horizontally to a velocity of about 8000 m/s and enter a parking orbit at a height of about 200 km. The plane of the parking orbit will be inclined to the equator at an angle not less than the latitude of the launch site. The most efficient way of getting from the parking orbit to a circular equatorial orbit is to convert the parking orbit into an elliptical orbit in the same plane, with the perigee at the height of the parking orbit and the apogee at about 36,000 km and then to convert the transfer ellipse to the GEO.

Thus, the third stage is fired as the satellite crosses the equator, which ensures that the apogee of the Geostationary Transfer Orbit (GTO) is in the equatorial plane. When the satellite is placed in the GTO, the third stage has completed its mission and is jettisoned. The final phase of the Hohmann transfer three-stage launch sequence is carried out by means of AKM built into the satellite. The propulsion of this motor is required to provide at the apogee of the GTO a velocity increment of such a magnitude and in such a direction as to reduce the orbit to zero and make the orbit circular. Once the satellite is in the GEO trajectory, the attitude is corrected, the antennas and solar panels are deployed and the satellite is drifted to the correct longitude (apogee position) for operation.

2. **Expendable Direct-Injection Vehicles**—Typical models of direct-injection launchers are the USA-based Titan IV and the Russian-based Proton, illustrated in Fig. 2.8 (a—Left) and (a—Right), respectively, and also Zenit (Ukraine). Otherwise, these types of vehicles do not need an AKM because direct-injection

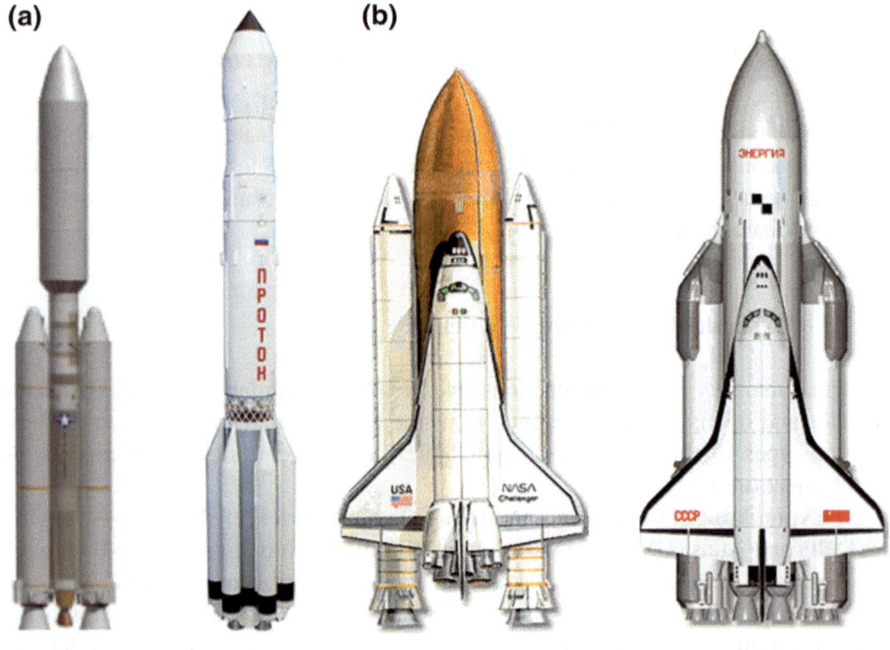

Fig. 2.8 Types of launch vehicles—Courtesy of Book: by Pascall

launchers have a fourth stage, which converts directly from GTO to GEO constellation. The Proton rocket is one of the most capable and reliable heavy lift launch vehicles in operation today. Proton D-1 and D-1-E launcher variants have three and four stages, respectively. At liftoff, the total weight of Proton is about 688 tons and this vehicle has the capability of placing a maximum of 4,500 and 2,600 kg into GTO and GEO, respectively.

2.2.2.2 Reusable Launch Vehicles

Reusable launch vehicles have already been developed in the USA (Space Shuttle) and former USSR (Energia/Buran), which have as their aim the development of vehicles that could journey into space and return, all or much of their structure being reusable, and thus, the satellite launching will cost less. Moreover, in using these launchers there will be less burnt-out upper stages than with expendable vehicles. What remains in space, the small pieces in transfer orbits for many years and much small debris, remains in LEO for a long time, adding to the growing space junk hazard for operational satellites and future space operations. There are other projects for the development of similar vehicles such as a small manned reusable Space Shuttle called Hermes (Europe) and Hope (Japan), unmanned space plane Hotol in UK is proposed, while in Germany and the USA two similar vehicles are projected: TAV (Trans Atmospheric Vehicle) and Sanger Space plane, respectively. Thus, in the development of these small vehicles, it is important to realize whether any of them could carry sufficient weight and be able to put communications satellites into the desired orbits.

1. **Space Shuttle**—The US-based NASA developed a fleet of manned reusable vehicles of Space Transportation System (STS) called Space Shuttle, which are capable of lifting a satellite of up to 29.5 tons into a parking orbit, inclined at 28.5°, with an altitude of up to 431 km, shown in Fig. 2.8 (b—Left). A Shuttle has three main elements: (1) the orbiter for carrying the satellite and crew; (2) a very large external tank containing propellant for the main engine of the orbiter; and (3) two solid-propellant boosters. The reusable Space Shuttle plane is 37.2 m long, the fuselage is 4.5 m in diameter, the wingspan is 23.8 m, and the mass is about 84.8 tons. This STS is designed to accommodate in total 7 crewmembers and passengers on board plane. The system came into service in 1981 and made over twenty successful operational flights until January 1986, when the Shuttle Challenger was destroyed by a fault in the solid-propellant booster and all the crew were killed in a tragic accident. Following this disaster, NASA redesigned the booster but decided to use STS only for regular launch program of government and scientific vehicles.

2. **Energia/Buran Space plane**—The launcher Energia is the most powerful operational reusable vehicle in the world, capable of carrying about 100 tons into space, whose four first-stage booster units are recoverable for reuse. In particular, it can launch the Buran space plane, enabling it to acquire a LEO and to land with the aid of its own rocket engine, shown in Fig. 2.8 (b—Right). The main purpose for which those very heavy lift vehicles were developed was to ferry personnel and

supplies for the Russian space station Mir and also to retrieve or repair satellites already in orbit.

The Energia vehicle can also carry into space a side-mounted canister containing an upper stage and a payload compartment suitable, for example, for a large heavy spacecraft or group of communications satellites to be placed in orbit. Energia flew for the first time on May 15, 1987, carrying a spacecraft mock-up and later on November 18, 1988, carrying an unmanned version of Buran space plane. The reusable Buran space plane is 36.3 m long, the fuselage is 5.6 m in diameter, the wingspan is 24 m, and the mass is about 100 tons. It can be flown in automatic configuration or under the control of a pilot to place satellites in LEO or to retrieve them and come back to base for the next use. Up to ten people, crew and passengers, can be accommodated and it can carry in the cargo bay up to 30 tons into an orbit of 200 km altitude and 51.6° inclination. In fact, this plane enables large satellites to be put into orbit and construction of space stations to be considered for both for telecommunication purposes and for scientific missions.

The Energia Launch Vehicle was also the successor to the N-1 Moon Rockets, except that Buran was also used to launch Polyus from Baikonur Cosmodrome in Kazakhstan (former Soviet Union). Energia was 60 m high and 18 m in diameter, consisting of a central core and four strap-on boosters, while the core was 58.1 m high and 7.7 m diameter. It used 4 RD-0120 rocket engines. The propellants were liquid hydrogen and oxygen. The strap-on boosters were then 38.3 m high and 3.9 m in diameter, with a single four-chamber RD-170 kerosene/liquid oxygen rocket engine.

In 1992, the Russian Space Agency (Roscosmos) decided to terminate the Energia/Buran Program due to Russia's economic difficulties after disintegration of former Soviet Union. At that stage, the second Orbiter had been assembled and assembly of the third Orbiter with improved performance was nearing completion.

Although the Energia project has been abandoned, it may return to service if a market is found, or adequate partners. Consideration is being given in Russia to the development of a more compact winged space plane designed to ferry personnel and their luggage into space, such as new developed space plane Kliper by NPO Energia, see Fig. 2.9. This compact shuttlecraft could be placed atop of a Proton, Soyuz, Angara or any other launchers.

2.2.2.3 Land-Based Launching Systems

Most satellite launches have taken place from the following launch facilities:

1. **US-Based Launch Centers**—The USA launches satellites from two main locations, in Florida Cape Canaveral, suitable for direct equatorial orbit, and the Vandenberg Air Force Base in California, suitable for polar orbit missions.
2. **Russian Launch Centers**—Russian satellites are launched from two main launch centers named Baikonur and Northern Cosmodrome. Baikonur lies north of Tyuratam in Kazakhstan, with the all launching support infrastructure for launching Proton and Energia heavy launchers. The Northern Cosmodrome is

Fig. 2.9 Russian space plane Kliper—Courtesy of Brochure: by Zak

located near Plesetsk, south of the town Archangelsk, suitable for launching satellites for all purposes in high-inclination orbits. This Cosmodrome is the world's busiest launch site.

3. **European Launch Centers**—The main European launch Cosmodrome is the Guiana Space Center in French Guiana, using Ariane vehicles. The position of this Cosmodrome enables the best advantage to be taken of the Earth's rotation for direct equatorial orbit.
4. **Chinese Launch Centers**—The principal launch sites in China are Jiuquan and Xi Chang, for launching Long March vehicles. In the meantime, the Xi Chang launch center has also become most used for launches into the GEO for the international market.
5. **Japanese Launch Centers**—The Japan's Tanegashima Space Center is situated in the prefecture of Kagashima. The facilities include the Takesaki Range for small rockets and the Osaki Range was used for the launch of H-I vehicles until the termination of program in 1992. After renovation, the Osaki Range will be used as the launching for the next generation of J-I Japanese vehicles. The new Yoshinobu launch complex has been constructed next to the Osaki center to satisfy the requirements of the new H-II launcher.

2.2.2.4 Sea-Based Launch Systems

The Sea Launch Multinational Organization was developed in March 1996 to overcome the cost of land-based launch infrastructure duplication around the world. The newly formed Sea Launch system is owned by the Sea Launch Partnership Limited in collaboration with international partners such as US Boeing Commercial

Space Company, Russian RSC Energia, Ukrainian KB Yuzhnoye/PO Yuzhmash, Shipping Anglo-Norwegian Kvaerner Group, and Sea Launch Company, LLC.

The Sea Launch Company, partner locations and operating centers, has US-based headquarters in Long Beach, California, and is manned by selected representatives of each of the partner companies.

The Sea Launch Partners have the following responsibilities and tasks:

1. Boeing responsibilities include designing and manufacturing the payload fairing and adapter, developing and operating the Home Port facility in Long Beach, integrating the spacecraft with the payload unit and the Sea Launch system, performing mission analysis and analytical integration, leading operations, securing launch licensing documents, and providing range services.
2. RSC Energia is responsible for developing and qualifying the Block DM-SL design modifications, manufacturing the Block DM-SL upper stage, developing and operating the automated ground support infrastructure and equipment, integrating the Block DM-SL with Zenit-2S and launch support equipment, planning and designing the CIS portion of launch operations, developing flight design documentation for the flight of the upper stage, and performing launch operations and range services.
3. KB Yuzhnoye/PO Yuzhmash are responsible for developing and qualifying Zenit-2S vehicle design modifications, integrating the launch vehicle flight hardware, developing flight design documentation for launch with respect to the first two stages, supporting Zenit processing and launch operations. Several significant configuration modifications have been made to allow the basic Zenit design to meet Sea Launch's unique requirements.
4. The Anglo-Norwegian Kvaerner Group is responsible for designing and modifying the Assembly and Command Vessel (ACV), designing and modifying the Launch Platform (LP), and integrating the marine elements. Furthermore, Barber Moss Marine Management is responsible for marine operations and maintenance of both vessels.

The partner team of contractors has developed an innovative approach to establishing Sea Launch as a reliable, cost-effective and flexible commercial launch

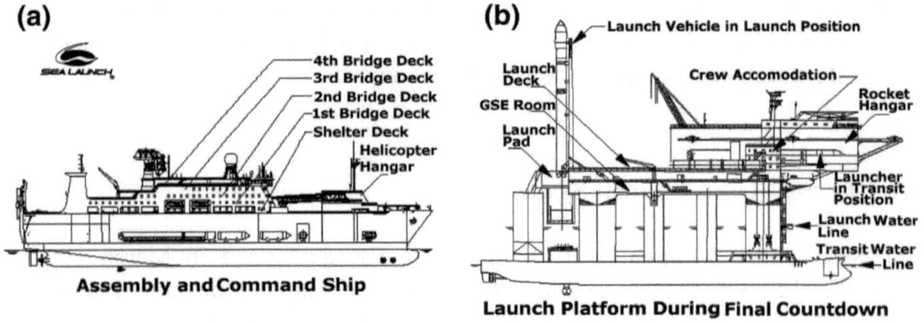

Fig. 2.10 Sea launch modules—Courtesy of Manual: by Sea Launch

system. Each partner is also a supplier to the venture, capitalizing on the strengths of these industry leaders. The system consists of two main modules: Assembly (Command and Control Ship) and Launch Platform, both illustrated in Fig. 2.10a, b, respectively. However, transit for the ACV and the LP from Home Port in Long Beach to the launch site on the equator takes 10–12 days, based on a speed of 10.1 knots.

The Sea Launch Home Port complex is located in Long Beach, California. The Home Port site provides the facilities, equipment, supplies, personnel, and other procedures necessary to receive, transport, process, test, and integrate the spacecraft and its associated support equipment with the Sea Launch system. The Home Port also serves as the marine base of operations for both of the Sea Launch vessels. The personnel providing the day-to-day support and service during prelaunch processing and launch conduct to Sea Launch and its customers are located at the Home Port. The ACV performs four important functions for Sea Launch operations: (1) It serves as the facility for assembly, processing, and checkout of the launch vehicle; (2) it houses the Launch Control Center (LCC), which monitors and controls all operations at the launch site; (3) it acts as the base for tracking the initial ascent of the launch vehicle; and (4) it provides accommodation for the marine and launch crews during transit to and from the launch site. Therefore, the ACV is designed and constructed specifically to suit the unique requirements of Sea Launch. The ship's overall dimensions are nearly 200 m in length, 32 m in beam, and a displacement of 34,000 tons.

Major features of the ACV include: a rocket assembly compartment; the LCC with viewing room; helicopter capability; customer work areas; spacecraft contractor; and customer accommodation. The rocket assembly compartment, which is located on the main deck of the ACV, hosts the final assembly and processing of the launch vehicle. This activity is conducted while the vessels are at the Home Port and typically in parallel with spacecraft processing. The bow of the main deck is dedicated to processing and fueling the Block DM-SL of the Zenit launch vehicle. After the completion of spacecraft processing and encapsulation, the encapsulated payload is transferred into the rocket assembly compartment, where it is integrated with the Zenit-2S and Block DM. The launchers and the satellite are assembled horizontally in the ACV before sailing from the port of Long Beach to the designated launch site. A launcher with a payload will then be transferred in the horizontal position to the launch pad on LP and raised to a vertical position for fueling and launching.

During the launch sequence, the crew of the LP will be transferred to the ACV, which will initiate and control the launch from a position about 3 miles away from the LP pad. The LP is an extremely stable sea platform from which to conduct the launch, control, and other operations. The LP rides catamaran-style on a pair of large pontoons and is self-propelled by a four-screw propulsion system (two in each lower hull, aft), which is powered by four direct-current double armature-type motors, each of which are rated at 3000 hp. The LP in navigation has normal draft at sea water level but once at the launch location, the pontoons are submerged to a depth of 22.5 m to achieve a very stable launch position, level to within approximately 1°. The ballast tanks are located in the pontoons and in the lower part of the columns. Six ballast pumps, three in each pontoon, serve them. The LP has an overall length of approximately 133 m at the pontoons and the launch deck is 78 by 66.8 m. The Zenit-3SL launcher is a two-stage liquid propellant launch vehicle solution capable of transporting a spacecraft to a variety of orbits.

The original two-stage Zenit was designed by KB Yuzhnoye quickly to reconstitute former Soviet military satellite constellations. The design emphasizes robustness, ease of operation, and fast reaction times. The result is a highly automated launch capability using a minimum complement of launch personnel. The launcher as an integrated part of the Sea Launch system is designed to place spacecraft into a variety of orbits and is capable of putting 5250 kg of payload into GEO.

The Sea Launch mission provides a number of technical support systems that are available for the customer's use in support of the launch process, including most importantly the following:

1. **Communications**—Internal communication systems are distributed between the ACV and LP. This system includes CCTV, telephones, intercom, video teleconferencing, public address, and vessel-to-vessel radiocommunications, known as the line-of-sight (LOS) direct system.

This system links with the external communication system and provides a worldwide network that interconnects the various segments of the Sea Launch program. The external communication system includes Intelsat and two ground stations. The LES are located in Brewster, Washington, and Eik, Norway, and provide the primary distribution gateways to the other communication nodes. Customers can connect to the Sea Launch communication network through the convenient Brewster site. The Intelsat system ties in with the ACV and launch platform PABX systems to provide telephone connectivity. Additionally, critical Voice, Fax, Tlx, or data capability can be ensured by the Inmarsat satellite SES service.

2. **Tracking and Data-Relay Satellite System (TDRSS)**—The Sea Launch system uses a unique dual telemetry stream with the TDRSS. Telemetry is simultaneously received from the Zenit stages, the Block DM upper stage, and the payload unit during certain portions of the flight. The Block DM upper stage and payload unit data are combined but the Zenit data are sent to a separate TDRSS receiver. Zenit data are received shortly after liftoff at approximately 9 s and continue until Zenit Stage 2/Block DM separation, at around 9 min. These data are routed from the NASA White Sands LES to the Sea Launch Brewster LES and to the ACV. Otherwise, the data are also recorded at White Sands and Brewster for later playback to the KB Yuzhnoye design center.

When the payload fairing separates, the payload unit transmitter shifts from sending high-rate payload accommodation data by LOS to sending combined payload unit/Block DM by TDRSS. The combined data are again routed from White Sands to Brewster, where it is separated into Block DM and payload unit data and then sent on to the ACV. The data are received on board the ship through the Intelsat communications terminal and are routed to Room 15 for upper-stage data and Room 94 for PLU data. Simultaneously, Brewster routes Block DM data to the Energia Moscow control center station. However, the TDRSS coverage continues until after playback of the recorded Block DM data.

3. **Telemetry System**—Sea Launch uses LOS telemetry systems for the initial flight phase, as well as the TDRSS for later phases. The LOS system, which includes the Proton antenna and the S-band system, is located on the ACV. Other telemetry assets include Russian ground tracking stations and the Energia Moscow control center. The following subsections apply to launch vehicle and payload unit telemetry reception and routing.

4. **Weather (WX) Data System and Forecast**—The ACV unit has a self-contained WX station, which includes a motion-stabilized C-band Doppler radar equipment, surface wind instruments, wave radar, upper-atmospheric balloon release station, ambient condition sensors, and access to satellite imagery and information from an on-site buoy.

2.3 Types of Orbits for Mobile and Other Satellite Systems

An orbit is the circular or elliptical path that the satellite traverses through space. This path appears in the chosen orbital plane in the same or different angle to the equatorial plane. All communications satellites always remain near the Earth and keep going around the same orbit, directed by centrifugal and centripetal forces. Each orbit has certain advantages in terms of launching (getting satellite into position), station keeping (keeping the satellite in place), roaming (providing adequate coverage), and maintaining necessary quality of communication services, such as continuous availability, reliability, power requirements, time delay, propagation loss, and network stability.

There is a large range of satellite orbits but not all of them are useful for fixed and Mobile Satellite Communications systems. In general, the one of most commonly used orbits for satellite communications is GEO constellations, after which HEO and latterly GIO, PEO, LEO, and MEO, shown in Fig. 2.11a.

Otherwise, it is essential to consider that satellites can serve all communication, navigation, meteorological, and observation systems for which they cannot have an attribute such as fixed or mobile satellites and the only common difference is which type of payload or transponder they carry on board. For example, its name can be satellite-specified for fixed communications but in effect it can carry major transponders for fixed communications and others for mobile or other purposes and vice versa.

After many years of research and experiments spent on finding the global standardization for spatial communications, satellites remained the only means of providing near-global coverage, even in those parts which other communications systems are not able to reach. There is always doubt about the best orbital

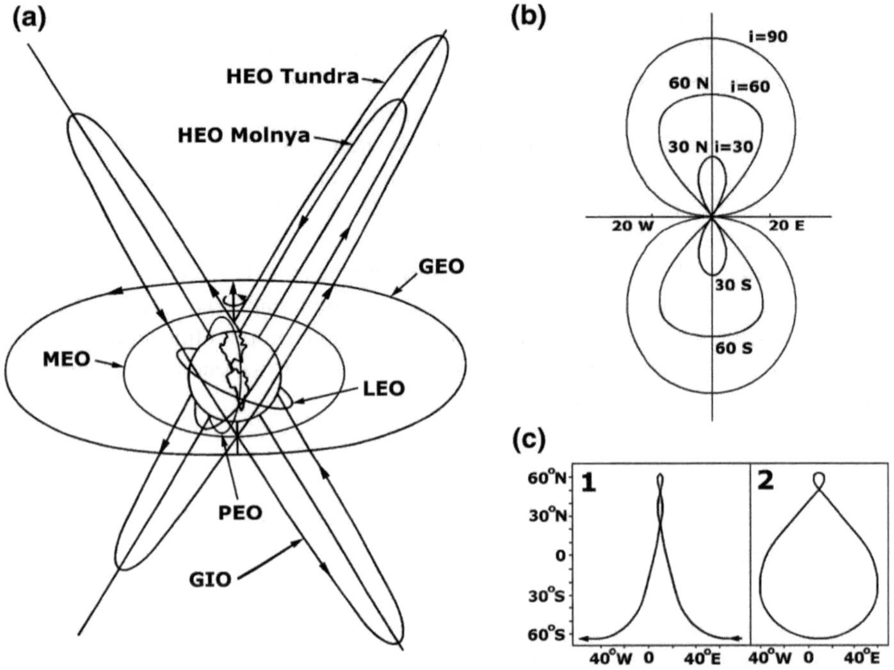

Fig. 2.11 Type of satellite orbits and tracks—Courtesy of Book: by Evans

constellation that can realize an appropriate global coverage and a reliable communications solution.

Unfortunately, there is no perfect system today; all systems have some advantages or disadvantages. The best conclusion is to abridge the story and to say briefly that today the GEO system is the best solution and has only congestion as a more serious problem. The extensive use of GEO shows that it provides something good. Inmarsat is the biggest GEO operator whose service and revenue confirm this point of view. The advantages of Inmarsat MSC solutions can be realized if someone uses them such as operators on board mobiles and finds out how powerful they are. Most of other regional GEO worldwide networks, such as ACeS, Optus and Thuraya, are also more successful than other non-GEO constellations.

In particular, Big LEO and ICO systems or hybrid constellations such as Ellipso have had several years of serious economical and concept difficulties. It is sufficient to see Table 2.3 to understand that the major reasons for LEO problems are more satellite cost, complex network, and short satellite visibility and lifetime. The LEO/PEO constellations are the same or similar and because of differences in inclination angle of orbital plane and type of coverage, they will be considered separately.

The track of the satellite varies from 0 to 360°, see Fig. 2.11b. The track of the GEO satellite is at a point in the center of the coordinate system; two tracks are

Table 2.3 The properties of four major orbits

Orbital properties	LEO	MEO	HEO	GEO
Development period	Long	Short	Medium	Long
Launch & satellite cost	Maximum	Maximum	Medium	Medium
Satellite life (Years)	3–7	10–15	2–4	10–15
Congestion	Low	Low	Low	High
Radiation damage	Zero	Small	Big	Small
Orbital period	<100 min	8-12 h	½ Sidereal Day	1 Sidereal Day
Inclination	90°	45°	63.4°	Zero
Coverage	Global	Global	Near global	Near global
Altitude range (km^{-3})	0.5–1.5	8–20	40/A − 1/P	40 (i = 0)
Satellite visibility	Short	Medium	Medium	Continuous
Handover	Very Much	Medium	No	No
Elevation variations	Rapid	Slow	Zero	Zero
Eccentricity	0 to High	High	High	Zero
Handheld terminal	Possible	Possible	Possible	Possible
Network complexity	Complex	Medium	Simple	Simple
Tx Power/Antenna	Low	Low	Low/High	Low/High
Gain	Short	Medium	Large	Large
Propagation delay	Low	Medium	High	High
Propagation loss	High	Medium	Low	Zero

apparent movements of the GIO satellite with respect to the ascending node of both 30° and 60° inclination angles and the last is the track of the PEO satellite with an inclined orbit plane to the equator of 90°. The tracks of HEO Molniya (part of the track) and Tundra (complete track) orbits are shown in Fig. 2.11 (c-1/c-2), respectively. These two tracks pass over the African Continent and almost all of Europe.

This is very important for MSC systems that the orbit used can provide satellite view during 24 h with less handovers and network difficulties. However, for other types of broadcasting a communications satellite must be visible from the region concerned during the periods when it is desired to provide a communication service, which can vary from a few hours to 24 h a day. When the service is not continuous, it is desirable that the intervals during which the service is available repeat each day at the same time.

2.3.1 Low Earth Orbits (LEO)

The LEO systems are either elliptical or more usually circular satellite orbits between 500 and 2000 km above the surface of the Earth and below the Inner Van Allen Belt. The orbit period at these altitudes varies between ninety minutes and two hours. The radius of the footprint of a communications satellite in LEO varies from 3000 to 4000 km.

Therefore, the maximum time during which a satellite in LEO orbit is above the local horizon for an observer on the Earth is up to 20 min. Earth is up to 20 minutes, which is shown in Fig. 2.11a. In this case, the traffic to a LEO satellite has to be handed over much more frequently than all other types of orbit. At this point, when a satellite, which is serving a particular user, moves below the local horizon, it needs to be able to quickly handover the service to a succeeding one in the same or adjacent orbit.

Due to the relatively large movement of a satellite in LEO constellation with respect to an observer on the Earth, satellite systems using this type of orbit need to be able to cope with large Doppler shifts. In fact, satellites in LEO are not affected at all by radiation damage but are affected by atmospheric drag, which causes the orbit to gradually deteriorate. Satellites in LEO and MEO constellations are subjected to orbital perturbation. For very LEO satellites, the aerodynamic drag is likely to be significant and in general, some of the other perturbations, such as precession of the argument of the perigee, resolve to zero in the orbit is circular or polar. On the other hand, a perturbation is unlikely to have a serious effect on the operation of a multisatellite constellation since it will usually affect all satellites of the configuration in equal measure.

The major advantages of LEO are as follows:

(a) The LEO system may become important in the field of MSC using handheld terminals with global roaming and to be exceedingly useful in areas not covered by cellular systems. The LEO constellations cover almost the entire Earth's surface and some of them provide polar coverage and show promise in the fields of mobile data and Internet and FSS networks for broadband data transmission and communications.
(b) High Doppler shift allows the LEO system to be used for satellite positioning, tracking, and determination.
(c) The relatively small distance between LES and LEO results in much lower power and smaller user terminals. Furthermore, the one-way speed-of-light propagation delay of at least 0.25 s using GEO is obviated with LEO, in which effect can be annoying in two-way voice transmission. For example, for two-way voice via a satellite at an altitude of about 1000 km, the delay is only 13 ms in total for uplink and downlink.
(d) Satellite path diversity eliminates signal interruption due to path obstruction. In Fig. 2.12, handover from satellite A to satellite B is demonstrated, as well as path diversity between satellites B and C. This figure illustrates the LEO MSS space and ground architecture with utilization of handheld personal terminals (PES). On the other hand, the Satellite Access Node (SAN) is the LES providing a link between PES terminals through satellites and ground telecommunications infrastructures.

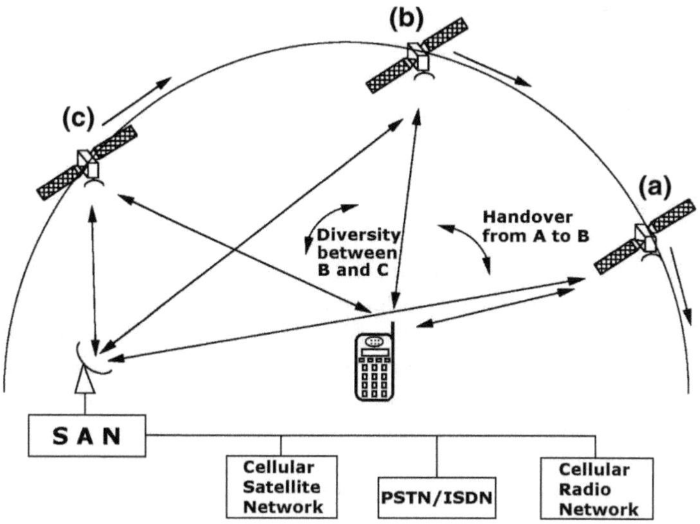

Fig. 2.12 LEO MSS diversity and handover—Courtesy of Book: by Huurdeman

The disadvantages of LEO are as follows:

(a) The orbit period at about 1000 km altitude is in the order of approximately 100 min and the visibility at a point on the Earth is only some 10 min, requiring 40–80 satellites in six to seven planes for global coverage. Thus, in reality a GMSC system using this type of orbit requires a large number of satellites, in a number of different inclined orbits, which increases the total cost of the network.
(b) Frequently handover is necessary for uninterrupted communications. Satellite visibility for MES could be improved by using more satellites. In such a way, the optimum number of satellites of about 48 inclined in the constellation in a carefully optimized pattern of orbit planes will provide continuous visibility of one or other of the satellites at any location on Earth surface.
(c) During times of the year that the orbital plane is in the direction of the Sun, a satellite in LEO is eclipsed for almost one-third of the orbit period. Consequently, there is a significant demand on battery power, with up to 5000 charge/discharge cycles per year, which, with existing NiCd types of batteries, reduces satellite lifetimes to 3–7 years.
(d) The launch cost is low, with direct injection into the orbit of several satellites, but the total cost is very high, with a minimum of 40 satellites being produced.

The first-generation LEO satellites were used for military communications because single GEO could be an easy target for an opponent. The large number of LEO satellites will reduce enormously the risk of vulnerability if someone wishes to destroy only one satellite.

Because this orbit configuration is in the initial phase of its exploitation, it is still free of congestion problems. There are two types of LEO constellations known as Little LEO, useful for messaging and satellite tracking systems and Big LEO, suitable for voice, video, and data communications.

2.3.1.1 Little LEO

The Little LEO mobile satellite systems are a category of LEO solutions that utilize birds of small size and low mass for low-bit-rate transmission under 1 Kb/s. Thus, the Little LEO systems are the constellations of very small non-GEO satellites, which operate in LEO orbits, providing mainly mobile data messaging and tracking services for vehicles and ships and other FSS and broadcasting services. The FCC has allocated a frequency band of 137–138 MHz for the downlink and 148–149.9 MHz for the uplink to Little LEO systems.

The Little LEO satellite frequency spectrum is a heavily utilized for different private and government services worldwide, such as Orbcomm, Falsat, VITASat, Starnet, and other systems. The mass of satellites in these solutions ranges from 40 kg in Orbcomm to 150 kg in the Starnet system. These systems prefer a spectrum below 1 GHz, because it enables the use of cost-effective equipment for non-voice two-way messaging and positioning satellite transceiver, which would be equipped with an alphanumerical display.

2.3.1.2 Big LEO

The Big LEO is a larger non-GEO satellite system, which operates in LEO constellations and provides mainly mobile telephony, Fax, data, and RDSS services. Compared to the Little LEO systems, satellites in Big LEO systems are expected to be bigger in body and to have more power and bandwidth to provide a different service to their subscribers. This system will use the underutilized spectrum available in the L-band because of the commercial failure of proposed RDSS service. Currently, the frequency spectrum of 1610–1626.5 MHz for uplinks and 2483.5–2500 MHz for downlinks is assigned to these MSC systems. It is interesting to note that although the names of these systems include LEO, their frequencies are the ones usually utilized in MEO and GEO satellite systems. For this reason, a new ICO system is systematized in the category of Big LEO constellations together with real Big LEO systems, such as Iridium and Globalstar, which are located at a lower altitude than ICO, at about 700–1500 km from the Earth's surface. Thus, all of the new proposed Big LEO systems would offer global handheld telephone service by means of satellites on lower altitudes moving very fast, instead of fixed GEO relays. The bigger size of the satellites enables them to carry a transponder on board with more complex data processing facilities than the simple store-and-forward feature of the Little LEO satellite configuration. Hence, an important fact is that these systems are networking with cellular and spreading their roaming and billing capabilities in real global coverage.

2.3.2 *Circular Orbits*

The GEO satellite constellation has great advantages for MSS communication applications where polar coverage is not required but there are solutions for providing polar roaming. Satellite orbits in 63.4° inclined high-apogee HEO have some advantages from GEO also providing polar coverage. The most popular circular equatorial orbit with zero inclination is the GEO satellite constellation. The period of rotation is equal to that of the Earth and has the same direction. However, both of these satellite orbits exhibit high LOS loss and long transmission times and delays. Using new technology, these problems can be solved, or as an alternative to these orbits there are LEO and MEO constellations with their good and bad characteristics. The choice of orbit depends on the nature of the MSC mission, the acceptable interference in an adequate environment and the performance of the launchers.

2.3.2.1 Medium Earth Orbits (MEO)

The MEO satellite constellations, known also as Intermediate Circular Orbits (ICO), are circular orbits located at an altitude of around 10,000–20,0000 km between the Van Allen Belts. The MEO satellites are operated in a similar way to Big LEO systems providing global coverage, which orbit is shown in Fig. 2.11a.

However, compared to a LEO system a MEO constellation can only be in circular orbit; Doppler effect and handover is less frequent; propagation delay is about 70 ms and free-space loss is greater; satellites are affected by radiation damages from the Inner Van Allen Belt only during the launching period; fewer eclipse cycles means that battery lifetime will be more than 7 years; cosmic radiation is lower, with subsequently longer life expectancy for the complete MEO configuration; higher average elevation angle from users to satellite minimizes probability of LOS blockage and higher RF output power required for both indoor and handheld terminals.

A LEO constellation for MSS global coverage requires around 10 satellites in two or three orbital planes, each plane inclining 45° to the equator. Their orbit period measures about 6–8 h, providing slightly over 1 h local visibility above the horizon for an observer on the Earth and handover from one to the next satellite is every 6 h minimum.

There is in exploitation a special model of MEO constellation known in practice as Highly Inclined Orbit. This particular orbit is of interest because it has been chosen for existing and proposed GNSS systems such as Navstar (GPS), Navsat, GLONASS, and the newly developed Galileo. In all, complete implementation of this orbit configuration would have 24 satellites in 3 orbital planes equidistant from each other, at an altitude of 20,000 km and at an inclination of 55°. In comparison with existing GNSS, the new Galileo system will have 30 satellites in high MEO of about 28,000 km and at a similar inclination of 56°. At this point, its interest to polar MSC would be the eventual prospect of satellite sharing with navigation services, in a similar fashion to a high PEO with minimum of 3 satellites in the same orbital plane.

2.3.2.2 Geostationary Earth Orbits (GEO)

A GEO has a circular orbit in the equatorial plane, with an orbital period equal to the rotation of the Earth of 1 sidereal day, which is achieved with an orbital radius of 66,107 (Equatorial) Earth Radii, or an orbital height of 35,786 km. Otherwise, a satellite in a GEO will appear fixed above the surface of the Earth and remain in a stationary position relative to the Earth itself. Theoretically, this orbit is with zero inclination and track as a point but in practice, the orbit has small nonzero values for inclination and eccentricity, causing the satellite to trace out a small figure eight in the sky, which is shown in Fig. 2.11a.

The footprint or service area of a GEO satellite covers almost one-third of the Earth's surface or 120° in longitude direction and up to 75°–78° latitude North and South of the Equator but cannot cover the polar regions. In this way, near-global coverage can be achieved with a minimum of three satellites in orbit moved apart by 120°, although the best solution is to employ four GEO satellites for better overlapping. This type of orbit is essentially used for commercial communication services for both FSS and MSS with the following advantages:

(a) The satellite remains stationary with respect to one point on the Earth's surface and so the LES antennas can be beamed exactly toward the focus of the GEO satellite without periodical tracking. Only mobile high-gain antennas need auto-tracking systems, while low-gain omnidirectional antennas are free of tracking systems.
(b) The new Inmarsat GEO space constellation consisting of four satellites can cover all three-ocean regions with four overlapping longitudes, except for the polar regions beyond latitudes of 75° North and South. Otherwise, the polar regions can be covered for maritime and aeronautical MSS applications with current HF radio systems or in combination with PEO or HEO satellite constellations.
(c) The Doppler shift, affecting synchronous digital systems caused by satellites to drift in orbit (affected by the gravitation of the Moon and to a lesser extent of the Sun), is small for all LES and MES within satellite coverage.

The disadvantages of GEO compared with LEO and MEO operation are as follows:

(a) The long signal delay is due to the large distance of about 35,800 km if the satellite is in zenith for MES and about 41,000 km at the minimum elevation angle of about 5°. For the EM waves traveling at the speed of light, this causes a round-trip signal delay of 240–270 ms and full duplex delay of 480–540 ms. Thus, the voice used via satellite can experience some disturbance but echo cancelation devices developed in the 1980s can reduce the problem. Besides, for data transmission equipment, especially when using error-correction protocols that require retransmission of blocks with detected errors, complex circuitry with special high-capacity buffer devices is required to overcome delay problems. In addition, practical experience has shown that given good control of the echo, a telephone connection which includes one hop in each direction via a GEO satellite is acceptable to public users.

(b) The required higher RF output power and the use of directional antennas aggravate GEO operation slightly for use with handheld terminals, although it is not critical, because some GEO operators provide this service, such as Thuraya and ACeS.

(c) The launch procedure to put a satellite in GEO is expensive but the total cost of 4 satellites is less than the cost of a minimum of 12 or 40 for MEO and LEO, respectively.

As stated earlier, the major disadvantage of a GEO satellite in a voice transmission is the round-trip delay between satellite and LES of approximately 2.5 s, which can be successfully solved with current and newly advertised echo cancelation circuits. Because of the enormous use of the GEO constellation for many space applications, some parts of the GEO are becoming congested, owing to only one radius and latitude. This orbit is geostationary and so its track is one point called the subsatellite point and obviously, handover and Doppler effect does not apply to GEO. A GEO satellite is at essentially fixed latitude and longitude, so even a narrow-beam Earth antenna can remain fixed. Satellites in GEO can use high- and recently low-gain antennas, which helps to overcome the great distances in achieving the required Effective Radiated Isotropic Power (EIRP) at ground level. On the other hand, using satellite spot beam antennas, coverage can be confined to smaller spot areas, bigger power and higher speed of transmission, such as new generation of Inmarsat-3 spacecraft. Furthermore, a variety of perturbing forces causes the GEO satellite to drift out of its path and assigned position toward so-called inclined orbit (GIO). By far the most important perturbations are the lunar and at a lesser degree the solar gravitational forces, which cause the satellite to drift in latitude or North–South direction. The longitudinal drift in East–West direction is caused by fluctuations in the gravitational forces from the Earth, due to its non-spherical shape and by fluctuations in solar radiation pressure. To counteract these perturbations, the spacecraft needs station-keeping devices. The GEO satellites pass through both Van Allen Belts only on launch, so their effect is insignificant. After reaching the end of operational life, a satellite has to be removed from its orbital slots into a graveyard orbit some 200 km above the GEO plane.

Thus, the GEO satellite constellation seems likely to continue to dominate in the satellite communications world, especially in MSS, providing near-global coverage with low- and high-power transmission. In particular attractive is the reliable and economical use of the Inmarsat standard-C low-power transceiver and low-gain omnidirectional antenna for maritime, land vehicles and aeronautical two-way data/messaging and telex and one-way e-mail service.

The major existing GEO mobile systems in the world are Inmarsat and GEOSAT of the COSPAS–SARSAT system as global solutions and ACeS, AMSC, MSAT, Artemis, Emsat, Optus, N-Star, Solidaridad, and Thuraya as regional networks. Some of these systems, such as ACeS and Thuraya, also provide a service for handheld and mini indoor terminals, which makes it obvious that some authors made the mistake of assuming that for GEO it is very difficult to provide a handheld service and that Inmarsat mini-M is the smallest terminal for GEO, as is mentioned on page 13 of "Low Earth Orbital Satellites for Personal Communication Networks" written by A Jamalipour.

Early in 1995, Pasifik Satelit Nusantara of Indonesia along with Philippine Long Distance Telephone and Jasmine International Public Co Ltd of Thailand came together and formed a joint venture for MSS today known as Asia Cellular System (ACeS). The ACeS handheld dual mode (GSM/ACeS) terminal is manufactured by Ericsson.

Thuraya is Private Joint Stock Company registered on April 26, 1997, in UAE under Federal Law No. 8 of 1984 as a Regional GEO Mobile Satellite Communication System Operator providing voice, low-bit-rate data and facsimile services. The two prototypes of Thuraya handheld terminals are being manufactured by Hughes Network Systems (USA) and ASCOM (Switzerland).

2.3.2.3 Geosynchronous Inclined Orbits (GIO)

This system would consist in four satellites at six-hour intervals around the Earth orbit at an inclination of 45° to the equatorial plane, which is shown in Fig. 2.11a. The satellites provide polar coverage for six hours either side of their most northerly and southerly movement. Special LES with full-tracking antennas are needed; therefore, this system in general must be considered complex and expensive for a polar communication system.

Otherwise, a GIO satellite has a period of orbit equal to or very little different from a sidereal day (23 h 56 min and 4.1 s), which is time for one complete revolution of the Earth. The satellite movement speed has only very little difference from the angular velocity of the Earth, so this movement also has constant angular velocity. Otherwise, the projection of this movement on the equatorial plane is not at a constant velocity. There is an apparent movement of the satellite with respect to the reference meridian on the surface of the Earth and that of the satellite on passing through the nodes. The orbit may be inclined at any angle, which produces a repeating ground track. Figure 2.11b presents tracks of 30° and 60° inclined orbits. The coupled N–S and E–W motion of GIO satellites is shown as a figure eight pattern, while the patterns could also be distorted circles. Depending on the inclination angle, the GIO satellite shows points on the equator at various longitudes.

A satellite may operate in this orbit for several reasons. First, it is often desirable to save the inclination control fuel required for GEO circle. Sometimes there is no need to control inclination because tracking LES antennas are required for other reasons, while mobiles such as ships and aircraft require tracking antennas. Some GEO satellites may last beyond their planned lifetime if run low on fuel and cease inclination control. In effect, the GIO constellation with nonzero inclination can be chosen because of easy launching and placing of the satellite into orbit. This satellite must move with an angular velocity equal to the Earth and be in a prograde orbit, that is, revolving eastward in the same direction as the Earth rotates. Otherwise, the only requirements for a GIO constellation are the right period and direction of rotation.

2.3.3 *Highly Elliptical Orbits (HEO)*

Using inclined HEO configuration, both polar areas can be effectively covered with four satellites; two in each polar orbit. The elevation angle to the HEO satellites remains high for most of the 12-h period of visibility, which is especially required for continuous Euro-Asian regional coverage providing land MSC service. At this point, blocking of the beam due to occlusion of the satellite by buildings, mountains, hills, and trees is minimized. Beside, multiple trajectories caused by successive reflection of various obstacles are also reduced in comparison with systems operating with low-elevation angles, such as GEO.

The apogee altitude combines polar coverages with nearly synchronous advantages. Thus, minimum two special LES in both northern and southern polar regions are required to serve MES terminals. The LES tracking can be reached by a fairly directive fixed antenna while the satellite is in its slow apogee sector, the HEO space constellation is namely designed to cover the area under the apogee. Tracking of the satellite is facilitated on account of the small apparent movement and the long visibility duration. Otherwise, it is even possible to use antennas whose 3 dB bandwidth is a few tens of degrees, with fixed pointing toward the zenith, which permits the complexity and cost of the terminal to be reduced while retaining a high gain. A satellite in HEO constellation near the apogee can also use a high-gain antenna to overcome the great distances in achieving the required EIRP values. The noise captured by the LES antenna, from the ground or due to interference from other radio systems and atmosphere, is also minimized due to the high-elevation angles. At any rate, these advantages have led the former USSR to use these orbits for a long time in order to provide coverage of high-latitude territories for mobile systems.

The HEO satellite two-way voice transmission has a similar delay as a GEO at the apogee of about 0.25 s. Therefore, free-space loss and propagation delay for HEO is comparable to that of the GEO constellation. Compared with GEO, the launch and satellite cost of the HEO constellation is reasonably low; this constellation is free of congestion because of only a few current and projected new HEO systems and provides high-elevation angles for LES, which reduces atmospheric losses. Due to the relatively large movement of a satellite in HEO with respect to an observer on Earth, satellite systems using this model of orbit need to be able to cope with large Doppler shifts, 14 kHz for Molniya and 6 kHz for Tundra orbits in L-band 1.6 GHz. However, as the former USSR's experience has shown, satellites in this orbit tend to have rather a short lifetime due to the repetitive crossing of both Van Allen Belts. The rest of the disadvantages are the necessity of constant satellite tracking at the MES, compensation of signal loss variation, long eclipse periods, and complex control system of MES and spacecraft.

The HEO satellite typically has a perigee at about 500 km above the Earth's surface and an apogee as high as 50,000 km. The orbits are inclined at 63.4° in order to provide services to locations at high northern latitudes. The particular inclination value is selected in order to avoid rotation of the apses, i.e., the intersection of a line from the Earth's center to the apogee, and the Earth's surface will always occur at latitude of 63.4°N. Orbit period varies from eight to 24 h. Owing to the high eccentricity of the orbit, a satellite will spend about two-thirds of the orbital period near apogee and during that time it appears to be almost stationary for an observer on Earth (this is referred to as apogee dwell). After this period, a switchover needs to occur to another satellite in the same orbit in order to avoid loss of communications. There have to be at least three HEO satellites in orbit, with traffic being handed over from one to the next every eight hours at a minimum.

When there is an orbit in HEO plane of nonzero inclination, the satellite passes over the region situated on each side of the equator and will possibly cover the polar regions if the inclination of the orbit is close to 90°. By orienting the apsidal line, namely the line between perigee and apogee, in the vicinity of the perpendicular to the line of nodes (when ω is close to 90° or 270°), the HEO satellite at the apogee systematically returns above the regions of a given hemisphere. In this way, it is possible to establish satellite links with LES or MES located at high latitudes. Although the satellite remains for several hours in the vicinity of the apogee, it does move with respect to the Earth and after a time dependent on the position of the MES, the satellite disappears over the horizon as seen from the mobiles. However, to establish permanent links it is necessary to provide several suitably phased satellites in similar orbits, which are spaced around the Earth (with different right ascensions of the ascending node and regularly distributed between 0 and 2π) in such a way that the satellite moving away from the apogee is replaced (handover) by another satellite in the same area of the sky as seen from the MES. However, the problems of satellite acquisition and tracking by the MES are simplified. Finally, there only remains the problem of handover and switching the links from one satellite to other, so the RF link frequencies of the various satellites can be different in order to avoid interference.

Examples of HEO systems are Molniya, Tundra, Loopus, Borealis of Ellipso system and Archimedes, which obits are shown in Fig. 2.11a. The ESA proposed Archimedes system employs a so-called "M-HEO" 8-hour orbit. This produces three apogees spaced at 120°. Each apogee corresponds to a service area, which could cover a major population centre, for example the whole European continent, the Far East and North America.

2.3.3.1 Molniya Orbit

The first prototype HEO Molniya satellite was launched in 1964 and to date more than 150 have been deployed, primarily produced by the Applied Mechanics NPO in Krasnoyarsk, former USSR. The HEO Molniya satellites weigh approximately 1.6 metric tons at launch and stand 4.4 m tall, with a base diameter of 1.4 m. Electrical energy is provided by 6 windmill-type solar panels, producing up to 1 kW of power. A liquid propellant attitude control and orbital correction configuration maintains satellite stability and performs orbital maneuvers, although the latter usage is rarely needed. Sun and Earth sensors are used to determine proper spacecraft attitude and antenna pointing. The first Molniya 3 spacecraft appeared in 1974, primarily to support civil communications (domestic and international), with a slightly enhanced electrical power system and a communications payload of three 6/4 GHz transponders with power outputs of 40/80 W.

The second stratum of the Russian space-based communications system consists of 16 HEO Molniya-class spacecraft in highly inclined 63° semi-synchronous orbit planes, which initial perigees between 450 and 600 km fixed deep in the Southern Hemisphere and apogees near 40,000 km in the Northern Hemisphere. In fact, Molniya satellites are synchronized with the Earth's rotation, making two complete revolutions each day with orbital period of 718 min. The laws of orbital mechanics dictate that the spacecraft orbital velocity is greatly reduced near apogee, allowing broad visibility of the Northern Hemisphere for periods up to eight hours at a time. Thus, by carefully spacing 3 or 4 Molniya spacecraft, continuous communications can be maintained. This type of orbit was pioneered by the USSR and is particularly suited to high-latitude regions, which are difficult or impossible to service with GEO satellites.

The 16 operational Molniya satellites are divided into two types and four distinct groups. Namely, eight Molniya 1 satellites were divided into two constellations of four vehicles each. Both constellations consist of four orbital planes spaced 90° apart, but the ascending node of one constellation is shifted 90° from the other; that is, the Eastern Hemisphere ascending nodes are approximately 65° and 155°E, respectively. Although the system was designed to support the Russian Orbita TV network, a principal function was to service government and military communications traffic via a single 40 W 1.0/0.8 GHz satellite transponder.

The hypothetical Russian Molniya network can employ minimum 3 HEO satellites in three 12-h orbits separated by 120° around the Earth, with apogee distance at 39,354 km and perigee at 1000 km. This orbit takes the name from the communication system installed by the former USSR, whose territories are situated in the Northern Hemisphere at high latitudes. The orbital period (t) is equal to (t_E /2), or about 12 h. The characteristics of an example Molniya orbit are given in Table 2.4.

Table 2.4 Molniya and Tundra orbit parameters

Characteristics	Molniya orbit	Tundra orbit
Orbital period (t)	12 h	24 h
Sidereal period	11 h 58 min 2 s (half day)	23 h 56 m 4 s (full day)
Semi-major axis (a)	26,556 km	42,164 km
Inclination (i)	63.4°	63.4°
Eccentricity (e)	0.6–0.75	0.25–0.4
Perigee altitude (h_p) (e.g.,: e = 0.71)	a(1 − e) − R 1250 km	a(1 − e) − R 25,231 km
Apogee altitude (h_a) (e.g.,: e = 0.71)	a(1 + e) − R 39,105 km	a(1 + e) − R 46,340 km

The only one-track cycles of a total of two satellite tracks on the surface of the Earth is shown in Fig. 2.11 (c-1) for a perigee argument equal to 270°. The shape of this track is cycles of one orbit only near Greenwich Meridian, so the center of the next identical track is around 180° westward. Therefore, the satellite at apogee passes successively on each orbit above two points separated by 180° in longitude. The apogee is situated above regions of 63° latitude (the altitude of the vertex is equal to the value of the inclination and the apogee coincides with the vertex of the track when the argument of the perigee is equal to 270°). The large ellipticity of the orbit results in a transit time for the period of the orbit situated in the Northern Hemisphere greater than that in the Southern Hemisphere. The value of inclination, which makes the drift of the argument of the perigee equal to zero, is 63.45°. A value different from this leads to a drift, which is nonzero but remains small for value of inclination, which does not deviate too greatly from the nominal value. By the way of example, for an inclination i = 65°, that is variation of 1.55°, the drift of argument of the perigee has a value of around 6.5° per annum.

It is evident that the Molniya HEO satellite has the advantage of high-elevation-angle coverage of the Northern Hemisphere because of a need to completely cover a great part of the Russian territory. Three satellites in this orbit and phasing are chosen so that at least one satellite is available at any time over the horizon. Thus, with three satellites, each satellite is used (or handover is) 8 h per day, while with four satellites handover is every 6 h. The LES must use tracking antenna systems, so a terminal with only one antenna will have an outage during handover (switching) from one satellite to another.

The disadvantages of the Molniya orbit include the need for multiple satellites (which the system does not need), the poor, virtually useless coverage of the Southern Hemisphere and the need for tracking antennas at each LES. Since the distance from terminal to satellite is continually changing, the received power and frequency vary (Doppler effect). The former may require automatic uplink power control and scheduling is needed to allow LES to switch satellites simultaneously. As the satellite altitude varies, the beam also coverage changes, so the satellite carries a tracking antenna that must be kept continuously pointed at operating LES.

2.3.3.2 Tundra Orbit

The Russian Tundra HEO system employs 2 satellites in two 24-h orbits separated by 180° around the Earth, with apogee distance at 53,622 km and perigee at 17,951 km, which provides visibility duration of more than 12 h with high-elevation angles. The Tundra orbit can be useful for regional coverage for both FSS and MSS applications. Similar to the Molniya orbit, this orbit is particularly useful for LMSS, where the masking effects caused by surrounding obstacles and multiple path are pronounced at low-elevation angles, (>30°).

The period (t) of the orbit is equal to t_E, which is around 24 h. The characteristics of an example orbit of this type are given in Table 2.4. This orbit has only one track on the Earth's surface, as shown in Fig. 2.11 (c-2), for a perigee argument equal to 270°, inclination i = 63.4° and eccentricity e = 0.35. The latter parameter can have three values of eccentricity e = 15, e = 25, and e = 45.

According to the value of orbital eccentricity, the loop above the Northern Hemisphere is accentuated to a greater or lesser extent. For eccentricity equal to zero, the track has a form of Fig. 8, with loops of the same size and symmetrical with respect to the equator. When the eccentricity increases, the upper loop decreases, while the lower loop increases and the crossover point of the track is displaced toward the North. This loop disappears for a value of eccentricity of the order of e = 0.37 and the lower loop becomes its maximum size. The transit time of the loop represents a substantial part of the orbital period and varies with the eccentricity. The position of the loop can be displaced toward the East or West, with respect to the point of maximum latitude, by changing the value of argument of the perigee (ω) and the eccentricity.

2.3.3.3 Loopus Orbit

The proposed Loopus system, which employs 3 satellites in three 8-h orbits separated by 120° around the Earth, has an apogee distance at 39,117 km and perigee at 1238 km. This orbit has similar advantages and disadvantages as for the Molniya orbit. One of the problems encountered by the LES is that of repointing the antenna during the handover (changeover) from one satellite to another. With orbits whose track contains a loop, it is possible to use only the loop as the useful part of the track in the trajectory. Handover between two satellites is performed at the crossover point of the track. At this instant, the two satellites are seen from the LES in exactly the same direction and it is not necessary to repoint the antenna. To achieve continuous coverage of the region situated under the loop, the transmit time of the loop must be a submultiple of the orbit period and the number of satellites. Hence, the coverage can be extended to one part of the hemisphere by increasing the number of satellites in orbit regularly spaced about the globe.

2.3.4 Polar Earth Orbits (PEO)

The PEO constellation is today a synonym for providing coverage of both polar regions for different types of meteorological observation and satellite determination services. Namely, a satellite in this orbit travels its course over the geographical North and South Poles and will effectively follow a line of longitude. Certainly, this orbit may be virtually circular or elliptical depending upon requirements of the program and is inclined at about 90° to the equatorial plane, covering both poles. The orbit is fixed in space while the Earth rotates underneath and consequently, the satellite, over a number of orbits determined by its specific orbit line, will pass over any given point on the Earth's surface. Therefore, a single satellite in a PEO provides in principle coverage to the entire globe, although there are long periods during which the satellite is out of view of a particular ground station. Accessibility can of course be improved by deploying more than one satellite in different orbital planes. If two PEO satellite orbits are spaced at 90° to each other, the time between satellites passes over any given point will be halved, which orbit is shown in Fig. 2.11 (A).

The PEO system is rarely used for communication purposes because the satellite is in view of a specific point on the Earth's surface for only a short period of time. Any complex steerable antenna systems would also need to follow the satellite as it passes overhead. At any rate, this satellite orbit may well be acceptable for a processing store-and-forward type of communications system and for satellite determination and navigation.

There are four primary requirements for PEO systems as follows:

(1) To provide total global satellite visibility for worldwide LEOSAR COSPAS–SARSAT distress and safety satellite beacons EPIRB, PLB, and ELT applications;
(2) To provide global continuous coverage for current or newly developed and forthcoming satellite navigation systems;
(3) To provide at L-band or any convenient spectrum the communication requirements of ships and aircraft in the polar regions not covered by the Inmarsat system; and
(4) To provide global coverage for meteorological and synoptic observation stations.

The Inmarsat team has studied two broad ranges of orbit altitude of PEO for both distress and communication purposes: first, low altitudes up to 1400 km and second, high altitudes above 11,000 km. In reality, these two orbit ranges are separated by the Inner Van Allen radiation belt. In the regions of the radiation belt, the radiation level increases roughly exponentially with height at around 1000 km, reaching a peak at about 5000 km altitude. Therefore, a critical requirement to reduce high-energy proton damage to the solar cell arrays of the satellite system constrains the PEO to low and high altitudes. As is evident, another Outer Van Allen Belt has no negative influence on these two PEO constellations because it lies far a way between MEO and GEO satellite planes.

These two specific systems studied by Inmarsat are COSPAS–SARSAT low PEO at 1000 km altitude and high PEO at 12,000 km altitude, similar to that studied by ERNO, named SERES (Search and Rescue Satellite) system. Thus, it is considered that these two systems demonstrate clearly the solution tradeoff and constraints on a joint PEO distress, SAR, and communication mission. Other possible orbits for polar coverage can be an inclined HEO Molniya constellation of four satellites; GIO 45° inclined orbit of four satellites and 55° inclined circular MEO at 20,000 km altitude for GPS and GLONASS satellite navigation systems. In the meantime, the COSPAS–SARSAT system has developed a special GEOSAR system using three GEO satellites for global distress communications satellite beacons in combination with already-developed LEOSAR systems using four PEO satellites.

For both the low and high PEO systems, the number of operational satellites required to provide adequate Earth coverage needs to be minimized in order to achieve minimum system costs. An IMO and ICAO requirement for the GMDSS/COSPAS–SARSATmission is that there should be no time delay in distress alerting anywhere in the globe.

2.3.4.1 Low PEO

The low PEO satellite constellation similar to the LEO satellite constellation mostly employs both polar and near-polar orbits for communications and navigation utilities. Thus, a particular example of a system that uses this type of orbit is the COSPAS–SARSAT SAR system for maritime, land, and aeronautical applications. This system uses 8 satellites in 4 near-polar orbits: four US-based SARSAT satellite constellations at 860 km orbits, inclined at 99°, which makes them Sun-synchronous and four COSPAS satellite configurations at 1000 km orbits, inclined at 82°. However, this orbit was also suitable for the first satellite navigation systems Transit and Cicada, developed by the USA and the former USSR, respectively.

Otherwise, with a limited number of low-altitude PEO satellites, it is impossible to provide continuous coverage to polar region, because the view of individual spacecraft is relatively small and their transit time is short. However, because the time for a single orbit is low, less than two hours, and a different section of the polar region is covered at each orbit due to Earth rotation, this drawback is somewhat offset. For a given number of satellites, preferably about eight, it is possible to optimize the constellation that maximizes total system coverage, to improve handover and minimize waiting time between transits. This problem can be solved with additional GES terminals over pole area.

Figure 2.13 illustrates the Earth track of ten successive orbits of satellite in low PEO with an altitude of 1000 km. The MES in shaded area A (4200 km in diameter) would see the satellite, in the absence of environmental screening, at an angle of elevation not less than 10°, while the satellite was passing through the equatorial plane. The coverage area has the same size and shape wherever the

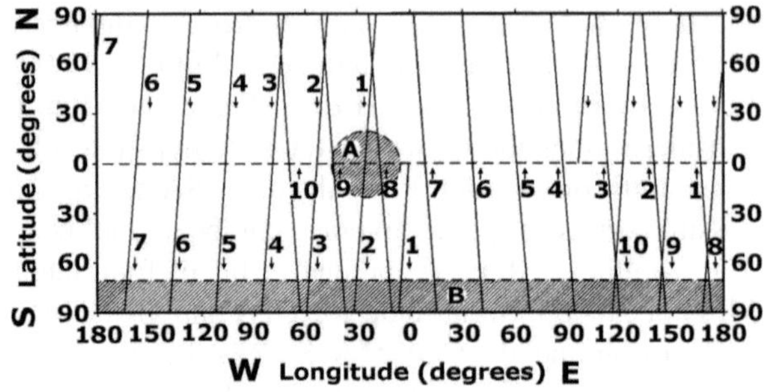

Fig. 2.13 Type of satellite orbits and tracks—Courtesy of Book: by Pascall

satellite is in the orbit but its apparent size and shape would change with latitude, being distorted by the map projection used in the figure. Thus, the South Pole coverage area at a single pass of the satellite is shown in figure by shaded area B.

The same figure shows that a single PEO satellite in a polar orbit will have a brief sighting of every part of the Earth's surface everyday. There will be 2 or 3 of these glimpses per day near the equator, the number increasing as the poles approach.

The period of visibility was seen from the MES range from about 10 min, the satellite passing overhead, down to a few seconds when the satellite appears briefly above the horizon. If the orbital plane of the satellite is given an angle of inclination differing from 90° of the PEO, a similar Earth track is obtained but the geographical distribution of the satellite visibility changes. One LEO satellite with an orbital inclination of 50° would have better visibility between 60° N and 60° S latitude than a PEO satellite but it would have no visibility at all of the polar regions.

The low PEO configuration is attractive for mobile distress communications for two reasons. First, the transmission path loss is relatively low, allowing reliable communication with a low-powered satellite beacon and PEO spacecraft. An altitude of about 1000 km is the upper limit for good reception of signals at 243/406 MHz sent from emergency distress beacons. Secondly, the Doppler shift is high, approximately 30 kHz at 1.6 GHz, allowing accurate location of the distress transmitter. On the other hand, there are several significant disadvantages. However, as mentioned earlier, PEO coverage is not continuous unless there is simultaneous communication between a distress buoy and a ground terminal because of the small footprint of each individual satellite. Accordingly, storage and retransmission of distress messages onboard processing would be necessarily adding to the distress alert delay time and also to satellite mass and complexity.

The short visibility period during a transit and the uneconomic need for large numbers of satellites for continuous coverage makes a low PEO unattractive for communications considerations. If this orbit configured well as an economic solution for distress coverage in polar regions to be used for communications purposes, users would have to operate with the following restrictions: (1) only burst mode, non-simultaneous data communication would be possible; (2) transmission time and/or bit rate would be limited by satellite message storage capability;

(3) replies to the message would require an interrogation or polling system from the MES expecting a reply; and (4) depending on the PEO constellation and MES position, a reply could take some hours.

However, many of these PEO communication limitations would be removed if a system of intersatellite links, possibly in addition to inter-GEO infrastructure, were used to provide a near-continuous, simultaneous two-way communication system. The complexity and likely cost of such system would almost certainly not be justified by the expected low level of polar communication traffic. Thus, in considering the possible integration of PEO and GEO for communication purposes, it is necessary to determine the additional requirements and constraints arising from polar operation. In this context, for reliable communications the number of additional LES required for operation to PEO is a significant element of the overall system. For example, a constellation of eight low PEO would require about six LES worldwide for polar coverage assuming message storing and forwarding techniques, where a high PEO would require a minimum of two LES located in North and South polar latitudes for continuous polar coverage with simultaneous two-way communications. In addition, it would be necessary to obtain reliable terrestrial links between the LES of each system, as well as intersatellite links between the PEO and the GEO satellites.

In any case, by using the store and transmit method, a low PEO system could effectively be served for the relay of mobile distress, safety, and urgency messages, for maritime, land, and aeronautical applications via satellite beacons to receive-only terminals onshore.

2.3.4.2 High PEO

The high PEO constellation would consist of three satellites separated by 120° in the same circular orbit of 12,000 km altitude, geometrically similar to the GEO and as orbit similar to MEO configuration. This orbit can provide continuous coverage to both polar regions above 59° latitude. In such a way, six satellites (in two orbital planes of three satellites each) would provide continuous and real global coverage if that was required, which GEO constellation alone cannot obtain.

By comparison with low PEO systems transmission path losses are higher at an altitude of 12,000 km but not to the extent that a distress beacon need be especially high powered to transmit successfully to a high PEO satellite. Reception of the COSPAS–SARSAT existing two very low-powered distress frequencies will be interfered, but not impossible. The Doppler shift is lower (about 10 kHz at 1.6 GHz), not allowing very accurate area location of the distress transmitters. Single high-latitude LES in both Arctic and Antarctic polar regions allows reception with no delay of all distress messages transmitted from above 59° latitude. Furthermore, using these two LES positioned at high latitude with continuous visibility of at least one of the three satellites and collocated or linked with an Inmarsat LES, can offer a full range of near continuous communication services to the polar regions.

2.3.5 *Hybrid Satellite Orbits (HSO)*

The hybrid satellite constellation can be configured by several types of combinations between existing orbital solutions today. Namely, any of these combinations can provide better global coverage for both hemispheres, including both polar regions. In this context will be introduced shortly five hybrid constellation systems, which are currently using or developing MSC and navigation systems as follows:

1. **Combination of GEO and HEO Constellations**—The development of a MSS that would provide reliable communications with MES terminals, such as oceangoing vessels, land mobiles (road vehicles and trains) and aircraft, rural areas and remote terminals, illustrated in Fig. 2.14. This MSC system, called Marathon, includes five GEO Arcos-type satellites and four Mayak-type satellites in a HEO, as well as a ground segment that is composed of base stations and terminals installed at fixed or mobile users premises.

This MSC system, called Marathon, includes five GEO Arcos-type satellites and four Mayak-type satellites in a HEO, as well as a ground segment that is composed of base stations and terminals installed at fixed or mobile users premises. Therefore, the combination of GEO and non-GEO satellite constellations makes it possible to render GMSC services, including those at high latitudes and in the both polar areas; this is especially important for Russia, with its vast northern Eurasian territories and to provide the most reliable satellite communication between the territories of the Western and Eastern Hemispheres.

This hybrid constellation can be useful for the Alaska, Greenland, and northern territory of Canada as well. The similar hybrid constellation with 2 HEO satellites can be configured for complete coverage of Southern Hemisphere with apogees on opposite side as shown in the same figure. This HSO configuration is the best solution for providing complete global coverage for aeronautical applications.

2. **Combination of GEO and PEO Constellations**—This current combination of orbits has been developed by the efforts of the COSPAS–SARSAT organization, with the assistance of IMO, Inmarsat, and other international and regional contributors. At the other words, the COSPAS–SARSAT space segment is a combination of three GEO operational satellites of the subsystem called GEOSAR and four PEO operational satellites of the subsystem called LEOSAR, with spare

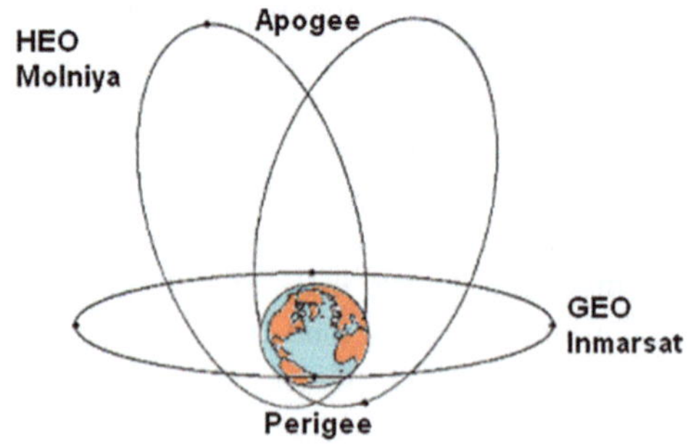

Fig. 2.14 Combination of GEO–HEO constellations—Courtesy of Paper: by Ilcev

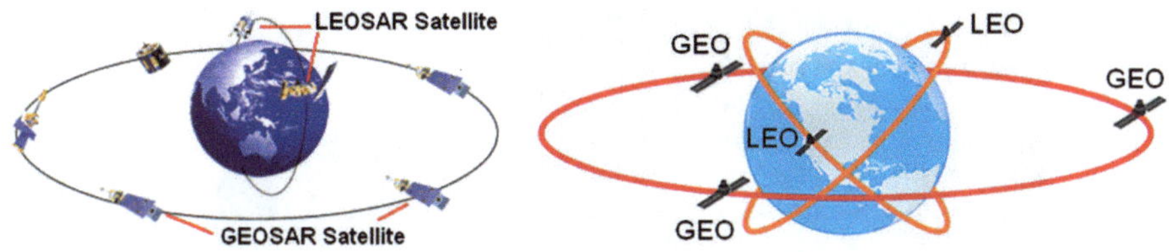

Fig. 2.15 Combination of GEO–PEO and GEO–LEO constellations—Courtesy of Paper: by Ilcev

spacecraft for all participants. The GEOSAR employs one satellite type of INSAT-2A and two GOES-type GOES-E and GOES-W, while the LEOSAR configuration provides two COSPAS and two SARSAT spacecrafts. Otherwise, the GEOSAR project in the future has to include the European MSC and two Russian Luch-M spacecraft. This system is responsible for providing distress alert via special radio beacons (EPIRB, PLB and ELT) and to help SAR forces on-scene determination for maritime, land, and aeronautical applications, in which constellation is shown in Fig. 2.15 (Left).

3. **Combination of GEO and LEO Constellations**—Celestri is the Motorola trademark name for a proposed GEO and LEO satellite hybrid communication network, shown in Fig. 2.15 (Right). The network will combine 9 GEO and 63 LEO satellites in 7 planes with Earth-based control equipment and provide interfaces to existing telecommunication infrastructures, the Internet, and corporate and personal networks. The system will offer a 64 Kb/s voice circuit from anywhere in the world. The architecture is not limited to fixed-sized channels but permits dynamic bandwidth assignment based on application demand. Business users will benefit by Celestri's to provide remote access to LAN infrastructures.

4. **Combination of MEO and HEO Constellations**—The newly proposed MSS Ellipso is developing in combination with an initial complement of seven Concordia satellites deployed in a circular equatorial MEO at an altitude of 8050 km and ten Borealis satellites in two HEO planes inclined at 116.6°. They have apogees of 7605 km and perigees of 633 km and a three-hour orbital period.

This combination of two constellations, shown in Fig. 2.16 (Left), would provide coverage of the entire Northern Hemisphere including North Pole areas and part of the Southern Hemisphere up to 50° latitude South. The HEO satellites can spend a greater proportion of their orbital periods over the northern latitudes and, together with the MEO constellation, the Ellipso hybrid system will provide voice, data, and Fax communication and navigation RDSS services to areas with large landmasses, enormous populations with a large density of users and potentially widespread markets. This system is also planned to cooperate with the terrestrial PSTN and other services.

5. **Combination of MEO and LEO Constellations**—The Kompomash consortium for space systems in Russia have prepared the Gostelesat satellite system for MSS, shown in Fig. 2.16 (Right), using 24 satellites in MEO and 91 in LEO

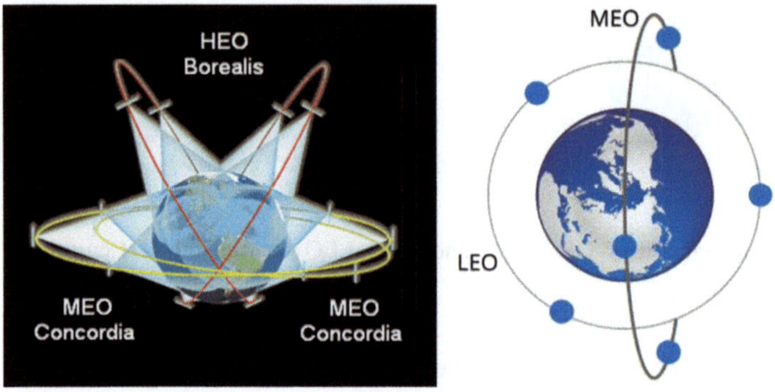

Fig. 2.16 Combination of MEO and HEO and MEO and LEO constellations—Courtesy of Paper: by Ilcev

satellite constellation. Thus, this satellite project is provided for future global MSC and navigation applications with possibility to cover both polar regions.

2.4 Spacecraft Subsystems

A communications satellite essentially consists of two major functional units: payload and bus. The primary function of the payload is to provide communication between LES and MES, while the bus provides all the necessary electrical and mechanical support to the payload and all satellite missions illustrated in Fig. 2.17

The payload is made up of the multipurpose repeaters and antenna systems. The repeater performs the required processing of the signal, and the antenna system is used to receive signals from LES and to transmit signals to MES in the coverage area and vice versa.

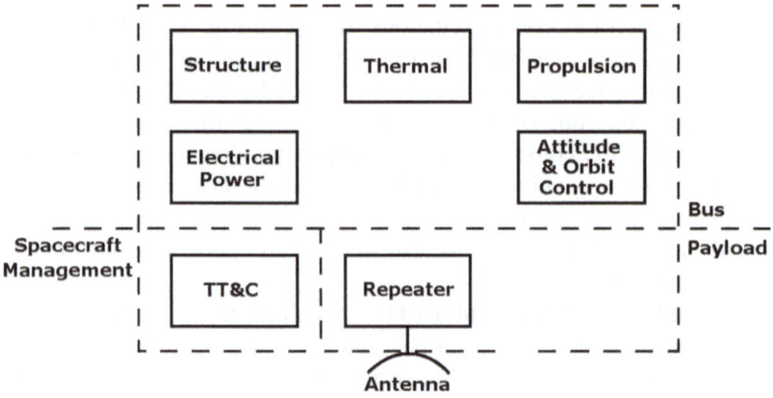

Fig. 2.17 Spacecraft subsystem—Courtesy of Book: by Richharia

2.4.1 *Satellite Repeaters for Mobile Satellite Communications*

The function of a satellite repeater is to receive the uplink RF signals from either ground segment service or feeder links, then to convert these signals to the appropriate downlink frequency and power, and to retransmit them toward the service or feeder links ground segment. Two main types of repeaters are possible for onboard utilization: Transparent and regenerative transponders, however, are developed many other types for different satellite applications.

2.4.1.1 Transparent or Bent-Pipe Transponder

The basic function of the satellite transponder is to isolate individual carriers or groups of carriers of signals and to boost their power level before they are retransmitted to the ground stations. The carrier frequencies are also altered as the carriers pass through the satellite. Satellite repeaters that process the carrier in this way are typically referred to as transparent or bent-pipe satellite transponders, illustrated in Fig. 2.18. Only the basic RF characteristics of the carrier (amplitude and frequency) are altered by the satellite. The detailed signal carrier format, such as the modulation characteristics and the spectral shape, remains completely unchanged.

Transmission via a transparent satellite transponder is often likened to a bent-pipe because the satellite simply channels the information back to the ground stations. A bent-pipe is a commonly used satellite link when the satellite transponder simply converts the uplink RF into a downlink RF, with its power

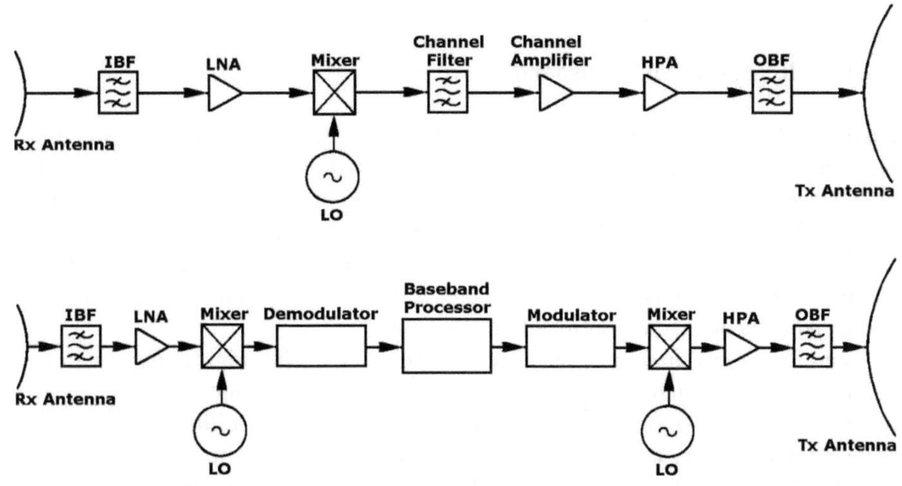

Fig. 2.18 Configuration of spacecraft Transponders—Courtesy of Handbook: by ITU

amplification. Initially, the received uplink signals from LES or MES by Rx antenna are filtered in an Input Bandpass Filter (IBF) prior to amplification in a low-noise amplifier (LNA).

In addition, the output of the LNA is then fed into a local oscillator (LO), which performs the required frequency shift from uplink to downlink RF and the bandpass channel filter after the mixer removes unwanted image frequencies resulting in the downconversion, prior to undergoing two amplification stages of signals in the channel and high-power amplifier (HPA). Finally, the output signal of the HPA is then filtered in the output bandpass filter (OBF) prior to transmission through Tx antenna to the ground. The IFB is a bandpass filter which blocks out all other RF used in satellite communications. After that, the receiver converts the incoming signal into a lower frequency, using an LO which is controlled to provide a very stable frequency source. This is needed to reduce all noises to facilitate processing of the incoming signal and to enable the downlink frequencies to be established.

The channel filter isolates the various communications channels contained in the waveband allowed through by the input filter. Filtering often leads to large power losses, creating a need for extra amplification, usually followed by a main amplifier. In order to attain the required gain of HPA, this segment may employ either a solid-state power amplifier (SSPA) or a traveling wave tube amplifier (TWTA). In a more complex transponder design, in order to achieve higher RF power, it may be possible to combine the output of several amplifiers.

Therefore, to do this the incoming signal must be divided in such a way so as to provide separate identical input to each amplifier, see 6 TWTA presented in Fig. 2.19. A power combiner then recombines the RF signals from the amplifiers to produce a single RF output. The output filter removes all unwanted signals from the transmitted downlink returning to the Earth stations. High reliability throughout the lifetime of the satellite is achieved by duplicating critical units in the receiver, such as TWTA.

2.4.1.2 Regenerative Transponder

Other satellite system designs go through a more complex onboard process to manipulate the carrier's formats, by using onboard processing architecture. This payload architecture offers advantages over the transparent alternative, including improved transmission quality and the prospect of compact and inexpensive MES and handheld user terminals. A typical onboard processing system will implement some or all of the functions that are performed by the ground-based transmitter and/or receiver in a transparent satellite system.

Therefore, these functions may include recovery of the original information on board the satellite and the processing of this information into a different carrier format for transmission to the ground stations. In fact, any satellite transponder that recreates the signals carrier in this way is usually referred to as a regenerative transponder, illustrated in Fig. 2.18. This type of satellite transponder provides demodulation and modulation capacity completely on board the satellite.

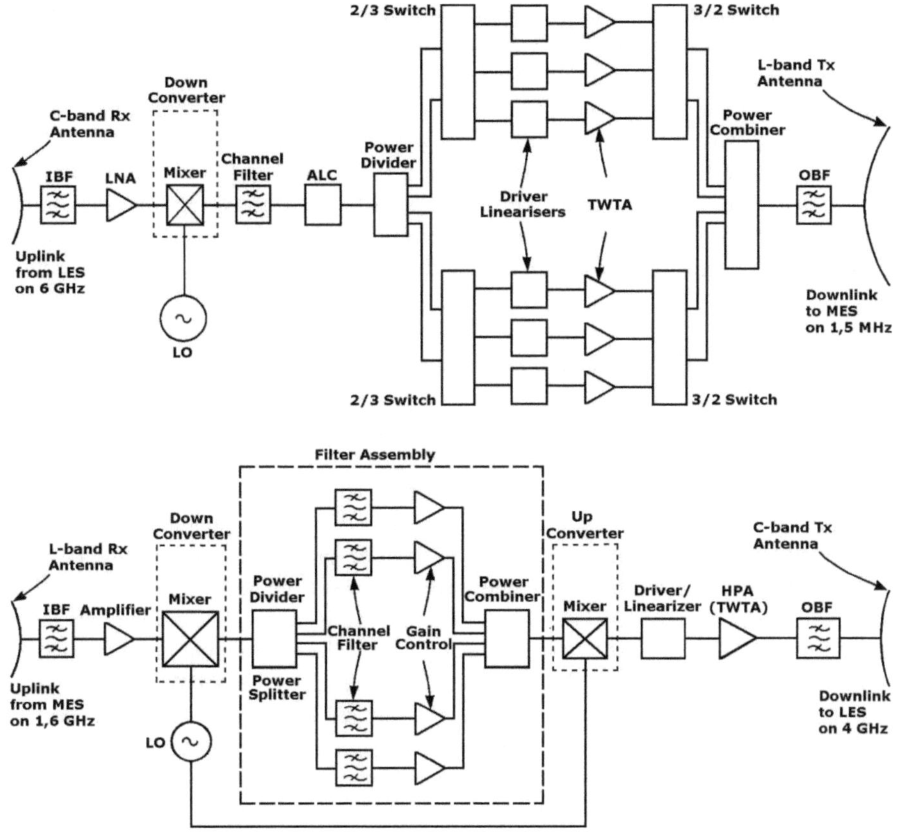

Fig. 2.19 Diagram of spacecraft C/L and L/C-band Transponders—Courtesy of Book: by Gallagher

The received uplink signal goes along the downconverter segment prior to coming into the onboard demodulator, where it is demodulated and processed in the base band processor.

This technology provides flexible functions, such as switching and routings. The downlink signal generated by an onboard modulator passes along the upconverter segment and is transmitted via the antenna. For this type of system link design can be separately conducted for the uplink and downlink because link degradation factors are decoupled between the uplink and the downlink by the onboard demodulator and modulator, supported by the base band processor.

A regenerative transponder with base band processing permits reformatting of data without limitation to MES Rx, while the bent-pipe system requires a satellite link design for the entire link, involving both uplink and downlink, but the forward link burst rate is limited by the MES G/T and demodulation performance. Advanced MSS include intersatellite links to establish a direct connection between satellite transponders and in this way to enlarge system coverage and help the reception of signals from other GEO satellite coverage not visible for particular LES and MES terminals. Moreover, an intersatellite link can help to solve the problems associated with some TTN infrastructures or to reduce landline charges.

In the same way, intersatellite links can also provide a connection from a satellite to neighboring satellites in a constellation of non-GEO space segment and so, they are beneficial in reducing the number of satellite hops when Earth coverage of each satellite is limited. Intersatellite links are usually implemented with regenerative satellite transponder systems, mainly due to the flexible connection to intersatellite links.

2.4.1.3 Satellite Transponders for Inmarsat-2 MSC

A transponder as the nucleus of the Inmarsat network receives information from LES in a directed beam of energy at 6 GHz in the C-band and converted information broadcasts at 1.5 GHz in the L-band. The link from MES follows the reciprocal path. The Inmarsat-2 payload consists of two transparent transponders: the C/L-band, illustrated in Fig. 2.19, and the L/C-band, shown in Fig. 2.19.

1. **Inmarsat-2 C/L-band Transponder**—This transponder receives uplink signals in the C-band of 6.4 GHz from LES and retransmits downlink signals in the L-band of 1.5 GHz to MES, after frequency conversion and signal amplification by a HPA. The signals received by a C-band antenna are fed via IBF and LNA to a downconverter section. A signal channel is followed by an automatic level control (ALC) device, which limits the level of the signal to the amplifier. The HPA consists of six TWTA and their associated power supplies. In front of each TWTA is a driver/linearizer, predisposed to compensate the nonlinear RF properties of the TWTA. The signal driver supplies an equal drive signal to each of the four TWTAs that are active at any given time and the other two can be activated for backup if the operating TWTA malfunctions. For this reason, the signal driver is preceded by an amplitude equalizer. However, the active TWTA are selected by 2/3 and 3/2 switches and their output powers are combined by a power combiner. The total power is fed to an L-band transmission antenna via OBF.

2. **Inmarsat-2 L/C-band Transponder**—This transponder receives uplink signals in the L-band of 1.6 GHz from MES and retransmits downlink signals in the C-band of 3.6 GHz to LES, after frequency conversion and signal amplification by the HPA. The signals received by an L-band antenna are fed to a downconverter via IBF and LNA. At the down-converter, signals are converted into 60 MHz IF by LO. A filter assembly then provides the required characteristics divided into four channels. Following upconversion, the signal passes to an ALC unit and the power for four channels is combined and signals are upconverted from 60 to 3.8 GHz by activated TWTA. The amplified signal in HPA then goes through bandpass and harmonic filters in OBF before being distributed among the 7 cup-dipole elements of the C-band transmit antenna for radiation to the Earth's surface.

2.4.2 *Satellite Repeaters for COSPAS–SARSAT System*

The COSPAS–SARSAT organization initially developed the first generation of LEO Search and Rescue (LEOSAR) systems, such as COSPAS and SARSAT LEO spacecraft. COSPAS–SARSAT is also employing four GEO satellites for its new GEO Search and Rescue (GEOSAR) 406.05 MHz mission, such as the Russian Luch-M, the US Geostationary Operational Environmental Satellite (GOES), Indian National Satellite System (INSAT), and the European Meteosat Second Generation (MSG). In the following context will be introduced Luch-M and GOES transponders of the GEOSAR subsystem including COSPAS transponder and SARSAT Rx/Tx configurations of the LEOSAR subsystem.

2.4.2.1 COSPAS GEOSAR Luch-M Transponder

The Luch-M spacecraft is developed by Russia and is serving for COSPAS–SARSAT and also for communication applications. The COSPAS GEO Luch-M East bird is at 95°E covering Indian Ocean and Luch-M West at 16°W is covering Pacific Ocean.

A functional diagram of the Luch-M GEOSAR transponder is shown in Fig. 2.20. The repeater is redundantly configured and is comprised of the following subsystems:

(a) The UHF 406 MHz receive antenna, LNA, and IF1 Amplifier as main components of SAR 406 MHz Receiver (Rx);
(b) The IF1 Amplifier, Power Amplifier (PA), and SHF 11,381 MHz transmit antenna as main parts of SAR 11,381 MHz Transmitter (Tx); and
(c) The intermediate components of Rx and Tx are local oscillator and synthesizer.

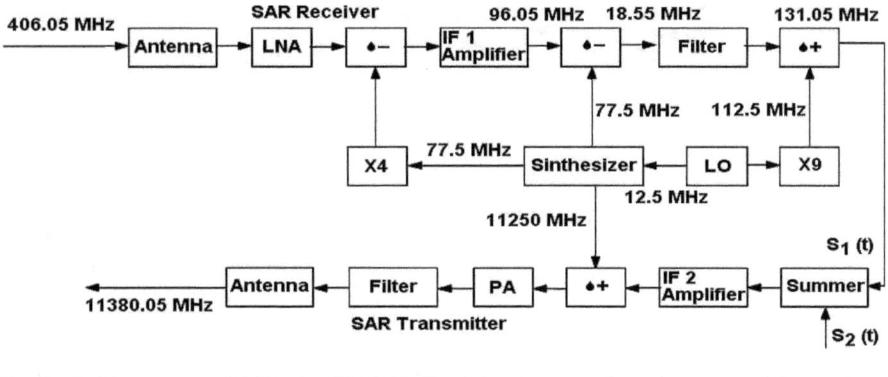

Fig. 2.20 Diagram of COSPAS GEOSAR Luch-M Transponder—Courtesy of Manual: by COSPAS–SARSAT

The 406.05 MHz signals from COSPAS–SARSAT distress beacons are received by the Luch-M UHF antenna. The signal is downconverted twice to an Intermediate Frequency (IF) of 18.55 MHz after which it is filtered. The 3 dB beamwidth of this filter is 600 kHz. This IF signal (S1(t)) is upconverted and then combined with signals from other instruments onboard the satellite (S2(t)). This composite signal is amplified, upconverted to 11381.05 MHz and amplified to a power of 3.75 W. The composite amplified signal is then filtered before being transmitted via the satellite .6 M parabolic antenna.

2.4.2.2 SARSAT GEOSAR GOES Transponder

The GOES spacecraft is developed by the USA and is serving for COSPAS–SARSAT and also for meteorological applications. The SARSAT GEO GOES East bird is at 75°W covering Atlantic Ocean and GOES West at 135°W is covering almost entire Pacific Ocean.

A functional diagram of the GOES SAR repeater is illustrated in Fig. 2.21. The repeater is redundantly configured and consists of the following units:

(a) One UHF 406 MHz receive antenna, Switch (SW) dividing two 406 MHz LNA shared with another satellite subsystem and two dual-conversion 406 MHz receivers;
(b) Two 3 W (watt) phase-modulated L-band transmitters synchronously switched and one SHF 1544.5 MHz transmit antenna; and
(c) Telemetry and Command (T&C) points interfaced with the spacecraft Command and Telemetry subsystem.

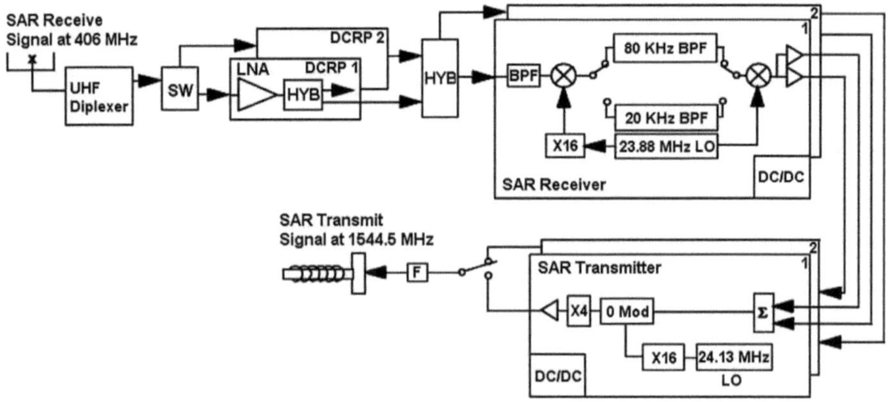

Fig. 2.21 Diagram of SARSAT GEOSAR GOES Transponders—Courtesy of Manual: by COSPAS–SARSAT

The 406 MHz SAR signals from COSPAS–SARSAT distress beacons are received on the UHF antenna and fed through the antenna diplexer and switch to an LNA module in one of the redundant pairs of Data Collection Platform Repeaters (DCPR). The DCPR LNA modules are used as a part of the SAR implementation to accommodate circuit efficiency on the spacecraft.

The LNA module outputs are connected to the redundant pair of SAR receivers. The signal applied to the selected receiver is downconverted for bandpass filtering in accordance with one of two commandable bandwidth modes; a narrow band mode of 20 kHz or a wide band mode of 80 kHz. The filtered output signal is further downconverted to near baseband and fed through amplifiers to the SAR transmitter. However, the overall gain of the SAR receiver can be command-selected into a fixed gain or ALC mode.

The outputs of the receivers are provided to the redundant pair of SAR transmitters. The selected SAR transmitter phase modulates the signal, multiplies the signal to 1544.5 MHz, and amplifies the modulated carrier to 3 W. The phase-modulated signal has the nominal modulation index set such that the carrier suppression is 3 dB with the receiver in the ALC mode or with the receiver in the fixed gain mode operating with two nominal beacon signals plus the noise. A baseband limiter restricts the modulation index from exceeding 2 radians. The transmitter output is applied through a 4 MHz bandwidth filter to the helical antenna and radiated with an EIRP of +15.0 dBW.

The GOES repeater has redundant LNA modules, receivers, and transmitters that can be selected to define a complete repeater configuration. At this point, a specific repeater configuration can be operated in the modes described in Table 2.5.

ALC mode is concerning two equal test tones each at 7 dB above the receiver noise applied to the receiver input will not produce intermodulation products within the transponder bandwidth greater than 30 dB below the test tone output level.

The GOES SAR Receiver and Transmitter parameters are shown in Table 2.6.

Note 1 in Table of Rx Parameters is nominal input level at antenna from 5 W beacons located at 45± Elevation angle to the satellite, which includes 4.1 dB polarization losses.

Note 2 in Table of Tx Parameters is concerning Fixed Gain Mode, which is presenting two equal test tones each at 2 dB above the Rx noise applied to the repeater receiver input will not produce intermodulation products within the transponder bandwidth greater than 20 dB below the test tone output level.

Table 2.5 GOES repeater operating mode

Mode	Band center frequency (MHz)	Receiver 3 dB bandwidth (kHz)
Narrow band with ALC	406.025	20
Narrow band fixed gain	406.025	20
Wide band with ALC	406.050	80
Wide band fixed gain	406.050	80

Table 2.6 GOES SAR receiver and transmitter parameters

Receiver (Rx) parameters	Unit	Values	Transmitter (Tx) parameters	Unit	Values
Nominal input level antenna[1]	dBW	−173.1	Center frequency	MHz	1544.5
System noise temperature	K	359	Output power transmitter	W	3.0
G/T	dB/K	−18.5	Repeater EIRP	dBW	+15.0
Rx bandpass characteristics: Narrow band mode (to 406.025 MHz)	KHz	±6.0(1 dB BW) ±10.0(3 dB BW) ±20.0(20 dB BW)	Phase Jitter (50 Hz Bandwidth)	deg. (rms)	≤10
Narrow band mode (to 406.05 MHz)	KHz	±30.0(1 dB BW) ±40.0(3 dB BW) ±50.0(20 dB BW)			
Dynamic range	dB	≤15	Modulation type	Type	Linear phase
Group delay (over 4 kHz) in KHz	μs	≤13	Tx Nominal Nodulation Index Modulation Index Limit	Radians peak	1.0 2.0
Image rejection	dB	60	Downlink frequency stability	N/A	$\pm 2.5 \times 10^{-6}$
AGC time constant	ms	≤40	Amplitude ripple (over 24 h)	dB	±1
Frequency (RF) stability (<0.5 s) RF conversion oscillator	N/A	$\pm 1 \times 10^{-9}$	Linearity	N/A	See Note 2

2.4.2.3 COSPAS LEOSAR Transponder

The COSPAS LEOSAR payload system functional diagram is illustrated in Fig. 2.22.

The uplink signal from SARP and SARR receive antenna (SPA) is entering via filter and RF switch in one of two receivers. From each Rx, the 75 kHz signal by speed of 2.4 Kb/s is passing formatters, T&C, and entering in one of two Tx units. The downlink signal from the SAR L-band transmit antenna (SLA) can be detected by any COSPAS–SARSAT ground Local User Terminal (LUT) in the LEOSAR satellite system known as LEOLUT.

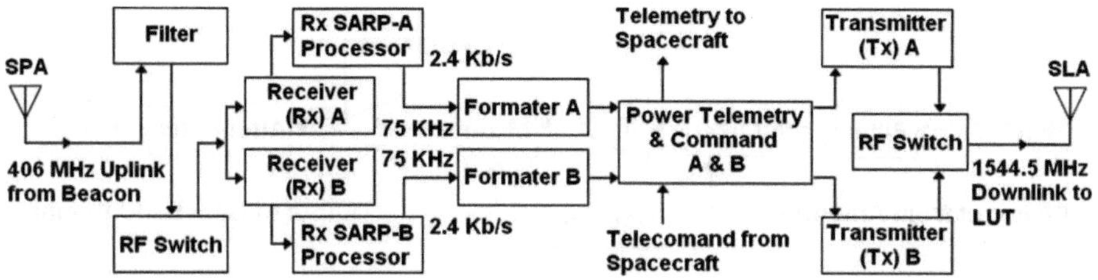

Fig. 2.22 Diagram of COSPAS LEOSAR Transponders—Courtesy of Manual: by COSPAS–SARSAT

The COSPAS LEOSAR payload of the COSPAS–SARSAT system is composed of:

(a) SAR Repeater (SARR) is getting SAR distress signals from ground beacons and without processing is retransmitting them to the LUT land infrastructures;
(b) SAR processor (SARP) is processing receiving SAR signals and retransmit them to the first-available LUT station; and
(c) Uplink UHF 406.05 MHz and downlink SHF 1544.5 MHz antennas.

The SARR provides local mode coverage for the 406 MHz band without processing distress and SAR signals. The SARP provides both local mode and global mode coverage for the UHF 406 MHz band via COSPAS satellites, which have an improved SARP with memory known as SARP-2, shown in Fig. 2.20 as Rx SARP A and B processors. Processed data are transmitted to the ground stations via the downlink transmitter. In addition, this transponder is also providing routine Telecommand (TC) from spacecraft and Telemetry (TM) to spacecraft for better control and management of spacecraft.

2.4.2.4 SARSAT LEOSAR Spacecraft Receiver and Transmitter

The LEOSAR receiver is an integrated component of SARR without processing elements and can be also integrated in SARP configuration with processing and formatting modes. As shown in Fig. 2.23, the two 406 MHz receivers are getting RF signal from beacons via antenna and via mixers is transforming it in IF signal. Each receiver contains automatic gain control (AGC) and provides two outputs to drive the two transmitters, while via mixers they share the same oscillator and are coherent in a ratio of 1:2.

As illustrated in Fig. 2.23, each one of the transmitters has four inputs in Baseband Summer: one for each of the two 406 MHz receivers, one for the Processed Data Stream (PDS) channel, and one spare port. The signals are then entering in Phase Modulator where the receiver input is illuminated by a sinusoidal

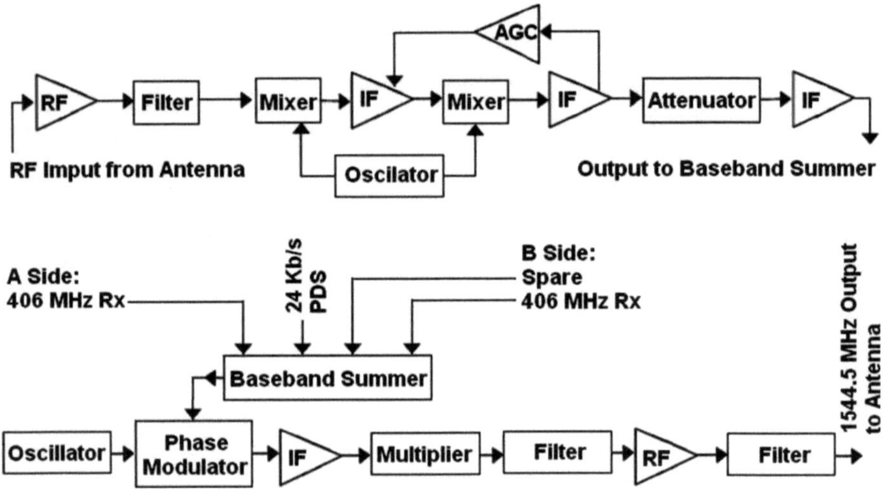

Fig. 2.23 Diagram of Inmarsat-3 transponders—Courtesy of Book: by Martin

signal from oscillator at the maximum frequency and level of IF. The IF signal is passing multiplier and filtering process and as RF is transmitted at 1554.5 MHz via antenna to Local User Terminal (LUT) on the ground.

2.4.3 Satellite Repeaters for New Generation of GEO and non-GEO MSC

The new generations of GEO and non-GEO satellites for MSC are developed by Inmarsat GEO satellite operator, while non-GEO satellites, such as Big LEO satellite constellations, are developed by Iridium and Globalstar, and Little LEO satellite constellation is developed by Orbcomm.

2.4.3.1 Inmarsat-3 GEO Satellite Communication Repeater

The Inmarsat-3 development was not paced by a lifetime limit of Inmarsat-2, but by its capacity limitation. The number of ships using Inmarsat increases every year, both by choice and gradually because of requirements that ships be equipped for MSC. In addition, since 1990, Inmarsat has provided service to airplanes as well as ships. Besides requiring increased satellite capacity, the airplanes also require more of the satellite power because of their smaller antennas.

Planning of the Inmarsat-3 series began about 1988. Proposals were received early in 1990; that summer, negotiations with a contractor began in parallel with a

5-month technology validation program. Thus, the technology program concentrated on the L-band multibeam antennas and the L-band transmitter power amplifier.

The final contract was signed early in 1991; it included four satellites and options for more. In March 1994, one option was changed to an order for a fifth satellite. The capacity increase relative to the foregoing Inmarsat satellites is achieved by the use of spot beams at the L-band transmit-and-receive frequencies used between the satellites and the mobile terminals. The main requirements on the Inmarsat-3 payload illustrated in Fig. 2.24 are as follows:

1. Reconfigurable spot beams so that each satellite can provide the desired coverage from any of the five operating locations;
2. Flexible allocation of the total L-band power among the spot beams and the global beam to adapt to changing traffic loading; and
3. Channelization matching the international L-band allocations and switchable to adapt to Inmarsat needs and to ensure compatibility with other L-band satellites.

The Inmarsat-3 payload consists of five sections. The C-band-to-L-band forward channel is for MSC from fixed terminals to mobiles and is matched to the L-band-to-C-band return channel for communications in the opposite direction. These two channels are the main part of the payload. There is also a C-band-to-C-band channel for administrative traffic between mobile terminals in limited circumstances such as during search and rescue efforts. The final section of the payload is a navigation channel that augments the US GPS and the Russian GLONASS system for new communication, navigation, and surveillance (CNS) service known as Global Satellite Augmentation System (GSAS). When Inmarsat-3 was developed, the L-band-to-L-band and navigation capabilities were new features in MSS.

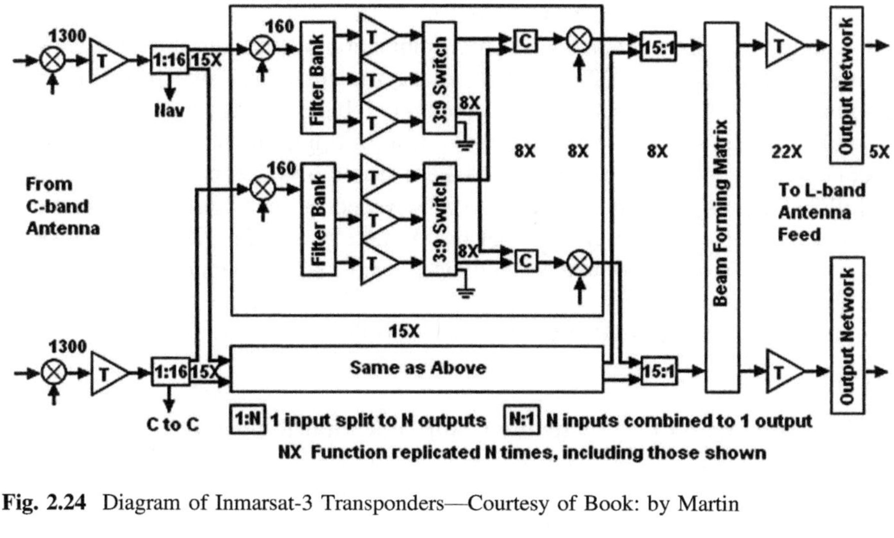

Fig. 2.24 Diagram of Inmarsat-3 Transponders—Courtesy of Book: by Martin

Inmarsat-3 spot beams in coverage area are quite small to allow frequency reuse and large enough to moderate the payload complex, which grows with the number of coverage areas. Many of these coverage areas have the same geometry at more than one location, or differ only by a rotation of the coverage areas about the subsatellite point. As a result, the Inmarsat-3 design has seven spot beams for the L-band transmission, each generated by four- or six-element subarrays of the 22-element feed array, while global coverage beams are used at C-band. The feed array can be rotated 21° about its axis to change the coverage geometry at the time of satellite relocation. Adjacent beams overlap, but some separated beam combinations have sufficient isolation to allow frequency reuse, namely to use of the same frequencies in two different beams.

The limited L-band spectrum available for use with mobile terminals is reused twice in the Inmarsat-3 design. When Inmarsat-3 was developed, the L-band allocation was subdivided for Maritime, Land, Aeronautical, and general mobile applications. To match this, the Inmarsat-3 frequency plan has 10 bands with 1–4.5 MHz bandwidths. All except the narrowest band have switchable Surface Acoustic Wave (SAW) filters that can divide the band into two or three sub-bands separated by 200 kHz guard bands for use in different antenna beams, or combine the sub-bands with the intervening guard bands for use in the same beam. The frequencies of the guard bands are different in the two reuses of the spectrum in order to maximize the ability to adjust spectrum usage. More recently, most of the allocations have been generalized to allow any application. Nevertheless, the Inmarsat-3 channelization is still good for efficient spectrum utilization.

The combination of many frequency bands and multiple satellite antenna beams would typically imply a very complex switching network on the satellite to provide flexible interconnections; this has been avoided by using a multiport amplifier. This amplifier is composed of a beam-forming matrix followed by an output filtering and routing network. The beam-forming matrix has eight inputs, corresponding to seven spot beams plus one global beam. The output network has 22 output ports, corresponding to the 22 antenna feeds. The combination of the matrix and the network allows the power of all 22 amplifiers to be routed to any one beam or to be distributed among many beams, both spot and global. The distribution of power is controlled by the signal phasing set in the beam-forming matrix. All the amplifiers operate at high efficiency regardless of the distribution of signals among the spot and global beams. Even if the total power is routed to one beam, it is spread over the four or six feeds that generate that beam, thereby avoiding excessive power at any one feed. Two global channels, and the L-band-to-L-band channel (when used), are always transmitted through global beams; all other channels in the L-band frequency plan may be routed to either spot or global beams.

2.4.3.2 Inmarsat-4 GEO Satellite Communication Repeater

The next question Inmarsat faced was whether to use a bent-pipe (transparent) transponder as on Inmarsat-3 before or to design a new regenerative payload on the last generation of the Inmarsat-4 payloads. In the end, notwithstanding the better characteristics offered by a regenerative solution, Inmarsat decided to also implement its fourth-generation navigation payload as a bent-pipe repeater.

The company has negotiated European spacecraft manufacturer Astrium a 700 million US $ contract to build three Inmarsat I-4 satellites. The contract was awarded in May 2000 with a planned service start sometimes in 2004. However, satellite development difficulties and slower than expected demand delayed the start of service to 2005. Astrium includes the former Matra Marconi Space, which built the Inmarsat-2 satellites and the payload for the Inmarsat-3s. Two Inmarsat-4 satellites were placed in two of Inmarsat's existing IOR and AOR-W orbits in 2005. The third Inmarsat-4 satellite was launched later to extend the coverage area so as to make it global, based on business and operational considerations. The satellite locations are subjected to review and may change.

The focus for the Inmarsat 4 satellite transponders is what Inmarsat calls the Broadband Global Area Network (BGAN) with communication speeds to 432 Kb/s for Internet access and multimedia transmissions, as well as traditional voice and data services. As a prelude, Inmarsat started Regional BGAN at the end of 2002 with transponder capacity leased from Thuraya (described below). This service has data rates to 144 Kb/s and is available in the coverage area of Thuraya spacecraft. The communication payload requires a powerful satellite platform, capable of providing the required DC power (10–12 kW) and other resources. Inmarsat 4, like prior Inmarsat generations, uses L-band for links between the satellite and large, fixed gateway stations. The satellite has a 9-m diameter L-band reflector used for both transmission and reception. This antenna consists of 120-element feed on the adjacent face of the satellite body, to form up to 228 spot beams that cover Earth. In addition, for compatibility with Inmarsat 3 satellites there is a global coverage beam and a 19-beam array of wide spots that cover Earth, both in L-band. The satellite also has a dual-polarization Earth coverage C-band antenna beam. Each satellite has the capacity of 630 channels each with a data rate up to 432 Kb/s. Onboard processing can dynamically reassign channels among the beams to satisfy user requests.

Inmarsat 4 has a two-channel navigation payload, each with wider bandwidth than the single channel on Inmarsat 3. Navigation data are transmitted to the satellite at C-band and transmitted from it on two frequencies. These signals augment GPS and other navigation satellite systems by transmitting integrity and correction data that can improve the accuracy of users' position determination. Other information about Inmarsat 4 is as follows.

2.4.3.3 Projected Inmarsat-5 GEO Satellite Communication Repeater

Inmarsat's first wholly owned satellites, the Inmarsat-2, were launched in the early 1990 s, and the Inmarsat-3 as the first generation to use spot beam technology followed later in the decade, both of these spacecrafts are spare or still operational. The new constellation of the Inmarsat-4 set a new benchmark for MSC in terms of their power, capacity, and flexibility. In fact, one Inmarsat-4 satellite is 60 times more powerful than an Inmarsat-3 and its constellation series is expected to continue in commercial operation until about 2020.

In the meantime, Inmarsat has entered into agreement with ESA to become the commercial operator of a new satellite called Alphasat. The satellite is part of an ESA initiative to develop a new spacecraft platform capable of carrying onboard a large communications payload. Accordingly, Alphasat is scheduled for completion in 2012 and to supplement the existing I-4 satellites. It will provide service over Europe, the Middle East, and Africa.

On the other side, in 2010 Inmarsat satellite organization has contracted the US aerospace manufacturer Boeing to build a new constellation of Inmarsat-5 satellites as a part of a new 1.2 billion US $ worldwide wireless broadband network called Inmarsat Global Xpress™. Boeing will construct three Inmarsat-5 (I-5) satellites based on its 702HP spacecraft platform. The first is scheduled for completion in 2013, with full global coverage expected by the end of 2014.

The spacecraft will break new ground by transmitting in a portion of the radio spectrum never before utilized by the commercial operator of a global satellite system and with the extremely high-frequency Ka-band transponder for the first time. Accordingly, each fifth generation of Inmarsat-5 will carry a payload of 89 Ka-band beams capable of flexing capacity across the globe and enabling Inmarsat to adapt to shifting subscriber usage patterns over their projected lifetime of 15 years.

2.4.3.4 Globalstar Big LEO Satellite Communication Repeater

The Globalstar satellite uses a simple transparent repeater rather than a digital onboard processor such as that being flown on Iridium. Figure 2.25 shows a block diagram of the Globalstar communication payload and a phased array antenna with 16 spot beams, which provides gain and a degree of frequency reuse. This spacecraft is a 3-axis stabilized platform with the Earth-facing panel always parallel to the orbit tangent. All transmissions are relayed at C-band (7/5 GHz) through GES terminals that track the satellites as they pass near by in LOS. However, this condition places a constraint that a given LEO satellite must simultaneously see both the mobile user at L/S-bands and the GES at C-band.

Through a relationship with Qualcomm, Inc., Globalstar provides service using of the CDMA cellular standard. A GPS onboard receiver is used to accurately determine the orbit parameters and also to supply accurate time and frequency to the satellite systems. The attitude control system uses small, one Newton, thrusters for onboard attitude control. At this point, yaw steering is employed to provide sufficient solar array power during all phases of mission. Solar panel and a large nickel–hydrogen battery provide power for all phases of the mission. However battery recharge takes place where satellite is moving over the oceans, there is less traffic with mobile users.

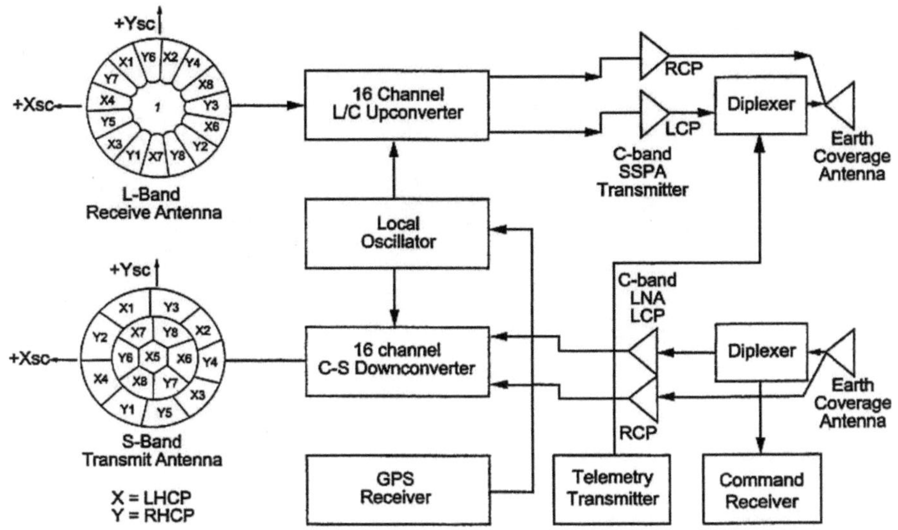

Fig. 2.25 Diagram of Globalstar spacecraft transponders—Courtesy of Paper: by Dietrich

2.4.3.5 Iridium Big LEO Satellite Communication Repeater

Iridium, such as Globalstar, as Big LEO satellite system, provides too much useless coverage of the ocean regions, resulting in less effective capacity over land areas than can be obtained with either a MEO or a GEO strategy. Without intersatellite links, LEO satellite cannot serve users out on the oceans since an operating GES must simultaneously be in view. Globalstar has not this facility, but this is in contrast to Iridium's ability to serve users no matter where on or above the planet they may be located by employing intersatellite links.

The Iridium satellite bus presented in Fig. 2.26 is adapted on the satellite's Earth face, which integrates communication and other types of antennas, crosslink arrays, secondary payload, and other parts. The secondary payload envelope reserved is 300 × 400 × 700 mm, the two larger dimensions being aligned with the Earth face, the single largest dimension being aligned with the y-axis. While traveling on its orbit, the satellite keeps its z-axis pointed toward the center of the Earth and its velocity vector is aligned with the x-axis. The field of view of the secondary payload is a 75° half-cone opening at its top and center, and pointing toward the nadir (+Z). It is clear of any obstruction.

2.4.3.6 Diagram of VSAT GEO Satellite Communication Repeater

As stated earlier, a transparent payload makes no distinction between uplink carrier and uplink noise, and both signals are forwarded to the downlink. Therefore, at the

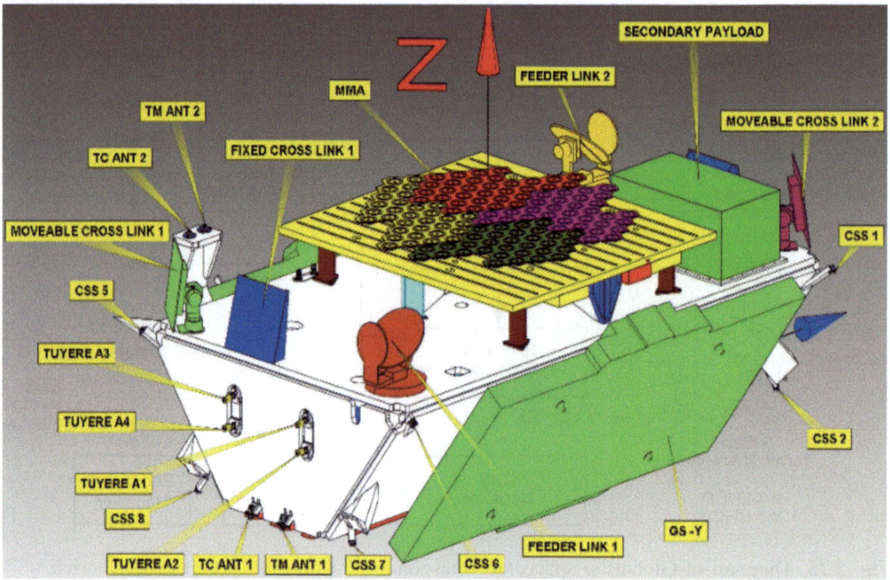

Fig. 2.26 Components of Iridium transponders—Courtesy of Brochure by Thales Alenia

Earth station receiver, one gets the downlink noise together with the uplink-retransmitted noise.

However, a regenerative payload entails onboard demodulation to process of the uplink carriers. Onboard regeneration is most conveniently performed on digital carriers. The bit stream obtained from demodulation of a given uplink carrier is then used to modulate a new carrier at downlink frequency. This carrier is noise-free; hence, a regenerative payload does not retransmit the uplink noise on the downlink. The overall link quality is therefore improved.

Moreover, intermodulation noise can be avoided, as the satellite Channel Amplifier (CA) is no longer requested to operate in a multicarrier mode. Indeed, several bit streams at the output of various demodulators can be combined into a Time Division Multiplex (TDM), which modulates a single high-rate downlink carrier. However, this carrier is amplified by the CA section, which can be operated at saturation without generating intermodulation noise, as the carrier it amplifies is unique. This concept of VSAT Satellite transponder is illustrated in Fig. 2.27.

It should be emphasized that today's commercial satellites that can be used for VSAT services are not equipped with regenerative payloads but only with transparent ones. Only a few experimental satellites such as NASA's Advanced Communications Technology Satellite (ACTS) and the Italian ITALSAT satellite have incorporated a regenerative payload, but they are no longer in operation. Some satellites of the EUTELSAT fleet are equipped with a regenerative payload (Skyplex) but can be used only by Earth stations operating according to the DVB-S standard.

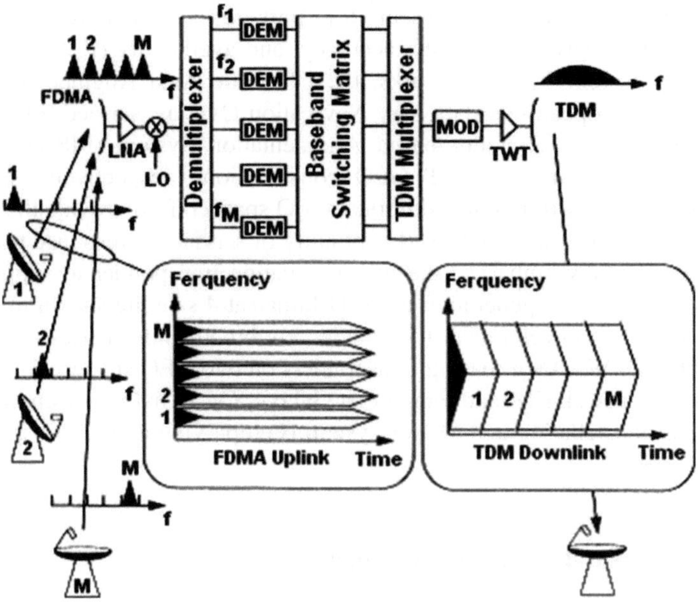

Fig. 2.27 Diagram of VSAT Spacecraft transponders—Courtesy of Book: by Maral

2.4.4 Satellite Navigation Repeaters for GNSS

The new generations of modern GEO satellites can augment the performance of the US GPS and Russia GLONASS by providing a separate ranging channel to transmit integrity and correction data. In fact, this concept dates back to the late eighties and has evolved to its current form known as Regional Satellite Augmentation System (RSAS).

In this book, the use of the ICAO known nomination Satellite-based Augmentation System (SBAS), which appear in the world classification of the acronyms, will be replaced by the Regional Satellite Augmentation System (RSAS) as more convenient nomenclature.

The RSAS complementary information processed in any Master station is diffused by GEO satellite by means of a pseudo-GPS or GLONASS signals and covers a wide regional or geographical area integrated in Global Satellite Augmentation System (GSAS). The RSAS data will allow satellite navigation to meet the stringent reliability and other requirements set by authorities for Air Traffic Control (ATC) and Maritime Traffic Control (MTC). Land users, such as road and railways, will also be able to take advantage of the improvements in positioning accuracy. However, augmented GNSS-1 solutions of RSAS were recently developed to improve the mentioned deficiencies of current GPS and GLONASS military systems and to meet the present transportation civilian requirements for high-operating integrity, continuity, accuracy, and availability (ICAA).

The three current RSAS networks, the US Wide Area Augmentation System (WAAS), the European Geostationary Navigation Overlay Service (EGNOS), and the Japanese MTSAT Satellite-based Augmentation System (MSAS) are using Inmarsat, Artemis, and MTSAT GEO constellations, respectively. The GSAS network worldwide will require additional GEO spacecraft to provide the necessary coverage, availability, and improved continuity of service. In response to this need, Inmarsat has already embarked the new navigation transponder to support RSAS functions on its current generation of GEO Inmarsat-4 satellite shown in Fig. 2.28, which is a modern development to provide new broadband services.

In additional, WAAS is leasing transponders on two GEO spacecraft, such as the Canadian Telsat Anik F1R and PanAmSat Galaxy XV, EGNOS is using Inmarsat-4 and Artemis constellation and MSAT system is employing own GEO spacecraft of MTSAT series.

2.4.4.1 Inmarsat-4 Navigation Payload

The first GNSS transponder was installed in the payload of Inmarsat-3 spacecraft providing RSAS service until 2005.

The current EGNOS space constellation is composed of two leased transponder hosts of the Inmarsat-4 AORE and IOR, and one ESA Artemis GEO spacecraft. The Inmarsat-4 hosts a navigation payload capable of transmitting RSAS information on both the GPS L1 and the L5 frequencies via GES to the mobile users. Therefore, this payload will transmit satellite navigation signals and allow the real-time relay from a single-ground-monitoring network of integrity and accuracy augmentation data for orbiting GNSS. A functional diagram of the Inmarsat-4 satellite is shown in

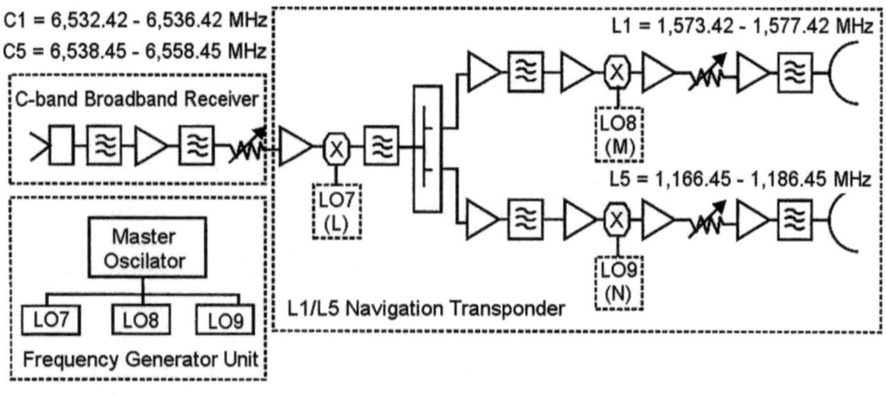

Fig. 2.28 Diagram of Inmarsat-4 Navigation (GNSS) transponder—Courtesy of Article: by Razumovsky

Fig. 2.26. This payload is a dual-channel bent-pipe transponder that converts two C-band (C1 and C5) uplink signals from one GES to two downlink signals in two separate bands with the following characteristics:

1. **Bandwidth Limitations**—A 4.0 MHz-wide C1 band uplink channel will be relayed in the L1 downlink channel and allow the transmission of the RSAS L1 signal. A 20.0 MHz-wide C5 band uplink channel will be relayed in the L5 downlink channel and allow the transmission of the newly defined L5 RSAS signal. The system has not allocated 24 MHz of bandwidth at both frequencies, but the design included extra spectrum necessary to uplink the newly defined signal at L5 and that left only 4 MHz available at L1, still more than what was provided by Inmarsat-3.

2. **Transponders**—The Inmarsat-4 communication and navigation transponders share both the C-band broadband receiver and the Frequency Generator Unit, while the two channels (C1-L1 and C5-L5) share hardware until after filtering at IF. The uplink channels are frequency-translated, using coherent translation oscillators (derived from a common master oscillator), to the corresponding L1 and L5 downlink channels. The navigation uplink signals are downconverted with the same local oscillator (LO7) to a suitable IF and then split into two signal paths where the appropriate channel bandwidths are applied. The two signals are then upconverted into the final output frequency (L1 and L5) and fed to the L-band HPA unit before transmission to the dedicated navigation antenna providing global coverage. Both channels have independent telecommandable gain adjustment. All signal paths through the navigation transponder are cold redundant.

3. **Broadcast Power**—The payload broadcast power will be 28.1 dBW at L1 and 26.2 dBW at L5 in a global coverage antenna beam. The output power at the L1 frequency has been designed to assure the payload would be capable of delivering the necessary power as required by the ICAO Standards and Recommended Practices (SARP) for an RSAS signal. The EIRP output of 26.2 dBW at L5 will permit the user and receiver to receive the same level of radiated power it receives at L1, taking into consideration the smaller free-space loss at L5 frequency. The transponder is expected to accommodate only one signal per channel, which means that the specified broadcast powers are saturated values.

4. **Navigation Performance**—The overall ranging and navigation performance of the RSAS signal through a bent-pipe repeater is influenced not only by the satellite repeater characteristics, but also by the supporting ground signal generator and timing system. Unlike GPS satellites, Inmarsat navigation payloads do not generate the signal. Rather, their "clock" is on the ground and the signal is generated there. Hence, a closed-loop control system is necessary in order to make the ranging signal appear as if it is originated on board the GEO satellite by compensating for the uplink delay contribution due to the ionosphere, Doppler shift and the frequency offset induced by the GEO transponder itself.

In other words, the final aim was to have the signal broadcast by the transponder resemble a GPS signal. In particular, the closed-loop control system must be capable of maintaining signal code/carrier coherence; that is, the code-chipping rate (1.023 MHz of GPS C/A code) and carrier frequency (L1) must be kept in the constant ratio of 1:1540 as it happens in the GPS signals. This requirement allows GPS receivers to improve the accuracy of their ranging measurements and ICAO has included it in the SBAS signal specification.

5. **Transponder Control**—In the past, Inmarsat sponsored the development of signal generators and receivers to demonstrate achievable performance and to perform validation testing of the Inmarsat-3 payloads, both prior to and after launch. Inmarsat-4 navigation signals, like those for Inmarsat-3 satellite, will require a proper control of the signal code and carrier: This time the challenge will be to maintain a proper code/carrier coherence on both L1 and L5 signals. The navigation satellite transponder control will be realized through the Ground Uplink System (GUS) at a GNSS GES terminal. Inmarsat plans to design and develop prototype onboard equipment for proper navigation signal generation and control that will be used for the ground and in-orbit test campaign in order to conduct end-to-end system tests.

2.4.4.2 Artemis Navigation Payload

The aim of the Artemis payload illustrated in Fig. 2.29 was the same as GNSS payload of Artemis spacecraft, namely to provide an overlay function with supplementary services to the current operating GPS and GLONASS constellation. This kind of services uses existing GEO communications satellite to transmit overlay signals to provide improvement of accuracy, system's integrity data, and wide area differential corrections.

The Artemis Navigation (GNSS) satellite payload is caring transparent transponder capable to receive a signal from a European coverage and to translate it onto two separate downlink carriers, the first operating at L-band (L1) and the second one at Ku-band (channel F2), which is illustrated in Fig. 2.30.

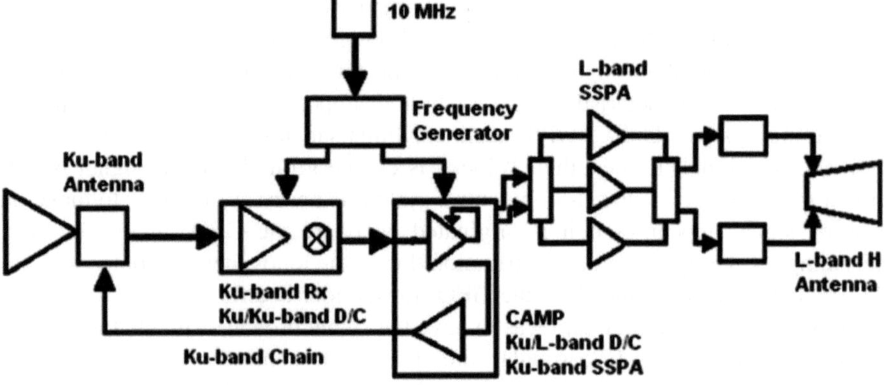

Fig. 2.29 Diagram of Artemis Navigation Transponder—Courtesy of Paper: by Comparini

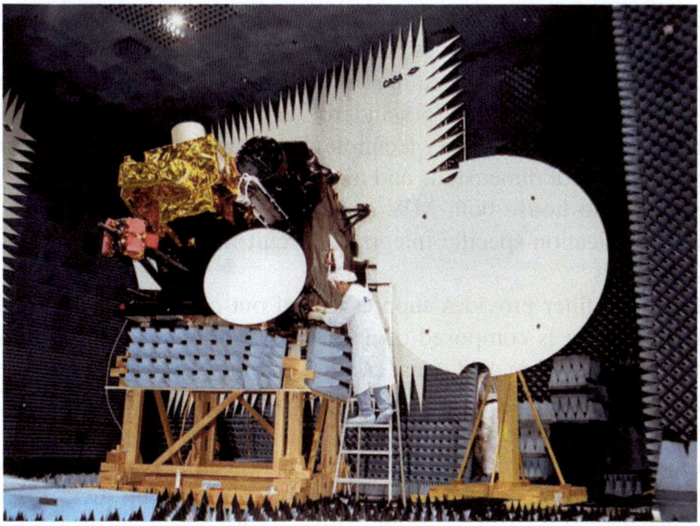

Fig. 2.30 Artemis Communication and GNSS Payload—Courtesy of Brochure: by ESA

The signal transmitted from the ground is a single 500 b/s carrier modulated with timing, integrity, ranging, and other information and spread with a PN code, a type compatible to those used for GPS. The code orthogonality properties assure the possibility to transmit the GNSS signal on the same nominal carrier frequency of the GPS and GLONASS.

The uplink signal is received at 13.875 GHz by a Ku-band antenna and filtered by the input multiplexer, which separates the incoming communication signal from the NOS signal and to reduce interference. This filter is mounted directly on the spacecraft top floor. The input signal is divided into two paths and translated; therefore, the heart of the payload performing this function is the CA module. At this point, the low-noise amplification mode is mainly performed by the Ku-band receiver (KRx), which provides a noise figure of 2.3 dB and a gain of about 53 dB. The signal is then fed into the Ku-band channel filter. Realized using high Q waveguide dielectric resonator, it gives only 10 nSec of group delay variation into the useful bandwidth of 4 MHz. The signal feeds the CA element at the output of the channel filter. The unit splits the signal into two paths: One feed the Ku-band transmit section without RF conversion while the other is downconverted to the GPS frequency. The two CA output signals are amplified by two SSPA components providing 16 watts end of life. All the RF units use latest technologies (hybrids, MMIC, MCM) both for the RF portion and for the control and interface functions.

1. **Ku-band Rx**—The KRx unit provides low-noise amplification and down-conversion of signals at RF of 13.855-13.895 GHz down to 12.728-12.768 GHz, by means of a net subtraction of an externally generated LO frequency. The equipment is provided with an internal power supply and a command and telemetry interface. The signal is then amplified in a linear amplifier. Thus, group delay,

phase, AM/PM conversion, and gain are specified as required by GNSS payload performances analysis. The unit is internally redundant and includes a redundant DC/DC converter to derive the required secondary voltages from the satellite main bus. The input interface is WR75 standard, and the LO input and the RF output interfaces are SMA 50 O. The key technologies characterizing the units are: MMIC to achieve reduction of dimensions and mass and a high level of reproducibility; RF hybrid technique to house both MW and low frequency analog/digital integrated circuits; and application-specific integrated circuit (ASIC) for control and TM/TC interface functions.

A IF bandpass filter provides suppression of out-of-band mixing products. The rest of the IF section is composed of a cascade of different MMIC circuits which provide the required gain and gain control dynamic. The first gain stage is an MMIC LNA as the one above described: The large passband of the MMIC allows use on the IF section as well, in order for it to have negligible contribution to the overall noise figure. The MESFET process has been used for the MMIC units following in the IF chain. The Flatness Corrector (FC) circuit gives the possibility to control the gain slope variation consequent to the complete system assembling and possible slope variations over temperature. The gain control block is the variable gain amplifier (VGA), which is composed of three self-biased stages of amplification with an embedded analog attenuator, allowing compensation of gain drift over temperature of the complete assembly. The analog attenuator design was based on a broadband-extended T configuration using cold FET units. This approach was chosen because of its minimal VSWR variation and flat gain response over the attenuation range. A medium power amplifier (MPA) is used as the output stage in order to have higher output level and better linearity. The circuit design is single-ended and consists of a two-stage amplifier. Output FET gate periphery was selected to achieve a 32 dBm third-order intercept point while keeping the channel temperature below 110 °C when in maximum environmental temperature.

2. **Channel Amplifier (CA)**—This unit is used to amplify the signal to a level suitable to drive the power amplifier. It can be set into two operational modes: **a)** Nominal operation of amplifier level control (ALC) mode is settable by means of ground command to different levels. In fact, the ALC loop maintains the output power level within the specified limits recovering uplink variations and input signal dynamic; and fixed gain mode, when ALC is off, is the amplifier provides a fixed gain set by ground command.

The CA unit is in general redundant and equipped with internal DC/DC converter and provides two different outputs: **L1** at a frequency of **1575.42 MHz**, generated by means an external local oscillator, and **F2** at a frequency of **12748 MHz** (no conversion is required). As in the Ku-band RX, also in CA hybrid circuits are used to realize the RF channels and the digital control circuits. The following modules compose the unit: two complete Ku-band CA modules (main and redundant); two Ku/L-band RF downconverter including CA module functionality (main and redundant); redundant DC/DC converter; and couplers for redundancy and channels splitting.

3. **Solid-State Amplifier (SSA)**—The SSA module is used for high-power amplification of L-band downlink signal. This unit, manufactured by the UK MMS company, now Astrium UK), consists of: an amplifier module in which high-power RF amplification and power generation occur; an electronic power conditioner (EPC) which accepts primary bus from the spacecraft and generates the required secondary stabilized voltages.

4. **RF Generator (RFG)**—This unit generates two frequencies at L- and X-bands using a 10.0 MHz reference signal coming from the external USO. In the Artemis implementation, the onboard USO has been used. The unit consisting of the navigation RFG plus the 10 MHz reference has been specified as a whole, in terms of performance; however, the RFG and the 10 MHz USO are mechanically separated in order to maintain the qualification of the assembly achieved on Artemis for the RFG, on SICRAL, and on UHF spacecraft update programs for the 10 MHz ultra-stable oscillator (USO). The RFG output frequencies are 1127 MHz and 11172.58 MHz. Direct synthesis approach is used: Multiplication is obtained using active stages and step recovery diodes. At this point, quartz and helical filters are used to reduce spurs and phase noise to levels within the specification.

2.4.4.3 MTSAT Navigation Payload

The Japanese GNSS payload is integrated together with meteorological payload onboard series of GEO Multifunctional Transport Satellite (MTSAT). The MTSAT GNSS payload is the forerunner of a family of GNSS payloads able to fulfill augmentation system for Japan and surrounded areas allowing interoperability among the different systems.

The MTSAT-1R spacecraft launched in February 2005 replaced unsuccessful MTSAT-1, which was destroyed in a launch failure in 1999. Air traffic communications will use the L-band and the AMSS protocol, which is specific to mobile satellite communications. Links between the Earth segment and satellite segment will use the Ku- and Ka-band.

This advanced GEO satellite has rideshare payloads that play a critical role in the next generation of the Japanese satellite-enhanced air traffic management systems and weather forecasting. The aeronautical payload is a natural improvement on the previously designed MTSAT-1 payload. A new state-of-the-art digital processor facilitates communication link frequency adjustments that may be required by international coordination agreements after MTSAT-1R is put into service. It also facilitates efficient use of the allocated frequency spectrum and improves transponder performance.

2.4.5 Repeaters for Stratospheric Platform Systems (SPS)

The SPS or high-altitude platform (HPA) transponder shown in Fig. 2.31 illustrates a block diagram of a typical onboard architecture, which together with its associated

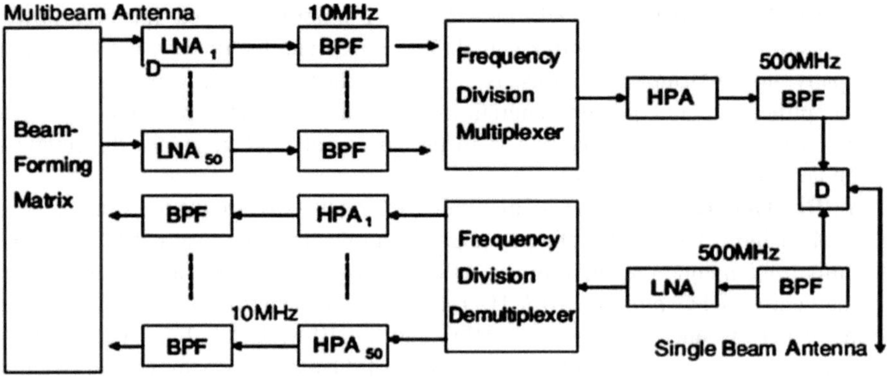

Fig. 2.31 SPS transponder—Courtesy of Book: by Aragon-Zavala

antenna subsystem would make a complete communications payload. In this case, SPS transponder has a bandwidth of 500 MHz using Code Division Multiple Access (CDMA) scheme. This transponder can accommodate up to 50 antenna beams with eight wideband carriers (assuming a carrier bandwidth of 1.25 MHz each). For the uplink, the carriers are received by the platform antenna subsystem and amplified by a LNA.

The bandwidth of these carriers is limited to only 10 MHz using passband filters, and multiplexed using Frequency Division Multiplexing (FDM). Before these multiplexed signals are transmitted to the ground station, they are amplified further by a HPA stage, filtered and multiplexed. For the downlink, the process is similar but in the reverse direction, and instead of a multiplexer, a demultiplexer is used [ITU-F1500, 00].

Other transponder components play a very important role in the system performance, and their design and selection should not be underestimated. Thus, the LNA element having low-noise temperature and sufficient gain, ensure that noise contributions from the succeeding stages are kept small. Frequency converters serve as frequency-change devices (also known as mixers), which either upconvert or downconvert the signals in the SCP to differentiate between uplink and downlink. Noise performance is also important for these frequency converters, although not as critical as the noise figure for the LNA, since this is the first element in the frontend and therefore, is the one that contributes most to the overall noise. IF processors have the main functions of providing most of transponder gain, defining a frequency response of each channel and, whenever required, performing beam-to-beam routing functions. Filters are in charge of limiting adjacent spurious signals as well as noise that can be generated by the payload itself. Their design is crucial for the performance of the communications payload. Transmitters are in charge of amplifying the signals to the level required for downlink transmissions. Examples of well-known transmitters used for satellite communications are the TWT and power transistors, the latter used for specific applications in lower frequency bands. Payload system performance or performance parameters of individual equipments need to be specified so that the required SCP payload performance is achieved by the following important parameters:

1. **Antenna Coverage Area**—Since antenna gain and SCP coverage are related to antenna dimensions, antenna electrical efficiency and antenna feeder losses;
2. **Figure of Merit G/T**—Since this parameter as ratio of the receive antenna gain to system noise temperature (gain-to-noise temperature) affects link performance;
3. **EIRP**—Is determined by the power capability of the SCP;
4. **Power per Backhaul Carrier**—Is the user link;
5. **Isolation Between Channels**—It reduces potential adjacent channel interference issues;
6. **Spurious Outputs**—Reduce interference levels to other contiguous wireless systems;
7. **Amplifier Linearity**—Since nonlinearities produce intermodulation products that can leak into the passband of other equipment, causing interference; and
8. **Group Delay Variation**—Is the time delay experienced by the modulation waveform in passing through the equipment, causing signal dispersion and performance degradation.

2.4.6 Satellite Antenna System for MSC

The antenna array system of Inmarsat-2 satellite for MSC is shown in Fig. 2.32 (Left). The antenna system mounted on the spacecraft structure, similar to the transponders, is composed of two main integrated elements: the C/L-band and the L/C-band antenna.

1. **Inmarsat-2 C/L-band Arrays**—This uplink is actually the feeder link, which operates in the 6 GHz RF range. The signals sent by LES are detected by a C-band receiving array, comprising seven cup-dipole elements in the smallest circle. On the other hand, the L-band transmit antenna is the biggest segment of the whole system, consisting of 43 individual dipole elements, arranged in three rings around a single central element. Thus, this antenna is providing near-global coverage service downlink for MES in the 1.5 GHz RF spectrums.

2. **Inmarsat-2 L/C-band Arrays**—These arrays are actually the service uplink and operates in the 1.6 GHz RF range. The signals sent by MES in adjacent global coverage region are detected by L-band receiving array, comprising nine cup-dipole elements arranged in a circle. Finally, the C-band transmit antenna consists of seven cup-dipoles for radiation of the feeder downlink to LES in the 3.6 GHz RF spectrum.

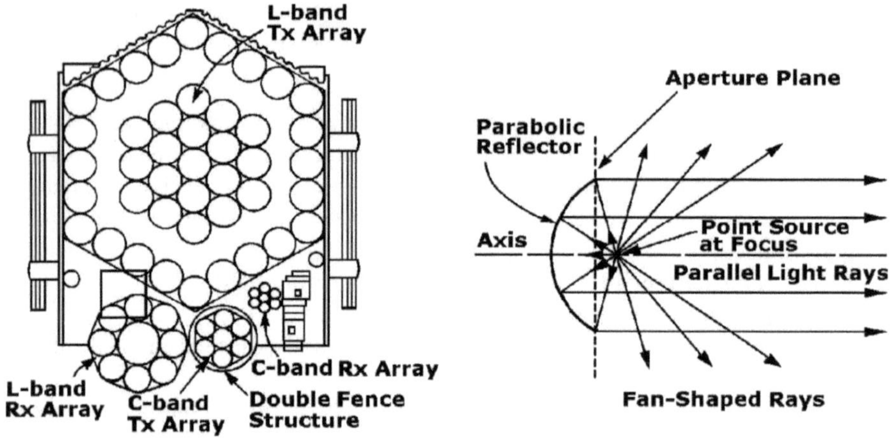

Fig. 2.32 Spacecraft antenna systems—Courtesy of Book: by Gallagher

2.4.6.1 Characteristics of Satellite Antennas

Both transmit antenna array systems are providing a global (wide) footprint on the Earth's surface. However, narrow circular beams from GEO or non-GEO can be used to provide spot beam coverage. For instance, from GEO the Earth subtends an angle of 17.4°. Antenna beams 5.8° wide can reuse three frequency bands twice in providing Earth disk coverage. The directional properties of antenna arrays can be exploited to permit RF reuse in space communications, which is similar to several radio stations using the same RF being geographically far apart.

Earth coverage by seven spot beams (six spots are set out around one spot in the center) can be arranged by three pairs of beams: 1 and 4, 2 and 5, and 3 and 6, operating on frequencies f2, f3, and f4, respectively. Mutual interference within pairs is avoided by pointing one beam as far away from the other as possible. Satellite antenna coverage of the center of the disk is provided by a single-beam operating on frequency f1. Thus, the main advantage with this satellite spot footprint that is specific Earth areas can be covered more accurately than with wide beams.

Furthermore, a greater power density per unit area for a given input power can be achieved very well, when compared with that produced by a global circular beam, leading to the use of much smaller receiving MES antennas. The equation that determines received power (P_R) is proportional to the power transmitted (P_T) separated by a distance (R), with gain of transmit antenna (G_T) and effective area of receiving antenna (A_R) and inversely proportional to 4π and square of distance. The relations for P_R and G_T are represented as follows:

$$\begin{aligned} P_R &= P_T G_T A_R / 4\pi R^2 \\ G_T &= 4\pi\, A_T / \lambda^2 \end{aligned} \tag{2.46}$$

where G_T = effective area of transmit antenna and λ = wavelength. The product of P_T and G_T is gain, generally as an increase in signal power, known as an EIRP. Signal or carrier power received in a link is proportional to the gain of the transmit and receive antennas (G_R) represented as:

$$P_R = P_T G_T G_R \lambda^2/(4\pi R)^2 \text{ or } P_R = P_T G_T G_R / L_P L_K [W] \tag{2.47}$$

The last relation can be derived with the density of noise power giving:

$$P_R/N = P_T G_T (G_R/T_R)(1/K\, L_P L_K) \tag{2.48}$$

where L_P = coefficient of energy loss in free space, L_K = coefficient of EMW energy absorption in satellite channels, T_R = temperature noise of receiver, G_R/T_R is the figure of merit and K = Boltzmann's Constant (1.38×10^{-23} J/K or its alternatively value is –228.6 dBW/K/Hz).

At any rate, P_R has a minimum allowable value compared with system noise power (N); that is, the Carrier and Noise (C/N) or Signal and Noise (S/N) ratio must exceed a certain value. This may be achieved by a trade-off between EIRP (P_T G_T) and received antenna gain (G_R). If the receive antenna on the satellite is very efficient, the demands on the LES/MES are minimized. Similarly, on the satellite-to-Earth link, the higher the gain of the satellite transmit antenna, the greater the EIRP for a given transmitter power. Satellites often have parabolic dish antennas, though there are also other types, such as phased arrays. The principal property of a parabolic reflector is its ability to turn light from a point source placed at its focus into a parallel beam, mostly as illustrated in Fig. 2.32 (Right). In practice the beam can never be truly parallel, because rays can also be fan-shaped; i.e., a car headlamp is a typical example. In a microwave antenna, the light source is replaced by the antenna feed, which directs waves toward the reflector. The length of all paths from feed to aperture plane via the reflector is constant, irrespective of their angle of parabolic axis. The phase of the wave in the aperture plane is constant, resulting in maximum efficiency and gain. The gain of an aperture (G_a) and parabolic (G_p) type of antennas are as follows:

$$\begin{aligned} G_a &= \eta\left(4\pi A/\lambda^2\right) = 4\pi A_E/\lambda^2 \\ G_{yps} &= \eta(\pi D)^2/\lambda^2 \end{aligned} \tag{2.49}$$

where η = efficiency factor, A = projected aperture area of antenna, $A_E = \eta^A$ is the effective collecting area, and D = parabolic antenna diameter. Thus, owing to correlation between frequency and wavelength, $f = c/\lambda$ is given in the following relations:

$$G_p = \eta(\pi D f/c)^2 = 60,7(D f)^2 \tag{2.50}$$

where the second relation comes from considering that $\eta \approx 0.55$ of numerical value. If this value is presented in decibels, then the gain of antenna will be calculated as follows:

$$G_T = 10 \log G_p \tag{2.51}$$

For example, a parabolic antenna of 2 m in diameter has a gain of 36 dB for a frequency at 4 GHz and a gain of 38 dB for a frequency at 6 GHz. Parabolic antennas can have aperture planes that are circular, elliptical, or rectangular in shape. Thus, antenna with circular shape and homogeneous illumination of aperture with a gain of –3 dB has about 47.5 % of effective radiation, and the rest of the power is lost. To find out the ideal characteristics, it is necessary to determine the function diagram of radiation in the following way:

$$F(\delta_o) = s(\delta_o)/s(\delta_o = 0) \tag{2.52}$$

where parameter s (δ_o) = flow density of radiation in the hypothetical satellite angle (δ_o) and s $(\delta_o = 0)$ = flow density in the middle of the coverage area. Looking the geometrical relations in Fig. 2.2a follows the relation:

$$F(\delta_o) = d_o/h = \cos \delta \sqrt{(k^2 - \sin^2 \delta_o)}/1 - k \tag{2.53}$$

where, as mentioned, $k = R/(R + h) = \sin \delta$ and if $\delta_o = \delta$, the relation is defined by the following equation:

$$F(\delta) = k \cos \delta \tag{2.54}$$

For GEO satellite, the value of ΔL is given as a function of angle δ, which is the distance from the center of the coverage area, where the function diagram of the radiation is as follows:

$$F(\delta) = \Delta L = 20 \log = 20 \log R/(R + h) \cos \delta = 10 \log R/(1 + 2R/h)[dB] \tag{2.55}$$

Therefore, in the case of GEO satellites the losses of antenna propagation are greater around the periphery than in the center of the coverage area for about 1.32 dB. The free-space propagation loss (L_P) and the input level of received signals (L_K) are given by the equations:

$$\begin{aligned} L_P &= (4\pi d/\lambda)^2 \\ P_R/S &= P_T G_T / 4\pi d^2 L_K \end{aligned} \tag{2.56}$$

The free-space propagation loss is caused by geometrical attenuation during propagation from the transmitter to the receiver.

2.4.6.2 Link Performance with Monobeam and Multibeam Antenna Coverage

As stated earlier, the most important parameter of spacecraft transponder and the overall RF link quality depends on the gain of the satellite antenna. From Eq. (2.47), it can be seen that the satellite antenna gain is constrained by its beamwidth, whatever the frequency at which the link is operated. So the antenna gain is imposed by the angular width of the antenna beam covering the zone to be served. If the service zone is covered using a single-antenna beam, this is referred to as single or monobeam beam coverage, which displays one of these characteristics:

- The satellite may provide coverage of the whole region of the Earth, which is visible from the satellite as a global coverage and thus permits long-distance links to be established, for example from one continent to another with 20 dB bandwidth. In this case, the gain of the satellite antenna is limited by its beamwidth as imposed by the coverage.
- The satellite may provide coverage of only part of the Earth (a region or country) by means of a narrow beam (a zone or spot beam), with 3 dB beamwidth of the order of 1° to a few degrees.

With a single-beam satellite antenna coverage, it will be therefore necessary to choose between either extended coverage providing service with reduced quality of service to geographically dispersed GES terminals, or with reduced coverage providing service with improved quality of transmission and reception to selected geographically concentrated GES terminals.

Multibeam satellite antenna coverage allows these two alternatives to be reconciled. Thus, satellite extended coverage may be achieved by means of the juxtaposition of several narrow-beam antenna coverages, so in such a way, each beam provides an antenna gain that increases as the satellite antenna beamwidth decreases (reduced coverage per beam). The satellite link performance improves as the number of beams increases, which limit can be determined by the antenna technology, whose complexity increases with the number of beams and the mass.

The complexity originates in the more elaborate satellite antenna technology and the requirement to provide onboard interconnection of the coverage areas, so as to ensure within the satellite payload routing of the various carriers that are unlinked in different beams to any wanted destination beam.

In Fig. 2.33 (Left) is illustrated satellite that provides global coverage with a single beam (monobeam) of beamwidth and in Fig. 2.33 (Right) is illustrated that satellite supports spot beams with beamwidth of a consequently reduced coverage, known as multibeam satellite coverage. In both cases, all GES terminals in the satellite network are within the correspondent satellite coverage or in LOS with satellite. Multibeam coverage is providing the following advantages:

1. **Impact on Earth Segment**—The satellite communication link performance is evaluated as the ratio of the received carrier power C to the noise power special

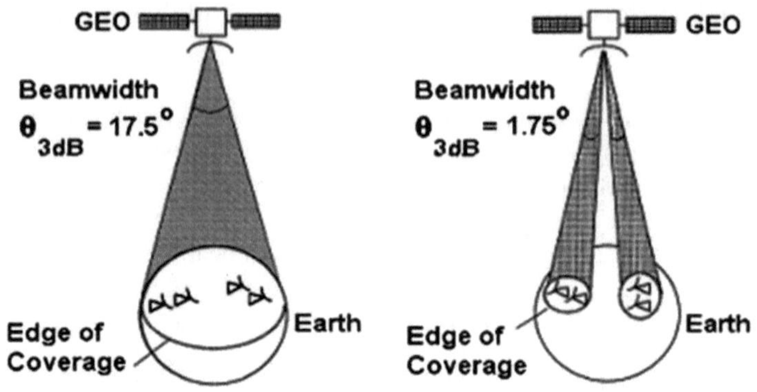

Fig. 2.33 Global monobeam and multibeam antenna coverage—Courtesy of Book: by Maral

density N_0 and is quoted as the C/N_0 ratio, expressed in Hz. The expression for $(C/N_0)_U$ for the uplink (U) is given by the following equation:

$$(C/N_0)_U = (EIRP)_{station}(1/L_U)(G/T)_{satellite}(1/k)\ [Hz] \qquad (2.57)$$

Assuming that the noise temperature at the satellite receiver input is $T_{satellite}$ = 800 K = 29 dBK and is independent of the beam coverage (this is not rigorously true but satisfies a first approximation), let L_U = 200 dB and neglect the implementation losses. At this point, this equation becomes (all terms in dB):

$$\begin{aligned}(C/N_0)_U &= (EIRP)_{station} - 200 + (G_R)_{satellite} - 29 + 228.6 \\ &= (EIRP)_{station} + (G_R)_{satellite} - 0.4\ [dBHz]\end{aligned} \qquad (2.58)$$

where $(G_R)_{satellite}$ is the gain of the receiving satellite antenna in the direction of the GES transmitting terminals. This relation is represented by the two cases considered receiver:

- Global coverage ($\theta_{3\ dB}$ = 17.5°), which implies $(G_R)_{satellite}$ = 29 000/$(\theta_{3\ dB})^2 \approx$ 20 dBi.
- Spot beam coverage ($\theta_{3\ dB}$ = 1.75°), which implies $(G_R)_{satellite}$ = 29 000/$(\theta_{3\ dB})^2 \approx$ 40 dBi.

The expression for $(C/N_0)_D$ for the downlink (D) is given by:

$$(C/N_0)_D = (EIRP)_{satellite}(1/L_D)(G/T)_{station}(1/k)\ [Hz] \qquad (2.59)$$

Assume that the power of the carrier transmitted by the satellite is P_T = 10 W = 10 dBW. Let L_U = 200 dB and neglect the implementation losses. Thus, this equation becomes (all terms in dB):

$$
\begin{aligned}
(C/N_0)_D &= 10-200+(G_T)_{satellite}+(G/T)_{station}+228.6 \\
&= (G_T)_{satellite}+(G/T)_{station}+38.6\,[\text{dBHz}]
\end{aligned}
\tag{2.60}
$$

This relation is represented for the two cases considered transmitter:

- Global coverage ($\theta_{3\ dB} = 17.5°$), which implies $(G_T)_{satellite} = 29\ 000/(\theta_{3\ dB})^2 \approx 20$ dBi.
- Spot beam coverage ($\theta_{3\ dB} = 1.75°$), which implies $(G_T)_{satellite} = 29\ 000/(\theta_{3\ dB})^2 \approx 40$ dBi.

In case that values indicate the reduction in $(EIRP)_{station}$ and $(G/T)_{station,}$ the transmission system is changing from a satellite with global coverage to a multibeam satellite with coverage by several spot beams. In this case, the multibeam satellite permits an economy of size and, hence, cost of the Earth segment. For instance, a 20 dB reduction of $(EIRP)_{station}$ and $(G/T)_{station}$ may result in a tenfold reduction of the antenna size (perhaps from 30 m to 3 m) with a cost reduction for the GES terminal for more than 100 times. If an identical GES is retained (a vertical displacement toward the top), an increase of C/N_0 is achieved which can be transferred to an increase of capacity, if sufficient bandwidth is available, at constant signal quality in terms of Bit Error Rate (BER).

2. **Frequency Reuse**—Frequency reuse consists of using the same frequency band several times in such a way as to increase the total capacity of the network without increasing the allocated bandwidth (B). In the case of a multibeam satellite, the isolation resulting from antenna directivity can be exploited to reuse the same frequency band in separate beam coverages. The frequency reuse factor is defined as the number of times that the bandwidth is used. In theory, a multibeam satellite with M single-polarization antenna beams, each being allocated the bandwidth, combines reuse by angular separation and reuse by orthogonal polarization may have a frequency reuse factor equal to 2 M. This signifies that it can claim the capacity which would be offered by a single-beam satellite with single polarization using a bandwidth of $M \times B$. In practice, the frequency reuse factor depends on the configuration of the service area which determines the coverage before it is provided by the satellite. If the service area consists of several widely separated regions (for example, urban areas separated by extensive rural areas), it is possible to reuse the same band in all beams. The frequency reuse factor can then attain the theoretical value of *M*. Figure 2.31 shows an example of multibeam coverage.

On the other hand, multibeam coverage is providing the following disadvantages:

1. **Interference Between Beams**—In practical reality, the interference generation within a multibeam satellite system is called self-interference. Thus, the effect of self-interference appears as an increase in thermal noise under the same conditions as interference noise between systems. At this point, it must be included the term $(C/N_0)_1$, which expresses the signal power in relation to the spectral density interference. Taking account of the multiplicity of sources of interference, which become more numerous as the number of beams increases, relatively low values of

$(C/N_O)_I$ may be achieved and the contribution of this term impairs the performance in terms of $(C/N_O)_T$ of the total link. As modern satellite systems tend to re-use frequency as much as possible to increase capacity, self-interference noise in a multibeam satellite link may contribute up to 50 % of the total noise.

2. **Interference Between Coverage Areas**—A satellite payload using multibeam coverage must be in a position to interconnect all network Earth stations and consequently must provide interconnection of coverage areas. The complexity of the payload is added to that of the multibeam satellite antenna subsystem, which is already much more complex than that of a single-beam satellite. Different techniques, depending on the onboard processing capability (no processing, transparent processing, regenerative processing, etc.) and on the network layer, are considered for interconnection of coverage:

- Interconnection by transponder hopping (no onboard processing);
- Interconnection by onboard switching (transparent and regenerative processing); and
- Interconnection by beam scanning.

Multibeam satellite systems make it possible to reduce the size of GES terminal and hence the cost of the Earth segment infrastructure. Frequency reuse from one satellite beams to another permits an increase in capacity without increasing the bandwidth allocated to the system. However, interference between adjacent satellite channels, which occurs between beams using the same frequencies, limits the potential capacity increase, particularly as interference is greater with Earth stations equipped with small antennas.

The simplest form of a payload with multibeam antenna radiation is shown in Fig. 2.34. A three-transponder payload uses one transponder per coverage circle. At this point, there is not connectivity between coverage areas in this simple transponder. However, this payload could be designed so each transponder antenna illuminates three coverage circles and provide connectivity, but would cover three times area with just one-third the gain.

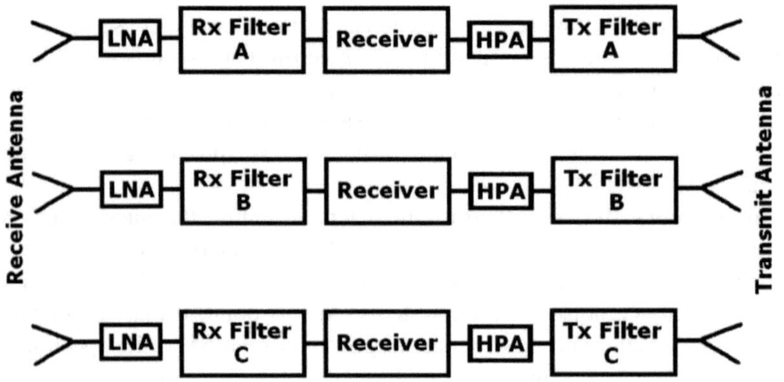

Fig. 2.34 Multibeam antenna coverage transponder—Courtesy of Book: by Swan

The relatively simple changes to multiple small satellite antenna beams have the following significant consequences as follows:

(a) The area covered by each beam is much smaller, increasing the antenna gain and allowing smaller, less expensive ground terminals;
(b) The same total RF power can be used to carry more traffic, and/or reduce the RF power;
(c) The same transponder bandwidth can be used multiple beam antenna, greatly increasing the available bandwidth of the satellite:

    1. This allow the same bandwidth to be reused, increasing the amount that can be accommodated within the bandwidth; and
    2. A terminal has to be tuned to the correct RF to function and retuned if it is moved.

(d) Connectivity between beams, if required for the mission, must be provided by additional hardware on the satellite since a single uplink does not encompass the entire coverage area.

The beams may be formed by individual feeds (circles) and by mechanical or electronic beam former. Mechanical beam formers use fixed wave-guide components to control the RF phase and amplitude. Usually there is one amplifier for each transponder in each composite beam. Electronic beam formers use electronic RF phase shifters, and sometimes electronic amplitude control, to produce the multi-beam. There are many radiating elements and each usually has its own amplifier.

Antenna radiators fall into two categories, reflector and direct radiating antennas. Reflector antenna uses a feed that indirectly radiates the energy toward the illuminated area of users, while direct radiating antenna radiates the energy direct to the coverage area. The spot beam antenna can be pointed in various directions within a cone characterized by the scan angle, shown in Fig. 2.35. In this way, a direct radiating array can radiate at large scan as illustrated in Fig. 2.35. Such arrays are

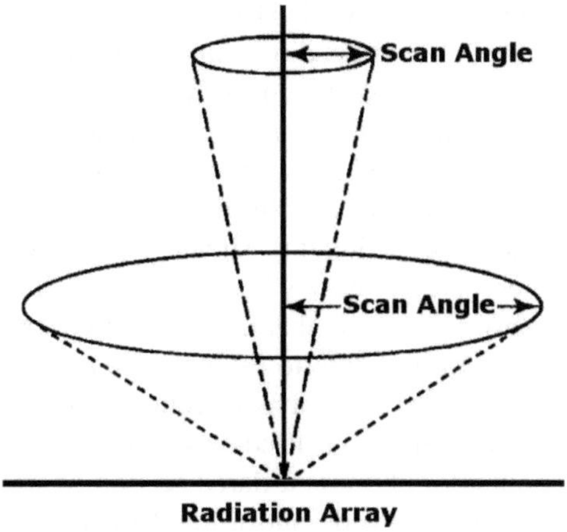

Fig. 2.35 Antenna scan angles—Courtesy of Book: by Swan

attractive for LEO communications satellites because they operate over large scan angle than reflectors antenna, so they require a scan angle of 63° to cover its field of view. In contrary, a GEO communications satellite requires a scan angle of 7° to cover the entire visible circle on the Earth.

In addition to the fact that satellite antenna gain decreases as the scan angle increases, and polarization purity decreases as well. Thus, the LEO satellites require a direct radiating antenna, while commercial GEO satellite do not use this type of antenna, but directional reflector antennas. At MEO, the situation is less clear and both direct radiating and reflector antennas have been proposed for this orbit.

2.4.6.3 Basic Types of Antenna Onboard Spacecraft

Four main types of antennas are used on satellites: 1. Wire antennas: monopoles and dipoles; 2. Horn antennas; 3. Reflector antennas; and 4. Array antennas.

Wire antennas are used primarily at VHF and UHF to provide communications for the Telemetry, Tracking, and Command (TT&C) systems. They are positioned with great care on the body of the satellite in an attempt to provide omnidirectional coverage. Most satellites measure only a few wavelengths at VHF frequencies, which make it difficult to get the required antenna patterns, and there tends to be some orientations of the satellite in which the sensitivity of the TT&C system is reduced by nulls in the antenna pattern.

An antenna pattern is a plot of the field strength in the far field of the antenna when a transmitter drives the antenna. The gain of an antenna is a measure in dB of the antenna's capability to direct energy in one direction, rather than all around. A useful principle in antenna theory is reciprocity, which means that an antenna has the same gain and pattern at any given frequency whether it transmits or receives. An antenna pattern measured when receiving is identical to the pattern when transmitting.

As stated earlier, the antenna is providing global, spot, and multiple beam coverages, but it can provide scanning and orthogonally polarized beams or coverage zones as well. The pattern is frequently specified by its 3-dB beamwidth, the angle between the directions in which the radiated (or received) field falls to half the power in the direction of maximum field strength. However, a satellite antenna is used to provide coverage of a certain area or zone on the Earth's surface, and it is more useful to have contours of antenna gain with maximum strengths of the signal in the middle of the coverage area and with decreasing of signals to the peripheries. When computing the signal power received by a GES from the satellite, it is important to know where the station lies relative to the satellite transmit antenna contour pattern so that the exact EIRP can be calculated. If the pattern is not known, it may be possible to estimate the antenna gain in a given direction if the antenna boresight or beam axis direction and its beamwidth are known.

Horn antenna is used at MW when for global coverage relatively wide beams are required. A horn is a flared section of waveguide that provides an aperture several wavelengths wide and a good match between the waveguide impedance and the free space. It is also used as feeds for reflectors, either singly or in clusters. Horns and reflectors are examples of aperture antennas that launch a wave into free space from a waveguide. It is difficult to obtain gains much greater than 23 dB or beamwidths narrower than about 10° with horn antennas. For higher gains or narrow beamwidths, a reflector antenna or array must be used.

Reflector antennas are usually illuminated by one or more horns and provide a larger aperture than can be achieved with a horn alone. For maximum gain, it is necessary to generate a plane wave in the aperture of the reflector. This is achieved by choosing a reflector profile that has equal path lengths from the feed to the aperture so that all the energy radiated by the feed and reflected by the reflector reaches the aperture with the same phase angle and creates a uniform phase front. One reflector shape that achieves this with a point source of radiation is the paraboloid, with a feed placed at its focus. The paraboloid is the basic shape for most reflector antennas and is commonly used for Earth station antennas. Satellite antennas often use modified paraboloidal reflector profiles to tailor the beam pattern to a particular coverage zone. Phased array antennas are also used on satellites to create multiple beams from a single aperture and have been used by Iridium and Globalstar to generate up to 16 beams from a single aperture for their LEO system.

2.4.7 Satellite Bus

The satellite bus is usually called a platform and consists of several sections, shown in Fig. 2.17. The function of the satellite platform is to support the payload operation reliably throughout the mission of primary construction section, such as Structure Platform (SP), Electric Power (EP), Thermal Control (TC), Attitude and Orbit Control (AOC), Telemetry, Tracking and Command (TT&C) and Propulsion Engine.

2.4.7.1 Structure Platform (SP)

The structure has to house and keep together all components of bus and communications modules, enable protection from the environment, and facilitate connection of the satellite to the launcher. It comprises a skeleton on which the equipment modules are mounted and a panel, which covers and provides protection for sensitive parts during the operational phase from micrometers and helps to shield the equipment from extremes of heat, coldness, vacuum, and weightlessness, including the relatively small dynamic forces produced by the station-keeping, attitude control engines, and inertial momentum devices.

The spacecraft is protected during the launch phase with an enclosure, or nose cone. At the end, the nose cone is jettisoned, at which time the spacecraft must survive the inertial and thermal stress of an additional propulsion stage until it is inserted into orbit. In this sense, a spacecraft is virtually free of gravitational stress

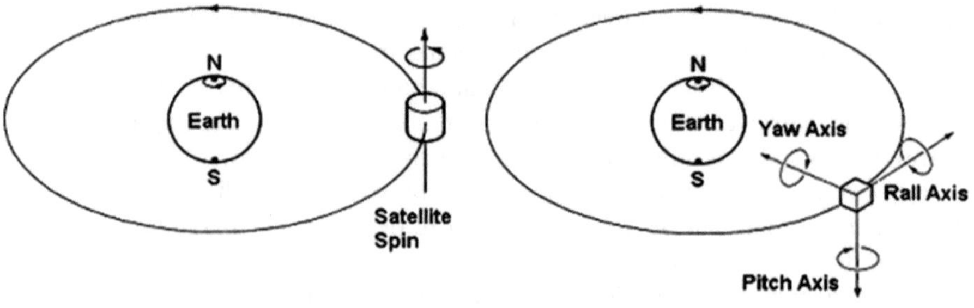

Fig. 2.36 Satellite spin RPY 3-axes stabilization—Courtesy of Book: by Roddy

when in orbit, which allows the use of very large deployable arrays, which would collapse under their own weight on the Earth's surface without problems.

Thus, large stresses are developed during launch as a result of massive acceleration and intense vibration, so the SP body must be sufficiently strong to withstand all external forces. On the other hand, all large structures such as antenna and solar arrays have to be folded and protected during a launch sequence and must have a deployable mechanism. The deployment of structures requires a special technique in the vacuum of space because of the lack of a damping medium, such as air.

Most satellites are either cylindrically shaped and are stabilized by spinning the whole of main body shown in Fig. 2.36 (Left) or box-shaped three-axis body stabilized illustrated in Fig. 2.36 (Right). Spin stabilized structures have a cylindrical part, which rotates at a speed of 50–100 rpm and the despun stabilized part has mounted antennas always facing to the Earth. The spin axis lies along the pitch axis parallel to the Earth's N–S axis. Thus, spin is initiated during the launch phase by means of small gas jet.

The spinning part of the cylinder is covered with solar cells and its spin axis is oriented perpendicularly toward the Sun. Body-stabilized structures rotate once for every rotation of the Earth so that the side with mounted antenna will always face the Earth. This platform utilizes a deployed set of solar panels with solar cells mounted on one side of the panel surface relative to the Sun.

The box-shaped body-stabilized satellite has three-axis known as Roll, Pitch, and Yaw (RPY), which define the attitude of satellite. Thus, all three-axes pass through the center of satellite gravity. The Yaw axis is directed toward the Earth's center, the Pitch axis is normal to the orbital plane, and Roll axis is perpendicular to the other two. In fact, the Roll axis is tangential to the orbit and lies along the satellite velocity vector.

Materials in space are not subjected to gravitational stress or atmospheric corrosion and the effects of the space environment are not all benign by any means. The high vacuum causes some materials to sublime or evaporate and some to weld together on contact. The latter behavior means that special attention has to be given to the materials used for bearings. The basic materials for the main frame are aluminum or magnesium alloys and special plastic or fiber materials and for other components carbon fiber, epoxy resins, and carbon nanotube filaments are used.

2.4.7.2 Electric Power (EP)

The primary source of power for a communications satellite is the Sun. Hence, solar cells are used to convert energy received from the Sun into an electrical source. The principal components of the power supply system include: (1) The power electric generator, usually solar cell arrays, located on the spinning body of a spin-stabilized satellite or on the paddles for a three-axis stabilized satellite; (2) reliable electrical storage devices, such as batteries, for operating during periods of solar eclipses; (3) the electrical harness for conducting electricity to all of the devices demanding power; (4) the special converters and regulators delivering regulated voltage and currents to the devices on board the spacecraft; and (5) the electrical control and protection section is associated with the remote monitoring TT&C satellite system.

The solar arrays are the motor during entire life of the satellite, providing sufficient power to all active components. Occasionally, if the Earth or Moon blocks the Sun, then researchable batteries provide secondary electrical power.

A typical satellite power subsystem is shown in Fig. 2.37. Most components are duplicated for redundancy. On spinning satellites, the rings are between the power subsystem, on the rotating part, and the satellite bus.

Each cell delivers about 150 mA at a few hundred millivolts and an array of cells must be connected in series or in parallel together to give the required voltage and current for operating the equipment until the end of its life and to recharge the batteries when the satellite moves out of an eclipse. Charge is applied via the main electric power bus or a small section of the solar cell. In the course of exploitation, batteries are sometimes reconditioned by intentionally discharging them to a low charge level and recharging again, which usually prolongs their life.

The operational status of batteries including recharge, in-service, or reconditioning is remotely controlled by a special ground segment. The mass of a battery constitutes a significant portion of the total satellite mass. Therefore, a useful figure of merit to evaluate the performance of a battery is capacity in W/h per unit weight taken at the end of its life. Until recently virtually all satellites used Ni-Cd (Nickel-Cadmium) batteries because of their high reliability and long lifetime. These batteries provide a low specific energy of about 30–40 Wh/kg. The latest type of Ni-H (Nickel-Hydrogen) batteries can store at least 50 % more energy per kilogram.

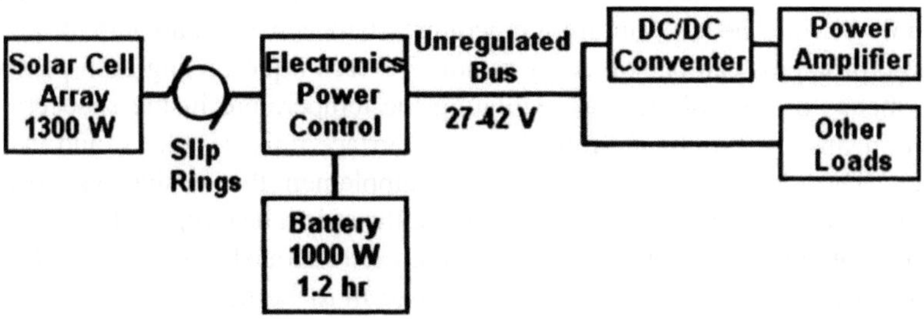

Fig. 2.37 Satellite electric power subsystem—Courtesy of Book: by Gordon

When a satellite passes through the Earth's shadow, the solar arrays stop producing power and the satellite structures use the energy from batteries. The GEO satellite undergoes around 84 eclipses in a year, with a maximum duration of 70 min. Thus, the eclipse occurs twice a year for 42 consecutive days each time. The percentage of eclipses' duration for GEO and HEO is much less than for lower satellite orbits. The LEO satellites can undergo several thousand eclipses in a year. For example, a LEO satellite in equatorial orbit at an altitude of 780 km can remain in the Earth's shadow for 35 % of the orbital period. For a MEO under similar conditions, the maximum eclipse duration would be about 12.5 % of the orbital period and a total duration of about 3 h a day, with about 4 eclipses per day. Otherwise, the Sun can also be sometimes eclipsed by the Moon's shadow, which is less predictable.

2.4.7.3 Thermal Control (TC)

There is not air at satellite orbit surrounded by very harsh space environment. The average satellite temperature is determined by the absorbed solar energy, different thermal radiation into space, and internal electric dissipation, what depends on the satellite shape and surface.

Thermal control of a communications satellite is very important factor during entire satellite lifetime, which is necessary to achieve normal temperature balance and proper performance of all subsystem. Thermal stress results in high-temperature effects from the Sun and low temperatures occurring during eclipse period. The obvious objective of the TC is to assure that the spacecraft structure and all equipment is maintained within temperatures that will provide successful operations. A satellite undergoes different thermal and other conditions during the launch and operational phase. The vacuum in space limits all heat transfer mechanisms to and from a spacecraft and its external environment to that of radiation. However, some main parts are usually in direct sunlight with a flux density of over 1 kW/m^2, while other parts are facing the shadow side at a temperature of about −270 °C. In addition, an eclipse causes temperature variation from around −180° to +60 °C, when the ambient temperature falls well below 0 °C and rises rapidly from the moment the satellite emerges from the eclipse. All these extremes have to be eliminated or moderated for normal satellite operations, especially because all electronic devices need optimum temperatures between −5° and +45 °C.

These problems can be solved by remote TC techniques, using both passive and active means of controlling and regulating the temperature inside spacecraft. The passive means are simple and reliable, using surface finishes, filters, and insulation blankets. The active means are necessary to supplement the passive systems, which include louvers and blinds operated by bimetallic strips, heat pipes, thermal louvers, and different electrical heaters. Heat pipes are used to transfer heat from internal hot spots or devices to remote radiator surfaces or must be transported to the outside surface where it can be dissipated. On the other hand, special electric heaters are used to maintain minimum component or structure temperatures during cold conditions. Accordingly, the TC subsystem ensures temperature regulation for optimum efficiency and satellite performance.

2.4.7.4 Attitude and Orbit Control (AOC)

The attitude and orbital control subsystem checks that a spacecraft is placed in its precise orbital position, and maintains, thereafter, the required attitude throughout its mission. Control is achieved by employing momentum wheels, which produce gyroscopic torques, combined with an auxiliary reaction control gas thruster system. Various sensors are employed to detect attitude errors, including Sun's initial orientation purposes.

The satellite antennas require AOC system that will keep them pointed always at the Earth, frequently within 0.1° and 0.01°. Figure 2.38 shows a block diagram of AOC system. The sensors direct any pointing errors and correct them by changing the speed or direction of a rotating wheel. The main performance specification of an AOC system is determined by the disturbance torques and the required pointing accuracy.

The AOC system performs satellite orientation and accurate orbital positioning throughout its lifetime, because loss of attitude renders a spacecraft useless. There are in use two common AOS, such as attitude control and orbit or station-keeping control systems. The objective of attitude control was to keep the antenna RF beam pointing at the intended areas on the Earth, which procedure involves as follows: (1) measuring the attitude of the satellite by sensors; (2) comparing the results of measurements with the required values; (3) calculating the corrections to reduce eventual errors; and (4) introducing these corrections by operating the appropriate torque units.

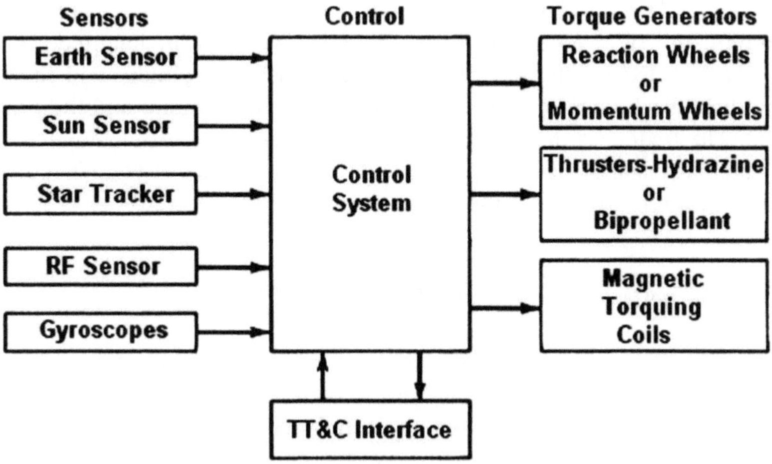

Fig. 2.38 Satellite AOC subsystem—Courtesy of Book: by Gordon

1. **Attitude Control**—Currently, all types of attitude stabilization systems have relied on the conservation of angular momentum in a spinning element, which can be classified into the two categories already mentioned, such as spin-stabilized and three-axis stabilization. The satellite is rapidly spun around one of its principal axes of inertia. Thus, in the absence of any perturbing torque, the satellite attains an angular momentum in a fixed direction in an absolute frame of reference. For the GEO satellite, the spin (pitch direction) axis must be parallel to the axis of the Earth's rotation. The perturbation torques reduce the spin of the satellite and they affect the orientation of the spin axis. The second system of attitude control is a body-stabilized design in a three-axis stabilized satellite, whose body remains fixed in space. This solution is the simplest method of attitude control using a momentum wheel, which simultaneously acts as a gyroscope, in combination with spin and drive stabilization. Certain perturbing torques can be resisted by changing its spin speed and the resulting angular momentum of the satellite.

2. **Orbit or Station-keeping Control**—Onboard propulsion requirements for both GEO and non-GEO are important to keep a satellite in the correct orbital attitude and position. For this reason, several types of propulsion systems are used, such as arc jet thrusters, ion and solar electrical propulsion, pulsed plasma thrusters, and iridium-coated rhenium chambers for chemical propellants. In order that the appropriate station-keeping corrections can be applied, it is essential that the orbit and position of a satellite are accurately determined. This may be done by making measurements of the angular direction and distance of the satellite from the Earth station, or a number of LES. When the orbit and position of the satellite have been determined, it is possible to calculate the velocity increments required to keep the N–S and E–W excursion of the satellite within the tolerated limits. The frequency with which N–S correction must be made depend on the maximum allowable value of the orbital inclination but the total increment required each year to cancel out the attraction of the Sun and Moon is 40–50 m/s. Otherwise, E–W station-keeping is usually achieved by allowing the satellite to drift toward the nearest point of equilibrium until it reaches the maximum tolerable error in longitude, then the process is repeated on the other side of the nominal longitude, and finally, the satellite drifts back once more toward the point of equilibrium, and the process is repeated. The frequency and magnitude of the velocity increments required depend on the angular distance between the satellite and the points of equilibrium and on the tolerable error, which is a maximum of about 2 m/s.

2.4.7.5 Telemetry, Tracking, and Command (TT&C)

The telemetry, tracking, command, and communication equipment enables data to be send continuously to the Earth stations, received from these stations and allows ground control stations to track the spacecraft and to monitor the health of the spacecraft and also to send commands to carry out various tasks such as switching the transponders in and out of service, and switching between redundant units.

The TT&C subsystem monitors and controls the satellite functions right from the liftoff stage to the end of its operational life in space. The TT&C subsystem is therefore very important, not only during orbital injection and the positioning phase, but also throughout the operational fife of the satellite. During the orbital injection and positioning phase, the telemetry link is primarily used by the tracking system to establish the satellite-to-Earth station communication channel. After the satellite is put into the desired slot in its intended orbit, its mission is to monitor the health of various subsystems onboard satellite.

In Fig. 2.39 is shown the block schematic arrangement of the basic TT&C subsystem.

The TT&C system supports the function of spacecraft management for successful operation of payload and bus subsystems. The main functions of a TT&C are as follows:

1. **Telemetry Subsystem**—The function of telemetry is to monitor various spacecraft parameters and performances such as voltage, current, temperature, output from attitude sensors, reaction wheel speed, pressure of propulsion tanks and equipment status and to transmit the monitored data to the Satellite Control Center (SCC) on the Earth. At this point, the telemetered data are analyzed at the SCC and used for routine operational and failure diagnosis purposes, to provide data about the amount of fuel remaining, to support determination of orbital parameters, etc.

2. **Tracking Subsystem**—The function of tracking is to provide necessary sources to Earth stations for the tracking and determination of orbital parameters. In such a way, to maintain a satellite in its assigned orbital slot and provide look angle information to LES in the network, it is necessary to estimate the orbital parameters regularly. These parameters can be obtained by tracking the communications satellite from the ground and measuring its angular position and range. Most SCC employ angular and range or range-rate tracking to control satellite orbits.

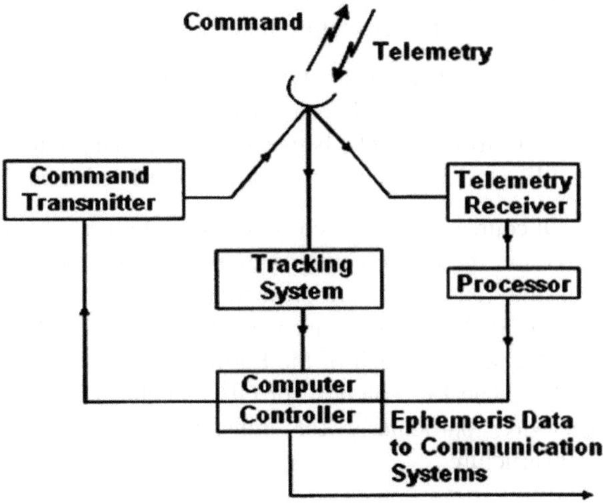

Fig. 2.39 Satellite TT&C—Courtesy of Book: by Maini

3. **Command Subsystem**—This subsystem receives commands transmitted from the ground SCC, verifies reception, and executes commands to perform various functions of the satellite during its operational mission, such as Satellite transponder and beacon switching, Antenna pointing control, Switch matrix reconfiguration, Controlling direction and speed of solar arrays drive, Battery reconditioning, Thruster firing, and Switching heaters of the various systems.

2.4.7.6 Propulsion Engine (PE)

The functions of the propulsion motors are to generate the thrust required for the attitude and orbital control of errors caused by solar and lunar gravity and other influences, or possibly the adequate assistance of the satellite into its final orbit.

Hence, these errors are normally corrected at set intervals in response to commands from SCC. The necessary impulse is provided by thrusters, which operate by ejecting hot or cold gas under pressure. The thrust requirements for orbital control are provided by mono- or bi-propellant fuels. The attitude control thrusters are positioned away from the center of the mass to achieve the maximum thrust, the thrust being applied perpendicular to the direction of a spacecraft's center of mass. The orbit control thrusters are mounted so that the thrust vector passes through the center of mass. The relocation of a satellite from transfer orbit into GEO may be performed by apogee boost motor. In some satellite configurations, this is achieved by a solid or liquid fuel engine. Moreover, the choice between these two motors has a significant effect on the internal arrangements of the satellite.

2.5 Intersatellite Links (ISL)

Intersatellite or cross-links are communication links between satellite constellations. Three types of intersatellite links can be considered:

1. GEO to LEO links between geostationary (GEO) satellites and low Earth orbit (LEO) satellites also called Interorbital Links (IOL);
2. GEO to GEO links between geostationary satellites; and
3. LEO to LEO links between low Earth orbit satellites.

Of course, one could consider intersatellite links between satellites in any type of orbit, but the above configurations are those most considered in practice. Only the transmission aspects are presented here.

An GEO satellite can cover about 40 % of the Earth's surface. Because each satellite sees so much of the Earth, no many commercial GEO satellite to date has an ISL. The ISL solution for GEO satellites is being considered for network connectivity, particularly for transfer of data, and generally to reduce ground infrastructure.

However, if connectivity over a large geographical area is required, the smaller footprint of LEO satellite needs that they have GES or Gateway in the coverage area of each satellite, with onboard store and forward capacity or ISL service. The Big LEO Globalstar has not ISL system, while Big LEO Iridium system using ISL optical links can offer service anywhere and anytime.

The ISL configuration may use either RF or optical technology for interconnections. Thus, a requirement for an optical or RF ISL to operate through solar conjunction, when the Sun is within the field of view of an optical or RF antenna, is a major ISL design driver. Operation through solar conjunction is possible with both microwave and optical systems, but is actually easier with an optical system.

Optical ISL has very high antenna gain because of their short wavelength. Laser diodes have a frequency of the order of $1–6 \times 10^{15}$, or about 105 higher than a 30 GHz RF wave. The optical antenna gain is 100 dB more than a RF antenna of the same size and efficiency. Moreover, a 10-cm-diameter optical antenna has 80 dB more gain than a 1 m RF antenna. This allows a 1 W laser diode to provide the same flux density as a 100 MW RF source. Optical detection is less efficient than RF detection so the above overstates the actual comparison, but gives a clear idea of the advantages of an optical system. Thus, to appreciate the narrowness of the optical beam, a 1-micron laser using a 10-cm antenna would have a bandwidth of about 1 milliradian. The very narrow bandwidth requires a closed-loop tracking system, where each ISL tracks the signal from the communicating satellite.

The choice between radio and optical links depends on the mass and power consumed. In general terms, it can be said that the advantage is with radio links for low throughputs (less than 1 Mb/s). For high-capacity links (several tens of Mb/s) optical links command attention. For a link involving one uplink, one or more optical intersatellite links and one downlink, the overall station-to-station link performance should be established in the same way as the overall link performance for regenerative satellites. Indeed, the implementation of intersatellite links, be it at radio frequency or with optical technology, is mainly of interest when onboard demodulation is available, so as to provide flexible onboard switching.

2.5.1 Direct ISL Data Transmission Over GEO Satellite

In space activities, there is a stronger need of exploring and making usage of advanced technologies to transmit data directly in cross-link from one satellite to another, both conventionally using radio waves or in a revolutionary and new way using laser beams.

Earth observation satellites, for instance, cannot send their data to just one Earth station, because they fly at low altitude of about 800 km. As they scan the Earth's surface, they disappear behind the horizon after a few minutes at the most. The data must therefore be stored in onboard memories, which are far from fail-safe or else transmitted to a large number of distributed Earth stations, which are costly to operate.

Fig. 2.40 ISL of GEO spacecraft Artemis—Courtesy of Webpage: by ESA

From its vantage point in GEO at 36,000 km above the Earth such as the ESA Artemis spacecraft can be in constant communication with satellites in LEO orbit for long periods of each orbit and can beam data from those satellites directly to the users, illustrated in Fig. 2.40. To demonstrate the reception of data from other satellites and their onward transmission to users, Artemis is carrying an advanced radio data-relay payload and a revolutionary laser data-relay payload called Semiconductor Laser Intersatellite Link Experiment (SILEX). SILEX is the world's first civil intersatellite data-relay system using lasers as carriers for the signal transmission.

Using laser communications, very high data rates can be achieved via small terminals on LEO satellites, such as existing Iridium. The radio data-relay payload SKDR (S/Ka-band Data Relay) features the use of two frequencies for relaying data at low, medium, or high rates again both ways. The first users of SKDR include ESA's environmental ENVISAT-1 satellite scheduled for launch in 1999. Under an agreement with NASDA, the Japanese space agency, the ARTEMIS satellite, will also serve several Japanese spacecraft and the Japanese module of the International Space.

2.5.2 Radio Frequency (RF) ISL

Table 2.7 indicates the frequency bands allocated to ISL by the radiocommunication regulations. These frequencies correspond to strong absorption by the atmosphere and have been chosen to provide protection against interference between intersatellite links and terrestrial systems. However, these bands are shared with

Table 2.7 Frequency bands for ISL

Intersatellite service	Frequency bands
Radio frequency	22.55–23.5 GHz
	24.45–24.75 GHz
	32–33 GHz
	54.25–58.2 GHz
Optical	0.8–0.9 μm (AlGaAs laser diode)
	1.06 μm (Nd:YAG laser diode)
	0.532 μm (Nd:YAG laser diode)
	10.6 μm (CO_2laser)

other space services and the limitation on interference level is likely to impose constraints on the choice of the defining parameters of intersatellite links (CCIR Reports 451, 465, 874, 951). The same table also indicates the wavelengths envisaged for optical links. These result in the transmission characteristics of the components.

Propagation losses reduce to free-space losses since there is no passage through the atmosphere. Antenna pointing error can be maintained at around a tenth of the beamwidth and this leads to a pointing error loss of the order of 0.5 dB. The antenna temperature in the case of a GEO–GEO link, in the absence of solar conjunction, is of the order of 10 K. Table 2.8 indicates typical values for the terminal equipment. For practical applications, antenna dimensions are of the order of 1–2 m. In such a way, considering a frequency of 60 GHz and transmission and reception losses of 1 dB leads to:

- A receiver figure of merit G/T of the order of 25–29 dBK^{-1}; and
- A transmitter EIRP of the order of 72–78 dBW.

Because of the relatively wide beamwidth of the antenna (0.2° at 60 GHz for a 2 m antenna), establishing the link is not a problem. Each satellite orientates its receiving antenna in the direction of the transmitting satellite with a precision of the order of 0.1° to acquire a beacon signal, which is subsequently used for tracking. The development of high capacity RF ISL between GEO satellite systems implies reuse of frequencies from one beam to another. In view of the small angular separation of the satellites, it is preferable to use narrow-beam antennas with reduced side lobes in order to avoid interference between systems. Consequently, and in view of the limited antenna size imposed by the launcher and the technical complexity of the deployable antennas, the use of high frequencies is indicated. The use of optical links may be usefully considered in this context.

Table 2.8 Typical values for terminal equipment of an ISC

Receiver noise factor (dB)	Transmitter power (W)
3–4.5	150
4.5	75
9	30

2.5.3 Optical ISL

Except RF links, ISL can use optical or laser links as well.

In comparison with radio links, optical links have specific characteristics, which are important for establishment a link.

2.5.3.1 Establishing an Optical ISL

Two aspects should be indicated to establish an optical ISL:

1. The small diameter of the telescope is typically of the order 0.3 m. In this way, one is freed from congestion problems and aperture blocking of other antennas in the satellite payload.

2. The narrowness of the optical beam is typically 5 microradians. Notice that this width is of several orders of magnitude less than that of a radio beam, and this is an advantage for protection against interference between systems. But it is also a disadvantage since the beamwidth is much less than the precision of satellite attitude control (typically 0.1° or 1.75 mrad). Consequently, an advanced pointing device is necessary; this is probably the most difficult technical problem.

There are three basic phases to optical communications:

1. Acquisition: The beam must be as wide as possible in order to reduce the acquisition time. But this requires a high-power laser transmitter. A laser of lower mean power can be used which emits pulses of high peak power with a low duty cycle. The beam scans the region of space where the receiver is expected to be located. When the receiver receives the signal, it enters a tracking phase and transmits in the direction of the received signal. On receiving the return signal from the receiver, the transmitter also enters the tracking phase. The typical duration of this phase is 10 s.

2. Tracking: The beams are reduced to their nominal width. Laser transmission becomes continuous. In this phase, which extends throughout the following, the pointing error control device must allow for movements of the platform and relative movements of the two satellites. In addition, since the relative velocity of the two satellites is not zero, a lead-ahead angle exists between the receiver and the transmitter LOS. On the other hand, as demonstrated below, the lead-ahead angle is larger than the beamwidth and must be accurately determined.

3. Communications: Information is exchanged between the two ends.

Otherwise, the narrow optical beam gives rise to several problems for optical cross-links as:

1. **Difficult acquisition**—The very narrow beam requires very precise knowledge of the location of the two satellites establishing the link and the special acquisition mode.

2. **Sensitivity to spacecraft vibration**—A fast-tracking system, of the order of 100 Hz, is necessary so that naturally occurring vibrations, from thruster valves, momentum wheels, solar array devices and so on, do not degrade performance.

3. **Point ahead**—The finite speed of light requires that the tracking system point to where the communicating satellite will be when its signal reaches it, not where it was when it received the signal. This is particularly true for GEO satellite.

4. **Optical cleanliness**—The satellite antenna gain is so high that even small amounts of contamination can significantly degrade it. The optical system must be kept clean through ground test and launch procedure, including from satellite thrusters used for station-keeping system.

2.5.3.2 ISL Relations Between Two GEO Satellites

Consider two satellites, S1 and S2, respectively, moving with velocity vectors V_{S1} and V_{S2}, whose components orthogonal to the line joining S1 and S2 at time t are, respectively, the two vectors represented in Fig. 2.41 (Left) by V_{T1} and V_{T2}. In fact, Lead-ahead angle for link between two satellites S1 and S2 with velocity vector components V_{T1} and V_{T2} in a plane perpendicular to the line joining S1 and S2 at time t; t_p is propagation time of a photon from S1 to S2.

The propagation time of a photon from S1 to S2 is $t_p = d/c$, where d is the distance between the two satellites at time t and c is the speed of light ($c = 3 \times 10^8$ m/s).

The lead-ahead angle *B* is given by:

$$\beta = 2/V_{T1} - V_{T2}/c[\text{rad}] \qquad (2.61)$$

where $|V_{T1} - V_{T2}|$ is the modulus of the difference vector $V_{T1}–V_{T2}$.

Two situations are now considered: ISL between two GEO satellites; interorbital links between a GEO and an LEO satellite.

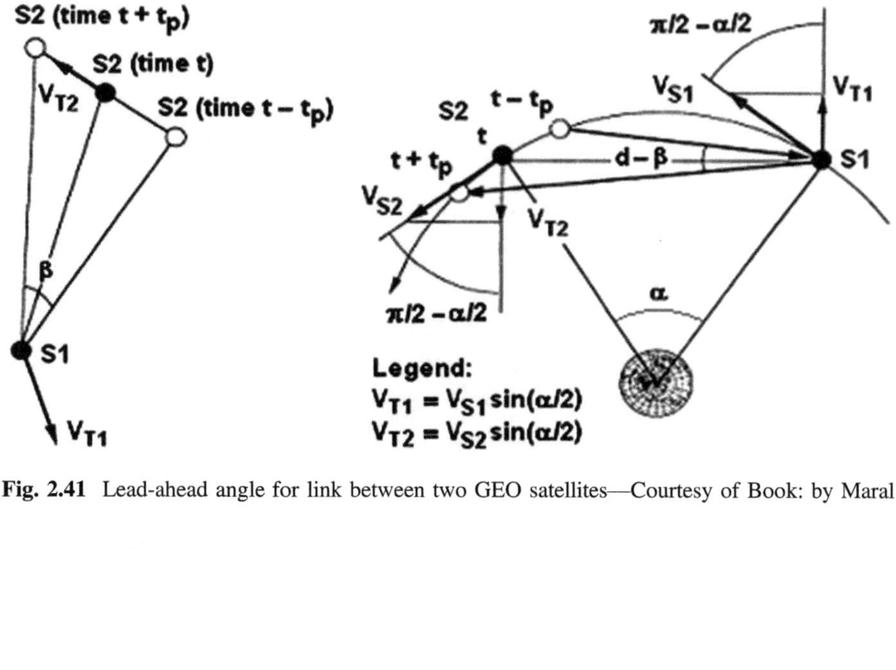

Fig. 2.41 Lead-ahead angle for link between two GEO satellites—Courtesy of Book: by Maral

As both satellites are on the same circular orbit, shown in Fig. 2.39 (Right), the velocity vectors V_{S1} and V_{S2}, which are tangential to the orbit, have equal modulus:

$$|V_{S1}| = |V_{S2}| = \omega(R_O + R_E) = 3075\,\text{m/s} \tag{2.62}$$

where ω is the angular velocity of a GEO = 7.293 × 10^{-5} rad/s; R_O is the altitude of a GEO satellite = 35,786 km; and R_E is the Earth radius = 6378 km.

The component vectors V_{T1} and V_{T2}, perpendicular to the line joining S1 and S2 at time t, both lie in the plane of the orbit and are opposite. They are at an angle (π/2 − α/2) with respect to vectors V_{S1} and V_{S2}. Therefore, following these values is giving relation:

$$|V_{T1} - V_{T2}| = 2\omega(R_O + R_E)\cos(\pi/2 - \alpha/2) = 2\omega(R_O + R_E)\sin(\alpha/2)\ [\text{m/s}] \tag{2.63}$$

This equation is possible to express the following relation:

$$\beta = 2/V_{T1} - V_{T2}/c = 4\omega(R_O + R_E)\sin(\alpha/2)/c\ [\text{rad}] \tag{2.64}$$

At this point, the lead-ahead angle β as a function of the separation angle α between the two GEO satellites. Here is important to note that for a separation angle larger than 15°, the lead-ahead angle is larger than the beamwidth (typically 5 microradians). For instance, β = 10.6 microradians for α = 30°, β = 20.5 microradians for α = 60°, and β = 35.5 microradians for α = 120°.

2.5.3.3 ISL Relations Between GEO and LEO Satellites

The relative velocity of the two satellites varies with time and so does the value of the lead-ahead angle, as shown in Fig. 2.42. Its maximum value is obtained when the LEO satellite crosses the equatorial plane. Denoting as i the LEO satellite orbit inclination, then:

$$|V_{T1} - V_{T2}| = \left\{|V_{S1}|^2 + |V_{S2}|^2 - 2|V_{S1}||V_{S2}|\cos i\right\} \tag{2.65}$$

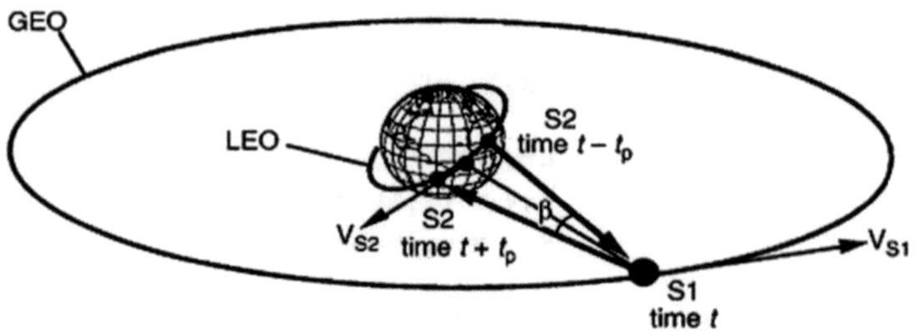

Fig. 2.42 Lead-ahead angle for link between GEO and LEO satellites—Courtesy of Book: by Maral

Using this equation is possible to determine both values with the following relations:

$$\begin{aligned} |V_{S1}| &= \omega_{GEO}(R_O + R_E) = 3,075\ [m/s] \\ |V_{S2}| &= \omega_{LEO}(h + R_E) \end{aligned} \tag{2.66}$$

where h is the LEO satellite altitude and ω_{LEO} + $\mu^{1/2}$ $(h + R_E)^{-3/2}$ is the LEO satellite angular rate ($\mu = 3986 \times 10^{14}\ m^3/s^2$). From Eq. 2.56, the lead-ahead angle is given by:

$$\beta = 2/V_{T1} - (2/c)\left\{|V_{S1}|^2 + |V_{S2}|^2 - 2|V_{S1}||V_{S2}| \cos I\ [rad]\right. \tag{2.67}$$

The lead-ahead angle is the same for the two satellites. Considering i = 98.5° and h = 800 km, β = 57 microradians. Note this value is even larger than for intersatellite links between two geostationary satellites.

2.5.4 *Transmission and Reception of Optical Sources*

The ISL relay link from a LEO satellite to the ground via high-altitude platforms (HAP) relay and from a LEO satellite over a GEO satellite and HAP is presented in Fig. 2.43.

The "last mile" downlink from the HAP or stratospheric communication platform (SCP) to GES would be either bridged by a high-bandwidth optical or microwave link.

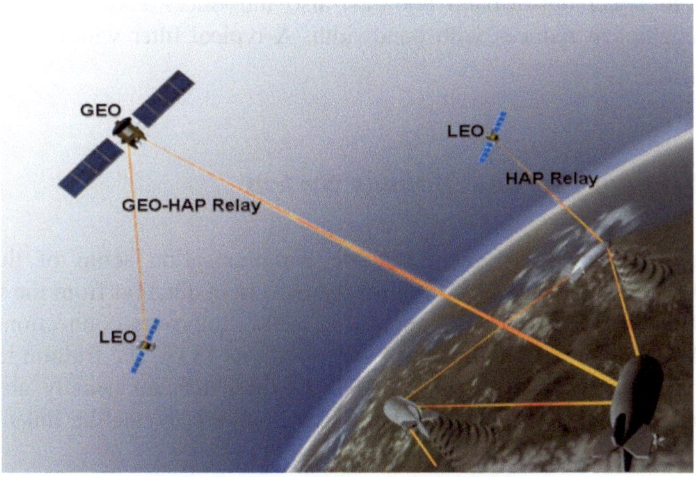

Fig. 2.43 ISL between GEO, LEO, and HAP—Courtesy of Paper: by Knapek

Optical or laser sources operate in single- and multifrequency modes. In single frequency mode, spectral width varies between 10 kHz and 10 MHz. In multifrequency mode, it is from 1.5 to 10 nm. The power emitted depends on the type of laser. Modulation can be internal or external. Internal modulation implies direct modification of the operation of the laser. External modulation is a modification of the light beam after its emission by the laser. The intensity, the frequency, the phase, and the polarization can be modulated. Phase and polarization modulation are external. Intensity and frequency modulation can be internal or external. Polarization modulation requires the presence of two detectors in the receiver, one for each polarization. Because of this, it is preferable to reserve polarization for multiplexing of two channels. The intensity distribution of a laser beam, as a function of angle with respect to the maximum intensity, follows a Gaussian law. If total beamwidth is reduced, there is benefit in gain but the pointing error loss increases. In general, for a pointing error of any kind, the beamwidth may be adapted to the pointing error. In addition to losses due to pointing error, transmission losses and degradation of the wavefront in the emitting optics occur.

The receiver can be of a direct detection or a coherent detection receiver. With direct detection, the incident photons are converted into electrons by a photodetector. The subsequent baseband electric current at the photodetected output is amplified then detected by a matched filter. With coherent detection, the optical signal field associated with the incident photons is mixed with the signal from a local laser. The resulting optical field is converted into a bandpass electric current by a photodetector and is subsequently amplified by an intermediate frequency amplifier. The demodulator detects the useful signal either by envelope detection or by coherent demodulation. The receiving losses include optical transmission losses and, for coherent detection, losses associated with the degradation of the wavefront. The quality of the wavefront is an important characteristic for optimum mixing of the received signal field and the local oscillator field at the photodetector frontend. Filtering, to reject out-of-band photons, also introduces losses, since the transmission coefficient reduces with bandwidth. A typical filter width is from 0.1 to 100 nm.

2.5.5 Iridium ISL and Mobility System

Figure 2.44 illustrates a typical satellite communication setup of the Iridium Big LEO constellation ISL and mobility system. In distinction from the Globalstar Big LEO, Iridium is providing optical ISL, which improves connections between users and GES terminals, although there is not direct LOS. The Iridium ISL system employs frequencies of 22.55–23.55 GHz as these RF are heavily absorbed by water and hence are not useful for establishment Earth to satellite links.

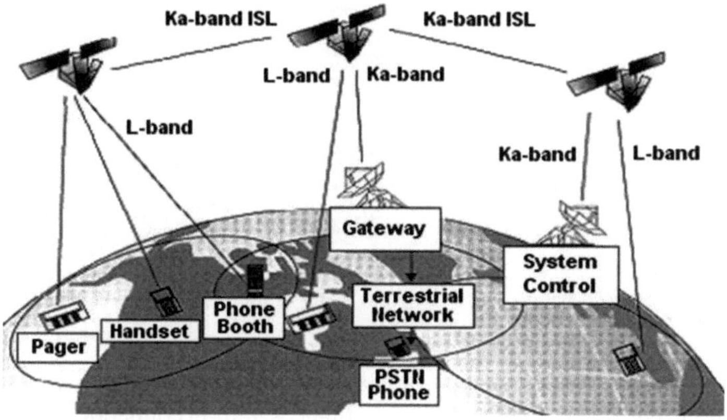

Fig. 2.44 Iridium big LEO ISL system—Courtesy of Book: by Maini

2.5.5.1 LEO Satellite ISL

The LEO network topology is constructed by the ISL, which establish a network between satellites. In order to establish ISL network, each satellite needs to have extra equipment, including transceivers and antennas. This additional equipment will increase the weight and the cost of the satellite, but on the other hand a satellite system that has this ISL features does not require a GES to establish a long distance connection. This advantage reduces the dependency of the satellite network on terrestrial systems. As stated earlier, in the satellite constellation, there are two types of ISL solutions.

The permanent ISL between satellites in the same plane (intraplane satellite link) and semi permanent between satellites in the neighboring planes (interplane satellite link) will function as edges in LEO satellite constellation with LEO satellites as the nodes.

As illustrated in Fig. 2.45, ISL between satellites 1 (S1), satellite 2 (S2), and ISL between satellite 1 (S1) and satellite 3 (S3) in orbital plane 1 are interplane ISL. The ISL between S1 in orbital plane 2, S4 in orbital plane 3 and the ISL between S1 and S5 in orbital plane 1 are intraplane ISL. Iridium is the first satellite communication system that provides ISL. The ISL system uses radio or laser media for their direct communication for mobile and semi-fixed systems. Iridium uses a GSM-based telephony architecture, and a geographically controlled system access process. Each satellite is connected to its four neighboring satellites through their ISL network. Connections between the IRIDIUM network and the Public Switched Telephone Network (PSTN) are provided via GES installations or Gateway. By having ISL the location of the GES can be flexibly located.

As stated, satellite links in the LEO constellation are constructed from two different types of ISL solutions. The first one is the interplane satellite connection, which is constructed between two satellites in the same plane, whereas interplane satellite connection is an ISL which is constructed between two neighboring

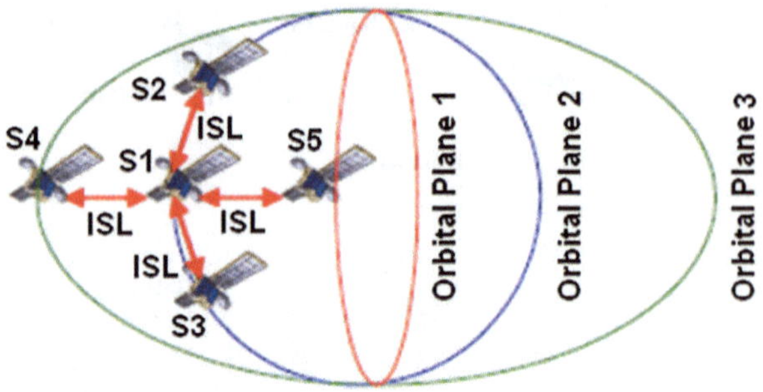

Fig. 2.45 Iridium big LEO ISL system—Courtesy of Theses: by Septiawan

satellites in a deferent plane. The intraplane satellite connection remains continuous for the whole period of time, while interplane satellite connection in some LEO topology will be turned on for a period of time. For example, in the near-polar constellation, intraplane satellite connection is switched on in the polar region, to reduce the signal interference. In the case of Iridium intraplane, ISL will only be maintained between latitudes of approximately 60° North or South of the equator. A first-degree ISL only provides a connection between satellites with its direct neighbors. A second-degree ISL will provide a connection between satellites with not only its direct neighbor satellites, but also with the next neighbor satellites.

Besides intersatellite handover, there is another type of satellite handover, which is called intrasatellite handover. This is a handover between spot beams in one satellite or handover occurs between satellites in the same plane. Interplane satellite handover: This type of handover occurs when a connection is handed over to another satellite in another plane, due to the unavailability of a satellite in the same plane, or due to the unavailability of satellite channels to accommodate this connection. The last type of handover is a handover that occurs once the ISL systems between satellites (either on the seam region or near polar region) are turned off.

2.5.5.2 LEO Satellite Handover

The mobility management in LEO satellite network is specific because a LEO satellites movement is relatively higher than the movement of any object on Earth, when an LEO satellite becomes the moving object in this context, while the mobile user on Earth has a relatively fixed position. The LEO mobility aspect is more concerned with the movement of the satellite footprint on Earth, in which an on-going connection of MES users has to be handed over from one satellite to another. Thus, these transitions have to be smooth and seamless and mobility management has to be able to maintain any on-going network connections and perform a handover process if needed. There is an advantage in a LEO satellites environment, in which the movement of the LEO satellite is roughly predictable.

Prior to the disappearance of viewed satellite, a new connection needs to be established. A handover procedure to the new approaching satellite needs to be

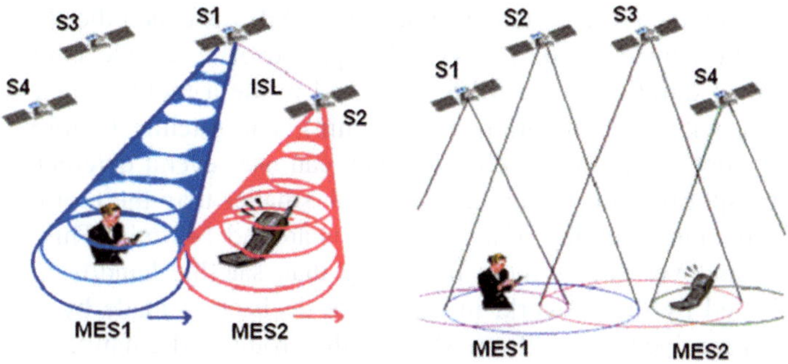

Fig. 2.46 Handover between neighboring satellites and soft handover procedure—Courtesy of Theses: by Septiawan

constructed. In case there is only one visible satellite at a time, the handover procedure needs to be initiated before the currently used satellite becomes invisible. This case is shown in Fig. 2.46 (Left), where a connection between MES1 and MES2 is constructed using satellite 1 (S1) and satellite 2(S2), via ISL between these two satellites. The MES1 unit is currently inside the service coverage of S1, while MES2 is located inside S2 service coverage. Both satellites move in clockwise direction, so MES2 starts losing service coverage from S2 and starts to obtain S1 coverage. Since currently MES2 is only covered by S2, it needs to start the handover procedure to allocate the connection to the next available satellite S1.

In the case that two or more LEO satellites are visible at the same time, a soft handover can be used to allocate the ongoing call into another visible satellite, which has a better signal performance or better visibility than the current one. This is shown in Fig. 2.46 (Right), when MES2 is currently in the coverage of S3 and S4. Soon, S3 and S2 will cover MES2. Every time there should be two visible satellites for each user. MES2 compares signal strength between these two satellites and chooses the best satellite to build a connection.

Therefore, due to the rapid movement of satellites, MES terminals can only use the same LEO satellite for short periods of times and before they lose their LOS to the current satellite, their connection must be handed over to the next available satellite.

It is necessary to conclude that a handover is initiated from one satellite to another satellite. The term handover is used in US cellular standard documents, while in ITU documents the term handover is used as well. Both terms have the same meaning. Handover in LEO satellite systems differs from the one in the cellular terrestrial system in term of the mobile and fixed units. In the LEO satellite systems, handover means a procedure of changing the assignment of a fixed unit, a MES terminal on Earth, from one mobile unit, a LEO satellite, to another as the LEO satellites move, while in the terrestrial cellular system a handover means a procedure of changing the assignment of a mobile unit from one fixed unit (Base Station) to another, as the mobile unit moves permanently.

The time in which a satellite is visible from a MES is called the sliding window of this satellite constellation. The visibility period of a satellite is defined as the maximum time duration that a MES is located inside a footprint and can directly communicate with that satellite (about 15 min). The satellite footprint is divided into several spot beams, and each spot beam can use several deferent frequencies. Because the spot beam coverage area is much smaller than the footprint coverage area, the maximum visibility of a spot beam is around 1–2 min. This means that a spot beam handover occurs more frequently than satellite handover. A handover can be initiated by the network, in which the decision is made by network measurements of received signals from MES on the ground. Alternatively, MES on the ground can provide a feedback to the LEO satellite concerning the signal received at the MES user. Various performance metrics may be used in making a decision, such as call dropping probability due to the handover itself, probability of unsuccessful handover due to an execution of a handover with an inadequate reception condition, call blocking probability due to unavailable capacity, rate of handover the maximum number of handovers per unit time, and handover delay due to the distance of the point that handover should occur and the point that the handover does occur.

Stallings outlines several handover strategies, which firstly suggests a handover can be initiated by measuring relative signal strength between two LEO satellites and the MES user on the ground. The second strategy depends on threshold signal strength. A handover is initiated when the signal strength between the LEO satellite and the MES user is lower than a certain threshold value. In the third strategy, a handover can only occur if the signal strength of the new LEO satellite is stronger by a margin than the current signal strength. The last strategy uses by using prediction techniques. The handover decision is based on the expected future value of the received signal depending on the movement of LEO satellites. A satellite handover takes place when an ongoing connection needs to be handed over from one satellite to another satellite. There are three types of satellite handovers:

1. **Intra-satellite Handover**—Handover occurs between satellites in the same plane. The Gateway monitors the signal strength and the MES unit position relative to the satellite. Since the Gateway has information about the MES unit positions and the satellite positions, when the currently used satellite moves away from the MES unit, the Gateway will contact the next available satellite in the same plane, to replace the currently used satellite.

2. **Interplane Satellite Handover**—It occurs when a connection is handed over to another satellite in another plane, due to the unavailability of a satellite in the same plane, or due to the unavailability of satellite channels to accommodate this connection.

3. **Interbeam Handover**—This handover occurs while the MES decides to use another frequency in the adjacent candidate beams, due to the weakness of RF signal power from the used frequency. Once the RF power strength in the current beam is lower than the MES initiates, a handover requests to hand over a user to the new beam.

Chapter 3
Transmission Techniques

Early satellite communication systems, primarily fixed, and later mobile networks used the analog transmission technique. Although modern MSC networks are still using the analog technique, the rapid development of high-speed digital mobile equipment is fostering a trend toward completely digital MSC. This chapter includes techniques and technologies that enable signals to be sent from one user to another and vice versa via satellites and LES.

There are several methods of modulation, multiplexing, and multiple access techniques used in MSC and their reverse processes, with some overlapping. Modulation is the process by which the baseband signal in EM form can be impressed upon a carrier, so phase (PM) and frequency (FM) modulations are used heavily in satellite communications because of their positive ability to deal with nonlinear distortion, noise, and interference. The amplitude of the PM and FM carrier is held constant, so there is no apparent change in the power level. Most nonlinear distortion is the result of amplitude variations on the carrier, while PM and FM are able to perform better in this environment than amplitude modulation (AM), so because of that problem there is a major limitation in directly using AM for MSC.

Multiplexing and multiple access require sharing the resources of the satellite. Facilities are shared based on spectrum assignment (frequencies), by time-sharing (time domain) and by spatial separation (antenna beam and polarization). Theoretically, any method can be used for the transmission of analog and digital signals. In practice, frequency is easier used with analog signals, whereas time division is easier to use with digital signals. Multiplexing consists of combining the signals from several users into a single signal, which then forms the signal used for modulation of the carrier. After demodulation, the individual signals are separated by an inverse operation called demultiplexing. All types of coding, decoding, and error corrections are presented. Finally, up-to-date technologies of Internet and broadband with conclusive explanations of the major protocols used in MSS are discussed.

S.D. Ilčev, *Global Mobile Satellite Communications Theory*,
DOI 10.1007/978-3-319-39171-7_3

3.1 Baseband Signals

The information transmitted over an RF communications link consists of signals conveyed from one user to another via telephone and video telephone; via facsimile, telex, data, packets, radio, and PC terminals; and via television cameras and receivers. Such signals are called baseband signals, which consist of basic electrical impulses carried by the radio path of a satellite or via other communication networks. If the baseband signal is analog, the voltage which represents it can take any value within a given range, but if the signal is digital, the voltage takes discrete values within a given range.

The types of baseband signals are determined by the communication requirements of the final users and the nature of the network. The main types of baseband used in MSS are voice (Tel), fax, and Tlx signals on Tel channels and then data, video, and image signals. The baseband signals of mobile broadcast satellite services consist of direct TV (television) and sound transmissions, while in a broadband wireless or satellite service baseband, the signal is ICT via Internet and multimedia, such as data, video retrieval, and image transfer.

Each of these radio signals has to be arranged in a form suitable for transmission over some physical layer (air or wire), and such a technique is called baseband signal processing. The modified baseband signal is then superimposed onto a higher frequency carrier wave, when the signal modulates the carrier to a value suitable for propagation over the many different transmission links, such as radio and satellite. The analog process at the transmit end of the link is called modulation and the circuit that performs modulation is the modulator, while at the receive end of the link, it is called demodulation and the circuit that recovers the baseband information is the demodulator. Digital systems use a modulator for transmission and demodulator for reception of the RF carriers within one unit, known as a modem.

The process of modulation/demodulation applies to radio, satellite, and terrestrial links using either analog or digital transmission methods. Where GMSC are concerned, there could also be a terrestrial RF link involved between service subscribers' terminals on shore and LES for maritime, land, and aeronautical services. The RF terrestrial link may be radio and relay microwave link or landline (wire and optical fiber) or a combination of all three mediums. In both government and E-commerce, the need for communications to be secure is of vital importance. Thus, by attaching an encryption device to the transceivers, MSC customers can be confident that their baseband voice, data, or video signals are authentic, not tampered with and secure from eavesdroppers almost anywhere in the world.

The bandwidth classifiers and a brief description of system characteristics are as follows:

1. **Narrowband**—It uses low data of 10 Kb/s by channel bandwidths of 12.5–25 kHz;
2. **Wideband**—It is developed to provide higher data rate services by channel bandwidths of 50, 100, or 150 kHz and provides data rates that are measured in 100 Kb/s.

3. **Broadband**—It provides megabits per second of data. The current allocation is 50 MHz of spectrum at 4.9 GHz band.

In this chapter, they are considered as follows: telephone or voice signal; data and multimedia signals; sound signals; and television or video signals.

3.1.1 Voice (Telephone) Signals

The first commercial voice service via satellites was established by Intelsat at a time when underwater cables could not keep pace with the rapid growth of telephone traffic. In spite of data traffic growth in recent years, voice traffic has become dominant and will remain in the future integrated with videophones. In general, signals from the most telephone sets are analog, but these signals can be transformed into digital mode and multiplexed at a local telephone exchange for transmission on trunk routes. In addition, satellites are effective for the broadcast distribution of voice, data, and video (VDV) signals anywhere and anytime, including voice (Tel) modems that are able to send digital data over satellite voice channels or via an 3.3 kHz analog voice channel interfaced to the landline voice equipment.

Based on the extensive studies of voice signals, it is now well known that most of the speech energy during normal conversation lies between 0 and 3400 Hz and the range of voice frequency that can be registered by the human ear is up to about 20 kHz. Therefore, a bandwidth of 3–4 kHz is allocated to each voice channel. The CCITT recommends the range of telephony signals as 300–3400 Hz. The maximum energy of a signal representing speech is in the range of 800 Hz, and about 99 % of the energy is situated below 3000 Hz. The signal power of an average talker implied to a zero relative level point which is given by:

$$P_m = P_a + 0.115\sigma^2 + 10 \log \tau \, [\mathrm{dBm_o}] \tag{3.1}$$

where $P_a = -12.9$ $\mathrm{dBm_o}$ represents the average power of the speech signal; $\sigma = 5.8$ dB is the standard deviation of the normal distribution of the active speech power; $\tau = 0.25$ is the active factor of a talker. Thus, in total, the value of $P_m = -15$ $\mathrm{dBm_o}$ (Article G.223 of CCITT Recommendation).

The quality of a received analog voice signal has been specified by the CCITT to give a worst-case baseband signal-to-noise ratio for voice (Tel) signals, for transmission over a long distance, as 50 dB and with the maximum allowable noise in the baseband of 10,000 picowatts. Speech is characterized by having a large, dynamic range of up to 50 dB to accommodate the volume difference between a whisper and a shout. Speakers also tend to pause often, and on average, a speaker will talk for only 40 % of the period available, while the remainder of the time the link is idle.

In recent years, considerable effort has been spent in digitizing radio voice and telephone systems. Some of these techniques are applicable to transfer any type of analog into digital signal and vice versa, using various analog-to-digital conversion techniques. The main need has arisen from the advent of MSS and cellular systems, which both operate on a RF bands that have great demand; hence, considerable effort is being spent on improving the spectral efficiency of transmission as the spectrum becomes scarcer. The use of spectrally efficient voice coding coupled with an effective combination of modulation and multiple access techniques provides a viable transmission solution, so Inmarsat-M mobile standard enables voice communications using voice codes operating at about 6 Kb/sec. The quality of the reconstituted speech for digital transmission at the receive end, among other factors, will depend on the number of bits transmitted per second and the number of bits received in error. This is BER, which value is considered to be about 10^{-4} (1 bit error per 10,000 bits) to provide good speech quality. The BER can be used as a design threshold, although some systems commonly have values superior to 10^{-5} or even more.

Today, many modern transmission techniques are in use for the transfer of analog and digital interactive voice signals between fixed and mobile transmitters and receivers and for direct transmitting of voice/fax signals to the FES and MES. Despite the rapid technological advances and the diverse range of data capabilities available, voice is very important for quick, instant, and easy communication, meeting the needs of different markets worldwide. Therefore, Inmarsat mobile users can rely on clear digital voice radiocommunications up to full broadcast quality. Voice GMSC terminals onboard ships, in the cabin of vehicles or in the aircraft cockpits, enable direct dealing with landline telephone subscribers via GEO or non-GEO satellites, LES, and TTN.

The voice service of Globalstar GMPSC infrastructure is based on a code-excited linear prediction (CELP) variable rate vocoder. The voice quality has to meet the value provided by IS-96, which is the terrestrial CDMA standard. The superior voice quality can be offered at low data rates in large part due to the adaptable vocoder. In marginal areas, where the user terminal cannot generate sufficient power to close the link, the peak data rate is reduced to 4.8 or 2.4 Kb/s.

Hence, this will provide intelligible voice service in areas that otherwise could not be served. The vocoder will incorporate echo cancelation, which can be disabled if this function is provided by the satellite network. On the other hand, the forthcoming project if ICO Global Communications has a technical solution that can also enable voice roaming inside buildings or tunnels, which will eliminate the blocking effect of objects, which is the current situation in personal satellite communications, which is a very serious problem of GSM or other cellular systems.

The radio broadcasting program occupies a spectrum from 40 Hz to 15 kHz, enabling a high quality of voice signal. The test signal is a pure sinusoid at a frequency of 1 kHz. When the broadcast radio program is transmitted by communication satellite in digital form, the performance objective is stipulated in terms of error probability. These errors have the effect of generating audible "clicks." To limit their frequency to about 1 per hour, a binary error rate in the order of 10^{-9} is required by CCIR Report 648. The present broadcasting satellite system operates at 12 GHz and is designed for community fixed stations with large antennas, while Direct Broadcasting Satellite (DBS) is serving the individual reception of signals by fixed or semi-fixed terminals with small antennas called VSAT. The present system is broadcasting both radio and TV programs only for fixed terminals. Although some developed countries have started to use Direct Audio Broadcasting (DAB) satellite systems in the L-, S- and Ka-band for MSS. At any rate, these models of broadcasting systems are suitable for mobile utilization on ships, vehicles, and aircraft.

3.1.2 Data and Multimedia Signals

Data transmission is composed using digital signals consisting of a series of bits. Actually, bits are bistate pulses with low state called "0" (–V) and high state "1" (+V). Information is superimposed on a digital stream by arranging groups of bits called words, which can be used for the transmission of analog signals from a telephone or alphanumeric characters from a PC keypad. Therefore, facsimile, telex, data, and PC e-mail networking terminals are used for the transfer of data through the medium of different transmission techniques and applications that are expanding rapidly, including GMSC systems.

Data signals used by Inmarsat MSS can be broadly classified into three ranges:

(1) Narrowband data of about 300 b/s can be transmitted via MES TDMA 24 Kb/s Tlx channel with possibly 16 data bursts directed to LES or other MES terminals and PSTN via a modem.
(2) Full-duplex voice and 9.6 Kb/s data signals can be transmitted in the single channel per carrier (SCPC) data channel with a rate of 24 Kb/s in O-QPSK modulation scheme. In fact, the data channel enables packet data transfer using the CCITT X.25 recommendation to interface data terminal equipment (DTE) and data circuit-terminating equipment (DCTE) operating in packet mode to the PSDN by dedicated circuits. This channel also supports CCITT Group-3 Fax, which is also available in the SCPC voice channel, using a 2.4 b/s data rate and APC voice codes.
(3) Wideband one-way high-speed data (HSD) transmission can be supported with a rate of 64/56 Kb/s via voice channels on a dedicated frequency with a special type of V.32 (9600 b/s) modem. This service can be used for the transfer of PC data, high-quality digital audio, and compressed video. The duplex HSD will offer two-way 492 Kb/s via standard IP service for aircraft in both directions. The two-way HSD can be also used for digital multiplexing data channels and video conferencing between ships and shore.

There are several known techniques in transmitting digital data, facsimile, e-mail, and PC networking data applications, whose utility is also expanding rapidly for fixed and mobile systems. The new satellite global area network (GAN) portfolio of the HSD service developed by Inmarsat offers a cost-effective extension to corporate LAN and WAN even in the world's most rugged regions. Moreover, mobile packet data allows users to pay according to the amount of data they send rather than how long they spend online suited to Internet Web-based applications such as intranet access and e-commerce via satellite.

Meanwhile, the quality of data service provided by GMPSC systems enables transmission speeds up to 7.2 Kb/s. The BER contributed for instance by the Globalstar system is less than 1×10^{-6}. Hence, a higher terminal rate of about 9.6 Kb/s can be processed if the equipment incorporates an elastic buffer to accommodate the required flow control.

Therefore, data is becoming the most common vehicle for information transfer related to a large variety of services, including voice telephony, video, and PC-generated information exchange. One of the most appealing aspects of data is the ability to combine, onto a single transmission, support data generated by a number of individual sources, which resulting in a single data stream called aggregate traffic. This is paramount for transfer of multimedia traffic integrating VDV signals that often displays a reduced bit-rate variance versus traffic generated by the individual sources due to the embedded statistical multiplexing.

A network operator uses this opportunity to dimension its links (and particularly satellite links) with a capacity that is less than the sum of the peak bit rate of the individual sources. This dimensioning considers both the burstiness of the data traffic and the multiplexing techniques in use. The traffic is typically transported by packets, so the data structure (transport stream) developed for conveying TV program components could be used to carry any type of data, taking advantage of the well-recognized standard and mass-market production of equipment.

For example, the Moving Picture Experts Group (MPEG-2) transport stream (MPEG-TS) packets are used with DVB-S data satellite transmission. As for video program, MPEG-2 transport stream (MPEG-TS) packet has a fixed size of 188 bytes of which 4 bytes are used for the packet header and 144 bytes for the payload. The header consists of a synchronization (sync) byte, a packet identifier (PID), transport error indication, and a field of adaptation options.

The asynchronous transfer mode (ATM) data format is used in high data rate, terrestrial networks. The ATM packet, called an "ATM cell," is a fixed size of 53 bytes, of which 5 bytes are used for the header and 48 bytes for the payload. The header consists of a virtual channel identifier (VCI), a virtual path identifier (VPI), a payload type, and priority and header error check (HEC) fields.

3.1.3 Video Signals

Prior to the advent of satellite communications, it was very impracticable to relay video TV programs across oceans because of the limited bandwidth of submarine coaxial cables. Today, optical fiber cables have enough channels but cannot, like satellite systems, transmit DBS TV video to many widely separated fixed or VSAT stations and especially to mobile terminals. Direct broadcast by satellite has been in use almost 3 decades, because the first such European satellite was launched in 1987. The demand for this type of satellite service continues to gain interest in the distribution of TV programs direct to homes (households) or other fixed stations and, later, to provide a service for mobile stations.

There are various inexpensive ways in which it is possible to add data transmissions to the DBS and to provide cheap information services and e-mail for both domestic and business purposes, and to onshore or offshore TV subscribers.

More than two decades ago, Inmarsat started to provide transmission of still picture mode from the oil fields to any office using a wide range of analog and digital portable devices for transmitting, receiving, and processing high-resolution color photographs in less than 5 min.

In addition, Inmarsat developed the HSD store-and-forward video system. Namely, with the aid of advanced video codecs, it is possible to digitize and compress video material and then transmit it at 492 Kb/s to a receiver, which decompresses and buffers the data in almost full motion video. The store-and-forward technique also ensures that the received material is error-free, since the data transmission is achieved by using error detection and repeat transmission mode.

There is a new possibility today for Inmarsat MES to also use integrated services digital network (ISDN) interfaces for the transfer of intensive data interactive applications such as video conferencing (VC) and digital image transfer (DIT). Therefore, utility of standard ISDN interfaces will enable one to easily connect mobile fleet offices with corporate applications. Video baseband allows Inmarsat users to relay video from almost anywhere on the globe. This provides an effective flexible solution for media companies and the growing number of corporation who also use the VC system to facilitate "real-time" discussion on a regular basis. Users benefit from ISDN speeds and easy-to-use portable equipment, which delivers immediate live action, training, or medical and equipment diagnostic support, as required.

3.1.4 Basic Concept of Modulation

Modulation is when the signal message that is to be conveyed is transposed in baseband signals, such as voice or raw PC data, into a suitable form so that it can be transmitted over the media involved, usually plain air with absorbing gases as described earlier. The raw data or voice signal can be modulated onto a carrier frequency as illustrated in Fig. 3.1 (left), and demodulated (Right). Thus, a frequency carrier is selected to suit propagation conditions, and then spectrum and frequency management constraints and ITU radio regulations and efficient and reliable transportation of the information across the media. There are mathematical equations that describe this modulation process. The reverse process demodulation is when the useful information is extracted from the unuseful carrier.

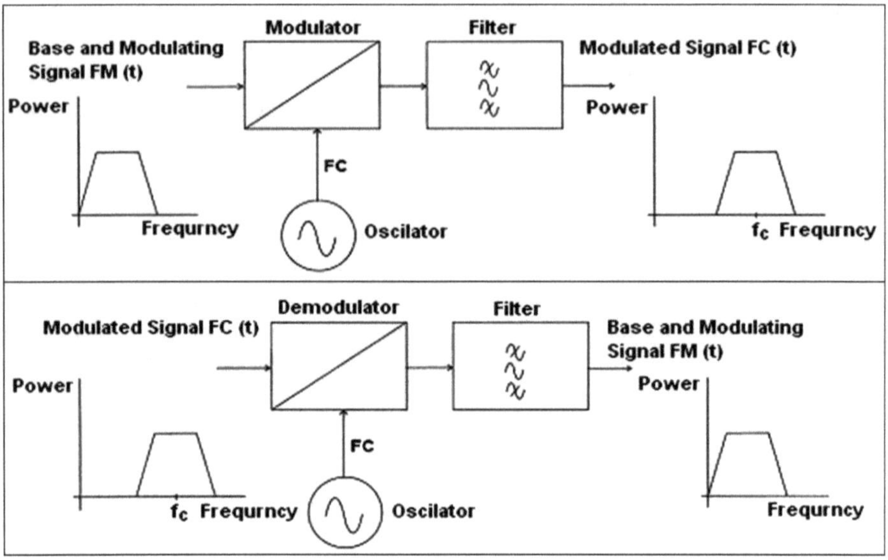

Fig. 3.1 Basic modulator and demodulator—Courtesy of Book: by Stacey

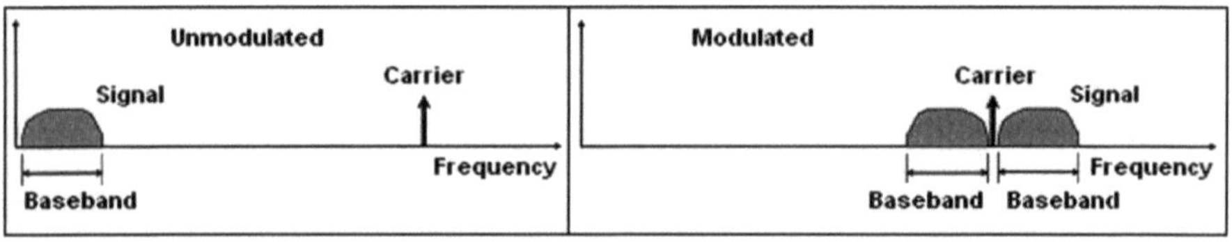

Fig. 3.2 Frequency domain—Courtesy of Book: by Stacey

There are many different types of modulation schemes available to the radio designers. In principle, these can be broken down into analog modulation such as amplitude or frequency modulation (AM or FM) schemes or variants on these, as can be used to modulate voice signals. More recently, digital modulation such as frequency-shift keying (FSK), phase-shift keying (PSK), or advanced variants of these can be used to convey digital messages and/or digitized voice streams, or of course both.

The selections of appropriate modulation schemes for an application are usually dependent on, but not necessarily limited to, the following factors: 1. required ranges of propagation; 2. frequency of operation and propagation properties; 3. spectral efficiency; 4. equipment complexity, reliability, size weight, and cost; and 5. required data or voice throughput rate and regulatory constraints.

The modulation conundrum of unmodulated signal is shown in Fig. 3.2 (left) and modulated signal (right), how baseband unmodulated signal is transforming into modulated baseband signal in frequency domain. At this point, there is a relationship between all of these aspects; mathematically, it is Shannon's law; and it is usually necessary to trade off one requirement against the others to achieve the best overall engineering solution. In reality, for low-specification systems where performance is not so critical, rugged and rudimentary low-order modulation schemes can work very efficiently, and this keeps cost and complexity down. For some of the more sophisticated systems where data throughput in a limited bandwidth is premium, some of the more sophisticated and elaborate modulation schemes pushing the limits of Shannon's law become applicable if not essential.

3.1.5 Analog and Digital Domains

Before coming into the details of the modulation schemes employed by the aeronautical systems, it is worth momentarily looking at the analog-to-digital (A/D) aspects of signal modulation. Modern communication systems generally digitize voice, data, and images and transmit them via a digital medium, where quality can generally be kept, before putting them back in their original analog domains. With data, there is never any A/D and it stays entirely in the digital domain. In such a way, the modulation process can be considered as the inverse of this process; i.e., data is modulated onto a carrier conversion, which is akin to digital-to-analog (D/A) version, and then, it is transmitted over the media and converted back to digital via A/D conversion at the end, illustrated in Table 3.1.

With voice over legacy radio systems, it is never even coming into the digital domain. It is analog when it starts; it goes through an analog-to-analog conversion as it is modulated; and then, at the end, it is an analog-to-analog demodulation process.

Analog modulation is the supporting emissions of double-sideband amplitude modulation (DSB-AM), single-sideband AM (SSB-AM), and suppressed carrier AM (SC-AM), and the frequency modulation (FM) scheme is the supporting emissions of phase-shift keying (PSK), frequency-shift keying (FSK), differential phase-shift keying (DPSK), differential frequency-shift keying (DFSK), quadrature amplitude modulation (QAM), and trellis coded modulation (TCM).

Table 3.1 Analog and digital domains

Analog-to-analog conversion (analog speech to analog radio)	Digital-to-analog conversion (data or digitized speech to radio)
Amplitude modulation Frequency modulation	Amplitude-shift keying Frequency-shift keying Phase-shift keying

3.2 Analog Transmission

Analog transmission is characterized by processing performed on the baseband signal before and after modulation in order to improve the quality of the link. The carrier can support only one or few channels for the transmission of baseband signals. In the case of a carrier transmitted from the station representing only a single-user channel, this is SCPC transmission. On the other hand, if the carrier represents a number of multiplexed users, it is designated multiple channel per carrier (MCPC) and also are using several channels of frequency division multiplexing (FDM) transfer.

A signal needs to be impressed on an RF carrier for transmission through the satellite, and for this purpose, it uses a process known as modulation. The objective of any communications is to transmit the modulated carrier to Rx as reliably as possible, so that the demodulated signals can be satisfactorily recovered. In analog transmission systems, the information waveform in the form of voice, data, or video signals are modulated directly from the source onto the carrier at the modulator of Tx, by using the methods of amplitude modulation (AM), frequency modulation (FM), and phase modulation (PM). Frequency and phase modulations are used most widely for direct analog transmission in satellite communications, while amplitude modulation is a process used indirectly in the satellite link.

Modulation may also be used at very low frequencies, like the more common and various forms of PSK modulation scheme. In addition, modulation is often done on a carrier with an RF of about 70 MHz lower than the transmission RF. This RF is then upconverted to the transponder frequencies on 6/4 GHz for amplification and retransmission. However, the previous types of satellite do not change the received modulation before retransmission. Satellites are now being designed to allow only one modulation method to be employed in the uplink and another for the downlink; each link can be optimized. The space link between two LES can generally be accomplished by the combination of modulation and multiplexing techniques.

In the context of analog modulation will be included specific schemes as well as DSB-AM, SSB-AM, and SC-AM.

3.2.1 Baseband Processing

The purpose of baseband processing is to improve the quality of the space (satellite) link using different methods whose cost is less than that arising from modification of one of the parameters involved in the link budget. The principal methods of baseband processing for telephone transmission are speech activation, pre- and de-emphasis, and companding and for TV transmission pre- and de-emphasis only.

3.2.1.1 Speech Activation

The principle of speech activation is to establish the space link only when the subscriber is actually speaking. As the activity factor is $\tau = 0.25$, its application to a multicarrier SCPC system should permit a reduction of the power required on the satellite by about 6 dB.

In practice, the reduction is only in the order of 4 dB, allowing guard times for activation and deactivation of the carrier and the sensitivity of S/N spikes. Therefore, the activation threshold parameter has to be from –30 to –40 dBm_o, the carrier activation time can be 6–10 ms, and deactivation time can be 150–200 ms.

3.2.1.2 Pre- and De-emphasis

The noise at the output of the demodulator of a FM transmission has a parabolic spectral density, when the high-frequency components of the signal are more affected by noise than the low frequencies. Figure 3.3 shows the principle of the pre- and de-emphasis effects. The pre-emphasis filter prior to modulation of the signal increases the amplitude of the high-frequency components.

The gain of the crossover frequency in the pre-emphasis filter is 0 dB. After demodulation, the de-emphasis filter is performing inverse effects, namely reducing the amplitude of the high-frequency signal and noise power. In such a way, the signal is restored without spectral distortion and the S/N is improved. The improvement can be 4–5 dB for telephony and around 13 dB for TV signals.

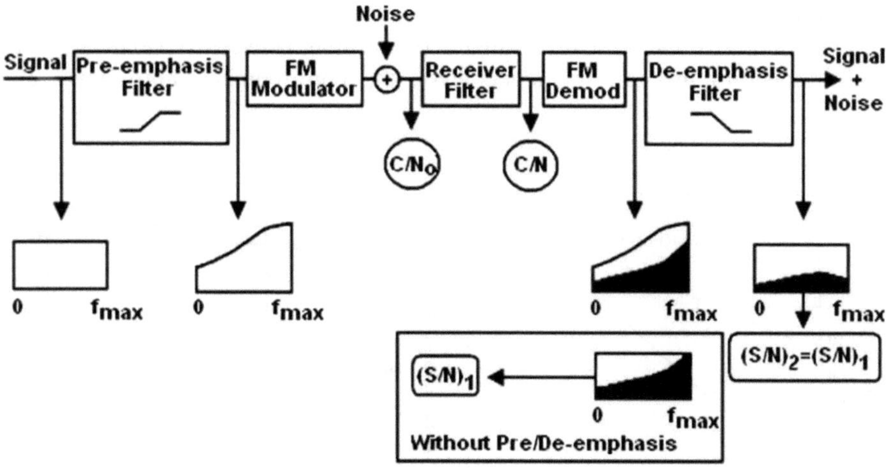

Fig. 3.3 The principle of pre- and de-emphasis—Courtesy of Book: by Maral

Companding Process

Companding is a process of compression and expansion for reducing the effect of noise on speech channels in accordance with the specification of CCITT Recommendations G.162 and article G.166. In effect, a compandor comprises a compressor and an expander, and an improvement in the S/N ratio at the output of the demodulator is obtained by reducing the dynamic range of the signal before modulation (compression) and performing the inverse operation after demodulation (expansion) in order to restore the original speech signal to its correct relative level.

These circuits perform the task of modifying the speech signal in analog voice channels; thus, when the gain of the compressor and expander is controlled by the speech power at a syllabic rate, the technique is referred to as syllabic companding.

The characteristics of a typical compressor and expander are shown in Fig. 3.4 (left). The compressor varies in such a way that an input level of x dB results in an output level of x/2 dB. For instance, input levels to the compressor of +10 by 0 (the reference level) and –60 dB result in output levels from the compressor of +5 in relation to the 0 and –30 dB, respectively. The signal characteristic of the expander is inverse to that of the compressor. It is not practicable to use compressors and expanders with a ratio of greater than 2:1.

The effect of companding on the relative levels of a signal and the noise added during transmission, i.e., after the compression and before the expander, is shown in Fig. 3.4 (right). The compressor raises the power of weak signals relative to the noise. In this way, for an input signal at a level of –40 dB, the effect of the compressor is to raise the S/N power ratio from 0 to 20 dB. Otherwise, in the absence of companding, a signal would be at the same level as the noise. The expander restores the original level of the speech signal, and it reduces the level of the noise.

For example, if the noise level at the receiver input between syllables is –40 dB with respect to the zero reference level, the corresponding noise level after

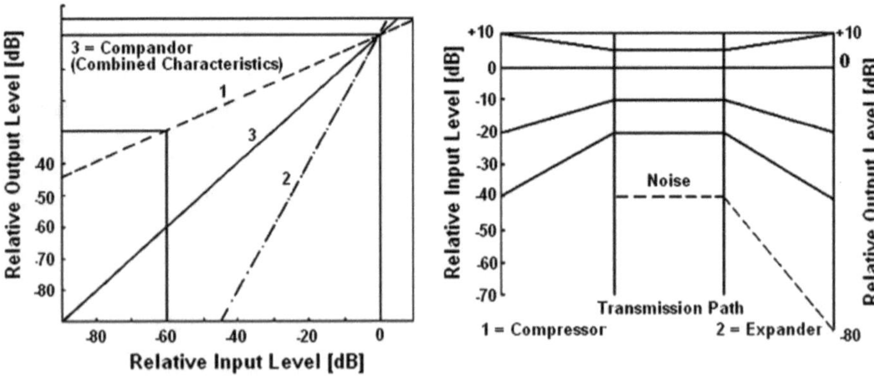

Fig. 3.4 Characteristics of compandor and companding effect—Courtesy of Book: by Dalgleish

expansion is reduced from –40 to –80 dB. The improvement in S/N ratio is the result of attenuation and is subjective, since it is associated with the absence of perceived noise during silences in the conversation. It is considered to be in the order of 15 dB.

3.2.2 Analog Modulation and Multiplexing

Source waveforms of analog signals are directly modulated onto RF or IF carriers at the transmitter, using any form of three types of modulation: AM, FM, or PM.

Analog modulation is a process in which some characteristics of a HF carrier are varied in accordance with the baseband signal, which is the creation, transmission, and reception of EM fields. In reality, EM fields and waves can be used to communicate all kinds of information from place to place. Therefore, an analog modulated signal can be represented as the following sinusoidal wave:

$$c(t) = A \cos [2\pi f t + \varphi] \tag{3.2}$$

where A = amplitude, $2\pi f$ = form of ω_c = angular frequency of the carrier, f = frequency, and φ = carrier phase. Modulation can be achieved by altering the amplitude, frequency, or phase of the wave in accordance with the information signals. Consequently, AM, FM, and PM analog carriers are simply forms of modulated carriers in which either amplitude, frequency, or phase is modulated by the information waveform.

Thus, by changing the amplitude, frequency, or phase is resulting in AM, FM, or PM mode, respectively, which is illustrated in Fig. 3.5. In fact, since EM fields are vector fields, it is also possible to modulate the polarization of a radio wave to communicate information. However, polarization modulation is only normally used for special purposes, as the polarization state of a wave can easily be "scrambled" during free-space transmission by effects such as unwanted reflections from buildings.

The AM system is a type of linear modulation in which the baseband signal is linearly related to the modulated signal, while FM and PM are kinds of angle modulation in which the baseband signal is angularly related to the modulated signal. However, angle-modulated FM and PM signals require more carrier bandwidth than AM but achieve a higher demodulated C/N ratio for the same carrier C/N value. The reverse process of recovering the information signal from the modulated carrier at the demodulator of the receiver is known as demodulation. Multichannel operation of FM for analog high-capacity satellite telephone transmission is normally accomplished by a system called FDM scheme. In such a system, different channels are separated from one another by being assigned to different subscribers, which are then combined to fill the total bandwidth of the transponder.

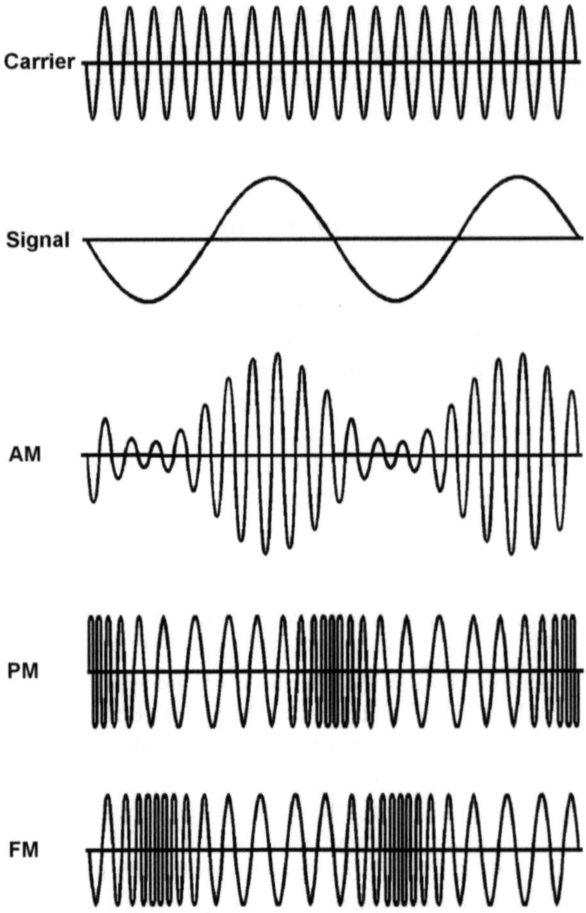

Fig. 3.5 Modulation options —Courtesy of Book: by Huurdeman

3.2.2.1 Amplitude Modulation (AM)

Amplitude modulation is a type of linear modulation and is not used as a modulating process in the satellite link. However, it can be used to modulate individual voice channels before combining them using FDM technique. In AM, the amplitude of the carrier frequency is modified by the amplitude of a modulating signal; namely, a carrier has to be amplitude modulated when the amplitude of the carrier varies in accordance with the signals. The sidebands are displaced at either side of the carrier frequency by the magnitude of the modulation frequency, and the amplitude depends on the modulation amplitude. The fundamental equation of AM modulation signals is as follows:

$$c(t) = A\left[1 + \Delta_a\, m_{AM}(t)\right] \cos\left(\omega_c t + \varphi\right) \tag{3.3}$$

where m_{AM} (t) = AM signal represents the information waveform to be transmitted and Δa = AM index (coefficient giving the degree of modulation of angular frequency of the carrier).

The AM method is employed usually in AM radio broadcasting and radiocommunication transmissions. In these systems, the intensity or amplitude of the carrier wave varies in accordance with the modulating signal. When the carrier is thus modulated, a fraction of the power is converted to sideband extending above and below the carrier frequency by an amount equal to the highest modulating frequency. If the modulated carrier is rectified and the carrier frequency filtered out, the modulating signal can be recovered. This form of AM is not a very efficient way to send information because the power required is quite large and because the carrier, which contains no information, is sent along with the information. In AM, the information is carried out only in the sidebands, and therefore, power in the carrier remains unutilized. In a double-sideband suppressed carrier (DSB-SC), modulation of the carrier is suppressed and only sidebands are used. The amplitude of this wave does not follow the signal amplitude, and consequently, the inherent simplicity of using envelope detection is lost. Therefore, this modulation is not used in satellite communications.

In a variant of amplitude modulation, called single sideband (SSB), the modulated signal contains only one sideband and no carrier. The information can be demodulated only if the carrier is used as a reference. This is normally accomplished by generating a wave in the receiver at the carrier frequency. The SSB is used for long-distance HF radiotelephony and telegraphy over land, such as in the maritime HF radio frequency bands and submarine cables. In the other words, the SSB scheme is called an SSB suppressed carrier (SSB-SC) when the carrier is suppressed. The most common application of SSB modulation in satellite communications is to multiplex voice carriers into a composite baseband signal. The required C/N ratio and the occupied bandwidths are two aspects considered in assessing the suitability of SSB for satellite transmission. The occupied RF bandwidth of an SSB transmission is the same as the baseband bandwidth, which is 4–5 kHz for a single telephone channel transmission. Typical bandwidth of – 30 kHz is necessary for FM and 20 kHz for the O-QPSK, so both schemes are widely used in satellite communications.

A typical satellite link can economically provide C/N ratios in the order of 10–12 dB, making SSB transmission inefficient from power considerations. This disadvantage can be offset to a large extent by the use of compandors, which offer an S/N ratio advantage of about 15–20 dB, and in such a manner, SSB transmission appears attractive. It is called Amplitude Companded SSB (ACSSB). Bandwidth efficiency is essential for MSC service, and for this reason, ACSSB has been considered favorably for such applications.

3.2.2.2 Frequency Modulation (FM)

In FM scheme, the amplitude of the carrier frequency is modified by the frequency deviation of a modulating signal. The sidebands are displaced at either side of the carrier frequency with an infinite number of separate bands, most resolving near zero. The amplitude depends on the modulation frequency change, and the frequency depends on the rate of frequency change of the modulation. Therefore, the fundamental equations of FM modulation are given below:

$$c(t) = A\cos\left[\omega_c t\, 2\pi\Delta_f \int m_{FM}(t)dt + \varphi\right] \quad (3.4)$$

where m_{FM} (t) = FM signal representing the information waveform to be transmitted and Δf = FM index (coefficient giving the degree of modulation angular frequency of the carrier).

Using FM, the frequency of the carrier wave is varied in such a way that the change in frequency at any instant is proportional to another signal that varies with time. The FM band has become the choice of music listeners and for the audio portion of TV broadcasting because of its low-noise and wide-bandwidth qualities; however, the FM schemes are also extensively used in satellite communications. Examples of FM applications are multiplexed telephony, SCPC systems, and TV broadcasting.

Thus, FM systems are well suited for those cases where the baseband signal is in analog form. An example is FM with companding used in Rx and Tx voice channels of the analog Inmarsat-A system of the SES transceiver. Furthermore, this scheme also offers advantages for transmission of digital data in applications where simple receivers are essential, such as the Inmarsat paging system. At any rate, an important requirement of a paging system is the need for simple, low cost, and rugged receivers.

3.2.2.3 Phase Modulation (PM)

Phase modulation (PM) like frequency modulation is a form of angle modulation so-called because the angle of the sine wave carrier is changed by the modulating wave. The two methods are very similar in the sense that any attempt to shift the frequency or phase is accomplished by a change in the other. The PM relation can be expressed with:

$$c(t) = A\cos\left[\omega_c t + \Delta_p m_{PM}(t) + \varphi\right] \quad (3.5)$$

where m_{PM} (t) = PM signal representing the information waveform to be transmitted and Δp = PM index, in which coefficient is giving the degree of modulation angular frequency of the carrier).

3.2.3 Double-Sideband Amplitude Modulation (DSB-AM)

As is known AM is when the amplitude of the carrier is directly proportional to the modulating signal. Probably, the simplest form of AM is DSB-AM mode, whose diagram is illustrated in Fig. 3.6 (left). The DSB-AM emission is described mathematically with the following equation:

$$f_c(t) = A\cos(2\pi f_c t + \varphi)\,\pi \tag{3.6}$$

where $f_{oci}(t)$ is the modulated signal output. Thus, from this relation, it is possible to derive the following equation:

$$A = K + f_m(t) \tag{3.7}$$

where *K* is the unmodulated carrier amplitude and f_m (t) is the baseband signal, which can be defined by the following equation:

$$f_m(t) = a\cos(2\pi f_m t) \tag{3.8}$$

The DSB-AM emission can be also presented as follows:

$$f_c(t) = [K + a\cos(2\pi f_m t)]\cos(2\pi f_c t + \Phi) \tag{3.9}$$

When Φ = 0, then multiplying out the cos function:

$$f_c(t) = K\{\cos(2\pi f_c t) + 0.5\,m\cos[2\pi(f_c - f_m)\,t] + 0.5\,m\cos[2\pi(f_c + f_m)t]\} \tag{3.10}$$

At this point, depth of modulation (m) = modulating signal amplitude (a)/unmodulated carrier amplitude (K). In the frequency domain, these discrete components are shown in Fig. 3.6 (right) and are called upper and lower sidebands, respectively.

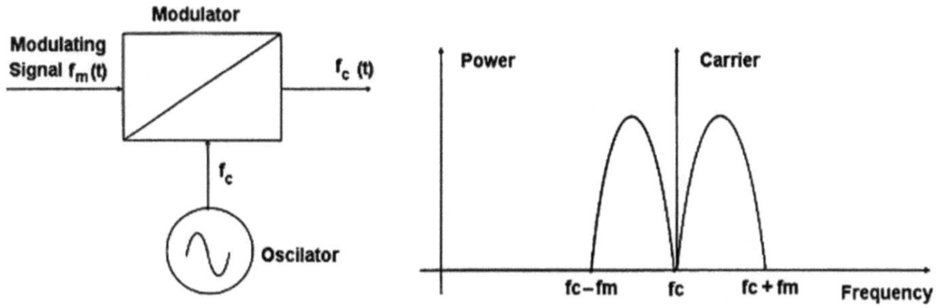

Fig. 3.6 Basic AM modulator and power spectrum for DSB-AM—Courtesy of Book: by Stacey

It is interesting to note that the modulated bandwidth of DSB-AM is twice the size of the minimum baseband bandwidth or the frequency component. It is independent of depth of modulation.

Following are the advantages of DSB-AM:

1. It is very easy to synthesize DSB-AM in nonlinear device such as diode or transistor;
2. It is very easy to demodulate DSB-AM in non-coherent detection, which uses a simple diode, capacitor, and resistor circuit or can use a coherent detector;
3. It is incredibly simple to work with, troubleshoot, and see the waveforms at every stage of the process;
4. There is always a carrier present, even when no information is being sent; this means a receiver can easily tune to this carrier frequency and keep locked to it; and
5. In the amplitude domain, signals cannot get distorted or shifted by Doppler shifts of receivers and transmitters that are physically moving with relation to one another.

Some of the relative disadvantages of DSB-AM when comparing this modulation to other schemes can be the following:

1. At its best performance, only one-third of power being transmitted is the information signal, and the other two-thirds of power is absorbed by the carrier and is not useful. This is with a modulation index of M = 100 %, which in theory can never be fully reached, and so, the practical realization of this is even worse, with typically always less than 25 % power transfer efficiency.
2. DSB-AM uses more RF bandwidth than is actually needed. As was previously pointed out, the RF bandwidth required is double the maximum baseband frequency component. More elaborate modulation schemes can half this or reduce it even more; and
3. There is no encryption of the signal, so it is easy to interfere with (security, etc.).

3.2.4 Single-Sideband Amplitude Modulation (SSB-AM)

With DSB-AM practical work, it is found that the upper sideband is a perfect reflection of the lower sideband; i.e., the same information is carried by both. By removing one of the sidebands and by removing the carrier, therefore, no information is lost and the S/N (signal-to-noise) ratio increases for a given transmit power as all the transmitted energy is concentrated in the information signal. This is the basis for SSB-AM block diagram and power spectrum illustrated in Fig. 3.7.

The construction of an SSB is fractionally more complicated than that of DSB-AM. It can be synthesized in two different ways.

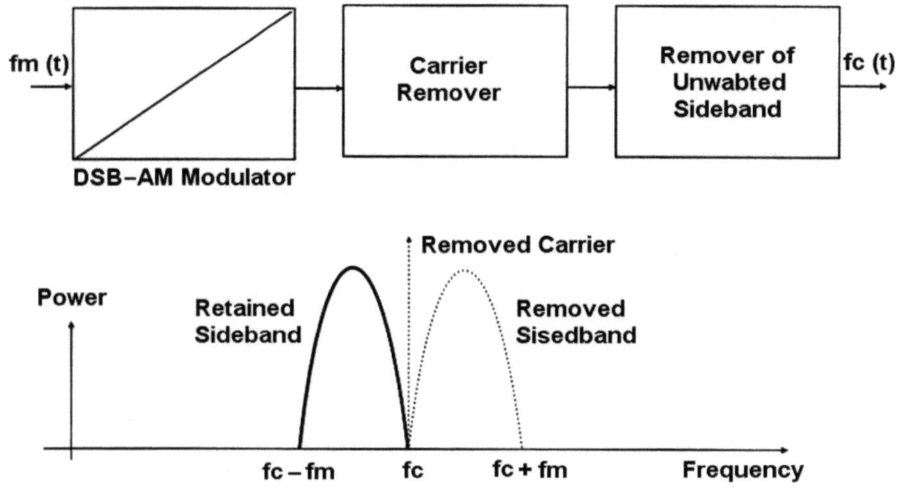

Fig. 3.7 SSB modulation block diagram and power spectrum for SSB-AM—d. Courtesy of Book: by Stacey

Method 1—As SSB indirect synthesis and filtering is shown in Fig. 3.8, a SSB signal can be synthesized either by using basic DSB-AM or DSB-SC-AM modulator and filtering out the unwanted carrier and sideband (usually done at a IF).

Method 2—Arguably, a more elegant way of modulation is to use Hilbert modulator as SSB direct synthesis, which essentially cancels out the unwanted sideband and carrier.

Mathematically, this is described with the following equation:

$$V_{out} = 0.5\,k\,\cos[2\pi\,(2f \pm f_m)\,t + \varphi] \pm 0.5\,k\,\cos\,(2\pi\,f_m t - \varphi) \tag{3.11}$$

The HF component is filtered out and expressed as follows:

$$V_{out} = 0.5\,k\,\cos\,(2\pi\,f_m t - \varphi) = 0.5\,k\,\cos\,(2\pi\,f_m)t \;\; \cos\,\Phi \pm 0.5\,k\,\sin\,(2\pi\,f_m)\,\sin\,\varphi \tag{3.12}$$

For demodulation, a Costas loop is required to present SSB demodulation as shown in Fig. 3.8.

Following are the advantages of SSB-AM:

1. As already stated, all transferred power is concentrated into the useful sideband carrying information. This gives transmitters efficiency theoretically up to 100 %; in reality, a more practical figure about 70 % is very useful for HF SSB telephony;
2. It is twice as bandwidth efficient as the DSB-AM system.

Disadvantages of SSB-AM are as follows:

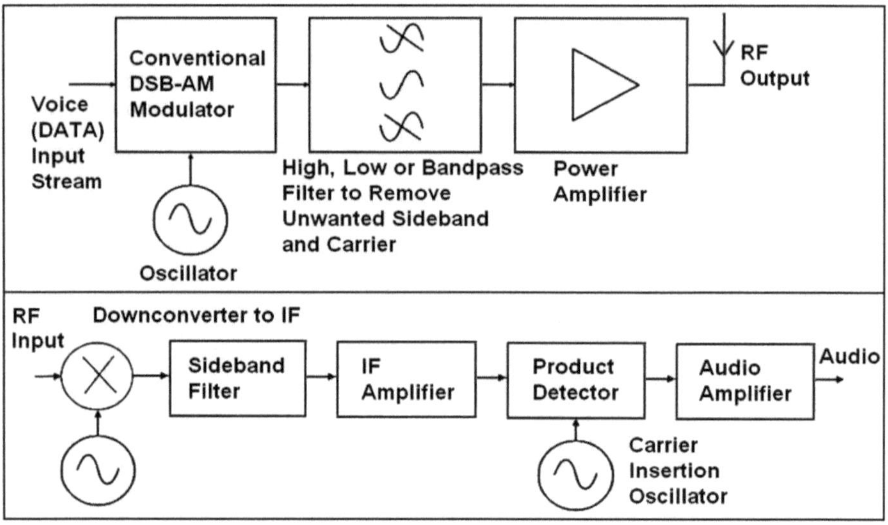

Fig. 3.8 SSB modulator and demodulator—Courtesy of Book: by Stacey

More sophisticated demodulation is required (coherent detector with PLL, Costas loop).

It can be prone to phase distortions (due to Tx and Rx oscillators beating and Φ changing), so high oscillator specification is required at both ends, not really heard by listeners (quality not high).

The aeronautical HF system and other SSB systems are using SSB modulation for the reasons of spectral efficiency and a limited resource, which will be discussed further in Chap. 9. Also, a number of earlier military systems are known to have used SSB as a basis.

Suppressed carrier double-sideband AM (SCDSB-AM) is modulation technique also used in aeronautical radio traffic. Worth mentioning in passing, there is an intermediate version of modulation called suppressed carrier double-sideband AM, which lies between DSB-AM and SSB-AM.

In its basic form, it is same as DSB-AM, with the carrier filtered out. It has the advantage of concentrating more power into the useful sidebands, so there is an S/N improvement; however, there is no bandwidth saving and it requires a more complex synthesis. It is rarely used, as most applications will do a full conversion to the SSB, where power and spectral efficiency become important and there are no mainstream applications of this in the aeronautical bands. However, there are some specialist military systems using this for aeronautical communication.

3.2.5 Frequency Division Multiplexing (FDM)

The process of combining baseband signals and sharing the communication channels is known as multiplexing, while the reverse process of extracting individual baseband signals is called demultiplexing. The CCITT proposed all multiplexing standards including FDM, which is applicable only to telephony baseband signals, whereas the digital time division multiplexing (TDM) standard is applicable to all types of baseband signals.

Each type of MES usually transmits and receives many kinds of signal transmissions to and from spacecraft. Multiplexing enables the division of channels or the combination of two or more input signals into a single output for transmission. In effect, the most common and known analog multiplexing method is FDM, which is used in satellite communication transmissions. The simplest approach to multiplexing is to assign a specific part of the available frequency or bandwidth spectrum to each signal. If two signals initially have the same spectrum, the frequency of one or both is shifted, so in such a way, they will not overlap. Therefore, the FDM is a multiplexing solution where signals occupy the channel at the same time but on different frequencies; namely, different message signals are separated from each other in frequency. The multiplexing scenario of FDM is shown in Fig. 3.9, as a concept of FDM, which illustrates simultaneous transmission of three message signals over a common communication channel. It is clear from the block schematic arrangement shown that each of the three messages signals modulates a different carrier.

Thus, the FM mode is used for some specialist military systems, in aeronautical telemetry and finally in the adjacent band to the VHF aeronautical communications for broadcasting information. It is mainly used by broadcasting in the band between 88 and 108 MHz, just below the VHF navigation and communication bands

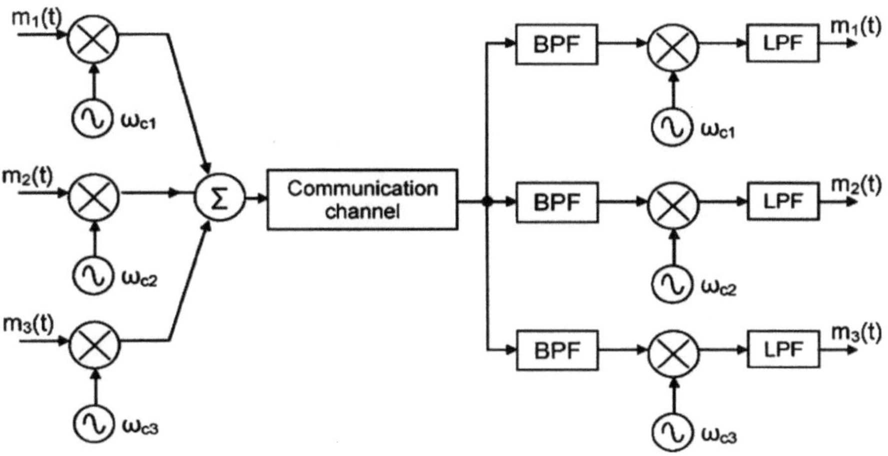

Fig. 3.9 Frequency division multiplexer—Courtesy of Book: by Maini

108.000–117.975 and 117.975–137.000 MHz, respectively. However, a clear understanding of FM is potentially useful for adjacent channel compatibility work. In addition, it is synthesized using a voltage-controlled oscillator. Also, it is more susceptible to Doppler shifts, which even though these have been shown to be very minor in the previous section, make it impractical for mobile aeronautical communications. Historically, FM gives a better quality of voice service. The FM modulation is used by entertainment systems on board aircraft where channels are FDM into time slots. Otherwise, the most common used modulation technique in aeronautical telephony is SSB.

3.3 Digital Transmission

Before providing any explanation about solutions of digital transmission will be necessary to introduce some relations:

1. **Shannon's Theory**—Shannon's theory provides a theoretical relationship between bits per second, baud, and S/N for error-free transmission presented in Table 3.2.

$$\mathrm{R\,(b/s) = M\,(number\ of\ signal\ states)\ x\ r\,(baud\ or\ signal\ rate),\ and} \tag{3.13}$$

$$\mathrm{C = B\ \log_2[(1 + S/N)]} \tag{3.14}$$

where C is channel capacity in b/s, B is bandwidth in Hz, S/N is signal-to-noise ratio, and R is the data rate and is always $<C$ for errorless transmission. If R > C, errors will occur. Theoretical relationship between M and S/N is as follows:

$$\mathrm{M = [(1 + S/N)]^{0.5}\ or\ S/N = M^2 - 1} \tag{3.15}$$

2. **Non-Errorless Transmission**—Of course, Shannon's law does not apply if it is decided to push past this theoretical limit of data throughput if errors can be tolerated. Actually, it is sometimes a prudent approach to tolerate the errored environment and to make facilities for correcting them by forward error

Table 3.2 Shannon's Law

M	2^M	Minimum S/N required	Maximum spectrum efficiency (maximum capacity) (bits/Hz)
1	2	3	1
2	4	15	2
3	8	63	4

correction (FEC) and cyclic redundancy coding (CRC); of course, FEC and CRC too are overhead and not useful data payload. However, this can be a more efficient way to operate and maximize data throughput. Alternatively, the errors can just be tolerated, and the application layer of the data system can be designed to tolerate this or to correct it. Also in a number of digital systems, errors will be present from the propagation problems already present, and these have to be tolerated. These too can be corrected by FEC or CRC checks, or other mechanisms discussed later.

Digital transmission relates to the link for which the MES is designed to produce the digital signals by PC or modem and to send them through Tx. It is possible to transmit analog signals (voice or broadcast) in digital form. Although this choice implies an increased baseband, it permits signals from diverse origins to be transmitted on the same satellite channels and the satellite link to be incorporated in ISDN, which implies the use of TDM. The digitization of analog signals implies the stages of sampling, quantization, and coding. The simple digital transmission chain includes a transmitting and receiving segment. The first unit of the transmitting segment is TDM with input signals from digital (direct) and analog (via encoder) sources and after multiplexed signals pass devices such as data encryption, channel encoding, scrambling, and digital modulation, where the digital signal is transmitted through up or down satellite links. In the receiving segment of reverse mode, incoming signal goes through devices such as demodulator, descrambling, channel decoding, data decryption, and TDM demultiplexing in its direction to different users.

Digital signals use the same principle to modulate a carrier as analog signals. The AM, FM, and PM schemes are all applicable to digital modulation; digital equivalents are ASK, FSK, and PSK; and special hybrid modulation solutions have also been developed to optimize digital modulation, such as QAM. The pioneering generation of MSS of Marisat and the first Inmarsat-A SES used analog FM for the voice circuits and digital PSK for the telex and signaling traffic. However, the current generation of Inmarsat standards for all mobile applications uses a combination of analog and digital transmission techniques and/or full digital modulation and multiplexing. In addition, newly developed MSS non-GEO systems such as Iridium, Globalstar, and others usually employ digital solutions.

3.3.1 Delta Modulation (DM)

Analog signals, such as speech and video signals, generally have a considerable amount of redundancy; namely, there is a significant correlation between successive samples. Thus, when these correlated samples are coded as in the pulse code modulation (PCM) system, the resulting digital stream contains redundant information. The redundancy in these analog signals makes it possible to predict a sample value from the preceding sample values and to transmit the difference between the actual sample value and the predicted sample value estimated from the past samples. This results in a technique called difference encoding. One of the simplest forms of this is DM, which provides a staircase approximation of the sampled version of the analog input signal. This type of modulation is a way of digitizing a voice waveform, transmitting the digits and reconstructing the original analog waveform that avoids the quantizer and the A/D and D/A converters employed in PCM. In linear delta modulation (LDM), a circuit of DM determines the difference between an incoming waveform or message signal m(t) and estimated waveform or error signal e(t), where a difference of signal error voltage is as follows:

$$\Delta m(t) = m(t) - e(t) \tag{3.16}$$

The quantizer output is a positive constant when Δm(t) is positive and vice versa. Namely, the difference between the input and the approximation is quantized into two levels, +Δ and –Δ, corresponding to a positive and a negative difference, respectively. Moreover, at any sampling instant, the approximation can be increased or decreased by step size (Δ) value, depending on whether it is below or above the analog input signal.

Furthermore, a digital output of 1 or 0 can be generated according to whether the difference is +Δ or –Δ. These pulses go to a conventional PSK digital modulator for transmission. Increasing the step size (Δ) will result in poor resolution, and increasing the sampling rate will lead to a higher digital bit rate. In LDM, the step size is fixed at a value that provides performance near the peak, while better final performance may be achieved through a scheme of adaptive delta modulation (ADM) in which the value of step size is varied during the modulation process. On the other hand, it is difficult to make comparisons between the performance of PCM and DM because the latter is continually improving. Thus, most commercial satellite links to date use PCM for digital transmission of speech, while DM has been used for military satellite applications.

The DM mode has various forms, so in one of the simplest form, only one bit is transmitted per sample just to indicate whether the amplitude of the current sample is greater or smaller than the amplitude of the immediately preceding sample.

It has extremely simple encoding and decoding processes, but then it may result in tremendous quantizing noise in case of rapidly varying signal. In Fig. 3.10 is illustrated a simple delta modulator (DM) of DM system, where m(t) is added to a reference signal with the polarity shown. However, in Fig. 3.10 is shown the DM waveform of DM system. The reference signal is integral part of the delta-modulated signal. The error signal is fed to comparator and at output of comparator is (+Δ) for $e(t) > 0$ and (−Δ) for $e(t) < 0$.

Thus, the output of DM is a series of impulses with the polarity of each impulse depending upon the signal of e(t) at the sampling instants of time. A DM signal can be demodulated by integrating the modulated signal to obtain the staircase approximation and then passing it through lower filter. The smaller the Δ value, the better is the reproduction of the message signal.

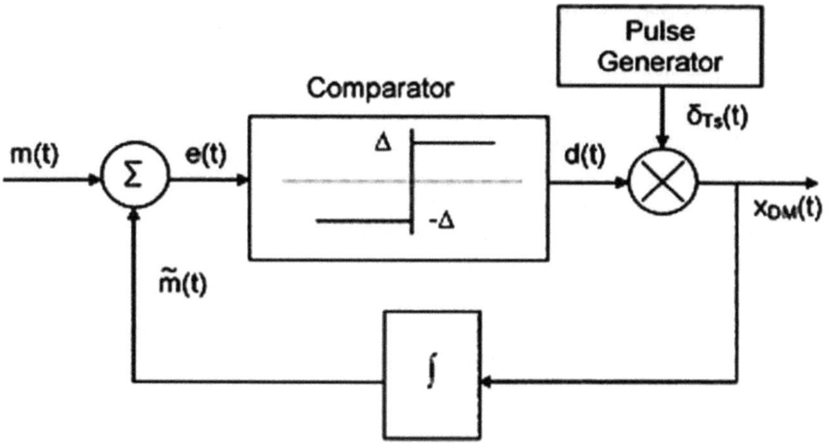

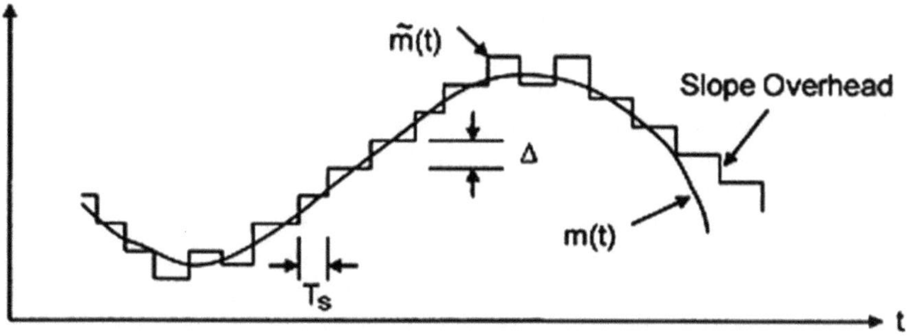

Fig. 3.10 DM and output waveform—Courtesy of Book: by Maini

3.3.2 Coded Modulation (CM)

The CM is a combination of modulation and error correction codes without degrading the power of bandwidth efficiency. Thus, using FEC, such as block and convolutional codes, the bit error performance is improved by expanding the required bandwidth. Obtaining the power efficiency requires twice the bandwidth of the original uncoded signal because of the increase in the symbol rate of modulation and complex implementation. This can be done by increasing the number of phases in PSK modulation without expanding signal bandwidth. In this case, the 8-PSK signals with 2/3 rate of convolutional code have the same bandwidth as uncoded 4-PSK. However, the bit error performance degrades by about 4 dB due to an increase in phase but will be referable if the coding gain becomes more than 4 dB. There are two practical CM in use for MSS: the trellis coded modulation (TCM) and the block coded modulation (BCM).

3.3.2.1 Trellis Coded Modulation (TCM)

The TCM scheme uses the combination of convolutional coding and expanded signal sets of 8-PSK to transmit two information bits per symbol. The modulation signals in TCM are assigned to each one- or two-satellite trellis branch, although binary code symbols are assigned in the convolutional codes. Here is a very important definition of measuring the distance between modulation signals assigned to each trellis branch. The modulation signal assignment in TCM can be designed either by the Euclidean or Hamming distances.

Initially, the modulator states are based on a trellis illustrated in Fig. 3.11 (left). This scheme is intelligent and involves more signals processing theory than the simpler prior modulation schemes. This sample works on the principle of putting states that are likely to occur simultaneously as far apart on the constellation diagram shown in Fig. 3.11 (right), with maximum distance of separation employing Euclidean distance. This sample is showing variable phase and amplitudes determining Euclidean distance as the distance between two points in constellation. Similarly, a binary sequence that is unlikely to occur can be put together on the constellation diagram with minimum Euclidean distance. Thus, this offers an effective coding gain over the simpler QAM for a given BER and data throughput would consequently require a lower SNR than QAM

The performance of TCM can be improved by increasing the number of states or by modifying the signal constellation. At this point, one solution is provided by multidimensional signals and another is performed by multiple coded modulations, known as multiple TCM (MTCM). A TCM 8-DPSK modem with a rate of 2/3 and 16 states for 4800 b/s has been implemented in the NASA MSAT-X experimental program for LMSC.

3.3.2.2 Block Coded Modulation (BCM)

Instead of convolutional coding, this scheme uses short binary block codes, which could be simpler and faster to decode. Similar to TCM, this type of modulation can improve performance by using multiple BCM (MBCM). Thus, MBCM with two symbols per branch has a coding gain of 3 dB relative to conventional BCM. The two symbols per branch MBCM has a performance for coding gains of 1.1 and

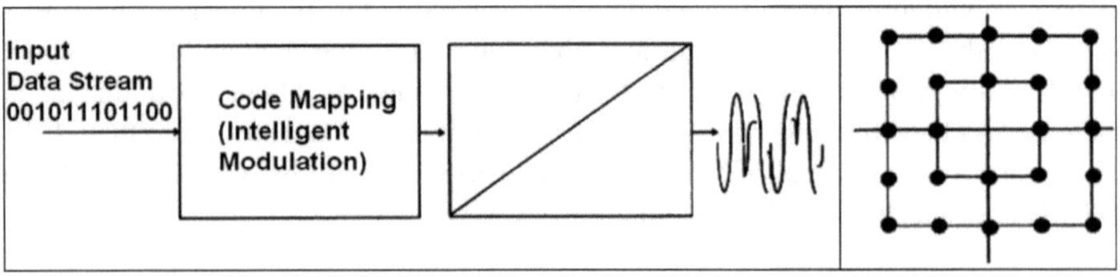

Fig. 3.11 TCM and constellation diagrams—Courtesy of Book: by Stacey

2.2 dB at BER = 10^{-3} relative to the BCM 8-PSK and the uncoded QPSK, respectively.

3.3.3 Pulse Code Modulation (PCM)

The digitization starts with the conversion of analog voice signals into a digital format. An analog can be converted into a digital signal of equal quality if the analog signal is sampled at a rate that corresponds to at least twice the signal's maximum frequency.

The peak-to-peak amplitude rate of the modulating signal in PCM is divided into a number of standard levels, which in the case of binary systems is an integral power of 2. The amplitude of the signal to be sent at any sampling instant is the nearest standard level. So, if at a particular sampling instant, the signal amplitude is 3.2 V, it will not be sent as a 3.2 V pulse, as can be the case of pulse amplitude modulation (PAM); instead, it will be sent as the digit 3, if 3 V is the nearest standard amplitude. Thus, where the signal range has been divided into 28 levels, it will be transmitted as 0000011, which coded waveform is shown in Fig. 3.12. This scheme is known as quantizing process in PCM shown in Fig. 3.12, where the number of bits for 2^n chosen standard levels per code group is n + 1. In fact, it is evident that the quantizing process distorts the signals. This distortion is referred to

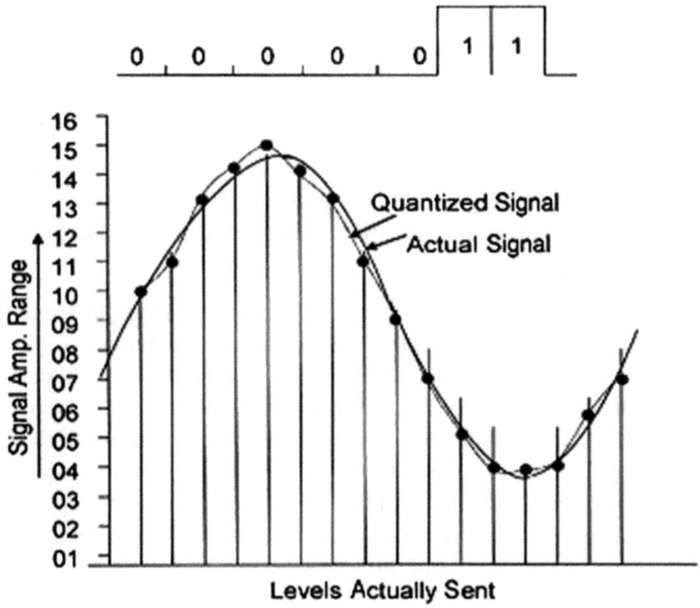

Fig. 3.12 Quantizing process in PCM. Courtesy of Book: by Maini

a quantizing noise, which is random in nature as the error in the signal's amplitude and that actually sent after quantizing is random. The maximum error can be as high as half of sampling interval, in which number of level is 16 or 1/32 of total signal amplitude range. A technique for converting an analog signal to a digital form is PCM, which requires three operations:

1. **Sampling**—This operation converts the continuous analog signal into a set of periodic pulses, the amplitudes of which represent the instantaneous amplitudes of the analog signal at the sampling instant. Thus, the process of sampling involves reading of input signals at discrete points in time. Hence, the sampled signals consist of electrical pulses, which vary in accordance with the amplitude of the input signal. In accordance with the Nyquist sampling rate, an analog signal of bandwidth B Hz must be sampled at a rate of at least 1/2B to preserve its wave shape when reconstructed.
2. **Quantizing**—This technique is the process of representing the continuous amplitude of the samples by a finite set of levels. If V quantizer levels are employed to represent the amplitude range, it take the $\log_2$ V bit to code each sample. In voice transmission, 256 quantized levels are employed; hence, each sample is coded using $\log_2 256 = 8$ bits, and thus, the digital bit rate is $8000 \times 8 = 64{,}000$ b/s.

The process of quantization introduces distortion into the signal, making the received voice signals raspy and hoarse. This type of distortion is known as quantization noise, which is only present during speech. However, when a large number of quantization steps, each of ΔS volts, are used to quantize a signal having an rms signal level S_{rms}, the signal-to-quantization noise ratio is given by the following equation:

$$S_{qn} = (S_{rms})^2/(\Delta S)^2/12 \tag{3.17}$$

A large number of bits are necessary to provide an acceptable signal-to-quantization noise ratio throughout the dynamic amplitude range. Some analysis of speech signals shows that smaller amplitude levels have a much higher probability of occurrence than high levels.

3. **Coding**—This solution protects message signals from impairment by adding redundancy to the message signal. Another important approach in digital coding of analog signals is differential PCM. This is basically a modification of DM where the difference between the analog input signals and their approximation at the sampling instant is quantized into V levels and the output of the encoder is coded into $\log_2$ V bits. In such a way, it combines the simplicity of DM and the multilevel quantizing feature of PCM mode and in many applications can provide good reproduction of analog signals comparable to PCM, with a considerable reduction in the digital bit rate.

3.3.4 Quadrature Amplitude Modulation (QAM)

Each higher PSK modulation requires a better S/N ratio performance, which is difficult to achieve without special schemes. Instead of 16 or higher PSK modulation, QAM is used, which is a combination of amplitude and phase modulation. In effect, modulation can be achieved in a similar manner to that of QPSK, by which the in-phase and quadrature carrier components are independently amplitude modulated by the incoming data streams. The incoming signals are detected at the receiver using matched filters. In terms of bandwidth, it is a highly efficient method for transmitting data flow. However, the sensitivity of the QAM method to variation in amplitude limits its applicability to satellite communication systems in practice, where nonlinear payload characteristics may distort the waveform, resulting in the reception of erroneous messages.

The QAM scheme is a combination of the phase domain. The number of possible states = 360/number of phase changes used. The functional block diagram of modulator for QAM is shown in Fig. 3.13 (left). The demodulation and detection process for QAM signals shown in Fig. 3.13 (right) becomes quite complex, but is not unachievable with modern technology. It is used frequently in high-capacity data point-to-point radio links. 64QAM, 128QAM, 256QAM, and even 518QAM are not uncommon. Obviously, these have high spectral efficiency in the amount of data per unit RF bandwidth they are able to pass. But consequently, a very good SNR is needed to retain a high BER. In Fig. 3.18d is shown a 16-QAM signal space diagram with a 16-state grid that permits both carrier amplitude and phase change. This scheme is not yet considered favorable for satellite communication.

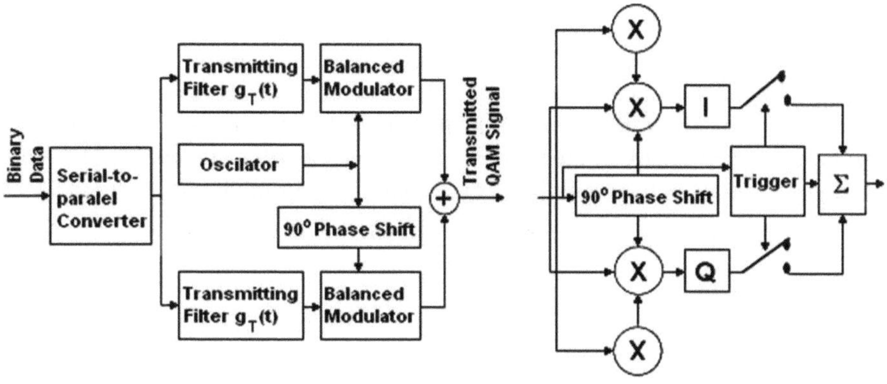

Fig. 3.13 Modulator and demodulator of QAM—Courtesy of Book: by Stacey

3.3.5 *Time Division Multiplexing (TDM)*

Satellite links normally relay many signals from many MES, but to avoid interfering with each other, it is necessary for some kind of separation or division. This separation is known as multiplexing, and its common forms are FDM (already explained) and TDM. The TDM is easier to implement with digital modulation and to form hybrid solutions applicable to all type of baseband signals.

The TDM is a time multiplexing solution where a group of various transmission signals on the same frequency at different times take turns using a channel. In this way, a group of pulses from a number of channels may be interleaved to form a single high-rate bit stream of multiplexed assembly directly modulated onto the RF carrier. Since digital signals are precisely timed and consist of short pulse groups with relatively long intervals between them, TDM is the only natural way to combine digital signals for transmission. This system has the advantage that less equipment is required than is needed to modulate each channel onto a separate carrier and that the transmission efficiency of a satellite is usually better when it is carrying a few transmissions on many channels and vice versa. Accurate timing is essential to the correct operation of digital systems. For that reason, the TDM system uses a synchronous clock, which controls the timing of all slave clocks and plesiochronous independent clocks with very good accuracy. If the transmission is without errors and breaks, in the synchronous system it would be only necessary to provide single markers at the beginning of transmission where the decoder could identify all streams of bits.

The TDM is using when there is one RF channel shared in time between the different users. If a suitable TDM repetition frequency is chosen, the user does not need to be aware of the discontinuity shown in Fig. 3.14. This is the principle deployed in modern ground cable and fiber networks to which the radio systems described in this book need to interface.

However, in any practical communication system, regular markers must be provided in the bit stream so that the decoder can extract groups of digitally representative samples, identify the channels of a TDM assembly, and resynchronize the system after errors, breaks of transmission, or drift of the clock. Therefore, when plesiochronous time multiplexes are combined to form higher order of TDM,

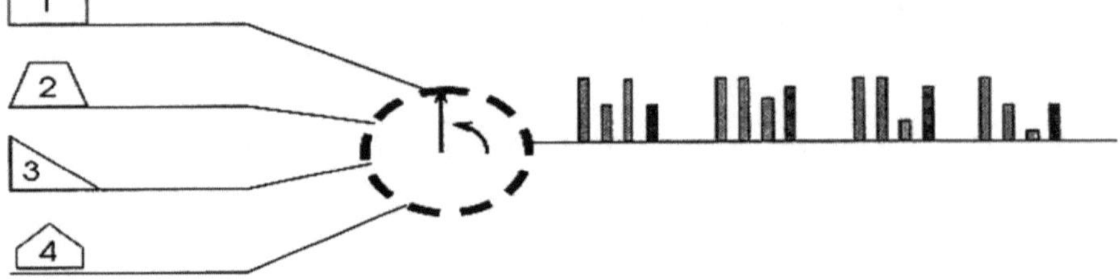

Fig. 3.14 Modulator and demodulator of QAM—Courtesy of Book: by Stacey

it is necessary, in order to preserve synchronism over the long term, to allow for the occasional addition of dummy or padding bits.

3.3.6 Types of Digital Shift Keying

Digital signals can be used to modulate the amplitude, frequency, and phase, and therefore, the solutions of shift keying available for digital modulation are amplitude-shift keying (ASK), frequency-shift keying (FSK), and minimum-shift keying (MSK) as applications of FSK and phase-shift keying (PSK).

In terms of performance, ASK and FSK, both illustrated in Fig. 3.15a, b, respectively, require twice as much power to attain the same BER performance as PSK, shown in Fig. 3.15c. At the top of the same figure, a stream of a digital signal with 1 and 0 binary state is presented. Consequently, the vast majority of MSS employs a method of phase modulation known as PSK.

3.3.6.1 Amplitude-Shift Keying (ASK)

This scheme can be accomplished simply by the on–off gating of a continuous carrier. The simplest ASK technique is to represent one binary level (binary 1) by a single signal of fixed amplitude and the other level (binary 0) by switching off the signal. The absence of the signal for one of the binary levels has the disadvantage that if fault conditions exist, it could be misinterpreted as received data. Waveform for ASK, using different amplitude signals for the logic levels, are an alternative method to prevent this disadvantage. As with speech telephony circuits, the upper sideband and carrier may be suppressed to reduce the bandwidth requirement and concentrate the available power on the signal containing the information.

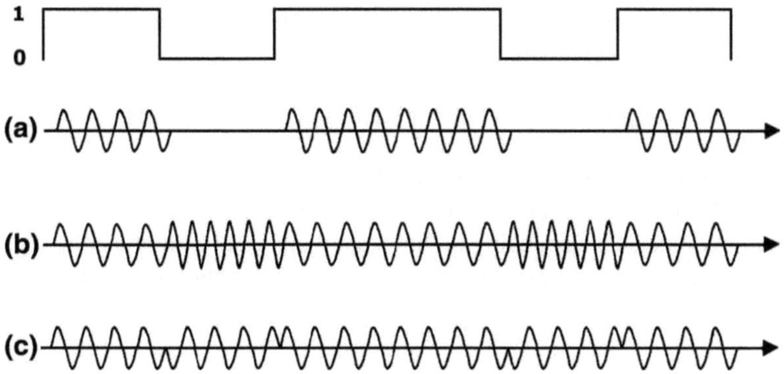

Fig. 3.15 Comparison of: **a** ASK, **b** FSK, and **c** PSK—Courtesy of Book: by Sheriff

3.3.6.2 Frequency-Shift Keying (FSK)

This solution may be used whereby the carrier frequency has one value for a 1 bit and another for a 0 bit. The main difficulty in the use of this FM technique is that the gap between the frequencies used must be increased as the modulation rate increases.

More exactly, for a restricted channel bandwidth, especially using in-band supervisory signaling, there is a limit to the maximum bit rate that is possible with this technique.

3.3.6.3 Minimum-Shift Keying (MSK)

The MSK is a binary form of continuous-phase frequency-shift keying (CPFSK), where the frequency deviation (Δf) from the carrier is set at half the reciprocal data rate of 1/2T. The MSK scheme may also be viewed as a special form of offset QPSK, consisting of two sinusoidal envelope carriers, employing modulation at half the bit rate. It is for this reason that the MSK demodulator is usually a coherent quadrature detector, similar to that for QPSK. The error rate performance is the same as that of BPSK and QPSK. Similarly, differentially encoded data has the same error performance as D-PSK. This solution can also be received as an FSK signal using coherent or non-coherent methods; however, this will degrade the performance of the link. At any rate, the side lobes of MSK are usually suppressed, using Gaussian filters and the modulation method scheme adopted by GSM cellular systems.

3.3.6.4 Phase-Shift Keying (PSK)

The PSK scheme is a technique using a multistate signaling stream in which the rate of data transmission can be increased without having to increase the bandwidth. In this shift keying system, the phase of the carrier changes in accordance with the baseband digital stream or information content. Hence, a general form of a PSK scheme is given by expression:

$$s_m(t) = A\cos(\omega t + \phi) \text{ and } \phi = (2m + 1)\pi/M \tag{3.18}$$

where A = amplitude; ω = frequency angle; ϕ = phase angle varied in accordance with the information signal; m = integer in the range from 0 to (M – 1); and M = number of states. Depending on how many bits can be combined in a group of information as a symbol, there are a number of combination possibilities for PSK digital carriers.

3.3.7 *Combinations of PSK Digital Carriers*

There are several types of hybrid solutions used for the combination of PSK digital carriers. The PSK family is most popular for satellite communications and especially for MSC. A real-valued band bass signal s(t), which has a common form for these types of PSK methods, is expressed as follows:

$$s(t) = A(t)\cos[2\pi f_c t + \phi(t)] = A(t)\cos[2\omega_c t + \phi(t)]\,[V] \tag{3.19}$$

where A (t) = amplitude; f_c = carrier frequency of signal s(t); ϕ (t) = phase angle; and ω_c = frequency angle varied in accordance with the signal s(t). Including information waveform m(t), the previous relation can be determined as follows:

$$s(t) = \mathrm{Re}\,[m(t)]\,e^{j2\pi fct} \tag{3.20}$$

In linear modulation methods such as pulse, PAM and PSK can be expressed by:

$$m(t) = A(t)\exp[j\phi(t)] = \sum_{n=0}^{\infty} a_n p(t)\,(t-nT) \tag{3.21}$$

where Re = real part of the complex in the next bracket; a_n = information carrying symbols; p(t) = signal pulse; and T = time interval of symbol.

3.3.7.1 Binary PSK (BPSK)

The simplest form of PSK is binary PSK (BPSK), where the digital information modulates a sinusoidal carrier. For a general case of M-ary PSK (number of states) in BPSK, M = 2, so the baseband bit rate and the symbol rates are the same. Binary data is expressed by $a_n = \exp(j\phi_n)$ for $\phi_n = 0$ or π, and the phase changes every data bit of duration period T_b. When p(t) is a rectangular pulse over symbol duration $T = T_b$, the BPSK signal is expressed by the following equation:

$$s(t) = Aa_n \cos 2\pi f_c t = A\cos(2\omega_c t + \phi_n) \text{ for } nT \le t < (n+1)T \tag{3.22}$$

In Fig. 3.16a, a diagram is illustrated of how, theoretically, the phase of the carrier changes instantaneously by 180° when the baseband signal switches from 0 to 1.

In addition, Fig. 3.18a represents the two states of the carrier by two vectors with a phase difference of 180°. In this sense, ϕ = 0° and ϕ = 180° when the baseband signal is 0 and 1, respectively, and A = + V and A = –V when the baseband signal is 0 and 1, respectively.

Except for standard-A, the TDM/BPSK scheme is used by almost all Inmarsat standards for process: Forwarding signaling/assignment channels, return request

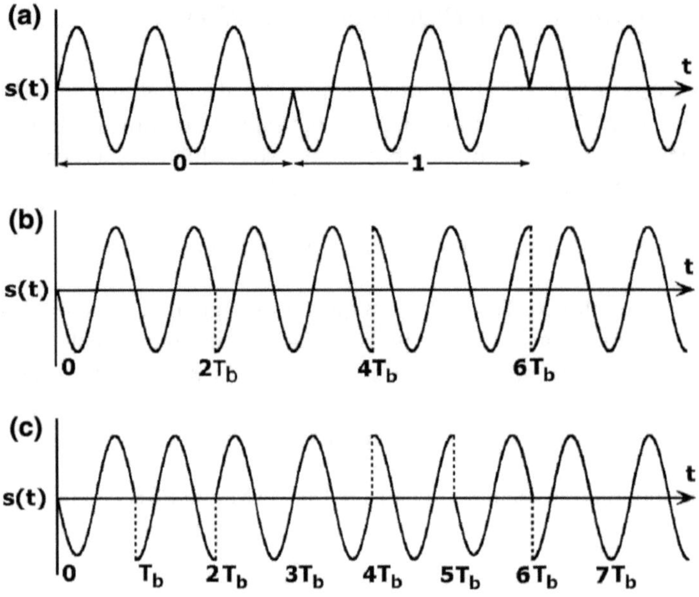

Fig. 3.16 Hybrid PSK modulations: **a** BPSK, **b** QPSK, and **c** O-QPSK—Courtesy of Book: by Ohmori

channels use Aloha BPSK in Inmarsat standard-A, standard-C and Aero, while Slotted Aloha BPSK is dedicated for Inmarsat standard-M. On the other hand, Tlx message channel modulation in the forward direction uses TDM BPSK for Inmarsat standard-A and standard-B and in the return link serves only for Inmarsat standard-A. In the similar manner, Inmarsat standard-C uses the BPSK (1/2-FEC) scheme in both directions.

3.3.7.2 Quadrature PSK (QPSK)

For M-ary, PSK is selected from M signals such as exp [j2π (m–1)/M], where m = 1, 2, …, M, and the resulting signal s(t) is written as follows:

$$s(t) = A \cos[2\pi f_c t + 2\pi/M(m-1)] \text{ for } m = 1, 2, \ldots, M \tag{3.23}$$

A slightly more complex form PSK is QPSK or 4-PSK, for which ϕ_n is a set of 0, π/2, π, and 3/2π. Then, the signal s(t) is given by:

$$s(t) = A/\sqrt{2}\,(a_n^I \cos(2\pi f_c t + \pi/4) + A/\sqrt{2}\,(a_n^Q \cos(2\pi f_c t + \pi/4) \tag{3.24}$$

where a_n^I and a_n^Q are the ±1 value data, which are converted from input data sequence a_n into the in-phase channel (I channel) and quadrature channel

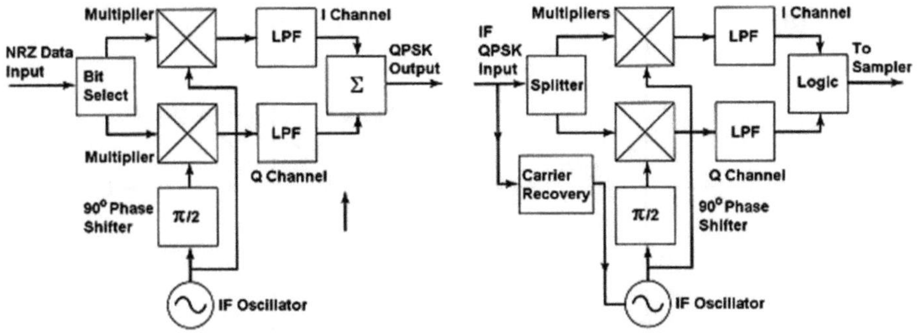

Fig. 3.17 QPSK modulator and demodulator—Courtesy of Book: by Pratt

(Q channel), respectively. In such a way, the relation (a_n^I, a_n^Q) is (1, 1) for $\phi_n = 0$; (1, −1) for $\phi_n = \pi/2$; (−1, −1) for $\phi_n = \pi$; and (−1, 1) for $\phi_n = 3\pi/2$.

In Fig. 3.16b, the shape of QPSK-modulated signals is presented and a QPSK scheme in the I-Q plane is shown in Fig. 3.18b.

Only two binary digits are needed to describe four possible states as follows: The ϕ values of 45°, 135°, 225°, and 315° correspond to 00, 01, 11, and 10, respectively. In fact, each state of the signal carries two bits of information. Thus, a combination of two bits or more, which corresponds to a discrete state of a signal, is called a symbol, in the case of QPSK; the symbol rate is half the bit rate. Namely, a bit rate of 120 Mb/s corresponds, with QPSK, to a symbol rate of 60 Mb/s or Mbauds.

The QPSK schemes of modulators are basically dual-channel BPSK schemes shown in Fig. 3.17 (left) and (right), respectively. At this point, one channel processes the u_i (I) bits and uses the reference carrier, and the other processes the u_q (Q) bits using a 90° phase-shifted version of the reference. The demodulation signals are passing out via low-power filter (LPF) and logic circuit. The bits values u_i and u_q are selected alternately from the input bit stream. For example, u_i may represent the odd numbers bits and u_q the even. In this case, one binary data channel enters the QPSK modulator, and the outgoing symbol rate is equal to half of the incoming bit rate. The octal phase modulation known as 8-phase PSK is a constant amplitude scheme with a higher bandwidth efficiency of 3 b/s/Hz, shown in Fig. 3.18c. The demands of high bit-rate applications are related to images, TV, and high-definition television (HDTV) transmission.

3.3.7.3 Offset QPSK (O-QPSK)

The O-QPSK scheme delays the quadrature bit stream by T sec relative to the in-phase bit stream to restrict the phase transition to phase changes of 0 or $\pi/2$ every T sec. Using a_n^I and a_n^Q, the equivalent low-pass and resulting signals are expressed by:

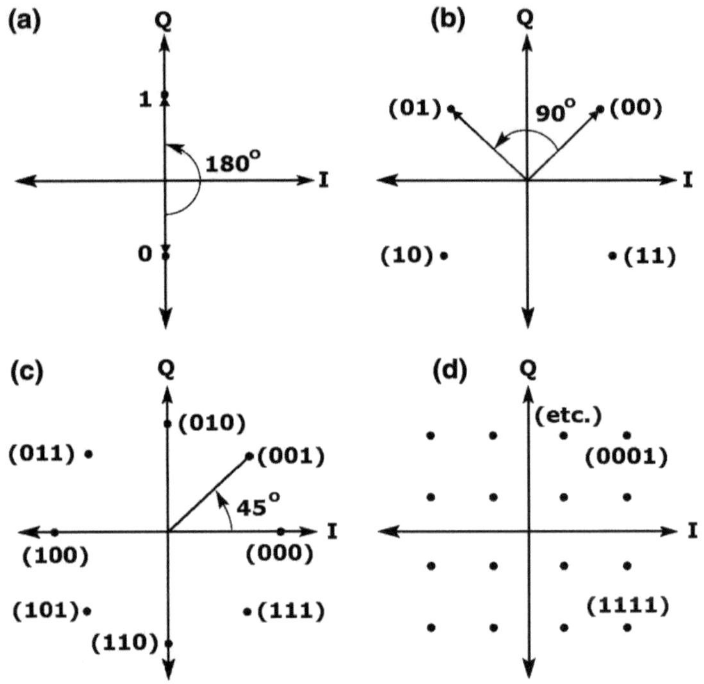

Fig. 3.18 Representation of PSK signals in I-Q plane—Courtesy of Book: by Richharia. **a** BPSK. **b** QPSK. **c** 8-phase PSK. **d** 16-QAM

$$u(t) = \sum_{n=0}^{\infty} a_n^I p(t)\,(t-2nT) - j a_n^Q p(t)\,(t-2nT-T)\, e^{j\pi/4} \tag{3.25}$$

$$s(t) = \left[\sum_n a_n^I p(t)\,(t-2nT)\right] \cos\left(2\pi f_c t + \pi/4\right) + \left[\sum_n a_n^Q p(t)\,(t-2nT-T)\right] \sin\left(2\pi f_c t + \pi/4\right)$$

The data transmission occurs in the conventional QPSK at the same time in both I and Q channels. This scheme has larger phase changes than O-QPSK, which has phase changes of at most $\pm\pi/2$ data transmission, and large envelope fluctuations do not occur as they do with Π-phase changes in QPSK. Figure 3.16c shows the O-QPSK-modulated signal's shape. The Aloha O-QPSK (1/2-FEC) serves in return request channels by a transmission speed of 24 Kb/s for Inmarsat-B and 600 b/s for Aero standards. The Inmarsat-B and Aero standards use 16 Kb/s or 9.6 Kb/s, respectively, of the APC O-QPSK scheme in voice channel coding/modulation, while Inmarsat standard-M uses 4.8 Kb/s IMBE in coding O-QPSK. The Inmarsat-B in Tlx return link uses the TDMA O-QPSK (1/2-FEC) modulation scheme.

3.3.7.4 Differential PSK (DPSK)

Channel conditions in MSC are more severe than additive white Gaussian noise channels due to multipath fading, shadowing, and Doppler effects. These problems can be solved by using differentially coherent detection of encoded signals. The received signal phase is generally an 180° ambiguity sign, which cannot be resolved unless some known reference signal is transmitted and a comparison is made. In the worst case, if such case left unresolved, the received signal could end up being the complement of the transmitted signal.

When the phase condition is constant for 2T seconds, the DPSK demodulator can obtain the optimum a posteriori probability. The DPSK scheme can also be used to remove sign ambiguity at the receiver. Differentially encoding data prior to modulation occurs when a binary 1 is used to indicate that the current message bit and prior code bit are of the same polarity and 0 to represent the condition when the two pulses are of opposite polarity.

The equivalent low-pass signal in the interval of m is given by:

$$v_m = \alpha e^{j(\phi_m - \phi)} u_m(t) + n_m(t) \tag{3.26}$$

where α = loss factor; ϕ_m = phase difference in m interval; $u_m(t)$ = signal in m interval; and $n_m(t)$ = Gaussian noise in m interval. The non-coherent detection technique is useful when the carrier phase is difficult to estimate at the receiver. In MSC channels, the transmitting oscillator that generates the carrier cannot be completely stabilized and an MSC link has a low S/N ratio. Besides, for coherent detection, the bandwidth of the carrier-tracking loop at the Rx must be decreased in the proper manner to increase the S/N and obtain a good phase reference.

Since propagation characteristics in MSC such as multipath fading and the Doppler effect cause rapid phase variation, the carrier phase of the Rx signal does not remain fixed long enough to be estimated. The problem can be solved with non-coherent detection, which is equivalent low-pass signal in relation to random phase value (ϕ_n), and Gaussian noise is as follows:

$$v_m = \alpha e^{-j\phi_n} u_n(t) + n(t) \tag{3.27}$$

Therefore, noncoherent detection results in the degradation of the bit error performance with respect to coherent detection. In this case, the BER probability (P_b) for binary DPSK is determined with the following relation:

$$P_b = 1/2 \exp(-\gamma_b) \tag{3.28}$$

where $\gamma_b = E_b/N_0$ is the value of the S/N density ratio per bit signal. In this sense, when considering the use of DPSK schemes, a trade-off between simplified receiver complexities against reduced performance characteristics in the presence of noise, particularly when employing higher order modulation techniques, needs to be made.

3.3.7.5 π/4-QPSK

Recently, π/4-QPSK, or π/4 shift QPSK, has become very popular for MSC as well as for cellular systems because it has a compact spectrum with small spectrum restoration due to nonlinear amplification and can perform differential detection.

Thus, the phase point of this scheme always shifts its phase over successive time intervals by $\pm\pi/4$ or $\pm 3\pi/4$. Therefore, the spectrum of this scheme is the same as that of a QPSK that undergoes an instantaneous $\pm\pi/24$ or $\pm\pi$ phase transition. In such a way, the π/4-QPSK signal can reduce the envelope fluctuation due to band-limited filtering or nonlinear amplification more than can QPSK. In a more general sense, this is because differential detection can be used since π/4-QPSK is not an offset scheme.

For MSC channels, a strong line-of-sight signal can be expected even in the Rician fading channel, so coherent demodulation is desirable for improved power efficiency. The bit error performance of coherently detected π/4-QPSK is the same as that of QPSK. On the other hand, differential detection is also desirable for simple hardware implementation. The bit error probability for differential coherer detected π/4-QPSK is given by:

$$P_b = e^{-2\gamma b} \sum_{k=0}^{\infty} (\sqrt{2}-1)^k I_k(\sqrt{2}\gamma_b) - 1/2\, I_0(\sqrt{2}\gamma_b)\, e^{-2\gamma b} \tag{3.29}$$

where $\gamma_b = E_b/N_0$ and I_k = modified Bessel function of the first kind in k interval order. The differential detection is about 2–3 dB inferior to the coherent detection in AWGN and fading channels, and non-coherent detection is also applicable. Because of these advantages, π/4-QPSK has been chosen as the standard modulation technique for several systems of cellular and satellite mobile communications. Some experimental results show that in fully saturated amplifier systems, π/4-QPSK still has significant spectral restoration. Thus, to reduce this restoration, π/4-controlled transition QPSK (CTPSK) uses both sinusoidal shaping pulses and timing offsets of the phase transition between I and Q channels.

3.3.8 Digital Voice Coding

The MSC transmission applications need the utilization of digital voice (speech) coding technique, which require high-quality speech coding at bit rates as low as possible, and that digital voice coding techniques can be classified in groups of waveform and parametric coding. The waveform coding group is achieved by matching as closely as possible the waveform of the original and the reconstructed signals, while the parametric coding group represents the speech signal using a model for speech production, such as time-varying linear predictive filters and transmission parameters of the filter only. At the decoder, the speech signal is reconstructed by the inverse prediction filters, using received parameters. Several hybrid-coding methods are proposed as a combination of waveform and parametric coding. In the hybrid coding, both the filter parameters and the quantized residual sample, which is the signal that remains after filtering the speech signal, are transmitted.

3.3.8.1 Adaptive Predictive Coding (APC)

The APC digital technique is based on an initial estimation and prediction of the input speech (voice) waveform samples, plus a residual (error) signal derived by comparing the estimated samples with the actual speech input. This initial estimation and residual signal are quantized and transmitted to the receiver, where an inverse quantized residual signal is fed into a synthesized filter in order to reconstruct the speech signal.

Coding efficiency is achieved by removing the waveform redundancy in the input signals. Furthermore, a quantizer operating only on the residual signal requires fewer bits per sample. The effect of quantization is to distort the reconstructed output speech signals. Maximum likelihood quantization (MLQ) techniques are used to minimize quantization noise. On the other hand, an adaptive quantization scheme with a noise-shaping filter is used to provide the subjective performance. A combination of short-term and long-term predictors reduces the variance of the residual signal and the subsequent quantization error.

The Inmarsat digital MSS utilizes the APC/O-QPSK speech (voice) coding technique, in which major parameters are systematized within the framework of the APC system for both standard-B and Aero of 16 and 9.6 Kb/s, respectively. The typical APC digital codec configuration consists of transmitting and receiving sides, both including two main components. The Tx side is an APC coder containing an inverse filter with predictor of negative predisposition and quantizer (Q), which produces output signals, while the Rx side is an APC decoder containing a synthesis filter with an inverse quantizer (Q^{-1}), which takes input signals and then forwards them in the predictor with positive predisposition.

3.3.8.2 IMBE Coding

After extensive testing in trials during which seven candidates' voice codecs (coder and decoder) were evaluated, a codec manufactured by Digital Voice Systems Incorporated (DVSI) was selected: the improved multiband excitation (IMBE) model of voice coding algorithm. Namely, it was judged to provide reasonable speech (voice) quality and speaker recognition; excellent intelligibility; acceptable tolerance to burst and random errors up to 4 % error rate; good performance in the presence of background acoustic and landline noise; relative immunity to the effect of nonlinear microphones; and relatively straightforward implementation.

In use, the IMBE codec speech is sampled and digitized, and then, fast Fourier transform is performed to determine a smoothed frequency spectrum (narrow bandwidth). Thus, where the detected energy is above a predetermined threshold, the harmonic is declared voiced and where the energy is declared unvoiced; then, the amplitudes are measured in both cases. The smoothed spectral envelope, the voice/unvoiced status of each harmonic, and the amplitudes of voice/unvoiced harmonics are transmitted to the speech synthesizer, located at the receiver. Accordingly, at the receiver, the pitch period and voiced harmonics are reconstructed in the speech synthesizer, using sinusoidal oscillators whose amplitudes are controlled by voiced amplitude information from the transmitter. Besides, the unvoiced harmonics are simulated by band-limited noise from a noise source in the receiver, the amplitude of the noise at each harmonic being controlled by the unvoiced amplitude information transmitted over the communication channel. The sampled smoothed spectrum information is multiplied with the synthesized harmonic spectrum, and an inverse fast Fourier transform performed to construct a digitized and synthesized voice signal.

At any rate, by employing speech analyses and coding at the transmitter, together with decoding and speech syntheses in the receiver, a large reduction in bandwidth and hence the transmitted bit rate can be achieved. The HF speech components no longer have to be sent over the communication channel but can be regenerated in the synthesizer, which is the only mathematical model that includes frequency sources and modulators, thereby reducing occupied channel bandwidth. More exactly, to reproduce the speaker's voice, it is only necessary to send the receiver a coded set of instructions, which instruct the speech synthesizer how to: **1**. simulate the distinguishing characteristics of the speaker's voice, namely to set the pitch frequency and selected harmonics, and **2**. modulate these to reproduce the speaker's message.

Both types of information change only slowly during speech and can be transmitted in a narrow bandwidth. Inmarsat-M uses digital voice coding at 6.4 Kb/s speech codec rates, including error correction coding to protect the most critical speech syntheses parameters. The coding rate has been optimized to preserve voice (speech) quality in a typical mobile environment where high background acoustic noise levels, multipath distortion (maritime), and blockage of the line-of-sight path to the satellite (land) are often encountered.

3.3.8.3 Adaptive Differential PCM (ADPCM)

The PCM scheme is the simplest method of waveform coding, and the voice signal s(t) quantized to one of many (2^k) amplitude levels, however, where (k) is the number of binary digits only to represent each sample.

When the sampling rate is chosen to be several times the Nyquist rate, the adjacent samples become highly correlated and the signal does not change rapidly from sample to sample. At this point, using this correlation between adjacent speech samples, differential PCM (DPCM) quantizes the differences between the input sample and the predicted value, which is estimated by a linear predictor. Since such differences are smaller than the sampled amplitude themselves, fewer bits are required to represent the Tel signal.

The DPCM scheme with fixed predictors can provide from 4 to 11 dB improvement of gain over direct quantized PCM. The ADPCM method uses both adaptive quantization and adaptive prediction to reduce coding errors.

In this sense, the typical configuration of an ADPCM encoder is composed of sample, quantizer, and predictor, providing feed-forward quantization where the quantizer step size is proportional to the variance of the input of the quantizer. This type of ADPCM provides about 10–11 dB improvement of S/N ratio over PCM. Several adaptation algorithms have been proposed for this scheme. Since ADPCM uses a scalar quantization, it is difficult to reduce the bit rate to less than 8 Kb/s.

3.3.8.4 Linear Predictive Coding (LPC)

The LPC method synthesizes the speech using a linear predictive filter, which is excited by appropriate signals such as an impulse sequence for voiced speech or random noise for unvoiced speech. The LPC codec transmits only the parameters of the linear predictive filter and the index of its selected excitation signals. The basic configuration of LPC codec is composed of excitation generator, long-term predictor, short-term predictor, and weighting filter. The input speech signals come into the weighting filter and are assumed to be produced by an all-pole filter and returned in the minimization procedure.

The short-term predictor removes the redundancy in the speech signal by the predicted value, using the past (p) sample. However, some periodicity, which is related to the pitch period of the original signal in the 50–400 Hz pitch frequency, still remains. This residual signal is removed by the long-term predictor (pitch predictor) and turns into a noise-link signal. The pitch predictor is not essential for medium bit-rate LPC codecs, although it can improve their performance, but it is very essential for low bit-rate codecs such as CELP, which use the excitation signal modeled by a random Gaussian noise process.

3.3.8.5 Multipulse Excited LPC (MELPC)

The configuration of the MELPC encoder is the same as CELP, the only difference being the excitation generator. In the MELPC mode, the excitation generator produces a sequence of pulses located at nonuniformly spaced intervals with different amplitudes. Both the amplitudes and positions of these pulses are determined using a closed-loop analysis-by-synthesis method. Thus, the synthesized signal is reconstructed using a sequence of pulses produced by the excitation generator and the long-term and short-term predictor.

While MELPC assumes that both amplitudes and positions of excitation pulses are initially unknown, the regular pulse-excited LPC assumes that the pulses are regularly spaced, but the amplitudes are unknown. Using a suitable error criterion, the error between the original and the synthesized signal is minimized. Two types of codecs need similar bit rates for the same speech quality, because MELPC needs less excitation pulses due to the optimization of the pulse position and needs both the transmission of pulse positions and its amplitudes.

3.3.8.6 Code-Excited Linear Prediction (CELP)

The LPC and MELPC can produce good voice quality at bit rates as low as 9.6 Kb/s, but they cannot maintain their quality below this rate, because they have to expend a large number of bits for encoding the excitation pulse. Besides, as excitation signals, the CELP voice coder uses the collection of code vectors, which are previously produced using vector quantization techniques based on Gaussian processes and are stored as a large codebook.

Using each code vector and the predictors, synthesized speech is produced and then the most suitable code vector that produces the lowest error between the original and the reconstructed signal is selected. The index is assigned to the code vector and the voice gain, and the values of the parameters for short- and long-term predictors are transmitted.

3.4 Channel Coding and Decoding

Voice, video, data, and telex information are transmitted in digital form through a channel that can cause degradation of these transmission signals. The noise, interference, fading, and other obstacle factors experienced during transmission could increase the probability of bit error at the receiver. Differently to say, the data signal may be encoded in such a way as to reduce the likelihood of bit error. Anyway, the coding process uses redundant bits, which contain no information to assist in the detection and correction of errors. The subject of coding emerged following the fundamental concepts of information theory laid down by Shannon in 1948, which is the relationship between communication channel and the rate at which information can be transmitted over it. Basically, the theorems laying down the fundamental limits on the amount of information flow through a channel are given.

3.4.1 Channel Processing

Channel processing is composed of special activities, which can improve the transmission techniques throughout satellite channels in connection with gain, errors, noise, interference, concentration, and authenticity.

3.4.1.1 Digital Speech Concentration and Channel Multiplication

The system for digital speech concentration (interpolation) uses the activity factor of telephone channels in order to reduce the number of satellite channels required to transmit a given number of terrestrial channels. The digital speech interpolation (DSI) technique is based on the fact that in a normal telephone conversation, each participant monopolizes the circuit for only around half the time. As the silence between syllables, words and phrases increases so does the unoccupied time. Hence, on average, the activity time of a circuit is from 35 to 40 % of the connection time.

By making use of the actual activity of the channels, several users can be permitted to share the same telephone circuit. Certain numbers of terrestrial satellite channels require only half the satellite channels and the gain is about 2. By adding a low-rate encoder to the digital speech concentrator, the gain can be further increased. For example, with encoding at 32 Kb/s, a gain increases by a factor of 2 can be obtained in voice channels used alternately for speech or data transmission. The theoretical DSI gain is defined by the ratio between the actual number of speakers (input trunks) and the number of transmission channels (bearers) required to service them.

On the other hand, the function of the digital circuit multiplication (DCM) equipment is to concentrate a number of input digital lines (trunks) onto a smaller number of digital output channels (bearers), thereby achieving a higher digital efficiency of the link or channels. This technique is qualified by the circuit multiplication gain, which is defined as the ratio of the input channels number over the number of DCM output channels. It is used in digital circuit multiplication equipment of the Intelsat/Eutelsat system.

3.4.1.2 Channel Encoding

The two fundamental problems related to reliable transmission of information via channels were identified by C.E. Shannon as follows:

(1) The use of minimal numbers of bits to represent the information given by a source in accordance with a fidelity criterion. In reality, this issue is usually identified as a problem of inefficient MSC, to which the source coding provides most practical solutions.
(2) The recovery as exactly as possible of the information after its transmission through a communication channel in the presence of noise and other interference. This is a problem of unreliable MSC, to which channel (error) coding is the basic solution.

In such a way, Shannon proved that by proper encoding, these two objectives can always be achieved, provided that the transmission rate (R_b) verifies the fundamental expression $H < R_b < C$, where H = source entropy and C = channel capacity. The BER of a digital system may be improved either by increasing E_b/N_0 or by detecting and correcting some of the errors in the received data. Therefore, for the additive white Gaussian noise (AWGN) channel, the Shannon Hartley law states that capacity of a channel is given by the following relation:

$$C = B\ \log(1 + C/N)\ [b/s] \tag{3.30}$$

where B = channel bandwidth in Hz and S/N = signal-to-noise ratio at the receiver. Thus, the channel capacity is the measure rate of the maximum information quantity that two parties can communicate without error via a probabilistically modeled channel. Namely, this chain of channels is composed of information data on input rate (R_b), channel encoder with redundancy data (r) and encoded data symbols on output rate (R_c). A reverse channel model contains input encoded data symbols, channel decoder, and output information data symbols. According to Shannon, if information is provided at rate R, which is less than the capacity of the channel, then a means of coding can be applied such that the probability of error of the received signal is arbitrarily small. If this rate is greater than the channel capacity, then it will not be possible to improve the link quality by means of coding techniques. Indeed, its application could have a detrimental effect on the link. Rearranging the above equation in terms of energy-per-bit and information rate, where the information rate is equal to the channel capacity, results in the following:

$$C/B = \log\left(1 + E_b\, C/N_0 B\right) \tag{3.31}$$

where E_b = information bit rate; E_b = related to the carrier power and the information bit rate; and C/N_0 = carrier-to-noise density ratio. Moreover, the above expression can be utilized to derive the Shannon limit, the minimum value of Eb/N_0 below which there can be no error-free transmission of information. As C/B tends to zero, this can be shown to be equal to –1.59 dB ($1/\log_2 e$). The code rate and input rate can be defined as follows:

$$c = n/n + r \quad \text{and} \quad R_c = R_b/n \quad [b/s] \tag{3.32}$$

The capacity of the channel is independent of the coding/modulation scheme used. Hence, Shannon's channel coding theorem exactly stated that, for a given carrier-to-noise ratio, the error probability could be made as small as desirable, provided that the information rate (R_b) is less than the capacity (C) and a suitable coding is used. In MSC systems, channel coding is especially interesting because of the severe power, bandwidth, and propagation limitations. Moreover, the considerable progresses in multiple access modulation schemes, resource assignment algorithms, signal-processing techniques, and advanced error-control coding provide the most efficient means to realize highly reliable information transmission.

3.4.1.3 Digital Compression

Digital transmission in general uses compression techniques for data and video signals. The effective data transfer via the Inmarsat MSC system can be significantly increased by using data compression software. Essential results were provided on PC by the PKZIP/PKUNZIP program developed by US-based PKWARE, which in a fraction of a second gives a 2–3 times reduction in size of ASCII files and 1.5 times for many types of binary files. The ARJ compression software from Robert K. Jung is slower but more effective than PKZIP. It can also be recommended for the compression of data files containing graphic information.

Thus, the real-time data compression incorporated into the most advanced modems can also increase the effective data rate of ASCII files transmission, but for transmission of already compressed files with information, it is better not to use the compression in the modem.

Their use in compressed video systems, where a TV receive only (TVRO) can also receive many channels of video from one transponder (about 6–8), has become very widespread, first in the USA and then in Europe. The compression system that has now become standard refers to a Moving Picture Experts Group (MPEG), formed under the auspices of the ISO and the International Electrotechnical Commission (IEC). In such a system, a number of digitized videos are combined into a single bit stream in a source coder. That bit stream is then sent to a channel coder for FEC and then to a QPSK modulator, an upconverter, amplifier, and an antenna for uplinking to the satellite transponder. Since only one signal is present in the transponder at any on time, there is no need for back off and full transponder power is used.

At the reverse side, the compressed video downlink comprises a line in the chain of an antenna, tuner, downconverter and QPSK demodulator, FEC detector, demultiplexer, and MPEG decoder. The MPEG is determined to provide standard compression that allows video and accompanying audio signals to be compressed in channel width. The packetizer function is to enter a suitable code in the bit stream for the individual digitized TV program so that it can be separated in the receiving chain, allowing the enabled user to select the desired program. The BER is determined from the (E_b/N_0) obtained for a combination of whatever transponder EIRP, FEC coding, transmission symbol rate, and receiver system are used. If a plot of that FEC system is not available, then the Viterbi mode FEC coding performance could be used for a good estimate of results. This type of compression has effects such as: MPEG-2 compression results in the removal of most audio and video redundancy; the FES utilization scheme results in a rapid BER increase; and the resultant (E_b/N_0) should be high enough to achieve a BER of 10^{-6} for a TVRO.

3.4.1.4 Voice Encryption

Encryption is used when it is wished to prevent exploitation or tampering with transmitted messages or voice conversation by unauthorized users, in the form of algorithmic operation in real time, bit-by-bit, on the binary stream. Thus, the set of parameters, which defines the transformation, is called a key. Its use is often associated with military communications, but commercial satellite systems are increasingly induced by customers to propose encrypted links, particularly for administrative and government sectors.

In fact, due to the extended coverage of satellite networks and the easy access to them by small MES, eavesdropping and message falsification are potentially within the reach of a large number of agents of modest means. The encryption transmission chain is composed of an encryption unit with plain text input, satellite channel with intruder and key distribution for retransmission of cipher (encrypted) text, and de-encryption with a unit for production of plain output text. The encryption and de-encryption units operate with a key provided by the key generation unit. Acquisition of a common key implies a secure method of key distribution. This key is entered into the encryption unit through a key injector about the size of a matchbox. Without this key, a potential eavesdropper attempting to listen in on the conversation would hear nothing but a noise made up of digital signals.

The technique used for most voice encryption consists of speech compressing and digitizing, using a very complicated coding process. In such a way, the voice signal is sliced into small bits, which are processed by an algorithm into bits of voice with a very complex structure. At the other end of the process, using the same key pattern, the voice is reproduced as it was before the encryption.

A typical example of voice encryption for MSC is Satsec A1 for secure voice transmission, of Inmarsat standard-A MES. This unit is housed in a modern smart telephone, which uses a very sophisticated Swiss encryption technique. The Satsec A1 features include full digital voice and facsimile encryption and LPC voice compression, using full-duplex operations CCITT V.22bis and V.27 Modem with a rate of 2400 b/s, giving business users a security level comparable to that used by government agencies. In case of transmission only in half-duplex mode, the unit automatically falls back to the built-in VOX-controlled quasi-duplex operational mode. Hence, this device should not be confused with a voice scrambler, even a digital one, which is no longer a competitive alternative to high-level encryption. The unit digitizer uses linear predictive coding and has, unlike self-synchronizing stream ciphers, no bit error multiplication. In a more general sense, under the same circumstances, this device may even have an enhanced effect on the channel that is quality of transmission.

The aspects of encryption are confidential to avoid exploitation of the voice/message by unauthorized persons and to provide authentic protection against any modification of the message by an intruder. This system uses the following technique:

1. **Online encryption (stream cipher)**—Each bit of the original binary message stream (plain text) is combined using a simple operation, for example, modulo-2 addition, with each bit of a binary stream (keystream) generated by a key device. Otherwise, the latter could be a pseudorandom sequence generator whose structure is defined by the key.

2. **Encryption by Block (block cipher)**—The transmission of the original binary stream message into an encrypted stream is performed simply block-by-block, according to the logic defined by the key.

Besides, encryption is commonly used in direct TV broadcasting to avoid illegal reception and military applications to minimize the probability of message interception. The principle of encryption and decryption unit for text is that both operate with a key generation unit. At this point, key generation unit is giving necessary key outputs to encryption and decryption units. In the same time, plain text is entering in encryption unit, coming out as a cipher text, then via intruder entering in decryption, and is transforming again in the same plain text.

3.4.2 Coding

As is known, satellite communication systems are generally limited by the available power and bandwidth. Thus, it is of interest if the signal power can be reduced while maintaining the same grade of service (BER). As mentioned, this can be achieved by adding extra or redundant bits to the information content by using a channel coder. Otherwise, excepting several main classes of channel coder, the three most widely used in MSC are block, cyclic, and convolutional encoders.

3.4.2.1 Block Codes

Binary linear block codes are expressed in the (n, k) form, where (k) is the information bits number that is converted into (n) code word bits. There are (n, k) party bits in each encoded block, where the difference between (n) and (k) bits are added by the coder as a number of redundancy bits (r). In the other words, a coded block comprising (n) bits consists of (k) information and (r) redundant bits expressed as follows:

$$n = k + r \tag{3.33}$$

Such a code is designated as a (n, k) code, where the code rate or code efficiency is given by the ratio of (k/n). Mapping between message sequences and code words can be achieved using look-up tables although as the size of the code block increases, such an approach becomes impractical. However, this is not such a problem as linear code words can be generated using some form of linear transformation of the message sequence. In this sense, a code sequence (c) comprising of the row vector elements (c_1, c_2, …, c_n) is generated from a message sequence (m), comprising the row vector elements (m_1, m_2, …, m_k) by the following linear operation:

$$c = m\,G \tag{3.34}$$

where G = generator matrix. Thus, in general, all (c) code bits are generated from linear combinations of the (k) message bits.

A special category known as a systematic code occurs when the first (k) digits of the code are the same as the first (k) message bits, namely if input message bits appear as part of the output code bits. The remaining n-k code bits are then generated from the (k) message bits using a form of linear combination, and they are termed the party data bits.

The generator matrix for a linear block code is one of the main bases for the vector space of valid codewords. The generator matrix defines the length of each codeword (n), the number of information bits (k) in each codeword, and the type of redundancy that is added; however, the code is completely defined by its generator matrix. The generator matrix is a (k • n) matrix that is the row space of V_k value. In such a way, one possible generator matrix for a typical (7, 4) linear block code has to be simply presented in the following four rows: 01000/0110100/1110010/1010001.

However, the distance between two coded words (e.g., first 2 and second 2 digits) in a block is defined as the number of bits in which the words differ and is called the Hamming distance (d_h). The Hamming distance has the capability to detect all coded words having errors (e_d), where $e_d < (d_h - 1)$; to detect and correct (e_{dc}) bits, where $e_{dc} = (d_h - 1)/2$; and to correct t and detect (e) errors, where the Hamming distance as a minimum space between two coded blocks is given by:

$$d_h = t + e + 1 \tag{3.35}$$

In the detection process, two coded words separated by (d_h) are most likely to be mistaken for each other. The extended Golay code offers superior performance to Hamming codes but at a cost of increased receiver complexity. In practice, code words are conveniently generated using a series of simple shift registers and modulo-2 adders.

In Fig. 3.19 (left) is shown the concept of block codes and their rate, which operate on groups of bits organized as blocks, namely information bits assembled as blocks before coding, while in Fig. 3.19 (right) is shown cyclic coder.

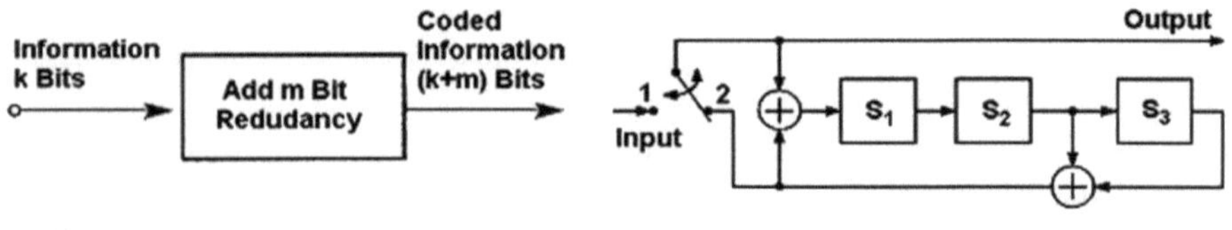

Fig. 3.19 Block and cyclic coders—Courtesy of Books: by Richharia/Calcutt

3.4.2.2 Cyclic Codes

These code methods are a subclass of linear codes, where a code word is generated simply by performing a cyclic shift of its predecessor. In other words, each bit in a code sequence generation is shifted by one place to the right and the end bit is fed back to the start of the sequence, hence the term cyclic. Both the linear Hamming and extended Golay codes have equivalent cyclic code generators. Thus, non-systematic cyclic codes are generated using a unique generator polynomial g(p) and message polynomials in the forms as follows:

$$g(p) = p^{n-k} + g_{n-k}p^{k-1} + \ldots\ g_1p + 1 \text{ and} \tag{3.36}$$

$$m(p) = m_{k-1}p^{k-1} + m_{k-2}p^{k-2} + \ldots + m_1p + m_0 \tag{3.37}$$

where the generator polynomial is a factor of p^{n+1} and the value ($m_{k-1} \ldots m_0$). When this is multiplied by the generator polynomial, it results in the generation of a code word by:

$$c(p) = (m_{k-1}p^{k-1} + m_{k-2}p^{k-2} + \ldots + m_1p + m_0)\, g(p) \tag{3.38}$$

Thus, an alternative to this approach is to generate systematic cyclic codes, which can be generated in three steps, involving the use of feedback shift register: (**a**) The message polynomial is multiplied by p^{n-k}, which is equivalent to shifting the message sequence by (n – k) bits. This is necessary to make space for the insertion of the party bits; (**b**) the product of step 1, $p^{n-k}m(p)$, is divided by the generator polynomial, g(p); and (**c**) The remainder from step 2 is the party bit sequence, which is then added to the message sequence prior to transmissions. The cyclic code scheme has two methods used in MSC systems such as: Bose–Chadhuri–Hocquenghem (BCH) and Reed–Solomon (RS).

1. **BCH Codes**—The BCH codes are the most powerful of all cyclic codes with a large range of block length, code rates, alphabets, and error correction capability. They are superior to all other codes because of similar block length and code rate. Most commonly used BCH codes have a code word block length as $n = 2^m - 1$, where (m = 3, 4 …). Inmarsat standard-A uses 57 bits plus 6 party bits encoded with BCH (63, 57) code in TDM channels and for the return request channel burst employs Aloha BPSK (BCH) 4800 b/s.
2. **RS Codes**—The RS codes are a subset of the BCH codes specially suited for correcting the effect of the burst errors. The latter consideration is particularly important in the context of the MSC channels, and hence, RS codes are usually incorporated into the system design. This set of codes has the largest possible code minimum distance of any linear code with the same encoder input and output block length. Thus, the RS codes are specified using the convention RS (n, k), where n = number of code symbols word length per block; k = data

symbols encoded; and the difference between (n) and (k) is the number of parity symbols added to the data. The code minimum distance is given by:

$$d_{min} = n-k+1 \tag{3.39}$$

The code is capable of correcting errors such as: $e = 1/2\ (d_{min}\ -\ 1)$ and $e = (n-k)/2$, or to use an alphabet of 2^m symbols with: $n = 2^m - 1$ and $k = 2^m -1-2e$, where m = 2,3 ... and so on. The advantage of RS codes is the reduction in the number of words (n) symbols, which are code words, producing a possibly large value of minimum distance (d_{min}).

3.4.2.3 Convolutional Codes

The second family of commonly used codes is known as convolution codes. Unlike block codes, which operate on each block independently, these codes retain several previous bits in memory, which are all used in the coding process. They are generated by a typed-shift register and two or more modulo-2 adders connected to particular stage of the register. The number of bits stored in the shift register is termed the constraint length (K). Bits within the register are shifted by (k) input bits. Each new input generates (n) output bits, which are obtained by sampling the outputs of the modulo-2 adders. The ratio of (k) to (n) is known as the code rate.

These codes are usually classified according to the following convention: (n, k, K), for example (2, 1, 7), refers to a half-rate encoder of constraint length 7. It is important to know what sequence of output code bits will be generated for a particular input stream. There are several techniques available to assist with this question, the most popular being connection pictorial, state diagram, tree diagram, and trellis diagram.

However, to illustrate how these methods are applied, the simple example of half-rate (1/2) encoder will be considered with constraint length k = 3. The system has two modulo-2 adders, so that the code rate is 1/2. The input bit (m) placed into the first of the shift register causes the bits in the register to be moved one place to the right. The output switch samples the output of each modulo-2 adder, one after the other, to form a bit pair for the bit just entered. The connections from the register to the adders could be one, two, or three interfaces for either adder. The choice depends on the requirement to produce a code with good distance properties. A similar encoder used by the Inmarsat standard-A is a half-rate convolutional encoder. Therefore, in terms of connections to the modulo-2, adders can be defined using generator polynomials in the encoder configuration.

Thus, convolutional codes are forming in convolutional coder by convolving information bits R with the impulse response of a shift register encoder, whose block diagram is shown in Fig. 3.20 (Left). These types of codes use previous information bits in memory (v) and continuously produce coded bits. The constraint length of convolutional code defines the number of information bits, which influence the encoder output. In such a way, the constraint length is decided by the

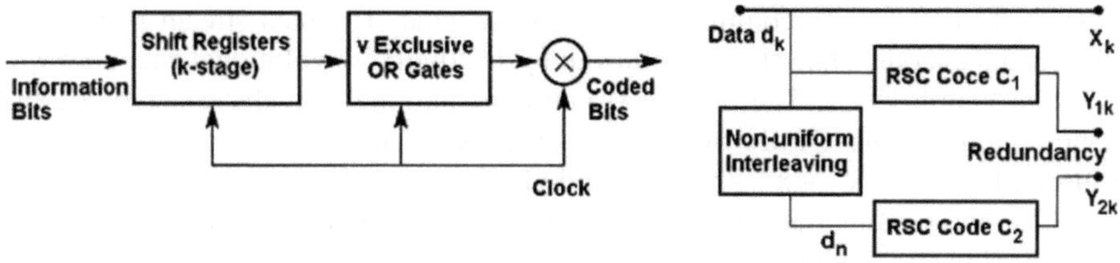

Fig. 3.20 Convolutional and turbo coders—Courtesy of Book: by Richharia

number of shift registers or code memory. The error correcting property of the convolutional code depends on the constraint length and its value improves as code memory is increased, and in such a way, decoding complexity increases.

The polynomial for the generating arm (n) of the encoder $g_n(p)$ and the generator polynomials representing encoder $g_1(p)$ can have the following relations:

$$g_n(p) = g_0(p) + g_1 p^1 + \ldots g_n p^n \quad \text{and} \tag{3.40}$$

$$g_1(p) = 1 + p + p^2 = 1 + p^2 \tag{3.41}$$

where the value of g_1 takes on the value of 0 or 1 and 1 is used to indicate that there is a connection between a particular element of the shift register and the modulo-2 adder. Thus, to provide a simple representation of the encoder, generator polynomials are used to predict the output coded message sequences for a given input sequences. For instance, the input sequence 10110 can be represented by the polynomial relation:

$$m(p) = 1 + p^2 + p^3 \tag{3.42}$$

Combining this with the respective generator polynomials and using the rules of module-2 arithmetic results in the following:

$$m(p)g_1(p) = (1 + p^2 + p^3)(1 + p + p^2) = 1 + p + p^5 \quad \text{and} \tag{3.43}$$

$$m(p)g_2(p) = (1 + p^2 + p^3)(1 + p^2) = 1 + p^3 + p^4 + p^5 \tag{3.44}$$

The output code sequence c(p) obtains by interleaving the above two products with the following relations:

$$c(p) = [1, 1]p^0 + [1, 0]p^1 + [0, 0]p^2 + [0, 1]p^3 + [0, 1]p^4 + [1, 1]p^5 \tag{3.45}$$

Here, the number between brackets represents the output code sequence.

The Inmarsat analog standard-A uses a HSD channel encoding configuration for the information data stream at 56 Kb/s. The scrambling sequence on the input data stream shall be provided by the scrambler before the convolutional encoder described in CCITT Recommendation V.35 scheme. The data stream then passes differential encoder state stage 1 followed by 1/2 (half) convolutional encoding with constant length k = 7. The half (1/2) rate convolutional encoder can provide two data streams to the QPSK modulator using two generator polynomials rates as follows: $G_1 = 1 + x^2 + x^3 + x^5 + x^6$ and $G_2 = 1 + x + x^2 + x^3 + x^6$. The encoder provides two parallel data streams to the modulator: I and Q, while (Q) should lag (I) by 90° in the modulator.

The Inmarsat digital B and aero standards for transmission and out-of-band signaling channels use digital modulation and FEC in order to efficiently utilize satellite power and bandwidth. The basic modulation and coding techniques are filtered by 60 % roll-off O-QPSK and 40 % roll-off BPSK, both with convolutional coding at either rate: 1/2 or 3/4 FEC and 8-level soft decision Viterbi decoding (constraint length = 7).

Hence, punctured coding is used to derive 3/4 and 1/2 rates. All BPSK channels are differentially encoded outside the FEC. The Inmarsat maritime and aero standard-M for all transmissions, with the exception of those fields carrying digitally coded voice, employs FEC with convolutional encoding of constraint length k = 7 and 8-level soft decision Viterbi decoding.

There are two generator polynomials rates: G_1 (133 octal) and G_2 (171 octal). The transmitted bit is nominated by 1 and deleted by 0. However, the first bit in each transmission frame is the output from the G_1 polynomial, and all bits are transmitted at the rate of 1/2 code. Finally, the output data from the bit selector are punctured coded data of 3/4 rate.

3.4.2.4 Concatenated Codes

These codes were originally developed for deep space communications and occur when two separate coding techniques are combined to form a large code. The inner decoder is used to correct most of the errors introduced by the channel, the output of which is then fed into the outer decoder, which further reduces the BER to the target level. That is to say, a typical concatenated coding scheme would employ half-rate convolutional encoding of constraint length 7 (2, 1, 7)—Viterbi decoding as the inner scheme and RS (255, 223) block encoding and decoding as the outer scheme. Interleaving between the inner and outer coders can be used to further improve the performance.

3.4.2.5 Turbo Codes

Turbo codes are a new class of error correction codes that were introduced in 1993 by a group of researchers from France, along with a practical decoding algorithm. In Fig. 3.20 (right) is shown a basic rate 1/3 turbo coder. At this point, it feeds data stream d_k directly into a recursive systematic convolution (RCS) coder c_1 and after interleaving into another RCS coder C_2, which is not necessarily identical to C_1. In addition, the transmitting bit stream compresses symbol X_k and redundancies Y_{1k} and Y_{2k} and is therefore a rate 1/3 code or it may be punctured to give higher code rate. The major importance of these codes is that they enable reliable transmission with power efficiencies close to the theoretical limit predicted by Claude Shannon.

Since their introduction, turbo codes have been proposed for low-power applications such as deep space and satellite communications, as well as for interference limited applications such as third-generation cellular/personal communication services. Due to the use of a pseudorandom interleaver, turbo codes appear randomly to the channel, yet possess enough structure so that decoding can be physically realized. Developed for deep space and satellite communication applications, turbo codes offer a performance significantly better than concatenated codes. For instance, they are generated using two or more recursive systematic convolutional code generators concatenated in parallel. Here, the term recursive implies that some of the output bits of the convolutional encoder are fed back and applied to the input bit sequence, and in sense, the first encoder takes the information bits as input. The key to the turbo code generation is the presence of a permuter, which performs a function similar to an interleaver, with the only difference that here the output sequence is pseudorandom.

The permuter takes a block of information bits, which should be large to increase performance, for example, more than 1000 bits and produces a random, delayed sequence of output bits, which is then fed into the second encoder. In such a manner, the outputs of the two encoders are partly bits transmitted along with the original information bits. In order to reduce the number of transmitted bits, the party bits are punctured prior to transmission.

From various simulation results, it is recognized that turbo codes are capable of achieving an arbitrarily low BER of 10-5 at an Eb/No ratio of just 0.7 dB. For instance, in order to achieve this level of performance, large block sizes of 65,532 data bits are required. Because of this prohibitively enlarged block size, an original turbo code is not well suited for real-time Tel communication systems such as IS-95 CDMA cellular standard. For that reason, the work on this problem has focused on the design of short block length codes, compatible with IS-95 standard.

3.4.3 Decoding

The complete transmission loop requires any type of encoder followed by modulation and transmitter via transmission channel to receiver, namely to demodulator and decoder. In such a manner, decoding is the reverse method of coding and every type of decoding on the transmit side needs the same convenient decoding method on the receive side.

3.4.3.1 Block Decoding

The simplest means of decoding block codes is by a method of correlation whereby the decoder makes a comparison between the received code word and all permissible code words, and however, it is selecting that word that gives the nearest match. This decoding will also depend on whether error detection or error correction is required, so decoders generally cannot use soft decision outputs from the demodulator, unlike the decoders for convolutional codes.

3.4.3.2 Convolutional Decoding

The effect of the transmission channel on the signal and the probability of detection of a 1 or 0 in the presence of Gaussian noise are important factors during detection. In such a manner, an output from the demodulator can be configured to give a correct decision regarding whether the incoming signal is 1 or 0. The process of decoding then depends on the two state inputs it receives.

An alternative demodulator configuration allows quantization of the predicted level which gives the decoder more necessary information regarding the probable state of the demodulator output. For example, if 3 bit ($2^3 = 8$ levels) quantization occurs then 0 0 0 would suggest a firm valuation of the level received as a 0. On the other hand, an 0 0 1 scheme suggests the 0 is received close to the threshold and this valuation as a 0 is made with less certainly. The reason for quantization is to provide the convolutional decoder with more information in order to correctly recover the transmitted information with better error performance probability.

3.4.3.3 Turbo Decoding

The turbo decoder operates by performing an interactive decoding algorithm, resulting in the partial transfer of an a priori likelihood estimate of the decoded bit sequence between the constituent decoders. Initially, the received information bits, which may be in some error due to the influence of the channel, are used to perform a priori likelihood estimates by the respective decoders.

In a more precise sense, the decoders employ so-called maximum a posteriori (MAP) algorithm to perform converge on the likely sequence of data transfer, after which the interaction between decoders ceases and the output sequence is obtained from one of the decoders.

An interleaver can be placed between the output of Decoder 1 and the input of Decoder 2, to provide an additional weighted decision input into Decoder 2; similarly, a de-interleaver is placed at the output of Decoder 2, to provide feedback to Decoder 1. The decoding time is proportional to the number of interactions between decoders.

3.4.3.4 Sequential Decoding

A sequential decoder may be used for convolutional decoding, and it operates in a similar manner to the Viterbi decoder. On receipt of the incoming code word sequence, this decoder will penetrate into the tree according to a decision made regarding the best path to follow. For that reason, using a trial and error technique, the decoder will progress as long as the chosen path appears correct; otherwise, it will backtrack to try a different route. At this point, either soft decision or hard decision decoding is possible with the sequential decoder, although soft decision would considerably increase the computational time and storage space required.

A major advantage of sequential decoding is that the number of states examined is independent of constraint length, allowing the use of large constraint lengths and low error probability. A disadvantage is the need to store input sequences, while the decoder searches for its preferred route through the tree. If the average decodes rate falls below that of the average symbol arrival rate, there is a danger that the decoder cannot cope, causing a loss of input information.

3.4.3.5 Viterbi Decoding

Viterbi decoding was developed by Andrew J. Viterbi and published in April 1967. A Viterbi decoder uses the Viterbi algorithm for decoding a bitstream that has been encoded using convolutional or trellis code. The Viterbi algorithm is the most resource-consuming, but it does the maximum likelihood decoding, mostly used for decoding convolutional codes with constraint lengths $k \leq 10$, but values up to $k = 15$ are used in practice.

Viterbi maximum likelihood decoding of convolutional codes provides the best possible results in the presence of random errors. Thus, in an attempt to match the output sequence received by the decoder, Viterbi's algorithm models the possible state transition through a trellis identical to that used by the encoder.

Accordingly, the Viterbi decoding algorithm is a maximum likelihood path algorithm that takes advantage of the remaining path structure of convolutional codes. This method works by modeling the possible state transitions of the encoder and finding the output sequence that matches most closely to that received by the decoder. Its task is to realize that not all paths through the encoder states can contribute to the final decoded output and that many paths can be rejected after each frame is received, which keeps the problem to manageable proportions.

If the encoder remembers (v) bits, then there are 2^v possible memory states to be modeled by the decoder. Hence, this term dominates expressions for speed, complexity, and cost of the decoder and currently imposes an upper limit of 8–10 on constraint length. By path, maximum likelihood decoding means that of all the possible paths through the trellis, a Viterbi decoder chooses the convenient path, most likely in the probabilistic sense to have been transmitted. Viterbi decoders easily make use of either hard or soft decision making. This decoding can incorporate soft decisions very simply, which will almost double the error correction power of the code and this can provide an additional gain of up to 3 dB. Otherwise, the procedure for choosing the best Viterbi scheme is to maximize constraint length within the limits of cost and speed, to find a nonsymmetrical code with the best value of d_{∞} and to use soft decisions. The maritime Inmarsat standard-B and multipurpose M utilize an 8-level soft decision Viterbi decoding in their channels (constraint length = 7).

3.4.4 Error Correction

There are several methods (such as ADPCM) that reduce the number of redundant bits in speech, audio, and visual signals in order to make more economic use of bandwidth. First of all, it is necessary to consider methods that require the deliberate addition of redundant bits to messages. The added bits are very carefully chosen, and error correction systems make it possible to achieve large savings in the power required to realize low BER. At the receiver, the additional bits are used to detect any errors introduced by channels. To achieve this technique in MSC, FEC, ARQ, and pseudo-noise and interleaving are employed. In such a way, it is also possible to combine FEC and ARQ in an integration form known as a hybrid error correction (HEC) transmitting scheme. At this point, however, the HEC method is used to reduce BER and the number of retranslated blocks. Such an arrangement could also be used to provide feedback information to the transmitter regarding slow variations, such as a fading.

3.4.4.1 Forward Error Correction (FEC)

The FEC is a technique where errors are detected and corrected at the receiver. Thus, this scheme requires only a one-way transmission link, since the message contains parity bits used for detection and correction of errors. In such a way, it is working only on receiving Tlx mode in radio and satellite one-way transmissions. The basic FEC technique used in MSC can be classified into two major (already explained) categories such as convolutional and block codes. The FEC coding as a result of convolutional coding is used in Inmarsat standards for some voice, telex, and signaling channels. For example, Inmarsat standard-B uses convolutional encoder of constraint length 7- and 8-level soft decision Viterbi decoder. The coding rate is either 3/4 or 1/2, while for voice channel, the code rate 3/4 is used and is derived by puncturing the rate 1/2 with k = 7 convolutional code.

On the other hand, the association of both basic coding techniques results in an even more powerful FES scheme known as the concatenated coding system. This powerful FEC scheme has been introduced in recent years, for a considerable

Table 3.3 Performances of FEC Techniques

Code	Decoding Mode	Gain (BER = 10^{-5})	Gain (BER = 10^{-8})	Bit rate	Complexity
Convolutional	Threshold	1.5–3.0	2.5–4.0	Very High	Low
Convolutional	Viterbi (soft decision)	4.0–5.0	5.0–6.5	High	High
Convolutional	Sequential (hard decision)	4.0–5.0	6.0–7.0	High	Low
Convolutional	Sequential (soft decision)	6.0–7.0	8.0–9.0	Medium	Low
Concatenated (Convol./RS)	Viterbi inner and algebraic outer	6.5–7.5	8.5–9.5	High	Medium
Concatenated (short block/RS)	Soft inner and algebraic outer	4.5–5.5	6.5–7.5	Medium	High
Short block linear	Soft decision	5.0–6.0	6.5–7.5	Medium	High
Block (BSH/RS)	Algebraic (hard decision)	3.0–4.0	4.5–5.5	High	Medium

increase of the service quality without appreciable expansion of bandwidth. While the inner code, with Viterbi decoding, can correct a large part of the random errors and very short error bursts, the residual errors at the outputs of the Viterbi decoder tend to be grouped in bursts. Thus, using a properly chosen interleaving that cuts the error bursts into shorter ones, a high-rate Reed–Solomon code can be used as the outer code in order to correct most of these dispersed errors bursts to achieve a very low BER. Thus, the introduction of concatenated coding and trellis coded modulation into MSC is the most remarkable event in the domain. Table 3.3 shows a list of FEC techniques along with their performance.

An FEC scheme can improve the quality of a digital transmission link by the following two aspects: (**a**) an BER reduction, closely related to the service quality criterion, and (**b**) a saving in the E_b/N_0 or C/N_0 to be considered in the link budget. The E_b/N_0 or C/N_0 saving is often called the coding gain, expressed in dB as a difference at certain BER values, of the coded system and the reference noncoded one. Thus, in the comparison between different transmission schemes, E_b/N_0 is usually used because it is independent of the coding scheme, where the gain is given as follows:

$$G = (E_b/N_0)\,\text{ref} - (E_b/N_0)\,\text{cod} \quad [\text{dB}] \tag{3.46}$$

The merit of a coding system can also be appreciated in terms of the savings in C/N_0 and C/N, considering information rate (R_b) and information transmission bandwidth, so values for carrier-to-noise density ratio (C/N_0) and the carrier-to-noise ratio (C/N) are as follows:

$$(C/N_0) = (E_b/N_0) + \log R_b \text{ and } (C/N) = (E_b/N_0) + \log R_b - 10 \log W \,[\text{dB}] \tag{3.47}$$

A coding gain in E_b/N_0 means in general a gain in C/N, but the coding in C/N depends on the bandwidth expansion with respect to the reference system. It is however possible to have a coding gain without bandwidth expansions using trellis coded modulation (TCM).

3.4.4.2 Automatic Request Repeat (ARQ)

The Automatic Repeat ReQuest (ARQ), also known as Automatic Repeat Query, is an error-control method for data transmission that uses acknowledgments (messages sent by the receiver indicating that it has correctly received a data frame or packet) and time-outs (specified periods of time allowed to elapse before an acknowledgment is to be received) to achieve reliable data transmission over an unreliable service. If the sender does not receive an acknowledgment before the time-out, the message will be retransmitted until is received.

The ARQ is a transmission technique with which a high degree of data integrity is required and latency is not a significant factor. In reality, the ARQ scheme, based on error detection coding and a retransmission protocol, is well adapted to the situation where a two-way channel is available. Typical examples of such systems can be encountered in a computer data network using satellite links. It is worthy of notice that the ARQ and improved ARQ, as well as HEC techniques, are widely used in modern digital communications and storage systems. The ARQ method requires a two-way link, since a receiver, detecting an error, does not attempt to correct it but simply requests the transmitter to retransmit the message. Thus, the ARQ scheme basically works with the following modus:

1. **Stop and Wait ARQ**—After each message block is sent via satellite link, the transmitter waits for acknowledgment. If the message block received is in error, the transmitter will retransmit that block, but if this message is correctly received, the next message block is transmitted. A half-duplex link is required, transmission on the link is possible in both directions but not at the same time.
2. **Continuous ARQ with Repeat**—The transmitter sends and the receiver acknowledges message blocks continuously. Hence, any message block not correctly received causes the transmitter to return to the block in question (incorrect received block) and recommence continuous retransmission from there. A full-duplex link is necessary for transmission in both directions simultaneously.
3. **Continuous ARQ with Selective Repeat**—In this ARQ arrangement, only the block received in error is retransmitted and the transmitter continues from where

it left off at the last block, instead of repeating all the subsequent even correctly received messages. In such a manner, however, full-duplex link is also necessary for transmission in both directions simultaneously.

A major advantage of ARQ compared with FEC is that decoding equipment for error corrections can be simpler and the redundancy in the total message stream is less. The ARQ efficiency is good for low error ratios, but for high ratios requiring retransmission of a large number of message blocks, the system becomes inefficient. A disadvantage of ARQ is the variability of the delays experienced from end to end of the link and so the possible requirement for large data stores of incoming data blocks.

The Inmarsat standard-C for maritime and aeronautical applications uses packets of data, and each one transmitted contains a 16-bit checksum field. After that, the receiver completes an expected checksum for each packet and compares this with the actual packet received in order to verify that the packet has been correctly received. The ARQ method is used if the packet received is in error.

3.4.4.3 Pseudo-Noise (PN)

The PN generator will produce a set of cyclic codes with good distance properties. Thus, the name of the sequence is given, because the sequence, although deterministic, appears to have the properties of sampled white noise. Furthermore, a PN sequence is easily generated using shift registers and has a correlation function that is high packet for zero delay and approximates to zero for other delays. The PN sequence, being deterministic, is usually for synchronization purposes between a transmitter and receiver.

Some Inmarsat standards use a scrambler circuit before FEC encoding and a descrambler at the receive end following FEC decoding. For Inmarsat standard-B and M, for instance, the scrambler/descrambler circuits are PN generators using 15 stages.

The scrambler/descrambler circuits are clocked at the rate of one shift per information bit. The first bit into the scrambler at the beginning of a frame is modulo-2, added with the output of the scrambler shift generator, corresponding to the initial state-scrambling vector. The initial state of the shift register is located at the beginning of a burst and a frame.

Considering the Inmarsat standard-M, the initial state of the scrambler shift registers sent by MES is received at the LES (for SCPC channel operating in voice mode) at the start of a call as part of a call set-up sequence. The MES chooses any initial state (except all zeros) on a random basis for each call and signals this scrambling vector message (8D in hexadecimal form or 10001101) for implementation at the LES with the least significant bit (LSB) in shift register No 1 and the most significant bit (MSB) in shift register No 15 of the scrambler. The MES simultaneously sets the descrambler shift register with the same scrambling vector. Otherwise, for MES-to-LES channels, a fixed initial state default value of 6959 in hexadecimal form or 110100101011001 is used in MES scramblers and LES descramblers.

3.4.4.4 Interleaving

As is well known, the MSC transmission channel introduces errors of a bursting nature. Hence, in the short term, the errors introduced by the channel cannot be considered to be statistically independent or memory less, the criterion upon which most coders (block and convolutional) optimally operate. In order to mimic a statistically independent channel, a technique known as interleaving is incorporated into the transmitter chain after the output of the encoder and from the interleaver the input signal passes via the modulator. In reverse mode, the output signal goes through the demodulator, deinterleaver, and channel decoder. This circle presents the interleaver/deinterleaver segment within the transmission/reception chain in the satellite link. Hence, the role of the interleaver is to reorder the transmission sequence of the bits that make up the code words in some predetermined fashion, such that the effect of an error burst is minimized.

Interleaving can be performed for both block and convolutional codes. Block interleaving is achieved by firstly storing the output code words of the encoder into a two-dimensional array. Consider the case of an (m, n) array, where (m) is the number of code words to be interleaved and (n) is the number of code word bits.

Thus, each row of the array comprises a generated code word. Once the array is full, the contents are then output to the transmitter, but in this case, data is read out on a column-by-column basis. Generally speaking, the transmission of each symbol of a particular code word will be non-sequential. Namely, the input signal goes via the input sequence into the interleaver block and, after processing the output sequence, would correspond to the chain starting with C_{11}, C_{21}, C_{31}, C_{41}, C_{51}, C_{61}, C_{71}, C_{12}–C_{72}, … until C_{18}–C_{78}. At this point, the effect of any error bursts will have been dispersed in time throughout the transmitted code words. Convolutional interleavers work along similar lines, achieving performance characteristics similar to block interleaving.

At the receiver, the inverse of the interleaving function is performed by a deinterleaver and the original code words are reconstituted prior to feeding into the encoder. Namely, a burst of error affecting the transmitted bits indicated by the chain coming from interleaver block would be dispersed among the code words at the receiver.

3.5 Multiple Access Technique

In satellite communication systems, as a rule, many users are active at the same time. The problem of simultaneous communications between many single or multipoint satellite users, however, can be solved by using multiple access

(MA) technique. Since the resources of the systems such as the transmitting power and the bandwidth are limited, it is advisable to use the channels with complete charge and to create a different MA to the channel.

This generates a problem of summation and separation of signals in the transmission and reception parts, respectively. Deciding this problem consists of the development of orthogonal channels of transmission in order to divide signals from various users unambiguously on the reception part. There are five the following principal forms of MA techniques:

(1) Frequency division multiple access (FDMA) is a scheme where each concerned LES or MES is assigned its own different working carrier radio frequency inside the spacecraft transponder bandwidth.
(2) Time division multiple access (TDMA) is a scheme where all concerned Earth stations use the same carrier frequency and bandwidth with time-sharing, non-overlapping intervals.
(3) Code division multiple access (CDMA) is a scheme where all concerned Earth stations simultaneously share the same bandwidth and recognize the signals by various processes, such as code identification. Actually, they share the resources of both frequency and time using a set of mutually orthogonal codes, such as a pseudorandom noise (PN) sequence.
(4) Space division multiple access (SDMA) is a scheme where all concerned Earth stations can use the same frequency at the same time within a separate space available for each link.
(5) Random (packet) division multiple access (RDMA) is a scheme where a large number of satellite users share asynchronously the same transponder by randomly transmitting short burst or packet divisions.

Currently, these methods of multiple access are widely in use with many advantages and disadvantages, together with their combination of hybrid schemes or with other types of modulations. Hence, multiple access technique assignment strategy can be classified into three methods as follows: (**1**) preassignment or fixed assignment, (**2**) demand assignment (DA), and (**3**) random Access (RA); the bits make up the code words in some predetermined fashion, such that the effect of an error burst is minimized. In the preassignment method, channel plans are previously determined for chairing the system resources, regardless of traffic fluctuations. This scheme is suitable for communication links with a large amount of steady traffic. However, since most mobile users in MSC do not communicate continuously, the preassignment method is wasteful of the satellite resources. In demand assignment multiple access (DAMA), satellite channels are dynamically assigned to users according to the traffic requirements. Due to high efficiency and system flexibility, DAMA schemes are suited to MSC systems. In RA, a large number of mobile users use the satellite resources in bursts, with long inactive intervals. In effect, to increase the system throughout, several mobile Aloha methods have been proposed.

Therefore, the MA techniques permit more than two Earth stations to use the same satellite network for interchanging information. Several transponders in the satellite payload share the frequency bands in use, and each transponder will act independently of the others to filter out its own allocated frequency and further process that signal for transmission.

Thus, this feature allows any LES located in the corresponding coverage area to receive carriers originating from several MES and vice versa, and carriers transmitted by one MES can be received by any LES. This enables a transmitting Earth station to group several signals into a single, multidestination carrier. Access to a transponder may be limited to single carrier, or many carriers may exist simultaneously. The baseband information to be transmitted is impressed on the carrier by the single process of multichannel modulation.

3.5.1 Frequency Division Multiple Access (FDMA)

The most common and first employed MA scheme for satellite communication systems is FDMA concept illustrated in Fig. 3.21 (**FDMA**), where transmitting signals occupy non-overlapping frequency bands with guard bands between signals to avoid interchannel interference. The bandwidth of a repeater channel is therefore divided into many sub-bands each assigned to the carrier transmitted by an Earth station. The MES transmits continuously, and the channel transmits several carriers simultaneously at a series of different frequency bands. Because of interchannel interference, it is necessary to provide guard intervals between each band occupied by a carrier to allow for the imperfections of oscillators and filters. The downlink Rx selects the required carrier in accordance with the appropriate frequency.

When the satellite transponder is operating close to its saturation, nonlinear amplification produces intermodulation (IM) products, which may cause interference in the signals of other users. In order to reduce IM, it is necessary to operate the transponder by reducing the total input power according to input back off and that the IF amplifier provides adequate filtering.

Therefore, FDMA allocates a single satellite channel to one user at once. In fact, if the transmission path deteriorates, the controller switches the system to another channel. Although technically simple to implement, FDMA is wasteful of

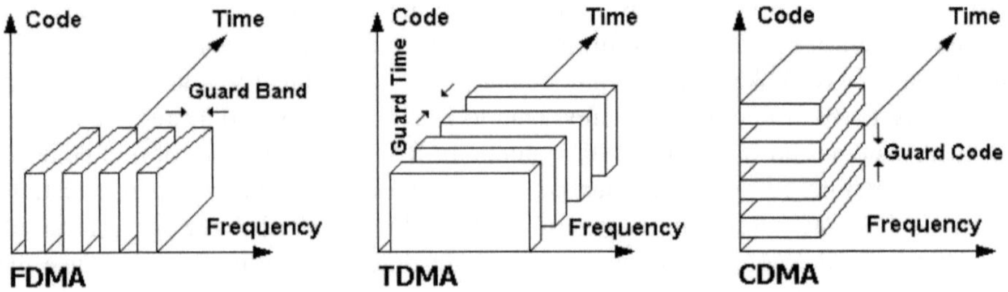

Fig. 3.21 Multiple access techniques—Courtesy of Web: by Google

bandwidth because the voice channel is assigned to a single conversation, whether or not somebody is speaking. Moreover, it cannot handle alternate forms of data, only voice transmissions. This system's advantages are that it is simple technique using equipment proven over decades to be reliable and it will remain very commonly in use because of its simplicity and flexibility.

It does have some disadvantages however:

(1) A FDMA method is the relatively inflexible system, and if there are changes in the required capacity, then the frequency plan has to change and, thus, involves many LES.
(2) Multiple carriers cause IM in both the MES HPA and in the transponder HPA. Reducing IM requires back off of the HPA power, so it cannot be exploited at full capacity.
(3) As the number of carriers increase, the IM products between carriers also increase and more HPA back off is needed to optimize the system. The throughput decreases relatively rapidly with the number of transmission carriers; therefore, for 25 carriers, it is about 40 % less than with 1 carrier.
(4) The FM system can suffer from what is known as a capture effect, where if two received signals are very close in frequency, but of different strengths, the stronger one tends to suppress the weaker one. For this reason, the carrier power has to be controlled carefully.

Therefore, with the FDMA technique, the signals from the various users are amplified by the satellite transponder in a given allocated bandwidth at the same time but at different frequencies. Depending on the multiplexing and modulation techniques employed, several transmission hybrid schemes can be considered and in general may be divided into two categories, based on the traffic demands of Earth stations on MCPC and SCPC.

3.5.1.1 Multiple Channels Per Carrier (MCPC)

The main elements of the MCPC are multiplexer, modulator, and transmitter using a satellite uplink, when LES-multiplexed baseband data is received from a terrestrial network and destined for various MES.

Moreover, the multiplexed data are modulated and transmitted to the allocated frequency segment, when the bandwidth of the transponder is shared among several MES, each with different traffic requirements. The transponder bandwidth is divided into several fixed segments, with several time frequency divisions allocated to these MES terminals. Namely, between each band segment is a guard band, which reduces the bandwidth utilization efficiency, and the loss is directly related to the number of accessing MES in the network as shown in Fig. 3.21 (**FDMA**). Depending on the number of receiving MES, a total number of carriers will pass through the satellite transponder.

On the other hand, the signals received from different MES extract the carrier containing traffic addressed to LES by using an appropriate RF filter, demodulator, baseband filter, and demultiplexer. The output of the demodulator consists of multiplexed telephone channels for a few MES together with the channels addressed to them. A baseband filter is used to filter out the desired baseband frequency segment, and finally, a demultiplexer retrieves individual telephone channels and feeds them into the terrestrial network for onward transmission.

Each baseband filter of LES receive stations in this scheme corresponds to a specific one in the LES transmitting station. However, any change in channel capacity requires the return of this filter, which is difficult to implement. Thus, many schemes may be categorized according to the type of baseband signal.

3.5.1.2 Single Channel Per Carrier (SCPC)

For certain applications, such as the provision of MSC service to remote areas or individual MES, traffic requirements are low. In reality, assigning multiple channels to each MES is wasteful of bandwidth because most channels remain unutilized for a significant part of the day. For this type of application, the SCPC type of FDMA is used. In the SCPC system, each carrier is modulated by only one voice or by low to medium bit-rate data channel. Some old analog systems use companded FM, but most new systems are digital PSK modulated. In the SCPC scheme, each carrier transmits a single carrier.

The assignment of transponder channels to each MES may be fixed preassigned multiple access (PAMA) or variable Demand Assignment Multiple Access (DAMA); the channel slots of the transponder are assigned to different MES units according to their instantaneous needs. In the case of PAMA, a few SCPC channels, about 5–10, are permanently assigned to each MES. In case of DAMA, a pool of frequency is shared by many MES terminals. When necessary, each MES requests a channel from frequency management of the network control station (NCS), which may always attempt to choose the best available channel or a lower quality one until an unoccupied channel has been found. The allocation is then announced on a signaling channel known as a broadcast channel, which is received by the calling and called MES terminal, which then tune to the allocated channel. The communication takes place on the allocated channel, and the end of call is announced by a signaling message, following which the NCS returns the channel to the common pool.

The SCPS solution requires an Automatic Frequency Control (AFC) pilot to maintain the spectrum centering on a channel-by-channel basis. This is usually achieved by transmitting a pilot tone in the center of the transponder bandwidth. It is transmitted by designated reference LES, and all the MESs use this reference to correct their transmission frequency. A receiving MES uses the pilot tone to produce a local AFC system which is able to control the frequency of the individual carriers by controlling the frequency of the LO.

Hence, drift in MSC translation frequency and frequency variations caused by the Doppler effect and the carriers retain their designated frequencies relative to each other. This feature is essential, because if uncorrected, the sum of the total frequency error can cause carrier overlapping, as carrier bandwidths are small. Thus, a stable receive frequency permits the LES demodulator design to be simplified. Centrally controlled networks, such as Inmarsat, are simple to manage missions, because they provide a higher usage of channels and can use simple demand assignment equipment. The SCPS scheme is cost-effective for networks consisting of a significant number of LES with a small number of channels.

3.5.2 *Forms of FDMA Operations*

There are several hybrid schemes of multiplexed FDMA in combination with SCPS, PSK, TDM, and TDMA techniques.

3.5.2.1 SCPC/FM/FDMA

The baseband signals from the network or users each modulate a carrier directly, in either analog or digital form according to the nature of SCPC signal in question. Therefore, each carrier accesses the satellite on its particular frequency at the same time as other carriers on the different frequencies from the same or other station terminals. Information routing is, thus, performed according to the principle of one carrier per link. The Inmarsat-A standard use SCPS, utilizing analog transmission with FM for telephone channels.

Thus, in calculating the channel capacity of the SCPC/FM system, it is necessary to ensure that the noise level does not exceed specified defined values. The CCIR Recommendations for an analog channel state that the relative level of noise power at a point of zero should not exceed 10,000 WOP with a 50 dB test tone, namely the noise ratio. In this way, it is assumed that the minimum required carrier-to-noise ratio per channel is at least 10 dB.

3.5.2.2 SCPC/PSK/FDMA

In this arrangement, each voice or data channel is modulated onto its own RF carrier. The only multiplexing occurs in the transponder bandwidth, where frequency division produces individual channels within the bandwidth, so Inmarsat is using various channel types of this multiplex scheme. The satellite transponder carrier frequencies may be PAMA or DAMA. For PAMA carriers, the RF is assigned to a channel unit and the PSK modem requires a fixed-frequency LO input. For DAMA, the channels may be connected according to the availability of particular carrier frequencies within the transponder RF bandwidth. For this arrangement, the SCPC channel frequency required is produced by a frequency synthesizer.

The forward link assigned by TDM in shore-to-ship direction uses the SCPC/DA/FDMA solution for Inmarsat standard-B voice/data transmission. This standard in the return link for channel request employs Aloha O-QPSK and for low speed data/telex uses the TDMA scheme in ship-to-shore direction. The Inmarsat-Aero in forward ground-to-aircraft direction uses packet mode TDM for network broadcasting, signaling and data, and the circuit mode of SCPS/DA/FDMA with distribution channel management for service communication links. Thus, the request for channel assignment, signaling and data in the return aircraft-to-ground direction the Slotted Aloha BPSK (1/2 – FES) of 600 b/s is employed and consequently, the TDMA scheme is reserved for data messages.

3.5.2.3 TDM/FDMA

This arrangement allows the use of TDM groups to be assembled at the satellite in FDMA, while the PSK is used as a modulation process at the Earth station. Systems such as this are compatible with FDM/FDMA carriers sharing the same transponders, and the terminal requirements are simple and easily incorporated. The Inmarsat standard-B system for telex low speed data uses this scheme in the shore-to-ship direction only and in the ship-to-shore direction uses TDMA/FDMA. The CES TDM and SES TDMA carrier frequencies are preallocated by Inmarsat. Each CES is allocated at least one forward CES TDM carrier frequency and a return SES TDMA frequency. Thus, additional allocations can be made depending on the traffic requirements. The channel unit associated with the CES TDM channel for transmission consists of a multiplexer, different encoder, frame transmit synchronizer, and modulator. So at the SES, the receive path of the channel has the corresponding functions to the transmitted end. The CES TDM channels use BPSK with differential coding, which is used for phase ambiguity resolution at the receive end.

3.5.2.4 TDMA/FDMA

As previously stated, however, the TDMA signals could occupy the complete transponder bandwidth. In fact, a better variation of this is where the TDMA signals are transmitted as a sub-band of transponder bandwidth, the remainder of which being available, for example, for SCPC/FDMA signals. Thus, the use of a narrowband TDMA arrangement is well suited for a system requiring only a few channels and has the all advantages of satellite digital transmission but can suffer from intermodulation with the adjacent FDMA satellite channels. Accordingly, the practical example of this scheme is the Tlx service of the Inmarsat-B system in ship-to-shore direction, which, depending on the transmission traffic, offers a flexible allocation of capacity for satellite communication and signaling slots.

3.5.3 Time Division Multiple Access (TDMA)

The TDMA application is a digital MA technique that permits individual Earth station transmissions to be received by the satellite in separate, non-overlapping time slots, called bursts, which contain buffered information. The satellite transponder receives these bursts sequentially, without overlapping interference and is then able to retransmit them to the MES terminal. Synchronization is necessary and is achieved using a reference station from which burst position and timing information can be used as a reference by all other stations. Each MES terminal must determine the satellite system in use, time, and range so that the transmitted signal bursts, typically QPSK modulated, are timed to arrive at the satellite in the proper time slots.

The offset QPSK modulation is used by Inmarsat-B SES. So as to ensure the timing of the bursts from multiple MES, TDMA systems use a frame structure arrangement to support telex in the ship-to-shore direction. Therefore, a reference burst is transmitted periodically by a reference station to indicate the start of each frame to control the transmission timing of all data bursts. A second reference burst may also follow the first in order to provide a means of redundancy. In the proper manner, to improve the imperfect timing of TDMA bursts, several synchronization methods of random access, open loop, and closed loop have been proposed.

In Fig. 3.21 (**TDMA**), a concept of TDMA is illustrated, where each mobile terminal transmits a data burst with a guard time to avoid overlaps. Since only one TDMA burst occupies the full bandwidth of the satellite transponder at a time, input back off, which is needed to reduce IM interference in FDMA, is not necessary in TDMA. At any instant in time, the transponder receives and amplifies only a single carrier. Thus, there can be no IM, which permits the satellite amplifier to be operated in full HPA saturation and the transmitter carrier power need not be controlled. Because all MES transmit and receive at the same frequency, tuning is simplified. This results in a significant increase in channel capacity. Another advantage over FDMA is its flexibility, and time-slot assignments are easier to adjust than frequency channel assignments. The transmission rate of TDMA bursts is about 4800 b/s, while the frame length is about 1.74 s and the optimal guard time is approximately 40 ms, using the open-loop burst synchronization method.

There are some disadvantages because TDMA is more complex than FDMA:

(1) Two reference stations and complex computer procedures are needed, for automated synchronizations between MES terminals.
(2) Peak power and bandwidth of individual MES terminals need to be larger than with FDMA, owing to high burst bit rate.

Therefore, in the TDMA scheme, the transmission signals from various users are amplified at different times but at the same nominal frequency, being spread by the modulation in a given bandwidth. Depending on the multiplexing techniques employed, two transmission hybrid schemes can be introduced for use in MSS.

3.5.3.1 TDM/TDMA

The Inmarsat analog standard-A uses the TDM/TDMA arrangement for telex transmission. Each SES has at least one TDM carrier, and each of the carriers has 20 telex channels of 50 bauds and a signaling channel. Moreover, there is also a common TDM carrier continuously transmitted on the selected idle listening frequency by the NCS for out-of-band signaling. The SES remains tuned to the common TDM carrier to receive signaling messages when the ship is idle or engaged in a telephone call. When an SES is involved in a telex forward call, it is tuned to the TDM/TDMA frequency pair associated with the corresponding CES to send messages in shore-to-ship direction. Telex transmissions in the return ship-to-shore direction form a TDMA assembly at the satellite transponder. Each frame of the return TDMA telex carrier has 22 time slots, while each of these slots is paired with a slot on the TDM carrier. The allocation of a pair of time slots to complete the link is received by the SES on receipt of a request for a telex call. Otherwise, the Inmarsat-A uses for forward signaling a telex mode, while all other MSS Inmarsat standards for forward signaling and assignment channels use the TDM BPSK scheme.

The new-generation Inmarsat digital standard-B (inheritor of standard-A) uses the same modulation TDM/TDMA technique, but instead of Aloha BPSK (BCH) at a data rate of 4800 b/s for the return request channel used by Inmarsat-A, new standard-B is using Aloha O-QPSK (1/2 – FEC) at a data rate of 24 Kb/s. This MA technique is also useful for the Inmarsat standard-C terminal for maritime, land, and aeronautical applications. In this case, the forward signaling and sending of messages in ground-to-mobile direction use a fixed assigned TDM carrier. The return signaling channel uses hybrid, slotted Aloha BPSK (1/2 FEC) with a provision for receiving some capacity, and the return message channels in the mobile-to-ground direction are modulated by the TDMA system at a data rate of 600 b/s.

3.5.3.2 FDMA/TDMA

The Iridium system employs a hybrid FDMA/TDMA access scheme, which is achieved by dividing the available 10.5 MHz bandwidth into 150 channels introduced into the FDMA components. Each channel accommodates a TDMA frame comprising eight time slots, four for transmission and four for reception. Each slot lasts about 11.25 ms, during which time data are transmitted in a 50 Kb/s burst. Each frame lasts 90 ms, and a satellite is able to support 840 channels. Thus, a user is allocated a channel occupied for a short period of time, during which transmissions occur.

3.5.4 Code Division Multiple Access (CDMA)

The CDMA solution is based on the use the modulation technique also known as spread-spectrum multiple access (SSMA), which means that it spreads the information contained in a particular signal of interest over a much greater bandwidth than the original signal.

In this MA scheme, the resources of both frequency bandwidth and time are shared by all users employing orthogonal codes, shown in Fig. 3.21 (**CDMA**). Therefore, the CDMA is achieved by a PN (Pseudo-noise) sequence generated by irreducible polynomials, which is the most popular CDMA method. In this way, a SSMA method using low-rate error correcting codes, including orthogonal codes with Hadamard or waveform transformation, has also been proposed.

Concerning the specific encoding process, each user is actually assigned a signature sequence, with its own characteristic code, chosen from a set of codes assigned individually to the various users of the system. This code is mixed, as a supplementary modulation, with the useful information signal. On reception, from all the signals that are received, a given user is able to select and recognize, by its own code, the signal, which is intended for it, and then to extract useful information. The other received signal can be intended for other users, but they can also originate from unwanted emissions, which gives CDMA a certain antijamming capability. For this operation, where it is necessary to identify one CDMA transmission signal among several others sharing the same band at the same time, correlation techniques are generally employed. From a commercial and military perspective, this MA is still new and has significant advantages. Interference from adjacent satellite systems including jammers is better solved than with other systems.

This scheme is simple to operate as it requires no synchronization of the transmitter and is more suited for a military MES. Small mobile antennas can be very useful in these applications, without the interference caused by wide antenna bandwidths. Using multibeam satellites, frequency reuse with CDMA is very effective and allows good flexibility in the management of traffic and the orbit/spectrum resources. The power flux density (PFD) of the CDMA signal received in the service area is automatically limited, with no need for any other dispersal processes. It also provides a low probability of intercept of the users and some kind of privacy, due to individual characteristic codes. The main disadvantage of CDMA by satellite is that the bandwidth required for the space segment of the spread carrier is very large, compared to that of a single unspread carrier, so the throughput is somewhat lower than with other systems.

Therefore, in the CDMA scheme, the signals from various users operate simultaneously, at the same nominal frequency but are spread in the given allocated bandwidth by a special encoding process.

Depending on the multiplexing techniques employed the bandwidth may extend to the entire capacity of the transponder but is often restricted to its own part, so CDMA can possibly be combined in the hybrid scheme with FDMA and/or TDMA. The SSMA technique can be classified into two methods: direct sequence (DS) and frequency hopping (FH). A combined system of DC and FH is called a hybrid CDMA system, and the processing gain can be improved without increases of chip rate. The hybrid system has been used in the military Joint Tactical Information Distribution System (JTIDS) and OmniTRACS, which is Ku-band MSS, developed by the Qualcomm Company. In a more precise sense, the CDMA technique was developed and implemented by experts of the Qualcomm Company in 1987.

At present, the CDMA system advantages are practically effective in new satellite systems, such as Globalstar, also developed by Qualcomm, which is devoted to MSS handheld terminals and Skybridge, involved in FSS. This type of MA is therefore attractive for handheld and portable MSS equipment with a wide antenna pattern.

Antennas with large beam widths can otherwise create or be subjected to interference with adjacent satellites. This MA technique is very attractive for commercial, military and even TT&C communications because some Russian satellites use CDMA for command and telemetry purposes.

The synchronous CDMA (S-CDMA) scheme proves efficiently to eliminate interference arising from other users sharing the same carrier and the same spot beam. Interference from other spot beams that overlap the coverage of the intended spot is still considerable. This process to ensure orthogonality between all links requires signaling to adjust transmission in time and frequency domains for every user independently.

3.5.4.1 Direct Sequence CDMA (DS-CDMA)

This dominant DS-CDMA technique is also called pseudo-noise (PN) modulation, where the modulated signal is multiplied by a PN code generator, which generates a pseudorandom binary sequence of length (N) at a chip rate (R_c), much larger than information bit rate (R_b), with a relation as follows:

$$R_c = N \cdot R_b \tag{3.48}$$

This sequence is combined with the information signal cut into small chip rates (R_c), thus speeding the combined signal in a much larger bandwidth ($W \sim R_c$); namely, the resulting signal has wider frequency bandwidth than the originally modulated signal. The transmitting signal can be expressed in the following way:

$$s(t) = m(t)\,p(t)\,\cos(2\pi f_c t) = m(t)\,p(t)\,\cos\omega_c t \tag{3.49}$$

where m(t) = binary message to be transmitted and p(t) = spreading NP binary sequence. At the receiver, the signal is coherently demodulated by multiplying the received signal by a replica of the carrier.

Neglecting thermal noise, the receiving signal at the input of the detector of low-pass filter (LPF) is given by:

$$r(t) = m(t)\,p(t)\,\cos\omega_c t\,(2\cos\omega_c t) = m(t)\,p(t) + m(t)\,p(t)\,\cos 2\omega_c t \tag{3.50}$$

The detector LLF eliminates the HF components and retains only the LW components, such as u(t) = m(t) p(t). This component is then multiplied by the local code [p(t)] in phase with the received code, where the product $p(t)^2 = 1$. At the output of the multiplier this gives:

$$x(t) = m(t)\,p(t)\,p(t) = m(t)\,p(t)^2 = m(t) \quad [\mathrm{V}] \tag{3.51}$$

The signal is then integrated over one bit period to filter the noise. The transmitted message is recovered at the integrator output, so in fact, only the same PN code can achieve the despreading of the received signal bandwidth. In this process, the interference or jamming spectrum is spread by the PN codes, while other user's signals, spared by different PN codes, are not despread.

Interference or jamming power density in the bandwidth of the received signal decreases from their original power. Otherwise, the most widely accepted measure of interference rejection is the processing gain (G_p), which is given by the ratio $R_c/R_{b\ \mathrm{and}}$ value of G_p = 20–60 dB. The input and output signal-to-noise ratios are related as follows:

$$(S/N)_{\mathrm{Output}} = G_p(S/N)_{\mathrm{Input}} \tag{3.52}$$

Therefore, for DS-CDMA, a transmitter will assemble a bit stream consisting of message and code information as a function of time.

In the forward link, the LES (HUB station) transmits the spread-spectrum signals, which are spread with synchronized PN sequence to different MSC users. Since orthogonal codes can be used, the mutual interference in the network is negligible and the channel capacity is close to that of FDMA. Conversely, in the return link, the signals transmitted from different MES users are not synchronized and they are not orthogonal. The first case is referred to as synchronous and the second case as asynchronous SSMA.

However, the nonorthogonality causes interference due to the transmission of other MES in the satellite network, and as the number of simultaneously accessing users increases, the communication quality gradually degrades in a process called graceful degradation.

3.5.4.2 Frequency Hopping CDMA (FH-CDMA)

The FH-CDMA system works similar to the DS system, since a correlation process of dehopping is also performed at the receiver. The difference is that here the pseudorandom sequence is used to control a frequency synthesizer, which results in the transmission of each information bit rate in the form of (N) multiple pulses at different frequencies in an extended bandwidth. The transmitted and received signals of FH-CDMA system have the following forms:

$$s(t) = m(t) \cos \omega_c(t)\, t$$

$$r(t) = m(t) \cos \omega_c(t)\, t \cdot 2 \cos \omega_c(t)\, t = m(t) + m(t) \cos 2\omega_c(t)\, t \qquad (3.53)$$

Thus, at the receiver, the carrier is multiplied by an unmodulated carrier generated under the same conditions as at the transmitter. The second term in the receiver is eliminated by the LPF of the demodulator. The relation of processing gain for FH is as follows:

$$G_p = W/\Delta f \qquad (3.54)$$

where W = frequency bandwidth and Δf = bandwidth of the originally modulated signal. At this point, coherent demodulation is difficult to implement in FH receivers because it is a problem to maintain phase relation between the frequency steps. Due to the relatively slow operation of the frequency synthesizer, DS schemes permit higher code rates than FH radio.

3.5.5 Space Division Multiple Access (SDMA)

The previous systems have used frequency, time, and code domains to achieve multiple access schemes, while SDMA uses spatial separation.

The significant factor in the performance of MA in a satellite communications system is interference caused by different factors and other users. In the other words, the most usual types of interference are cochannel and adjacent channel interference. The cochannel interference can be caused by transmissions from non-adjacent cells or spot beams using the same set of frequencies, where there is minimal physical separation from neighboring cells using the same frequencies, while the adjacent channel interference is caused by RF leakage on the subscriber's channel from a neighboring cell using an adjacent frequency.

This can occur when the user's signal is much weaker than that of the adjacent channel user. Signal-to-interference ratio (SIR) is an important indicator of call quality; it is a measure of the ratio between the mobile phone signal (the carrier signal) and an interfering signal. A higher SIR ratio means increasing overall system capacity.

Taking into account that within the systems of satellite communications, every user has their own unique spatial position, and this fact may be used for the separation of channels in space and, as a consequence, to increase he SIR ratio by using SDMA. In effect, this method is physically making the separation of paths available for each satellite link.

Terrestrial telecommunication networks can use separate cables or radio links, but on a single satellite, independent transmission paths are required. Thus, this MA control radiates energy into space and transmission can be on the same frequency, such as TDMA or CDMA, and on different frequencies, such as FDMA.

3.5.5.1 Special Effects of SDMA in Wireless Systems

The new technologies are recently implemented for the third-generation wireless (3G) with future enhancements. Some of the key enhancements to wireless technology include SDMA, the introduction of 3G wireless technologies and integration with personal and mobile applications including aeronautical. The SDMA method is a special system access technology that allows a single transmitter location to provide multiple communication channels by dividing the radio coverage into focused radio beams that reuse the same frequency. To allow multiple accesses, each mobile radio is assigned to a focused radio beam. These radio beams may dynamically change with the location of the mobile radio.

Analogously, in reception, the mobile antenna picks up signals coming from all directions, including noise and interference. These considerations have lead to the development of the SDMA technique, which is based on deriving and exploiting information on the spatial position of mobile terminals. The use of an adaptive antenna array at the base station thus allows introducing the SDMA technique, whose main advantage is the capability to increase system capacity, i.e., the number of users it can handle.

This increase can be obtained in two technical ways for the following applications:

1. **Reduction in Cochannel Interference**—The reduction in the level of cochannel interference between the different cells using the same group of radio channels is obtained, as above seen, by minimizing the gain in the direction of interfering mobile units. This technique, indicated with the acronym spatial filtering for interference reduction (SFIR), allows reducing frequency reuse distance and cluster size. In this way, each transmission cell can be assigned a higher number of channels by phase array antenna, as shown in Fig. 3.22 (left).
2. **Spatial Orthogonality**—In conventional multiple access techniques, orthogonality between signals associated with different users is obtained by transmitting them in different frequency bands of FDMA, in different time slots of TDMA, or using different code sequences of CDMA. However, using an antenna array, it is possible to create an additional degree of orthogonality between signals transmitted to and from different directions. It is thus possible to assign the same

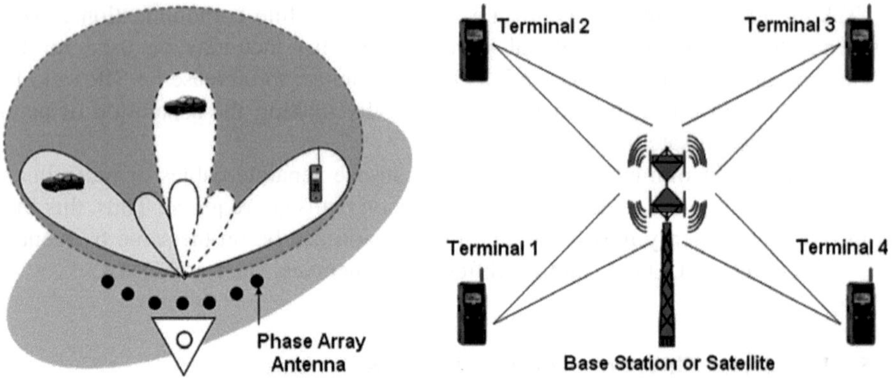

Fig. 3.22 Phased-array antenna and base station for SDMA mode—Courtesy of Web: by Google

physical channel to several mobile units, as depicted in Fig. 3.22 (right), when the angles at which they are seen by the base station or satellite transponder are sufficiently separated. The result is an increase in the number of available channels, since the same physical channel, for example, the same carrier in a FDMA or the same time slot in a TDMA system, can be subdivided into multiple spatial channels, each of which is assigned to a different user. So, the multiple users belonging to the same cell use the same channel. In addition, the same frequency can be also available in same cell or satellite spot coverage by using the more improved adaptive array system (AAS). Thus, the advantage of this system is providing better control of radiated energy in space; it is employing usage of spot beams antennas, may be served different areas by using same frequency for TDMA and CDMA, or can be used different frequencies for FDMA and can use adaptive antennas that dynamically adopt to number of users. Disadvantages of this system are possible problems with reverse links, limitation of battery capability for some handheld units, and interference impairments.

Therefore, the SDMA mode in MSC optimizes the use of radio spectrum and minimizes system cost by taking advantage of the directional properties of dish antennas. In SDMA, satellite dish antennas transmit signals to numerous zones on the Earth's surface. Moreover, the mobile satellite antennas are highly directional, allowing duplicate frequencies to be used for multiple surface zones.

3.5.5.2 Special Effects of SDMA in Mobile Radio Systems

In a FDMA mobile radio system, however, capacity is limited by two different factors. On the one hand, a limited number of radio channels, carriers, and time slots are available, and they must be subdivided among beams making up a cluster. On the other hand, cochannel interference limits channel reuse. The SDMA technique allows to expand both these limits and to enhance system capacity. As already

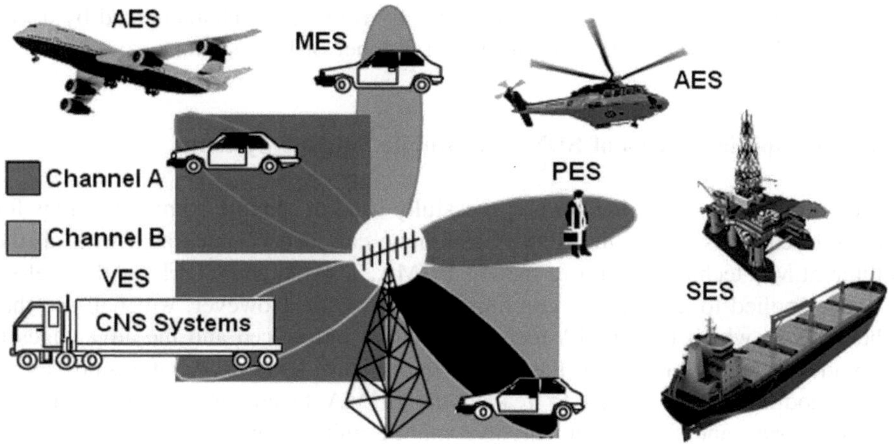

Fig. 3.23 Hypothetical scenario of SDMA for mobile radio applications—Courtesy of Paper: by Buracchini

described, this can occur in two different ways: with the SFIR technique and spatial orthogonality. In such a way, interference level is reduced and channel reuse distance is decreased, whereas the actual SDMA technique assigns the same channel to multiple, spatially separated users. In Fig. 3.23 is shown a SDMA system, which diagram shows a single tower that is serving many different users from the same radio tower on the same frequency using independent beams of radio energy. In practice, this is not possible to achieve, because each type of mobile system has different frequency band, but they can be seen in separate way. On the other hand, the SDMA technique requires an array composed of more antennas than the SFIR technique does. In fact, spatial orthogonality is exploited by eliminating, through the use of spatial filtering, intra-cell cochannel interference, which is due to mobile units belonging to the same cells and using the same channel as the mobile unit in question.

Differently to say, in traditional radio systems, the base station, having no information on the position of mobile units, is forced to radiate the signal in all direction, in order to cover the entire area of the cell. This entails both a waste of power and the transmission, in the directions where there are no mobile terminals to reach, of a signal which will be seen as interfering for cochannel cells, i.e., those cells using the same group of radio channels.

It must be noted that the term SDMA refers, strictly speaking, only to the latter application, in which SDMA is actually accomplished. In spite of this fact, the SFIR technique is also considered within the SDMA technique, since it is based on the same principles. In addition to the opportunity to increase system capacity, the SDMA technique has additional characteristics making its introduction in a mobile radio system advantageous. In particular, it is possible to exploit the higher receive gain offered by an antenna array with respect to an omnidirectional case, to allow mobile units to transmit at reduced power, and therefore lower consumption. At equal power, gain can be exploited to extend beam size. This is useful when it is necessary to cover vast surface areas, typically rural areas, characterized by a low mobile radio traffic density, with a limited number of base stations.

3.5.5.3 Special Effects of SDMA in Mobile Satellite Systems

The SDMA technology has been successfully used in satellite communications for several years. As stated, the SDMA technique can also be integrated with all the different MA techniques in use, such as FDMA, TDMA, and CDMA and therefore can be applied to any mobile communication system. However, we shall see that the ways in which the SDMA technique can be introduced and the advantages it provides differ depending on the system under consideration. As aforesaid described, modifications required to realize the SDMA technique are limited to the satellite array and thus do not involve mobile units. However, this allows introducing this technique in existing mobile satellite systems, with no need to modify their characteristics.

The ability to reject jammer of adaptive arrays can be ensured for performing SDMA mode where several mobiles are allowed to share the same classical access in a cell, leading to a capacity increase. In Fig. 3.24 is shown this sharing with possibility through the use of adaptive beam forming and interference rejection on the satellite uplink and downlink communication for mobiles, which are located at different angular sectors.

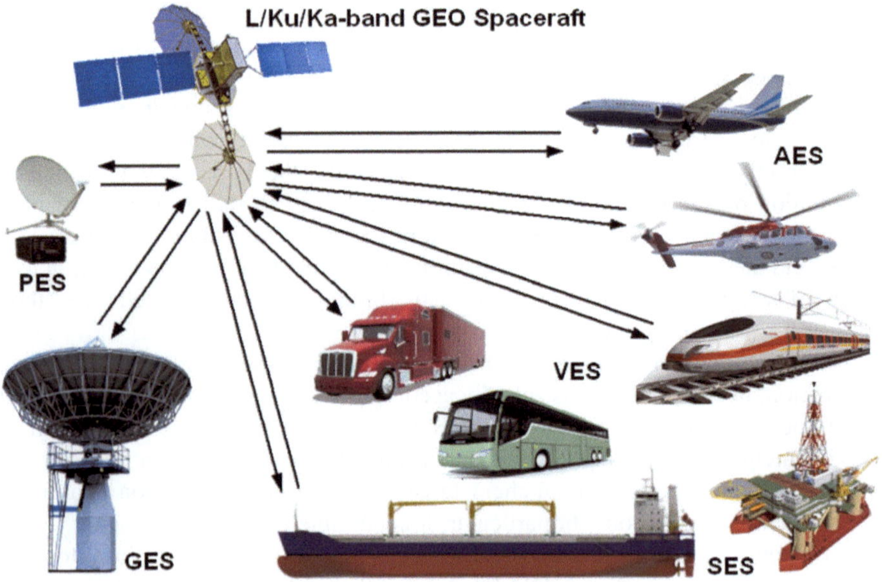

Fig. 3.24 SDMA for MSC applications—Courtesy of Manual: by Ilcev

In using SDMA, either FDMA or TDMA is needed to allow GES to roam in the same satellite beam or for polarization to enter the repeater. Thus, the frequency reuse technique of same frequency is effectively a form of SDMA scheme, which depends upon achieving adequate beam-to-beam and polarization isolation. Using this system, reverse line means that interference may be a problem and the capacity of the battery is limited.

On the other hand, a single satellite may achieve spatial separation by using beams with horizontal and vertical polarization or left-hand and right-hand circular polarization. This could allow two beams to cover the same Earth surface area, being separated by the polarization. Thus, the satellite could also have multiple beams using separate antennas or using a single antenna with multiple feeds.

In case of multiple satellite communication systems, spatial separation can be achieved with orbital longitude or latitude and for intersatellite links, by using different planes. Except for frequency reuse, this system provides onboard switching techniques, which, in turn, enhance channel capacity. Additionally, the use of narrow beams from the satellite allows the Earth station to operate with smaller antennas and so produce a higher power density per unit area for a given transmitter power.

Therefore, through the careful use of polarization, beams (SDMA) or orthogonal (CDMA), the same frequency spectrum may be reused several times, with limited interference among mobile users.

The more detailed benefits of an SDMA system include the following:

(1) The number of cells required to cover a given area can be substantially reduced.
(2) Interference from other systems and from users in other cells is significantly reduced.
(3) The destructive effects of multipath signals, copies of the desired signal that have arrived at the antenna after bouncing from objects between the signal source and the antenna, can often be mitigated.
(4) Channel reuse patterns of the systems can be significantly tighter because the average interference resulting from cochannel signals in other cells is markedly reduced.
(5) Separate spatial channels can be created in each cell on the same conventional channel. In other words, intra-cell reuse of conventional channels is possible.
(6) The SDMA station radiates much less total power than a conventional station. One result is a reduction in network-wide RF pollution. Another is a reduction in power amplifier size.
(7) The direction of each spatial channel is known and can be used to accurately establish the position of the signal source.
(8) The SDMA technique is compatible with almost any modulation method, bandwidth, or frequency band, including GSM, PHP, DECT, IS-54, IS-95, and other formats. The SDMA solution can be implemented with a broad range of array geometry and antenna types.

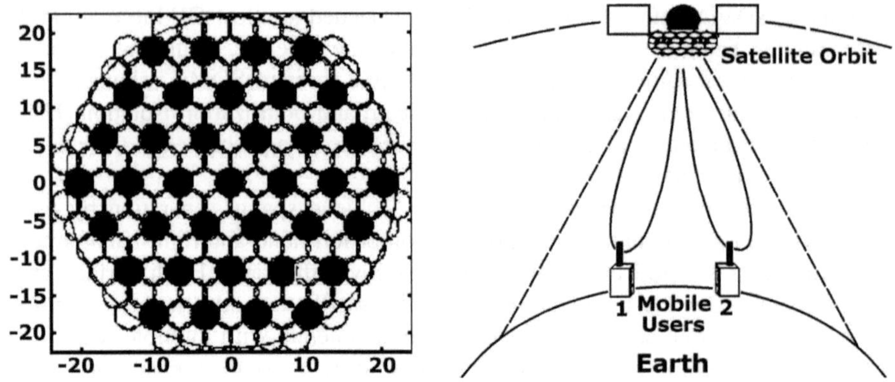

Fig. 3.25 Beam patterns and adaptive antenna solutions for SDMA—Courtesy of Paper: by Zaharov

Another perspective of the realization of SDMA systems is the application of smart antenna arrays with different levels of intelligence consisting of the antenna array and digital processor. Since the frequency of transmission for satellite communications is high enough (mostly 6 or 14 GHz), the dimensions of an array placed in orbit are commensurable with the dimensions of the parabolic antenna, which is a necessary condition to put such systems into orbit.

Thus, the SDMA scheme mostly responds to the demands of LEO and MEO constellations, when the signals of users achieve the satellite antenna under different angles ($\pm$ 22° for the MEO). In this instance, ground level may be split into the number of zones of service coverage determined by switched multiple beam pattern lobes in different satellite detections, or by adaptive antenna separations, shown in Fig. 3.25 (left). There are two different beam-forming approaches in SDMA satellite communications: (1) The multiple spot beam antennas are the fundamental way of applying SDMA in large fixed and mobile satellite systems and (2) adaptive array antennas dynamically adapt to the number of users.

3.5.5.4 Switched Spot Beam Antenna

Switched multibeam antennas are designed to track each subscriber of a given cell with an individual beam pattern as the target subscriber moves within the cell (spot). Therefore, it is possible to use array antennas and to create a group of overlapping beams that together result in omnidirectional coverage. This is the simplest technique comprising only a basic switching function between separate directive antennas or predefined beams of an array. Beam-switching algorithms and RF signal-processing software are incorporated into smart antenna designs. For each call, software algorithms determine the beams that maintain the highest quality signal and the system continuously updates beam selection, ensuring that customers get optimal quality for the duration of their call. One might design overlapping beam patterns pointing in slightly different directions, similar to the ones shown in Fig. 3.25 (left).

Every so often, the system scans the outputs of each beam and selects the beam with the largest output power. The black cells reuse the frequencies currently assigned to the mobile terminals, so they are potential sources of interference. In fact, the use of a narrow beam reduces the number of interfering sources seen at the base station. Namely, as the mobile moves, the smart antenna system continuously monitors the signal quality to determine when a particular beam should be selected.

Switched-beam antennas are normally used only for the reception of signals, since there can be ambiguity in the system's perception of the location of the received signal. In fact, these antennas give the best performance, usually in terms of received power, but they also suppress interference arriving from directions away from the active antenna beam's center, and because of the higher directivity, compared to a conventional antenna, some gain is achieved. In high-interference areas, switched-beam antennas are further limited since their pattern is fixed and they lack the ability to adaptively reject interference. Such an antenna will be easier to implement in existing cell structures than the more sophisticated adaptive arrays, but it gives only limited improvement.

3.5.5.5 Adaptive Array Antenna Systems

Adaptive array antenna systems select one beam pattern for each user out of a number of preset fixed beam patterns, depending on the location of the subscribers. At all events, these systems continually monitor their coverage areas, attempting to adapt to their changing radio environment, which consists of (often mobile) users and interferers. Thus, in the simplest scenario, that of a single user and no interferers, the system adapts to the user's motion by providing an effective antenna system pattern that follows the mobile user, always providing maximum gain in the user's direction.

The principle of SDMA with adaptive antenna system application is quite different from the beam-forming approaches described in Fig. 3.25 (right).

The events processed in SDMA adaptive array antenna systems are as follows:

(1) A "snapshot," or sample, is taken of the transmission signals coming from all of the antenna elements, converted into digital form and stored in memory.
(2) The SDMA digital processor analyzes the sample to estimate the radio environment at this point, identifying users and interferers and their locations.
(3) The processor calculates the combining strategy for the antenna signals that optimally recovers the user's signals. With this strategy, each user's signal is received with as much gain as possible and with the other users/interferers signals rejected as much as possible.
(4) An analogous calculation is done to allow spatially selective transmission from the array. Each user's signal is now effectively delivered through a separate spatial channel.

(5) The system now has the ability to both transmit and receive information on each of the spatial channels, making them two-way channels.

As a result, the SDMA adaptive array antenna system can create a number of two-way spatial channels on a single conventional channel, be it frequency, time, or code. Of course, each of these spatial channels enjoys the full gain and interference rejection capabilities of the antenna array. In theory, an antenna array with (n) elements can support (n) spatial channels per conventional channel. In practice, the number is somewhat less because the received multipath signals, which can be combined to direct received signals, takes place. In addition, by using special algorithms and space diversity techniques, the radiation pattern can be adapted to receive multipath signals, which can be combined. Hence, these techniques will maximize the SIR or signal-to-interference and noise ratio (SINR).

3.5.5.6 SDMA/FDMA

This modulation arrangement uses filters and fixed links within the satellite transceiver to route an incoming uplink frequency to a particular downlink transmission antenna, shown in Fig. 3.26 (left). A basic configuration of fixed links may be set up using a switch that is selected only occasionally. Thus, an alternative solution allows the filter to be switched using a switch matrix, which is controlled by a command link. Because of the term SS (switching satellite), this scheme would be classified as SDMA/SS/FDMA, whose block diagram is shown in Fig. 3.26 (right).

The satellite switches are changed only rarely, only when it is desired to reconfigure the satellite, to take account of possible traffic changes. The main disadvantage of this solution is the need for filters, which increase the mass of the payload.

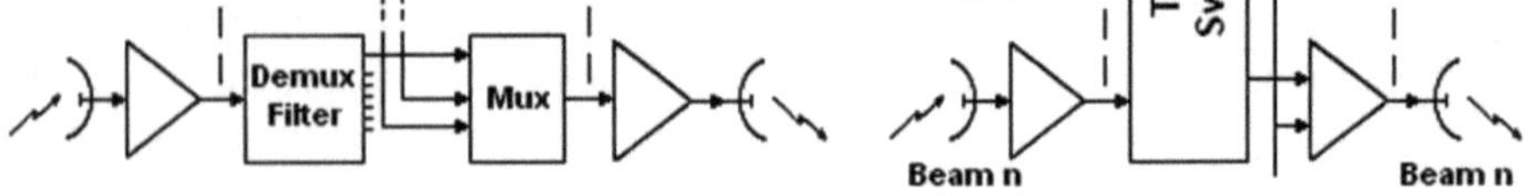

Fig. 3.26 Block diagram of SDMA/FDMA and SDMA/SS/FDMA-SDMA/SS/TDMA—Courtesy of Book: by Calcutt

3.5.5.7 SDMA/TDMA

This solution is similar to the previously explained SDMA/SS/FDMA in that a switch system allows a TDMA receiver to be connected to a single beam. Switching again is only carried out when it is required to reconfigure the satellite. Under normal conditions, a link between beam pairs is maintained and operated under TDMA conditions.

The utilization of time slots may be arranged on an organized or contention basis, and switching is achieved by using the RF signal. Thus, onboard processing is likely to be used in the future, allowing switching to take place by the utilization of baseband signals. The signal could be restored in quality and even stored to allow transmission in a new time slot in the outgoing TDMA frame.

This scheme is providing up and downlinks for the later Intelsat VI spacecraft, known as SDMA/SS/TDMA, whose block diagram is shown in Fig. 3.26b. This MA is using to allow TDMA traffic from the uplink beams to be switched to downlink beams during the course of TDMA frame. At this point, the connection exists at a specific time for the burst duration within the frame time before the next connection is made and so on.

3.5.5.8 SDMA/CDMA

This arrangement allows access to a common frequency band and may be used to provide the MA to the satellite, when each stream is decoded on the satellite in order to obtain the destination addresses. Thus, onboard circuitry must be capable of determining different destination addresses, which may arrive simultaneously, while also denying invalid users access to the downlink. However, onboard processors allow the CDMA bit stream to be retimed, regenerated, and stored on the satellite. Because of this possibility, the downlink CDMA configurations need not be the same as for uplink and the Earth link may thus be optimized.

3.5.6 Random Division Multiple Access (RDMA)

For data transmission, a bit stream may be sent continuously over an established channel without the need to provide addresses or unique words if the channel is not charred. In fact, where charring is implemented, data are sent in bursts, which, thus, requires unique words or synchronization signals to enable time-sharing with other users, to be affected in the division of channels. Each burst may consist of one or more packets comprising data from one or more sources that have been assembled over time, processed and made ready for transmission. However, this type of multiplex scheme is also known as Packet MA. Packet access can be used in special RDMA solutions, such as Aloha, where retransmission of blocked packets may be required.

Random access can be achieved to the satellite link by contention, and for that reason, it is called a contention access scheme. This type of access is well-suited to satellite networks containing a large number of stations, such as MES, where each station is required to transmit short randomly generated messages with long dead times between messages.

The principle of RDMA is to permit the transmission of messages almost without restriction, in the form of limited duration bursts, which occupy all the bandwidth of the transmission channel. Therefore, in other words, this is MA with time division and random transmission, and an attribute for the synonym random division multiple access is quite assessable. A user transmits a message irrespective of the fact that there may be other users equally in connection. The probability of collisions between bursts at the satellite is accepted, causing the data to be blocked from receipt by the Earth station. In case of collision, the destination Earth station receiver will be confronted with interference noise, which can compromise message identification and retransmission after a random delay period. The retransmissions can occur as many times as probably are carried out, using random time delays.

Therefore, such a scheme implies that the transmitter vies for satellite resources on a per-demand basis. It will provide that no other transmitter is attempting to access the same resources during the transmission burst period, when an error-free transmission can occur. The types of random protocols are distinguished by the means provided to overcome this disadvantage. The performance of these protocols is measured in terms of the throughput and the mean transmission delay. Throughput is the ratio of the volume of traffic delivered at the destination to the maximum capacity of the transmission channels. The transmission time, i.e., delay, is a random variable. Its mean value indicates the mean time between the generation of a message and its correct reception by the destination station.

3.5.6.1 Aloha

The most widely used contention access scheme is Aloha and its associated derivatives. This solution was developed in the late 1960s by the University of Hawaii and allows usage of small and inexpensive Earth stations (including MES) to communicate with a minimum of protocols and no network supervision. This is the simplest mode of operation, in which time shares a single RF divided among multiple users and consists of stations randomly accessing a particular resource that is used to transmit packets. When an Aloha station has something to transmit, it immediately sends a burst of data pulses and can detect whether its transmission has been correctly received at the satellite by either monitoring the retransmission from the satellite or by receiving an acknowledgment message from the receiving party. Should a collision with another transmitting station occur, resulting in the incorrect reception of a packet at the satellite, the transmitting station waits for a random period of time, prior to retransmitting the packet. Otherwise, a remote station (MES) uses Aloha to get a HUB station (LES) terminal's attention. Namely, the MES terminal sends a brief burst requesting a frequency or time slot assignment for the main transmission. Thus, once the assignment for MES is made, there is no further need for the Aloha channel, which becomes available for other stations to use. After that, the main transmissions are then made on the assigned channels. At the end, the Aloha channel might be used again to drop the main channel assignments after the transmission is completed. The advantages of Aloha are the lack of any centralized control, giving simple, low-cost stations and the ability to transmit at any time, without having to consider other users.

In the case where the user population is homogenous, so that the packet duration and message generation rate are constant, it can be shown that the traffic carried S (packet correctly interpreted by the receiver), as a function of total traffic G (original and retransmitted message), is given by the relation:

$$S = G\ \exp(-2G) \quad [\text{packet/time slot}] \tag{3.55}$$

where S = transmission throughput and G are expressed as a number of packets per time slot equal to the common packet duration. The Aloha protocol cannot exceed a throughput of 18 %, and the mean transmission time increases very rapidly as the traffic increases due to an increasing number of collisions and packet retransmissions.

The Aloha mode is relatively inefficient with a maximum throughput of only 18.4 % (1/2). However, this has to be counterweight against the gains in simple network complexity, since no-coordination or complex timing properties are required at the transmitting MES.

3.5.6.2 Slotted Aloha

This form of Aloha or S-Aloha, where the time domain is divided into slots equivalent to a single packet burst time; there will be no overlap, as is the case with ordinary Aloha. In Fig. 3.27 is shown a simplified arrangement that presents this point. The transmissions from different stations are now synchronized in such a way that packets are located at the satellite in time slots defined by the network clocks and equal to the common packet duration. Hence, there cannot be partial collisions; every collision arises from complete superposition of packets. In effect, the timescale of collision is thus reduced to the duration of a packet, whereas with the Aloha protocol, this timescale is equal to the duration of both packets. At this point, this situation divides the probability of collision by two and the throughput becomes:

$$S = G\ \exp(-G) \quad [\text{packet/time slot}] \tag{3.56}$$

This protocol enables collisions between new messages and retransmission to be avoided and increases the throughput of S-Aloha in the order of 50–60 % by introducing a frame structure, which permits the numbering of time slots. Each packet incorporates additional information indicating the slot number reserved for retransmission in case of collision.

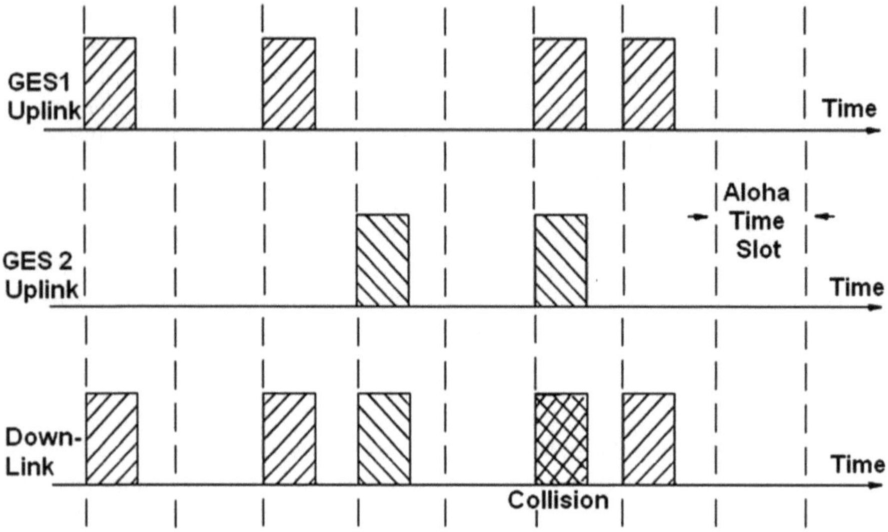

Fig. 3.27 Time slot organization for Slotted Aloha—Courtesy of Book: by Calcutt

For the same value of utilization as basic Aloha, the time delay and probability of packet loss are both improved. The major disadvantages of S-Aloha are that more complex equipment in the Earth station is necessary, and because of the timing requirement and because there are fixed time slots, customers with a small transmission requirement are wasting capacity by not using the time slot to its full availability.

3.5.6.3 Slot Reservation Aloha

This solution of an extension for the slotted Aloha scheme allows time slots to be reserved for transmission by an Earth station. In general terms, this mode of operation is termed as a Packed Reserved Multiple Access (PRMA). Slot reservation basically takes the following two forms:

1. **Implicit**—When a station acquires a slot and successfully transmits, the slot is reserved for that station for as long as it takes the station to complete its transmission. The network controller then informs all stations on the network that the slot is available for contention once more. There is only the problem that a station with much data to transmit could block the system to other users.
2. **Explicit**—Every user station may send a request for the reservation of a time slot prior to transmission of data. A record of all time slot occupation and reservation requests is kept. Actually, a free time slot could be allocated on a priority basis. Some kind of control for the reservation of slots is necessary, and

this could be accomplished by a single or all stations being informed of slot occupancy and reservation requests.

3.6 Mobile Broadband and Internet Protocols

Under the third- and fourth-generation IMT-2000 services, high-speed and large-capacity multimedia data and Internet will be delivered through the satellite and radio networks, and a wide variety of network services such as global roaming will be provided at the same time. In such a way, many solutions and protocols will be employed in terrestrial radio and telecommunication systems, including ISDN, IP, Transmission Control Protocol (TCP), IP/ATM, Unspecified Bit Rate (UBR), and Available Bit Rate (ABR) broadband and Internet solutions and protocols.

With its new advanced service menu, the IMT-2000 network is required to provide not only mobility management for individual terminals but also various service control functions. With capabilities such as QoS, in which the desired communications quality is individually established for each session, the IP/ATM switching system provides for a sophisticated and economic network configuration.

Usage of different satellite fixed and mobile solutions has increased dramatically in past years. Along with that trend, there has been an increasing demand for higher transmission rates. Hence, to cope with this demand, new technologies have emerged onto the radio and telecommunications stage. The IP/ATM over satellite is one fine example, which is creating waves in the global telecommunication world today and is ideal for a wide range of satellite applications, including data transfer, voice, imaging, full motion video, and multimedias. Its characteristics include scalability, which make it ideal for the traffic demands of broadband multimedia expected on data networks of the future. Just as the Internet revolutionized worldwide communications, ATM brings new meaning to high-speed satellite networking.

The development of the Internet technologies is the concern of the Internet Engineering Task Force (IETF), which publishes its recommendations on the Web site under a series of Request for Comments (RFC) documents, with a specific identification number for each RFC. These documents are available for downloading from the IETF Web site in the form of information FYI RFC documents or those that specify Internet Standards (STD RFC).

3.6.1 Mobile Internet Protocol (IP)

The drive toward the establishment of an information society will bring together the two most successful solutions of all of the technological advances of the latter quarter of the twentieth century. This is the integration of Internet and mobile communications to deliver Mobile IP to all mobile customers. In this respect, fourth-generation (4G) mobile networks will be based upon an IP-environment.

Significant effort around the world is now underway toward standardizing such a mobile environment through such organizations as the Mobile Wireless Internet Forum (MWIF), IETF, 3GPP, and 3GPP2. Recently, two different approaches to the network architecture by 3GPP and 3GPP2 have been proposed. First, the 3GPP solution for the W-CDMA radio interface is based on an evolution of the GMPSC network, with enhancements to the call control functionalities obtained though the introduction of a new network element, Called State Control Function (CSCF), to allow the provision of voice over IP (VoIP) services. Second, the 3GPP2 solution for the CDMA2000 radio interface system has adopted a forthcoming packet network architecture incorporating mobile IP functionalities in support of packet data mobility. Eventually, it is hoped that a single, harmonized solution will emerge from the two distinct approaches. In the future mobile network, therefore, where satellites will operate alongside terrestrial mobile networks, various categories of service will coexist, where "best effort," which is presently available over the Internet, will operate with guaranteed QoS classes.

Moreover, the move toward an all IP network should facilitate the networking between satellite and terrestrial mobile networks, since the problem of providing mobile link connectivity across the different networks reduces to the level of attaining the appropriate route to direct the packets of information to the appropriate terminal. In the mobile IP environment, the mobile terminal has a temporary IP address, known as the Care-of Address (CoA), which is associated with a correspondent node of a particular access network. The CoA is in addition to the IP address that is permanently assigned to the mobile terminal, which is its home address, which is stored at its home subnetwork. Thus, the mobile node's CoA is made known to the home agent. Packets addressed to the mobile node are thus tunneled by the home agent to the correspondent node, identified by the CoA. Namely, when a mobile node moves from one access network to another, a new CoA will be assigned by the visiting network operator corresponding to the new correspondent node.

The mobile node then makes the home agent aware of the change in the CoA, through the transmission of a binding update message. Mobile IP and the mobile router will play a major role in NASA's aeronautics programs, including AATT, WINCOMM, and SATS. Each program requires continuous connectivity, while dynamically traversing various subnetworks. Figure 3.28 illustrates what a potential aeronautical network could entail in 2005. Although only the USA is shown, connectivity will however be global. In addition, the future aeronautical networks will use whatever wireless networking is available at reasonable cost. The LEO, MEO, HEO, or GEO satellite constellations may be used, as will VHF data link (VDL), future stratospheric communication platforms (SCP), and perhaps next-generation cellular, if it survives the competition previously mentioned.

Unique, cost-effective satellite techniques will also be deployed using GEO satellite direct broadcasting, known as Digital Video Broadcasting (DVB). Digital TV satellites with data embedded in the Moving Picture Expert Group 2

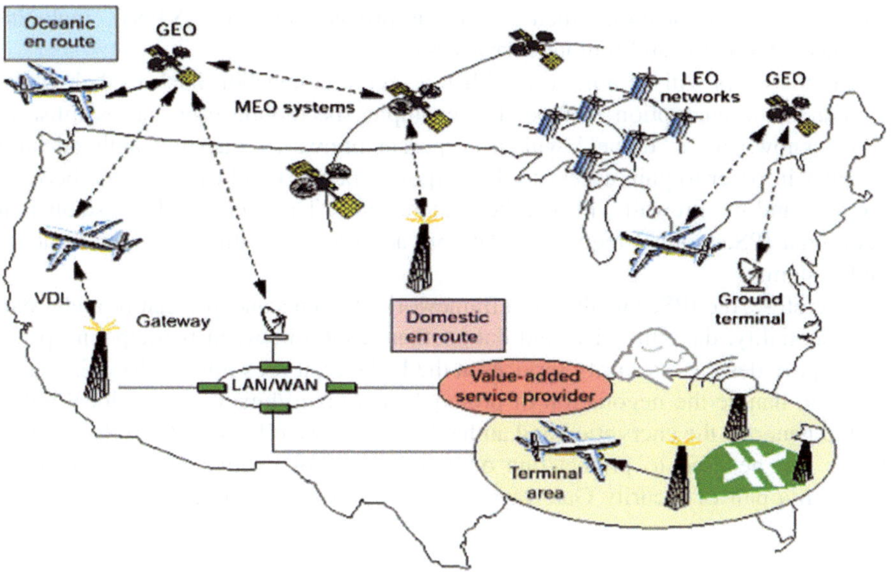

Fig. 3.28 WAAS CNS over IP—Courtesy of Paper: by NASA

(MPEG-2) transport streams to transmit data to the aircraft. The return path would be an inexpensive, low-bandwidth, duplex channel. Under the AATT program, NASA is working in alliance with the FAA to enable an increase in capacity, flexibility, and efficiency, while maintaining flight safety of aircraft operations within the USA and the global airspace system. Similar to a WAAS, the goals are to increase terminals throughput by 40 % and to increase en route throughput by 20 %. Current operations in the National Air System (NAS) are severely constrained by an antiquated ATM, supported by a decades-old CNS infrastructure, using primarily analog radio voice, radar surveillance, and ground-based navigational aids. Extensive ATC system delays and gridlock have resulted. A digital airspace WAAS/IP infrastructure is needed to meet the bandwidth, latency, security, and integrity requirements of the future free-flight ATM system. The mobile router is an enabling technology for such a system.

3.6.1.1 IP Security Protocol (IPSec)

The IPSec protocol is a framework of open standards developed by the IETF, which can provide security for the transmission of sensitive information over unprotected networks such as the Internet. In this sense, the IPSec protocol acts at the network layer, protecting and authenticating IP packets between participating IPSec devices, such as Cisco or other IP routers. With IPSec, data can be transmitted across a public network without fear of observation, modification, or spoofing. This mode enables fixed applications such as virtual private networks (VPNs), including intranets, extranets, and remote user access.

Mobile users will be able to establish a secure connection back to their office, similar to the encryption method. For example, the mobile user can establish an IPSec known as a "tunnel" with a corporate firewall, requesting authentication services in order to gain access to the corporate network; all of the traffic between the user and the firewall will then be authenticated. The user can then establish an additional IPSec tunnel requesting data privacy services with an internal router or end system.

Therefore, the IPSec method is a framework of open standards that provides data confidentiality, data integrity, and data authentication between participating peers. IPSec provides these security services at the IP-layer; it uses Internet Key Exchange (IKE) to handle the negotiation of protocols and algorithms based on local policy and to generate the encryption and authentication keys to be used by IPSec. In fact, IPSec can be used to protect one or more data flows between a pair of hosts, between a pair of security Gateways, or between a security Gateway and a host.

3.6.1.2 Mobile Transmissions Over IP (MToIP)

The IP data link can be used for transmissions such as regularly made long-distance phone calls, videoconferencing connections, or sending fax messages. In such a way, IP telephony is known as a Voice over IP (VoIP), while similar to this, Videophone can also be established over IP (VPoIP) and fax over IP (FoIP). Actually, these services are the transmission of telephone, video, or fax calls over a data network like one of the many terrestrial networks that make up the Internet or will be possible through Mobile Satellite Internet (MSI) as future MSC standards.

On the other hand, a VoIP application meets the challenges of combining legacy voice networks and packet networks by allowing both voice and signaling information to be transported over the packet network. A FoIP application enables the networking of standard fax machines with packet networks. It accomplishes this by extracting the fax image from an analog signal and carrying it as digital data over the packet network. The VPoIP mode is a similar application, which can enable the mobile and semi-fixed networking of standard video cameras, videoconferencing, or videophones with mobile packet network or via mobile ISDN links.

However, the TDM over IP (TDMoIP) is a transport technology that extends T1, E1, T3, E3, serial data, or analog voice circuits transparently across IP or Ethernet networks. When used for voice mode, these TDM over IP circuits are transparent to signaling and provide superior voice quality with much lower latency than VoIP. The TDM scheme over IP supports all PBX features and all modem and fax rates. In other words, TDMoIP is not limited to voice and can extend to circuits carrying any protocol over IP including Frame Relay, ATM, ISDN, SS7, SNA, HDLC, Asynchronic, Synchronic, and X.25, as well as H.320 and H.324 video over IP (VPoIP). The TDM scheme over IP was recently developed by the RAD Company to provide a simple, inexpensive migration strategy to IP-based networks.

3.6.1.3 Mobile IP Version 6 (MIPv6)

The IP was introduced in the Arpanet in the mid-1970s; thus, the version in common used today is IP version 4 (IPv4). Although several protocol suites (including Open System Interconnection) have been proposed over the years to replace IPv4, none have succeeded because of IPv4's large and continually growing, installed base.

Nevertheless, IPv4 was never intended for the current Internet, either in terms of the number of hosts, types of applications, or security concerns. However, in the early 1990s, IETF recognized that the only way to cope with these changes was to design a new version of IP as a successor to IPv4. In such a way, the IETF formed the next IP generation (IPgen) Working Group to define this transitional protocol to ensure long-term compatibility between the current and new IP versions and support for current and emerging IP-based applications.

The result of this effort was IP version 6 (IPv6), described in RFC 1883–1886; these four RFC were officially entered into the Internet Standard Track in December 1995. IPv6 is designed as an evolution from IPv4 rather than as a radical change. Useful features of IPv4 were carried over in IPv6 and less useful features were dropped. Mobile IPv6 is an IP-layer mobility protocol for the IPv6 Internet, being standardized by the IETF. Thus, the idea is that when mobility, like any other functionality, is implemented in the network layer, it needs to be implemented only once and will then be transparently available for all higher layer protocols. It remains to be seen how well this promise is fulfilled in practice. There are, however, some applications such as mobile VPN access, for which Mobile IP is clearly a good solution.

The mobile network is configured in the way that the first half of an IPv6 address indicates the subnet to which the address belongs and it is used for routing IP packets across the Internet. When a mobile Internet host, known as a mobile node (MN) in the Mobile IPv6 terminology, moves to a different place in the network topology, its subnet and IP address necessarily change. This creates two kinds of problems: existing connections (e.g., TCP connections and IPSec security associations) between the mobile and other hosts known as a correspondent nodes (CN)s become invalid and the mobile is no longer reachable at its old address for new connections. The former problem is important in stateful protocols and has little effect on stateless protocols, such as Hypertext Transfer Protocol (HTTP). The latter problem typically concerns servers and not client computers.

Mobile IPv6 protocol has two basic goals: All transport-layer and higher layer connections and security associations between the mobile and its correspondents should survive the address change, and the mobile host should be reachable as long as it is connected to the Internet somewhere in the world. Mobile IP makes some quite strong assumptions about the environments in which it is used. First, that all mobile hosts have a home network and a Home Address (HoA) on that network.

This is a reflection from a time when mobility was an exception: Few Internet nodes would be mobile and even they would for most of the time remains stationary at home.

In any case, Mobile IP solves the reachability problem by ensuring that the mobile is always able to receive packets sent to its home address. The IP address of a stationary IP node normally serves two purposes: It is both an identifier for the node and an address that is used for routing messages to the node. Mobile IP preserves this dual use of home addresses, which are an identifier for the mobile, as well as an address to which correspondents can send packets.

The mobile's current location, called Care-of Address (CoA), on the other hand, is a pure address and serves no identification purposes. Any IPv6 address can be or become mobile, and there is no way of distinguishing a mobile and stationary host by just looking at its address. This is because the Mobile IP protocol was originally designed to be transparent to the mobile's correspondents, and the correspondent did not need to know that the mobile, in fact, was a mobile.

An additional technical advantage in the adoption of the MIPv6 IP solution as a final stage of standardization should be determined to provide the opportunity to perform seamless communications handover between satellite and TTN. Recent trials involving handover between TTN using MIPv6 have demonstrated the feasibility of such an approach. At any rate, researchers are now addressing the needs of integrated space/terrestrial mobile communication networks, based upon packet-oriented service delivery.

3.6.2 Transmission Control Protocol (TCP)

The transmission of Internet packets in Tx-Rx direction is primarily achieved using the TCP solution. The TCP mode is a connection-oriented transport protocol that sends data as an unstructured stream of bytes and provides the functionality to ensure that the transmission rate of data over the network is appropriate for the capabilities of the Rx device, as well as the devices that are used to route the data from the Tx to the Rx. Thus, by using sequence numbers and acknowledgment messages, TCP can provide a sending node with delivery information about packets transmitted to a destination node. Where data has been lost in transit from source to destination, TCP can retransmit the data until either a time-out condition is reached or until successful delivery has been achieved or can also recognize duplicate messages and will discard them appropriately.

If the sending PC is transmitting too fast for the receiving computer, TCP can employ flow control mechanisms to slow data transfer. The TCP method can also communicate delivery information to the upper-layer protocols and the applications it supports and is responsible for ensuring that the network's resources are divided in an equitable manner among all users of the network. Applications such as file transfer protocol (FTP) and HTTP (the language of the Web) rely on TCP to transport their data over the network as quickly as the network will allow.

3.6.2.1 TCP/IP Over Satellite

As far a TCP/IP connection is concerned, a mobile satellite network should be viewed as any other network connection. Given the way that TCP operates, the transmission of IP packets over the MSC link poses several problems that need to be overcome if services are to be delivered efficiently. In such a manner, the major difficulty is due to the latency of the satellite link, which when combined with a burst error channel and the characteristics of the TCP protocol itself can result in an inefficient means of transmission. This is because TCP operates using a conservative congestion control mechanism, whereby new data can be transmitted only when an ACK (acknowledgment) from the previous transmission has been received.

Therefore, with this in mind, the need for a high-quality link between the satellite and MES is re-emphasized, since packets in error are presently deemed to be due to congestion on the satellite network. Hence, in this case, TCP responds by reducing its transmission rate accordingly. It can be seen that TCP operates on the basis of "best effort," founded on the available resources of the network.

When starting transmission, TCP enters the network in a restrained manner, whereby the initial rate of transmission is carefully controlled to avoid overloading the network with traffic. The TCP method achieves this by employing a congestion control mechanism such as slow start, congestion avoidance, fast retransmit, and recovery.

Slow start is used, as its name suggests, at the start of transmission or after congestion of the network has been detected and the data transmission is reduced. On the other hand, congestion control is used to gradually increase the transmission rate once the initial data rate has been ramped-up using the slow start algorithms. In this sense, the fast transmit and fast recovery algorithms are used to speed up the recovery of the transmission rate after significant congestion in the communication network has been detected.

3.6.2.2 TCP Intertwined Algorithms

Modern implementations of TCP contain four intertwined algorithms that have never been fully documented as Internet control mechanism standards, such as slow start, congestion avoidance, fast retransmit (FRet), and fast recovery (FRec).

1. **Slow Start**—The old TCP mode would start a connection with the sender injecting multiple segments into the network, up to the window size advertised by the receiver. While this works when the two hosts are on the same LAN, if there are routers and slower links between the sender and the receiver, problems can arise. Some intermediate router must queue the packets and it is possible for that router to run out of space. Hence, the algorithm to avoid this is called slow start. It operates by observing that the rate at which new packets should be

injected into the network is the rate at which the acknowledgments are returned by the other end.

Slow start adds a window to the sender's TCP known as Congestion Window (CWnd), which is initialized to one segment when a new connection is established with a host on another network. Each time an ACK is received, the CWnd is increased by one segment. The Receiver Advertised Window (RWnd) is the maximum amount of data that can be buffered at the receiver, which is determined by the minimum of CWnd and RWnd and ensuring that the Rx is not overloaded with data. The sender can transmit up to the minimum of the CWnd and the RWnd. The CWnd is flow control imposed by the sender, while the RWnd is flow control imposed by the Rx. The former is based on the sender's assessment of perceived network congestion; the latter is related to the amount of available buffer space at the Rx for this connection. The sender starts by transmitting one segment and waiting for its ACK. When that ACK is received, the CWnd is incremented from one to two and two segments can be sent. When each of those two segments is acknowledged, the CWnd is increased to four. This provides an exponential growth, although it is not exactly exponential because the receiver may delay its ACK, typically sending one ACK for every two segments that it receives. At some point, the capacity of the Internet may be reached, and an intermediate routers will start discarding packets. This tells the sender that its congestion window has become too large. However, it can be seen that after transmitting the first segment, the data sender remains idle until an ACK message has been received. For a GEO satellite, this would take in the region of roughly 500–570 ms.

2. **Congestion Avoidance**—Congestion can occur when data arrives on a big pipe (a fast LAN) and gets sent out via a smaller pipe (a slower WAN). It can also occur when multiple input streams arrive at a router whose output capacity is less than the sum of the inputs. Therefore, congestion avoidance is a way to deal with lost packets of data. The assumption of the algorithm is that packet loss caused by damage is very small (much less than 1 %); therefore, the loss of a packet signals congestion somewhere in the network between the source and destination. There are two indications of packet loss: a time-out occurring and the receipt of duplicate ACK.

The congestion avoidance and slow start are independent algorithms with different objectives. But when congestion occurs, TCP must slow down its transmission rate of packets into the network and then invoke slow start to get things going again. Hence, in practice, they are implemented together. They require that two variables be maintained for each connection: a CWnd and a Slow Start Threshold (SSThres) size. If CWnd is less than or equal to SSThres, TCP is in slow start, while TCP is performing congestion avoidance. Slow start continues until TCP is halfway to where it was when congestion occurred and then congestion avoidance takes over. Slow start has CWnd begin at one segment and be incremented by one segment every time an ACK is received.

3. **Fast Retransmit**—Before describing the change, realize that TCP may generate an immediate ACK (a duplicate ACK) when an out-of-order segment is received, with a note that one reason for doing so was for the experimental fast retransmit algorithm. Thus, this duplicate ACK should not be delayed, and the purpose is to let the other end know that a segment was received out of order and to tell it what sequence number is expected. Since TCP mode does not know whether a duplicate ACK is caused by a lost segment or just a reordering of segments, it waits for a small number of duplicate ACK to be received. It is assumed that if there is just a reordering of the segments, there will be only one or two duplicate ACK before the reordered segment is processed, which will then generate a new ACK. If three or more duplicate ACK are received in a row, it is a strong indication that a segment has been lost. The TCP mode then performs a retransmission of what appears to be the missing segment, without waiting for a retransmission timer to expire.
4. **Fast Recovery**—After fast retransmit sends what appears to be the missing segment, the congestion avoidance but not slow start is performed. That is to say, this is practically the fast recovery algorithm. It is an improvement that allows high throughput under moderate congestion, especially for large windows. The reason for not performing slow start in this case is that the receipt of the duplicate ACK tells TCP that more than just a packet has been lost. Since the receiver can only generate the duplicate ACK when another segment is received, that segment has left the network and is in the receiver's buffer. That is, there is still data flowing between the two ends and TCP does not want to reduce the flow abruptly by going into slow start.

3.6.3 Mobile Asynchronous Transfer Mode (ATM)

The ATM solution is a network switching technology used by Broadband Integrated Service Digital Network (BISDN). It uses a technique called cell switching, breaking all data into cells, or packets and transmits them from one location on the network to another, connected by switches. So, the small, constant cell size allows ATM equipment to transmit audio, video, and computer data over the same network and assure that no single type of data hogs the line.

The latest implementations of ATM support data transfer rates from 25 to 622 Mb/s, which compares to a maximum of 100 Mb/s for Ethernet. It is a protocol designed to handle isochronous (time critical) data such as video and telephony (audio), in addition to more conventional data communications between computers. These protocols are capable of providing a homogeneous network for all traffic types.

The same protocols are used regardless of whether the application is to carry conventional telephony (VoIP), fax over IP (FoIP), entertainment video over IP (VPoIP), or computer network traffic over LAN, MAN, WAN, or satellite networks.

Small and constant packet size allows switching to be implemented in hardware, rather than have routing done in the software. This makes ATM switches sufficiently rapid that multiple isochronous data can be statistically multiplexed together. The significance of this is that ATM protocols provide bandwidth on demand. Network management software will allow small amounts of bandwidth to be set aside for simple transactions, such as e-mails, while allowing more bandwidth for resource intensive multimedia applications. This sort of technology means that a level of bandwidth can be guaranteed and that a connection is not delayed or interrupted by network traffic, which in ATM parlance is known as QoS.

The ATM transport protocol is universal since it can be used for all kinds of networks be it physical networks, (twisted pair, coax, and fiber optics) or virtual networks, such as radio and satellite systems. Today's emphasis on multimedia presentations, videoconferencing, remote lecturing, etc., has made ATM more attractive to network administrators. Simply throwing more bandwidth onto the network is not the solution to the problem, since not all demands on the network are equal. The ATM's ability to provide the bandwidth and QoS guarantees makes it the obvious upgrade choice. Other areas where ATM is used include Digital TV, HDTV, and video vending. A public ATM network would offer some entirely new services to people in their homes. The most significant type of new services will be the possible video services. What will be so unique about these services is the opportunity for a viewer to interact more with the programs they receive. At this point, the use of the ATM solution would reduce costs considerably, making ATM a very attractive technology for mobile applications, such as oil companies, shipping, and airlines. Many designers of satellite systems are thinking about implementing the ATM protocol, which transmits data that have been placed in cells of a constant length (53 bytes).

Thus, the ATM guarantees data transmission at a rate ranging between 2 Mb/s and 2.4 Gb/s. The protocol acts on the principle that a virtual channel should be set up between two points whenever such a need appears. This is what makes the ATM protocol different from the TCP/IP protocol, in which messages are transmitted in packet form, where each packet may reach the recipient via a different route. The ATM protocol enables data transmission through various media. Taking into account the header of the cell (cell-tax), which takes 5 bytes, the application of the ATM protocol may not appear to be so cost-effective when the rate of transmission is low and the capacity of the link (e.g., in two-way modem channels) becomes a basic limitation.

As noted, the ATM technology is expected to provide QoS-based networks that support voice, video, and data applications. Initially, the ATM protocol was originally designed for fiber-based terrestrial networks that exhibit low latencies and error rates. With the increasing demand for electronic connectivity across the world, satellite networks play an indispensable role in the deployment of global networks. The Ka-band communication link using the GHz frequency satellites spectrum can reach user terminals across most of the populated world. Thus, the ATM-based satellite networks can effectively provide real-time as well as non-real-time communications services to remote areas.

There are four different types of ATM service.

1. **Constant Bit Rate (CBR)**—The CBR ATM service specifies a fixed bit rate so that data is sent in a steady stream. This is analogous to a leased line.
2. **Variable Bit Rate (VBR)**—This ATM service provides a specified throughput capacity but data is not sent evenly. This is a popular choice for voice and videoconferencing data.
3. **Unspecified/Undefined Bit Rate (UBR)**—It does not guarantee any throughput levels. This is used for applications, such as file transfer, that can tolerate delays.
4. **Available Bit Rate (ABR)**—It provides a guaranteed minimum capacity but allows data to be burst at higher capacities when the network is free.

However, there is opinion that ATM holds the answer to the Internet bandwidth problem, but others are skeptical. Thus, ATM creates a fixed channel, or route, between two points whenever a data transfer begins. This differs from TCP/IP, in which messages are divided into packets and each packet can take a different route from source to destination. This difference makes it easier to track and bill data usage across an ATM.

3.6.3.1 IP/ATM Over Satellite

As mentioned, the ATM terrestrial technology can be successfully used in satellite communication networks to enhance the overall performance of the network. This network as usual can comprise corresponding satellite configuration and several ground stations communicating via satellite on the one hand and ATM Switch TTN infrastructure on the other, interconnected with the main component of the network, known as ATM Satellite Internetworking Unit (ASIU). The ASIU is responsible for management and control of system resources as well as the overall system administrative functions, such as real-time bandwidth allocation, network access control, system timing, and traffic control. This fixed satellite configuration can be very easily implemented in MSC with connecting MES terminals via satellite with LES and ASIU to the ATM Switching TTN.

The internal architecture of the ASIU in the satellite direction stage is to extract the ATM cells from the various transport methods. After the ATM cells are extracted, they are error corrected and then buffered according to their type. After that, the cells are encoded and transmitted through the modem to the satellite. Hence, looking at the various components in a bit more detail, the first stage will be the extraction of ATM cells from the transport protocols, using the following methods:

1. **Plesiochronous Digital Hierarchy (PDH)**—The PDH protocol was developed to carry digitized voice efficiently. At any rate, it mainly operates by multiplexing various rates of bit streams at the highest allowed clock speed. When necessary, it adds a stuffing bit which the demultiplexer can later remove, and

therefore, the stuffing procedure is inefficient. Also, rerouting signals after network failures and managing remote network elements are extremely difficult.
2. **Physical Layer Convergence Protocol (PLCP)**—On the other hand, PLCP mode works differently. The PLCP combines 12 ATM cells in one frame format with a header in front. Consequently, this method is not very good for satellite transmission as it is very sensitive to burst errors, which may result in the loss of a whole frame and loss of synchronization of the PLCP device.
3. **Synchronous Optical Network (SONET) or Synchronous Digital Hierarchy (SDH)**—For the reasons already explained, PDH and PLCP are not suitable to be used with ATM and engineers have been turning to SDH as the protocol for ATM. The SONET, similar to the SDH protocol, uses pointer bytes to indicate the location of the first byte in the payload of the SDH frame. Thus, the SDH protocol incorporates a cell delineation mechanism for the acquisition and synchronization of the ATM cells on the receiver side of the network and has some considerable advantages over PDH/PLCP, the main being flexibility of service, allowing the network operator to respond quickly to customers' requirements, improved quality and supervision, permitting the operator to increase the quality of the offered services.

Furthermore, the SDH protocol reduces operations costs, by using efficient network management technology, which makes the remote control of the network possible mostly without on-site activities. Then, the higher rates of transfer are well defined and direct multiplexing is possible without an intermediate multiplexing stage. In fact, despite these advantages, SDH has an important problem in that incorrect pointer detection may produce an incorrect payload extraction and an error block. As a result, all received blocks may be incorrect and severely errored. To overcome this problem, an efficient error mechanism, error detection, and recovery must be used in the next stage of the ASIU architecture.

However, satellite systems have several inherent constraints. The resources of the satellite network, especially the satellite and the Earth station, are expensive and typically have low redundancy; these must be robust and be used efficiently. Thus, the large delays in GEO and delay variations in LEO systems affect both real-time and non-real-time applications. In an acknowledgment and time-out-based congestion control mechanism (like TCP), performance is inherently related to the delay–bandwidth product of the connection. Moreover, TCP round-trip time (RTT) measurements are sensitive to delay variations that may cause false timeouts and retransmissions. As a result, the congestion control issues for broadband satellite networks are somewhat different from those of low-latency terrestrial networks. Both interoperability issues as well as performance issues need to be addressed before a transport-layer protocol like the TCP model can satisfactorily work over long latency satellite ATM networks.

Satellite ATM networks can be used to provide broadband access to remote locations, as well as to serve as an alternative to fiber-based backbone networks. In either case, a single satellite is designed to support thousands of Earth terminals.

The Earth terminals set up virtual channels (VC) through the onboard satellite switches to transfer ATM cells among one another.

Because of the limited capacity of a satellite switch, each MES has a limited number of VC it can use for TCP/IP data transport. In backbone networks, these MES are IP/ATM edge devices that terminate ATM connections and route IP traffic in and out of the ATM network. Namely, these high-capacity backbone routers must handle thousands of simultaneous IP flows. As a result, the routers must be able to aggregate multiple IP flows onto individual VC. Flow classification may be done by means of a QoS that can use IP source–destination address pairs, as well as transport-layer port numbers. Therefore, the QoS manager can further classify IP packets into flows based on the differentiated services code points in the TOS byte of the IP header.

In addition to flow and VC management, all MES terminals also provide a means for congestion control between the IP and ATM networks. The onboard ATM switches must perform traffic management at the cell and VC levels. Hence, TCP hosts implement various TCP flow and congestion control mechanisms for effective network bandwidth utilization. The enhancements that perform intelligent buffer management policies at the switches can be developed for UBR service to improve transport-layer throughput and fairness. A policy for selective cell drop based on per-VC accounting can be used to improve fairness.

Providing a minimum guaranteed rate (GR) to UBR traffic has been discussed as possible candidate to improve TCP performance over UBR. The goal of providing GR is to protect the UBR service category from total bandwidth starvation and provide a continuous minimum bandwidth guarantee. It has been shown that in the presence of high loads of higher priority constant bit rate (CBR), variable bit rate (VBR), and available bit rate (ABR) traffic, TCP congestion control mechanisms benefit from a minimum GR.

Moreover, guaranteed frame rate (GFR) has recently been proposed in the ATM Forum as an enhancement to the UBR service category. GFR will provide a minimum rate guarantee to VC at the frame level.

The GFR service also allows for the fair usage of any extra network bandwidth. GFR is likely to be used by applications that can neither specify the traffic parameters needed for a VBR VC, nor have the capability for ABR (for rate-based feedback control). Current Internetworking applications fall into this category and are not designed to run over QoS-based networks. Routers separated by satellite ATM networks can use the GFR service to establish VC between one another. GFR can be implemented using per-VC queuing or buffer management.

3.6.3.2 UBR Over Satellite

The Unspecified Bit Rate (UBR) service class is intended for delay-tolerant or non-real-time applications, that is, those that do not require tightly constrained delay and delay variation, such as traditional computer communications applications. Sources are expected to transmit non-continuous bursts of cells. Namely, the

UBR service supports a high degree of statistical multiplexing among sources and includes no notion of a per-VC allocated bandwidth resource. Thus, transport of cells in UBR service is not necessarily guaranteed by mechanisms operating at the cell level. However, it is expected that resources will be provisioned for UBR service in such a way as to make it usable for some sets of applications. The UBR service may be considered as an interpretation of the common term "best effort service." The UBR service is typically used for data transmission applications such as file transfer and e-mail. Neither an ATM-attached router nor an ATM switch provides traffic or QoS guarantees to a UBR virtual circuit.

As a result, UBR VC can experience a large number of cell drops or a high cell transfer delay as cells move from the source to the destination device. In its simplest form, an ATM switch implements a tail drop policy for the UBR service category. Hence, if cells are dropped, the TCP source loses time waiting for the retransmission timeout. Even though TCP congestion mechanisms effectively recover from loss, the link efficiency can be very low, especially for large delay–bandwidth networks. In general, link efficiency typically increases with increasing buffer size. In fact, the performance of TCP over UBR can be improved by using buffer management policies. In addition, TCP performance is also affected by TCP congestion control mechanisms and TCP parameters such as segment size, timer granularity, receiver window size, slow start threshold, and initial window size.

The TCP Reno implements the fast retransmit and recovery algorithms that enable the connection to quickly recover from isolated segment losses. However, fast retransmit and recovery cannot efficiently recover from multiple packet losses within the same window. A modification to Reno TCP is proposed, so that the sender can recover from multiple packet losses without having to timeout. The TCP solutions with selective acknowledgment (SACK) are designed to efficiently recover from multiple segment losses. With SACK, the sender can recover from multiple dropped segments in about one round trip. The studies show that in low-delay networks, the effect of network-based buffer management policies is very important and can dominate the effect of SACK. The throughput improvement provided by SACK is very significant for long latency connections. When the propagation delay is large, time-out results in the loss of a significant amount of time during slow start from a window of one segment.

Reno TCP (with fast retransmit and recovery) results in the worst performance (for multiple packet losses) because time-out occurs at a much lower window than in Vanilla TCP. With SACK TCP, a time-out is avoided most of the time and recovery is complete within a small number of round trips. The NewReno modification to the fast retransmit and fast recovery algorithms (TCP Reno) has been therefore proposed to counteract multiple packet drops where the SACK option is not available. For lower delay satellite networks (LEO), both NewReno and SACK TCP provide high throughput but as the latency increases, SACK significantly outperforms NewReno, Reno, and Vanilla.

3.6.3.3 ABR Over Satellite

The ABR service category is another option to implement TCP/IP over ATM. This service category is specified by a Peak Cell Rate (PCR) and a Minimum Cell Rate (MCR), which is guaranteed by the network. ABR connections use a rate-based closed-loop end-to-end feedback control mechanism for congestion control. The network tries to maintain a low cell loss ratio by changing the Allowed Cell Rate (ACR) at which a source can send. Switches can also use the virtual source/virtual destination (VS/VD) feature to segment the ABR control loop into smaller loops. Studies have indicated that ABR with VS/VD can effectively reduce the buffer requirement for TCP over ATM, especially for long delay paths. In addition to network-based drop policies, end-to-end flow control and congestion control policies can be effective in improving TCP performance over UBR. The fast retransmit and recovery mechanism can be used in addition to slow start and congestion avoidance to quickly recover from isolated segment losses. Otherwise, the SACK option has been proposed to recover quickly from multiple segment losses.

The TCP performance over ABR service is important for the satellite IP network. Namely, a key ABR feature, VS/VD, highlights its relevance to long delay paths. Most of the consideration assumes that the switches implement a rate-based switch algorithm such as ERICA+. Credit-based congestion control for satellite networks has also been suggested, in long-latency satellite configurations; therefore, the feedback delay is the dominant factor in determining the maximum queue length. A feedback delay of 10 ms corresponds to about 3670 cells (at OC-3) of queue for TCP over ERICA+, while a feedback delay of 550 ms corresponds to 201,850 cells. This indicates that satellite switches need to provide at least one feedback delay of buffering to avoid loss on these high delay paths. A point to consider is that these large queues should not be seen in downstream workgroup or WAN switches, because they will not provide as much buffering. Satellite switches can isolate downstream switches from such large queues by implementing the VS/VD option, while VS/VD can effectively isolate nodes in different VS/VD loops. As a result, the buffer requirements of a node are bound by the feedback delay–bandwidth product of the upstream VS/VD loop. At this point, VS/VD helps to reduce the buffer requirements of terrestrial switches connected to Gateway terminals. The feedback delay–bandwidth products of the satellite hop are about 160,000 cells and dominates the feedback delay–bandwidth product of the terrestrial hop (about 3000 cells). Without VS/VD, terrestrial switch S, a bottleneck, must buffer cells up to the feedback delay–bandwidth product of the entire control loop (including the satellite hop). With a VS/VD loop between the satellite and the terrestrial switch, the queue accumulation due to the satellite feedback delay is confined to the satellite switch. The terrestrial switch only buffers cells that are accumulated due to the feedback delay of the terrestrial link to the satellite switch.

3.6.4 Fixed Digital Video Broadcasting-Return Channel via Satellite (DVB-RCS)

According to a recent Satellite Industry Association's report, over 130 M direct satellite TV subscribers are spread around the world. The key enablers attributed to this success include modern satellite systems capable of powerful transmissions over vast expanses, efficient transmission, highly integrated, low-cost receiver systems, and a vast variety of rich content at an affordable price. Current audio and video compression techniques dwell specifically on the MPEG multimedia standards, which constitute the industry's standards for compression and transport. At this point, a majority of direct broadcast satellite systems beaming standard-definition TV (SDTV) either use or can support the MPEG-2 standard, while HDTV and multimedia broadcast systems rely firmly on the standard's upgraded version, MPEG-4 mode. However, in last two decades is implemented the new standard of DVB-RCS system.

To encourage mass-scale acceptance, regulatory bodies promote standardization of the satellite broadcast service. At this point, ITU recognizes at least six direct-to home (DTH) broadcast systems that encompass television, data, and multimedia. Modern satellite and handheld receiver technologies enable television and radio broadcasts to handheld and portable personal devices. In the meantime, numerous commercial systems that offer sound and television services to the individual and groups have emerged in the last decade. The final developments of this technique offers the new personal and mobile solutions that will be in the future recognized as Digital Video Broadcast–Satellite Personal (DVB-SP) and Digital Video Broadcast–Satellite Mobile (DVB-SM) standards, both of which support commercial mobile television services.

The major development in satellite broadcasting technology was the standardization of the DVB-RCS, which allows the users within direct broadcast terrestrial network (DVB-T) to communicate directly with the broadcast satellite network (DVB-S) through an assigned return channel. The DVB-T cell can be comprised by UMTS/GPRS, ISDN, B-ISDN, and ATM broadband networks, and the DVB-S cells may include rural, consumer broadcasting SoHo/SME LAN, Corporate WAN, and multicasting networks. This greatly simplifies the overall network architecture and associated network management procedures, in that now all kind of communication solutions takes place over the same access network. Thus, the DVB-RCS system has been specified for indoor use only, so can be envisaged that very soon, research and standardization efforts will be directed toward establishing a suitable mobile standards. Will be enough to develop corresponding mobile satellite antenna, and this could open up significant new opportunities in the mobile satellite sector.

A sample of fixed DVB-RCS system is depicted in Fig. 3.29. The DVB-RCS network enables via HUB as a GES with C-, Ku-, or Ka-band antenna to interface the TTN (DVB-T) cell via corresponding GEO satellite connections (C-, Ku-, or

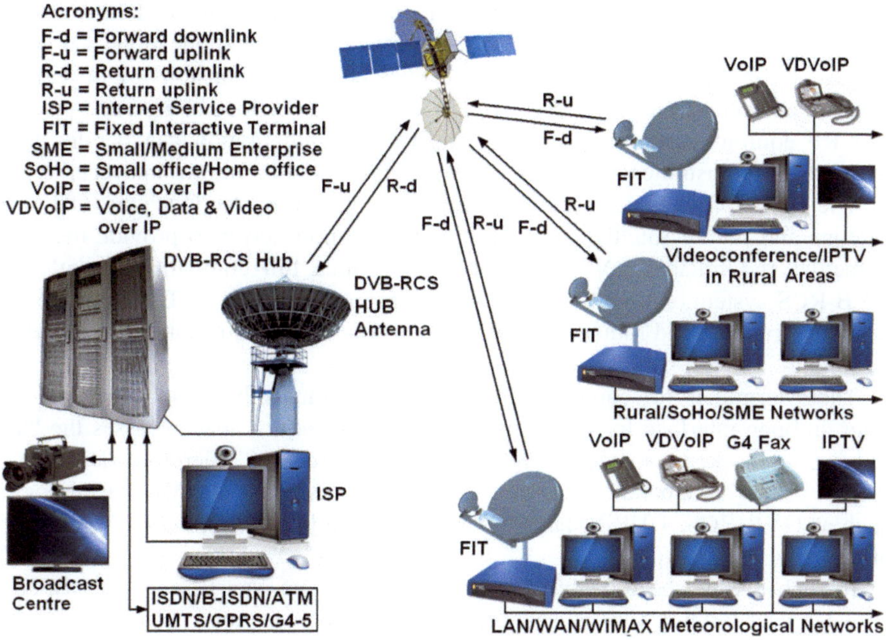

Fig. 3.29 Mixed DVB-RCS solutions—Courtesy of Manual: "Global Mobile CNS" by Ilcev

Ka-band GEO) to the Fixed Interactive Terminals (FIT) or Remote (DVB-S) cells for the following services:

(1) Regenerate rural communications: VoIP, IPTV, Internet access, and interactive TV/radio two separate way broadcasting (telephony/broadband/broadcast);
(2) Broadband access: Asymmetric Digital Subscriber Line (ADSL) anywhere and anytime, Internet access/e-mail, consumer, SoHo/SME LAN, Corporate WAN, Intranet/VPN, FTP, and HTTP; the FTP scheme is service for moving and copying an electronic file of any type from one computer to another over the Internet. For instance, it can be used both for downloads and uploads;
(3) Teleservice of all e-solutions, which include e-medicine and e-education and others: Videoconference, image/video/audio transfer and interactive distance learning; and
(4) Multicasting: Web casting, video streaming, satellite newsgathering, and push/pull data delivery and voice data and video over IP (VDVoIP).

The HUB terminal supports existing DVB-RCS compliant forward link system (FLS) and provides all the ground required interfaces and management functions necessary to set up a DVB-RCS service over 100 Mb/s downstreams. The HUB operator can interface DVB-RCS terminals to a terrestrial network or service provider and manage all the operational aspects of the system. The HUB station design optimized for robustness and stability in operations and can be delivered in a number of different configurations to suit customer precise applications. Its architecture is also designed to accommodate upgrades and expansions of DVB-RCS network and offer to DVB-RCS providers a choice of C-, Ku-, or Ka-band antenna and RF equipment, so one ordinary DVB-RCS configuration can handle 1–10 forward link transmitters and from one to several hundred return link receivers

The DVB-RCS FIT for rack mounting is composed by outdoor unit (ODU) of antenna and indoor satellite unit (ISU) of remote transceiver to provide the most attractive and reliable DVB solutions and ensures full compatibility with any DVB-RCS system or VSAT modem, ETSI/CE approval. The DVB-RCS remote is desktop terminals and router that provide two-way IP communications via satellite at C-, Ku-, and Ka-band frequencies.

These terminals for corporations, institutions, home offices, and householders offer an open-interface for high-capacity broadband access that bypasses the "last mile" bottleneck associated with terrestrial infrastructure. Therefore, DVB-RCS system offers broadband access to core IP networks using standard technologies such as DVB-RCS, DVB-S, latest DVB-S2, and IP interfacing DVB-T with fixed and mobile user terminals.

In such a way, the broadband network delivers two-way IP connectivity both between user terminals in the satellite system and between user terminals and the terrestrial network coming in a scalable choice of performance with a range of data IP throughputs from 4 to 30 Mb/s.

The broadcast MPEG2/DVB-S service is available in one-way (unidirectional) from the HUB to the users terminals. With very low cost per bit, service prices become comparable to those offered by terrestrial networks and can be delivered where other technologies cannot reach. This network can provide solutions in vas variety of user segments, such as enterprises and private networks, broadcasting and content distribution (BCD), satellite news gathering, satellite emergency and security management, and defense information management.

3.6.5 Mobile Digital Video Broadcasting-Return Channel Over Satellite (DVB-RCS)

The convergence of MSC and Internet technique has opened many opportunities to deliver new multimedia service over hybrid satellite systems to MES. The interactive nature of the Internet has paved the way for new-generation MSS to support interactivity. Apart from the convergence between mobile and Internet technologies, the other major technical driver is the convergence between mobile and fixed technologies.

By supplementing broadcasting systems with a narrowband uplink, new interactive services can be facilitated in DVB and DAB solutions. This is foreseen for fixed network operation and could equally be adopted onto a mobile network such

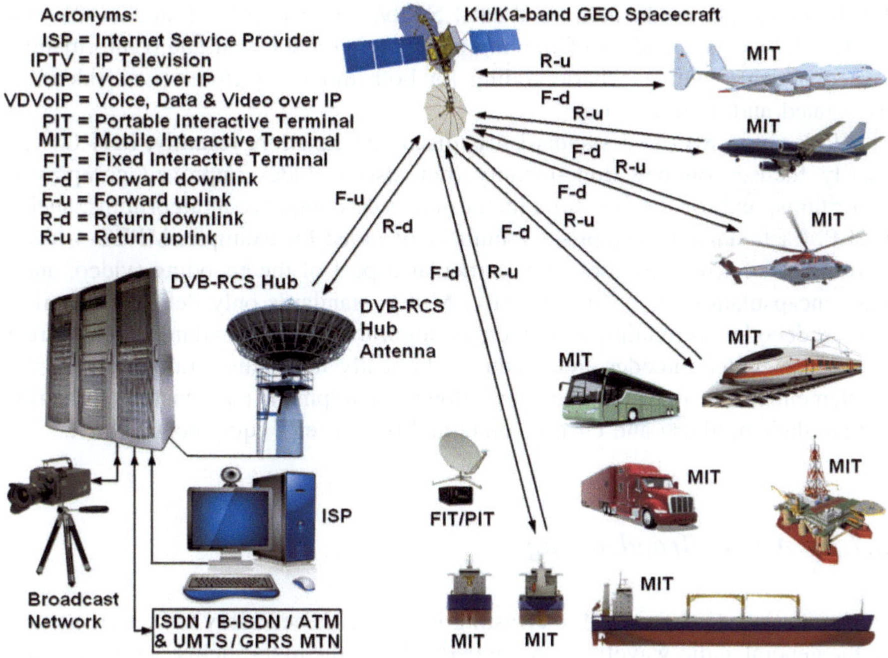

Fig. 3.30 Mobile DVB-RCS solutions—Courtesy of Manual: by Ilcev

as the UMTS, thus demonstrating the concept of convergence of personal and mobile communications, Internet, and broadcasting technology.

A configuration of mobile DVB-RCS system shown in Fig. 3.30 is designed by author in 2000. The DVB-RCS mobile network includes an HUB as a GES with C-, Ku-, or Ka-band antenna to interface the TTN (DVB-T) cell via corresponding satellite connections (C-, Ku-, or Ka-band GEO) to the Mobile Interactive Terminals (MIT) or Remote (DVB-S) cells for the following services: maritime and offshore installations, land (road and rail) solutions, and aeronautical service. This infrastructure is the best solution for establishment a network for connection of all seaports or airports in one hypothetical country or region. In any case, however, the modern aeronautical DVB-RCS network is the most reliable communication link between aircraft and airports for commercial and security purposes.

3.7 MPEG Multimedia Standards

The encoding of both high-fidelity (non-speech) audio signals and video signals today is dominated by the standards defined by the International Standards Organization (ISO) MPEG, subsequently adopted by both the ISO/IEC and the ITU. The first standard, MPEG-1, “Coding of moving pictures and associated audio for digital storage media at up to about 1.5 Mb/s," issued in 1992, and its successor MPEG-2, "Generic coding of moving pictures and associated audio information," issued in 1994, define generic coding for both moving pictures (i.e., video) and associated audio information.

Work on an MPEG-3 standard was discontinued. More recently, MPEG-4 primarily focuses on new functionality, but also includes improved compression algorithms, and for the first time includes speech compression algorithms, such as CELP. Each standard comprises a number of parts; for example, MPEG-4 has 23 parts, each of which focuses on a particular aspect of the encoding (video, audio, data encapsulation, etc.). Significantly, MPEG standards only define in detail the source decoder, providing a toolbox of algorithms, and standardized bit stream formats. Detailed encoder architecture is typically not defined (although example implementations are given), thereby allowing equipment manufacturers to differentiate their products and ensure compatibility of user's equipment.

3.7.1 Audio Broadcasting

With reference to the audio broadcasting in general is necessary to be turn attention to the general audio waveform and in particular to the high-fidelity audio, including stereo music, multichannel surround-sound content. and speech encoding.

3.7.1.1 MPEG-2 Audio Layer II (MP2) Encoding

MPEG-1 provided for three different types of audio encoding with sampling rates of 32 kHz (kSamples/s), 44.1 and 48 kHz for monophonic (mono), dual mono, and stereophonic (stereo) channels. MPEG-2 added support for up to five audio channels and sampling frequencies down to 16 kHz. Subsequently, MPEG-2 Advanced Audio Encoding (AAC) added sampling from 8 to 96 kHz with up to 48 audio channels (and 15 embedded data streams). MPEG-2 audio layer I is the simplest encoding scheme and is suited for encoded bit rates above 128 kb/s per channel (Pan 1995). Audio layer II, known as MP2, has an intermediate complexity and is targeted at bit rates around 128 kb/s. MP2, also known as Musicam, forms the basis of the Digital Audio Broadcasting (DAB) system and is incorporated into the DVB standard.

3.7.1.2 MPEG-2 Audio Layer III (MP3) Encoding

MPEG-2 audio layer III, known as MP3, provides increased compression (or, alternatively, improved quality at the same data rate) compared with MP2 at the expense of slightly increased complexity and computational effort. Today, MP3 is widely used for the storage and distribution of music on personal computers via the

Internet and for digital satellite broadcasting by 1 worldspace (see Chap. 10). The MP3 psychoacoustic model uses a finer-frequency resolution than the MP2 polyphase quadrature filter band provides (Pan 1995). MP3 divides the audio spectrum into 576 frequency bands and processes each band separately. It does this in two stages. First, the spectrum is divided into the normal 32 bands, using a polyphase quadrature filter, in order to ensure compatibility with audio layers I and II. In MP3, however, each band is further divided into 18 sub-bands using a modified discrete cosine transform (MDCT). The MP3 MDCT is a variant of the discrete cosine transform that reuses a fraction of the data from one sample to the next. We have learnt in Chap. 8 that following the transformation into the frequency domain, component frequencies can be allocated bits according to their audibility using the masking levels in each filter. We also noted that MP3 exploits interchannel redundancies, for example, in situations when the same information is transmitted on both stereo channels. Typically, MP3 permits compression of CD-quality sound by a factor of –12.

3.7.2 *Video Broadcasting*

As with audio encoding, the dominant video encoding standards are currently the MPEG standards, notably MPEG-2 for standard-definition video and, increasingly, MPEG-4 for high-definition video. The MPEG-2 video decoders contain a "toolbox" of compression algorithms grouped into a number of subsets called "profiles." Furthermore, profiles support a number of levels (combinations of image size, frame rate, etc.), and decoders may implement a subset of profiles.

3.7.2.1 MPEG-2 Video Encoding

The MPEG video encoding algorithms aim to transmit differences between frames where possible and use a DCT (a form of Fourier transform) to encode the difference information into the frequency domain, discarding insignificant high spatial frequency information that would not normally be noticeable. For this reason, they are sometimes referred to as hybrid block interframe differential PCM (DPCM)/discrete cosine transform (DCT) algorithms. A video buffer is used to ensure a constant bit stream on the user side. Typical bit rates for MPEG-2 encoded standard-definition video are in the region of 3–15 Mb/s, which is around 10–50:1 compression achieved over the raw PCM bit rate.

3.7.2.2 High-Definition TV and MPEG-4

Even greater compression ratios are required for the transmission of high-definition TV video. HDTV frame resolutions of up to 1920×1080 pixels are in use (at 25 Hz)—with up to 5 times as many pixels per frame compared with SDTV. While MPEG-2 high level can support resolutions of up to 1920 × 1080 (sometimes referred to as full-HD), HDTV generally requires the increased compression available in MPEG-4 to be viable.

The MPEG-4 Advanced Video Coding (MPEG-4 part 10 AVC), jointly developed with the ITU Video Coding Experts Group (as H.264), employs additional techniques to achieve compression ratios greater than those for MPEG-2 and is used by several satellite services for broadcasting HDTV. MPEG-4 AVC is utilized in BluRay videodiscs. Specifically, MPEG-4 AVC utilizes a number of features to achieve higher compression than MPEG-2. Some of the more significant features of MPEG-4 AVC are as follows:

1. Up to 32 reference pictures, however, may be used for motion compensation, rather than just 1 (I-Pictures) or 2 (B-Pictures);
2. The macroblock size used for motion compensation may be varied from 16 × 16 to 4 × 4 pixels with subpixel precision, and new 4 × 4 and 16 × 16 pixel block transforms; and
3. Improved nonlinear quantization size control and improved entropy encoding.

3.7.2.3 Multiplexing and Transporting

In Fig. 3.31 is illustrated packetization of encoded video and audio elementary streams to produce a packetized elementary stream (PES). The PES packets are next combined with system-level information to form transport streams (TS) or program streams (PS).

The systems part of the MPEG-2 specification (ISO/IEC 13818-1:2000(E)) specifies all syntactical and semantic rules to combine MPEG-2 video and audio elementary streams (output of an encoder), including other types of data content, into a single or multiple streams to enable storage or transmission.

The program stream consists of one or more streams of PES packets of common time base into a single stream. The stream is useful for operation in relatively error-free environments such as interactive multimedia applications.

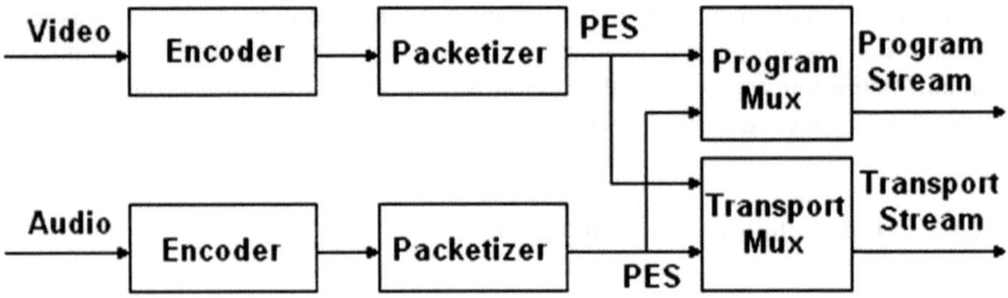

Fig. 3.31 MPEG Program and transport streams—Courtesy of Book: by Richharia

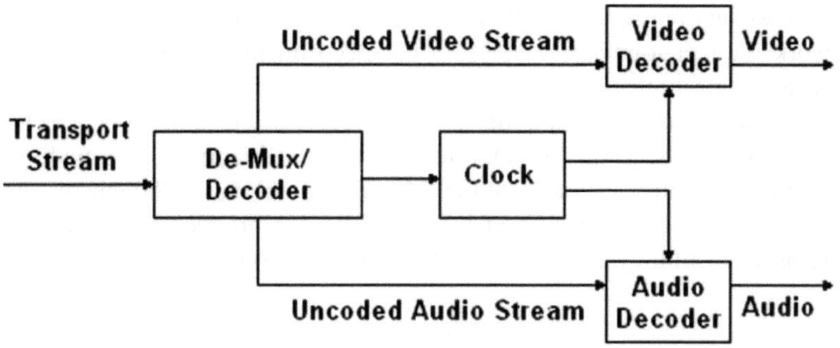

Fig. 3.32 Demultiplexing and decoding of MPEG—Courtesy of Book: by Richharia

The transport stream consists of one or more independent programs into a single stream. This type of stream is useful in error-prone environments as satellite broadcasts. Packets are 188 bytes in length. Transport stream rates may be fixed or variable. Values and locations of Program Clock Reference (PCR) fields define the stream rate. The transport stream design is such as to facilitate: retrieval and decoding of a single program within the transport stream, extraction of only one more program or contents from a transport stream, and conversion of a program into a transport stream including recovery at Rx.

Figure 3.32 illustrates the concept of demultiplexing and decoding of a single program from a received transport scheme containing one or more program streams. The input stream to the demultiplexer/decoder includes a system layer wrapped about a compression layer. The system layer facilitates the operation of the demultiplexer block, and the compression layer assists in video and audio decoding. Note that, although audio and video decoders are illustrated, these may not always be necessary when the transport stream carries other classes of data.

3.8 Direct-to-Home Broadcast System

Recognizing the advantages of digital television transmissions, and their potential to transport HDTV efficiently, considerable effort was directed in the 1980s and 1990s (notably in the USA, Europe, and Japan) toward the development of digital transmission systems, resulting in the design of several terrestrial and satellite systems. The convergence of computing, telecommunications and broadcast disciplines led the developers to adapt a generic architecture that would offer a variety of enabling services in addition to SDTV and HDTV.

Enabling technology and services include a digital video recorder to facilitate recording of program directly in a digital format, interactivity (e.g., multichannel display or multiple camera angle displays), receiver-enabled home networking, and reception onboard mobiles.

Broadcast satellite service (BSS) radio frequency plan of ITU provides a useful framework with guaranteed availability of spectrum in Ku-band to each member country, allowing high-power radio transmissions amenable for reception at home via small non-obtrusive antenna on low-cost receivers. It is necessary to highlight the technology of the DVB-S2 satellite system, as it incorporates a wide range of recent technology advances to provide a highly flexible and efficient medium, and moreover, it is uniquely identified by the ITU as a broadcast system for digital satellite broadcasting system with flexible configuration.

3.8.1 Transmission System Architecture

A typical direct-to-home system comprises incoming signals to the uplink facility from one or more transmission sources, such as a studio, local terrestrial broadcast signal feed, and prerecorded material. Occasionally, additional material such as advertisements may be added locally at preagreed points of the incoming program. The incoming signals are monitored, routed within the facility and, if necessary, readjusted and synchronized. The prerecorded material is checked for quality, edited, when necessary, and read into video file servers to be played at the broadcast time.

Commercial systems incorporate a facility known as conditional access to facilitate reception solely by the authorized users. Other functionalities include additions of service information (SI) and electronic program guide (EPG), analog-to-digital conversion, compression, multiplexing to create a suitable transport stream, error control, and modulation. The equipment of SI and EPG creates signals for displaying program-related information, for example, programs title, and start/end time.

The compression equipment is typically MPEG-2 standard, although migration to MPEG-4 is endemic because of its tighter encoding. At this point, several channels are multiplexed to provide a single high-rate channel for transmission. The stream is FEC mode, modulated and transmitted. The ITU BSS plan recommends uplink transmissions in the 17.5 GHz band and downlink transmissions to the user community in the 12.5 GHz band, but their transmission powers are restricted by the FSS regulations. The BSS satellites are placed in geostationary orbits. The satellites use transparent transponders capable of transmitting very high powers through spot and, often, shaped beams to be able to provide the high signal quality necessary for reception on small DTH receivers.

The DTH receivers typically use a 45–60 cm dish at Ku-band, depending on satellite EIRP. The relatively high gain of the receive antenna in conjunction with high-power satellite transmissions provides sufficient link margins to counter rain fades common in this band.

3.8.2 Generic Reference Integrated Receiver Decoder (IRD) Model

The indoor unit housing the electronics is known as the integrated receiver decoder (IRD). The core functions of all the direct-to-home television systems are nominally identical, and hence, a generic model of the IRD is feasible. Operators may tailor the remaining functions around these core functional entities. The ITU proposes reference architecture of the IRD on this premise. The model provides a structured definition of functionalities to facilitate a generic receiver design. As observed in the preceding section, the core functions relate generically to a transmission system. The additional essential functions relate to the service provision, operation of the system, and additional or complementary features, which may differ depending on implementation. These functions and units include a satellite tuner, output interfaces, an operating system and applications, EPG, service/system information (SI), conditional access (CA), a display, a remote control, read-only memory (ROM), random-access memory (RAM), flash memory, an interactive module, a microcontroller and units to support teletext, and subtitling.

Software-reconfigurable receivers are common as they simplify an upgrade to the receiver. The upgrade may, for example, become necessary to repair a software anomaly, add a new functionality, or reconfigure receiver subsystems when a new satellite transponder is deployed. Modern receivers typically include a digital video recorder permitting users to record a program directly in a digital format, or to store content, ready for an "instant" video-on-demand display, thereby avoiding the interactive delay. Often L-band IF signals from a single dish/LNB flow to one or more receivers in the customer's home. Home networking systems permit interworking between receivers, thereby permitting programs recorded on one set to be viewed by other sets elsewhere in the house and, in addition, support features such as security and quality of service management.

3.9 Transmission Standards

The development of digital television has evolved independently around the world, and hence, several types of transmission system are in use. The majority operates in the 11/12 GHz downlink band. Being digital, the systems can support numerous applications and services efficiently be it television, multimedia, data services, or audio. To assist in the selection of an appropriate system, ITU recommends four systems—system A, system B, system C, and system D.

3.9.1 Digital Video Broadcast Second-Generation (DVB-S2) Standard

The DVB transmission project, initiated in Europe but subsequently extended worldwide, defines digital broadcast standards by consensus among the participants representing the manufacturing industry, broadcasters, program providers, network operators, satellite operators, and regulatory bodies. The DVB standards embrace broadcast transmission technology across all media of cable, terrestrial, and satellite. The DVB-S specifications were standardized in 1993. A second-generation specification, DVB-S2, was produced in 2003 in response to a growing demand for more capacity, which was standardized by the European Telecommunication Standard Institute (ETSI) as EN 302 307 (ETSI Online). In August 2006, the ITU's study group on satellite delivery issued a recommendation (BO.1784) that DVB-S2 be the preferred option for a Digital Satellite Broadcasting System with flexible configuration (television, sound, and data), entitled system E.

The DVB-S standards support up to 12 categories of transmission medium encompassing a plethora of media contents, among others, standard- and high-definition television, radio, and data with or without user interactivity. The standard includes specifications for Internet protocol (IP) data, software downloads, and transmissions to handheld devices. The standards, namely DVB Satellite (DVB-S), DVB Return Channel Satellite (DVB-RCS), and DVB Second Generation (DVB-S2) apply to fixed user terminals. The DVB Satellite Handheld (DVB-SH) standard, discussed later in the chapter, applies to handheld terminals.

Many parts of the DVB-S specifications are shared between various transmission media. The source coding of video and audio and formation of the transport stream comprises MPEG-2 tailored for satellite systems (MPEG specifications are too generic). DVB-S also supports H.264/AVC video and MPEG-4 high-efficiency AAC audio and, additionally, audio formats such as Dolby AC-3 and DTS coded audio. Guidelines are also available for transporting IP content. The specifications support teletext and other data transmitted during the vertical blanking period, subtitles, graphical elements, service information, etc. DVB-S's enhanced version, DVB-S2, is based on three key concepts: 1. best transmission performance; 2. total flexibility; and 3. reasonable receiver complexity.

The specifications enable delivery of services that could never have been delivered by DVB-S. According to the developers, the DVB-S2 standard is not intended to replace DVB-S in the short term for conventional TV broadcasting applications but is rather aimed at new applications such as the delivery of HDTV and IP-based services, fly-away small DSNG stations, and low-cost Internet access to rural areas and in developing countries. The DVB-S2 specifications, in conjunction with recent advances in video compression, have enabled the widespread commercial launch of HDTV services. The supported applications are as follows: 1. standard and high-definition television broadcasts; 2. interactive services, including Internet access for consumer applications; 3. professional applications: digital TV contribution (DTVC) facilitating point-to-multipoint transmissions, and digital satellite news gathering (DSNG); 4. content distribution; and 5. Internet trunking.

In addition to MPEG-2 video and audio coding, DVB-S2 is designed to handle a variety of advanced audio and video formats, which the DVB Project is currently defining. In fact, DVB-S2 accommodates most common digital input stream format, including single or multiple MPEG transport streams, continuous bit streams, IP as well as ATM packets. It is 30 % spectrally more efficient than its predecessor, employing an adaptable modulation scheme consisting of QPSK and 8-PSK for broadcast applications and also 16-APSK and 32-APSK for professional applications such as newsgathering and interactive services.

In addition, the modulation and coding schemes may be dynamically adapted to variable channel condition on a frame-by-frame basis scheme. The coding arrangement consists of a Bose–Chaudhuri–Hocquenghem (BCH) outer code scheme and a low-density parity-check (LDPC) inner code. The communication performance lies within 0.7 dB of the theoretical limit. The flexibility offered by variable coding and modulation provides different levels of protection to services as needed.

The specifications support operation on any type of satellite transponder characteristics with a large variety of spectrum efficiencies and associated C/N requirements. Thus, the DVB-S2 broadcast services comprise a backward compatible mode and a more optimized version, which is not backward compatible. The system allows interactive services by return channels established either via satellite or another medium incorporating the added flexibility of adaptive coding and modulation (ACM) to the forward channel through feedback. Structured as a toolkit, DVB-S2, attempts to optimize transmission efficiency and flexibility, keeping receiver costs at an acceptable level.

3.9.2 DVB-S2 Architecture

Figure 3.33 shows a block schematic of functional components of a DVB-S2 transmission system. There are two levels of framing structures, one at baseband (BBFRAME) and the other at physical layer (PLFRAME).

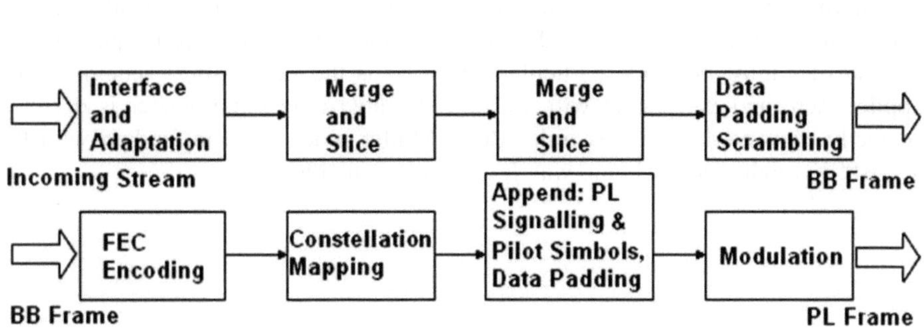

Fig. 3.33 DVB-S2 transmission system—Courtesy of Book: by Richharia

Thus, BBFRAME includes signaling to configure the receiver for the given specification and service. The PLFRAME comprises a regular sequence of frames, each coded and modulated for the given channel condition and application, containing a few signaling bits for synchronization and physical layer signaling.

Referring to BBFRAME shown in Fig. 3.33, the mode and stream adaptation block interfaces with the incoming stream, provides synchronization, and supports the adaptive coding–modulation schemes. It merges multiple streams and slices them into blocks, each of which are modulated and coded homogeneously. A header of 80 bits containing transmission characteristics is next appended to the baseband data block to assist reception. The information within the header, for example, informs the receiver as to whether the transmission stream is single or multiple, the type of coding modulation schemes, signal format, etc. When the volume of data is insufficient to fill the frame, data padding is introduced, and finally, the frame is scrambled and passed over for coding.

The FEC coding–modulation schemes are instrumental in high transmission efficiency. The FEC code comprises a low-density parity-check (LDPC) code that exhibits a low distance from the Shannon limit for the specified decoder complexity (equating to 14 mm^2 of silicon 0.13 μm technologies).

The LDPC codes are suitable for iterative decoding at reasonable complexity because of their easily parallelizable decoding algorithm, which can use simple operations resulting in reduced complexity and storage. To avoid error floors at low error rates, a BCH outer code of the same block length as the LPDC code is concatenated.

The coded payload modulates the carrier with a QPSK, 8-PSK, 16-APSK, or 32-APSK scheme, as required by the given application data rate and link conditions. In fact, the corresponding spectral efficiency ranges from 0.5 to 4.5 b/symbol. Typically, the QPSK and 8-PSK schemes are applied for broadcast applications through nonlinear transponders, whereas 16- and 32-APSK schemes are better suited for professional and other applications operating through transponders driven in the quasi-linear region. The DVB-S2 provides backward compatibility with DVB-S through a hierarchical modulation scheme.

The physical layer (PL) signals are composed of a regular sequence of frames, within each of which the modulation and coding schemes remain unchanged, that is, a homogeneous operation. A header of 90 bits preceding the payload assists in synchronization at the receiver, detection of modulation, and coding parameters. This header must be made particularly robust, as the LPDC/BCH scheme is not applied here and the packet will not be detectable unless the header is detected correctly. An interleaved first-order Reed–Muller block code in tandem with the $\pi/2$-BPSK modulation scheme was found to be suitable.

Chapter 4
Mobile Satellite Antenna Systems

In the beginning of radio development, mobile communication systems were conceived for the transmission and receiving of telegraphy and telephony signals via antenna from ships, cars, trains, and aircraft. The consideration of antenna transmission is inevitable, especially in Global Mobile Satellite Communications (GMSC) applications, where their propagation characteristics are much affected by different and changeable local environments during movement of mobile and differ greatly from those observed in fixed satellite systems. To create antenna hardware for GMSC systems, engineers have to consider all related factors in order to realize full electrical, mechanical, and transmission potentials.

This chapter describes antenna characteristics, requirements, and basic relations of antenna systems, and considerable antenna classification for mobile satellite applications in general and for aeronautical satellite applications in particular. The antenna systems can be installed on board ships, vehicles (tracks, buses, cars, and trains), aircraft (airplanes and helicopters), GNSS antennas, and integrated solutions with GSM antenna on the top of handheld phones. Thus, all basic mobile antenna configurations for GMSC, such as low-gain omnidirectional antennas, three principal divisions of medium-gain directional antennas, and two types of high-gain directional aperture antennas, are presented. Finally, all types of antennas for particular mobile systems including antenna mounting and tracking systems are included.

4.1 Evolution of Antenna Systems for Mobile Radio Communications (MRC)

The Russian professor of physics Popov designed his first world's radio receiver in 1895 with antenna in the shape of wire mounted on a balloon in the air and transmitter with a lightning conductor as an antenna, including a metal filings

S.D. Ilčev, *Global Mobile Satellite Communications Theory*,
DOI 10.1007/978-3-319-39171-7_4

coherer and a detector element with telegraph relay and a bell. Soon later, Marconi started commercially to deploy radio and antenna equipment on board different merchant ships and to establish his own company for the production of maritime radio and antenna equipment.

Since the initial use of mobile radio was for long-distance wireless communications at LF and the first shipboard antenna was all made of haphazard lengths of wire strung as high as possible above the ship's topside, evidently the thinking was that the longer and higher the wire, the better the results should be. After that, those different kinds of wire and whip antennas were developed for MF/HF/VHF maritime and aeronautical applications.

4.1.1 Development of Antennas for Mobile Satellite Communications (MSC)

The MSC systems introduced new complexities into the design of shipboard antennas. The direct line of sight between antenna and satellite requires the antenna to see from horizon to overhead (zenith—90°) in elevation and 360° in azimuth angle, with total hemispherical coverage. This is fulfilled in the case of transceiver antenna through the use of tracking rotatable, high-gain antennas often installed in pairs on board ship to attain full coverage, irrespective of blockage in the form of the funnel, musts, stacks, and other deck objects.

The ship platform itself imposes even more stringent requirements. Therefore, in spite of constant vibrating, pitching, rolling, and yawing during bad weather conditions, the MSC antenna's narrow radiation beam must be pointed accurately from any position on the high seas. The situation regarding land or aero antenna is less complicated. As the 1970s dawned, optimism and enthusiasm about satellite communications were so great with ideas to virtually replace HF radio in the navy with the new FltSatCom military mobile system. For instance, since this time, several types of UHF transceiving antennas developed for US naval shipboard services were used, such as the oldest crossed-dipole array, its improved version and the so-called wash tube similar to an SBF antenna and one type of SHF parabolic dish antenna. However, as discussed in Chap. 1, the first real global MSC system was the Marisat network, which employed Ship Earth Station (SES) and L-band antenna systems similar to the abandoned Inmarsat-A and current Inmarsat-B terminals.

4.1.2 Classification and Types of Mobile Satellite Antennas (MSA)

The general classifications of mobile satellite antenna (MSA) systems in connection with the service operators and providers are performed on the following major types: satellite communications antennas, satellite navigation and surveillance antennas, digital video broadcasting return channel via satellite (DVB-RCS) or broadband antennas, satellite meteorological antenna, and military satellite antennas. Owing to the fundamental values and characteristics, gain classification of MSA falls into three antenna groups: low-gain omnidirectional antennas, medium-gain directional antennas, and high-gain directional aperture antennas. In many respects, the mobile satellite antennas currently available for MSC applications constitute the weakest links of the system. If the mobile antenna has a high gain, it has to track the satellite, following both mobiles and satellite orbital motions. Thus, sometimes this is difficult and expensive to synchronize.

Therefore, if the vehicular antenna has low gain, it does not need to perform tracking, but the capacity of the communications link is limited. In general, according to the transmission direction, there are three types of MSA: (1) transmitting and receiving or so-called transceiving antennas; (2) receiving antennas only for GMSC and GNSS solutions; and (3) transmitting only built-in satellite beacon and for satellite asset tracking (SAT) devices. In particular, MSA antenna can be a shipborne, vehicleborne, personal, GNSS, broadband, and airborne antenna.

The shipborne MSA was developed for oceangoing ships and oilrigs, fishing, and inland vessels and boats, whose size and weight are not very critical. The vehicleborne antennas can be mounted on board vehicles and wagons, which need small size. The airborne antennas have very specific design for installations on board aircraft and helicopters. Because of the high speed of aircraft, the aerodynamic constraints are significant and antennas for both radio and satellite systems must conform to minimum drag and reliability requirements.

4.2 Antennas Requirements and Technical Characteristics

This section describes important general requirements for mobile antennas for maritime, land, and aeronautical applications, including for personal terminals. The MSA has to be compact, flexible, and lightweight and performs with good mechanical and electrical characteristics, especially for heavy mobiles such as ships and aircraft, owing to the special conditions of installation and the influence of changeable environmental conditions.

4.2.1 Mechanical Characteristics

Mobile antennas have to satisfy the requirements of mechanical characteristics in relation to construction strength and easy installation. Easy installation and appropriate physical shape are very important requirements in addition to compactness and lightweight. In the case of shipborne antennas, the installation requirements are not severe compared to that of aircraft and cars because even in small ships there is a comfortable space to install an antenna set. Otherwise, the only problem is that all types of ship satellite antennas are sometimes under stress from vibration and sloping caused by strong winds, ship's rolling, and pitch or are subject to corrosion by sea salt.

Owing to these problems, a ship's antenna has to be well protected by plastic radome and properly mounted on a strong mast, specially designed for a certain size of antenna. However, in the case of road vehicles, especially small cars, low-profile and lightweight equipment is required. The requirements are the same as that of aircraft, although more severe conditions are required to satisfy avionic standards. Namely, low air drag is one of the most important requirements for aircraft antennas. Vehicles and aircraft utilize smaller and more aerodynamic sizes of antenna.

4.2.2 Electrical Characteristics

Sometimes, the mechanical construction of antenna is perfect because of some functional or electrical characteristics; however, designers of antenna have to keep in mind that the compact design of antenna has two major disadvantages in electrical characteristics, such as low-gain and wide beam coverage. The gain is closely related to the beam width, and a low-gain antenna (LGA) should have a wide beam width. As the gain of antenna is theoretically determined by its physical dimensions, reducing the size of antenna means decreasing its gain.

Because of low-gain and limited power supply, it is very difficult for mobile antennas to have enough receiving capability (G/T) and transmission power (EIRP). The disadvantages of Mobile Earth Station (MES) can be compensated by large satellite antenna, high-power amplifier (HPA), and enough electrical power. A powerful satellite with high G/T and EIRP performance should permit the fabrication of MES with compact and lightweight antennas. The next disadvantage is that a wide beam antenna is likely to transmit undesired signals to and receive them from an undesired direction, which will cause interference in and from other systems. The wide beam is also responsible for several fading effects, such as that from sea surface reflections in MMSC and AMSC and multipath fading in LMSC, and so a compact mobile antenna system is required to prevent fading and interference.

Accordingly, it is inevitable for mobile antennas not to have enough performance, such as gain, radiation power, and receiving capability because of their small physical dimensions. Without taking this into consideration, the requirements of transmitting and receiving performance of mobile antennas mainly depend on the satellite transmission capability. The first and second generations of Inmarsat satellites have a global beam, and the third generation has spot and global beams to provide global coverage.

The regional or domestic MSS, such as AMSC, MSAT, and Optus, have spot beams. The spot gives higher satellite capacity than global beams although there are basically no big differences between requirements for mobile antennas in the global system, such as Inmarsat or the mentioned regional systems.

4.2.3 Basic Relations of Antennas

The basic relations of antenna systems are very important parameters to easily understand the mode of antenna functions in two-way (duplex) satellite transmission systems, such as MES transceiving antennas. Moreover, these characteristics of MES antenna systems are needed for link budget calculations and for good satellite up- and downlink design, which can provide reliable and acceptable quality satellite communications. At this point, this implies that the signal transmitted via the MES Tx antenna must reach the Rx antenna of other MES or LES at a carrier level sufficiently above the unwanted signals generated by various unavoidable sources of noise and interference.

4.2.3.1 Frequency and Bandwidth

Almost all present and forthcoming MSS use geostationary Earth orbit (GEO) satellites, at L-band (1.6/1.5 GHz) to link satellites and MES. The required frequency bandwidth in L-band MSS is about 8 % to cover transmission and receiving channels. In using a narrowband antenna, such as an omnidirectional patch antenna, some efforts have to be made to widen the bandwidth. The S- and L-bands are allocated in WARC-92 for the Big LEO Iridium and Globalstar systems, which require frequency bandwidths of about 5 %.

4.2.3.2 Gain and Directivity

The required antenna gain is determined by a link budget, which can be calculated by taking into consideration the required channel quality and the satellite capability. The channels are expressed as C/N_o and depend on the G/T and EIRP values of the satellite and MES. In the case of the present Inmarsat Aero system, a comparative high-gain antenna (HGA) of minimum 24 dBi is required, due to the difference in satellite capabilities. However, LGA of about 0–4 dBi are used in the Inmarsat-C and other omnidirectional systems to provide LSD of only about 600–1200 b/s. The GPS system has adopted LGA because of the extremely low data rate of 50 b/s from the satellites.

There are no exact definitions to differentiate between characteristics of low-, medium-, and high-gain antenna systems, except by the gain quantum, shape of the antenna, and type of service. In the present and upcoming L-band GMSC applications, classification of L-band MES antenna systems by their receiving and service capabilities is illustrated in Table 4.1.

The ideal antenna gain can be defined with an isotropic (hypothetical) antenna, which has an isotropic radiation pattern without any losses and therefore radiates power in all directions in uniform intensities. Thus, if input power (P_{in}) is put into an isotropic antenna, the power flux density per ideal unit area (P_{id}) at distance (r) from the antenna is given by the following relation:

$$P_{id} = P_{in}/4\pi r^2 [W/m^2] \tag{4.1}$$

However, if radiated power density is $P(\theta, \Phi)/r^2$ in directions (θ = angle between the considered direction and the one in which maximum power is radiated, known as boresight and Φ = phase) at distance (r) from the antenna under elevation, the gain of the antenna can be defined by the following equations:

$$G(\theta, \Phi) = P(\theta, \Phi)/r^2/P_{id} = P(\theta, \Phi)/r^2/P_{in}/4\pi r^2 = 4\pi P(\theta, \Phi)/P_{in} = P(\theta, \Phi)/P_{in}/4\pi [dBi] \tag{4.2}$$

The above-defined gain is called an absolute gain or directive gain, which is determined only by the directivity (radiation pattern) of the antenna without taking account of any losses in the antenna system, such as impedance mismatch loss or spillover loss. Thus, if direction is not specified and the gain is not given a function of (θ, Φ), it is assumed to be maximum gain. There is a general relationship between absolute gain and the physical dimensions of the antenna and this is given by the equation as follows:

$$G = 4\pi\lambda^2\eta a \tag{4.3}$$

where η = aperture efficiency and a = physical aperture, which will denote the effective aperture of the antenna. According to the above relation, it can be realized that compact antennas with small apertures must have low gain. If an antenna aperture is a dish, a known diameter (d) can be written in normal and in decibel expression as follows:

$$G = (\pi d/\lambda)^2\eta = 10 \log \eta(\pi d/\lambda)^2 [dBi] \tag{4.4}$$

Thus, it can be calculated that the gain in the Inmarsat shipborne antenna with a diameter of d = 1 m operated at 1.5 GHz is about 21 dBi.

The directivity of the antenna $D(\theta, \Phi)$ does not include dissipative losses and is defined as the ratio of $P(\theta, \Phi)$ to the power per unit solid angle from an isotropic antenna radiation, the same total antenna-radiated power (P_r). The antenna directivity can be expressed by:

$$D(\theta, \Phi) = P(\theta, \Phi)/P_r/4\pi \tag{4.5}$$

Table 4.1 Classification of L-band antenna systems in GMSC

Type of antenna	Gain class	Typical gain (dBi)	Typical G/T (dBK)	Typical antenna (dimension)	Typical GMSC services
Omnidirectional	Low	0–4	−27 to −23	Quadrifilar Drooping dipole Patch	LSD (messages) Ship (Inmarsat-C) Vehicles and Aircraft
Semi-directional (only in azimuth)	Medium	4–8 8–16	−23 to −18 −18 to −10	Array (2–4 elements) Helical, patch SBF (0,4 m *Φ*) Phased array (20 elements)	Voice/HSD Ship (Inmarsat-M) Vehicles Aircraft (Inmarsat Aero)
Directional	High	17–20 20–24	− 8 to −6 −4	Dish (0,8 m *Φ*) Dish (1 m *Φ*)	Voice/HSD Ship (Inmarsat-A, B)

The definition of antenna directivity does not take the efficiency of an antenna into account because $(P_r/4\pi)$ is related to the actual power launched into space. The ratio of $G(\theta, \Phi)$ to $D(\theta, \Phi)$ is termed the radiation efficiency of the antenna.

4.2.3.3 Radiation Pattern, Beam Width, and Sidelobes

Radiation calculation is possible in principle if the EM field can be described quantitatively at all points of the antenna surface whose boundaries are those of the apertures.

In this section, the radiation pattern from a circular aperture is considered as the aperture-type antenna that has generally been used in MSC, especially in MMSC. This simple problem will give an insight into the characteristics of mobile antennas. For an antenna that generates a single focused beam, the principal parameter affecting the antenna radiation pattern $E(\theta, \Phi)$, after the aperture size (a) is the aperture illumination distribution $E_a(r, \psi)$, which is the amplitude of the far-field radiation pattern E, at the point (θ, Φ), being essentially the Fourier transform of the illumination distribution and is given by:

$$E(\theta, \Phi) = 1/\pi a^2 \int_0^{2\pi} \int_0^{a} E_a(r, \psi) \exp[-jkr \sin\theta \cos(\Phi - \psi)] r dr \psi \tag{4.6}$$

An example considers an antenna employed in MSC, which utilizes a circular aperture, where for circularly symmetric aperture illumination distribution this relation reduces to:

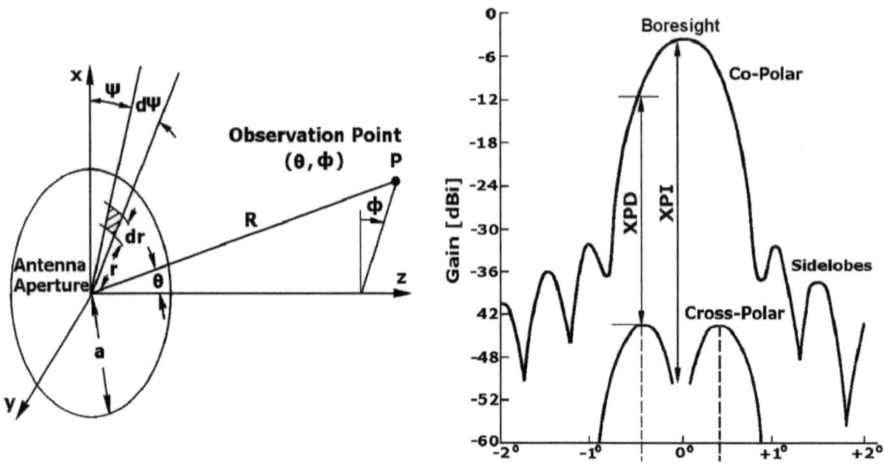

Fig. 4.1 Geometrical parameters of antenna pattern and gain characteristics—Courtesy of Book by Evans

$$E(\theta) = 2/a^2 \int_0^a E_a(r) J_0(kr \sin\theta \cos(\Phi - \psi)) r dr \quad (4.7)$$

where a = d/2 denotes the radius of antenna aperture; J_0 = first kind and order zero of the Bassel function and $k = 2\pi/2$ denotes the wave number. The other notations that denote distance and angles in coordinates are defined in the geometry illustrated in Fig. 4.1 (Left). Thus, the antenna radiation pattern is three-dimensional in nature, so it usually has to be represented from the point of view of a single-axis plot.

The characteristics of the MES antenna radiation pattern affect interference levels directly. Any improvement in the pattern will therefore be fully reflected in the interference level, and such improvement constitutes a very effective means of solving interference problems. To improve the pattern, one can either increase the antenna diameter or, with a constant diameter, use a specific technique for reducing the sidelobes. This method is therefore applicable when the MSC network is in the initial stages of development.

The antenna gain is normally calculated with reference to the boresight, i.e., the direction at which the maximum antenna gain occurs, in the case when $\theta, \Phi = 0°$. Gain is usually expressed in (dBi), where component (i) refers to the fact that it is relative to the isotropic gain. In this instance, the matter of moment in a dual polarization frequency reuse satellite communication system is polarization discrimination between the copolar and cross-polar signals, especially in the antenna main beam region, as illustrated in Fig. 4.1 (Right). An important parameter that is used in an antenna's specification is the beam width evaluated by half-power beam width (HPBW) $2\theta_{HP}$ where θ_{HP} is the half-power angle when radiated power becomes half the maximum level (−3 dB). The HPBW (θ_0) is given by the following equation:

$$\theta_0 = 65\,(\lambda/d) \tag{4.8}$$

Here, it is possible to realize that the half-power bandwidth is inversely proportional to the operating frequency and the diameter of the antenna. For example, a 1-m receiver antenna operating in the C-band (4 GHz) has a 3 dB bandwidth of roughly 4.9°, while the same antenna operating in the Ku-band (11 GHz) has a 3 dB bandwidth of approximately 1.8°.

The antenna systems have copolar and cross-polar gains, where the reception of unwanted, orthogonally polarized cross-polar signals will add as interference to the copolar signal. The ability of an antenna to discriminate between a wanted polarized waveform and its unwanted orthogonal component is termed as its cross-polar discrimination (XPD). When dual polarization is employed and the antenna's ability to differentiate between the wanted polarized waveform and the unwanted signal of the same polarization, introduced by the orthogonal polarized wave, it is termed as the cross-polar isolation (XPI). In this context, an antenna typically would have an XPI >30 dB.

The level of the antenna pattern's sidelobes is also important, as this tends to represent gain in an unwanted direction. For a transmitting gain, this leads to the transmission of unwanted power, resulting in interference to other systems, or in the case of a receiving antenna, the reception of unwanted signals, or noise. The sidelobe characteristic of MES is one of the main factors in determining the minimum spacing between satellites and therefore the orbit and spectrum utilization efficiency. The ITU-R S.465-5 recommendation gives a reference radiation diagram for use in coordination and interference assessment, which is defined by:

$$\begin{aligned} G &= 32 - 25\log\Phi\,[\text{dBi}] && \text{for } \Phi_{min} \le \Phi < 48° \\ &= 10\,\text{dBi} && \text{for } 48° \le \Phi \le 180° \end{aligned} \tag{4.9}$$

where G = gain relative to an isotropic antenna; Φ = off-axis angle referred to the main lobe axis and Φ_{min}=1° or $100(\lambda/d)$ degrees, whichever is the greater. In this context, most of the effective power radiated by an antenna is contained in the so-called main lobes of the radiation pattern, while some residual power is radiated in the sidelobes. Sidelobes are an intrinsic property of antenna radiation, and diffraction theory shows that they cannot be completely suppressed. However, sidelobes are also due partly to antenna defects which can be minimized by proper design. Conversely, due to the reciprocity theorem, the receive antenna gains and radiation patterns at the same frequency are identical to the transmit antenna gains and radiation patterns. Unlikely, unwanted power can also be picked up by the antenna sidelobes during reception.

For large satellite antennas, with a diameter over 100 λ (wavelengths), a reference radiation pattern is recommended by the CCIR for interference to and from other satellite and terrestrial communication systems. At this point, the diameters of the vehicle antennas under discussion are, in many cases, below five wavelengths in the L-band. Further CCIR action is expected to define a reference radiation pattern for mobile antennas in MSC.

4.2.3.4 Polarization and Axial Ratio

The antenna and the EM field received or transmitted have polarization properties. Thus, the polarization of an EM wave describes the shape and orientation of the locus of the extremities of the field vectors as a function of time. A wave may be described as linearly, circularly, or elliptically polarized.

Linear polarization is such that the electric E-field is oriented at a constant angle as it is propagated that can be either vertically or horizontally. If a plane wave is propagated along the (z)-axis and electric field (E) is on the (x-z) or (y-z) planes, relations for linear vertical and horizontal polarizations can be written as follows:

$$E_x = E_a e^{j(\omega t - kz + \Phi a)} \quad \text{and} \quad E_y = E_b e^{j(\omega t - kz + \Phi b)} \tag{4.10}$$

where E_a; ω; k and Φ_a denote the maximum amplitude of electric field, angular frequency (2πf), wave number, and initial phase, respectively, while E_b and Φ_b are the maximum amplitude and the initial phase of the wave.

Circular polarization is the superposition of two orthogonal linear polarizations, such as vertical and horizontal, with a 90° (π/2) phase difference. The tip of the resultant E-field vector may be imagined to rotate as it propagates in a helical path. There is a left-hand circularly polarized (LHCP) wave with anticlockwise rotation and a right-hand circularly polarized (RHCP) wave with clockwise rotation.

An elliptically polarized wave may be regarded as the result either of two linearly or of two circularly polarized waves with opposite directions. This type of polarization is the case when the amplitudes and phase difference between the two waves are not equal (π/2).

As discussed in a previous section, the signals fields can contain copolar and cross-polar components. In this way, the cross-polarization of a source becomes of increasing interest to MSC antenna designers. In the case of Tx or Rx antennas with a linearly polarized field, the cross-polar component is the field at right angles to this copolar component. Namely, if the copolar component is vertical, then the cross-polar component is horizontal.

Circular cross-polarization is that of the opposite hand to the desired principal or reference polarization. Impure circular polarization is in fact elliptical. The level of impurity is measured by the elliptical and known as the axial ratio (AR). The AR can be defined as the ratio of the major axis electric component to that of the minor axis by:

$$|AR| = |E_1/E_2|(1 \leq |AR| \leq \infty) \quad (4.11)$$

The signal for AR denotes the direction of rotation; however, an absolute value is usually used to evaluate circularly polarized radiated waves and can be expressed in decibels by the following equation:

$$|AR| = 20\log(|E_1/E_2|)[dB] \quad \text{for} \quad (0 \leq |AR| \leq \infty) \quad (4.12)$$

Accordingly, the AR is determined by the performance of the antenna, so the AR is one of the most important parameters of circularly polarized antennas. It can easily be understood that the AR depends on the direction with respect to the axis of the antenna. In general, the AR is best (smallest) in the boresight direction and is progressively worse further away from the boresight. Circularly polarized waves are used in order to eliminate the need for polarization tracking. RHCP has been used in the Inmarsat transmission system. In the case of aperture-type antennas, such as the parabolic reflector antenna, an axial ratio of below 1.5 dB in the boresight direction is so easy to achieve that polarization mismatch loss is almost negligible. However, in the case of phased-array antennas, a degradation of the axis ratio caused by beam scanning must be taken into account.

4.2.3.5 Figure of Merit (G/T) and EIRP

Although gain is an essential factor in considering antennas, the figure-of-merit ratio of a G/T is more commonly specified from the standpoint of MSS and FSS in general. The G/T for the receiving station is defined as the ratio of the gain of the antenna in the direction of the receiving signal to the receiving system noise temperature. The G/T is generally given for the maximum gain derived from gain formula (4.4) as follows:

$$G_{max} = P_{max}/P_0/4\pi = 10 \log G[dB] \quad (4.13)$$

The G_{max} is often called the antenna gain expressed in dB, where the total radiated power in all directions can be determined by the following integration:

$$P_0 = \int_0^{2\pi}\int_0^{\pi} P(\theta, \Phi) \sin\theta \, d\theta \, d\Phi \quad (4.14)$$

The G/T value is expressed in decibels per kelvin [dB(K^{-1})] by the following relation:

$$(G/T) = 10 \log G - \log T_{SA} = 10 \log G - \log T_S \left[dB\left(K^{-1}\right)\right] \quad (4.15)$$

The Earth station G/T typical values range from 35 dB(K^{-1}); for instance, an LES receive antenna with a 15–18 m diameter has some 15.5 dB(K^{-1}). The G/T is a very important parameter of an Earth station, so the methods used for its measurement and the contribution to the noise temperature are the subject of the ITU-R S.733 recommendation.

The noise temperature measured at the terminals of antenna pointed to the sky depends upon frequency of operation, elevation angle, and the antenna sidelobe structure. In more formal terms, the noise temperature will be derived from a complete solid angle integration of the noise power received from all noise sources (terrestrial and galactic) and determined for clear weather conditions by the following integral:

$$T_A = 1/4\pi \int_{\Omega} P(\theta, \Phi)T(\theta, \Phi)d\theta \, d\Phi \quad (4.16)$$

Thus, to produce a low-noise antenna, its sidelobes must be minimized, especially in the direction of the Earth's surface, where T = noise temperature. Total noise temperature of the system (T_{SR}) at an input port of receiver LNA or at the antenna output (T_A) is taking account of losses caused by tracking, feed lines, and a radome and is defined by:

$$T_{SR} = T_R + T_a(1-1/a) + T_A/a \quad \text{or} \quad T_{SA} = T_A + T_a(a-1) + a\,T_A \quad (4.17)$$

where T_R = noise temperature of the receiver (LNA) with a typical value of about 80 K to 100 K in the L-band; T_a = temperature of the environment of about 300 K; L_f = total loss of feed lines and components such as diplexer, cables, and phase shifters if a phased-array antenna is used; a = attenuation expressed as a power ratio (a ≥ 1 or in decibel $a_{dB} = 10 \log a$); TA = antenna noise temperature that comes from the ionosphere and the Earth, whose value of about 200 K depends on factors such as frequency and bandwidth; and T_{SA} = antenna with a noise temperature.

The noise of the antenna temperature must be kept as low as possible by proper design on in order to obtain a high figure of merit (G/T). With reference to the previously expressed formula ($P_{id} = P_{in}/4\pi r^2$) of transmitting antenna power density on the spherical surface if it has a transmitting gain (G_T) and where (P_{in}) is equal to the transmitted power (P_T), the power density (P_D) can be written as follows:

$$P_D = G_T P_T / 4\pi r^2 \left[W/m^2\right] \quad (4.18)$$

where (G_T P_T) related values are considered to be the radiation power transmitted by an ideal omnidirectional antenna. Therefore, this term is considered as an effective (or also equivalent) isotropically radiated power (EIRP), which can be expressed in antilogarithm and decibel expressions, respectively, as follows:

$$\mathrm{EIRP} = \mathrm{G_T P_T[W]} \quad \text{and} \quad \mathrm{EIRP} = [\mathrm{G_T}] + [\mathrm{P_T}][\mathrm{dBW}] \tag{4.19}$$

The EIRP value is an important parameter in evaluating the transmitting performance of an MES terminal including an antenna. However, the EIRP amount (dBW) is defined by the sum of the antenna gain (dB) and the output power of HPA (dBW), taking account of feed losses such as feed lines, cable, and a diplexer.

4.3 Classification of Mobile Satellite Antennas (MSA)

In many respects, the mobile satellite antennas currently available for MSC applications constitute the weakest links of the system. If the mobile antenna has a high gain, it has to track the satellite, following both mobiles and satellite orbital motions. Thus, sometimes, this is difficult and expensive to synchronize. Therefore, if the vehicular antenna has low gain, it does not need to perform tracking, but the capacity of the communications link is limited. In general, according to the transmission direction, there are three types of MSA: (1) transmitting and receiving or so-called transceiving, as a part of all types of MES; (2) only receiving is part of the special Inmarsat EGC receiver; and (3) only transmitting is built-in satellite beacon antennas for maritime, land, and aeronautical applications.

Existing and forthcoming GEO and non-GEO mobile satellite operators have conducted research on all network segments, including different types of MSA and their future development and improvements. The Engineering Test Satellite-V (ETS/V) experiments conducted in Japan for the transmission of voice, video, and different data rate digital communications between ships, land vehicles, and aircraft were successful. Moreover, a test of low-speed data transmission by using briefcase-size transportable equipment, onto which two small printed antennas were mounted, was among the experiments.

4.3.1 Shipborne MSA

The different types of shipborne satellite antenna systems were developed for installation on board oceangoing ships and inland sailing vessels, on sea platforms and other offshore infrastructures. In general, these antennas must have strong and rugged constructions, with corresponding mechanical and electrical particulars. The Inmarsat-A SES is the inheritor of the first generation of Marisat, similar to SES, and was the first Inmarsat operating standard of MMSC. In fact, this analog standard was started in 1982, using the Inmarsat-A standard transceiving antenna system known as above deck equipment (ADE).

In the meantime, Inmarsat-C and EGC were developed with small omnidirectional antennas. In addition, Inmarsat-B digital standard started to be in service on ships at the end of 1993, using the second generation of Inmarsat satellites. This standard, compatible with Inmarsat-A, uses the same antenna specification. After employing more powerful satellite constellations, the Inmarsat system developed Inmarsat-M, mini-M, and D+ standards with special shipborne antenna systems. All the above-mentioned MSA are transceiving antennas, except for the Inmarsat EGC receiver, which can also use a receiving antenna. At the same time, Inmarsat developed the Inmarsat-E L-band satellite EPIRB, which was decommissioned on December 1, 2006. On the other hand, the COSPAS–SARSAT system developed shipborne EPIRB, with small built-in VHF and UHF antennas, which is able to send distress alert for search and rescue (SAR) operations. Thus, rugged MSA with added-value system functions compatible with shipborne operation for Inmarsat, Iridium, Globalstar, and other systems, and compact antennas for small lifeboat operation are some of the fascinating requirements for antenna designers in the future.

4.3.2 Vehicleborne MSA

The vehicleborne MSA were developed for installation on road and rail vehicles, such as trucks, trailers, buses, cars, and trains. Thus, the transceiving Inmarsat-C, D, M, mini-M, and other MSA are used on board land vehicles. Similar to the maritime application, the COSPAS–SARSAT system has developed personal or vehicleborne PLB handheld satellite beacons with small built-in VHF- and UHF-band-transmitting antennas.

4.3.3 Airborne MSA

The airborne MSA were developed for installation on board aircraft, such as airplanes and helicopters. There are several types of low- and high-gain airborne satellite transceiving antennas for Inmarsat-C, H, H+, I, L, mini-M, and other MSS. Parallel to the maritime and land distress applications, the COSPAS–SARSAT system developed airborne ELT satellite beacons with small built-in VHF- and UHF-band-transmitting antennas. Because of the high speed of aircraft, the aerodynamic constraints are significant and antennas for both radio and satellite systems must conform to minimum drag and reliability requirements.

4.3.4 Transportable MSA

Inmarsat satellite operator developed special transceiving satellite antenna systems, which can be integrated into mobile Transportable Earth Station (TES) of Inmarsat-B, C, M, and mini-M terminals.

4.3.5 MSA for Personal Satellite Terminals

The new GEO and non-GEO MSC systems developed small satellite antennas for GMPSC terminals, such as handheld and roof antennas. These systems provide mobile tracking, so the tracking antenna has to be combined or integrated with the communications antenna.

4.3.6 Other Types of MSA

The other solutions for MSA are receiving broadcast radio and TV and GNSS, transceiving broadband and antennas for military satellite communications and navigation equipment.

4.4 Low-Gain Omnidirectional Antennas

As mentioned, the antenna systems for MSC are classified into omnidirectional and directional. The gain of omnidirectional antennas is low and generally from 0 to 4 dB in the L-band, which does not require the capability of satellite tracking. There are three types of low-gain omnidirectional antennas, which are very attractive for all mobile applications owing to the small size, lightweight, and circular polarization properties. These antennas are also used as elements of directional antennas for special configurations.

4.4.1 Quadrifilar Helix Antenna (QHA)

The QHA low-gain model is composed of four identical helixes wound, equally spaced, on a cylindrical surface. The helix elements are fed with signals equal in amplitude and 0, 90, 180, and 270° in relative phase. This antenna can easily generate circularly polarized waves without a balloon or a 3 dB power divider, which are required to excite a balanced fed dipole and circularly polarized

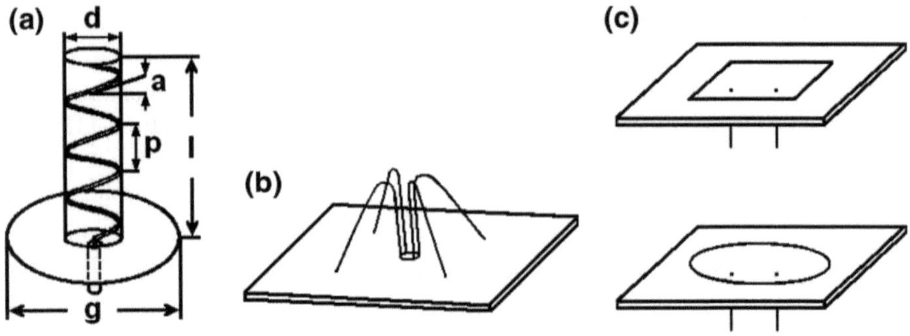

Fig. 4.2 Types of low-gain omnidirectional antennas—Courtesy of Book: by Ilcev

crossed-dipoles. It can also be operated on a wide frequency bandwidth of up to 200 % because it is a traveling-wave-type antenna. The components of QHA are ground plane **(g)**, pitch **(p)**, pitch angle **(a)**, length **(l)**, and diameter **(d)**, presented in Fig. 4.2a. The diameter of the ground plane is usually selected to be larger than one wavelength, and the number of turns is N = l/p. However, it is well known that the parameters for (a) are about 12–15 and the circumference of the helix (πd) is about 0.75–1.25 wavelengths. Circularly polarized waves with good axial ratios can be transmitted along the (z)-axis direction (axial mode). The gain of a helical antenna depends on the number of (N) turns, and typical gain and half-power beam width are about 8 dBi and 50° when N = 12 $\sim$ 12 but is usually about 3 dBi. This antenna is employed as a receiving antenna for GPS and as a transceiver antenna for L-band Inmarsat-C SES, Vehicle Earth Station (VES) and Aircraft Earth Station (AES), low air drag Aero-C applications covered by different kinds of radomes. All three types of antennas can also be combined with GPS receiving. This antenna is also a component of satellite terminals deployed by AMSC, MSAT, MSAT-x, Iridium, Globalstar, Mobilesat ETS-V, forthcoming ICO, and other MSC operators.

In general, QHA, as a mobile antenna, is the best solution and has two advantages over a conventional unifilar helical antenna. The first is an increase in bandwidth; namely, it can generate axial mode circularly polarized waves in the frequency range from 0.4 to 2.0 wavelengths of the helix circumference. The second is lowered frequency for axial mode operation. The principal disadvantage is an increase in the complexity of the feed system. The area of the ground plane is usually about 3 times the diameter of the helix.

4.4.2 Crossed-Drooping Dipole Antenna (CDDA)

A dipole antenna with a half wavelength ($\lambda/2$) is the most widely used, and it is also the most popular, having been used in antenna systems such as the parabolic antenna for MSC.

A half-wavelength dipole is a linear antenna whose current amplitude varies one-half of a sine wave, with a maximum at the center. As a dipole antenna radiates linearly polarized waves, two crossed-dipole antennas have been used in order to generate circularly polarized waves. The two dipoles are geometrically orthogonal, and equal amplitude signals are fed to them with $\pi/2$ in-phase difference. In order to optimize the radiation pattern, a set of dipole antennas is bent toward the ground, as shown in Fig. 4.2b, and for that reason, it is called a drooping dipole antenna. Otherwise, the CDDA serves as a transceiver antenna for L-band Inmarsat-C SES and VES applications mounted inside a radome.

The CDDA is the most interesting for LMSC, where required angular coverage is narrow in elevation and is almost constant in azimuth angle. By varying the separation between the dipole elements and the ground plane, the elevation pattern can be adjusted for optimum coverage for the region of interest. The general characteristics of this antenna are as follows: Gain is 4 dBi minimum, axial ratio is 6 dB maximum, and the height of the antenna is about 15 cm. This antenna has a maximum gain in the boresight direction.

4.4.3 Microstrip Patch Antenna (MPA)

A microstrip disk (patch) antenna is very low profile and has mechanical strength, so it is considered to be the best type for mobiles such as cars and especially in aircraft at the hybrid L to Ku-band, which requires low air drag. In general, a circular disk antenna element has a circular metallic disk supported by a dielectric substrate material and printed on a thin dielectric substrate with a ground plane. In order to produce a circularly polarized wave, a patch antenna is excited at two points orthogonal to each other and fed with signals equal in amplitude and 0 and 90° in relative phase. Thus, a higher-mode patch antenna can also be designed to have a similar radiation pattern to the drooping dipole. To produce conical radiation patterns (null on axis) suitable for land mobile satellite applications, the antenna is excited at higher-mode orders. In Fig. 4.2c is illustrated the basic configuration for a circular patch antenna (above it is shown square patch with the same characteristics), which has two feed points to generate circularly polarized waves. The resonant frequency is excited by basic mode and given as follows:

$$f = 1.84c/2\pi a\sqrt{\epsilon_r} \tag{4.20}$$

where (a), (c), and (ϵ_r) are the radius of circular disk, the velocity of light in free space, and the relative dielectric constant of the substrate, respectively.

In LMSC, a MPA antenna with higher-order excitation is considered better because it can optimize the gain in elevation angle to the satellite in the same way as a CDDA. The area of a higher-mode circular MPA is about 1.7 times larger in radius as the gain is about 6–8 dBi. The circular patch is also suitable as a satellite navigation-receiving antenna for GPS receivers. The MPA antenna with higher-order excitation is considered better because it can optimize the gain in elevation angle to the satellite in the same way as a CDDA.

In fact, the area of a higher-mode circular MPA is about 1.7 times larger in radius as the gain is about 6–8 dBi. The circular patch is also suitable as a satellite navigation-receiving antenna for GPS receivers.

4.5 Directional Medium-Gain Antennas (MGA)

The medium-gain directional MSA are solutions with a typical gain between 12 and 15 dBi, although some antennas can have even bigger gains. These MSC antenna systems can provide voice, fax, and HSD for Inmarsat-M MES applications and for Inmarsat airborne standards, including other systems developed by ESTEC.

4.5.1 Aperture Reflector Antennas

The aperture reflector antennas are good solutions with medium-gain characteristics used in MSC, with three basic representatives such as SBF, modified SBF, and improved SBF antennas, illustrated in Fig. 4.3 (Left), (Middle), and (Right), respectively. The main characteristics of these three antennas are shown in Table 4.2.

Moreover, due to the excellent radiation characteristics and voltage standing wave ratio (VSWR) of SBF antennas, all three types of aperture antenna with half-power beam width of about 34° have been in their time proposed for Inmarsat-M antenna. The SBF antennas consist of the stabilized platform with two gyroscopes for azimuth and elevation angles, diplexer, HPA, and LNA, which are

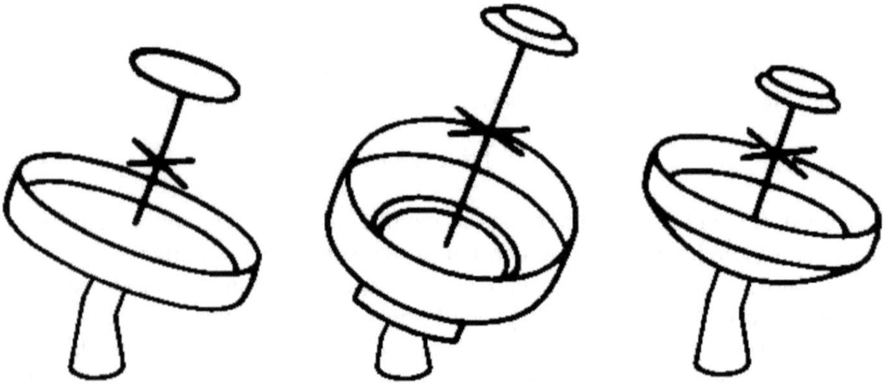

Fig. 4.3 Types of directional medium-gain aperture antennas—Courtesy of Book: by Fujimoto

Table 4.2 Particulars of aperture types of antennas

Characteristics	SBF antenna	Modified SBF antenna	Improved SBF antenna
Effective gain	14.5 dB	15 dB	15 dB
Half-power bandwidth	34°	34°	34°
Directive gain	14.8 dB	15.5 dB	15.5 dB
First sidelobe level	−21 dB	−22 dB	−22.5 dB
Axial ratio	−1.3 dB	−1.1 dB	−1.1 dB
Aperture efficiency			
Effective–directive gain	65–75 %	75–80 %	76–85 %
RF/VSWR bandwidth, under 1,5	3 %	7 %	9 %
Diameter of large reflector (D_R):			
Bigger (D_{R1})	40 cm (2.05λ)	40 cm (2.05λ)	40 cm (2.05λ)
Smaller (D_{R2})	–	27 cm (1.38λ)	–
Diameter of small reflector (D_r):		–	
Bigger (D_{r1})	9 cm (0.46λ)	9.5 cm (0.48λ)	9 cm (0.46λ)
Smaller (D_{r2})	–	8.5 cm (0.43λ)	8 cm (0.41λ)
Width of a rim	4.9 (0.25λ)		4.9 (0.25 λ)
Distance between (D_R) and (D_r)	9.7 cm (0.49λ)	19.5 cm (0.99λ)	12.9 cm (0.66λ)
Distance between exciter and D_r	4.9 cm (0.25λ)	–	5.7 cm (0.29λ)
Distance between (D_{r1}) and (D_{r2})	–	–	1.8 cm (0.09λ)
Slanting angle of a (D_R)	0°	–	15°

enclosed under the protective cupola of a radome. In order to stabilize the antenna, two gyro wheels rotate in opposite directions on a platform.

4.5.1.1 Short Backfire (SBF) Plane Reflector Antenna

The SBF plane reflector antenna that was developed experimentally by H.W. Ehrenspeck in the 1960s is well known as a highly efficient antenna of distinctly simple and compact construction. Its high directivity and low sidelobe characteristics make it a single antenna with high, even values, which are applicable to MSC, tracking, and telemetry. Therefore, an SBF antenna is very attractive for gains in the order of 13–15 dBi peak RHCP and can be mounted primarily on small but on any size of ships. This antenna consists of two circular planar reflectors of different diameter, separated generally by one-half wavelength, forming a shallow leaky cavity resonator with a radiation beam normal to the small reflector. The antenna is fed by a dipole at around the midpoint between two reflectors, and it has almost a quarter-wavelength rim on the larger reflector. It has the problem of a narrow bandwidth of about 3 % because of its leaky cavity operation.

The Rx terminal G/T is −12 dBK, and the EIRP of the Tx terminal is 28 dBW. The basic configuration of the SBF antenna consists of a crossed-dipole element, which is required to generate a circularly polarized wave, large and small reflectors, and a circular metallic rim. It has the strong directivity normal to the reflector, and its performance is superior to that of other types of mobile antennas with the same diameter; however, it has the problem of narrow frequency band characteristics. This antenna has many beneficial characteristics, such as efficiency and the simplicity of construction, and is also considered a favorite option for a compact and high-efficiency shipboard antenna. It is produced by many world manufacturers approved by Inmarsat or ESTEC.

4.5.1.2 Modified SBF Plane Reflector Antenna

A modified SBF antenna differs from the conventional SBF antenna in that there is either an additional step on the large reflector or a change in the shape of the large reflector from a circular to a conical plate in order to improve the gain characteristics and the frequency bandwidth of the VSWR.

The dual reflector improves the input impedance characteristics covering the frequency range between transmitting and receiving sides. The conventional SBF model is a resonant-type antenna, producing input impedance characteristics that are narrow in bandwidth, so wider bandwidth is required to cover the 1.6/1.5 GHz range for MES of the Inmarsat system. In effect, the improvement in the input impedance is greatly dependent on the size and the separation of the small reflectors. The VSWR can be reduced from 1.7 and 1.5 (at 1.54 and 1.64 GHz) to below 1.2 for each RF.

4.5.1.3 Improved SBF Conical Reflector Antenna

The main research activities of the ETS-V program in MSC have been focused on studying the reduction of fading, using compact and high-efficiency antennas with a gain of around 15 dBi, so the electrical characteristics of a simple SBF antenna have been improved by changing its main reflector from a flat disk to a conical or a step plate and by adding a second small reflector. The gain is improved by 1 dB without changing sidelobe levels. Comparisons of electrical and other parameters of three types of SBF antennas are shown in Table 4.2. Stabilization of the antenna is obtained by a two-axis stabilized method, and satellite pointing is carried out by a tracking program using output signals from the ship's gyroscope. It is also considered to be a suitable option for mounting aboard ships.

4.5.2 Wire Antennas

The wire antenna systems are monosyllabic construction or combinations of elements, such as different shapes of wire spirals and helixes, dipoles, and patches.

These types of antennas have a very simple construction, with any reflector specified for medium-gain directional antennas, and, with some modification, respond well to the demands of MSC applications.

4.5.2.1 Helical Wire Antennas

Since an axial mode helical antenna has good circular polarization characteristics over a wide frequency range, it has been put into practical use as a single wire antenna or as an array element. With respect to the structure, this antenna can be considered a compromise between the dipole and the loop antennas and the radiation mode varies with the pitch angle and the circumference of the helix. In particular, a helix with a pitch angle of 12–15° and a circumference of about 1λ has a sharp directivity toward the axial direction of the antenna. This radiation mode is called the axial mode, which is the most important mode in helical antennas. Several studies have been carried out on the properties of the axial mode helical antenna with a finite reflector. The current induced on the helix is composed of four major waves, which are two rapidly attenuating waves and two uniform waves along the helical wire. These waves include the traveling wave and the reflected wave. Thus, in a conventional helical antenna, the uniform traveling wave will be dominant when the antenna length is fairly large, with typical versions such as a conical helix shown in Fig. 4.4a and a cylindrical helix in Fig. 4.4b. A conical helix is interesting for L-band MSS enabling HPBW in the order of 100° and circular polarization without hybrid gain of 4–7 dBi. Cylindrical antennas can be monofilar or multifilar, also suited for L-band MSS, while in a shortcut cylindrical helix antenna, the rapidly attenuating traveling wave will be dominant, especially in a two-turn (N = 2) helical antenna.

1 **Conical Helix Antenna**—This antenna can be regarded as a low-gain development of the cylindrical helix antenna and is suitable for wide beam-width applications with good efficiency. Thus, with suitable choices of cone angle and turn spacing, it is possible to achieve a beam width in the order of 100°. This type of antenna can also achieve an input VSWR of 1.5:1 or better than 5 %

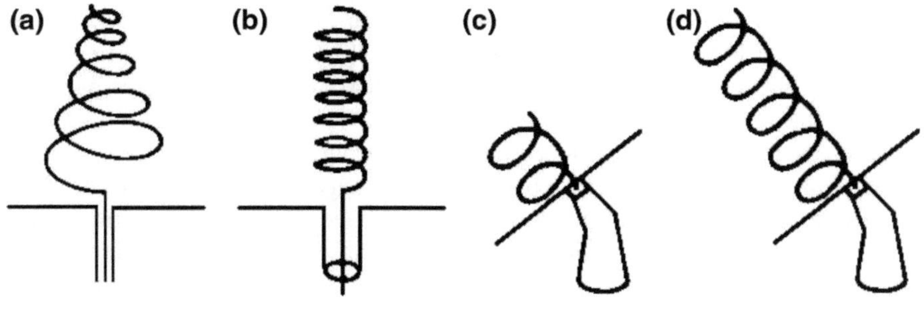

Fig. 4.4 Types of helical wire antennas—Courtesy of Book: by Fujimoto

frequency bandwidth merely by incorporating a simple quarter-wavelength transformer. The typical size of an L-band application is in the order of 15 cm in length, and the ground plane is about 20 cm in diameter. The resultant gain is approximately 4–7 dBi, which is between low- and medium-gain requirements.

2 **Two-Turn Cylindrical Helix Antenna**—This antenna has two turns of wires, forming a simple helical antenna with reflector, shown in Fig. 4.4c. This model has relatively high antenna gain and excellent polarization characteristics for its size. Radiation patterns characteristically are calculated with respect to (E_0) and (E_Φ) planes. Its gain is 9 dBi, and the axial ratio is about 1 dB, with reflector diameter (d) around 1λ. Such types of antenna have comparatively high performance in spite of their small size and compact construction. From the above-mentioned considerations, a highly efficient antenna for the Inmarsat-M MES can be realized by applying this antenna to elements of an array antenna.

3 **Five-Turn Cylindrical Helix Antenna**—This antenna solution is illustrated in Fig. 4.4d. The main electrical characteristics are as follows: Gain is 12.5 dBi of peak RHCP for Tx and 11.5 dBi for Rx; sidelobe level has value of about −13 dB; axial ratio is 3 dB; beam width of 3 dB has angle of −47°; terminal G/T has −16 dBK; and terminal EIRP has 29 dBW. This antenna solution is designed and developed by the European research institution ESTEC. Its stabilization is obtained by gravity elevation on double-gimbaled suspension. Thus, the pendulum aligns itself with the vertical when not subject to other acceleration. However, the center of rotation of the pendulum is distant from that of the ship pitch and roll movement induces horizontal acceleration to which the pendulum is sensitive. In order to limit perturbations, the resonant frequency of the pendulum must be low with respect to the excitation frequencies in pitch and roll and the damping (friction) must be minimum. Low resonance frequency is achieved by minimizing the distance between the center of gravity of the rotating part and its center of rotation. This also reduces torque due to horizontal acceleration but at the same time reduces the stabilizing torque due to gravity.

4.5.2.2 Inverted V-Type Crossed-Dipole Antenna

The inverted V-type crossed-dipole antenna is an advanced circularly polarized antenna with tick V-elements, illustrated in Fig. 4.5 (Left). The resonance of this antenna is obtained when the length is somewhat shorter than a free-space half wavelength.

Thus, as the thickness is increased, the resonant length is reduced. Circular polarization can be produced by a pair of orthogonally positioned dipoles driven in quadrature phase with equal amplitudes. The crossed-dipole antenna arrangement cannot provide a good axial ratio off boresight because the radiation patterns for the straight dipole are different in both principal planes, called the H and E planes. This

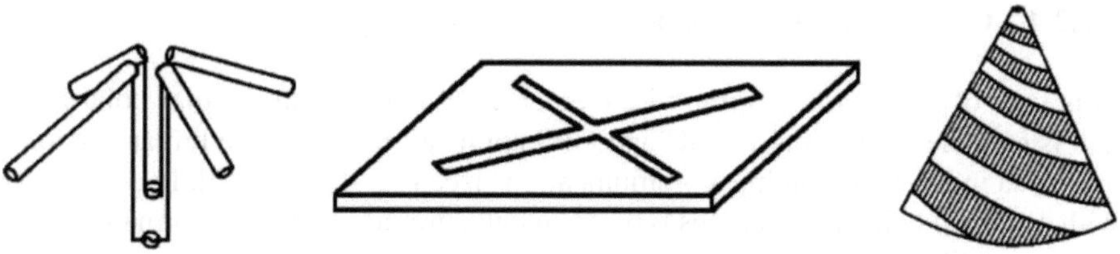

Fig. 4.5 Types of Crossed-dipole, crossed-slot, and conical wire antenna systems—Courtesy of Book: by Fujimoto

shortcoming can be improved by modifying the straight dipoles to a non-straight version, such as the V- and U-forms.

The improved dipoles are called V- and U-type dipoles. According to some conducted measurements, the U-type provides better electrical performance than the V-type, though the V-type is simpler in mechanical structure and is less complex. The crossed-dipole can also produce circular polarization without using any external circuits, such as the hybrid component.

Thus, the condition to excite the circularly polarized waves can be established by a balun and the self-phasing of four radiating elements. Two of the elements are at a 0° phase angle, and the other two are at an 180° phase angle. The desired 90° phase difference is obtained by designing the orthogonal elements such that one is larger to make it inductive, while the other is smaller to make it capacitive. This type of antenna is a good model for Ku-band aeronautical satellite communications.

4.5.2.3 Crossed-Slot Antenna

These antennas are useful for L-band aeronautical satellite communications on high-speed aircraft because they are very low profile in structure and suitable for a flush-mounting application, shown in Fig. 4.5 (Middle). The slot antenna is circularly polarized and is complementary with the corresponding dipole antenna, so that the radiation pattern is the same as that for the horizontal dipole.

There are only two differences: First is the property that the electric and magnetic fields are interchanged, and second is that the slot electric field component normal to the perfectly conducting sheet is discontinuous from one side of the sheet to the other because of the direction of the field reverses.

In this case, the tangential component of the magnetic field is, likewise, discontinuous. This antenna can be also complementary with the corresponding crossed-dipole antenna, although the feeding method for the circular polarization is more complicated. Thus, on a model of this antenna known as a cavity-backed, it needs one 90° hybrid to produce the circular polarization. This feed technique is effective not only to suppress undesired coupling between the cross-slots but also to match the input impedance over a wider frequency band.

4.5.2.4 Conical Spiral Antenna

This type of antenna has spiral wire elements on a cone with circular polarization and is suitable for L-band LMSC and GPS applications, while the bifilar version is also used in Ku-band satellite communications [see Fig. 4.5 (Right)]. In comparison with a conical helix antenna, this type of antenna provides better performance and is more versatile, though the geometry is somewhat complex. Such antenna is independent of frequency, and its geometry can be presented mathematically in spherical coordinates (r, θ, Φ) as follows:

$$r = e^{a\Phi} g(\theta) \tag{4.21}$$

where (a) and $g(\theta)$ are an arbitrary constant and angular function, respectively. Its radiation mechanism can be understood by regarding the two spirals as a transmission line. When two conductor arms are fed in antiphase at the cone apex, waves travel out from the feed point and propagate along the spirals without radiating until a resonant length has been traversed. Strong radiation occurs at that point, and very little energy is reflected by the outer limits of the spiral.

Conveniently, two conductor arms can also be fed directly at the center point or apex from a coaxial cable bonded to one of the spiral arms without any external baluns because the spiral arm can itself act as a balun. In this case, a dummy cable may be bonded to another arm to maintain the symmetrical performance. If the width of arm is decreased to a narrow constant value, the arms can be formed by the cable alone.

4.5.2.5 Planar Spiral Antennas

Cavity-backed planar spiral antennas are commonly divided into three main categories: equiangular, logarithmic, and Archimedean spiral antennas. These types of antennas are well suited for flush mounting on aircraft for L to Ku-band satellite communications. In general, this antenna has been fed by using the external balun, but it can also be fed at the center point, or apex, from a coaxial cable bonded to one of the arms, without any external baluns, like the conical spiral antenna.

1 **Equiangular Spiral Antenna**—The geometry of this antenna corresponds to the special case of the conical spiral antenna, bifilar with logarithmic period, cavity-backed and can be obtained by substituting a $\pi/2$ into θ_0 to give:

$$r_1 = r_0 e^{a\Phi}; r_2 = r_0 e^{a(\Phi-\Delta)}; r_3 = r_0 e^{a(\Phi\pm\pi)}; r_4 = r_0 e^{a(\Phi-\Delta\pm\pi)} \tag{4.22}$$

This antenna needs no external hybrid circuits to produce circular polarization, and the example shown in Fig. 4.6 (Left) can radiate LHCP waves outward from the page and RHCP waves into the page when the pair of spirals is excited in antiphase at the center. Otherwise, according to experimental measurements, the

axial ratio is near unity and the HPBW is in the order of 90° over a decade bandwidth or even more. As for the input impedance, the resistive part on the thickness of the antenna elements and thin elements lead to high impedance values. This implies that the impedance depends on the arm width when the structure is planar. If the angular extent (Δ) is chosen to be 90°, the geometries of the arm and the space between arms are identical, except for a rotation of 90° around an axis.
This structure is defined as self-complementary, just like the conical spiral antenna, but it should be noted that the planar spiral antenna has a constant impedance of $60\pi[\Omega]$ for the two arm configurations.

2. **Logarithmic Spiral Antenna**—This bifilar antenna design with logarithmic period and cavity-backed, shown in Fig. 4.6 (Middle), can be presented mathematically by the following equation:

$$r_1 = a^{\Phi}; r_2 = a^{(\Phi-\Delta)}; r_3 = a^{(\Phi\pm\pi)}; r_4 = a^{(\Phi-\Delta\pm\pi)} \tag{4.23}$$

This antenna can radiate RHCP waves outward from the page and LHCP waves into the page without any external hybrid circuits, as a pair of spirals is excited with an antiphase at the center.

3. **Archimedean Spiral Antenna**—The Archimedean spiral thin-wire bifilar cavity-backed antenna, which is illustrated in Fig. 4.6 (Right), is another geometry of the planar spiral antenna. This spiral antenna has superior bandwidth properties when fully optimized and typically consists of a pair of thin-wire arms, of which the geometry can be presented by the following relation:

$$r_1 = r_0\Phi; r_2 = r_0(\Phi-\pi) \tag{4.24}$$

This antenna also needs no external hybrid circuits to produce circular polarization and can radiate RHCP waves outward from the page and the LHCP waves into the page if the pair of thin-wire arms is excited in antiphase at the center. It is a

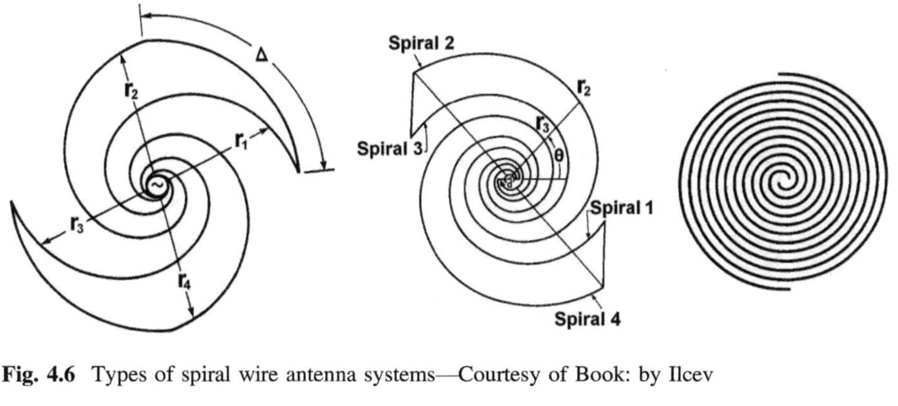

Fig. 4.6 Types of spiral wire antenna systems—Courtesy of Book: by Ilcev

broadband antenna and has properties similar to the standard planar spiral antenna, although it is not theoretically a frequency-independent structure. When placed in a quarter-wave cavity, this antenna can achieve near-octave bandwidth, even when the cavity consists in a metal-based cylinder without any absorber.

Thus, if an absorber-loaded cylinder is employed in the cavity, a greater-than-decade bandwidth may be achieved, although about half the power is dissipated into heat by the absorber. A typical Archimedean spiral antenna has an octave bandwidth for a VSWR less then 2, an axial ratio of less then 2 dB, and a beam width of about 70°, while a gain of 7–8 dBi is achieved without an absorber. The structure has several mechanical advantages: It is compact and fairly simple to construct, and the spiral arms can be easily fed, using a suitable impendence-transforming balun.

4.5.3 Array Antennas

Several different types of antenna can be arrayed in space to make a directional pattern or one with a desired radiation pattern. This type of integrated and combined antenna is called an array antenna consisting of more than two elements, such as microstrip, crossed-slot, crossed-dipole, helixes, or other wire elements, and is suitable for MSC. Each element of an array antenna is excited by equal amplitude and phase, and its radiation pattern is fixed.

4.5.3.1 Microstrip Array Antenna

The microstrip array antenna (MAA) is a nine-element flat antenna disposed in three lines spaced at 94 mm, namely about a half wavelength at 1.6/1.5 GHz and whose antenna volume is about 300 × 300 × 10 mm [see Fig. 4.7 (Left)]. As shown in this figure, the element arrangements of the MAA solutions are 3 × 3 rows square arrays in order to obtain similar radiation patterns in different cut planes.

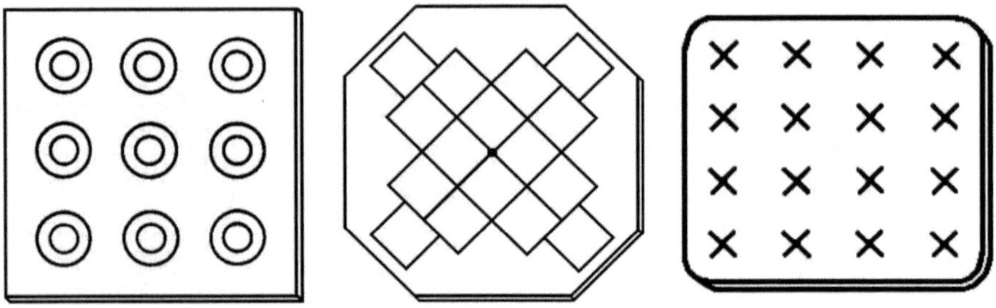

Fig. 4.7 Microstrip, crossed-slot, and crossed-dipole array antennas—Courtesy of Book: by Fujimoto

The MAA beam scanning is performed by controlling four-bit variable phase shifters attached to each antenna element. This type of antenna is very applicable for the MES Inmarsat-M and the Inmarsat Aero standard.

4.5.3.2 Crossed-Slot Array Antenna

The crossed-slot array antennas (XSA) are a 16-element solution with 97-mm spacing, and their volume is about 560 × 560 × 20 mm, shown in Fig. 4.7 (Middle). Evident is the element arrangement of the XSA, which is a modified 4 × 4 square array in order to obtain similar radiation patterns in different cut planes.

The XSA antenna beam scanning is carried out to control four-bit variable phase shifters associated with each antenna element. This antenna is suitable for the MES Inmarsat-M MES, including Inmarsat Aero solutions.

4.5.3.3 Crossed-Dipole Array Antenna

The crossed-dipole array antenna is composed of 16 crossed-dipoles fed in phase with a peak gain of 17 dBi and with the feeding circuit behind the radiating aperture, shown in Fig. 4.7 (Right). The main electrical characteristics of this antenna are as follows: Gain is 15–17 dBi with peak RHCP transmit; axial ratio has a value of 0.7 dB; beam width of 3 dB is -34°; terminal G/T is −9.5 dBK; and EIRP terminal value is 32 dBW.

Otherwise, the antenna system consists of a stabilization mechanism platform for tracking and satellite focusing, flat antenna array, diplexer, HPA, and LNA, which are all protected by a plastic radome. Stabilization of the antenna is obtained by a single-wheel gyroscope, when the azimuth pointing is controlled by the output from the ship's gyrocompass. Similar to the previous two models, this satellite antenna is also suitable for maritime, land, and aeronautical applications.

4.5.3.4 Four-Element Array Antennas

There have been several four-element antenna models developed, such as Yagi–Uda, quad-helix, and four-element SBF array.

1. **Yagi-Uda Crossed-Dipole Array Antenna**—This type of array antenna has been developed for use on board ships and is protected with a radome, shown in Fig. 4.8 (Left). The feeder of this antenna is a simple formation of four in-line crossed-dipoles fixed in the middle of the reflector. This end-fire array has circular polarization, and the gain is between 8 and 15 dBi.
2. **Quad-Helix Array Antenna**—The quad-helix array antenna solution is composed of four identical two-turn helical wire antennas in the shape of a square

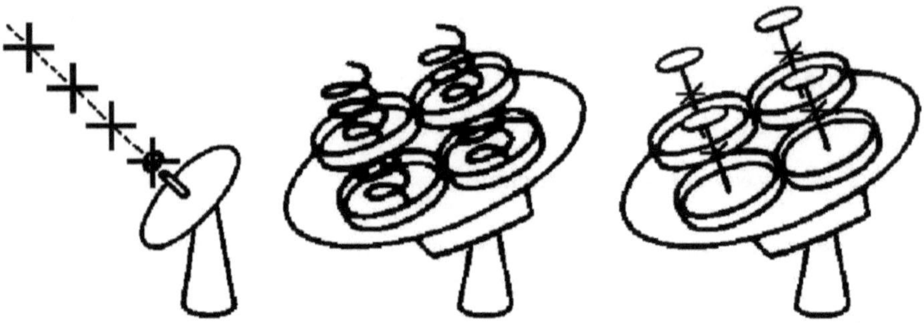

Fig. 4.8 Types of four-element array antennas—Courtesy of Book: by Fujimoto

and whose elements are oriented in the manner shown in Fig. 4.8 (Middle). According to the previous studies, the effect of mutual coupling between each element of this antenna is not negligible and this mutual coupling mainly degrades the axial ratio. The axial ratio of a single helical antenna is about 1 dB, but this value is degraded to about 4.5 dB in the case of the array antenna with an array spacing of 0.7λ. However, the best properties of antenna gain and axial ratio can be obtained at a rim height of about 0.25λ. The antenna gain is improved by 0.4 dB, and the axial ratio is also improved by 3.5 dB, compared to that of the quad-helix array antenna without rims. The performance characteristics of this small antenna are essentially, gain is about 13 dB (HPBW is 38°), and aperture efficiency is about 100 %. It appears that this type of helix-integrated antenna is also well suited for the shipborne Inmarsat-M standards.

3 **Four-Element SBF Array**—This antenna is developed on the basis of a conventional SBF antenna as an integrated array with four SBF elements [see Fig. 4.8 (Right)]. The antenna provides high aperture efficiency, circular polarization, and almost high-performance gain between 18 and 20 dBi. Because of the high-gain characteristics, this array is very suitable for maritime applications as a shipborne antenna.

4.5.3.5 Spiral Array Antenna

Directional antennas for LMSC have been expected to provide voice and HSD links not only for long-haul tracks but also for private cars. From that point of view, cost is an important factor to be taken into account in designing antenna systems. In the early stage of LMSC, a mechanical steering antenna system was considered the best candidate for vehicles; however, it will be replaced by a phased-array antenna in the near future because it has many attractive advantages, such as low profile, high-speed tracking, and potentially low cost.

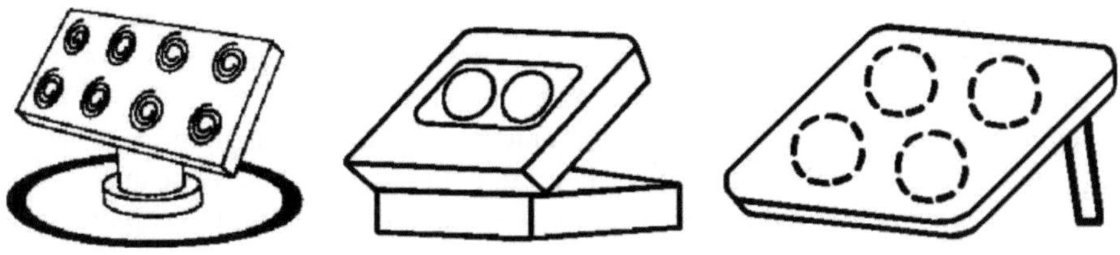

Fig. 4.9 Types of spiral and two- and four-patch array antennas—Courtesy of Book: by Fujimoto

The mechanical steering antenna with eight spiral elements and with adopted closed-loop tracking method gives about 15 dBi in system gain, shown in Fig. 4.9 (Left). The antenna is 30 cm in radius, 35 cm in height, and 1.5 kg in weight. The array consists in 2 × 4 spiral elements, and it forms a fan beam with a half-power beam width of 21° in the azimuth and 39° in the elevation plane at L-band. Its peak gain is about 15 dBi, including the feeder losses, and is suitable to track the satellite for MSC, because elevation angles to the satellite are not as varied as those of the azimuth angles. In effect, the antenna beam direction can be shifted in two azimuth directions, from the E or W side, by switching the pin diode phase shifters. Consequently, the difference between the received signals in both directions is used to drive the antenna system toward the satellite. The beam-shifting angle is set to approximately 4°.

4.5.3.6 Patch Array Antennas

The main feature of the future MSC systems will be portability and transportability, which means that a person can directly access the satellite to establish a link using a very small TES transceiver with antenna system. Even in the present Inmarsat L-band system, great efforts have been made to develop transportable and portable terminals with corresponding satellite antennas.

1 **Two-Patch Array Antenna**—This antenna for TES and briefcase portable terminals is developed in the Inmarsat and ETS-V programs, which is shown in Fig. 4.9 (Middle). This antenna has two microstrip patch elements (one for Rx and another for Tx), gain is 6 dBi, EIRP is 6 dBW, and G/T is −21 dBK. The reason for adopting separate Rx and Tx antennas is to eliminate a diplexer, which is too large and heavy for a compact and lightweight terminal. The antenna beam width on the lid is wide enough to point to a satellite by manual tracking. Two microstrip patch array antennas mounted on the lid of the briefcase TES transceiver for low-speed data transmission serve the Inmarsat and ETS-V TES mobile terminals.
2 **Four-Patch Array Antenna**—Several four-element patch array transportable antennas were developed for universal Inmarsat-C and other TES terminals. On the other hand, JPL and NASA designed similar L-band types of mobile antennas, mainly for LMSC regional utilities in the USA. These include a

mechanically steered, tilted 1 × 4 patch array and two electrically steered planar phased-array antennas. The mechanically steered four-square patch arrays can be fixed in one line, similar to two-patch array, or can have the shape of a four circular patch array manually steered antenna, whose arrangement is shown in Fig. 4.9 (Right). All three of these medium-gain antennas feature beams that are narrow in azimuth angle; hence, they require azimuth steering to keep the beam pointed toward the desired satellite as a mobile changes its azimuth orientation. They provide 9–12 dBi gain, reject multipath signals outside their beam pattern, and allow two communication satellites separated by 30° in a GEO arc to reuse the same frequency to cover the continental US region. A dither-tracking four-element, circular, polarized array for AMSC/MSAT terminals has been designed, which is 10.16 cm high and 50.8 cm in diameter, with 20–60° elevation coverage and with a minimum of 10 dBi gain. In fact, this antenna employs a kind of closed loop for tracking the satellite in azimuth. The rotating antenna platform is mounted on the fixed platform that includes the motor drive and pointing system hardware.

4.5.3.7 Phased-Array Antenna

A phase antenna array is a set of 2 or more antennas in one antenna complex, in which the relative phases of the respective signals feeding the antennas are set in such a way that the effective radiation pattern of the array is reinforced in a desired direction and suppressed in undesired directions. The signals from the antennas are combined or processed in order to achieve improved performance over that of a single antenna.

Phased-array antennas were developed for maritime, land, and aeronautical applications to provide a design for a thin satellite antenna that can be installed on board land vehicles and aircraft. Otherwise, these mobile antennas were developed initially for aircraft and are well known for their complexity and high cost. As a result, emphasis was placed on the selection of manufacturing techniques, materials, and component types, in addition to meeting the RF and pointing requirements and keeping the cost down.

As mentioned previously, the radiation pattern of an array antenna is fixed; however, the radiation pattern can be scanned in space by controlling the phase of the exciting electrical current in each element of the array. This type of antenna is called a phased-array antenna, which has many advantages in terms of MSC applications such as compactness, lightweight, high-speed tracking performance, and potentially low cost.

1. **Phased-Array Antenna for AMSC**—A directional medium-gain antenna is considered a key technology in aeronautical industry. This type of antenna was developed and tested by CRL for the ETS-V antenna program. At the same time, Inmarsat operator approved all aeronautical antenna standards for installations on board commercial and military aircraft as above cockpit equipment (ACE).

Fig. 4.10 Types of aeronautical and maritime phased-array antennas—Courtesy of Book: by Ilcev

In this sense, taking account of the electrical and mechanical requirements of AMSC, a phased array with low-profile antenna elements was chosen for a directional main antenna, while a microstrip antenna was chosen as an antenna element because of its very low profile, very lightweight, and mechanical strength, which satisfy the requirements of airborne antenna. However, one disadvantage is the very narrow frequency bandwidth, usually 2–3 %. The antenna adopted is a two-frequency resonant element because it provides a compact array and a simple feed line configuration.

On the other hand, this type of antenna has very poor axial ratio values. The problem was overcome by using the sequential array technique, where a thin substrate with high dielectric constant is used over a wide frequency bandwidth with excellent axial ratios. The microstrip phased-array antenna is mounted on top of the fuselage and has two planes with 16 circular patch elements, 2 of which are in elevation and 8 in azimuth, illustrated in Fig. 4.10 (Left). In practice, the required coverage angles are as narrow as +20°, so the beams are not steered in elevation directions and the array plane is set at 65° to the horizon in order to optimize the beam coverage on the flight routes. Thus, by controlling eight 4-bit digital phase shifters, the antenna beam scans in a 4° step with +60° with respect to a line perpendicular to the axis of the aircraft. In addition, a step track method was adopted to track the satellite. Therefore, this array has the following characteristics: 2 L-band Tx and Rx frequencies; polarization is LHCP; gain is 14.7 dBi for Tx and 13.5 dBi for Rx; EIRP is about 30 dBW, G/T −10.8 dBK; the axial ratio is about 2 dB; volume is 760 (l) × 320 (w) × 180 (h) mm; and the weight is 18 kg.

2 **Phased-Array Antenna for MMSC**—The new shipborne phased-array antenna has been developed for utilization on board big oceangoing vessels. Besides other directional types of shipborne antenna, this unit is designed to serve as above deck unit (ADE) of the Nera F77 transceiver or other brand SES for new Inmarsat Fleet 77 service. The Nera F77 antenna is a mechanically steered circular disk with 32 low-profile radiating elements arranged in two circles with 16 and 12 elements, and 4 elements are located in the middle of the disk, which is shown in Fig. 4.10 (Right). The ADE unit can be mounted on a mast or directly on the deck and is covered by a radome. The dimensions of antenna mast mount are 180 × 132 cm with a weight of 65 kg, and deck-mounted antennas are 108 × 91 cm and 50 kg in weight. The gain of the antenna is about 15 dBi in both transmission satellite links.

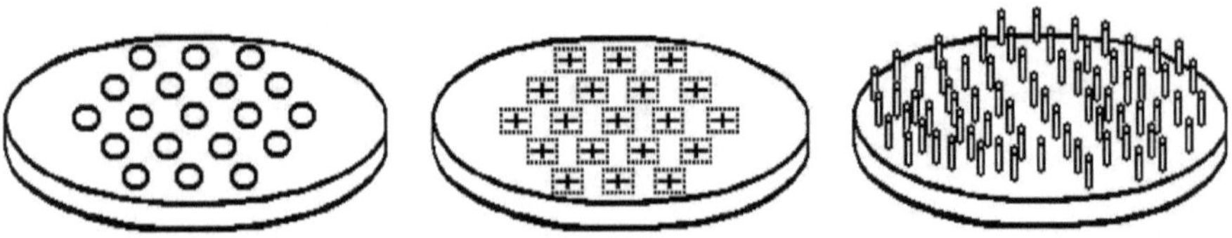

Fig. 4.11 Types of breadboard phased- and adaptive array antennas—Courtesy of Book: by Fujimoto

3. **Breadboard Phased-Array Antenna for LMSC**—The breadboard mobile phased-array antennas have been developed in the USA to meet the L-band frequency requirements, 20°–60° elevation and full 360° azimuth angles of special coverage, gain of 10 dBi above 30° in elevation and 8 dBi at 20° in elevation, half-power bandwidth of 25° in azimuth and 35° in elevation, inter-satellite signal isolation of 20 dB between two GEO separated by circa 35°, and beam-pointing accuracy of +5°. Two phased-array antennas were developed for LMSC: Ball and Teledyne designs shown in Fig. 4.11 (Left) and (Middle), respectively. Each antenna consists of 19 low-profile radiating elements, with 18 3-bit diode phase shifters. The Ball company model uses dual-resonant stacked circular microstrip elements to cover both the Tx- and Rx-bands, while the Teledyne company model employs stripling crossed-slot radiators. For the antenna's beam-pointing system, the initial acquisition of the satellite is accomplished by a full azimuth search for the strongest received signal. An angular rate sensor is used to establish an inertial reference point when the acquisition is performed while the mobile is turning.

Tests show that the antenna can acquire a reference pilot signal in two seconds from a random spatial position. After the desired satellite signal has been acquired, the antenna tracks the satellite by a closed-loop sequential lobing technique. In the event of severe signal fade due to shadowing, the sequential lobing can no longer function properly. In this case, the open-loop angular rate sensor takes over the pointing for a 10-s period, until the sensor drifts away. This antenna with all the radiated elements, hybrids, diode phase shifters, pin diode drivers, and power/combiner driver (except diplexer) can be mounted on the roof of a land vehicle cabin. Thus, based on the research experiments of the airborne phased-array antenna, a new phased-array car antenna has been proposed by CRL. The antenna was installed on a test van and tracks the ETS-V satellite at an elevation angle of 47°. The receiving signal from the satellite was almost constant except for shadowing and blocking effects. The gain of this antenna is 10 and 18 dBi for elevations of 30° and 90°, respectively, the system temperature is about 200 K, axial ratio is 4 dB for 30° of elevation, the volume is 60 × 4 cm, and weight is 5 kg.

4.5.3.8 Adaptive Array Antennas

The antenna for the MSAT program of LMSC was developed by the Canadian CRC. Linearly and circularly polarized arrays with gain values of 9–11 and 10–13 dBi, respectively, have been designed and evaluated in trials using the ex-Marecs-B satellite.

Both antennas use L-band with spatial coverage of 15°–50° elevation and 360° azimuth angles. The satellite tracking is initially acquired with a closed-loop method of stepping through 16 azimuth beam positions and selecting the beam with the strongest signal. In the event that the signal falls below a given threshold, the acquisition sequence is again initiated until the signal is required.

The speed of operation is determined by the terminal C/N_0 ratio and the value of S/N requirements in the control loop bandwidth, which takes 0.1 s to acquire the satellite after an initial phase look. Otherwise, tracking of the satellite is performed by periodically switching on either side of the current beam position and selecting the beam with the strongest signal. A number of algorithms have been devised to minimize any perturbation of the communications signal to less than 1 % of the time.

The main characteristics of this type of antenna are as follows: Operating RF is from 1.530 to 1.660 GHz; spatial coverage is 15°–50° elevation and 360° azimuth; gain is 9–11 dBi for linear and 10–13 dBi for circular model; and the size of both models is about 61 cm diameter, while the linear is 6.3 cm and circular is 20.3 cm high, respectively. Thus, the maximum phase transients in azimuth can be kept to less than ±10° over the required angular coverage and operating frequency band.

Here will be introduced two the following types of adaptive array antennas:

1. **Linearly Polarized Adaptive Array Antenna**—This antenna consists essentially of a driven quarter-wave monopole surrounded by concentric rings of parasitic elements all mounted on a ground plane of finite size. The parasitic elements are connected to the ground through pin diodes. With the application of suitable biasing voltages, the desired parasitic elements can be activated and made highly reflective. The directivity and pointing of the antenna beam can be controlled both in the elevation and in the azimuth planes using high-speed digital switching techniques. The use of a circular polarizer in the linearly polarized design can realize an increase in gain at the expense of an increase in antenna height. The polarizer has an elliptical cross section, a diameter of 40 cm, and a height of 20 cm. It consists of a number of conformal scattering matrices to achieve the 90° differential phase shift between two orthogonal polarizations. Part of a five-ring linearly polarized antenna is shown in Fig. 4.11c. The antenna incorporates sufficient electronics to control the radiation patterns and pointing on command. It is designed to serve MSAT and AMSC transceivers and to be mounted on the metallic roof of a vehicle, where the effective ground plane can significantly enhance antenna gain at low elevation angles.

2. **Circularly Polarized Adaptive Array Antenna**—This type of adaptive array is similar to linear adaptive array and hence is obtained by adding the linearly polarized array. With good cross-polarized isolation, it enables frequency reuse using polarization diversity.

4.6 High-Gain Directional Aperture Antennas

High-gain directional aperture antennas are more powerful transmission reflectors and panels used for Inmarsat maritime and transportable applications. The typical gain of these antennas is more than 20 dBi, EIRP is a maximum of 33 dBW, and G/T is about −4 dBK. There are few basic types of directional parabolic antennas: parabolic dish antenna, which uses parabolic reflector outside or inside radome, and umbrella antenna. The new antenna solution for Inmarsat transportable satellite units is known as a quad flat panel antenna designed by the South African-based company OmniPless, which is introduced in the next section.

1. **Parabolic Dish Antenna**—The first generation of parabolic dish antennas used a reflector in diameter of max 1.2 m, whereas on newer models it is likely to have reduced in size to approximately 0.7–0.8 m, which is illustrated in Fig. 4.12 (Left). Because a large proportion of the Rx signal gain and Tx EIRP is produced by the antenna, the area of the dish can only be reduced if the transmitting power from the satellite transponder is increased, when the receive preamplifier gain can be increased without an appreciable increase in noise.
The parabolic reflector is most often used for high directivity for radio signals traveling in straight lines, as do light rays. They can also be focused and reflected just as light rays can; namely, a microwave source can be placed at focal point of antenna reflector. The field leaves this antenna as a spherical wave front. As each part of the wave front reaches the reflecting surface, it is phase-shifted 180°. Each part is then sent outward at an angle that results in all parts of the field traveling in

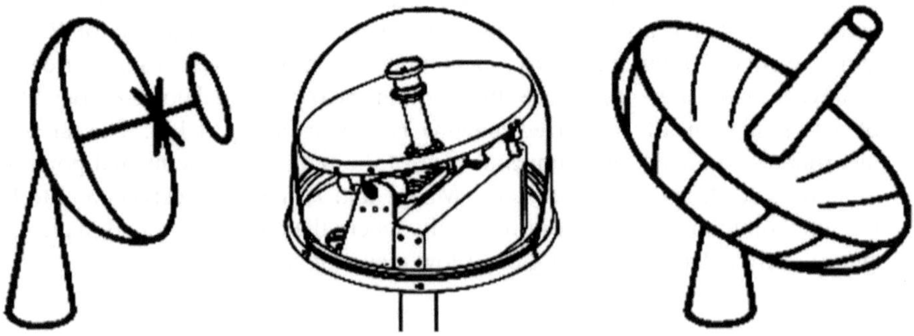

Fig. 4.12 High-gain directional aperture antennas—Courtesy of Book: by Ilcev

parallel paths. Due to the special shape of a parabolic surface, all paths from the focus to the reflector and back into space line are the same length. When the parts of the field are reflected from the parabolic surface, they travel to the space line in the same amount of time. This antenna is a large microwave parabolic consisting of the reflector of dish shape, feeder structure, waveguide assembly, servo and drive system, and protective radome.

Stable platform supports an antenna assembly, which must remain perfectly stable when MES is pitching and rolling in extremely bad weather conditions. Namely, it is essential that the stable platform holds the reflector in its A/E angular positions despite movement of the ship. The platform usually consists of a large solid bed mounted in such a way that four gyrocompasses are able to sense movement and correct any errors detected, holding the platform level. In practice, it is a form of electronic gimbal.

The antenna tracking system is controlled in A/E angles by stepping motors, which in turn are electronically controlled in a simple feedback system. This electromechanical antenna enables the dish to maintain a lock on a satellite despite navigation course changes. As the MES changes course, both A and E control corrections are made automatically.

The antenna unit processor controls all necessary functions, which include satellite tracking and electronic control. The RF electronics segment contains the Tx HPA and the Rx RF front-end LNA stage, plus all the critical bandpass signal filter stage. In modern equipment, it is common practice to reduce the number of cables between antenna and main unit. Hence, this is achieved by multiplexing up/down signals or commands between antenna and main unit onto the one coaxial feeder. This type of antenna can be used for aeronautical broadcasting and broadband mobile and fixed DVB units and for communications antennas between references with master stations of GNSS augmentation network.

2. **Parabolic Dish Antenna in Radome**—For MSC service, radomes are widely used to protect dish antennas, which are continually tracking satellites while the ship or aircraft experiences pitch, roll, and yaw movements. The sample of shipborne reflector dish antenna is shown in Fig. 4.12 (Middle), whose motorized elevation drive system is configured to selectively adjust an elevation of the satellite dish and a motorized azimuth drive system can be configured to selectively rotate the satellite dish. A control system is connected to the elevation and the azimuth drive systems managing automated operation of the satellite antenna system. Large ships may have radomes over 3 m in diameter covering antennas for broadband, TV, voice, data, video, and Internet. Recent developments allow similar services from smaller installations such as up to 85-cm motorized dish used in the maritime and land applications, while aircraftborne antennas are using more aerodynamically shaped radomes for installation atop fuselage or inside tail without radome. Small ships, fishing vessels, and private yachts may use radomes as small as 26 cm in diameter for voice and low-speed data. However, buses and trains can have low-profile parabolic TV antenna in radome.

3. **Parabolic Umbrella Antenna**—This type of portable antenna is a deployable, compact, and lightweight parabolic type suitable for portable and transportable Inmarsat transceivers, shown in Fig. 4.9 (Right). Otherwise, this antenna has almost all the same technical characteristics as a parabolic dish antenna.

4.7 Antenna Systems for Particular MSC

Typical MSC systems are Inmarsat, AMSC, MSAT, Optus, and others in which the L-band 1.6/1.5 GHz MSA are used for the service link. The Big LEO systems use S- and L-bands (2.5/1.6) and Little LEO uses VHF band. The only exceptions are US-based OmniTRACS and EutelTRACS in which the MSA antenna is serviced in the 14/12 GHz Ku-band, the Japanese N-Star service link uses 2.6/2.5 GHz S-band MSA, and some new advanced MSS will operate on 30/18 K- and Ka-bands. A typical satellite navigation system GPS uses 1.6 and 1.3 L-band for transmission in satellite-to-Earth direction. In addition, an antenna system for ships, land vehicles, and aircraft has been developed to receive TV programs. In general, the biggest mobile satellite operator, Inmarsat, developed four types of MSA for maritime, land, aeronautical, transportable, and fixed applications, while in particular, some maritime MSA are used or transformed for land vehicles and off/onshore installations.

4.7.1 Shipborne Satellite Antennas

The Ship Earth Station (SES) antennas are more sophisticated than other MSA owing to the huge ship structures, different ship's motions, and rugged environment. The MSA systems are designed by Inmarsat operator according to SES standards, which have been developed since 1981 (see four types of MSA in Fig. 4.13).

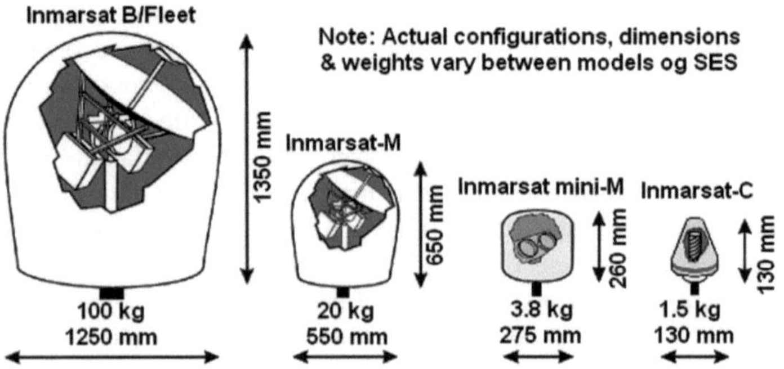

Fig. 4.13 Inmarsat antenna systems for SES—Courtesy of Manual: by Inmarsat

4.7.1.1 Previous Inmarsat Maritime Communication Antenna Standards

Inmarsat-A standard was used from 1982 to 2007 as a first operating analog SES, while existing Inmarsat-B, M, and mini-M will be in service until its end-of-life date, which is the December 30, 2016. In the meantime were developed Inmarsat standards Fleet 33, 55, and 77, which will be still in operation, while mini-M will be replaced by new developed Fleet One. These previous hugely popular Inmarsat maritime services are of course now being replaced by most shipping companies with the hugely popular FleetBroadband service, which offers quite greater speeds and much lower cost channels for both voice and data packages. All above-stated Inmarsat standards are using L-band frequency spectrum; however, new Inmarsat-5 satellite constellation is providing the next generation in maritime VSAT antenna with deployment of Inmarsat Global Xpress Ku- and Ka-band service.

1. **Inmarsat-B/Fleet MSA**—The Inmarsat-B SES was deployed in 1993 as first digital maritime standard using the second generation of Inmarsat-2 GEO satellites, which were decommissioned in December 2014. However, Inmarsat-B standard will be replaced by FleetBroadband or VSAT on December 30, 2016. The next standard Inmarsat Fleet 77 has been introduced in 2002 and Fleet 55 and 33 in 2003 using third generation of Inmarsat-3 satellites. Both Inmarsat-A and -B standards are compatible for mounting on board large ships, sea platforms, and onshore infrastructures. They use same high-gain parabolic steerable dish with the same features: Gain is from 20 to 24 dBi, and G/T is –4 dBK. The EIRP of Inmarsat-B is less than that of Inmarsat-A by 3 dB, due to the slightly shaped global beam antenna adopted in the Inmarsat-2 satellites. Considering the desired EIRP and the noise performance of current LNA, the Inmarsat-B EIRP of 33 dBW can be obtained by the antenna with a gain of about 20 dBi. For both standards is suitable an aperture antenna such as a directional parabolic dish with high aperture efficiency. In Fig. 4.14 (Left) is shown rear side and (Right) shows profile of the Inmarsat-B SES antenna elements with antenna control unit (ACU), pedestal control unit (PCU), and antenna control board (ACB), produced by Norwegian company Nera. In 2006, Nera merged with Danish Thrane and Thrane, and in 2012, both became part of UK Cobham.
   Inmarsat Fleet family of SES was announced by Inmarsat during 2002 as the first new dedicated Inmarsat SMC system since the development of Inmarsat-B in 1993, which is providing oceangoing vessels with comprehensive global voice, data, and fax service, whose comparison with A and B antenna standards is presented in Table 4.3. As with Inmarsat-B, the Inmarsat Fleet F77 service also provides GMDSS distress and safety communications.
   In Fig. 4.15 (Left) is illustrated 1-m parabolic stabilized reflector dish with the main components of Inmarsat-B antenna, while in Fig. 4.15 (Right) is shown Inmarsat-F or precisely Fleet 77 phased-array antennas with main components, both produced by Nera.

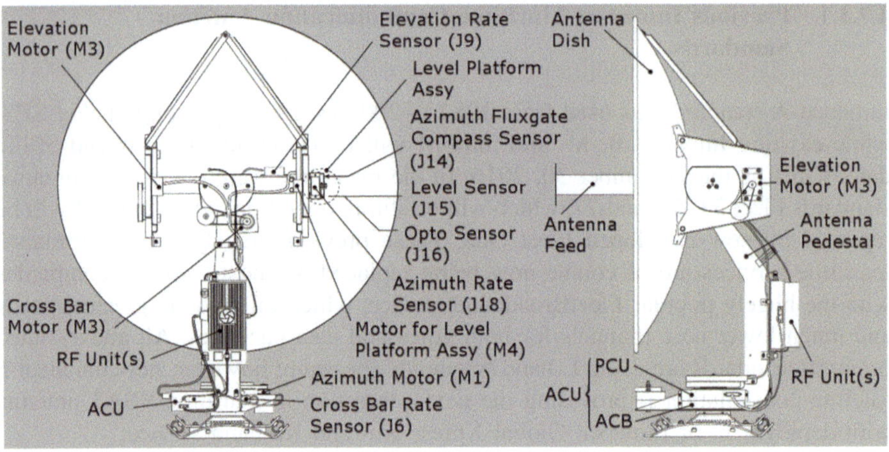

Fig. 4.14 Inmarsat antenna saturn Bm marine Class 2—Courtesy of Manual: by Nera

Table 4.3 Comparison of Inmarsat-A, -B, and -F antenna characteristics

Characteristics	Inmarsat-A	Inmarsat-B	Inmarsat-F
Antenna beam	Global	Global/spot	Global/spot
Antenna weight	110 kg	90 kg	65 kg
Voice	Available	Available	Available
HSD	4.8 Kb/s	64 Kb/s	64 Kb/s
MPDS	Nil	Nil	64 Kb/s

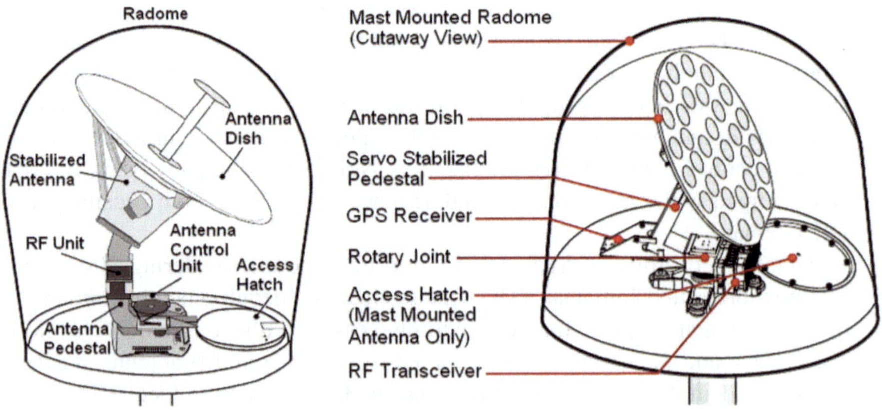

Fig. 4.15 Saturn Bm marine Class 2 and Fleet 77 Inmarsat antennas—Courtesy of Manual: by Nera

The Inmarsat-F 33, 55, and 77 provide further services within the Fleet family, each aimed at specific sectors of the Maritime industry serving different types of merchant fleet. The F77 SES antenna has the following specifications: G/T is −17 dB/K minimum, EIRP is 8–14 dBW in 2 dB steps, antenna cable has TNC/TNC female, and maximum cable loss is 15 dB at L-band and 1.8 O at DC. In addition, antenna is providing high-speed data (HSD) at a speed of about 128, 64, and 56 Kb/s data service. Speech is offering high-quality voice connections (it is the service simulated by Vox Maris), and 3.1 kHz audio for analogue signal transmission like fax or modems and also antenna provides transmission of mobile packed data service (MPDS) for the first time, whose tariff depends on the amount of data sent and received.

The B/Fleet family antenna stabilization is achieved by using the three-axis active method that detects the different ship's motions with three rate sensors and an inclinometer. These antennas are simple in their structure and can be designed flexibly with respect to the gain values larger than 20 dBi. Satellite tracking is an essential capability for ship's motions and the small half-power beam width of about 10°. The four-axis (X-Y-A-E) stabilizer is the one most commonly used, when a fixed horizontal plane is obtained by controlling motion about the x- and y-axes and an antenna system installed on the x-y plane can be stabilized by controlling the a- and e-axes. The antenna is directed to the satellite by controlling the a- and e-axes. It needs antenna pedestal control circuits with servomotors to control the axes and it also needs some sensors, such as an accelerometer, a rate sensor, and a level sensor, to provide information about ship motions. A flywheel stabilizer has sometimes been used in order to avoid the need for antenna control sensors and electronic circuits. This kind of stabilizer makes use of the inertial force generated by one or two rapidly rotating flywheels. Thus, a three-axis stabilizer was developed and used in some recent standards A and B, while the present standards have generally used a closed-loop tracking system because of their very simple configuration. The most popular open-loop tracking is a step-tracking method, which drives in elevation and azimuth directions alternatively by a step angle of 0.5° in such a way as to keep the received signal level as high as possible.

2. **Inmarsat-M MSA**—This standard was initiated in 1993 as a solution for small, light, and low-cost SES for installation on small ships, land vehicles, and rural infrastructures, whose antenna is shown in Fig. 4.16 (Left). The antenna applicable to this system can be a single medium-gain directional parabolic SBF dish or a phased array with an antenna gain that ranges from 13 to 16 dBi. For the medium-gain parabolic dish antenna, it will be difficult to illuminate the parabolic efficiency with the primary radiator because the diameter of this antenna is smaller than a high-gain antenna. In fact, this type of antenna is not adequate for the Inmarsat-M compared to other types of antennas with high aperture efficiency, such as SBF antenna. Traveling-wave antennas, such as the cross-Yagi/Uda array, a single helical antenna, and a log-periodic antenna, have high efficiency for medium-gain performances. Since those types of antennas are

comparatively long in their axial direction, the volume of the radome becomes fairly large and conflicts with the requirements of Inmarsat-M. This antenna needs a suitable mount/steering system on board ships. The SBF antenna, with typical gain of 15 dBi, is one of the favorite standard M antennas for SES. Although the SBF antenna is a compact, simple configuration, and high-efficiency shipborne antenna, it has a narrow frequency bandwidth of about 3 % (about 5 % less than required). The electrical characteristics of the conventional SBF have been improved in an upgraded SBF antenna by changing the main reflector from a flat disk to a conical or a step plate and by adding a small second reflector. These improvements give better performance with an aperture efficiency of about 80 % and an RF bandwidth of 20 % for VSWR under 1.5, and gain is also improved by about 1 dB without changing sidelobe levels. Inmarsat-M was replaced by FleetBroadband or VSAT on December 30, 2014.

3. **Inmarsat mini-M MSA**—The Inmarsat mini-M SES introduced in 1997 offers the same service as Inmarsat-M but in smaller, more lightweight and compact unit and antenna for mounting on small ships and land vehicles, whose antenna is shown in Fig. 4.16 (Right). This SES can be made smaller because it operates only in the spot beam coverage, whose typical representative is TT-3064A Capsat SES, designed by Thrane and Thrane. Its antenna is three-axis stabilized platform with a minimum of moveable parts, and all motors, sensors, and electronics are mounted in the bottom bowl. Antenna pointing is controlled by various sensors and a step-tracking algorithm for signal level peaking. It does not require cable unwrapping but can track the satellite in any position above the horizon, and it is connected to the transceiver through a single coaxial cable to ensure simple installation. Inmarsat mini-M will be replaced by FleetBroadband or IsatPhone Marine on December 30, 2016

Fig. 4.16 Inmarsat-M and mini-M antennas—Courtesy of Manual: by Nera/Thrane & Thrane

The sensors in the antenna are as follows:

(a) 3D compass (magnetometer) which measures the Earth's magnetic field vector;
(b) 2D inclinometer which measures the vessel's roll and pitch angles; and
(c) Gyro that measures the vessel's yaw, pitch, and roll rates.

The sensor stabilization of the antenna pointing compensates for the fast movements of the ship. In case of a no signal situation, the sensors are able to point the antenna independently, but in all other situations, a step-tracking algorithm maximizes the strength of the received signal. The antenna has step-tracking and sky-scan modes of normal operation:

1. During step-tracking mode, the platform tracks the satellite using the sensors and the step-tracking algorithm. If synchronism with the satellite is lost for more than 130 s, the sky-scan mode is entered to reacquire synchronism with the satellite.
2. The sky-scan mode is automatically entered during the initial acquisition of a satellite after power-up, when a different satellite is selected by the user or the signal has been lost for more than 130 s. During a sky-scan, the antenna pointing is moved over the sky from horizon to zenith in a smooth spiral movement and stabilized by the sensors, searching for a particular satellite. Therefore, when the satellite is found, the antenna exits the sky-scan mode and enters step-tracking mode. Sensor-stabilized platforms work with directional RHCP antennas. Maximum pointing error is $\pm 10°$ and beam width $\pm 30°$. The value of G/T is minimum -17 dBK, and EIRP is from 8 to 14 dBW in 2 dB steps.
3. **Inmarsat-C and mini-C MSA**—This MSA system is the simplest and has most compact configuration without mount of tracking and pointing systems, introduced in 1990 as the second Inmarsat SES standard, whose antenna shape is shown in Fig. 4.17 (Left). The omnidirectional low-gain antennas are the most suitable for this mobile standard, such as cross-drooping dipole, quadrifilar

Fig. 4.17 Inmarsat-C and IsatData Pro (former-D+)—Courtesy of Manual: by Thrane and Thrane/SkyWave

helix, and the microstrip patch antennas. In fact, the quadrifilar antenna is well known as a shipborne antenna because of its good performance and axial ratios in wide angular coverage, with about 4 dBi in gain, maximum 3 dB in axial ratio, −24 dBK in G/T, and 14 dBW EIRP. The Inmarsat-C MSA is suitable for installation on board medium and small ships, land vehicles, and rural infrastructures.

4. **Inmarsat-D and D+ MSA**—The Inmarsat-D (receiver), D+, and IsatData Pro (ISP) Inmarsat (transceivers) standards were developed in 1996, 1997, and 2011, respectively, whose omnidirectional antenna is illustrated in Fig. 4.17 (Right). These three standards use omnidirectional small low-gain antenna similar to that introduced for Inmarsat-C, with the only difference that is smaller and more compact and with the G/T value of −25 dBK.

Besides, this antenna can be a separate unit in the smallest radome for Inmarsat-D+ or ISP transceiver, or the antenna can be built-in a small radome together with the Inmarsat-D+ or ISP transceiver and GPS Rx antenna made by Canadian-based SkyWave. These antennas are serving equipment similar to paging systems for vessels or vehicle messaging and SCADA Rx/transceivers equipment.

Therefore, both IsatM2M and IsatData Pro Inmarsat solutions were developed from initial Inmarsat-D+ system and they can be used for all mobile and fixed SAT applications. The global IsatM2M satellite network is able to offer better service than Inmarsat-D+ system, whose examples for maritime applications include: tracks geographic location of fleets, finds deviation from predefined geofenced areas, monitors ship locations worldwide, sends PVT data to the special shore ship traffic control, and enables polling by request from shore to be obtained data of any ship in certain sailing area.

4.7.1.2 New Inmarsat Maritime Communication Antenna Standards

Inmarsat Maritime-B, M, and mini-M terminals are replaced by the new Inmarsat standards known as FleetBroadband (FB) and Fleet One (F1), which use Inmarsat-4 satellites, and last developed is Global Xpress (GX) that may use Inmarsat-5 satellite constellation.

1. **Inmarsat-FB MSA**—The FleetBroadband (FB) was launched by Inmarsat sometimes in 2007 providing high-speed Internet connectivity, crew IP solutions in parallel with e-mail, Web, data, VPN, and voice calling at sea. It enables tracking and telemetry functionality, so shippers on shore have all the information they need about voyage, ETA, cargo operations, real-time engine data for maintenance, drilling data for improving production or position data for fleet tracking and management and so on. There are two terminal types defined and type-approved for the FleetBroadband service utilizing Inmarsat-4 satellite constellation at L-band. They are referred to as Class 8 (high-gain antenna "FB500") and Class 9 (low-gain antenna "FB250" and "FB 150") terminals.

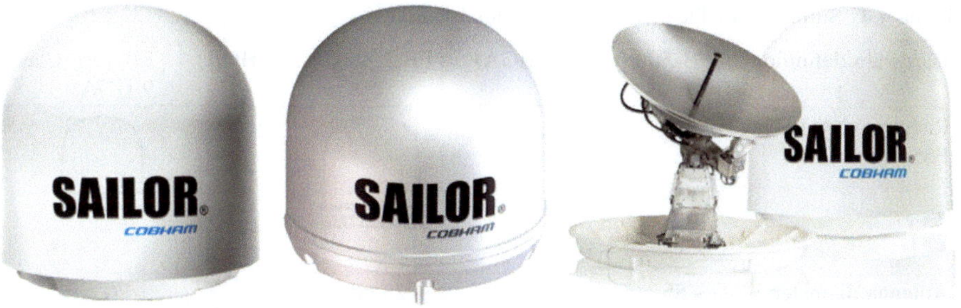

Fig. 4.18 Inmarsat-FB, fleet one, and global Xpress antennas—Courtesy of Manual: by Cobham

The key difference between Class 8 and Class 9 is the antenna. Therefore, for each type of FB terminal, Inmarsat defines:

1. The air interface at the output of the terminal's antenna;
2. The mandatory features and service types for each class of user terminal; and
3. The performance requirements of the user terminal.

The Sailor 500 FB transceiver uses the TT-3052A antenna, which is a maritime Broadband Global Area Network (BGAN) antenna that complies with Inmarsat's Class 8 definition for ship antenna illustrated in Fig. 4.18 (Left). This antenna is larger and provides more bandwidth than the TT-3050A used for the Sailor 250 FB system.

This antenna contains all functions for satellite tracking including a GPS system. A single coaxial cable carries all RF communication, supply voltage, and modem communication between the antenna and the terminal. Manufacturers must meet all of these requirements in order to obtain type-approved antennas. The definition of other equipment features such as physical connections, user interfaces, firewalling, routing, and control is determined by each manufacturer according to specific market-driven needs. Because of the possibility of different physical interfaces on terminals from different manufacturers, users should pay particular attention to the installation guidelines for different terminal configurations and features. A summary of the features of each of three FleetBroadband terminals is shown in Table 4.4, entitled "Summary of FleetBroadband Terminal Features."

From above table is evident that only FB 500 HGA provides more than 20 dBW EIRP and uses integrated services for digital network (ISDN) mode.

2. **Inmarsat-F1 MSA**—The Fleet One (F1) transceiver is complement to the new Inmarsat Fleet One service, which utilizes the existing Inmarsat-4 satellites at L-band RF to deliver voice and data connectivity ideal for use on board fishing vessels, boats, and luxury ships, whose antenna is shown in Fig. 4.18 (Middle). With its compact, lightweight antenna and simple installation, F1 is a low-cost solution for maritime satellite broadband that keeps oceangoing ships connected at all times. The Sailor F1 system uses small-size maritime 2-axis stabilized

Table 4.4 Summary of FleetBroadband terminal features

Hardware definition	FB 500 (Class 8 HGA)	FB 250 (Class 9 HGA)	FB 150 (Class 9 HGA)
Radome view (as per F55 & mini-M)			
Antenna diameter	~55 cm	~25 cm	~20 cm
Antenna G/T at 5° elevation	−7 dB/K	−15 dB/K	Same
Antenna EIRP	22 dBW	15.1 dBW	Same
Antenna type	Directional 3-axis stabilized	Directional 3-axis stabilized	Same
Antenna weight	15–20 kg	3–5 kg	Same
Voice simultaneously with data	4 KB/s	4 KB/s	4 KB/s
Contended standard IP T × /R × KB/s	Up to 432/432 kB/s	Up to 239/284 kB/s	Up to 150 kB/s
ISDN	Yes	3.1 kHz audio	None
IP-guaranteed streaming throughput	32, 64, 128, and 256 KB/s	32, 64 and 128 KB/s	None
Physical interfaces and ports	RJ11, Ethernet, RJ45 (ISDN), and L-band RF	RJ11, Ethernet, RJ45 (ISDN), and L-band RF	RJ11 and Ethernet

antenna with the following specifications: transmit at 1626.5–1660.5 MHz; receive at 1525.0–1559.0 MHz; channel spacing 1.25 kHz; Rx gain 11.0 dB min; G/T −15.5 dB/K; Min. EIRP 3.1 dBW and max. 16.1 dBW; return loss is <−15 dB (reflection loss <0.15 dB); cable loss RF attenuation is max. 20 dB; input voltage is 22.7 V; operational power max. 39 W; weight is 3.94 kg; and dimension 291.9 × 275.6 mm.

3. **Inmarsat GX MSA**—The Global Xpress (GX) is the newest Inmarsat standard that uses the last constellation of Inmarsat-5 satellites at Ka-band RF, whose Cobham Sailor 100 GX satellite reflector antenna with radome is illustrated in Fig. 4.18 (Right). This light and compact light antenna is an advanced 3-axis stabilized tracking system designed for the Inmarsat GX satellite network. It is a direct development from the immensely successful Sailor 900 VSAT antenna systems, which have created a new industry standard through innovative design and reliable operation. The GX antenna features advanced tracking receiver technology that enables it to verify the right satellite in less than a second. It uses a single cable between satellite antenna known as above deck equipment (ADE) and transceiver with peripherals known as below deck equipment (BDE) for RF, power, and data, while advanced features such as automatic azimuth calibration (home flag) and automatic cable calibration significantly

reduce installation time further. This GX antenna provides reliable access to the full range of high-throughput satellite broadband services for maritime business applications, ship operations, and crew welfare, including passenger multimedia services. The technical specifications of ADE are as follows: Antenna pedestal is 3-axis stabilized tracking with integrated GPS Rx; antenna reflector system is reflector/subreflector including ring focus; transmit gain is 47.5 dBi at 29.5 GHz (excl. radome); receive gain is 44.0 dBi at 19.7 GHz (excl. radome); system G/T is 20.1 dB/K at 19.7 GHz, at ≥10° elevation (ε), and clear sky (incl. radome); block-up converter (BUC) has output power 5 W GX BUC; EIRP ≥53.5 dBW (incl. radome) is MAX. 36.0 dBW/40 kHz; low-noise block (LNB) GX Ka is single-band LNB; tracking receiver internal is "all-band/modulation type" including power, DVB-S2, GSC, and modem RSSI; polarization is circular cross-pol (Inmarsat GX, TX: RHCP, RX: LHCP); elevation range is −25° to +125°; azimuth range is unlimited (rotary joint); ship motion (angular) is roll ±30°, pitch ±15°, yaw ±10°; ship's turning rate and acceleration are 15°/S2 and 15°/S2; ADE motion is of linear acceleration ±2.5 g max in any direction; satellite acquisition is automatic with or without gyro/GPS compass input; humidity tolerance is 100 %; rain/IP class considers EN60945 Exposed/IPX6; wind tolerance is about 80 kt for operational and 110 kt for survival conditions; ice survival tolerance is 25 mm/1″; solar radiation is 1120 W/m^2 to MIL-STD-810F 505.4; compass safe distance is 1 m/40″ to EN60945; dimension (overall) height: H 150 cm/58.9″; diameter: Ø 130 cm/51.3″; and Weight is 126 kg/276 lbs.

4.7.1.3 Inmarsat Maritime Multiband VSAT and TVRO Antenna Standards

Inmarsat as the biggest GEO mobile satellite operator developed the following new antenna standards for maritime applications, which includes multiband Ku and Ka-band VSAT and own TV Receive-Only (TVRO) antennas for TV programs on board ships:

1. **Multiband VSAT Antenna**—Intellian Company has launched in 2012 a new Ku-band VSAT antenna, which will be upgradeable for use with existing Inmarsat Global Xpress (GX) Ka-band after that service is launched. Thus, the 1 m, 3-Axis Ku-band v110GX antenna features an optimized Ku and Ka-band antenna reflector and radome which today can be "easily" changed to be used with the GX service, because Inmarsat's I-5 Ka-band satellites are operational since 2014. The modular GX Conversion Kit includes the GX BUC/LNB assembly, GX Feed Assembly and GX ACU, incorporating the modem. The pedestal and new mounting architecture has also been designed to support conversion to Ka-band by manually replacing the RF Module and Feedhorn inside the radome.

Fig. 4.19 Inmarsat multiband VSAT and TVRO—Courtesy of Manual: by Intellian/Inmarsat

The GX BUC/LNB assembly is attached to the rear side of the reflector, with no special requirement to rebalance the system. Once the ACU is replaced and power is supplied, the system will be immediately ready and operational for the GX service. A gyro-free satellite search function enables the v110GX to acquire and lock onto the satellite without requiring separate input from the ship's gyrocompass.

The Intellian brand of V100GX Ku/Ka-band convertible ship antenna is shown in Fig. 4.19 (Left). As already stated, this Ku-band antenna can be quickly converted in as little as 10 min without the need of a factory-trained technician to GX Ka-band antenna for Inmarsat broadband service. One would need to replace the RF module containing the BUC and LNB units, the feed, and the ACU to make the antenna immediately ready for Ka-band service. However, the V100GX is available with 4, 8, and 16 W BUC sizes. All models are standard with co-pol and cross-pol feed with Intellian's Global PLL LNB units. The local oscillator (LO) frequency of the Global PLL LNB unit can be remotely adjusted to receive any Ku-band VSAT satellite around the world. The antenna radome is optimized for minimal attenuation of both Ka- and Ku-band frequencies. The antenna will search and find the satellite without the need for an input from the ship's gyrocompass.

The ADE specifications of this antenna are as follows: radome (W × H) is 133 cm × 149 cm (52″ × 59″); dish diameter is 103 cm (41″); antenna weight is about 128 kg (282 lbs); platform provides azimuth, elevation, and cross-level; azimuth range is unlimited; elevation range is −20° to +115°; cross-level range is up to ±37°; RX frequency is 10.95–12.75 GHz at Ku-band; RX gain is 39.8 dBi at mid-band; TX frequency is 13.75–14.5 GHz at Ku-band; TX gain is 41.3 dBi at mid-band; cross-pol isolation minimum is 35 dB; G/T is >18.3 dB/K; polarization is cross-pol and co-pol; and power consumption is 100–240 V AC, 50/60 Hz, 4A.

2. **TVRO Antenna**—Inmarsat GEO satellite operator provides stabilized ship antennas that deliver high-quality TV reception or TV receive-only (TVRO) systems suitable for any region of the world. This TV antenna can be installed

on board any type and size of merchant ships, sea rigs or platforms, fishing boats, and luxury cruise vessels, which are designed to withstand the harshest marine conditions and to continue working in even the deep seas on the globe. Inmarsat has designed four types of shipborne stabilized TVRO antenna, such as 0.6-m ST 24 model, 1.2-m 5004 model, 1.25-m t130 W in 66.5″ radome, and 1.5-m 6004 in 76″ radome. The last 6004 type of TV antenna is illustrated in Fig. 4.19 (Right), whose total weight is approximately 159 kg and DAC 2202 is a control unit.

4.7.1.4 Maritime VSAT and TVRO Antennas Over GEO Satellites

The single-band and multiband VSAT antennas with transceivers provide fixed and mobile two-way or interactive satellite communications for new voice, data, and video over IP (VDVoverIP) and Internet service via GEO satellites. Thus, DVB-RCS VSAT standards in the maritime market are delivered as packaged solutions including GEO satellite space segment, ADE, and BDE with peripherals. The VSAT services historically have operated at Ku- and C-bands for the commercial and luxury maritime market, using satellites located at the GEO orbital location. Due to the location of the satellites in combination with the movement of vessels, stabilized antennas with tracking are required. A stabilized maritime VSAT antenna typically has a circular antenna (often concealed within a dome) that is 2.4 m or smaller in diameter. The majority of maritime VSAT antennas range from 60 cm to 1.5 m for Ku-band and 2.4 m for C-band. Data rates typically range from 64 Kb/s up to 8 Mb/s, but lower and much higher data rates are available. A remote maritime VSAT system typically consists of two parts: antenna and transceiver placed outdoors (typically inside a radome)—in maritime, this can be referred to as the above deck equipment (ADE). An indoor unit interacts with the outdoor unit and controls the antenna, which in maritime can be referred to as the below deck equipment (BDE).

The term “TVRO” has been in use on ships since it was introduced in the 1980s. One early provider of equipment was SeaTel with its first generation of stabilized satellite antennas that was launched in 1985, the TV-at-Sea 8885 system. Until this time, ships had not been able to receive TV signals from satellites due to their rocking motion rendering reception impossible. The SeaTel antenna was stabilized using electrically driven gyroscope and thus made it possible to point to the satellite accurate enough, that is to within 2°, in order to receive a signal. The successful implementation of stabilized TVRO systems on ships immediately led to the development of maritime VSAT systems. The second generation of SeaTel TVRO systems came in 1994 and was the 2494 antenna, which got its gyros signal from the ship rather than its own gyros, improving accuracy and reducing maintenance.

1. **Sea Tel VSAT 3011 Antenna**—The VSAT 3011 model is the latest 3-axis marine an 75-cm (30 in) stabilized antenna system for sub 1-m networks, whose reflector and antenna radome are shown in Fig. 4.20 (Left). It is compatible with

Fig. 4.20 Maritime VSAT for GEO satellites—Courtesy of Manual: by Cobham

Ku-band GEO satellites for maritime broadband connectivity, which incorporate the newest frequency-tuned radome design and electronics used in our industry-leading XX09 series systems. Its antenna platform provides stabilization accuracy of 0.2° (RMS); transmit gain is 39 dBi at 14.25 GHz; receive gain is 37 dBi at 11.85 GHz; G/T is 16 dB/K at 12.0 GHz (typical); elevation range of motion is −15° to +115°; and azimuth range of motion is unlimited.

2. **Sea Tel VSAT 6012 Antenna**—This model is an 3-axis marine stabilized VSAT antenna system compatible with C-band linear satellites and is powered by integrated marine antenna (IMA) software along with trusted XX09 pedestal design, which is shown in Fig. 4.20 (Middle). The 6012 C-band antennas are supplied in a 2.05-m (80.5 in) radome, which feature an integrated control unit (ICU) that offers a single box integrated electronic control to maintain the best and most efficient pointing accuracy. Specifications of this antenna are as follows: reflector antenna size is 1.5 m (58 in); Tx frequency is 5.850–6.725 GHz; Rx frequency is 3.400–4.200 GHz; Tx gain is 37.9 dBi at 6.20 GHz; Rx gain is 33.3 dBi at 3.70 GHz; BUC is 25 and 40 W standard; elevation joint angle is -15° to $+115^{\circ}$; azimuth is unlimited; and polarization is linearly cross.
3. **Sea Tel Multiband VSAT 9797B Antenna**—This radical offset VSAT antenna is most efficient ever multiband C-band, X-band, and Ku-band marine stabilized antenna system, which is illustrated in Fig. 4.20 (Right). Thus, it is both Intelsat and Defense Satellite Communications System (DSCS) compliant antenna model. Mariners get more throughputs using the same or lesser amount of valuable deck space and also are available as a dual-band selectable C- and Ku-band system for complete flexibility. This antenna utilizes the DAC 2202 ACU, which is 19″ rack mount unit (1.75″ tall) with the following features: Rx gain is 38.5 dBi at 3.95 GHz and 47.75 dBi at 11.85 GHz; Tx gain is 41.7 dBi at 6.18 GHz and 48.45 dBi at 14.25 GHz; G/T is calculated 20.2 dB/k at 3.95 GHz and 28.2 dB/k at 11.85 GHz (clear sky, 20° elevation); radome dimensions are 3.66 m (144 in) D × 4.17 m (164 in) H; sidelobes are 100 % below 29–25 log mask; and compliant to DSCS requirements is X-band G/T >22 db/k at 7.5 GHz.
4. **Orbit Multiband VSAT OceanTRx 7 Antenna**—This is an innovative ship antenna platform supporting a variety of 2.2 m stabilized maritime antenna

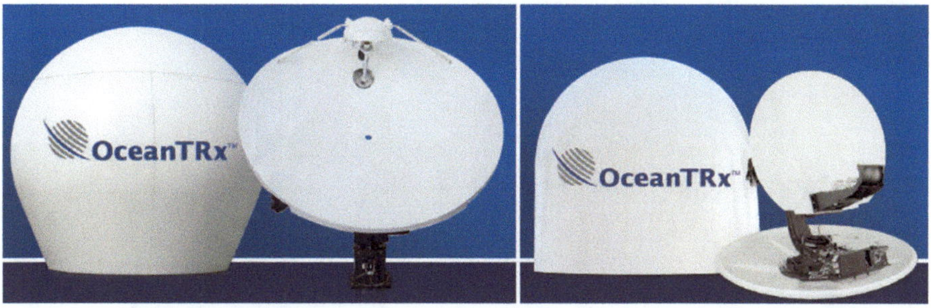

Fig. 4.21 Maritime multiband VSAT for GEO satellites—Courtesy of Manual: by Orbit

system configurations in C-, Ku-, and Ka-bands, shown in Fig. 4.21 (Left). As a common platform, it is inherently designed to accommodate the current and future broadband needs of the maritime market. This antenna is built for quick and easy installation, upgrade, and maintenance, combining exceptional RF performance and system availability with an extraordinarily small footprint. It supports the mission and business-critical broadband solution needs of commercial and naval vessels. This design comprises two product series: (a) OceanTRx 7-300 features multiband frequency support for C- and Ku-bands and (b) the OceanTRx 7-500 500 Series features built-in Ka-band fully compatible design to ensure smooth migration to future high-speed Ka-band services, for the entire Ka-band range using GEO and MEO satellites. The multiband 7-500 antennas provide multiband frequency support for Ku- and Ka-bands, based on field exchangeable kits.

5. **Orbit Multiband VSAT OceanTRx 4 Antenna**—OceanTRx 4 is an innovative antenna platform supporting a variety of 1.15 m stabilized maritime antenna system configurations in X-, Ku-, and Ka-bands, which is shown in Fig. 4.21 (Right). As a common platform, this antenna is inherently designed to accommodate commercial and military maritime applications. It features outstanding RF performance, system availability, and dynamic response under virtually any sea conditions. Its 500 Series features built-in Ka-band fully compatible design to ensure smooth migration to future high-speed Ka-band using GEO and MEO satellites
6. **Sea Tel ST144 Multiband TVRO Antenna**—This shipborne antenna is one of the top 3.6-m (144″) TV-at-Sea system, whose reflector, radome design, RF components, and the pedestal are based on Sea Tel well-proven 14400B system, which is shown in Fig. 4.22 (Left). This antenna is the preferred system for users who need the largest possible offshore operating area, for large vessels operating in deep oceans and compatible with all current and planned GEO satellites. The level cage used in the predecessor systems has also been replaced by high-accuracy accelerometers. Thus, this allows for continuous calculation of position, orientation, and velocity vector of a moving object without the need for external references. The micro electromechanical (MEM) sensors used in Sea

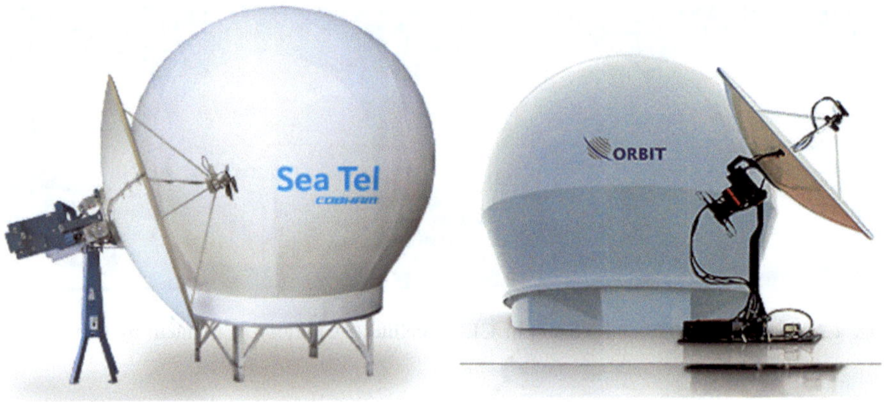

Fig. 4.22 Maritime TVRO for GEO satellites—Courtesy of Manual: by Sea Tel/Orbit

Tel's antennas are based on the same technology currently being used in missile and aerospace technology around the world. Much like its predecessor, the ST144 is available in C-band and Ku-band configurations or "dual-band" configuration that permits the same antenna to be used for both. This antenna has the following features: antenna gain at C-band is 42.2 dB at 4.2 GHz and at Ku-band is 49.6 dB at 12.5 GHz; minimum EIRP at C-band is 28 dBW and at Ku-band is 32–33 dBW; radome dimensions are 4.27 m (168 in) D × 4.21 m (165.7 in) H; antenna stabilization is 3-axis servo; built-in GPS is automatic satellite acquisition; ship's motion is ±15° roll or ±20° roll and ±15° pitch; full elevation range is −15° to +115°; azimuth range is unlimited; the entire frequency range for Ku-band satellite TV is 10.7–12.75 GHz; and the C-band frequency range is 3.7–4.2 GHz worldwide.

7. **Orbit AL-7205 Multiband TVRO Antenna**—This is 1.5-m (59″) quad Ku-band or dual C-band marine stabilized orbit TVRO antenna, which is illustrated in Fig. 4.22 (Right). It is designed to maintain constant TV reception on board all kind of oceangoing ships and sea platforms via both Ku-band and C-band satellites at RF of 3.7–4.2 and 10.7–12.75 GHZ. Its antenna specifications are as follows: antenna gain is 34 dB at 4.2 GHz and 43 dB at 12.75 GHz; system EIRP is 32 dBW at 4.2 GHz and 40 dBW at 12.75 GHz.

4.7.1.5 Maritime Antenna Standards Over MEO Satellites

The new OceanTRx 7-500 2.2-m Ka-band stabilized maritime VSAT antenna of orbit systems enables the most demanding maritime vessels and platforms to enjoy fiber-like broadband communications for very high-speed Internet services over O3b medium Earth orbit (MEO) satellites, which is shown in Fig. 4.2 (Left). In combination, the O3b satellites and OceanTRx 7-500 maritime VSAT features extremely high-speed Internet access over a cost-effective system and billing. O3b's MEO satellites will be deployed in a circular orbit along the equator at an altitude of 8000 km, less than one-fourth the altitude of GEO satellites. This lower orbit substantially lowers costs while maintaining low-latency fiber-like connections. No other service provider has the combination of fiber-like connections with the global reach of the O3b satellite network. Each O3b VSAT antenna incorporates ACU and central control unit (CCU) to manage its system operations. The controllers allocate antenna resources while managing constellation tracking of the MEO satellite constellation. Under normal conditions, the controllers receive and process satellite configuration and position updates from the O3b network operations center for a completely hands-off and lights-out operation. Thus, a pair of O3b stabilized maritime VSAT systems each with a Ka-band BUC keeps a continuous contact with the MEO satellite constellation. In such a way, continuous service is maintained by automatically transferring active links between setting and rising satellites when both are in view.

This compact maritime VSAT system offers industry-standard RF performance equivalent to a 2.4 m (95″) dish in just a 2.7 m (106″) footprint. The key to this antenna breakthrough is an extraordinarily small footprint with outstanding RF performance to its size, strict regulatory compliance, and support of multiple optional RF feeds. The key features of this antenna are as follows: highly efficient dual-offset Gregorian 2.2 m (87″) satellite antenna; superior stabilization and tracking under any severe sea conditions; four-axis pedestal with zero "keyholes" at zenith or horizon; supporting Ka/Ku-band frequencies, optional RF packages, and upgradeable BUC power levels; highly accessible pedestal design using small, light, and "easy-to-replace" field replaceable units (FRU); it does not require accurate or periodic balancing during installation and field upgrades (BUC, LNB, Feed); azimuth is continuous, and elevation is −30° to +120°; ship gyro interface is NMEA 0183, Step by Step, Synchro;

Operating frequency Rx is 17.8–19.3 GHz and is Tx 27.6–29.1 GHz; system G/T of (dB/°K as typical including all losses) is ≥24 dB/°K at mid-range at 20° elevation; system EIRP with multiple BUC options (dBW, typical including all losses) is 64 dBW with a 20 W BUC at mid-range and 67 dBW with a 40 W BUC at mid-range.

4.7.1.6 Maritime Antenna Standards Over LEO Satellites

Low Earth orbit (LEO) satellite systems employ a large fleet of spacecraft, each in a circular orbit at a constant altitude of a several hundred to few thousand kilometers. These orbits take the satellites over, or nearly over, the geographic poles. Each revolution takes approximately 90 min to a few hours. The fleet is arranged in such a way that, from any point on the surface at any time, at least one satellite is on a line of sight (LOS). The entire system operates in a manner similar to the way cellular functions. The main difference is that the transponders, or wireless receiver, and transmitters (transceiver) are moving rather than fixed and are in space rather than on the Earth. A well-designed LEO system makes it possible for anyone to access the voice, data, and Internet via wireless from any point on the planet, using an antenna no more sophisticated than old-fashioned television "rabbit ears."

The main LEO satellite orbits are divided into Big (Iridium and Globalstar) and Little LEO (Orbcomm). Iridium is the world's only truly global satellite communications system, with voice and data solutions covering entire Earth's surface with 66 cross-linked satellites. The Globalstar and Orbcomm have not intersatellite links and so are not providing both polar coverages. The LEO satellite network provides reliable and affordable maritime, land, and aeronautical communications enabling essential Internet, voice, data, and short burst data (SBD) communications using omnidirectional antennas. The antenna system for SBD units via LEO satellite is similar to the Inmarsat-C, mini-C, and IsatData Pro antennas.

1. **Iridium Pilot Antenna**—This omnidirectional antenna is ADE electronically switched and phase-steered six dual-element azimuth arrays and one zenith element connected to the Iridium BDE of Pilot shipborne transceiver, which is illustrated in Fig. 4.23 (Right). Small and easy to install, the Iridium Pilot unit is a cost-effective system for mini-VSAT maritime broadband maritime providing pole-to-pole global coverage for Internet, data, and voice transmission at standard IP of up to 128/134 Kb/s. The ADE provides the communication connection between the BDE and Iridium's LEO satellites. The ADE will automatically locate and track Iridium satellites overhead using a solid-state antenna array. Mounting bracket is a platform to attach mounting pole and ADE to flat surface, which is an optional accessory used to connect the ADE to the mast. This antenna has the following features: Antenna cable is either 20 m or 50 m length; dimensions (H × Ø) of ADU are 230 × 570 mm; weight of ADU is 11 kg; the AC power supply unit (PSU) provides power to the BDE; L-band RF for Tx and Rx is 1616–1626.5 MHz; gain is +8 dB; antenna beam width provides both hemispheric coverages (60° per element); steering is automatic solid state; and coverage is from horizon to horizon.

Fig. 4.23 Maritime VSAT for MEO and LEO satellites—Courtesy of Manual: by Orbit/KVH

2. **Iridium OpenPort Antenna**—The Iridium OpenPort low-profile satellite lightweight antenna provides pole-to-pole global coverage high-speed network specifically engineered for the ship applications, shown in Fig. 4.24 (Left). It offers true IP connectivity with up to three phone lines available for simultaneous use and high-speed data at a speed of 9.6–128/134 Kb/s. Characteristics of this antenna are as follows: height is 230 mm (9.06 in); diameter is 570 mm (22.44 in); and weight is 11 kg (24.25 Ib).
3. **Iridium Saylor Antenna**—The Sailor SC4000 omnidirectional helical antenna provides satellite phone system for all sizes and types of vessel, which is illustrated in Fig. 4.24 (Right). In spite of its very slim and delicate design, this antenna is very sturdy without any moving parts. Its helix construction is coated with glass-enforced epoxy, which is highly weather resistant and able to withstand the extreme conditions at sea. The characteristics technical of this antenna are as follows: helical antenna of 210° (3 dB bandwidth); dimensions Ø are 96 mm (3.8″), and height is 140 mm (5.5″); link margin is 16 dB; and elevation angle from any direction is 20° from horizontal.
4. **Globalstar Marine Helix Antenna**—For onboard ship installations, Globalstar is using the pole-mounted helix antenna GAT-17MR, which is shown in Fig. 4.25 (Left). This antenna provides superior reception and call retention while sturdily attached to recreation boats, small vessels, or vehicles inside of Globalstar satellite coverage. It also provides the new service known as Sat-Fi as a satellite hotspot-allowing operator to send and receive satellite calls through smartphones. Anyone on board ships of vehicles will be able to carry just a single phone and use that smartphone on both cellular and Globalstar networks.
5. **Orbcomm Marine Whip Antenna**—The Stellar Wip AT400-3 Eighth Wave (1/8) ship or vehicle VHF/GPS antenna is used with the DS400 satellite modems, which is illustrated in Fig. 4.25 (Right). This satellite antenna could be used in marine or land environments where small mounting plates could be used or for applications, which require a low-profile antenna. It has a fixed mount and

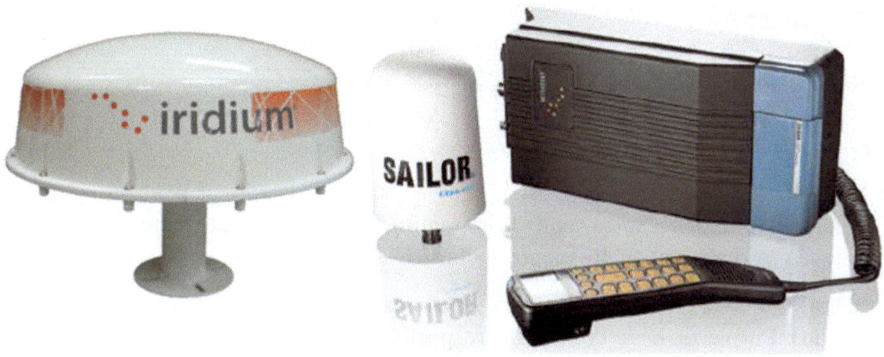

Fig. 4.24 Iridium maritime antennas—Courtesy of Manual: by Iridium

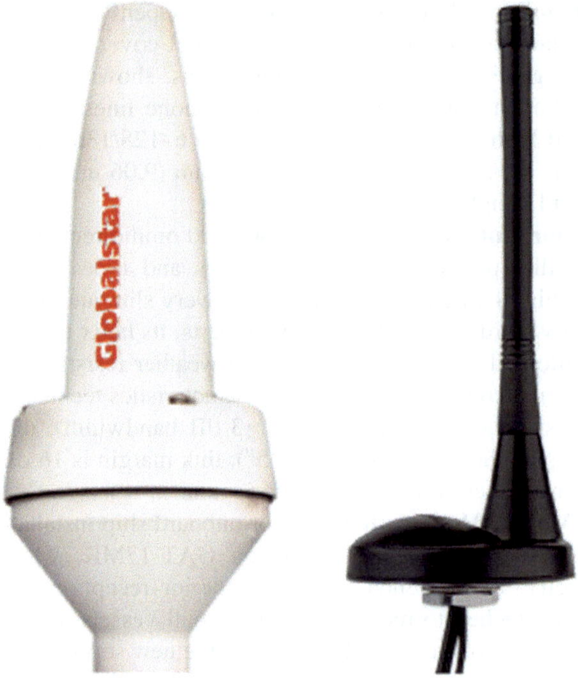

Fig. 4.25 Globalstar and Orbcomm maritime antennas—Courtesy of Manual: by Globalstar/Orbcomm

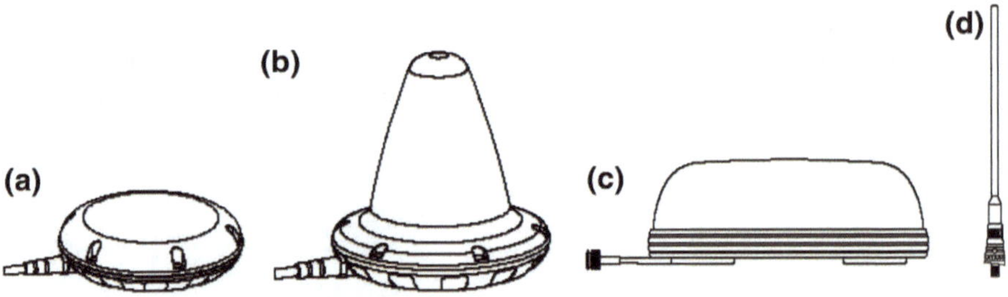

Fig. 4.26 Inmarsat antenna systems for VES—Courtesy of Manuals: by Thrane and Thrane

includes two RF cables, and its mounting surface ground plane of at least 6 inches in diameter or 36 square inches (6″ × 6″) is required (VHF/GPS).

The VHF and GPS cables are 15 feet long and are terminated with a FAKRA Bordeaux Purple or Blue connector, respectively. Characteristics of VHF antenna are as follows: RF is 137–150 MHz; gain is 2.0 dBi; maximum power is VSWR 200 W; and input impedance is 50 Ω (Fig. 4.26).

4.7.2 Vehicleborne Satellite Antennas

The vehicleborne satellite antenna system provides communications, broadband, and TVRO transmission of multimedia information such as audio and video images, voice and data by the integration of satellite communication, microwave-band transmission, wireless LAN, Internet, mobile public network, and audio and video compression. These antennas are serving road vehicles and rail wagons with Satcom on-the-move (SOTM) systems via Inmarsat and other GEO satellite constellations. However, the special vehicleborne satellite antenna system can quickly establish satellite broadband and TV transmissions.

4.7.2.1 Inmarsat Land Communication Antenna Standards

Land vehicles can generally use all shipborne terminals, except Inmarsat-A and -B because of the large size of their antennas and indoor equipment. The most popular standards for VES are Inmarsat-C, M, and mini-M sometimes with modified antennas and terminals. In the next section, three of the most known VES antennas, made by the Cobham manufacturer partner Thrane and Thrane, will be introduced.

1. **Inmarsat-C/GPS VES MSA**—The most suitable and smallest all-in-one antenna in the world for land vehicle mounting is the TT-3002L omnidirectional Inmarsat-C/GSP RHC polarized antenna with built-in LNA/HPA electronics designed to operate in an aggressive environment, which is illustrated in Fig. 4.14a. The antenna housing is sealed and contains no user serviceable parts, with dimensions of 122 × 35.5 mm and a weight of 0.5 kg. The antenna is very compact and is designed to operate in a corrosive environment and in extreme weather conditions without any service. It has a modular construction that allows easy exchange of antenna elements and is designed to work with the VES Capsat transceiver TT-3022C and GPS receiver in compliance with the Inmarsat-C specifications. Thus, this antenna can handle 32 Kbytes transmission length, with up to 100 m of coax cable, it is designed to operate when the satellite is visible over the horizon in coverage down to 20° of elevation, and no signal path blockage is present. The antenna gain is about 4 dBi, axial ratio is about 3 dB, G/T is −23 dBK, and EIRP is 14 dBK at a 5° elevation angle. This antenna is suitable for magnetic base mounting, especially on cars and on all land vehicles for message and data transmissions. The larger version of this antenna has only difference in dimension of 122 × 125 mm and in weight of 0.6 kg, which is shown in Fig. 4.14b. This antenna is designed to work with the Capsat Thrane and Thrane transceivers TT-3020C, TT-3022D, and TT-3022C and is suitable to be installed on any kind of land vehicle.
2. **Inmarsat mini-M VES Plate MSA**—The most suitable VES for installation in a large variety of vehicles, ranging from long-haul trucks to off-roaders, and train has the TT-3007F low-profile antenna built on an aluminum base plate and

covered by a radome, which is shown in Fig. 4.14c. mini-M was replaced by Explorer BGAN on June 30, 2015.

This satellite transceiver consists of an HPA/LNA board (TT-3010G), which amplifies received/transmitted voice and data radio signals. Otherwise, a controller board calculates tracking information based on input from a rate sensor and the received signal strength. Thus, the TT-3007F low-profile antenna is an automatic tracking antenna designed to work with the TT-3062D Capsat Compact Carphone. When it has locked onto a satellite, it tracks by turning the antenna element via a stepper motor. Elevation value is adjustable by switching between two different aerials; one helix aerial for low elevation and a patch for high elevation, both with an RHCP characteristic. The helix aerial has elevation coverage of 13°–40°; elevation beam of 37°; and azimuth beam of 73°, while the patch aerial has elevation coverage of 30°–70°; elevation beam of 62°; and azimuth beam of 67°. Thus, when driving in areas where the elevation is greater than 70°, the user might need to choose another satellite with a smaller elevation angle. To switch between the two aerials, a cable has to be manually moved from one connector to another, which requires the radome to be removed, and this is a disadvantage. Perhaps the problem can be solved with some kind of simple switch or diplexer.

Moreover, the antenna comes with magnetic feet for roof mounting on a vehicle. It is also possible to install the unit with bolts, for a more secure and permanent installation. The value of gain is 12–15 dBi, G/T is a minimum of −17 dBK, and EIRP is from 8 to 14 dBW in 2 dB steps. The dimensions of the antenna are 285 × 90 mm, and the weight is 2.7 kg.

3. **Inmarsat mini-M VES Whip MSA**—The TT-3007B rod antenna is a passive whip omnidirectional unit for use on cars and all kinds of land vehicles, which is illustrated in Fig. 4.14d. The only movable parts are the lower part of the unit, where adjustments for correct land zone are carried out. The rod is rigid but will bend down to an angle of approximately 30° at the base if necessary, e.g., when passing under obstacles. Thus, it must always be in its upright position when the system is in operational mode. To gain high receiver performance, the unit is connected to the antenna front-end unit via a low-loss coaxial cable. The elevation angle is set according to the geographic position and is determined by the maps provided. In fact, manual readjustment of the elevation is necessary when driving over long distances. Hence, the zones on the body of the antenna set the elevation angle as follows: Zone 1 = 25–32.5°; Zone 2 = 32.5–40°; Zone 3 = 40–47.5°; Zone 4 = 47.5–55°; Zone 5 = 55–62.5°; and Zone 6:62.5–70°. Tuned all the way down to maximum (beyond Zone 6), the antenna will work up to 90° because of the way the beam reforms. Tuned all the way up to the highest point, it will turn and the antenna will work down to 15°. The antenna has an elevation beam width of approx 15°. This relatively narrow angular aperture is necessary to obtain sufficient gain. Due to this, short signal dropouts might appear if driving in very rugged terrain. To avoid water ingression to the antenna, a rubber "raincoat," i.e., a rubber tube, is rolled down over the lower

part of the antenna. This antenna has the same electrical characteristics as the one previously discussed; dimensions are only 930 × 22 mm, and the weight is 0.63 kg.

The TT-3010A antenna front-end unit contains a high-power amplifier, a duplex circuitry, an LNA, and supervisory control circuitry. Data communication between the main unit and the antenna is accomplished by serial communication through the single coaxial cable, which also carries the RF signal.

4. **Inmarsat Vehicular BGAN MSA**—The vehicular BGAN transceiver Explorer 727 is a high-speed mobile broadband system, which is shown in Fig. 4.27 (Left). It consists of three fully integrated units: a transceiver, an IP handset, and a roof mountable antenna. The antenna automatically tracks satellite positions enabling high-speed connectivity while on the move providing simultaneous 432 Kb/s data, standard IP up to 492 Kb/s, ISDN up to 64 Kb/s, and phone calls, even at speeds in excess of 200 km/h. Placing the antenna on the roof of land vehicle, it is possible to turn it into a complete mobile communication HUB. It provides access to the Internet and phone networks, facilitates both live streaming and store and forward, and enables direct satellite transmissions from the hot spots of the world. Its specifications are as follows: RF is 1525.0–1559.0 MHz (Rx) and 1626.5–1660.5 MHz (Tx); antenna H/W is 152/Ø477 mm (6.3/Ø18.76 in), and weight is 6 kg (13.2 lbs); antenna cable loss is max/min 20 dB at 1.62 GHz and 1 Ω DC loop resistance, 3 dB at 36 MHz/4 dB at 54 MHz; EIRP is 20 dBW ±1 dB; and antenna connector is TNC Socket female.
5. **Inmarsat Portable BGAN Antenna**—The portable BGAN Explorer 700 inside mount antenna provides the same access to the highest bandwidth available via BGAN same as vehicular Explorer 727, thereby facilitating a platform for a wide range of applications with need of high performance, which is illustrated in Fig. 4.27 (Right). Thus, disregarding the environmental conditions, it is designed to last, and with the detachable antenna, it is ideally suited for temporary camps, vehicles, or semi-fixed installations. Antenna H/W/D is 297/99/51 mm (11.7/15.7/2.0 in), and weight is 2.1 kg (4.6 lbs).

4.7.2.2 Land VSAT and TVRO Antennas Over GEO Satellites

In addition to the Inmarsat satellites, there are other GEO satellite operators that provide innovative VSAT and TVRO transmissions for road vehicles and rail wagons. This GEO satellite constellations offer connectivity solutions via broadband voice, data, and video (VDV), Internet and TVRO video, data, TV, and/or IP-based service in a highly reliable and cost-effective packages or mobile maritime and land applications. The VSAT antenna is providing two-way or interactive broadband satellite communications, supporting VDV and Internet applications, while TVRO antennas provide one-way commercial on-the-move entertainment systems to individualized live satellite TV.

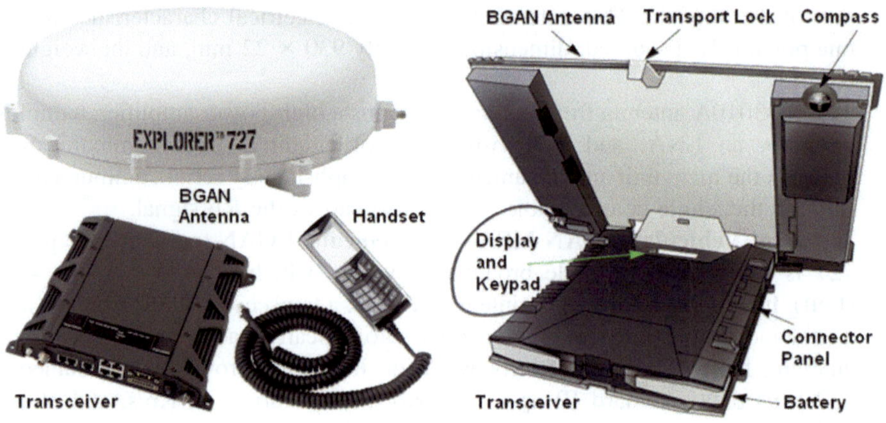

Fig. 4.27 Inmarsat vehicular and portable BGAN antennas—Courtesy of Manuals: by Cobham

Fig. 4.28 Land vehicular VSAT antennas for GEO satellites—Courtesy of Manuals: by KVH

1. **KVH Land Vehicular mini-VSAT Antenna**—The TracPhone V7-IP and V3-IP radome antenna units of KVH producer are mini-VSAT broadband communications system for vehicles on the move. The V7-IP radome and reflector antenna are shown in Fig. 4.28 (Left) and (Middle), respectively, while radome of V3-IP is shown in Fig. 4.28 (Right).
   Both IP antennas deliver a seamless and consistent Internet access with antennas that are smaller and lighter than traditional VSAT antennas. These VSAT systems consist of an antenna and Integrated CommBox Modem (ICM) that connect to a land-based HUB via a Ku-band GEO satellite constellation. The ground HUB is managed by a network operations center (NOC) that provides the link to the Internet and the terrestrial telephone network.
   The antenna transmits RF energy that is potentially harmful. Whenever the system is in use, it is necessary to make sure everyone stays more than 11 m

(36 feet) away from the radome antenna. However, no hazard exists directly below the antenna. Both antennas may accept an external GPS input as a backup to the antenna's internal GPS. In this sense, it will be more secure to connect a backup GPS and prevent a loss of service if the antenna's internal GPS fails. The specifications of both antennas are as follows: service coverage areas are seamless worldwide; RX receive/TX transmit RF band is 11.7–12.75 GHz/14.0–14.5 GHz; antenna gain is 35.6 dBi (RX-band, min) for V7-IP and 31.0 dBi (RX-band, min) for V3-IP; transmit power (BUC) is 4 W max for V7-IP and 3 W max for V3-IP; elevation range is −24° to +119° for V7-IP and 7.5°–75° for V3-IP; azimuth range is 710° rotation for V7-IP and 720° rotation max for V3-IP; maximum download speed is about 3 Mb/s for V7-IP and 2 Mb/s for V3-IP, and maximum upload speed is about 512 Kb/s for V7-IP and 128 Kb/s for V3-IP; antenna dish diameter is 60 cm (24″) for V7-IP and 36.8 cm (14.5″) for V3-IP; antenna diameter × height/weight is 66.3 cm (D) × 79.2 cm (h)/26.1 kg (26.1″ (D) × 31.2″ (h)/57.6 lbs) for V7-IP and 9.4 cm (D) × 44.7 cm (h)/11.3 kg (15.5″ (D) × 17.6″ (h)/25 lbs) for V3-IP.

2. **Gilat Vehicular Low-profile VSAT Antenna**—The Satcom on-the-move (SOTM) flat VSAT arrays are a product of Gilat Company with operations in Israel, and R&D office is in Sofia Bulgaria. SOTM is defined as a vehicle equipped with antenna that establishes and maintains communication while the vehicle is moving on the roads. There are different types of terminals, which meet customer-specific requirements varying in height from 60 down to 2.5 cm. In past years, requirements toward low-profile and smaller systems have prevailed. Key features of these antennas are as follows: It operates on any standard Ku-band and capacity (14–14.5 GHz); always on operation; automated signal acquisition and tracking; and immediate and rapid signal reacquisition following line-of-sight blockage.

   Figure 4.29 (Left) shows StealthRay 2000 vehicular VSAT antenna, whose specifications are as follows: Uplink EIRP is 41.7 dBW (with external 40 W BUC); G/T is 8 dB/°K at 30° elevation or 9 dB/°K at 45° elevation; Tx gain is 27 dBi; frequency band for Rx is high band 11.7–12.75 GHz/low b and 10.95–11.7 GHz and for Tx is 14.0–14.5 GHz; elevation look angle range is 20°–80°; dimensions W/L are 90/115 cm and 28 kg; and net height is 15 cm. Figure 4.29 (Right) shows SR-200 vehicular VSAT antenna, whose features are as follows: Uplink EIRP is 42 dBW (with external 40 W BUC); G/T is 5 dB/°K; Tx gain is 27.3 dBi; RF band for Rx is 10.9–12.75 GHz and for Tx is 14.0–14.5 GHz; elevation look angle range is 0°–90°; dimensions W/L are 49/54.5 cm and 12 kg; and net height is 21 cm.
3. **Orbit Rail Low-profile VSAT Antenna**—The Orbit RailTRx antenna systems are supporting a variety of stabilized train antenna system configurations in Ku- and Ka-bands. As a common platform, it is designed to accommodate the current and future needs of the train market. As such, these elliptic antennas are an optimal solution for the evolving rail broadband communications needs of intercity, regional, and high-speed trains.

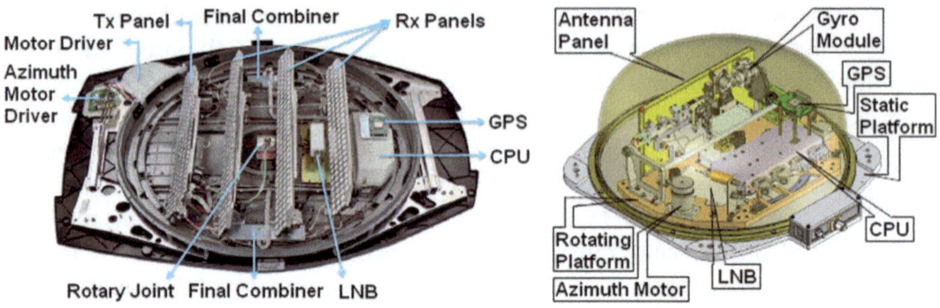

Fig. 4.29 Land vehicles VSAT antennas for GEO satellites—Courtesy of Manuals: by Gilat

The RailTRx antenna product line comprises two product series: RailTRx 2-300 features a low-profile high-gain Ku-band antenna, shown in Fig. 4.30 (Left), and RailTRx 2-500 features a low-profile high-gain Ka-band antenna, shown in Fig. 4.30 (Right). The 500 Series offers built-in Ka-band fully compatible design to ensure smooth migration to future high-speed Ka-band services. The values of 2-300 antennas are as follows: RF Ku-band for Tx is 13.75–14.50 and for Rx is 10.95–12.75 GHz; G/T is 13.5 dB/K$^{\circ}$; EIRP 44 dBW; travel azimuth is 360°; elevation is +5 to +50; antenna diameter is 105.8 cm (41.6 in); height is 48.8 cm (19.2 in); and weight is 96 kg (211 lbs). The values of 2-500 antennas are as follows: RF for Ka-band Tx is 29.00–31.00 and for Rx is 18.20–21.20 GHz; G/T is 14 dB/K$^{\circ}$; EIRP 44 dBW; 13.75–14.50 and for Rx is 10.95–12.75 GHz; G/T is 13.5 dB/K$^{\circ}$; EIRP 44 dBW; travel azimuth is 360°; elevation is +15 to +65; elevation is +5 to +50; antenna diameter is 105.8 cm (41.6 in); height is 41.8 cm (16.5 in); and weight is 94 kg (207 lbs).

4. **KVH Vehicular TVRO Antenna**—The KVH manufacturer of vehicular TV antennas has designed two different models: TracVision RV1 radome antenna shown in Fig. 4.31 (Left) and TracVision A9 low-profile antenna shown in Fig. 4.31 (Right). Both antennas are HDTV compatible, DVB-S2 acquisition and provide excellent performance and reliability for passengers in family cars, buses, and rail wagons with hundreds of channels of satellite TV entertainment in no time. In addition, high-performance TV antenna tracking with advanced algorithms provides crystal-clear television picture in extreme conditions.

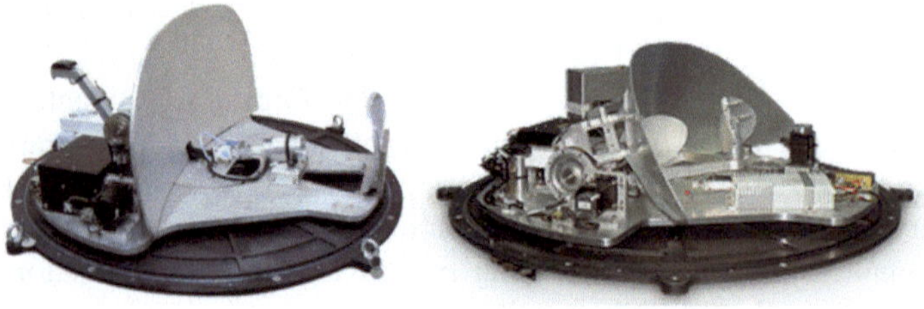

Fig. 4.30 Rail VSAT GEO antenna—Courtesy of Manuals: by Orbit

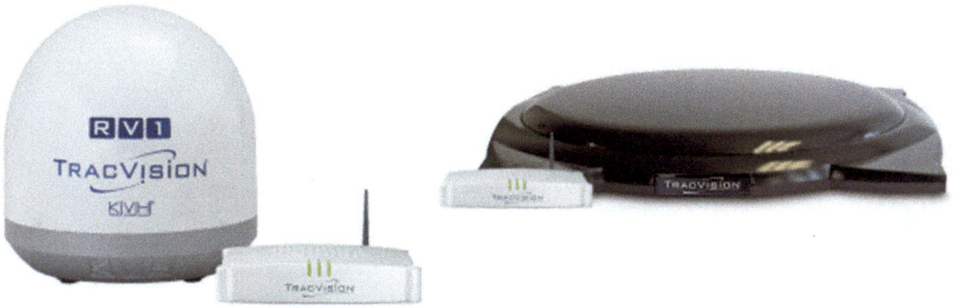

Fig. 4.31 Land vehicles TV antennas over GEO satellites—Courtesy of Manuals: by KVH

Antenna unit diameter × height and weight for RV1 are 34.3 cm × 33.7 cm (13.5″ × 13.3″) and 3.6 kg (8 lbs); for A9 are 81 cm × 12.7 cm (32″ × 5″) and 22.5 kg (49.5 lbs).

4.7.3 Transportable Earth Station (TES) Antennas

Recently, after the introduction of Inmarsat-A and -C standards, Inmarsat started to provide a new transportable and portable telephone and facsimile communication service in rural areas and in regions with out-of-range or damaged terrestrial telecommunication networks. At first, Inmarsat-A transportable transceivers were developed by Magnavox Corporation, EB Communications, JRC, and other manufacturers of MSC equipment, with directional aperture dish and umbrella parabolic antennas. Soon after Thrane and Thrane, OmniPless (South Africa), Toshiba, Nera, and other manufacturers of equipment designed the TES Inmarsat-C transceiver with small patch antennas.

1. **Inmarsat-C TES Patch Antenna**—This antenna was developed for the Inmarsat-C portable Capsat TT-3026A transceiver by Thrane and Thrane, which is shown in Fig. 4.32a. A transceiver unit was integrated into a small laptop PC as a portable terminal, providing the same service as other Inmarsat-C terminals in rural and remote areas.
2. **Inmarsat mini-M TES Patch Antenna**—The mini-M TT-3007A patch antenna resides on top of the Capsat TT3060A mini-M Electronics unit (EU), where a short coaxial cable of 0.15 m connects the antenna and the EU, which is illustrated in Fig. 4.32b. It is possible to place the patch antenna at a distance from the laptop-sized EU when the short antenna cable can be replaced with the supplied cable of 5 m or even longer up to 70 m. The patch antenna has a stand, which in unfolded condition can hold it at the correct angle for transmission; alternatively, when the stand is folded up, it can be fastened to a bracket. It is completely watertight and sealed for outdoor operation and has an acoustic

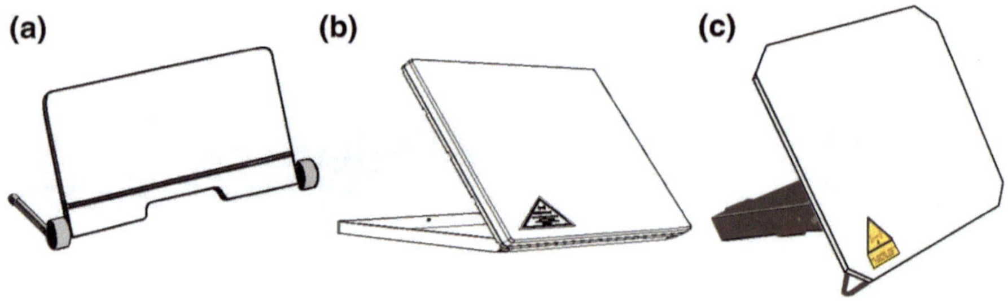

Fig. 4.32 Inmarsat small TES antennas—Courtesy of Manuals: by Thrane and Thrane

signal strength indicator, which beeps with an increased repetition rate as the signal received via the antenna becomes stronger. Generally, this TES antenna meets or exceeds current and proposed Inmarsat specifications for Inmarsat phone spot beam operations. It has a directional RHCP patch array with ±15° horizontal and ±15° vertical beam widths and electrical characteristics as follows: Gain is about 14–15 dBi, EIRP is from 11 to 17 dBW in 2 dB steps, and G/T is a minimum of −17 dBK. Dimensions of the antenna are 52 × 270 × 200 mm, and the weight is 2.2 kg (including handset, battery pack, and antenna). The label attached to the TT3007A antenna gives a warning that the antenna radiates microwave signals and that the operator has to keep a minimum of 0.5 m safety distance. All types of flat panel antennas have similar warning labels. The next version of this antenna has one panel or combination of two panels, which provided support to the M4 system multimedia service linking users to the GAN system via Inmarsat-3 satellites providing mobile HSD, ISDN line, voice, fax, or combinations of services from virtually anywhere and anytime. The Inmarsat GAN service was replaced by Explorer BGAN range on June 30, 2015.

3. **Inmarsat mini-M Small Dish TES Antenna**—The Small Dish TT-3008D antenna is just one of the many antennas available for the Thrane and Thrane M4 service, which is shown in Fig. 4.32c. The directional patch array antenna combined with Capsat messenger terminal is the perfect choice for a combined fixed and mobile solution at remote and rural sites. It offers both portability and an easy-to-setup fixed solution. The antenna has a robust frame mount construction that also gives the option of installation on a pole or tripod. Built-in compass, audible signal strength, and clear elevation indication aid in locking onto the satellite. The antenna can be stored in a separate and sturdy carry bag. The transceiver is placed in a bracket inside the antenna, making it one unit; the smallest ISDN units only weigh 5.2 kg. The 64 Kb/s mobile ISDN connections give the user access to the Internet, voice, fax, video conferencing, and many other applications. In addition, this antenna also enables access to the inexpensive low-speed mini-M services for voice, fax, and data. Should a 128-Kb/s speed solution be required, then this is made possible by combining two M4 terminals with two small dish antennas to the ISDN interfaces of the laptop. In

this case, it is possible to use the GAN service offered by Inmarsat, which covers both high-speed and low-speed data needs. The electric characteristics of the antennas are as follows: Gain is from 17 to 19 dBi, EIRP is 8–14 dBW for mini-M and 19–25 dBW for HSD, and G/T is −7 dBK. Dimensions of the antenna panel are 397 × 397 × 75 mm, and the weight is 3.1 kg. The new antenna version is combined with three connected (folded-over) panels.

4. **Inmarsat mini-M Big Dish TES Antenna**—The Inmarsat-M remote antenna is a proven product designed by OmniPless for use with satellite phones that do not have a remote function, which is shown in Fig. 4.33a. Due to its high gain, the antenna can be positioned more than 20 m from the terminal without any degradation of the signal, depending on the type of terminal and quality of cable used. Thus, this allows the transceiver to be used indoors even when no clear line of sight to the satellite is available. This antenna is fully sealed and can be mounted outdoors even in the harshest weather conditions. It can be mounted on a wall and remotely controlled from up to 70 m from the main unit. It is also available with a dedicated pole mount option, which allows simple installation for fixed or semi-fixed applications, such as rural public payphones, roadside emergency telephones, and SCADA applications. This kind of antenna supports Inmarsat GAN service with 64 Kb/s ISDN HSD and all Internet solutions. This antenna is also ideal for roof mounting of indoor transceivers in rural offices or remote business solutions in gas/oil pipe or mining industries. Besides, the antenna enables access to the inexpensive low-speed mini-M services for voice, fax, and data transmissions. The polarization of the dish antenna is RHCP with manual adjustable elevation setting and with a beam width of approximately 20°. The electrical characteristics of this antenna are as follows: Gain value is about 19 dBi, EIRP is from 8 to 14 dBW in 2 dB steps, G/T is a minimum of −7 dBK, and the axial ratio is 2 dB. The dimensions of the antenna are 558 × 550 × 63 mm, and the weight is 3.5 kg.

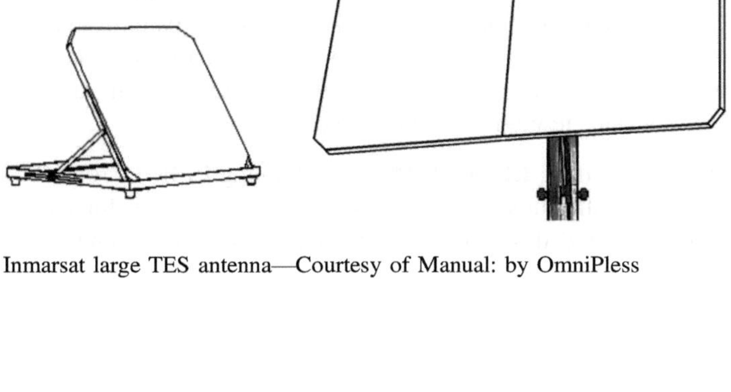

Fig. 4.33 Inmarsat large TES antenna—Courtesy of Manual: by OmniPless

5. **Inmarsat-B Quad Flat Panel TES Antenna**—The TES flat panel quadrant antenna is an innovative solution for new-generation transportable and portable Inmarsat-B terminals, which is shown in Fig. 4.33b. Designed by OmniPless for compact TES equipment, it provides ease of deployment and small stowage size. It consists in four separate identical panels and a base plate. Proprietary attachments allow the panels to be quickly mounted onto the base plate, ideal for integration with the devices of the deployment structure and RF electronics. A retaining clamp is available, which locks the panels to the base plate for permanent and semi-permanent installations. This antenna enables transmission of voice, data, and fax calls at speeds of 9.6 Kb/s; advanced ISDN capabilities for high-speed Internet/LAN access; and broadcast uplinks. By connecting a standard fax and a PC, one has a complete office at a remote or rural location, with optional HSD of 56/64 Kb/s. The electrical characteristics are as follows: Tx gain is 21.4 dBi and Rx gain >21.3 dBi, EIRP is 25–33 dBW, G/T is about −4 dBK, axial ratio is 1 dB, and antenna temperature is 114 °K and with RHCP polarization. The dimension of the antenna low panel stowage is 390 × 387 × 62 mm (4 panels), deployed panel size in total is 774 × 774 × 15.5 mm (about 0.6 m^2), and the weight is 2.4 kg.

4.7.4 Transmitting Antennas for the COSPAS–SARSAT System

The COSPAS–SARSAT system uses different types of antennas built-in or mounted atop of emergency position-indicating radio beacon (EPIRB), personal locator beacon (PLB), and emergency locator transmitter (ELT) units, for maritime, land, and aeronautical applications, respectively. The antenna for distress satellite beacons (DSB) may have been designed to transmit signals at RF of 406.025 MHz signals via LEOSAR, MEOSAR, and GEOSAR COSPAS–SARSAT satellites. Thus, the COSPAS–SARSAT system has ceased satellite processing of 121.5/243 MHz beacons on February 1, 2009. Some antennas for handheld DSB have to receive signals emitted on the same RF and to conduct power to a strobe light mounted above the antenna. It is possible that the radiated signal can be composed of an unknown ratio of vertical and horizontal polarizations. For this reason, some consideration shall be given to the type of antenna and its radiated field. The results shall encompass all wave polarizations, so the antenna pattern and field strength measurements should provide sufficient data to evaluate the antenna characteristics. There are many manufacturers of DSB for maritime, land, and aeronautical applications, but only few will be introduced.

1. **Antenna for Floating EPIRB/GPS**—The floating EPIRB Tron 60 GPS set is a GMDSS DSB with buoy shape and is designed by Jotron for manual or automatic operation and designated to enhance crew and passenger lifesaving on

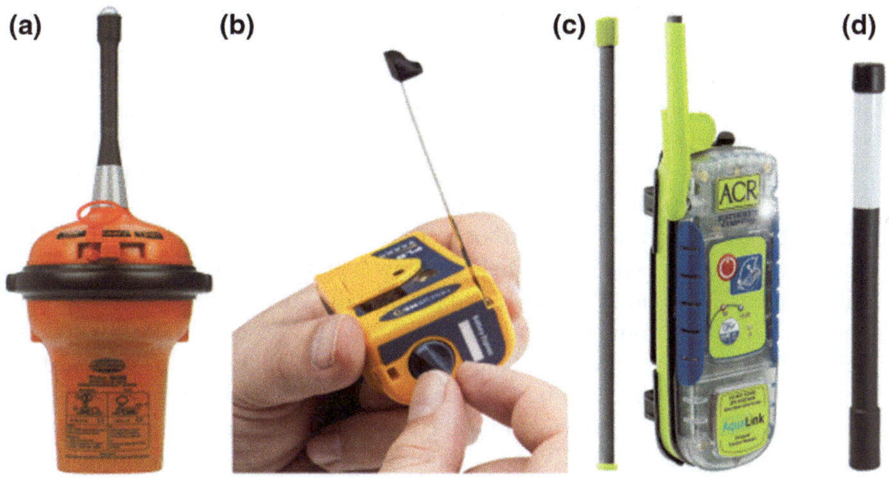

Fig. 4.34 Maritime EPIRB/PLB antennas—Courtesy of Manuals: by Jotron/Ocean Signal/ACR

board ocean vessels, as is shown in Fig. 4.34a. This DSB unit has two built-in antennas: The first is an omnidirectional 406.025 MHz antenna for Tx distress, and the second is a receiving GPS patch antenna providing accurate position improvement of the EPIRB from a radius of 5 km to an amazing 100 m. This unit can be used as ELT for aircraft in an emergency on the sea's surface. Similar top-mounted or built-in antennas have also been designed for ELT aeronautical applications with different body shapes.

2. **Antenna for Handheld PLB/GPS**—This is the smallest handheld DSP produced as PLB1 brand by Ocean Signal Company, whose unit with antenna is illustrated in Fig. 4.34b. The PLB1 unit is small enough to be easily carried in a pack or pocket with the following dimensions: Height/width/depth is 77/51/32.5 mm, and weight is 116 g. When it is activated, this beacon transmits via retractable antenna position of person wherever emergency situation is, at sea or on land. The PLB1 satellite transmitter is sending alert signals via antenna at UHF-band RF of 406.040 MHz and VHF homing beacon at 121.5 MHz. The built-in GPS receiver is using microstrip patch antenna.
3. **Antenna for Handheld–Floating PLB**—This is handheld and floating AquaLink PLB designed by ACR designed for manual activation during distress alert by crewmember or passenger on board ship, whose antenna and unit are illustrated in Fig. 4.34c. An omnidirectional whip transmitting 121.5 and 406 MHz antenna with a length of 180 mm is mounted on top of the electronic unit. Moreover, similar top-mounted or built-in antennas have also been designed for ELT and PLB beacon units for land and aeronautical mobile applications, respectively, with different body shapes. More exactly, with three levels of antenna types integrated in satellite beacon signal technology, such as: GPS Rx positioning, a powerful 406 MHz DSP Tx and 121.5 MHz homing Tx

capability, the AquaLink quickly and accurately relays ships or personal position in distress alerting the worldwide network for SAR operations.

4. **Replacement Antennas for EPIRB Transmitters**—This is ACR replacement antenna model 9368 for 406 and 121.5 MHz EPIRB transmissions, which is illustrated in Fig. 4.34d. The features of this antenna are as follows: Weight is 0.10 g; height is 18.77 cm (7.39″), and voltage standing wave ratio (VSWR) to measure transmission line imperfections is less than 1.5/1.

1. **ELT Blade Antenna**—The airborne ELT Rx blade antenna S65-1231-1 is product of sensor systems, which is shown in Fig. 4.35a. It is working on abandoned 121.5 and 243.0 MHz and the new digital satellite-aided tracking RF of 406.025 MHz, compatible with COSPAS–SARSAT and ELT systems. This aircraft antenna is one-piece aluminum casting construction designed for high-speed aircraft and can be fixed on rear side of fuselage of fixed and rotary wing aircraft. It has omni/azimuth and cos/elevation characteristics: Polarization is vertical; impendence is 50 ohms; lighting protection is DC grounded. Weight of antenna is 1.6 lbs, height is 9.8 inches, and it is built by 6061-T6 aluminum glass materials and is connected to the onboard unit via Bayonet Neill–Concelman (BNC) plug.
2. **Topside ELT Antennas**—This type of antenna is installing on topside of ELT and can be used in two types: the Kannad ELT with whip antenna, shown in Fig. 4.35b (Left), and the Kannad Rod ELT Rx antenna for business jet aircraft, shown in Fig. 4.35b (Right). Both antennas are compatible with COSPAS–SARSAT and other ELT standards.
3. **Antenna for Floating ELT**—This antenna is usually whip antenna installed on top of floating ELT, shown in Fig. 4.35c. The representative of this unit is product of the French company KANNAD 406 AS, an aeronautical survival beacon. Thanks to its small size and lightweight, this ELT fits easily inside a life raft. However, it is supplied with a floating collar and can be used for aircraft in an emergency on the sea's surface.

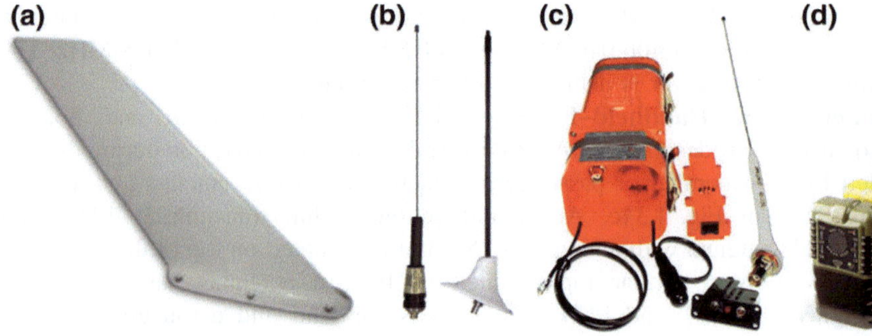

Fig. 4.35 Transmitting antennas for ELT—Courtesy of Manuals: by Sensor Systems, Kannad and Specmat Technologies

4. **Antennas for Handheld ELT**—This type of handheld ELT set is calling multifunction PLB with embedded GPS receiver with dual PLB/GPS antenna. These units are designed to survive for up to twenty years with the only maintenance being battery replacement every five years. Otherwise, there are various ELT antenna options available for most applications. The manufacturer of the US Specmat Technologies Inc different ELT Tx Antennas integrated with GPS Rx Antenna are shown in Fig. 4.35d.

4.7.5 Antenna Systems for GMPSC

Many emerging Global Mobile Personal Satellite Communication (GMPSC) networks are offering services based on small handheld terminals. The design of these terminals is based upon the services provided by the system, so the electronics and the antenna should follow certain guidelines to fulfill the system, network, service, and market requirements. Briefly, the requirements are for the GMPCS antenna system to have a shaped pattern in elevation (depending on the satellite antenna, orbital parameters, and satellite system statistics); circular polarization, an adequate G/T (depending on the link budget parameters, the mobile's environment, and handling of the handset by the user); an adequate bandwidth (depending on the system specifications); and finally, an appropriate size (dictated by the handset size and aesthetic constraints but also depending on all the other requirements). Hence, the QHA seems to be the most promising antenna for such an application because it offers an elevational shaped pattern, whose shape can be changed by small structural changes and circular polarization within the main lobe.

The G/T of these antennas has to be measured in conjunction with the environment and the operating system applied, while the instantaneous bandwidth for VSWR is better than that at 2:1–3 to 5 % for slim antennas and goes up to 10–15 % for wider structures. Otherwise, the pattern shape bandwidth goes up to 15 % for all structures. The required pattern shape was determined based on an analysis of the G/T for different antenna patterns in different environments and for a system with the constellation statistics of GMPSC, such as Iridium, Globalstar, and others. It is assumed that the user cooperates by pointing the antenna toward the zenith. The main lobe should have the maximum gain at about 60° above the horizon, with a half-transmission-power beam width starting from the maximum gain angle and ending at the minimum elevation angle.

The minimum elevation angle was defined by the system to be 10° above the horizon, but by taking into account the satellite statistics of the system, an 18° minimum elevation angle can be selected, offering a 100 % coverage of most of the Earth's surface and a 98 % coverage in latitudes between 3 and 5° and above 70° north and south. A small gain dip at the zenith is desirable, since this optimizes coverage at low elevation angles where extra path and fading losses are encountered. The antenna sizes for handheld terminals were designed to be as small as possible for the operational L- and S-bands. At this point, the antennas chosen are relatively short QHA with very good pattern shapes and can be produced using simple design instructions.

4.7.5.1 Wire Quadrifilar Helix Antenna (WQHA) for Handheld Terminals

The WQHA is a highly resonant antenna invented by Kilgus in the 1970s. This antenna consists of four helices placed at 90° to each other. The four helical elements are connected by radial parts and are fed in 90° phase differences (0°, 90°, 180°, and 270°).

The analysis by Kilgus for the resonant QHA (RQHA) is based on the assumption that the QHA consists of four helical and four radial parts. At this point, the current distribution on the helical elements is assumed to be sinusoidal, with magnitude maxima at the feed and the distal end for each l/2 length. The current in the radials is approximated by a uniform distribution. Another analysis approach to the RQHA is based on the assumption that a QHA consists of two bifilar helices (BH) placed at 90° angular distance and fed in phase quadrature. The radials in the distant end can be shorted or open-circuited giving only changes to the input impedance. One of the major disadvantages of the QHA is the complex feed network required. In addition, one approach is to feed each BH with the assistance of a balun. Most of these configurations need a 90° phase difference hybrid, and two baluns are needed to feed both BH. The exceptions to that rule are the self-phased configuration and the Keen balun, which can offer the phase difference without a hybrid and with the use of only one balun. The other way is to separately feed each one of the four helical elements, with 90° phase difference, using three hybrids. This method does not need a balun to feed the QHA but can be used only in an end-fire configuration, for constructional reasons.

Otherwise, the QHA can operate in both the satellite mode (hemispherical pattern) and terrestrial mode (toroidal pattern). There are three possible ways to accommodate a QHA on handheld terminals, illustrated in Fig. 4.36. The first one (a) is a bottom-fed, end-fire QHA with shorted-end radials and normal QHA. The second one (b) is the same antenna but uplifted to be above the head, known as a moved-up QHA, and finally, the third one (c) is a top-fed, backfire QHA with open bottom radials, i.e., with open radial QHA.

4.7.5.2 Antenna Solutions for Handheld Terminals

The new GEO and non-GEO MSC systems are employed as personal handheld terminals with small whip or telescope antennas similar to cellular systems. In this context, several different manufacturers in the world have designed types of antenna for top mount on the handheld terminals. These systems also provide service for all

Fig. 4.36 Three solutions of QHA for handheld phones—Courtesy of Paper: by Agius

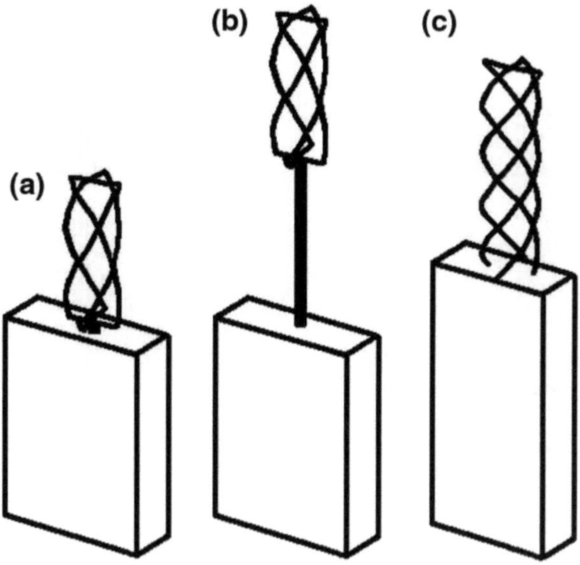

three mobile applications, with specific antenna solutions for installation atop mobile satellite phones.

1. **Iridium 9555 Satellite Handheld Phone**—This is compact, light, and easy-to-use with industrial-grade ruggedness and it is connected to the only truly global personal and mobile satellite communications network. The Iridium 9555 handset has an inbuilt transceiver and fully integrated internal omnidirectional whip antenna, which is illustrated in Fig. 4.37a (Left). This antenna is new and much more sensitive and effective for reliable satellite voice and data transmissions. During telephone call, internal antenna has to be extended, so if is not extended on the screen will be written alert message to extend the antenna. The antenna must be able to "see" the entire sky from approximately 8° above the horizon. The features of this handheld phone are as follows: Dimensions are 143 mm (L) × 55 mm (W) × 30 mm (D), and weight is 266 g (9.4 oz). This satellite phone is also designed for usage with an optional external (auxiliary) antenna without removing internal antenna, which is shown in Fig. 4.37a (Right). The optional auxiliary antenna adaptor allows to be easily attached the external antenna to the back of the Iridium 9555 handset. It allows for in-vehicle use of Iridium handset, such as on board ships, vehicles, and aircraft, and is calling satellite dock station. This antenna is fitted via an extension coaxial cable to the cradle that must be located more than >0.3 m (1 foot) from human body (person) when in operation.
2. **Inmarsat Handheld IsatPhone 2**—This Inmarsat handheld phone is designed for the most reliable voice and data satellite communications network in the world. It is designed to use whip internal active antenna that is able to receive incoming call alerts even with antenna stowed, which is illustrated in Fig. 4.37b.

Fig. 4.37 Antennas for handheld satellite phones—Courtesy of Manuals: by Iridium/Inmarsat/Globalstar

The features of this phone are as follows: Dimension is 16.9 cm, and weight is 0.318 kg. This phone has external antenna port for connection of external (auxiliary) antenna for all mobile applications. This external antenna is essential at sea, on the ground, and in the air to lock the geostationary satellite voice connections and also to make transfer of data without breaking. The docking stations can be used with a variety of antennas, including a fixed passive antenna, a marine pole mount mast antenna, and low-profile vehicular or aircraft antennas.

3. **Globalstar GSP-1700 Handheld Phone**—This is handheld portable satellite phone using Globalstar network. Rotate and extend the antenna located on the back of the phone, whose phone and antenna in stowed position is shown in Fig. 4.37c. The phone will ring with the antenna stowed, but cannot be answered until the antenna is extended. The optional external antenna id is designed for installation on board ships and vehicles.

4.7.6 Airborne Satellite Antennas

Typical medium and large jet antenna classification includes communication, VSAT, and TVRO antennas. The fully professional and more reliable GMSC system is Inmarsat GEO constellation serving AMSC in which the L-band 1.6/1.5 GHz MSA is used for the service link, such as low-gain antenna (LGA), intermediate-gain antenna (IGA), and high-gain antenna (HGA), which are shown in Fig. 4.38, LGA (a) produced by EMS Electronics and IGA (b) and HGA (c), both produced by Cobham. The Inmarsat and other GEO satellite operators also are deploying antennas for satellite data link (SDL) service.

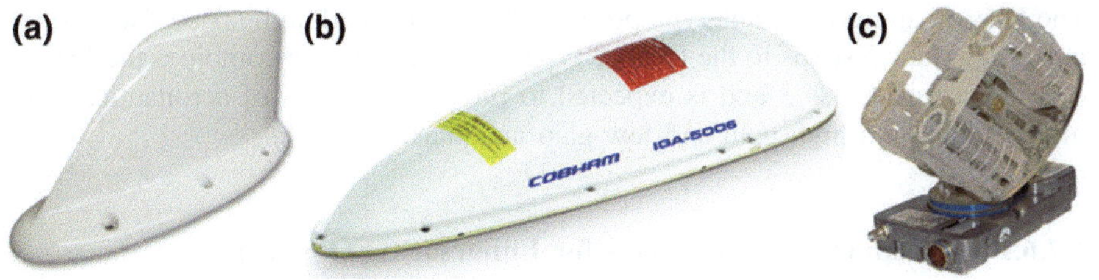

Fig. 4.38 Inmarsat LGA/IGA/HGA for AES—Courtesy of Manuals: by Cobham/EMS Electronics

The Inmarsat satellite communication system has been described in the aeronautical system definition manual (SDM), and the AES portion of the system for commercial aviation has been described in Aeronautical Radio Incorporated (ARINC) Characteristics 741. The ARINC 741 categorizes the operational types of the AES into four classes see Table 4.5.

The class 1 aircraft antenna type can provide only low-speed data services, which include aeronautical operational control (AOC) and aeronautical administrative communication (AAC) services by using a low-gain antenna (0 dBi). In the future, this type of station will be used for ATC. The class 2 aircraft antenna type provides only voice services, mainly for passengers in the cabin known as aeronautical passenger communication (APC), with a high-gain antenna (12 dBi).

Table 4.5 Operational types of AES for the ARINC 741 Standards

Type	Antenna	Voice/data	Service	User (air)	User (ground)
Class 1	Low-gain antenna (0 dBi)	Low-speed data	AOC AAC ATC [in the future]	Pilot Cabin crew Pilot	Airline company Airline company Control authority
Class 2	High-gain antenna (12 dBi)	Voice	APC	Passenger	Subscriber
Class 3	High-gain antenna (12 dBi)	Voice and high-speed data	AAC APC	Cabin crew Passenger	Airline company Subscriber
Class 4	Low-gain antenna (0 dBi) and high-gain antenna (12 dBi)	Voice and high- and low-speed data (Class 1 and Class 3)	AOC AAC APC ATC [in the future]	Pilot Cabin crew Passenger Pilot	Airline company Airline company Subscriber Control authority

Legend of Terms AOC Aeronautical Operational Control; *AAC* Aeronautical Administrative Communications; *ATC* Air Traffic Control; and *APC* Aeronautical Passenger Communication

The Class 3 aircraft antenna type provides both voice and high-speed data services by adding data systems to the Class 2 type. The Class 4 type combines the features of Class 1 and Class 3 and is expected to provide all kinds of aeronautical communications with high-gain and low-gain antennas.

4.7.6.1 Introduction of Antennas for Inmarsat GEO System

The commercial aeronautical MSC services worldwide have been provided by the Inmarsat network since 1990 with Aero-H and -L standards. Hence, aircraft can easily use Inmarsat-C and mini-M shipborne terminals, but because of the high speed of aircraft and small space in the cockpit, all AES terminals and antenna design are modified more aerodynamically and in a very flat shape for mounting on the aircraft fuselage. At this point, in 1992 and 1998, Aero-C and -L standards were developed, respectively, and finally, the aero mini-M standard for AES mounts was recently developed.

The biggest mobile satellite operator today Inmarsat Inc. developed five types of MSA for maritime, land, aeronautical, transportable (portable), and personal and semi-fixed applications, while in particular, some maritime MSA are used or transformed for land vehicles, aircraft, and off/onshore installations. There are many different categorizations of Aero standards, but as stated above, the most reasonable today is classified into three main groups: AES with LGA, IGA, and HGA, which all produce nominal gains of 0 and up to 12 dBi, respectively. The G/T of HGA and LGA is required to be over—13 dB/K (gain is about 12 dBi) and −26 dB/K (gain is about 0 dBi), respectively. The HGA is used for public voice/high-speed data (21 Kb/s) communications, which can be connected to terrestrial public telephone networks. On the other hand, the low-gain antenna is used for low-speed data (600, 1200, 2400 bps) communications.

The phased-array antennas are the best candidates for the airborne antenna because of such advantageous characteristics as low-profile and mechanical strength. Two types of phased-array antenna have been used; one is a conformal type, which has two sets of phased arrays on both sides of a fuselage, and the other is a top mount type, which has a set of phased arrays on the top of a fuselage.

As shown in Fig. 4.39, this is a top mount HGA phased-array antenna for the INMARSAT Aero terminal, which was improved for commercial aircraft use based on the research and development of the airborne phased-array antenna in the ETS-V program. It has four phased-array units with an A-class HPA, a DIP/LNA, and a beam steering unit (BSU). The nose and tail array units have been used to make keyhole areas narrow enough to keep communication links over 85 % in the upper hemisphere. All kinds of aircraft antenna provide simultaneous full duplex and two bands of transceiving operations. The main specifications and RF of LGA and HGA subsystems are shown in Table 4.6.

The AES terminal uses two types of aircraft phased-array antennas. The first is a top mount type, which is installed on the top of a fuselage, while the second is a side mount array type, which is installed on both aircraft sides, port and starboard

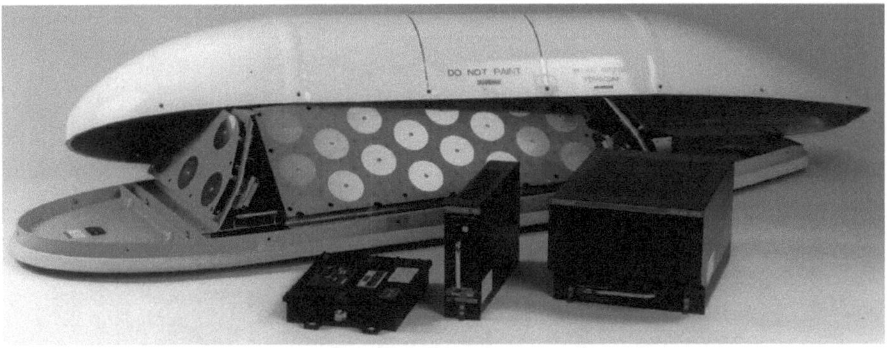

Fig. 4.39 Inmarsat HGA phased-array antenna—Courtesy of Book: by Ohmori

sides of the fuselage. Thus, the top mount type has the advantage of eliminating keyhole areas where the beam cannot be scanned, but it has the disadvantage of increasing air drag. The conformal type, on the other hand, has the advantage of low air drag, but has the disadvantage of keyholes. An HGA subsystem consists of a phased-array antenna, a DIP, an LNA, an A-class HPA, and a BSU, which steers the beam to track a satellite.

A program-tracking method carries out the satellite tracking. A BSU steers the beam by controlling digital phase shifters of the phased-array antenna; the information used to operate the BSU is calculated from the signals from the inertial navigation system (INS), which gives the position, heading direction, and altitude of the aircraft. To track the satellite, the HGA has to steer the beam to cover over 75 % of the hemisphere above an elevation angle of 5°. When a phased-array antenna is used, these are inevitably keyhole areas in the fore and aft directions shown in Fig. 4.40, because it is very hard for a phased-array antenna to scan the

Table 4.6 Specifications of LGA and HGA

	Low-gain antenna	High-gain antenna
Frequency	1530.0–1559.0 MHz (receive) 1626.5–1660.5 MHz (transmit)	
Polarization	Right-hand circular	
Axial ratio	Less than 6 dB	
Figure of merit (G/T)	Over −26 dBK	Over −13 dBK
Radiation power (EIRP)	Over 13.5 dBW	Over 25.5 dBW
Antenna gain	Over 0 dBi	Over 12 dBi
Coverage for semi-sphericity above elevation angle 5°	Over 85 %	Over 75 %
Tracking	None (omnidirectional)	Program tracking

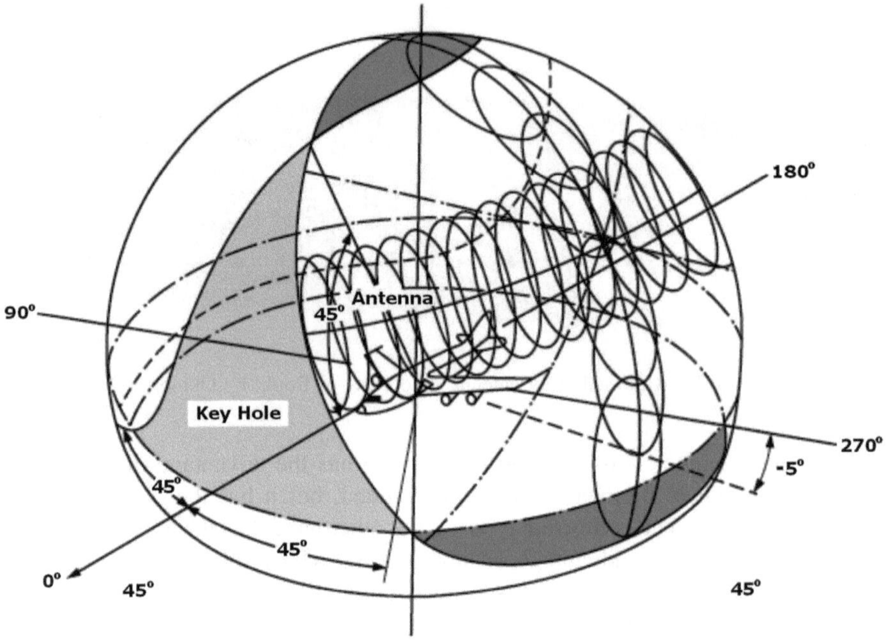

Fig. 4.40 Beam steering coverage and keyholes of phased-array antenna—Courtesy of Book: by Ohmori

beam to wide angular areas over 60°. An A-class HPA is used to avoid channel intermodulation in a multicarrier operation.

1. **LGA System**—The low-gain antenna (LGA) is a small omnidirectional and not a tracking aeronautical antenna designed to support both aircraft Inmarsat-C/GPS L-band transceivers and GPS Rx, whose shape is shown in Fig. 4.41a. It can be used for automatic low bit rate data reporting and message transfer of position reports, performance data, and operational messages on a global basis, from sea level to about 20 km and all the way from 70 °N to 70 °S. It consists in an antenna element, a diplexer, an LNA, and a C-class HPA. Its gain is 0 dBi, EPIRB is about 12 dBW, G/T is −32 dBW, and its radiation pattern is omnidirectional, to cover over 85 % of the upper hemisphere above an elevation angle of 5°. The main specifications of low-gain and high-gain antenna subsystems are listed in Table 4.6. In Fig. 4.43 is illustrated a photography of a low-gain antenna with a C-class HPA and a DIP/LNA.
   The sample of small blade omnidirectional antenna of EMS Electronics for Inmarsat Class 15 or SwiftBroadband LGA is illustrated in Fig. 4.38a. This antenna is all that is required to provide connectivity with new Inmarsat SB200 services. It can be upgraded to higher speeds of Swift64 and SwiftBroadband with an IGA and HGA up to 200 Kb/s.

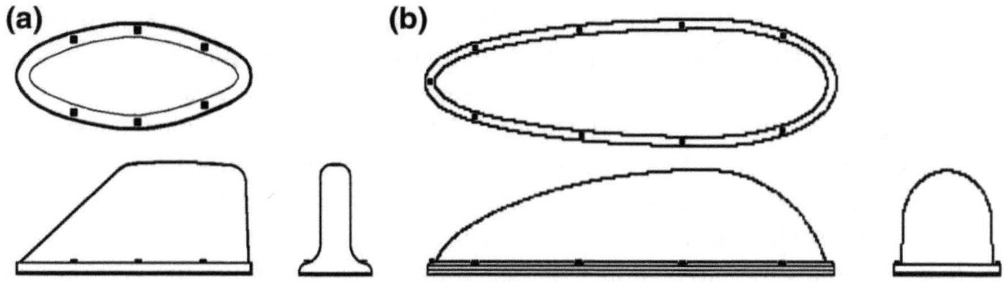

Fig. 4.41 Inmarsat LGA and IGA—Courtesy of Manuals: by Thrane and Thrane

A typical example of this antenna is the Capsat antenna of Sensor Systems Inc., consisting of a TT-3002A MSA (Jet Blade) and a TT-3001F HPA/LNA Pack. It is connected to the Thrane and Thrane TT-3024A Aero Inmarsat-C/GPS Capsat transceiver. Dimensions of the antenna are 116 × 297.4 × 108 mm, and the weight is 680 g. Working ambient temperature is between −55 °C and +85 °C at an altitude of about 21 km.

These particulars are for Inmarsat Aero-C standards, and the assembly drawing is presented in although some of the Aero-L standards have the same shape of antenna. The Aero-L antenna provides EIRP of minimum 16.5 dBW, G/T of −13 dBK, and 93 % coverage and can serve as the reversionary antenna below 7 dBW of EIRP for Aero-H configurations. The LGA system is suitable for aircraft that does not need voice or HSD communications. A LGA system consists of an antenna element, a DIP, an LNA, and a C-class HPA. Its gain is about 0 dBi, and its radiation pattern is omnidirectional to cover over 85 % of the upper hemisphere above an elevation angle of 5°.

2. **IGA System**—The intermediate-gain antenna (IGA) is a medium 2-axis mechanically steered phased-array IGA designed to support both aeronautical Inmarsat-I and -M. A typical example of IGA is the integrated TT-5006A innovative solution for Aero-I and -M installations on small- and medium-sized business aircraft, which can be used for both multichannel Aero-I and single channel Aero-M applications, shown in Fig. 4.41b. It is the smallest aero satellite antenna, which interfaces to the TT-5000 and TT-3000M series of Thrane and Thrane transceivers for Aero-I and -M, respectively, available for voice, faxes, prints, data transfer, or e-mail messages.

   This antenna contains its built-in navigational reference system, including the GPS antenna with magnetic and altitude sensing for stand-alone operations, angular sensors, and electronics for performing this task. All these functions are contained within the unit. The electrical characteristics of TT-5006A MSA are as follows: Gain is 6 dBi, EIRP is 14 dBW, G/T is −17 dBK, and axial ratio of minimum 6 dB with coverage volume of over 85 % above 5° elevation angles. Dimensions of this antenna system are 560 × 150 × 124 mm, and the weight is 2.2 kg. The antenna is the smallest combined NRS and MSC antenna in this class with an ultralight construction, which offers very low drag (less than 3 lb

drag at Mach 0.8 on a small jet). The latest state-of-the-art technology from the well-proven Thrane and Thrane IGA (Cobham) products provides ideal solutions for all short- and medium-haul business and military aviation.
The sample of the Thrane and Thrane TT-5006A IGA aeronautical subsystem, which can interface to HGA TT models, is an innovative solution for Aero-I and -M installations on small- to medium-sized business aircrafts, due to its integrated approach to functionality and installation, shown in Fig. 4.38b.
It incorporates the navigational reference functions, including the GPS aircraft antenna, magnetic, and attitude sensing. The antenna is mechanically steered with an ultralight construction, has a high-gain margin, and offers very low drag (less than 3 lb drag at Mach 0.8 on a small jet). With a short distance between the antenna and the LNA, an extremely high and superior coverage volume is achieved. However, the novel antenna pigtails interconnect approach used with the RF and GPS cables allows external antenna replacement without affecting the pressurization. Its features facilitate a single antenna installation, lower installation costs, installation flexibility, and transmitting data speed of support up to 300 Kb/s.

3. **HGA System**—The high-gain antenna (HGA) system is designed as a medium 2-axis electronically steered phased-array high-gain antenna. This antenna is designated to support aero Inmarsat-H, H+, HSD+, and Swift aeronautical standards. It may also provide service to new SwiftBroadband aeronautical standard. The two typical representatives of HGA products are as follows:
   (a) The solution of HGA configuration is designed to support Inmarsat Aero-H and H+ standards with which both provide channel rates up to 10.5 Kb/s. In addition, it supports multichannel voice, fax, and data transfer, while the Aero-H+ system optionally can also use the higher power of the Inmarsat-3 and Inmarsat-4 satellites when operating within the spot beam coverage area.
   (b) The combination of a HGA TT-5000HSD+ Thrane and Thrane, old Inmarsat GAN or new Inmarsat BGAN and Swift 64 advanced technology provides Aero-H+ capability together with a dedicated HSD satellite channel for up to 64 Kb/s of mobile packet data service (MPDS).

The HGA is usually a planar phased array of a number of identical crossed-dipole elements arranged in a grid. For instance, the CMA-2102 HGA of the Canadian Marconi Company has 39 crossed-dipole array elements, presented in Fig. 4.42a, and the T-4000 HGA array system manufactured by US-based Company Tecom has 44 crossed-dipoles, shown in Fig. 4.42b. Hence, both antennas provide simultaneous transmission and reception of full-duplex satellite signals, namely the Tx- and Rx-bands. Beam steering of the HGA is performed via serial transmission of phase shifter data from the beam steering unit (BSU).

The BSU converts tracking and pointing coordinates (relative azimuth and elevation of antenna) from the satellite data unit (SDU) into a signal needed to select the antenna array elements in combinations that point the antenna beam in the

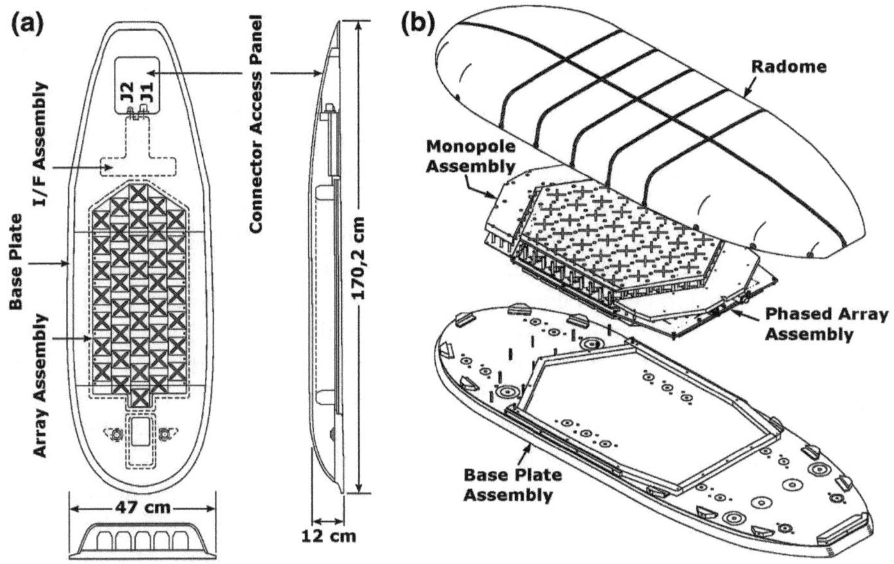

Fig. 4.42 Types of phased-array antenna for AES—Courtesy of Manuals: Canadian Marconi/Tecom

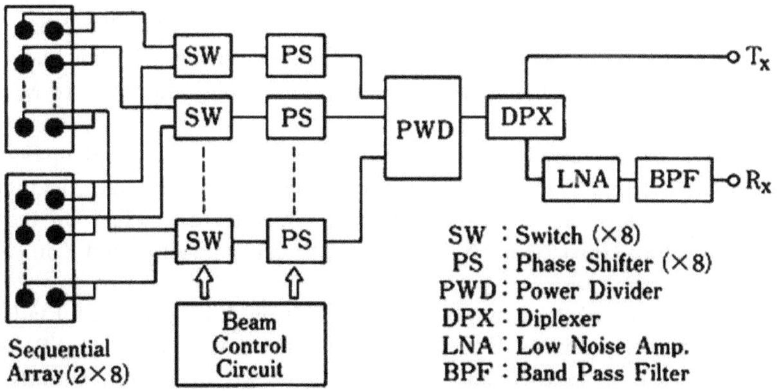

Fig. 4.43 Diagram of phased-array antenna—Courtesy of Book: by Fujimoto

desired direction. This form of beam steering is known as open-loop steering, which requires the SDU to receive accurate and timely navigation information. At any rate, all steering operations are transparent to the user since the antenna BSU/SDU combination ensures that the beam automatically points in the desired direction.

Diplexer/low-noise amplifier (DIP/LNA) enables an RF signal to be sent and received. The 3 RF ports antenna, transmit, and receive of the DIP/LNA provide signal routing and filtering functions. Signals in the Rx-band are routed from the antenna port to the Rx port, while Tx band signals are routed from the transmit port to the antenna port. The Rx path filters the Tx signal and other out-of-band signals to prevent the LNA and other Rx side components from being driven into nonlinear operation. The Tx path filters receive band signals so that noise and spurious signals from HPA do not increase the noise floor of the Rx. This unit establishes the noise floor by boosting the signals and noise received from the antenna to a level much greater than the noise level of subsequent components in the receive path.

The base plate of the antenna is a mount for the phased-array antenna and the radome and provides a mating interface to the aircraft airframe. The phased-array antenna is mounted to the antenna system base plate. The radome provides physical protection for the electrical components. It is of multilayer fiberglass construction and is painted with Teflon paint. This paint prevents ice buildup and resists common chemicals normally used around aircraft, such as de-icing fluid. The radome is outfitted with metallic strips, grounded to the base plate and fuselage, which serve to divert lighting and eliminate the disturbances of any antenna or avionics electrical components. Thus, the HGA low profile is aerodynamically designed radome that minimizes drag and eliminates icing. At this point, its design has been optimized for installation on long- and medium-haul passenger aircraft. Since it can be mounted anywhere along the top of the aircraft fuselage, it is suitable for a wide range of commercial and military airframe types. The typical performance characteristics of the HGA are as follows: Gain is between 12 and 17 dBi over 90 % of the Inmarsat hemisphere (IH), i.e., coverage or minimum of 9.5 dBi over 100 % of IH; EIRP is about 25.5 dBW over 92.6 % of IH; it assumes 40 W HPA and 2.5 dB of cable loss; G/T is about −13 dBK over 90 % of IH, and axial ratio is about 6 dB over 96.5 % of the industry (Inmarsat) hemisphere. Practically, HGA is known, as a high-gain antenna for aeronautical application, but actually, this type of MSA is a medium-gain antenna.

The system block diagram of phased-array antenna is illustrated in Fig. 4.43. Two array antennas were adopted to provide the required wide range coverage with high-gain beams, and ether one is selected to operate, depending on the flight direction. An array consists of 16 elements, with 2 in elevation and 8 in azimuth angles.

The sample of the EMS technologies AMT-700 HGA is designed for tail mount in housing radome, shown in Fig. 4.38c. This antenna is a complete ARINC 781 HGA and is used to receive and transmit voice and data from aircraft up to 432 Kb/s. It delivers the most gain of any high-gain Aero-H/H+, Swift64, or SwiftBroadband antenna currently on the market. The antenna weigh is less than 5 Pounds and achieves the maximum gain from its optimized array of helices, more than 13.5 dBic. The assembly of this mechanically steered antenna is made by two line replaceable units (LRU) and a diplexer/low-noise amplifier (DLNA) that meets stringent SwiftBroadband passive intermodulation (PIM) requirements. It operates with low-angle RF coverage >13.5 dBic gain of 100 % of Inmarsat coverage volume and up to −10° below horizon. Power supply options for antenna is +28 VDC, and nominal power consumption is 15 W (maximum is 30 W).

4.7.6.2 Classification of Inmarsat Aeronautical Communication Antenna

Airborne equipment for aeronautical satellite communication in the Inmarsat system is called the Inmarsat Aero terminal applications. As stated above, three types of antenna systems of LGA, IGA, and HGA are defined in the AES terminal systems. The G/T of aircraft antenna is required to be over −13 dB/K (gain is about 12 dBi) and −26 dB/K (gain is about 0 dBi). The HGA configuration is used for public voice, and high-speed data communication is about 21 Kb/s and up to 492 Kb/s, which can be connected to terrestrial public telephone networks. On the other hand, the LGA is used for low-speed data communications at 600, 1200, and 2400 b/s.

1. **Inmarsat Low-Gain Antenna (LGA)**—The LGA-3000 model of Inmarsat aero antenna produced by the US Cobham firm is a fuselage mount omnidirectional satellite antenna, designed to enable access to Inmarsat SB200 services, which is illustrated in Fig. 4.44a. The block diagram of the Inmarsat LGA-3000 antenna array is presented in Fig. 4.45a. A unique feature of this product is the integrated GPS receiving antenna that enables the Cobham SB200 solution to operate as a stand-alone system, without any inputs from aircraft navigation, or data control systems and the installations and connections of a third-party GPS antenna. In fact, this makes the SB200 system the most compact antenna solution for the provision of reliable and affordable voice and data connectivity and extends SwiftBroadband connectivity options right down to the smallest of airframes.
   It deals for Aero-C and other standards, has small footprint, aerodynamic profile, and simple installation, and has optimized coverage down to 20° elevation. It meets Inmarsat Class 15 technical specifications and is integrated GPS antenna for autonomous operation (no interface to aircraft navigation system required).
2. **Inmarsat Intermediate-Gain Antenna (IGA)**—The Cobham IGA-5001 antenna is a fuselage mount electronically steered, phased-array antenna, shown in Fig. 4.44b. The block diagram of this antenna is presented in Fig. 4.45b. It is a rugged and reliable product gaining from the field experience of the HGA-7000 and HGA-7001, the latter being the antenna of choice on new transport aircraft, fully compliant with the Inmarsat SDM, and is considered to

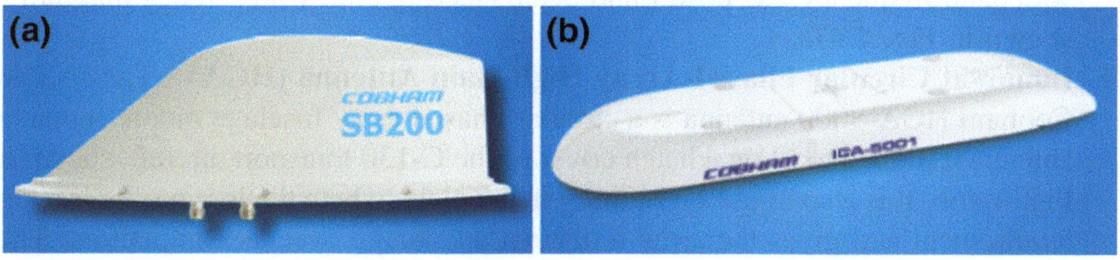

Fig. 4.44 Inmarsat LGA and IGA aircraft solutions—Courtesy of Manuals: by Cobham

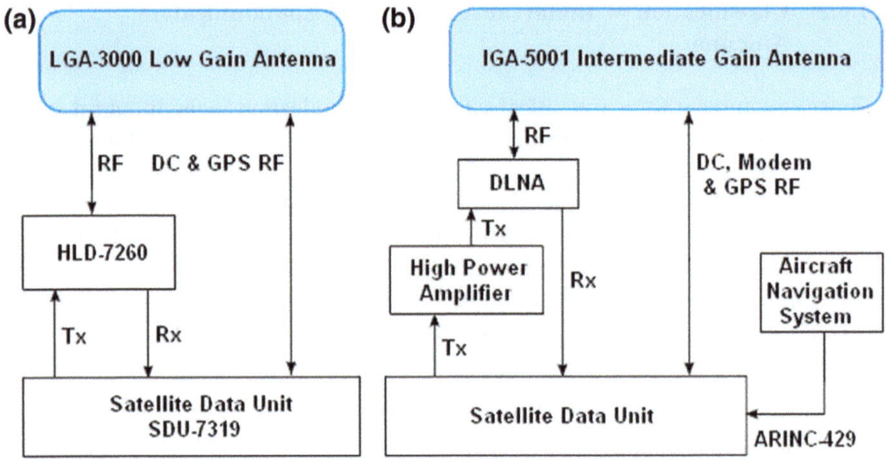

Fig. 4.45 Inmarsat LGA and IGA block diagram—Courtesy of Prospects: by Cobham

be the optimum antenna solution for Inmarsat's SwiftBroadband Class 7 service. It is the smallest IGA certified for both single- and dual-channel operations. Partnered with Cobham's SwiftBroadband systems, it makes a compelling case for moving Inmarsat voice and data services onto smaller aircraft types. This antenna has solid-state highly reliable design, low-profile, small-footprint, integrated GPS antenna, FAA TSO Certification, and Inmarsat approval service coverage up to 96.3 %.

3. **Inmarsat Phased-Array High-Gain Antenna (HGA)**—The Cobham HGA-7001 model is a fuselage mount low-profile antenna for Inmarsat (H/H+), Swift64, and SwiftBroadband systems, shown in Fig. 4.46a. This antenna complies and can operate with both ARINC 741 and ARINC 781 compliant avionics SDU. Its removable tail cap allows access to the connector compartment. In cooperation with various airframe manufacturers, a range of RF-matched mounting and adaptor plates have been designed for both fixed and flexible installations. It is Inmarsat SBB 7th-order PIM compliant, Inmarsat type-approved, is used on a range of aircraft, and has seamless coverage over Inmarsat-defined hemisphere up to 81.9 %. It uses DC power integrated with wideband frequency AC power and beam steering unit, whose block diagram is shown in Fig. 4.47a.
4. **Inmarsat Circular Phased-Array High-Gain Antenna (HGA)**—The specific Cobham HGA-8000 antenna is a circular phased-array fuselage mount Inmarsat antenna designed to fit on a hatch cover of the C-130 transport aircraft, shown in Fig. 4.46b. It is also in use on military helicopters, where the low profile is a key requirement and gives the array a distinct advantage over its competitors. This antenna is fully compatible with ARINC 741 standards and is compliant to

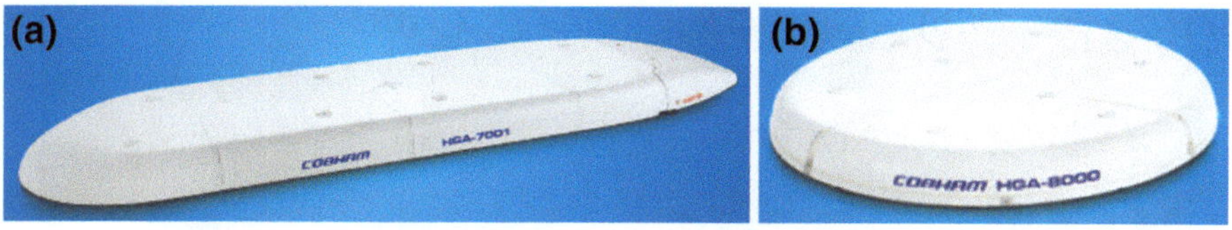

Fig. 4.46 Inmarsat HGA aircraft solutions—Courtesy of Prospects: by Cobham

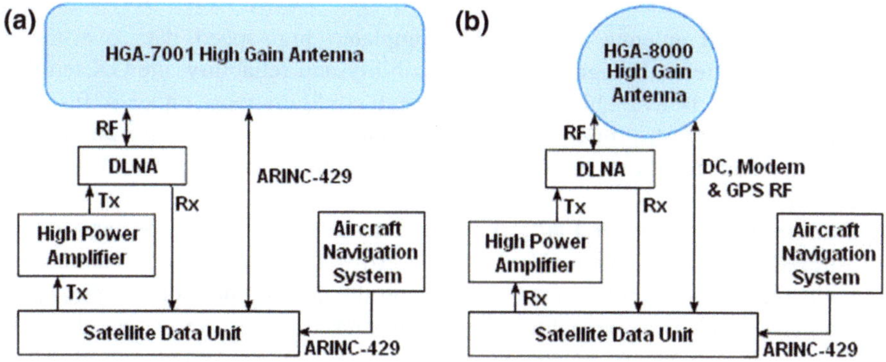

Fig. 4.47 Inmarsat HGA block diagram—Courtesy of Prospects: by Cobham

Classic Aero (H/H+), Swift64, and SwiftBroadband system specifications. It is certified for military applications and has Inmarsat approvals for seamless coverage over Inmarsat hemisphere up to 79.3 %. The block diagram of this antenna is presented in Fig. 4.47b.

5. **Inmarsat HGA ANT-50 for Helicopters**—The eNfusion Broadband of the Canadian EMS Electronics is a high-gain antenna used to receive and transmit voice and data from front view HELICOM with radome installed on board helicopters and rotor wing aircraft via the Inmarsat satellite network, which is illustrated in Fig. 4.48a, produced by EMS Electronics. The EMS antenna is Inmarsat SwiftBroadband Class 6 assessed and compatible with all ARINC 741 compliant satellite communications avionics. This antenna receives and transmits signals to and from Inmarsat satellites from up to 10° below horizon. This allows the helicopter to maintain a connection while maneuvering and near geographic obstacles. An approved ARINC Type F diplexer/DLNA unit provides an RF interface between the avionics and the antenna. An antenna control equipment (ACE) steers the antenna using ARINC 741 antenna-pointing vectors received from avionics.
6. **Inmarsat GX Airborne Fuselage Mount Antenna**—Similar to already described Inmarsat GX antenna for maritime applications, newly developed GX antenna by Honeywell for aeronautical applications uses the Inmarsat-5 satellite constellations at Ka-band RF inflight connectivity service, whose one-panel antenna is illustrated in Fig. 4.48b. In March 2015, Honeywell's GX Aviation

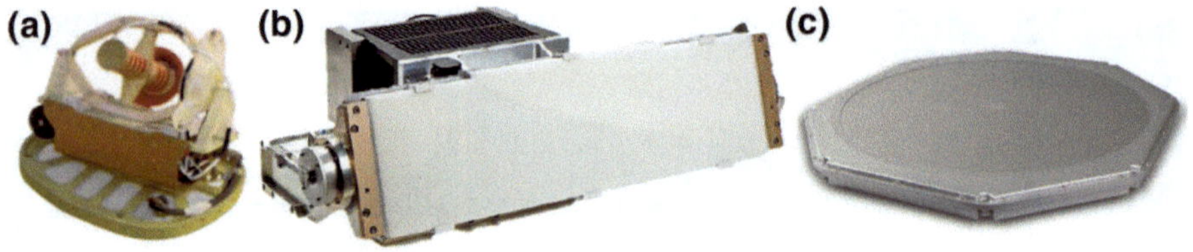

Fig. 4.48 Inmarsat GAHGA block diagram—Courtesy of Manuals: by EMS/Honeywell/Kymeta

hardware and antenna successfully completed high-speed data over-the-air performance tests. Designed to offer flexibility and reliability, the GX terminal with antenna is available in two different configurations, either a fuselage-mounted multilayer primary array aperture antenna system for larger aircraft or a tail-mounted reflector antenna system for smaller corporate and business aircraft.

7. **Inmarsat GX Airborne Flat Panel Antenna**—On the first day of exhibition at Aircraft Interiors Expo in Hamburg in 2015, Honeywell jointly announced collaboration with Inmarsat and Kymeta on the development of a higher-speed Ka-band wireless antenna for commercial aircraft and business jets. This airborne GX antenna complements Honeywell's JetWave aviation design that offers global broadband coverage powered by Inmarsat's GX Network and Inmarsat's new Ka-band i5 FX Aviation Network, which is illustrated in Fig. 4.48c. Thus, this GX airborne antenna also offers flexibility in design, such as a fuselage mount antenna system for larger business aircraft or a tail mount antenna system for smaller corporate and private aircraft.

4.7.6.3 Airborne Antenna Systems for VSAT Solutions

The DVB-RCS VSAT satellite technology is developed more than three decades ago and is taking big role in fixed satellite communications (FSC) and from recently in MSC applications. This service is providing broadcast, broadband, multimedia, and very fast Internet with IPTV via GEO satellites using different antenna solutions. The antenna serves just for transceiving facilities or can be integrated with IPTV reception. In general, the new VSAT broadcasting service provides digital VDVoIP. There is also possibility to install on board aircraft just antenna for receiving TV (TVRO) transmissions.

1. **VSAT Ku-band AL-1614 Tx/Rx Antenna**—This avionics DVB-RCS Tx/Rx antenna based upon a proven concept implemented on various applications of Orbit Technology Group is used over the last 15 years, whose antenna without radome is illustrated in Fig. 4.49a. Thus, this antenna is based on modular subassemblies such as follows: ACE, gear/motor/encoder assembly, and RF front-end, which are tested and proven to meet all necessary airborne

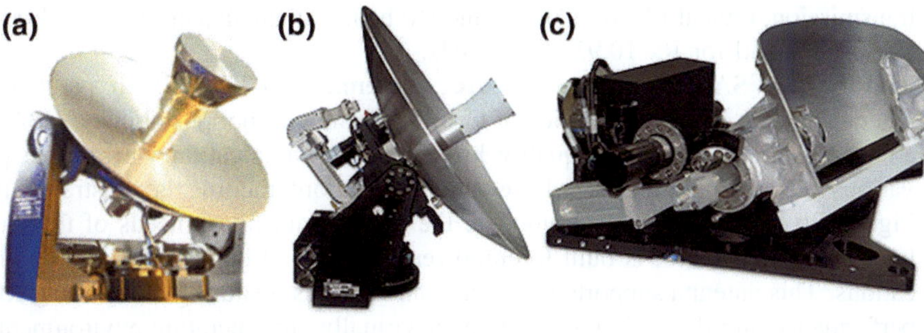

Fig. 4.49 Airborne DVB-RCS VSAT outdoor units—Courtesy of Manuals: by Orbit

environmental conditions. It provides access to satellite broadband networks anywhere, anytime under any weather condition while passengers are in the air, and has become essential. It is one of the first innovative stabilized VSAT Ku-band antenna solutions. Comprising a compact yet efficient dual-reflector antenna with an RF front-end delivering optimal nonstop quality connectivity with the selected satellite. The components of this antenna are assembled and integrated for this specific application in order to meet the required compact system design of airborne Ku-band AL-1614 Tx/Rx antenna, whose system layout of aircraft above and below fuselage is shown in Fig. 4.50.

The antenna unit complies with ARINC, ETSI, and FCC satellite regulations and provides the following key benefits and features: It is efficient "dual-reflector" antenna system, high EIRP of >44 dBW, G/T is >9.5 dB/°K, and the minimum dynamic tracking error meets aeronautic standard RTCA 160D and provides the typical data rate of 3 Mb/s in reception and 512 KB/s in

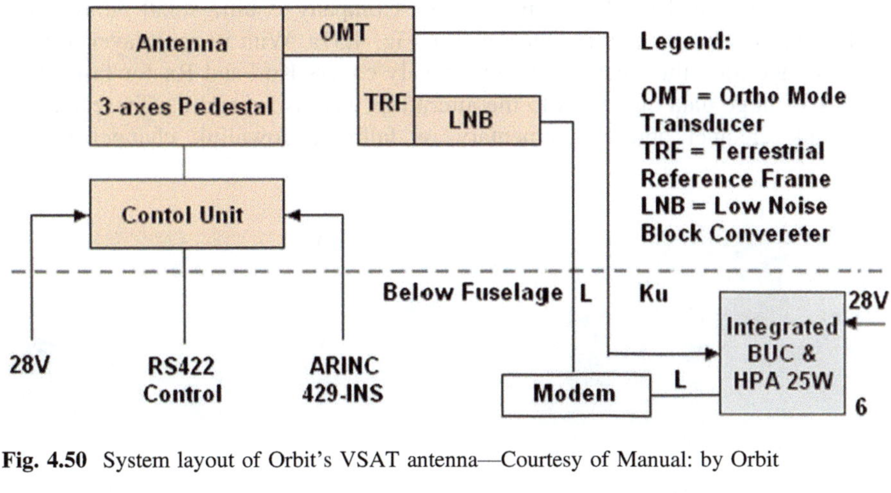

Fig. 4.50 System layout of Orbit's VSAT antenna—Courtesy of Manual: by Orbit

transmission, typical Eb/No is 6 dB, and RF bands of antenna are for Tx 14.00–14.50 GHz and for Rx 10.95–12.75 GHz.

2. **Multiband VSAT AirTRx 60 Tx/Rx Antenna**—AirTRx 60 cm parabolic antenna system is part of a series of Orbit innovative airborne stabilized VSAT antenna systems, providing quality broadband communication via satellite to various airborne platforms, whose antenna without radome is illustrated in Fig. 4.49b. Designed to accommodate the current and future needs of the airborne market, AirTRx is built to empower mission and business-critical applications. This antenna supports Ku-, Ka-, and X-bands, featuring outstanding RF performance and dynamic response under virtually any operating environment. Its features are as follows: Weight without radome and BUC is about 15 kg; swept volume is 70 cm (H) and 66 cm (D); G/T (typical, at mid-range, at 30° elevation, without radome) is 14.0 dB/°K (Ku-band), and 15.9 dB/°K (Ka-band); and EIRP (without radome) 52.7 dBW (Ku-band with 50 W BUC) and 48.5 dBW (Ka-band with 5 W BUC).
3. **Multiband VSAT AirTRx 25LP Tx/Rx Antenna**—AirTRx 25LP is a 25-cm elliptic antenna system, part of a family of innovative airborne stabilized VSAT antenna systems, providing quality broadband communication via satellite to various airborne platforms. Designed to accommodate the current and future needs of the airborne market, AirTRx is built to empower mission and business-critical applications. AirTRx™ supports Ku-, Ka-, and X-bands, or Ka/Ku auto switch, featuring outstanding RF performance and dynamic response under virtually any operating environment. Its features are as follows: Weight without radome and BUC is about 33 kg; swept volume is 29 cm (H) and 76 cm (D); G/T (typical, at mid-range, at 30° elevation, without radome) is 9.8 dB/°K (Ku-band) and 12.6 dB/°K (Ka-band); and EIRP (without radome) 43.8 dBW (Ku-band with 50 W BUC) and 43 dBW (Ka-band with 5 W BUC).
4. **Fuselage Mount Airborne Broadband Transceiving Antenna**—The modern antenna designed by the German QEST Company is ultra-small bidirectional Ka- and Ku-band antenna, illustrated in Fig. 4.51a. With its multilayer primary array aperture, this antenna simultaneously covers Ku-band Rx for fast downlink of large data quantities to the aircraft as well as L-band Rx/Tx services as uplink channel and supplementary or fallback downlink channel in one

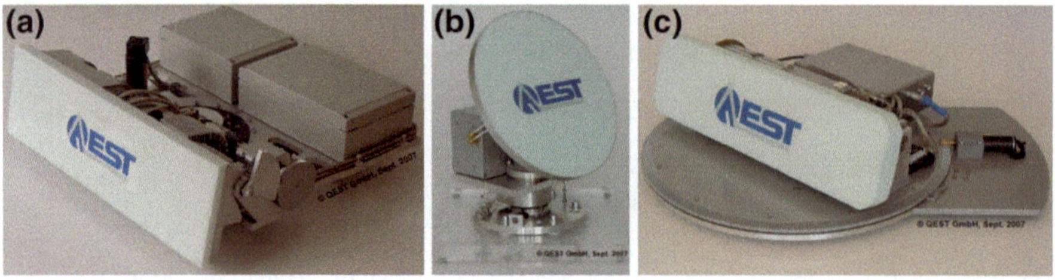

Fig. 4.51 Dual-band DVB-RCS transceiving and receiving antennas—Courtesy of Manuals: by QEST

integrated antenna. These result in numerous benefits, such as two antennas, smallest size, lightweight, including related services in one, physical shape without size increase, multisatellite operations and very low service and life cycle cost.

Antenna enables radically new service providing concepts: By combining Ku- and L-bands, a novel, hybrid data link can be established. In fact, Ku-band is used for a fast downlink of large data quantities over DVB-S and DVB-S2 quality to the aircraft, whereas L-band serves as uplink channel and supplementary or fallback downlink channel. However, this results in unprecedented connectivity options and maximum flexibility for owners and operators: combination of cost-efficient Ku-band downlink with globally available L-band connectivity. Asymmetric aircraft data link performs high-volume downlink combined with reasonable and affordable data rates in uplink, with avoidance of proprietary and expensive Ku-band uplink. This fuselage mount antenna is a real future-proof and secures your connectivity hardware investments, as it eliminates dependency of a proprietary Ku-band service provider or signal-encoding scheme. The special QEST technical solution is design of multilayer aperture for multiband antenna operations, which is shown in Fig. 4.52a. The QEST proposal of dual-band airborne broadband antenna coverage of Ku-band as well as L-band service simultaneously in one small antenna is shown in Fig. 4.6b.

The Ku-band frequency range is 10.7–12.75 GHz, antenna gain is 29.5 dBi, antenna G/T is >13 dB/K, data rates are up to 40 Mb/s per transponder Rx, coverage of more than 100 transponders simultaneously, and elevation coverage is 90°–0° (providing full performance for all elevation angles and regular flight maneuvers) including simultaneous Ku-band (Rx) and L-band (Tx + Rx) operation (multifrequency and multisatellite operation).

The L-band frequency range is 1.5–1.7 GHz, and thus, antenna gain and G/T comply with SwiftBroadband of Inmarsat specifications; data rates have SwiftBroadband capacity (432 Kb/s) per channel in Tx and Rx multichannel use possible, and elevation coverage is 90°–0° (full performance for all elevation angles and regular flight maneuvers) including simultaneous Ku-band (Rx) and L-band (Tx + Rx) operation (multifrequency and multisatellite operation).

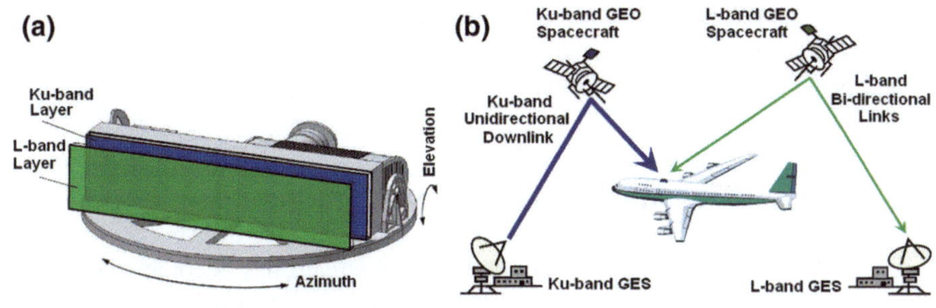

Fig. 4.52 Dual-band antennas and multiband GEO system—Courtesy of Manuals: by QEST

The diameter of this antenna is 64 cm (25.2 in), height is 14.5 cm (5.7 in), and weight is approximately 15 kg (33 lbs).

5. **Tail Mount Airborne Broadband Transceiving Antenna**—The specific design of the QEST Company is extremely small reflector bidirectional Ku-band satellite antenna, which is shown in Fig. 4.51b. The diameter of this antenna is 28 cm (11 in), height is 33 cm (13 in), and weight is approximately 10 kg (22 lbs).
6. **Fuselage Mount Airborne Broadband Receiving Antenna**—The modern design of the QEST Company ultra-small bidirectional Ku-band antenna is shown in Fig. 4.51c. It is dedicated for reception of large volume of data downlink and for high-performance airborne direct broadcasting satellite (DBS)/ Live TV receiving system.
   Based on similar core avionics satellite antenna technologies, it offers comparable benefits to the airlines such as ultra-small size, lightweight performance, efficient reception of high-volume DBS/Live TV (DSS, DVB-S, DVB-S2 up to HDTV quality), and low life cycle cost. The dimensions of this antenna are in diameter 40 cm (15.7 in), height is 10 cm (3.9 in), and weight is 12 kg (26.5 lbs). Antenna has RF range of 10.70–12.75 GHz, gain is 27.5 dBi, G/T is >11 dB/K, data rates are up to 40 Mb/s per Rx transponder with more than 100 transponders (>500 TV channels) simultaneously, and elevation coverage is 90°–0°.
7. **Multiband Airborne Broadband Transceiving Antennas**—The SOTM multiband transceiver airborne antenna of General Dynamics (GD) maintains the data link during dynamic platform motion for VDVoIP aeronautical onboard service in cockpit and cabins. This aircraft antenna provides reliable accuracy of satellite tracking via combination of on board aircraft integrated tracking Rx, gyro stabilization, and inertial measurement unit including attitude and heading reference system (AHRS). The model 13-13A series terminals, shown in Fig. 4.53a, provide high data rate at X-, Ku-, or Ka-bands (high- and low-band SOTM) over a wide range of operational conditions with an optional AHRS. It is lightweight and low-profile design with all RF components on elevation payload for high-efficiency RF Tx and Rx. This configuration includes the antenna positioner, tracking receiver, block-up and block-down converters, servo electronics, and high-efficiency solid-state power amplifier. Antenna is contained on the stabilized pedestal interfaced to input power, L-band Tx/Rx modem, and Ethernet. Another multiband antenna design of GD is model 13-17LPA, which is illustrated in Fig. 4.53b. It uses the same RF band 13-13A and provides the same performances and features as previous model.

4.7.6.4 Antennas for Aeronautical TV (TVRO) Broadcasting

Satellite broadcasting at the 12 GHz of Ku-band stream has become more widespread and there is a need for TVRO reception by relatively large- and medium-sized aircraft. This service is very attractive for both commercial and

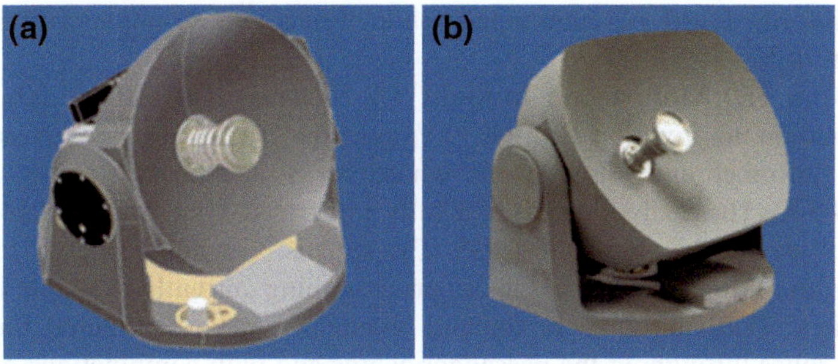

Fig. 4.53 Multiband aircraft antennas—Courtesy of Brochure: by GD

military applications including for entertainment of passengers in flight. The special digital TV and audio content distributed by GEO satellites are becoming the standard for TV reception on board aircraft worldwide. Since a satellite communication system covers a wide area of service, it is inherently suitable for mobile reception of two new transmission services known as direct audio broadcasting (DAB) and direct broadcasting satellite (DBS) in good propagation conditions compared to its terrestrial TV satellite broadcast counterparts, because the former suffers less interference due to multipath propagation as commonly experienced in practice. Thus, the TV satellite broadcasting service is deploying the following airborne satellite antenna systems:

1. **Tailwind 300 Multiregion In-flight TV System**—This antenna id designed by the Rockwell Collins Airborne TV ARINC as fully integrated system that enables passengers to receive Ku-band DBS TV program on board aircraft and access the spectrum of digital video and audio programming available from GEO DBS satellites, which antenna without radome is shown in Fig. 4.54a. The system takes these signals automatically, decodes them to the correct broadcast format, and then distributes them to the aircraft onboard audio–video or entertainment management systems for passengers in flight.
2. **Tailwind 500 Multiregion In-flight TV System**—This is the similar Ku-band ARINC antenna system for the executive jet business aircraft, which offers the kind of content-rich live programming from DBS satellites that allows you to

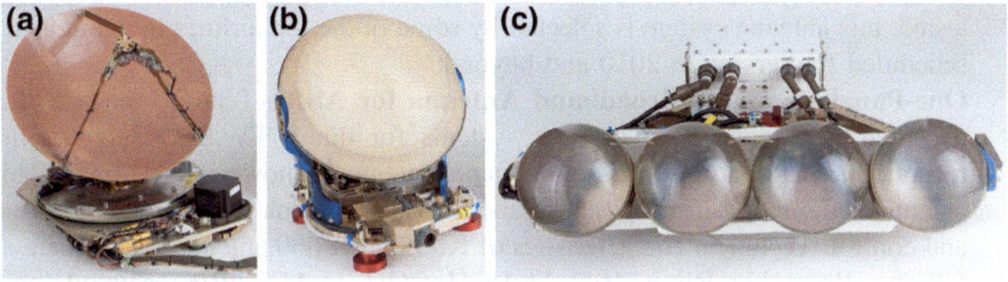

Fig. 4.54 Tailwind multiregion in-flight TV system—Courtesy of Manuals: by Rockwell Collins

stay informed and be entertained while flying, whose antenna in radome is shown in Fig. 4.54b.

3. **Tailwind 550 Multiregion In-flight TV System**—This robust airborne fuselage mount multiregion in-flight TV antenna system provides passengers in flight informed and entertained, whose array antenna system is shown in Fig. 4.54c. The Tailwind 550 is a fully integrated system that enables private aircraft to receive Ku-band DBS TV signals, providing access to the full spectrum of digital video and audio programming currently available from DBS satellites. These airborne broadband antennas can be configured for multiple flight zones throughout the passenger cabins and with combinations of up to 32 unique programming receivers providing flexibility to choose both personalized content and viewing location.
4. **Breakthrough Receiving Broadband Antenna for ASB**—The P&L International Inc. designed the next generation of airborne ASB antenna systems known as ZipPhaser-Air aviation antenna for all size aircraft broadband capability on board commercial airlines that need capability to connect over the ocean regions, which is illustrated in Fig. 4.55a. This antenna system being proposed offers the highest technology phased-array capability for DSB TV broadcasting onboard entertainment applications. The antenna system would install on the aircraft with a small low-profile footprint and radome with and astounding low weight. The modem router uses an iDirect X5 Evolution, and the antenna controller is included. The infrastructure on the aircraft can be wired or WiFi (wireless) and the choice is that of the airlines operator. The X5 router and HUB provide automatic antenna beam switching capability as the aircraft moves from one region to the next providing seamless transition for transatlantic flights. The TV broadcasting network ground stations are located in Italy, Florida, Singapore, and Hawaii. Moreover, the broadcasting satellite network is made up of a combination of Intelsat and Telesat Ku-band satellite transponders that downlink to the teleport infrastructures. Installation of this antenna would be simple with minimum effects to the aircraft.

   Today instant and uninterrupted airborne communications is a must globally employing the breakthrough technology. However, the ZipPhaser-Air aviation antenna system is the only low-profile, 2-way, Ku-band antenna system for all aircraft. Based on a unique flat panel technology, this antenna system has been especially designed to meet airlines expectation to provide reliable connectivity and global coverage at unprecedented speed and quality. Proved and flight tested, this antenna system is selected by some of the top airlines in the industry scheduled for service in 2010 and beyond.
5. **One-Panel Receiving Broadband Antenna for ASB**—Tecom Industries Inc designed KuStream 2000 antenna systems for direct TV broadcast satellite reception at Ku-band RF, whose one-panel antenna is shown in Fig. 4.55b. This antenna is mounting inside fuselage and provides antenna positioning command and control. Technical characteristics of KuStream 2000 antenna systems are as follows: Receiving RF is 10.7–11.4 GHz; G/T is 11.3 dB/K; azimuth gain

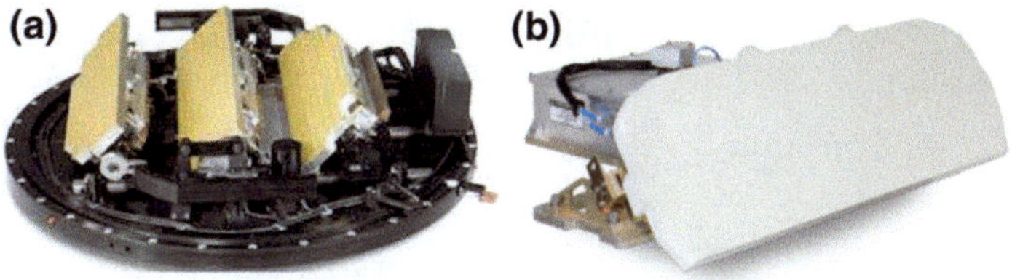

Fig. 4.55 Aircraft TVRO solutions—Courtesy of Prospect: by P&L/Tecom

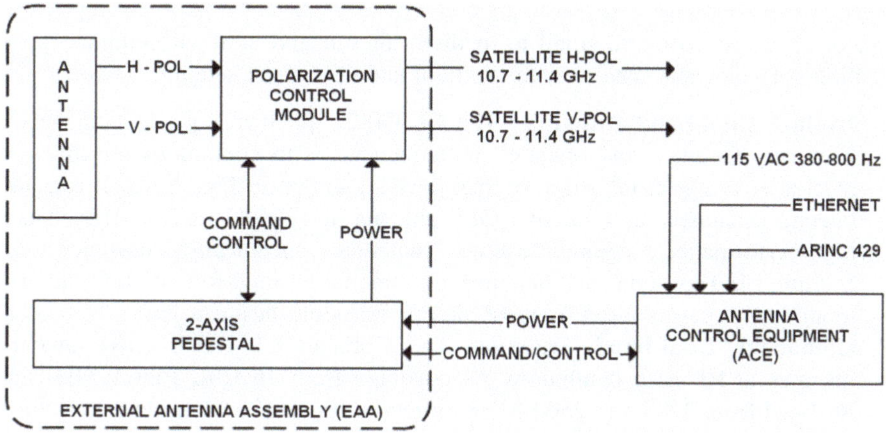

Fig. 4.56 Block diagram of KuStream 2000 antenna systems—Courtesy of Manual: by Tecom

pattern for first sidelobe is −17 dB at 5°, and grating lobe is −25 dB; antenna pedestal positioning range for elevation is 0°–90° and for Azimuth is 0°–360° continuous; satellite tracking accuracy is better than 0.48° combined with elevation and azimuth error; power supply of 115 VAC of 380–800 Hz; and its dimensions are 11.8 × 8.1 × 3.0 inches (W × L × H). This antenna is unique cavity array with suspended airstrip line feed network with superior G/T; it eliminates pointing error drifts and is designed for FAA certification. The block diagram of this antenna with external antenna assembly (EAA), and ACE is presented in Fig. 4.56.

4.7.6.5 Airborne Antennas for Non-GEO Satellite Systems

In this type of satellite antennas for aeronautical communication and tracking application can be included satellite antennas for non-GEO solutions such as Iridium, Globalstar, Orbcomm, and other satellite systems. These types of antennas

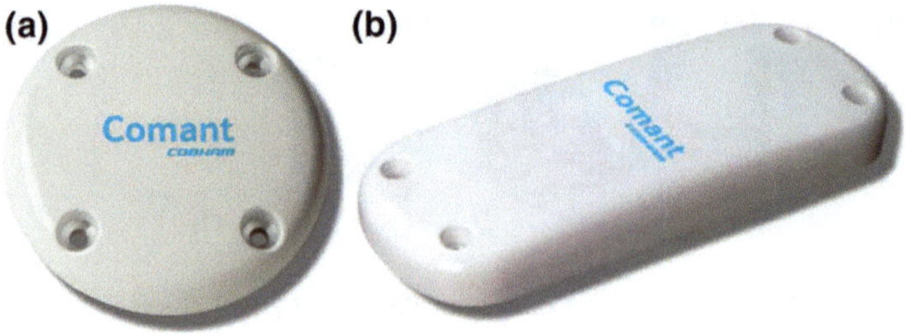

Fig. 4.57 Aircraft antennas for LEO satellite systems—Courtesy of Manuals: by Cobham

can be mounted on board small to medium aircraft and used for communication facilities of voice and data over the Iridium and Globalstar satellite networks.

1. **Iridium Dual-band Antenna**—The CI 490-22 antenna is product of the US Company Cobham and operates at frequencies with continuous transmit and receive coverage from 1616 to 1626.5 MHz, shown in Fig. 4.57a. It can also operate separately as a passive GPS antenna at 1575.42 ($\pm$10) MHz. It is a high-performance communications antenna specifically designed for Iridium LEO systems and features the standard round-format footprint and mounting. It can be used by most aircraft including business jets.
2. **Globalstar Dual-band Antenna**—The Cobham CI 480-1 active antenna operates at RF with continuous Tx coverage from 1610 to 1626.5 MHz and Rx-band from 2483.5 to 2500 MHz, shown in Fig. 4.57b. Receive side features a 29 dB gain amplifier with high-performance filtration and a built-in voltage regulator for DC bias operation of +3.3 and +28 VDC. This antenna is standard equipment with Northern Airborne Technology's STX 100 Airborne Satellite Communication System, capable of serving both voice and data operations. It can be installed on most aircraft including business jets.

4.7.6.6 Airborne Satellite Antennas for Unmanned Aircraft Systems (UAS)

Commercial and military Unmanned Aircraft Systems (UAS), Unmanned Aerial Vehicles (UAV), High-Altitude Platforms (HAP), or Stratospheric Platform Systems (SPS), either manned or unmanned, are often capable of long endurance time while carrying significant payload weight. Satellite communications fully exploit all platform capabilities, supporting intelligence, surveillance and reconnaissance (ISR), TT&C, communication, navigation and surveillance (CNS), Earth observation, remote sensing and other transmission links and missions beyond line of sight (LoS).

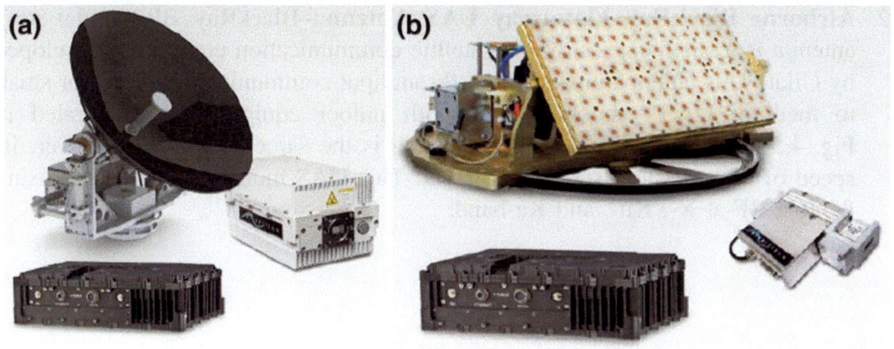

Fig. 4.58 Aircraft antennas for UAS applications—Courtesy of Manuals: by Gilat

These manned and unmanned terminals equipped with an adequate on board satellite and other electronic devices with antennas utilize commercial or government GEO satellite capacity to provide full-duplex satellite communication, linking customers via satellite and GES facilities. The forward link provides command and control capabilities, while the return link transfers sensor data. The antenna and modem may be installed as a unified unit or as separate components.

1. **Airborne BlackRay Parabolic UAS Antenna**—BlackRay satellite communication unit with parabolic reflector antenna is a compact lightweight airborne terminal, whose antenna with indoor equipment is illustrated in Fig. 4.58a. It comprises best-of-breed satellite technologies, all developed and manufactured by Gilat, which can be easily tailored to the UAS customer's needs. Main subassemblies are as follows: parabolic carbon fiber pointing antenna; high-performance satellite modem; and power-efficient BUC/SSPA. Thus, this system can transmit over 20 Mb/s from the UAS for any IP-based VDV beyond LoS applications. The BlackRay parabolic antenna provides spectrum-efficient IP connectivity, adaptive in real time to varying link conditions. Network implementation is provided straightforward by PAMA and DAMA, installing the GLT1000 (commercial grade) or MLT1000 (ruggedized military grade) modem with antenna in any teleport infrastructure or transportable HUBs.
   Antenna terminal is available in 2- or 3-axis gimbals, while the 3-axis mode may eliminate the keyhole effect when operating near the equator. The characteristics of Ku-band antenna are as follows: Antenna size is 60 cm; RF Tx is 13.75–14.5 GHz; RF Rx is 10.95–12.75 GHz; Tx gain is 36 dB; EIRP is 51 dBW (40W BUC); G/T is 14 dB/K; elevation is 0°–90° (operational elevation is $<80^{\circ}$; azimuth 360° continuous; and antenna size is 62L × 62 W × 62H cm (25L × 25 W × 25H in). The difference characteristics of Ka-band antenna are as follows: RF Tx is 29–31 GHz; RF Rx is 19.2–21.2 GHz; Tx gain is 43.5 dB; EIRP is 54 (12 W BUC); and G/T is 18 dB/K.

2. **Airborne BlackRay Flat-array UAS Antenna–**BlackRay BR300 flat-array antenna is a small-size, airborne satellite communication equipment developed by Gilat. This antenna enables high-throughput communication, even for small to medium UAS, whose antenna with indoor equipment is illustrated in Fig. 4.58b. The efficiency of this antenna is the same as above; however, its speed of transmission is 10 times lesser. This UAS multiband antenna is using flexible RF at X-, Ku-, and Ka-band.

Chapter 5
Propagation and Interference Consideration

Propagation and interference characteristics are very important for providing quality and reliability of MSC propagation channels. The quality of service (QoS) can be expressed in terms of the bit error rate (BER) performance, which depends on the carrier-to-noise density ratio C/N_0; meanwhile, the service reliability is manifested in the relation of service availability. The intervening medium between MES within the network and satellites is termed a transmission channel. The fixed satellite services have two constant channels between a minimum of two FES using the same spacecraft, while in the GMSC network there are two types of channels to be considered: the variable channel between the MES and satellite or service link and the constant channel between the LES (Gateway) and the satellite or feeder link. These two channels have many different characteristics, which need to be taken into account during the system design examination. The more critical are the variable channels used by MES, since transmitter power, receiver gain, and satellite visibility is sometimes quite restricted in comparison with the constant link. In this case, the propagation path profile between all kind of MES and satellites varies continuously whenever the mobiles (ships, land vehicles, or airplanes) are in motion. These variations are most significant and frequent in the case of maritime and aeronautical environments.

The common satellite channel environment affects radiowave propagation in changeless ways. The different parameters influenced are mainly path attenuation, polarization, and noise. The factors to be considered are gaseous absorption in the atmosphere, absorption, and scattering by clouds, fog, all precipitation, atmospheric turbulence, and ionospheric effects. In this sense, several measurement techniques serve to quantify these effects in order to improve reliability in the system design. Because these factors are random events, GMSC system designers usually use a statistical process in modeling their effects on radiowave propagation. To design an effective GMSC model, it is necessary to consider the quantum of all propagation characteristics, such as signal lost in normal environment, path depolarization causes, transionospheric contribution, propagation effects important for mobile

S.D. Ilčev, *Global Mobile Satellite Communications Theory*,
DOI 10.1007/978-3-319-39171-7_5

systems, including reflection from the Earth's surface, fading due to sea and land reflection, signal blockage, and to the different local environmental interferences for all mobile and handheld applications. At any rate, the local propagation characteristics on the determinate geographical position have very specific statistical proprieties and results for ships, vehicles, and aircraft.

On the other hand, for all navigation systems based on radio waves a sound knowledge of wave parameters is crucial to enable us to make an accurate position determination. The systems used for radio navigation operate at frequencies between 10 kHz and 10 GHz, and the presentation is therefore limited to this frequency range. As will be seen, there are wide propagation differences. Radio waves are often described as having the properties of light and can be described by traditional "wave" physics. Another important facet of radio engineering is the geometry and the relationship between the transmitters and receivers, and the Earth and space through which the radio waves are propagating.

5.1 Overview of Antennas for Radio and Satellite Communications

Antennas act as converters between conducted waves and electromagnetic (EM), electric, and magnetic waves propagating freely in space, which diagram is shown in Fig. 5.1. Thus, transmitter (Tx) is radiating EM via antenna and receiver (Rx) is detecting these EM affected by the propagation impairments. The name of antenna is borrowed from zoology, in which the Latin word antennae are used to describe the long, thin feelers possessed by many insects. In Fig. 5.2, it can be seen that the antenna is an important element in any radio or satellite system because it acts like a link of a chain. In such a way, the overall transmissions performances of radio waves in open space are significantly influenced by the performance of transmit and receive antennas.

At first glance, modern antennas may still look very similar to the ancient model developed by Hertz and Popov. However, they are nowadays optimized at great expense for their intended application. Communications antenna technology

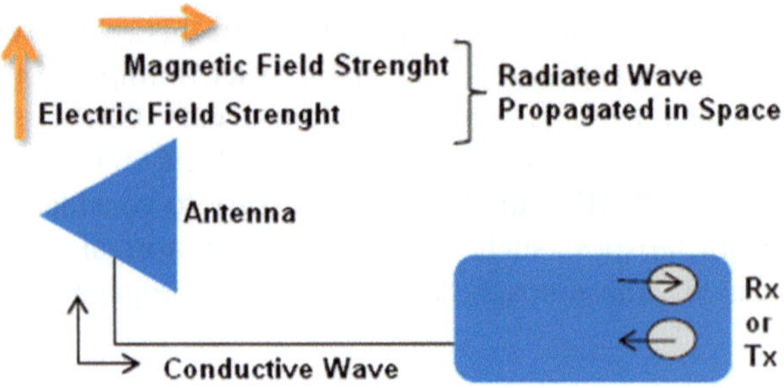

Fig. 5.1 Basic antenna functionality in propagation—Courtesy of Manual: by Rohde and Schwarz

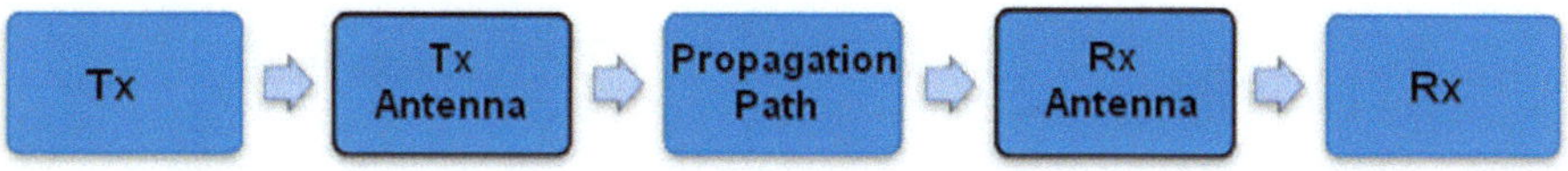

Fig. 5.2 Block diagram of radio link—Courtesy of Manual: by Rohde and Schwarz

primarily strives to transform one wave type into another with as little loss during propagation as possible.

This requirement is less important in the case of test antennas, which are intended to provide a precise measurement of the field strength at the installation site to a connected test receiver; instead, their physical properties need to be known with high accuracy. The explanation of the physical parameters by which the behavior of each communication antenna can be both described and evaluated is probably of wider general use.

The principal values used to determine the overall loss value for a wireless communication link include the free-space loss, in addition to any feed line loss at the transmitting and receiving ends of the link, which can be negligible when compared to the free-space segment losses.

In most cases, it is impractical to install an antenna that tracks the movements of these spacecraft, and in this point, omnidirectional antennas are normally used. Power levels for these types of installations must be carefully calculated so that a sufficient level of energy reaches the satellite regardless of its position in the sky at any given moment.

Communication with GEO satellite is almost always done using a directional parabolic reflector antenna. Since it is directional, the antenna must be accurately pointed toward the satellite and then fixed in place. Modern mobile satellite antennas have tracking mechanism that is constantly keeping GEI satellite in focus.

All GEO satellites rotate about the Earth above the equator, and at the same altitude, in the region of space known as the Clarke Belt. Before attempting to align an antenna with GEO satellite, it is useful to imagine what the Clarke Belt would look like if observer could see it in the sky from different locations on Earth.

The Clarke Belt forms an imaginary arc in the sky, which imagine in Fig. 5.3 is showing azimuth and elevation angles of antennas in Local Horizontal plane in the geographical north. If an observer is at the equator, the belt passes directly over his overhead. If an observer is in the Northern Hemisphere, the belt stretches from the horizon a little south of due east, rising to its maximum height directly toward the south, and then reaching the horizon a few degrees south of due west. If an observer is in the Southern hemisphere, the maximum height of the arc appears directly north of his location.

The true azimuths of all GEO satellites than are visible from a location in the Northern Hemisphere range from a little more than 90° to a little less than 270°. For locations in the Southern Hemisphere, the true azimuths are greater than 270° or less than 90°.

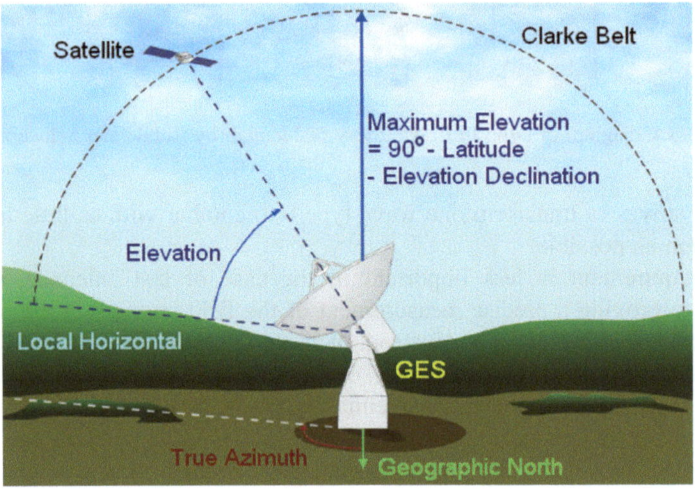

Fig. 5.3 Clarke Belt and antenna look angles in the Northern Hemisphere—Courtesy of Manual: by Rohde and Schwarz

The elevation of GEO satellite can range from a little more than 0° in local horizon to a maximum elevation that depends on the latitude of the Earth station antenna. For antenna at the equator, the maximum possible elevation is equal to 90° minus the latitude (φ) minus the apparent declination (d).

When the elevation angle to a GEO satellite is high, the path of the signal through the atmosphere is relatively short and degradation due to the atmosphere is low. However, when the elevation angle is low, the transmission signal follows a much longer path through the atmosphere and degradation is higher. The effect is significant below 10° elevation and very severe below 5°.

5.2 Propagation Fundamentals

Before to introduce propagation of electromagnetic (EM) waves, the best way to start is with providing elaboration of the following basic definitions:

1. **Radio Frequency (RF)**—This is a number of cycles per second or hertz (Hz) of a radio wave using in transmission systems.
2. **Transmitter (Tx)**—This device converts electrical signals into emitted RF energy. It is the source of a radiated wave or perturbation transmitted via antenna, which is illustrated in Fig. 5.4a.

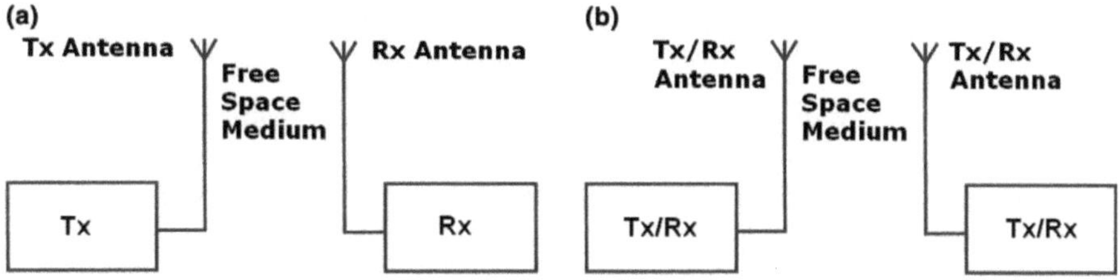

Fig. 5.4 Basic Tx, Rx, and Tx/Rx radio equipment—Courtesy of Book: by Stacey

3. **Receiver (Rx)**—It is equipment that receives RF energy via receiving antenna and converts it back into electrical signals, which is illustrated in Fig. 5.4a.
4. **Transmitter/Receiver (Tx/Rx)**—This is an integrated device containing transmitter (Tx) and receiver (Rx) and using common Tx/Rx antenna systems, which is shown in Fig. 5.4b.
5. **Propagation**—It is the ability of a radio wave to travel between a transmitter and a receiver in free space or in another medium such as air.
6. **Aerial or Antenna**—This can be considered as a part of the transmitter or receiver system (or it can be in both). It is usually a passive device that converts electrical signals straight into radio waves.

However, one of the main objectives in designing a communication system is to guarantee a specific quality of service, which depends on given user requirements. A key point for this design is the quality of the channel that is present in the link. This channel varies in time, and in order to perform an estimation of the losses that the signal will experiment, it is necessary to consider all the factors influencing the signal attenuation. Thus, a thorough understanding of the propagation mechanisms influencing radio waves propagating in the channel is therefore essential. The propagation mechanisms affecting the link produce various effects in the signal, which vary depending on the operating frequency. It is important to emphasize considerable differences in the models operating at low frequency bands in relation to the ones operating in upper bands.

Once these propagation effects have been studies, determined, and understood, an analysis of their nature and statistical behavior is performed, so that they can be represented and characterized through statistical models. In fact, by using these models, it is then possible to develop more realistic channel models, which are then applied when designing radio and satellite communication systems including navigation networks stratospheric platforms. For the case of communication systems in general, it is necessary to consider environmental conditions defined by the scenario, as well as the phenomena affecting the link. A model, which clearly characterizes signal affectations, will allow implementation of reliable and truthful evaluation schemes.

5.2.1 Electromagnetic Vectors and Polarization

From basic physics, radio waves are considered to be EM in origin. The EM waves have three perpendicular vectors describing their generation. An electrical vector (**E**), a magnetic vector (**H**), and the propagation vector (**V**) have to be considered as shown in Fig. 5.5.

A practical consideration here is the term polarity or polarization sense regarding radio wave propagation. A general convention used in radio engineering is when talking about polarization, where the electrical vector is being discussed. Vertical polarization is when the electrical **E** field is excited in a vertical direction relative to the Earth's surface, shown in Fig. 5.6a. Horizontal polarization is when the electrical **E** field is excited in a horizontal direction relative to the Earth's surface, shown in Fig. 5.6b. Circular polarization is a hybrid of both vertical and horizontal polarity where the direction of the electrical field excitation revolves between the two planes going out along the axis of propagation, where V changes with propagation, see Fig. 5.6c.

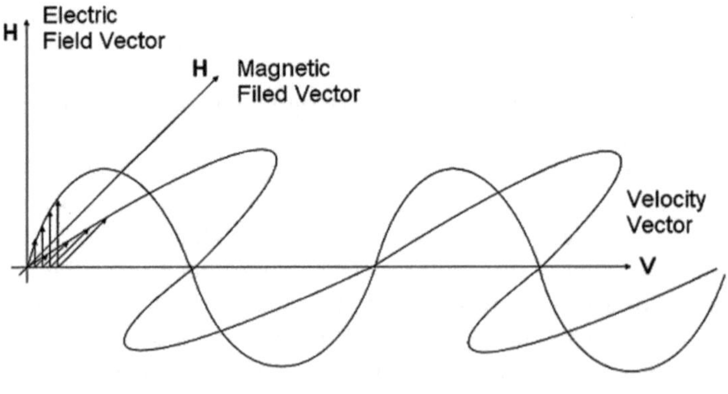

Fig. 5.5 EM wave propagation vectors—Courtesy of Book: by Stacey

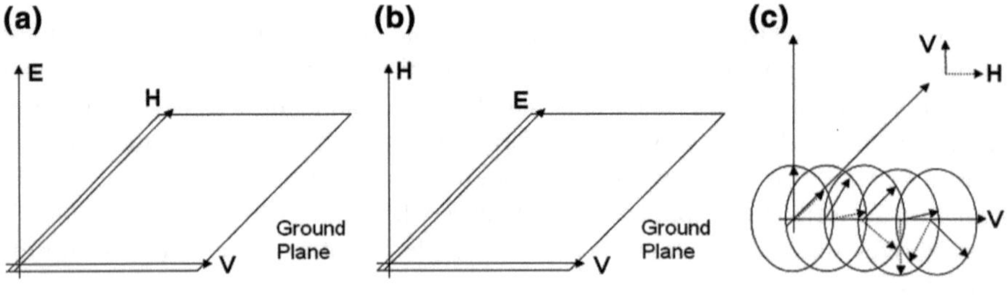

Fig. 5.6 Vertical, horizontal, and left-hand circular polarization conventions—Stacey Courtesy of Book: by Stacey

5.2.2 *Speed of Propagation and Relationship to Wavelength and Frequency*

Electromagnetic waves travel at the same speed as light in a vacuum expressed by:

$$v = f\lambda \tag{5.1}$$

where v is the velocity in meters per second (m/s), f is the frequency in hertz (Hz) or cycles per second, and λ is the wavelength in meters (m).

In particular, in free space with vacuum conditions, v has been found to be a constant denoted by *c*:

$$c = 3 \times 10^8 \mathrm{m/s} \tag{5.2}$$

5.2.3 *Radiowave Propagation*

Radio waves travel in space as electrons at the speed of light approximately 300,000,000 m/sec. When radio waves are propagated from a transmitter antenna, they form three modes of propagation path: surface, sky, and space wave propagation, shown in Fig. 5.7a. Radio waves are generally transmitted from an antenna omnidirectionally but transmission can be also modified in a directional path by using a directional dish or antenna arrays.

The surface wave is a radio wave that is modified by the nature of the terrain over which it travels. The surface wave propagation predominates at all radio

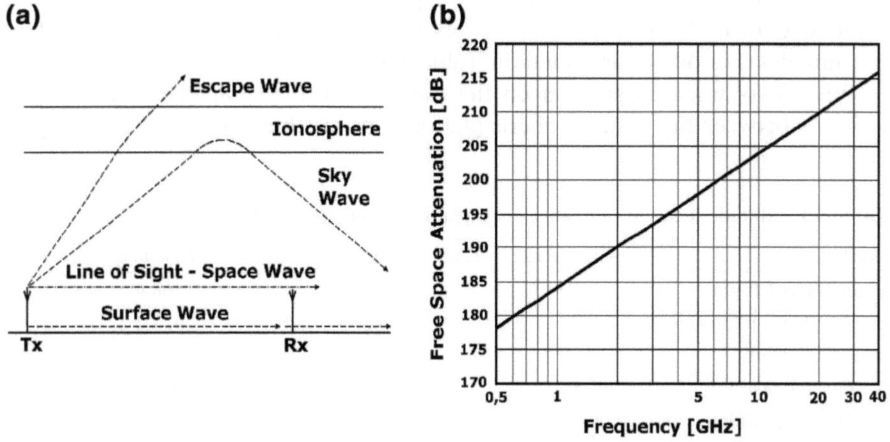

Fig. 5.7 Radiowave modes of propagation and free-space loss—Courtesy of Books: by Tetley/Ohmori

frequencies up to 3 MHz for VLF-, LF-, and MF-bands. Hence, sky waves are severely influenced by the action of free electrons, called ions, in the upper atmosphere, known as the ionosphere and are caused to be attenuated and reflected and possibly returned to Earth. In fact, sky wave propagation predominates in the HF-band between 3 and 30 MHz.

Above 30 MHz, the predominant mode of propagation of radio waves is by the space wave. This wave, when propagated between 30 and 300 MHz into the troposphere by an Earth VHF or UHF radio station, is subject to deflection by variations in the refractive index structure of the air through which it passes, which will cause the radio wave to follow the Earth's curvature for short distances up to about 100 km. However, space waves above 300 MHz propagated upwards, away from the troposphere, may be termed free-space waves and are primarily used for satellite communications.

While all transmitting antenna systems produce one or more of the three main modes of propagation, one of the modes will predominate. The predominant mode may be equated to the frequency used, as all other constraints remain constant. Therefore, for the purposes of this explanation of propagated radio waves, it is assumed that the mode of propagation is dependent upon the frequency used in an adequate system, for that is the only parameter, which may be changed by an operator.

The simplest situation involving the transmission of information by EM waves is when the Tx antenna transmits directly to the Rx antenna, without any obstacles in the path through which the EM waves must travel. This situation occurs when both antennas are located at a relatively short distance on the Earth's surface with full mutual visibility. As the distance between the two antennas increases due to the Earth's curvature, the EM waves are no longer in a direct path (line of sight).

In order to increase the path length, the antenna can be fitted on a higher mast but only to the extent that the costs do not become prohibitive. In this way, the problem can be solved by means of forthcoming stratospheric platforms or communications satellite payloads. Since these satellites are placed at very high altitudes, their reach is far superior to that of the higher towers that can be built on the Earth's surface, see altitudes and other parameters in Table 5.1.

When two antennas situated on the Earth are very far apart and there is no visibility between them, stratospheric platforms or satellites can be used to relay the

Table 5.1 Altitude location of stratospheric platforms and spacecraft

System	Altitude (h) in Km	Angle in degrees	Distance (d) in Km
International flight	10	3.313	368.4
Stratospheric platforms	30	5.549	617.1
Amateur satellites	150	12.320	136.2
LEO satellites/low	780	27.008	3003.6
LEO satellites/high	2000	40.438	4497.1
MEO satellites	10,000	67.095	7461.7
GEO satellites	35,600	81.268	9037.8

signals. There is otherwise visibility between the Tx antenna and the spacecraft (platforms) and vice versa, between spacecraft (platforms) and Rx antenna.

At this point, the ways these signals will travel on their line of sight depend on propagation characteristics and conditions of interference in the hypothetical environment. Because communications satellites are located at higher altitudes relative to the Earth's surface than stratospheric platforms and masts, they cover a wider area, which is beneficial for ocean spaces, continents and for countries with large dimensions.

5.2.4 Derivation of Free-Space Path Loss

It is necessary to define the loss between Tx and Rx antenna separated by a distance from the transmission medium, assumed to be a vacuum and the antenna is isotropic. It may serve either as Tx or Rx antenna in the radiation field where it is situated.

Considering that the isotropic antenna radiates signals in all directions of its spherical surface with total power flow P_T in Watts, including receiving antenna gain, at a sufficiently large distance (d) in meters away from the center to the surface of the sphere, the power flux per unit area through any portion of the spherical surface must be as follows:

$$P_{fd} = P_T/4\pi d^2 \tag{5.3}$$

To introduce free-space path loss has to be considered a radio link that starts at point P_{Tx} and is being received at point P_{Rx}, and assuming an isotropic transmitting source P_{Tx} (Watts) and, in addition, has to be considered the power passing through a unit aperture at distance *d* (meters) from the isotropic source (i.e., it needs to consider this area as separate on the surface of the propagation sphere).

The amount of power that the receiving antenna absorbs in relation to the RF power density of the EM field is determined by its effective aperture, which is defined as the area of the incident EM wave front that has a power flux equal to the power dissipated in the load connected to the receive antenna output terminals. Following the previous (Eq. 5.3) equation, the receiving power at a receiving antenna can be expressed by

$$P_{Rx} = P_T(\lambda/4\pi d)^2 = P_T/L_f \tag{5.4}$$

where λ = wavelength and L_f = free-space transmission loss, which can be given in dB as follows:

$$L_f = 10\log_{10}(P_T/P_{Rx} = 32.40 + 20\log f_{MHz} + 20\log d_{km} \tag{5.5}$$

where f = frequency of the emitted radio field in dB and d = distance in km between Tx and Rx antenna. Thus, free-space loss is related to operating frequency and transmission distance. Figure 5.7b shows free-space loss as a function of frequency in GHz for a distance of 36,000 km.

The power flux passing through this an area on the surface in respect of Tx power is as follows:

$$P_{fd} = P_{Tx}(\lambda/4\pi d)^2 \quad (5.6)$$

Power flux density can be also denoted as Ψ (W/m^2).

Now it has to consider an isotropic receiving antenna placed on this sphere. An isotropic receive antenna located on this sphere will absorb power from the radiation field it is situated in. The amount of power that the receiving antenna absorbs in relation to the RF power density of the field is determined by the effective aperture of the antenna in square meters. For an isotropic antenna, the effective aperture A_c is given by

$$A_c = \lambda^2/4\pi \quad (5.7)$$

Therefore, the received power into an isotropic antenna is

$$P_{Rx} = \lambda^2 P_f/4\pi \quad (5.8)$$

Substituting Eqs. (5.6) and (5.7) into (5.8) is giving the following relation:

$$P_{Rx} = \lambda^2 P_{Tx}/(4\pi d)^2 \quad (5.9)$$

Taking logs of each side of the equation is coming as follows:

$$10\log_{10}P_{Tx} - 10\log_{10}P_{Rx} = 20\log_{10}(4\pi) + 20\log_{10}(d) - 20\log_{10}(\lambda) \quad (5.10)$$

The expression $10\log_{10} P_{Tx}$–$10\log_{10} P_{Rx}$ is the difference in what is transmitted and what is received called path loss or sometimes called free-space path loss:

$$L_{fspl} = 10\log_{10}P_{Tx} - 10\log_{10}P_{Rx} \quad (5.11)$$

Or substituting Eqs. (2.11) into (2.12) can be presented as follows:

$$L_{fspl} = 20\log_{10}(4\pi) + 20\log_{10}(d) + 20\log_{10}(f) - 20\log(3 \times 10^8) \quad (5.12)$$

To have a more practical form, it has to be included distance in kilometers and frequency in MHz by the following equation now becomes

$$L_{fspl} = 20\log_{10}(4\pi) + 20\log_{10}(d_{km}) + 60 + 20\log_{10}(f_{MHz}) + 120 - 20\log(3 \times 10^8)$$
$$L_{fspl} = 32.44 + 20\log_{10}(d_{km}) + 20\log_{10}(f_{MHz}) \quad (5.13)$$

This is, one of the most important radio formulae, which is reproduced in the ITU Recommendation PN 525-2 "Calculation of free-space attenuation."

5.2.5 *Isotropic Power Source*

One of the most important concepts in radio engineering is that of an isotropic power source, which has the following definitions:

1. It is from a pinpoint source of infinitesimal low volume (i.e., volume $\approx 0\ m^3$);
2. It radiates radio power uniformly in all directions (i.e., the envelope of propagation is an outwardly expanding sphere); and
3. It has no loss.
   Obviously this is an ideal situation that cannot exist in practice. However, this concept is the central method of referencing nearly all radio calculations and for benchmarking most radio systems and antennas.

5.2.6 *Power Flux Density and Electric Field Strength*

The new concept of P_{fd}, Ψ, or S can be defined as the amount of power radiated per unit area from an isotropic source at a point some distanced from the isotropic source. From the geometry and the definition of isotropic:

$$\Psi \text{ or } P_{fd\,or\,(S)} = P_{Tx}/(4\pi d)^2 \quad (5.14)$$

From basic physics, remember (a) Ohm's law V = IR (where V is the voltage in Volts, I is the current in amperes and r is the resistance in Ohms (Ω), and (b) the electrical power equation P = I × V (where P is power in Watts). Using the two equations above can be produced new relations:

$$P = V^2/R \quad (5.15)$$

There is an equivalent analogous version for radio propagation:

$$P_{fd} = E^2/Z \quad (5.16)$$

where E is a new concept called field strength and is measured in volts per meter (V/m), and the impedance of free space (Z) is known to be a constant:

$$Z = 120\pi \text{ or } 380\Omega \tag{5.17}$$

However, now a relationship between Ψ, P_{fd} (S), and field strength E can be established. In logarithmic terms, taking 10 x logs of each side:

$$\begin{aligned} 10\log S &= 20\log E - 10\log(377) \\ \Psi \text{ or } P_{fd}(S)(dBW/m^2) &= E(dBV/m) - 10\log(377) \end{aligned} \tag{5.18}$$

Sometimes, it is more useful to consider field strength in dBμV/m. Therefore, the formula becomes with new relations:

$$\begin{aligned} \Psi \text{ or } S(dBW/m^2) &= E(dB\mu V/m) - 60 - 60 - 25.76 \\ \Psi \text{ or } S(dBW/m^2) &= E(dB\mu V/m) - 145.76 \end{aligned} \tag{5.19}$$

This is also a very useful formula giving a direct relationship between field strength and P_{fd} at the same geometrical point.

The relationship between field strength and transmitted power can be explained by the theme of the electrical field strength and combining Eqs. (5.15.) and (2.17.) in the following relation:

$$E^2/Z_0 = P_{Tx}/4\pi d^2 \tag{5.20}$$

Taking logs is getting the following relations:

$$\begin{aligned} 20\log E(V/m) &= 10\log P_{Tx} - 10\log 4\pi - 20\log(m) + 10\log Z_0 \\ E(dB\mu V/m) - 60 - 60 &= P_{Tx}(dBW) - 10.99 - 20\log d(km) - 30 - 30 + 25.76 \\ E(dB\mu V/m) &= P_{Tx}(dBW) - 20\log d(km) + 74.8 \end{aligned} \tag{5.21}$$

5.3 Refraction, Absorption, and Non-LOS Propagation

Discussion has so far been mainly concentrated on how radio waves propagate in free space with line-of-sight (LOS) conditions. This describes the majority case for mobile solutions and most situations can be approximated to this. Occasionally, however, mobiles and especially aircraft will not have LOS to the nearest ground node.

This is usually the case on a long approach, or approach in mountainous terrain or when the station is over the horizon. At such lower altitudes, other propagation mechanisms can come into effect, such as refraction, reflection, and ground wave/sky wave.

5.3.1 Refraction

Refraction effect as a propagation mechanism has two possible solutions: layer refraction and obstacle refraction.

1. **Layer Refraction**—The k factor starts to introduce the concept of non-straight-line radio beams. This is essentially incremental refraction occurring at every point along the beam. Much study and prediction of this has been carried out by ITU-R and archived in the P (propagation) archives (in particular 'ITU-R P-453-9[5]').

Thus, the radio refractive index: its formula and refractivity data are presented in Fig. 5.8a. Of course, in the limit, the phenomenon of total internal refraction or "ducting" occurs and a radio beam can extend well beyond its natural horizon limits, see Fig. 5.8b. In fact, this angle of incidence is equal to angle of refraction. These refractive layers sometimes are called ducts. Each layer has a different gas density and hence refractivity index.

2. **Obstacle Refraction**—A second type of refraction that occurs is that when the radio wave (showing its wave physics properties) is blocked by either a smooth obstacle that encroaches into its path or a knife-edge obstacle encroaching into its path. An attenuation of the beam occurs as a function of the Fresnel physics. In Fig. 5.9a is shown a smooth edge diffraction point, while in Fig. 5.9b is presented a knife-edge diffraction point.

Therefore, in both samples diffraction signal is attenuated as a function and Fresnel zone clearances over diffraction surface. In addition, it can be presented attenuation as a function of geometry by relative attenuation in free space and obstacle clearance.

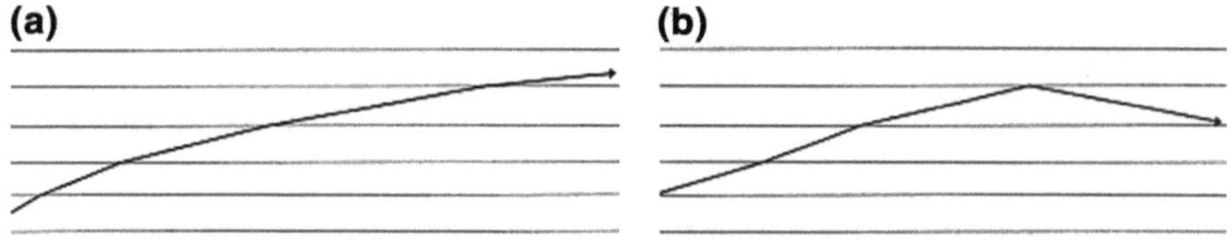

Fig. 5.8 Incremental radiation and total internal refraction of "ducting"—Courtesy of Book: by Stacey

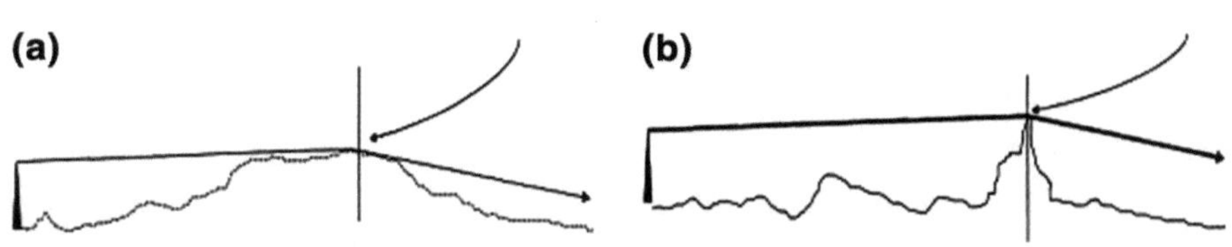

Fig. 5.9 Smooth edge and knife-edge diffraction points—Courtesy of Book: by Stacey

5.3.2 *Attenuation from Atmosphere Absorption*

Consider the molecular properties of radio waves and the concept that radio waves travel through gaseous medium. As the radio signal is promulgated through this medium, the gases take some of the energy out of the radio wave, particularly if the resonant frequency of the gas molecule under consideration is at or near the frequency of the radio wave. This phenomenon is called absorption.

1. **Water Absorption**—For terrestrial signal paths operating below about 5 GHz frequency multipaths, flat fading or selective fading (sometimes called fast fading) is the significant contributor to degradation from the free-space path loss condition (or fading as it is called). The gaseous absorption loss is usually negligible. Above the frequency of 5 GHz, water absorption particularly becomes more significant and usually is the dominant mechanism for fading on a terrestrial path. The proportion of the time for which the severe attenuation occurs is usually only when the path is subject to medium to heavy rain. The topographic rain profile of an area is an important feature to such paths. The ITU organization has studied these rain profiles and categorized each area of the world with a profile letter designation. (CCIR have rain zone tables and prediction methods; see rain zones described in ITU-CCIR Report 563-2.) In particular, the temperate climates are found to be less prone to rain fading than are the tropics, e.g., Northern Europe ranges from zone C to G, the subtropics are ranging typically to K and the tropics can go from M up to W and X. When designing links or paths, consideration of the rain zones should be given. For ICAO, the worst rain zone attenuations usually have to be considered when designing radio systems for operation in all regions of the world (Fig. 2.15).
2. **Oxygen Absorption and Other Gases**—Of lesser significance is absorption from other atmospheric gases, particularly oxygen. The frequencies of 6 and 13 GHz are where oxygen absorption peaks (See CCIR Report 719). At lower HF and VHF radio frequencies, gaseous absorption is negligible. In Table 5.2 are presented samples for CCIR rain regions.

5.3.3 *Non-LOS Propagation*

There are a number of other mechanisms for which propagation can happen in non-direct LOS. These are via ground wave propagation (which tends to be more significant at low or very low frequencies, in kilohertz or less), reflection, and refraction (which are both related and similar), which are prevalent in the HF-, VHF-, and UHF-bands. In the following, these will be considered in more detail.

1. **Ground Wave Propagation**—This is when radio waves follow the curvature of the Earth. This is particularly the case for low frequencies and very low frequencies in the kilohertz region. Ground waves that follow the Earth must be

Table 5.2 CCIR rain regions

CCIR rain region (Previous designation)	Example area	Worst hour (mm/hr)
A	Arctic/Antarctic	<10
B	Greenland	10
D	UK southeast	25
E	UK rest	30
K	Sydney	40
M	Brisbane	60
–	Rio de Janeiro	70.4
W	Cairns	80
–	Cameroon	100
–	Singapore	120

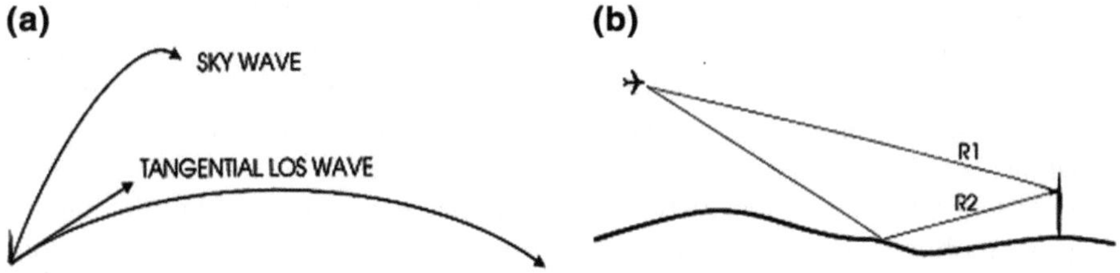

Fig. 5.10 Ground Wave Propagation and Two-Ray Model—Courtesy of Book: by Stacey

vertically polarized. The attenuation is a function of frequency and is lower at lower frequency. Generally, ground wave propagation is not appropriate for aeronautical purposes, either in frequency (the aeronautical bands are too high in frequency) or application. Fig. 5.10a shows the way that ground wave follows contour of Earth's surface.

2. **Reflection and Multipath**—The HF-, VHF-, and UHF-bands including microwave frequencies lend themselves easily to reflection, refraction, and multipath. This is when by the law of physics; a ray is incident to a "reflective surface" (and this can be, for example, still water, land, metal, or a highly conductive surface). In undergoing a reflection, usually some losses are incurred, but these can be small, in which case a reflected ray can be comparable with a direct ray in terms of incident power. From a practical perspective, this is the reason why these frequencies are often chosen for radar systems, which exploit this property.

For mobile communications, in some cases, this can be a favorable property, extending range and saturating coverage for certain systems (i.e., cellular mobiles) in a cluttered urban environment where frequencies of 400 and 800 MHz are well known for their good properties of bouncing off walls and propagating through windows and tunnels. Also for aviation, the property can be useful, particularly at HF where relatively low-loss reflections can be used to provide long-range radio.

However, in other aeronautical applications, particularly between low-altitude transceivers, usually it can also lead to severe attenuations even when direct LOS exists and is a very undesirable phenomenon. There are many models derived for terrestrial non-LOS, usually based on the two-ray model or the extended and advanced work on this theme by Hata and Okumura.

5.3.4 Two-Ray Model

Consider an instance where two signals (one usually direct and one indirect) from the same source arrive at a receiver. Constructive or destructive fading can take place, resulting in either good or bad signal reception, shown in Fig. 5.10b.

This phenomenon is called multipath and becomes very significant for marginal LOS cases or radio communications close to the Earth's surface, particularly over water or in metallic cluttered environments such as around airports. Of course the theme can be extended where multiple indirect rays can be arriving at the receiver with or without a direct ray, compounding the complexity and uncertainty of the situation.

Hata and Okamura extensively studied this phenomenon regarding its applications to mobile communications. It can even be exploited thoroughly in some instances, but generally, it is a nuisance and countermeasures are used to cancel the destructive interference aspects (using equalization techniques, delay equalization processing or data interleaving techniques, to name just a few of the system measures to be discussed later.) When multipath occurs and the delay period is less than a data period between two symbol states in the digital environment, intersymbol interference will occur. This can be seen when observing a data stream at the intermediate frequency (IF) level through an "eye diagram." The eye will close as a function of how much reflected signal is arriving at the receiver and the intersymbol interference builds. Fig. 5.11 is illustrated example transmitted and received data stream in a multipath environment.

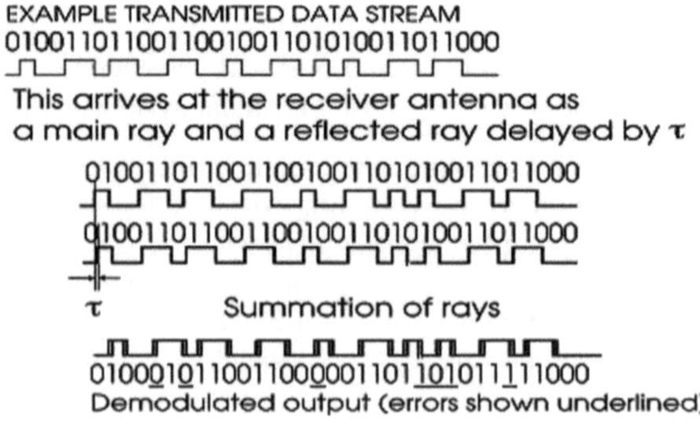

Fig. 5.11 Intersymbol interference—Courtesy of Book: by Stacey

5.4 Sky Wave Propagation

Sky wave propagation is an extension of the reflection theme. The waves that rely on the refractive properties of the ionosphere to propagate long distances are considered as sky waves, shown in Fig. 5.12a. In the limit, a refraction becomes a reflection and hence the relationship to reflections. This phenomenon is very frequent below 30 MHz. These properties are useful in that long-range communications can be provided well over the horizon. The HF-band (3–30 MHz) is well suited for this and aviation and other radio users exploit this band for this very purpose. Conversely, the reflective properties of the ionosphere can occur into the low VHF-band (above 30 MHz) and even beyond for a low proportion of time into the aeronautical communications band (currently 118–137 MHz). This is usually more of a problem where coverage can overreach past the horizon into an adjacent area sometimes up to 1000 km away using the same frequency for a different application of airspace and consequently causing interference. The HF-band traditionally is considered to extend from 3 to 30 MHz. The aeronautical HF-band, however, is considered to extend a little before this to start at 2 MHz and run up to 22 MHz. In the early days of radio, especially marine radio, HF was a favorite band because the propagation properties are such that under certain conditions the wave can totally internally reflect or "skip" its way right around the Earth. In addition, the antenna properties are very efficient and finally, the Tx/Rx was easily designed and economic. Today, this still holds true.

5.4.1 *Ionosphere*

The ionosphere extends nominally from 60 to 400 km above the Earth's surface. It is so-called as the particles in this region are easily ionized, giving its special electrical properties.

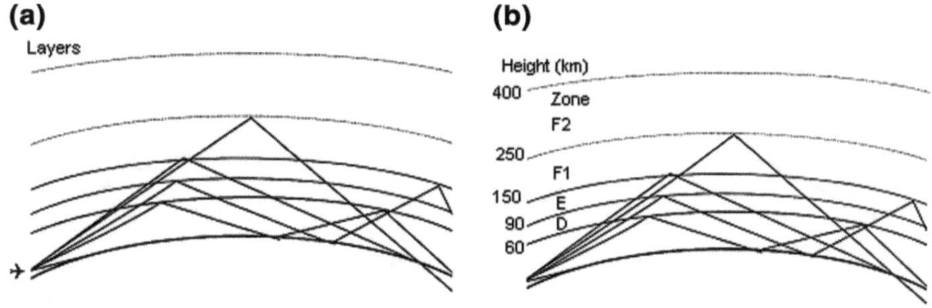

Fig. 5.12 Sky wave propagation and hf reflection from the different layers—Courtesy of Book: by Stacey

Under certain conditions, the HF radio waves will refract and in the limit totally internally reflect or "duct," shown in Fig. 5.12b. Theoretically, there are a number of layers that exist. The D layer runs from height 60 to 90 km and the E layer from 90 to 150 km. The F1 layer runs from 150 to 250 km, and the F2 layer from 250 to 400 km. The propagation conditions during day and nighttime are providing the different effects as follows:

1. **Daytime**—In general, the lower layers absorb lower frequencies typically up to 10 MHz during the daytime, and the higher frequencies typically 10–30 MHz are reflected off the F1 layer, shown in Fig. 5.13a.
2. **Nighttime**—In general, the D and E layers disappear and the F layer reflects frequencies below typically 10 MHz (the higher frequencies pass straight through the ionosphere without refracting/reflecting), shown in Fig. 5.13b.

Unfortunately, with HF there is an element of "luck" or randomness to these general rules. ITU recommendation P.533-7—"HF Propagation Prediction Method," gives an interesting insight into this. It provides a theoretical method to calculate which frequencies are most suitable at which times. It identifies a couple of the following concepts:

1. **Maximum Usable Frequency**—This is a median value, which is getting statistically. The chances at the time of validity of an HF channel frequency being "open" above this value are 50 %.
2. **Highest Possible Frequency**—This is the HF, above which statistically there is only a 10 % chance of the frequency being open.

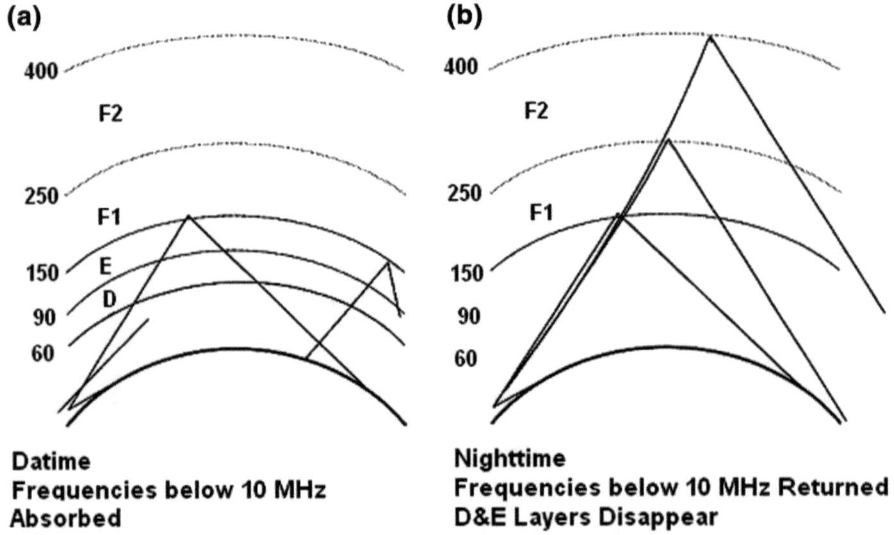

Fig. 5.13 Daytime and nighttime radio operations—Courtesy of Book: by Stacey

3. **Optimum Working Frequency**—This is the highest HF that gives an 85 % likelihood of the channel being open.
4. **Critical Frequency**—This is the highest radio frequency that is returned in ionosphere layer. Thus, this can be applied for all path lengths from hundreds of kilometers up to beyond 9000 km. (This is effectively half-way around the world, which is as far as you can practically go.)
   This theory also investigates the various modes discussed above in detail (e.g., E modes up to 4000 km range and F2 modes for all distances).
   There is a complex formula given in this recommendation:

$$\begin{aligned} L_t &= 32.45 + 20\log f + 30\log p\varphi - G_l + L_i + L_m + L_g + L_h + L_Z \\ P' &= 2R_0 \sum_{1}^{n} [\sin(d/2R_0)/\cos(\Delta + d/2R0)] \end{aligned} \tag{5.22}$$

   where L_i is the absorption loss (dB) for an *n*-hop mode, given by

$$L_i = n(1 + 0.0067R_{12})\sec i/(f + f_L)^2 1/k \sum_{j=1}^{k} AT_{noon}F(X_j)/F(X_{noon})\ \varphi(f_\upsilon/f_{oE})$$

   Note the resemblance to the free-space path loss equation, at this point, there is a time delay element as well.
   The time delay of an individual mode is given by:

$$\tau = (p'/c) \times 103\ [m/s] \tag{5.23}$$

   Where p' is virtual slant range (km) and c is the velocity of light (3×10^8 m/s).
5. **Practical Measurements (Ionosphere Sounding)**—This is a technique frequently used in practice to gauge the current ionosphere conditions and to be able to predict propagation. In its simplest form, consider sending a multifrequency signal straight up perpendicularly to Earth's surface and measure which frequency components are returned.
   A number of services send out "squitters" or multifrequency components like this in their transmissions, and receivers can look at what they receive and analyze, and from this evaluate what the current propagation conditions are and which frequencies are best to use at that moment in time. In reality, this practical approach is found to be a more powerful method than the theory.
6. **Solar Flares and Sunspot Activity**—Solar flares present total ionization of a specific frequency and thus can wipe out HF communication ability on a particular channel for minutes or hours. There is a random nature to their occurrences, although sunspot activity can be correlated with 11-year solar cycles.

5.4.2 Propagation Distance of Satellite and Atmospheric Losses

Probably, the most significant factor for satellite communications is the distance between the satellite and the Earth (or the distance between satellites for inter-satellite communication). This introduces significant delay time between transmitted and received signals. Also this has a direct bearing on the path attenuation (from the free-space path equation derived earlier, see Eq. (2.15)), which tends to be relatively large in satellite communications.

By contrast, satellite beams cut up through the atmosphere and quickly pass through the absorption/precipitation zone (usually this is less than 1 % of their path length) (the exception being at the poles to geostationary satellites (above the equator) where precipitation is much less anyway) and so rain and gaseous attenuation generally becomes insignificant.

For aeronautical applications, the main frequency bands of interest to aviation are L-band 1.4 GHz; C-band 4 and 6 GHz, and Ku-band 12–14 GHz, where the losses are minimal for satellite paths and international allocations have been made.

Doppler effects can be significant to the radio systems at the satellite end, as a satellite can be maneuvering at "galactic" speeds to maintain its orbital position.

There are generally three types of satellites:

High Earth orbit (HEO) or sometimes called GEO are the ones that are in stationary (relative to Earth movement) orbit at nominally 35,000 km above the equator. That is where the gravitational pull or centripetal force on the object balances or equals the centrifugal force of the object spinning with a period of one day. By the nature of physics, there is a time delay of 1/8 of a second in a radio beam traveling from the Earth's surface to such a satellite. These satellites are used extensively for a number of communications applications by civil aviation, military, and for entertainment broadcast and passenger services. These are the most common satellites presently used by civil aviation.

As discussed earlier, two other kinds of satellites are MEO and LEO satellites. These fly closer to the Earth and consequently are moving relative to the Earth's surface and will be perceived to go though a trajectory from one horizon to the other when being observed on the ground (sometimes within a few minutes). The lower the satellite, the shorter the trajectory period.

Also of consequence is that the latency or signal delay to these satellites is much reduced and the frequency reuse is much higher due to the ability of multiple satellites to be shielded from each other by the Earth itself. These satellites are not currently used by aeronautical applications (other than by exception the military ones); however, they could be used in future applications where they are more attractive for latency/voice critical applications.

5.4.3 Propagation Caused by Doppler Effect

In its simplest form, the Doppler effect is when an object with vector velocity relative to a receiver is transmitting a certain frequency that is perceived to be different due to either the waveforms being compressed (if the vector is toward the receiver) or elongated (if the vector is away from the receiver). Remember the school physics example of a fire engine coming toward a person with its siren on and as it passes and moves away, the note or tone apparently drops a pitch. The same applies to radio systems.

Considering Eqs. (5.1) and (5.2), the stationary receiver is receiving waves from a stationary source. Within a given second, the number of cycles received at the receiver would be received cycles = c/λ.

On the other hand, considering a source moving toward the receiver would be received cycles = $(c + y)/\lambda$, where y is the speed in meters per second of source. Therefore,

$$f_{apparent} = f_{carrier} \times (c + y)/\lambda \tag{5.24}$$

This formula is valid for transmitting sources traveling toward or way from the target. The argument can be further extended if both the transmitter and receiver are moving.

5.5 Atmospheric Effects on Propagation

This study deals with propagation effects in all regions of the atmosphere and free space, including the Earth's ionosphere. Namely, most of the Earth's weather precipitation and hydrometeors occur in the troposphere, which is the non-ionized region from the Earth's surface up to a height of about 15 km above the surface at the equator. The thickness of the troposphere decreases toward the poles. Therefore, propagation effects in the troposphere tend to increase in importance as the RF increases above 1 GHz. For GMSC systems, the effects of reflection from the Earth's surface are critically important at even lower RF.

At frequencies below about 1 GHz, the most important region of the Earth's atmosphere is the ionosphere, the ionized region above the stratosphere, within which low RF propagation effects are quite strong. Propagation effects within the ionosphere have an influence on the terrestrial and Earth space paths from VLF- to SHF-bands. However, it should be borne in mind that for an RF-band range below a certain number of GHz, the ionospheric propagation effects can be quite important and for frequencies above a certain number of GHz, tropospheric effects may be negligible. At this point, radiowave signals passing through the atmosphere or over the surface of the Earth begin to lose strength. The decrease of signal due to the medium through which it passes is called attenuation.

5.5.1 Propagation Effects of the Troposphere

The troposphere extends upwards from the Earth's surface to a height of approximately 15 km, where it meets the stratosphere. More exactly, at the boundary between the two spaces there is a region called the tropopause, which possesses a different refractive index to each neighboring layer. The effect exhibited by the tropopause on a space wave is to produce a downward bending action, causing it to follow the Earth's curvature. Hence, the bending radius of the radio wave is not as severe as the curvature of the Earth; nevertheless, the space wave will propagate beyond the visual horizon. In practice, the radio horizon exceeds the visual horizon by approximately 15 %. This effect is known to navigators whereby the surface radar range of a maritime PPI radar picture extends slightly beyond the horizon.

5.5.1.1 Attenuation Due to Atmospheric Gases

The different types of gases present in the atmosphere may attenuate the electromagnetic waves, which is caused by the molecular absorption of the atmospheric constituents and is strongly frequency dependent.

The main contributors to this attenuation below 70 GHz are water vapor and oxygen. Thus, absorption increases as elevation angle is reduced, which at any frequency is a function of temperature, pressure, humidity of the atmosphere, and the elevation angle of the satellite.

At sufficiently high frequencies, EM waves interact with the molecules of atmospheric gases to cause attenuation. These interactions occur at resonance radio frequencies and are apparent in plots of zenith (90° elevation angle) attenuation versus frequency, as shown in Fig. 5.14a. This is the theoretical total estimated one-way attenuation for vertical paths through the atmosphere, where solid curves are for moderately humid atmosphere, dashed curves represent the limits for 0 % and 100 % relative humidity, V is vertical polarization, and A denotes limits of uncertainty, for a vertical Earth space path as a function between 1 and 200 GHz at 45° north latitude using US standard atmosphere.

It may be noted that the specific frequency bands where the absorption is high are about 22.23 GHz caused by vapor (left peak in the curve) and centered between 53.50 and 65.20 GHz caused by oxygen (right peak in the curve). Meanwhile, in the frequency range of current interest about 1–18 GHz the zenith one-way absorption is in the range of 0.03–0.20 dB and 0.35–2.30 at 5° elevation. The corresponding upper limit for 100 % humidity is 0.7 dB at zenith and 8 dB at 5° elevation. Therefore, these effects are not very important for GMSC systems because the influence of attenuation due to atmospheric gases can be effectively ignored at frequencies below 10 GHz.

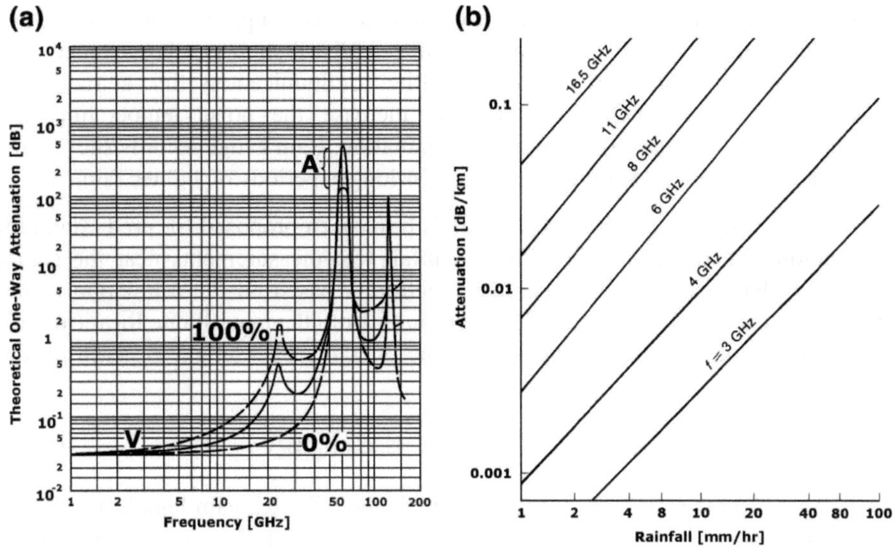

Fig. 5.14 Theoretical one-way and rainfall attenuations—Courtesy of Books: by Richharia/Gagliardi

5.5.1.2 Attenuation by Precipitation and Hydrometeors

Hydrometeors are condensed water vapors existing in the atmosphere, such as cloud, fog, rain, hail, and snow. The last three forms of atmospheric water are called precipitation. They produce transmission impairments and attenuation by absorbing and scattering the energy of radio waves.

1. **Clouds and Fog**—Cloud and fog may cause attenuation but much less than that caused by light rain of about 10 mm/h. Clouds and fog are suspended water droplets, usually less than about 0.10 mm in diameter, whose effect is significant only for systems operating above 10 GHz. It is important to note that attenuation to radio waves depends on the liquid water content of the atmosphere along the propagation path. Clouds have liquid contents from 0.05 to 5 g/m^3 and their formation shape can be due to a variety of atmospheric processes, which result in cloud layer types for each of three cloud heights: low, middle, and high. Thunderstorm cumulonimbus high clouds cause the maximum attenuation. The fog liquid content is in the order of about 0.40 g/m^3 and typically fog extends from 2 to 8 km. The attenuation from fog is negligible for satellite communications.
2. **Rain**—Rain precipitation is the most significant contribution to atmospheric attenuation, which is caused by radiowave absorption and scattering from raindrops. First of all, to evaluate this additional rain loss, it is necessary to obtain the expected rainfall rate in mm/hr for the region of the communication link. For this reason, in Fig. 5.14b are presented six curves, which can be used to read off dB losses per EM path length at the operating frequencies. Namely,

these curves are generated from a combination of empirical data and mathematical models that fit this data. Rainfall rate distribution is inhomogeneous in space and time and can impair a satellite communications link at radio frequencies above 10 GHz, as well as increase the noise temperature and impairment, or cross-polarization discrimination. Rain gauge records show short intervals of higher rain rates occurring in longer periods of lighter rain.

Therefore, various precipitation methods have been proposed for predicting rain attenuation statistics calculations from rainfall rate measurements near the signal path. The slant path rain attenuation prediction is based on the estimation of the attenuation exceeded at 0.01 % of the time $A_{0.01}$ from the rainfall rate $R_{0.01}$ (mm/h), exceeding at the same time percentage. The effective path length is given by the formula:

$$A_{0.01} =_{\gamma R} \cdot L_e =_{\gamma R} \cdot L_s \cdot r_{0.01} \text{ [dB]} \tag{5.25}$$

where γ_R (dB/km) = specific attenuation, L_e is the effective path length, L_s = slant path length and $r_{0.01}$ = path reduction factor. The spatial structure of rain can be modeled by an equivalent rain cell of uniform rainfall rate with a rectangular cross section of equivalent length L_O and effective height h_R–h_S in the plane of the path, see Fig. 5.15a.

The relation for slant path length for elevation angles above 5° is as follows:

$$L_s = h_R - h_S/\sin\theta \text{ [km]} \tag{5.26}$$

and for elevation angles less than 5° is

$$L_s = 2(h_R - h_S)/\left[\sin^2\theta + 2(h_R - h_S)/R_e\right]^{1/2} + \sin\theta \text{ [km]} \tag{5.27}$$

where θ = path elevation angle and R_e = effective radius of the Earth (8500 km). Thus, using simple trigonometry relations the horizontal projection is given by the following formula:

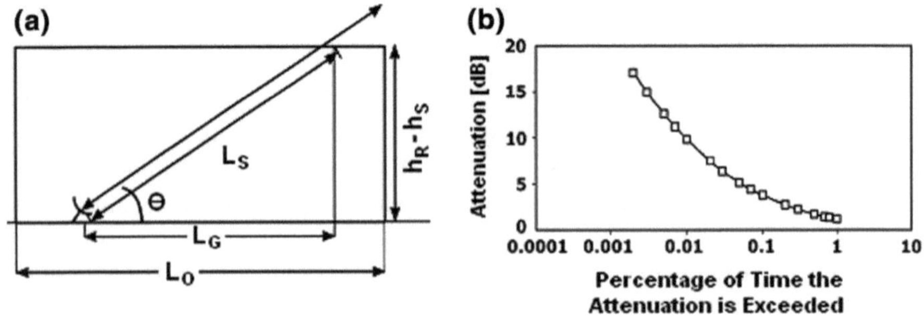

Fig. 5.15 Equivalent rain cell of rainfall rate and rain attenuation statistics—Courtesy of Handbook: by ITU

$$L_G = L_s \cos\theta \,[\text{km}] \tag{5.28}$$

The reduction factor has the following relation:

$$r_{0.01} = 1/1 + L_G/L_0 \tag{5.29}$$

For example, the specific attenuation using the frequency-dependent coefficients k and α gives the following expression:

$$\gamma_R = k(R_{0.01})^{\alpha} \,[\text{dB/km}] \tag{5.30}$$

Values of the frequency-dependent coefficients are shown in Table 5.3. Values not given there can be obtained by interpolation using logarithmic scales for frequency and k and a linear scale for α. For linear and circular polarization, these parameters can also be obtained by the following equations:

$$\begin{aligned} k &= \left[k_h + k_v + (k_h - k_v)\cos^2\theta\cos 2\tau\right]/2 \\ \alpha &= \left[k_h\alpha_h + k_v\alpha_v + (k_h\alpha_h + k_v\alpha_v)\cos^2\theta\cos 2\tau\right]/2k \end{aligned} \tag{5.31}$$

where θ = path elevation angle and τ = polarization tilt angle relative to the horizon (τ = 45° for circular polarization).

The attenuation value predicted for a percentage in the range of 0.001–1 % can be obtained by relation:

$$A_p = A_{0.01} 0.12 \cdot p^{-(0.546\,+\,0.043\log P)} \tag{5.32}$$

At this point, for percentages of the time outside the 0.001–0.1 % is a relatively complicated procedure, while the predicted attenuation exceeded for 0.01 % for an average year can be is $A_{0.01}$ = 9.8 dB.

Finally, the estimated attenuation exceeded for any other percentages of the time P may be calculated from $A_{0.01}$ by the expression:

$$A_p = A_{0.01}(P/0.01)^{-\alpha} \,[\text{dB}] \tag{5.33}$$

The cumulative distribution, for example, of the rain attenuation statistics is illustrated in Fig. 5.15b.

3. **Hail and Snow**—Hail depolarization often precedes or follows rain and the presence of hail above rain is one factor that causes wide variations in the

Table 5.3 Specific attenuation in dB/km for a visibility of 10 m at the given frequency

Frequency:	1 GHz		3 GHz		10 GHz		30 GHz	
g H_2O/g soil:	0.30 %	10 %	0.30 %	10 %	0.30 %	10 %	0.30 %	10 %
A_s [dB/km]	0.001	0.004	0.003	0.02	0.01	0.07	0.03	0.60

instantaneous value of the cross-polar discrimination (XPD) effect, which may accompany a given attenuation. Hail, with its ice meteors, causes a small loss due to scattering but not due to absorption.

Dry snow likewise does not absorb radiowave energy, however, the water content of wet snow dropping slowly through the ray path causes more absorption than would arise if the same amount of water were falling as rain.

5.5.1.3 Sand and Dust Effects

The Earth's atmosphere contains a variety of solid particles in suspension, whose size range from aerosols of approximately 50 μm to coarse dust about 1 mm. Thus, if the size of particles is larger, they fall out of suspension more quickly but aerosols can remain aloft for days and sometimes can be carried high up into the stratosphere. Aerosol and small solid particles do not significantly affect the propagation of EM until the optical frequencies are approached and their extinction cross section becomes appreciable. The effect of fine and coarse dust on the radiowave path can be noticeable at much lower frequencies and so in many cases, the impact is usually difficult to separate from the meteorological phenomena that often accompany the dust storms. Then, enhanced humidity or even rain that might accompany a severe convective activity in an arid region might cause the dust particles to attenuate more strongly than would be the case in a dry atmosphere due to water uptake into the crystal structure of the particles themselves. The strong winds that generate a dust storm may cause antenna depointing that will lead to a loss of radio signal strength, which is difficult to distinguish from the attenuating effect of dust particles.

An intensive study of the sand and dust particle effects on 4/6 and 11/14 GHz of satellite paths made a clear distinction between dust and sand. Sand particles have limits from 10 to 100 μm and are usually restricted to within 10 m of the Earth's surface and in this case their impact on the satellite link due to radio propagation effects is usually insignificant at elevation angles above 5°. On the other hand, dust particle sizes are below 10 μm and can come up to 1000 m in the atmosphere. Sand particles can absorb about 1 % and dust particles can absorb 6–9 % of air humidity. Thus, severe dust storms occur in eight desert areas with average wind speeds from 6 to 22 m/s, gust speed is from 15 to 50 m/s, width and length of affected area is maximum 8000 km; dust can spread up to 12 km high in the atmosphere, expected visibility is from 0 to 1000 m in a severe storm and the duration of events is from a few hours to 2 weeks. The location and direction of the dust storms are in the vicinity of big desert areas, such as the southwest part of the USA, central and southern part of the South American Continent, northern and southern African Continent, around the Caspian Sea and Persian Gulf including central Asia and regions of the northwest and east coast of Australia. The main sources of heavy dust storms are big deserts such as the Sahara and the Kalahari, including nearby areas affected by Haboob storms.

Namely, the Haboob storm is a direct result of a thunderstorm generated by a strong and cold wind, known as a cold dust storm. The severe turbulence and gusting in this storm area leads to very high suspended dust densities, which give rise to a significant reduction of optical visibility. For this reason, optical visibility has been used as a classification mechanism for dust storm effects. The principal criterion applied to defining the occurrence of a dust storm is when visibility drops below 1 km and when the duration of storm activities lasts from a few hours to 174 days in the various eight desert regions on Earth.

At this point, if parameters k_r and k_i are the real and imaginary parts of the complex relative dielectric constant of the dust particles, respectively, the specific attenuation (Flock 1987) can be given by

$$A_s = 189r/V\lambda\left[3k_i/(k_r+2)^2+k_i^2\right][dB/km] \tag{5.34}$$

where r = radius [m]; V = visibility 2 m above the ground [km], and λ = wavelength [m].

Table 5.3 gives typical results of attenuation predictions for nominated frequencies and where circular dust particles are assumed to have a uniform distribution. In any event, the specific attenuation is significant when Ka-band frequencies are employed on 20/30 GHz.

Even then, it is only in relatively humid air with about 10 % water uptake that the path attenuation approaches 2 dB. The value of XPD for depolarization is almost benign until the visibility becomes very small, about 2 m, and the elevation angle is low. However, if the air becomes more humid and the dust particles absorb the moisture, the XPD values are worsened by about 5 dB. On some paths, for elevation angles below 10° and for some frequencies below 10 GHz, a severe dust storm that contains a large proportion of moist particles could cause the XPD to fall below 12 dB.

In general, however, for satellite communications link paths, the radiowave impairments caused by sand and dust storms are significantly less of a problem than the mechanical difficulties they may cause, such as severe wind gusts and dust in the antenna feeds and auxiliary equipment.

5.5.1.4 Site Diversity Factors

Path diversity in satellite systems involves the provision of alternate propagation paths for signal transmission, with the capability to select the least-impaired path when conditions warrant. In this case, implementation of path diversity requires the deployment of two or more interconnected Earth terminals at spatially separated sites, hence the use of the term "site diversity." This combination is considerable when an LES is being used for a commercially or militarily important purpose, where an outage is considered more serious, e.g., as the head end of a cable system, the probability of an outage can be reduced by having a second LES a few kilometers away and using whichever has the better signal at moment. Therefore, this is because heavy rain, in particular, tends to be quite localized and would probably not impact both stations at the same time.

5.5.1.5 Other Diversity Schemes

Without an ability to increase the margin (e.g., through the application of power control via additional transmit gain and power, or increased resource allocation in the TDMA frame), there are basically three types of diversity schemes that can be used by satellite systems to overcome impairments at a given LES.

1. **Orbital Diversity**—This diversity is different from site diversity in that only one LES site is used. To achieve a measure of diversity, the LES uses two antennas that can access different satellites simultaneously and is not necessarily the diversity-interconnecting link between sites that is required for site diversity. However, to obtain significant decorrelation of concurrent attenuations along the two paths, the angle between the two paths at the LES must be large. Inasmuch as this angle is large, at least one or even both of the links will be at a relatively low elevation angle and therefore encounter a greater degree of impairment than at higher elevation angles. In general, the achievable diversity gain is very small, with values of about 2–3 dB in the 14/11 GHz bands.
2. **Frequency Diversity**—Path losses caused by particulates on the path increase as the frequency increases, particularly for rain. In effect, at 6/4 GHz, attenuation due to rain is negligible; at 14/11 GHz it can be significant in high rainfall regions of the world; at 30/20 GHz, it is the dominant link impairment nearly everywhere.
   At this point, if it is possible to switch communication transmissions to a lower-frequency band, significant increases in availability might be achieved. This capability requires that both frequency bands (the higher, impaired one and the lower one to which the communication channels are to be switched) be simultaneously available at the LES in question. Furthermore, there must be spare capacity available in the lower-frequency spectrum whenever needed, implying that significant spare capacity must be provided if the link is a high-capacity channel and that the complete network need to be under dynamic control. Namely, both elements require significant investment. Should such dynamic network control features be in place and the additional capacity in the lower-frequency band is available on-call, frequency diversity can therefore undoubtedly provide large increases in availability.
3. **Time Diversity**—Severe rain events do not usually last long at a given location. Thus, this characteristic can be used in any communication RF link that does not require interaction between the caller and the receiver. The service can be said to be acceptable if, for example, a fax is successfully sent without any error within a two-hour period. The delay in sending the fax can be considered a form of time diversity. This feature could also be used with advantage to determine the

capacity requirement of a given RF link for optimal economic performance. Thus, if a link is sized for a maximum anticipated capacity, it will have excess capacity for most of the time. If some transmissions can be delayed and sent, for example, at off-peak times, the capacity requirements can be reduced. The time delay could therefore be used either at times of peak capacity (i.e., the equivalent of call-blocking) or when the LES is undergoing a severe rain event.

5.5.2 Clear-Sky Effects on Atmospheric Propagation

Radio signals traveling through the atmosphere layers suffer attenuation even during fine weather. In any event, the clear-sky attenuation is mainly the result of absorption of energy from the transmission by water vapor and oxygen molecules, although there are other modes of clear-sky affect that have influences on propagation.

1. **Defocusing and Wave-Front Incoherence Contribution**—Several expressions have already been evaluated and are provided in No 2.3.2 of ITU-R P.618 recommendation to estimate the defocusing (beam-spreading) losses on paths at very low elevation angles. Otherwise, the loss is implicitly accounted for in the prediction methods for low-angle fading found in articles No 2.4.2 and 2.4.3 of the recommendation.
   Hence, small-scale irregularities of the refractive index structure of the atmosphere cause incoherence in the wave front at the receiving antenna. In any case, this will result in both rapid radio signal fluctuations and an antenna-to-medium coupling loss that can be described as a decrease of the antenna gain. In practice, signal loss due to wave-front incoherence is therefore probably only significant for large-aperture antennas, high frequencies, and elevation angles below 5°. Measurements made in Japan with 22 m antenna suggest that at 5° elevation angle the loss is about 0.2–0.4 dB at 6/4 GHz, while measurements with a 7 m antenna at 15.5 and 31.6 GHz gave losses of 0.3 and 0.6 dB, respectively, at a 5° elevation angle.
2. **Scintillation and Multipath Influence**—Small-scale irregularities in the atmospheric refractive index cause rapid amplitude variations. In such a way, different tropospheric effects in the absence of precipitation are unlikely to produce serious fading in radio and space telecommunication systems operating at a frequency about below 10 GHz and at elevation angles above 10°.
   Besides, at low elevation angles and at a frequency spectrum above about 10 GHz, tropospheric scintillations can, on occasion, cause serious degradations in performance. The atmosphere scintillation measure model includes frequency, elevation angle, and antenna diameter, but also including meteorological parameters, can be used to account for regional and seasonal dependencies.

3. **Propagation Delays**—Additional propagation delays superimposed on the delay due to free-space propagation are produced by refraction through the troposphere precipitation and the ionosphere. Therefore, at a frequency above 10 GHz, the ionospheric time delay is generally less than that for the troposphere.
4. **Angle-of-Arrival Values**—The gradient of the refractive index of the atmosphere causes a bending of the radio ray and the angle of arrival varies from that calculated based on the geometry of the path. Since the relative index varies largely with altitude, the angle-of-arrival variation is much greater in the elevation than in the azimuth angle. In addition, turbulent irregularities of the refractive index can give rise to angle-of-arrival scintillations. Both of these effects decrease markedly with elevation angle and are generally insignificant for elevation angles above 10°. The effects are independent of frequency.

5.5.3 *Transionospheric Propagation*

Radio waves at frequencies of VHF and above are capable of penetrating the ionosphere, and therefore, they provide transionospheric telecommunications. The ionosphere consists in a layer somewhere between 80 and 150 km altitude, where the density of the atmosphere is very low.

Radiation from the Sun ionizes some molecules and it takes a long time for them to be neutralized by other ions. The concentration of ions varies with height, time of day, the season, and in what part of its 11-year sunspot cycle the Sun happens to be.

5.5.3.1 Faraday Rotation and Group Delay

Ionospheric effects are significant for frequencies up to about 10 GHz and are particularly important for GEO and Non-GEO satellite constellations operating below 3 GHz. At the frequencies used for satellite transmission, signals pass right through and are subject to negligible refraction, less than 0.01° at 30° elevation.

The total electron content (TEC) accumulated through the transionospheric transmission path results in the rotation of the linear polarization of the signal carrier and a time delay in addition to the anticipated propagation path delay. Given knowledge of the TEC, Faraday Rotation, and group delay can be estimated for communication applications. Therefore, this delay is known as the group delay, while the rotation of the linear polarization of the carrier is known as Faraday Rotation. The TEC, denoted by $N_{T,}$ can be evaluated by the formula:

$$N_T = \int_s N_e(s)ds\,[\text{electrons}/\text{m}^2] \tag{5.35}$$

where N_e = electron density [electrons/m^2] and s = propagation path length through the ionosphere [m]. Typically, N_T varies from 1 to 200 TEC units (1 TEC unit = 10^{16} el/m^2). N_T has typical values in the range of 10^{16} and 10^{18} el/m^2. Even when the precise propagation path is known, the elevation of N_T is difficult to determine because N_e is highly variable in space and time. When propagating through the ionosphere, a linearly polarized wave will suffer a gradual rotation of its plane of polarization due to the presence of the geomagnetic field and the anisotropy of the plasma medium. Namely, this trend slows down the signal because the Earth's magnetic field penetrates the ionosphere when ions (charged particles), subject to the alternating electric field of a signal, tend to gyrate around the local line of force. The magnitude of Faraday Rotation will depend on the frequency of the radio wave, the geomagnetic field strength, and the electronic density (concentration) of the plasma as follows:

$$\Phi = N_T(KM/f^2) = 2.36 \times 10^2 B_E N_T f^2 \text{ [radians]} \tag{5.36}$$

where value K = 2.36×10^4 [MKS for units]; M = value of (B_E secϕ) at 420 km of height; B_E = longitudinal component of the Earth's magnetic induction along the ray path [Tesla]; ϕ = zenith angle of the ray and f = frequency [Hz]. Typical values of Φ as a function of frequency for representative TEC values are shown in Fig. 5.16.

Hence, the occurrence of Faraday is well understood and can be predicted with a high degree of accuracy and compensated for by adjusting the polarization tilt angle at the LES. The GPS and similar satellite navigation systems, which use the 1–2 GHz frequency spectrum and depend on measuring the travel time of EM signals, has to correct for this effect.

The presence of charged particles in the ionosphere slows down the propagation of radio signals along the path and produces a phase advance. Thus, the time delay

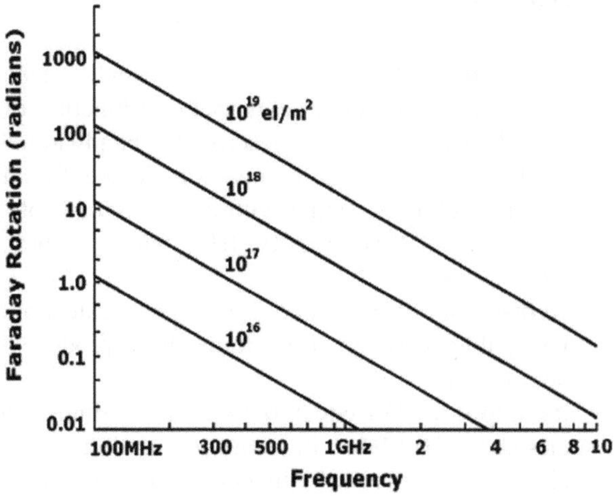

Fig. 5.16 Faraday rotation as a function of TEC and RF—Courtesy of Handbook: by ITU

in excess of the propagation time in free space is called the group delay and is given by

$$T_g = 1.34 \times 10^{-7} N_T / f^2 \text{ [sec]} \tag{5.37}$$

Accordingly, the time delay with reference to propagation in vacuum is an important factor to be considered for digital communication and navigation positioning systems.

5.5.3.2 Ionospheric Scintillation

Ionospheric effects are important at frequencies below 1 GHz, although they may even be important at frequencies above 1 GHz and are dependent on location, season, solar activity (sunspots), and local time. At this point, ionospheric scintillation occurs as short-term, rapid signal fluctuations and is mainly caused by irregularities in the ionosphere ranging from altitudes of 200–600 km. In fact, the frequency dependence depends on the ionospheric conditions, but the attenuation varies approximately at the same rate as the square of the wavelengths. The effect is greater for lower frequencies and at lower latitudes, while high-latitude areas near the Arctic polar region bounded between $\pm$ 20° are susceptible to intense scintillation activity. In the L- and S-band, this effect can be ignored at medium latitudes except during periods of solar activity. When the Sun is very active, L-band enhancement and fading of 6 and −36 dB, respectively, were observed even at 37° latitude. Scintillation activity is at a maximum during the night, lasting from 30 min to a number of hours.

5.5.3.3 Other Ionospheric Effects

1. **Dispersion**—When transionospheric radio signals occupy a significant bandwidth the propagation delay, being a function of frequency, introduces dispersion. The differential delay across the bandwidth is proportional to the integrated electron density along the ray path. Hence, for an integrated electron content of 5×10^{17} el/m^2, a signal with a pulse length of 1 μs will sustain a differential delay of 0.02 μs at 200 MHz, while at 600 MHz delay would be only 0.00074 μs.
2. **Refraction**—When radio waves propagate obliquely through the ionosphere layer, they undergo refraction, which produces a change in the direction of arrival of the ray.
3. **Absorption**—For equatorial and mid-latitude regions, radio waves of frequencies above 70 MHz will assure penetration of the ionosphere without significant absorption, while for frequencies below 70 MHz the ionospheric absorption loss is significant.

4. **Doppler Frequency Shift**—The special effects of frequency change due to the temporal variability of the ionosphere layer upon the apparent frequency of the carrier is the Doppler-shifted carrier. For example, at f = 1.6 GHz (GPS system), the observed frequency change Δf at high latitude is $\Delta f/f < 10^{-9}$.

5.6 Sky Noise Temperature Contributions

The mechanism that causes absorption of energy from a wave passing between space and the Earth also causes the emission of thermal noise at RF. In fact, some radio noise is added to the emission reaching the receiver, whereas the Earth itself radiates noise, which can enter the transmission path via satellite or the LES receiving antennas. Therefore, the radio noise emitted by all matter, while used as a source of information in radio astronomy and remote sensing, may be a limiting factor in communication services.

Otherwise, sources of radio noise of interest on Earth-to-space paths are the atmosphere, clouds, rain, extraterrestrial sources, and noise from the surface of the Earth. Prediction methods are given in the ITU-R P.372 Recommendation. The thermal noise power N, available from a blackbody having a noise temperature of the source T [K], measured in bandwidth B [Hz], is given by the form:

$$N = kTB\,[W] \tag{5.38}$$

where k = Boltzmann's constant. The special power density N_0 of noise from source is

$$N_0 = N/B = kT\,[WHz^{-1}] \tag{5.39}$$

Thus, in considering the level of noise received at an LES or satellite from sources external to the environment, it is convenient to identify a brightness temperature T_B for each separate source and a coefficient (η), which represents the efficiency, with which the receiving antenna captures noise from that source. Then t, the noise temperature component due to the identified source, is given by:

$$t = \eta T_B\,[K] \tag{5.40}$$

Therefore, the total noise entering the system from all of these sources, expressed as a noise temperature, can be obtained by summing all the component noise temperatures.

5.6.1 Environmental Noise Temperature Sources

The LES antennas in GEO satellite infrastructures are typically designed and sited so that the main lobe does not intersect the local terrain or obstructions, such as mountains or large buildings. Side lobes are also minimized to reduce the effect of the Earth's temperatures on the system performance. However, in LMSC systems, the antenna beam may pass through vegetation and be obstructed by buildings or mountain terrain. Measurements suggest that the impact of the additional terrestrial noise is greater when the antenna has a low internal noise temperature, that is, for less directive antennas. Although these obstructions will raise the noise temperature seen by the antenna, they will also cause shadowing or multipath effects, which are likely to be more significant in the total link performance.

Industrial man-made sources of noise affect VHF and UHF frequencies for all but the quietest rural areas. Unlike other noise effects, there is a polarization dependence in that the vertical component is higher that the horizontal. In general, the median level of noise will decrease linearly with log(f). There are significant variations with location and time and little data are available to develop models to predict levels.

Thus, considering the noise received at an LES from the ground, the sea, etc. and buildings nearby, T_B typically lies between 100 and 250 K, approaching the lower limit at sea at low angles of elevation angles and the upper limit on land. The relation for this noise temperature component t_{gr} is same as for t. If the angle of elevation of the main beam and the gain of the antenna are both high and the antenna design is good, with well-suppressed side lobes and little subreflector spillover, the corresponding value of η will be small and t_{gr} may be no more than 20 K. If, however, the gain of the antenna is very low, as is typical of LES, η may reach 0.5 and t_{gr} may exceed 100 K. Satellite antennas directed toward the Earth, having sufficient gain for the main lobe to be filled by the Earth, will also receive terrestrial noise with $\eta \approx 1$, while T_B will be about 210 K if land occupies a large fraction of the beam footprint. However, except in the atmospheric absorption bands, T will be somewhat less, perhaps as little as 160 K, if the sea occupies a large part of the footprint.

5.6.2 Atmospheric Noise Temperature Elements

The noise temperature of a satellite-based antenna is dominated by the high temperature emitted by the Earth, which fills, or mostly fills, the main beam of the antenna. Additional noise from precipitation or other variables is insignificant in this case. For a global beam, the noise temperatures are dependent both on frequency and on the position of the satellite with relation to the major landmasses of the Earth. The ground-based antenna observes the relatively cool sky and therefore, the presence of clouds and rain can significantly raise the noise temperature of the antenna. In general, the brightness temperature of the atmosphere due to the permanent gases and rain, if it is present and seen by an antenna, is given as follows:

$$T_B = T_m(1 - 10^{-A/10})\ [K] \tag{5.41}$$

where T_m = effective temperature of the attenuating medium (atmosphere, clouds, rain), typically about 270 K and A = total attenuation due to the medium. The effect of rain on a satellite downlink is not just the attenuation, but it decreases in C/N due to the higher noise temperature seen in rainy conditions, compared to clear-sky conditions. In some cases, the noise temperature increase can have more effect on the link than the attenuation itself.

5.6.3 Galactic and Other Interplanetary Noise Effects

Noise from interplanetary sources, particularly the Sun, the Moon, and from the galactic background, is well understood and the effect on the total extraterrestrial noise temperature of a system can be calculated with the following relation:

$$T_B = T_g \times 10^{-A/10}\ [K] \tag{5.42}$$

where T_g = temperature of any interplanetary radio sources, including background galactic noise (about 3 K above 3 GHz). Thus, the brightness temperature of the Sun decreases with increasing frequency, from about 10^6 at 30 MHz to 10^4 at 10 GHz under quiet conditions. At 20 GHz, an antenna of 2 m diameter and a beam width of about 0.5° would have an increase in noise temperature of about 8.100 K with a quiet Sun. The Sun and Moon each subtend an angle of about 0.5°, so that if the antenna beam is significantly larger than that, the effect t of the Sun or Moon is averaged with a larger portion of relatively cool sky.

5.7 Path Depolarization Causes

The atmosphere behaves as an anisotropic medium for radio propagation, which is the property of being directionally dependent. Consequently power from one polarization is coupled to its orthogonal component, causing interference between the channels of a dual polarized system.

In this sense, depolarization or cross-polarization may occur when EM waves propagate through media that are anisotropic, namely asymmetrical with respect to the incident of polarization. Meanwhile, depolarization in the form of Faraday Rotation of the plane of linear polarization occurs in the ionosphere because of presence of the Earth's magnetic fields. At this point, the resulting impairments are typically circumvented by using circular polarization at frequencies below 10 GHz, for which the effect can be significant. Depolarization is often the most significant path impairment for 6/4 GHz satellite systems and can be the limiting performance factor for some 14/11 GHz satellite paths, especially at lower path elevation angles in moderate rain climates.

On the other hand, depolarization in precipitation is caused by differential attenuation and phase shifts that are induced between orthogonal components of an incident wave by anisotropic hydrometeors. Orthogonally polarized radio waves propagating in a medium that causes only differential phase shift are depolarized but maintain orthogonality. If the medium induces differential attenuation, the waves are also deorthogonalized.

5.7.1 Depolarization and Polarization Components

The importance of depolarization for satellite communications systems depends on a few components: f = frequency signal, geometry of path (θ = elevation angle and τ = tilt angle of the received polarization), local climatic factors (severity of the rain), and sensitivity to cross-polar interference (whether the system employs frequency reuse).

The EM waves comprise both the electric and magnetic field vectors. Therefore, these two components travel in the direction of the transmission path and are orthogonal, while the orientation of the electric field vector defines the polarization of the transmitted waves. In general, as the wave progresses in time the tip of the electric vector traces an ellipse in a plane perpendicular to the propagation direction. A representative polarization ellipse of representative elliptically polarized radio wave is displayed in Fig. 5.17. Two important parameters in this figure are the axial ratio and τ = the inclination tilted angle with respect to the reference axis. The polarization ellipse may be tilted at an angle (τ) with respect to the particular coordinate frame. Thus, the general form of a polarized wave, when viewed perpendicular to the direction of travel, is elliptical in shape.

The polarization state of a wave is completely specified by its polarization ellipse, i.e., the amplitudes of the major axis (E_{max}), the minor axis (E_{min}), and the sense of rotation of the vector also defines the axial ratio by using the following expression:

$$A_R = 20\log\left(E_{max}/E_{min}\right)\ [\mathrm{dB}] \tag{5.43}$$

In satellite communications, four types of polarization are employed, shown in Fig. 5.17: (a) vertical linear polarization (VLP), (b) horizontal linear polarization (HLP), (c) left-hand circular polarization (LHCP), and (d) right-hand circular polarization (RHCP). The direction of the travel is symbolical “into the paper.” Horizontal and vertical polarizations are defined with respect to the horizon, LHCP

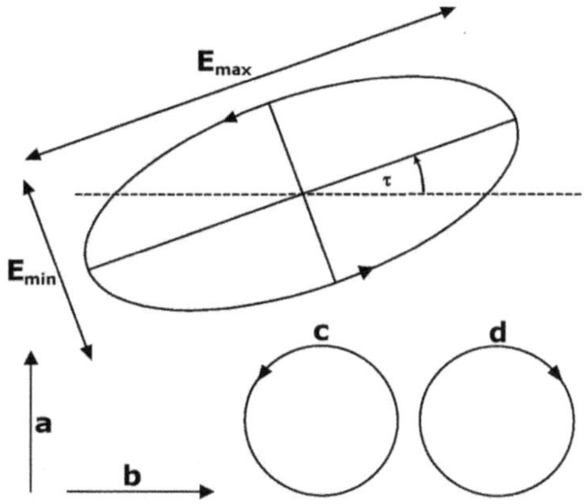

Fig. 5.17 Generalized elliptical waveform—Courtesy of Book: by Sheriff

has an anticlockwise and RHCP has a clockwise rotation when viewed from the antenna in the direction of travel.

Therefore, if E_{max} and E_{min} are equal in magnitude, the polarization state can be RHCP or LHCP, depending on the sense of rotation and in the case where E_{max} is nonzero and E_{min} is zero, the value of electric vector maintains a constant orientation defined by E_{max} and the polarization state is said to be linear.

The polarization quantity of interest for frequency reuse communication systems is the cross-polar isolation (XPI), defined as the decibel ratio of the (desired) copolar power received in a channel to the (undesired) cross-polar power received in that same channel. In practice, however, XPI is difficult to measure because the cross-polarized components cannot be distinguished from noise in the copolar channel.

The quantity usually measured is the XPD, defined as the ratio of the copolarized power received in one channel to the cross-polarized power detected in the orthogonal channel, both arising from the same transmitted signal. Moreover, theory predicts that XPD and XPI components are equivalent for most practical situations.

In fact, a polarized wave will comprise the wanted polarization together with some energy transmitted on the orthogonal polarization. The degree of the coupling of energy between polarizations is given by

$$\mathrm{XPD} = 20\log\left|E_{cpr}/E_{xpr}\right|\ [\mathrm{dB}] \tag{5.44}$$

where parameters E_{cpr} = received copolarized electric field strength and E_{xpr} = received cross-polarized electric field strength.

For the case of coexisting circular polarized waves, the XPD can be determined from the axial ratio by the expression:

$$\mathrm{XPD} = 20\log\left(A_R + 1/A_R - 1\right)\ [\mathrm{dB}] \tag{5.45}$$

5.7.2 *Relation Between Depolarization and Attenuation*

The ITU provides a step-by-step method for the calculation of hydrometer-induced cross-polarization, which is valid for frequencies within the range 8 GHz < f < 35 GHz and for elevation angles less than 60°. Also, the attenuation due to rain exceeded for the required percentage of time p and the polarization tilt angle τ, with respect to the horizontal, also needs to be known. The XPD due to rain is given by the following equation:

$$XPD_{rain} = C_f - C_A + C_\tau + C_\theta + C_\sigma \text{ [dB]} \quad (5.46)$$

where C_f = frequency-dependent term or 30 logf [GHz]; C_A = rain-dependent term or V(f) logA_p; C_τ = polarization improvement factor or −10log [1−0.484 (1 + cos4τ)]; C_θ = elevation angle-dependent term or −10log (cosθ) and C_σ = canting angle term or 0.0052σ. In the above, the canting angle refers to the angle at which a falling raindrop arrives at the Earth with respect to the local horizon. Terms τ, θ, and σ are expressed in degrees, while σ has value of 0°, 5°, 10°, and 15° for 1, 0.1, 0.01, and 0.001 % of the time, respectively. Taking ice into account, the XPD not exceeded for p% of time is given by

$$XPD_p = XPD_{rain} - C_{ice} \text{ [dB]} \quad (5.47)$$

where C_{ice} = ice depolarization or XPD_{rain} × (0.3 + 0.1 log p)/2.

5.8 Propagation Effects Important for GMSC Systems

The MES operates in a dynamic and often unsuitable environment in which propagation conditions are constantly changing. Namely, satellite transmission path profile in GMSC varies continuously while the mobiles are in motion. The MES use relatively broad beam antenna systems, which have only a limited discrimination against signals reflected from scattered objects and surfaces. Because of the inherent random nature of disturbances, radio signals are usually characterized statistically. In general, a signal arriving at the antenna of a MES consists in the vector sum of a direct component and diffused components arising from multipath reflection. The resultant effects of these additional considerations are as follows:

(1) Signals suffer attenuation whenever the satellite path to mobiles is shadowed.
(2) Signals fluctuate randomly because reflected and scattered random signal components arriving at the mobile antenna are picked up, which is known as the multipath.
(3) The power spectral density of multipath noise is a function of the mobile's speed and the environmental conditions.

Depending on the environment in which a mobile operates, the satellite channels in GMSC may be categorized as maritime, land, and aeronautically based. Each category has its typical channel characteristics because of the different propagation environments. Three different categories of MES: SES, VES, and AES, with their own distinctive channels, need to be considered separately. Therefore, the local operation variable environment has a significant impact on the achievable quality of transmission service in GMSC.

On the contrary, the LES or Gateway terminals serving in a mobile satellite network can be optimally located in fixed positions with constant channel characteristics to guarantee good visibility to the satellite at all times, eliminating or reducing the negative and undesirable effect of the local environment to a minimum.

5.8.1 Propagation in MMSC Systems

Maritime GMSC requirements have specific tasks and problems for shipping designers and operators. In fact, the modern ship is constructed of steel, which while floating on salt seawater becomes a very effective electromagnetic shield capable of rejecting and reflecting part of radio waves. Meanwhile, modern ocean-going vessels have reduced or removed many very heavy traditional deck structures, which can affect transmission and reception. Hence, ships are sailing sometimes on very rough seas, which cause the motion and vibration of satellite antennas. In this case, the mobile satellite antenna goes quickly out of satellite focus and causes interrupted or intermediate transmission.

Afterward, multipath fading caused by reflections from the sea surface can significantly impair MMSC channels, especially at the low elevation angles of an antenna system onboard ships. Fading characteristics depend on the antenna gain of the SES, elevation angle, sea surface conditions (smooth or rough), and so on. A number of multipath fading reduction techniques, such as polarization shaping, antenna pattern shaping, multielement antenna systems, and error-correction coding have to be considered.

Furthermore, MMSC systems may operate on a worldwide basis, including paths with low elevation angles. Due to the use of L-band frequencies for the Inmarsat system, the effect of ionospheric scintillation is not negligible, particularly in equatorial regions during years of high solar activity on an 11-year cycle. On the other hand, tropospheric effects, such as rain attenuation and scintillation, will be negligible for the said frequency bands. But signal-level attenuation due to blocking by the ship's superstructure is a considerable problem, for which there are some adequate solutions and modifications.

In any event, the necessity of communications to ships via satellite is greater every year. Due to safety regulations of the IMO and GMDSS, communication channels to vessels need to be specified to a high degree of reliability. A typical voyage onboard a ship comprises the following phases:

(1) The ship, after sailing, usually uses coastal navigation to a convenient position on the open sea ready to take over the shortest course or orthodrome toward the next port of call.
(2) Celestial and satellite navigation in a constant open sea environment.
(3) Going over to coastal navigation and maneuvering the ship at anchorage or in port.
(4) The ship is moored in port under merchant operations.

Each of the above phases can be considered to have particular channel characteristics. For instance, while the ship is in the port, an LMSC channel-type environment could be envisaged, where it is subject to sporadic shadowing due to buildings, cranes, other ships, and obstacles. However, during coastal navigation ship's satellite antennas could be under the influence of ground refraction and diffuse reflected components. With regard to effects of multipath reflection from the sea, this will be explained later in this chapter.

Clearly, the MMSC network offers a very different propagation scenario to that in the LMSC case. Whereas with the LMSC channel, the modeling of the specular ground reflection is largely ignored, in the instance of MMSC reflections from the surface of the sea provide the major propagation impairment. At this point, such impairments are especially severe when using antennas of wide beam width, while operating at a low elevation angle to the satellite. Such a scenario is not untypical in the MMSC environment.

5.8.2 Propagation in LMSC Systems

Land GMSC requirements have different problems to those for MMSC application because of a more specific local environment. Namely, when the direct satellite signal is obstructed by roadside trees, utility poles, buildings, tunnels, hilly terrain, mountains, or other kinds of blockage, communication outages will occur. If the disturbance lasts for a long period of time, the communication will be disconnected. At all events, it is necessary to know the statistical characteristics of signal attenuation and its frequency as well as the duration of outages in given service areas, such as urban, suburban, rural, vegetation, and mountain areas. However, in comparison with the modeling of the MMSC propagation channel, results for the LMSC channel are relatively few.

Radio waves reflected from mountains, hills, and artificial structures, such as buildings and bridges, cause interference to the direct signal from a satellite and cause amplitude and phase fluctuations in the received signal. Namely, such multipath fading may degrade the BER performance for digital modulation. The received LMSC signal is combined with three components: the direct line-of-sight wave, the diffuse wave, and the specular ground reflection. The direct wave arrives at the receiver without reflection from the surrounding environment.

Besides, the only L- and S-band propagation impairments that significantly affect the direct component are free-space loss and shadowing. The diffuse component comprises multipath reflected signals from the different nearby surrounding environment, such as buildings, mountains, trees, and telegraph poles. Without consideration of LMSC, networks rely on multipath propagation, where multipath has only a minor effect on mobile satellite links in most practical operating environments. The specular ground components are results of the reflected signal reception from the ground near to the mobile. Antenna of low gain and wide beam width operating through satellites with low elevation angle are particularly susceptible to this form of impairment. In fact, such a scenario could also include GMPSC systems operating via LEO spacecraft. The first step toward designing an LMSC channel is to define and divide certain transmission environments into typical categories as follows:

(1) Urban areas are characterized by almost complete obstruction of the direct wave.
(2) Suburban, mountains, and tree shadowing environments are affected when intermittent partial obstructions of the direct wave occur.
(3) Open and rural areas are not affected by obstruction of the direct wave.

The last two of the above environments are of particular interest for LMSS. However, in urban environments, it is difficult to guarantee visibility to the satellite, resulting in the multipath component dominating reception. Currently, it is not very economical and practical to use LMSC system in urban areas because it is possible to employ landline infrastructure or dual cellular/GMPSC networks, such as Globalstar and Iridium.

In open and rural areas, the multipath phenomenon is the most dominant link impairment. The multipath components can either add constructively, with enhancement of signal, or destructively, with fade caused to the direct wave component. This results in the received MES transmissions being subject to significant fluctuations in signal power.

In suburban areas, the major contribution to signal degradation is caused by buildings, trees, and other man-made obstacles. These obstacles manifest as shadowing of the direct signals, which results in attenuation of the received signal. The MES motion through suburban areas results in continuous variation of the received signal strength and variation in the received phase.

5.8.3 Propagation in AMSC Systems

Aeronautical GMSC requirements and propagation conditions are superior to those in MMSC and LMSC because there are no difficulties and obstacles between a satellite and AES. However, at low elevation angles and when a low-gain antenna is used, multipath fading caused by reflection from the sea or ground surface occurs, although it is less than in the maritime case. In an AMSC, an aircraft demodulator must track the received signal and remove the Doppler effect due to the high speed of flight, using digital signal processing or pilot signal. The ability of communications to aircraft via satellite is becoming increasingly important. Due to the ICAO safety regulations, communication channels to an aircraft need to be specified to a high degree of reliability. A typical flight has the following phases:

(1) The aircraft taxies to a position on the runway, ready to takeoff.
(2) Takeoff and ascension to cruise altitude at a constant height above the cloud layer.
(3) Descent from cruise altitude to a landing on the runway.
(4) Taxiing from the runway to a standby place in the airport for a new flight.

Each of the above phases can be considered to have particular channel characteristics. For example, while the aircraft is in the airport, an LMSC channel-type environment could be subject to sporadic shadowing due to adjacent buildings, airport structures, other aircraft, and close obstacles. In effect, the AMSC channel is further complicated by the maneuvers performed by an aircraft during the course of a flight, which could result in the aircraft's structure blocking the line of sight to the satellite. The body of the airplane is also a source of multipath reflection, which also needs to be considered. In fact, as was discussed earlier, the speed of an aircraft introduces large Doppler spreads as an additional effect.

The effect of multipath reflections from the sea for circular polarized L-band transmissions will be explained later, together with the maritime case. Thus, ITU provides a methodology similar to that for the maritime link, for determining the multipath power resulting from specular reflection from the sea. Thus, the ITU Recommendation includes a methodology to derive the mean multipath power as a function of elevation angle and mobile antenna gain. In this sense, by applying this method for an aircraft position 10 km above the sea and for a minimum elevation angle of 10°, the relative multipath power will be in the range of approximately −10 to −17 dB, for antenna gains varying from 0 to 18 dBi, respectively.

Presently, AMSC systems are served by the L-band. Due to the bandwidth restrictions at this spectra, services are limited to voice and low data rate applications. Inmarsat recently improved speed of transmission with a new generation of satellite constellation. The need to provide broadband multimedia satellite services, akin to those envisaged by new satellite UMTS/IMT-2000, will require the move up in frequency to the next suitable bandwidth, the K- and Ka-bands. At these frequencies, tropospheric effects will have an impact on link availability during the time when the aircraft is below the cloud layer. The channel characteristics for a K-band satellite system have been investigated in Europe and the USA. Thus, here a Rice factor of 34 dB is reported for line-of-sight operation, while shadowing introduced by the aircraft's wing during a turning maneuver resulted in a fade of 15 dB.

5.8.4 Surface Reflection and Local Environmental Effects

Surface reflections and local environmental effects are important for GMSC systems because such factors generally tend to impair the performance of satellite communications links, although signal enhancements are also occasionally observed.

Local environmental effects include shadowing and blockage from objects and vegetation near the MES. At this point, surface reflections are generated either in the immediate vicinity of the MES terminals or from distant reflectors, such as mountains and large industrial infrastructures. However, the reflected transmission signal can interfere with the direct signal from the satellite to produce unacceptable levels of signal degradation. In addition to the fading, signal degradations can include intersymbol interference, arising from delayed replicas.

The impact of the impairments depends on the specific application, namely in the case of typical LMSC links, all measurements and theoretical analysis indicate that the specular reflection component is usually negligible for path elevation angles above 20°.

Moreover, for handheld terminals, specular reflections may be important as the low antenna directivity increases the potential for significant specular reflection effects. For GMSC system links, design reflection multipath fading, in combination with possible shadowing and blockage of the direct signal from the satellite, is generally the dominant system impairment.

5.8.4.1 Reflection from the Earth's Surface

Prediction of the propagation impairments caused by reflections from the Earth's surface and from different objects (buildings, hills, vegetation) on the surface is difficult because the possible impairment scenarios are quite numerous, complex, and often cannot be easily quantified. For example, the degree of shadowing in LMSC satellite links frequently cannot be precisely specified.

Therefore, impairment prediction models for some complicated situations, especially for LMSC links, tend to be primarily empirical, while more analytical models, such as those used to predict sea-reflection fading, have restricted regions of applicability. Nevertheless, the basic features of surface reflections and the resultant effects on propagating signals can be understood in terms of the general theory of surface reflections, as summarized in the following classification:

1. **Specular Reflection from a Plane Earth**—Here, the specular reflection coefficient for vertical polarization is less than or equal to the coefficient for horizontal polarization. Thus, the polarization of the reflected waves will be different from the polarization of the incident wave if the incident polarization is not purely horizontal or purely vertical. For example, a circularly polarized incident wave becomes elliptically polarized after reflection.
2. **Specular Reflection from a Smooth Spherical Earth**—In this case, the incident grazing angle is equal to the angle of reflection. The amplitude of the

reflected signal is equal to the amplitude of the incident signal multiplied by the modules of the reflection coefficient.

3. **Divergence Factor**—When rays are specularly reflected from a spherical surface, there is an effective reduction in the reflection coefficient, which is actually a geometrical effect arising from the divergence of the rays.
4. **Reflection from Rough Surface**—In many practical cases, the surface of the Earth is not smooth. Namely, when the surface is rough, the reflected signal has two components: one is a specular component, which is coherent with the incident signal, while the other is a diffuse component, which fluctuates in amplitude and phase with a Rayleigh distribution.
5. **Total Reflected Field**—The total field above a reflecting surface is a result of the direct field, the coherent specular component, and the random diffuse component.
6. **Reflection Multipath**—Owing to the existence of surface reflection phenomena signals may arrive at a receiver from multiple apparent sources. Thus, the combination of the direct signal (line of sight) with specular and diffusely reflected waves causes signal fading at the receiver. The resultant multipath fading, in combination with varying levels of shadowing and blockage of the line-of-sight components, can cause the received signal power to fade severely and rapidly for MES and is really the dominant impairment in the GMSC service.

5.8.4.2 Fading in MMSC and AMSC Systems Due to Sea Surface Reflection

Multipath fading due to sea reflection of EM signals is caused by interference between direct and reflected radio waves. The reflected radio waves are composed of coherent and incoherent components, namely specular and diffuse reflections, respectively, that fluctuate with time due to the motion of sea waves. The coherent component is predominant under calm sea conditions and at low elevation angles, whereas the incoherent becomes significant in rough sea conditions. If the intensity of the coherent component and the variance of the incoherent component are both known, the cumulative time distribution of the signal intensity can be determined by statistical consideration.

In any event, a prediction model for multipath fading due to sea reflection, however, was first developed for MMSC systems at a frequency near 1.5 GHz. Although the mechanism of sea reflection is common for MMSC and AMSC systems, only with the difference that fading characteristics for AMSC are expected to differ from those for MMSC, this is because the speed and altitude of aircraft so much greater than those of ships. At this point, the effects of refractions and scattering by the sea surface become quite severe in case of MMSC and AMSC, particularly where antennas with wide beam widths are used.

The most common parameter used to describe sea condition is the significant wave height (H), defined as the average value of the peak-to-trough heights of the highest one-third of all waves. Empirically, H is related to the r.m.s. height (h_o) by

$$H = 4h_0 \tag{5.48}$$

Hence, at 1.5 GHz the smaller-scale waves can be neglected and the r.m.s. value of the sea surface slopes appear to fall between 0.04 to 0.07 in the case of wave heights less than 4 m.

Thus, with diminishing satellite elevation angle, the propagation path increases, causing a decrease of signal power at the Rx side. The noise level is initially constant; however, upon reaching some critical value of the elevation angle, sea-reflection signals appear at the Rx input, which begins to affect the C/N value. To include the effect of multipath interference caused by sea-refracted signals, the reception quality would be more properly described by the C/N plus M, where M is an interfering sea-reflected signal acting as a disturbance. Thus, sea-reflected signals differ in structure and can be divided into two categories:

1. Radio signals with the rapid continuous fluctuations of amplitudes and phases and with a possible frequency shift due to the motion of small portions of the specular cross section relative to the source of signals (noise or diffused components).
2. Radio waves with relatively slowly changing phase close to the phase of the basic signal and with amplitude correlating with that of the basic signal (specular component).

Consequently, within the overall specular cross section, an angle of arrival reflected radio signals relative to the horizontal plane may be regarded as constant and can be described by the following expression:

$$\alpha = 90^{\circ} - \gamma \tag{5.49}$$

where α = angle of radio signals arrival in accordance with Fig. 5.18 and γ = reflection angle. In this figure is presented satellite antenna onboard oceangoing ship, but can be also presented the similar antenna onboard aircraft instead. The modulus of sea-reflection factor for L-band signals is within 0.8 and 0.9, which means that the amplitude of the specular reflected signal is nearly the same as that of the direct signal.

As measurements have shown, the noise component depends only upon an elevation angle and a wave height. Decreasing the elevation angle and increasing the wave height result in an increase in the total amplitude of the noise, which includes the noise component. At elevation angles below 5°, the amplitude component reaches a peak value and is no longer affected by the wave height. Now an increase of the wave height causes primarily more frequent variations in the noise component. The corresponding deviation of C.N measured in 1 kHz bandwidth amounts from 4.5 to 5 dB.

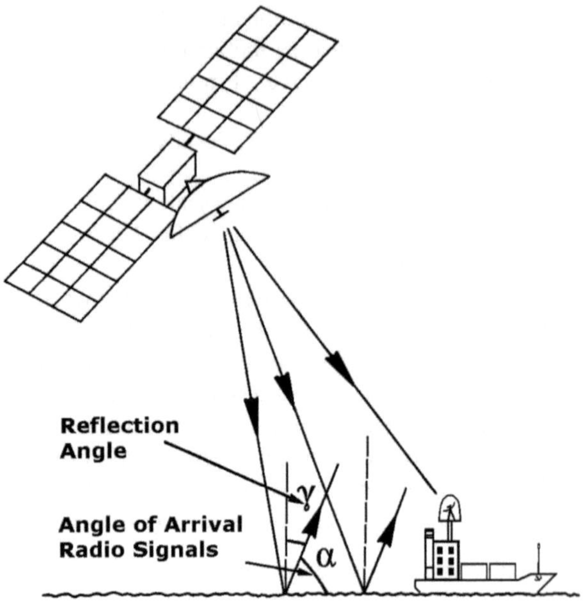

Fig. 5.18 Geometry of sea reflection of satellite radio signals—Courtesy of Paper: by Volodin

The specular component that appears at the Rx input together with the direct signal causes fading in the direct signal due to both the minor difference between their phases and the slow change of the parameters of the reflected signals. The ratio of the direct to specular reflected signal can be described as follows:

$$C/M = (C + G_\varepsilon) - [C - G_{(\alpha+\varepsilon)}] \quad (5.50)$$

where C = direct signal; M = specular reflected signal power from the sea; G_ε = maximum gain of the receive SES antenna pointing toward the satellite; ε = elevation angle and α, as is shown in Fig. 5.18. In addition, keeping accuracy sufficient for practical purposes, the previous relation gives:

$$C/M = \beta_{C/N} + [G_\varepsilon - G_{(\alpha+\varepsilon)}] \quad (5.51)$$

where $\beta_{C/N}$ = deviation of C/N ratio. With decreasing elevation angle, the C/M diminishes monotonically, except for the elevation angle range of 5°–8°, within which a rise in C/N is observed. This is obviously due to the fact that at the said angles the difference in path between the direct and the specular signal becomes negligible, so that conditions appear close to the summation of the similar signals at the receiver input. An increase of the C/N plus M ratio is observed simultaneously due to reaching a peak value of amplitude in the noise (diffused) component. In fact, experimental measurements show that as the elevation angle decreases from 10° to 1°, the mean C/N plus M diminishes from 22–24 dB to 17–18 dB, with the deviation increasing from 1.5–2 dB to 4.5–5 dB.

5.8.4.3 Multipath Fading Calculation Model for Reflection from the Sea

The amplitude of the resultant signal at the SES terminal, being the sum of the direct wave component, the coherent, and the incoherent reflection components, has a Nakagami–Rice distribution (see ITU-R P.1057). The cumulative distribution of fading depends on the coherent-to-incoherent signal intensities. For example, in the case of rough sea conditions at 1.5 GHz, the coherent reflection from the sea is virtually non-existent and the coherent signal is composed only of the direct component. Therefore, the fading is determined by the carrier-to-multipath ratio (C/M), i.e., the power ratio of the direct signal and multipath component caused by incoherent reflection. The maximum fade depth (Φ_{max}) occurs when the coherent multipath signal is in antiphase with the direct signal, given by:

$$\Phi_{max} = -20\log(1 - A_r)[dB] \tag{5.52}$$

where A_r = amplitude of the coherently reflected component. The value decreases rapidly with increasing wave height, elevation angle, and RF.

In practice, due to the vertical motion of the ship antenna relative to average sea surface height, the maximum fade value will seldom occur. By adding Φ_{max} and $\Phi_i(p)$ as signal fade due to the incoherent component in the function of time percentage (p), a practical estimate of the combined fading effects of the coherent and incoherent multipath signal for sea conditions is obtained:

$$\Phi_c = \Phi_{max} + \Phi_i(p) \tag{5.53}$$

The maximum fade value due to the coherent component will not occur constantly because of the vertical motion of the ship antenna relative to average sea surface height; therefore, the estimate using this equation seems to give the worst-case value. In practice, for low elevation angles (less than 10°) at around L-band frequencies, the maximum fading occurs when the significant sea wave height is between 1.5 and 3 m, where the coherent reflected component is negligible. Accordingly, the dependence of fading depth on wave height in this range is relatively small.

The amplitude level of the coherent component decreases rapidly with increasing sea wave height, elevation angle, and frequency. Figure 5.19a illustrates the relationship between coherent reflection and significant wave height. Namely, estimates of amplitude of the coherent component for an omnidirectional antenna as a function of a significant wave height for low elevation angles are illustrated; the frequency is 1.5 GHz and polarization is circular. Thus, the incoherent component is random in both amplitude and phase, since it originates from a large number of reflecting facets on the sea's waves. The amplitude of this component follows Rayleigh distribution and the phase has a uniform distribution.

Since the theoretical model concerning the incoherent components is not suitable for engineering computations using a small calculator, simpler prediction models

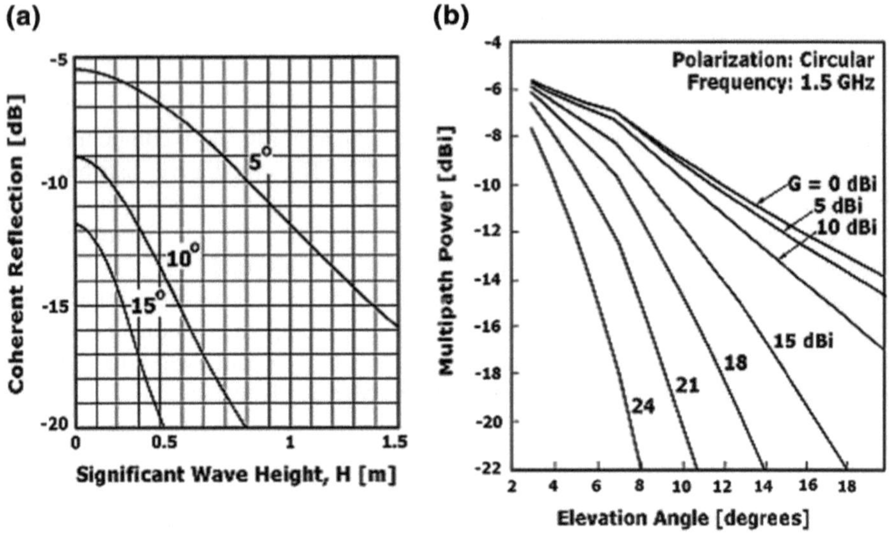

Fig. 5.19 Estimates of coherent reflection and multipath power—Courtesy of Book: by Fujimoto

are useful for the approximate calculation of fading. Such simple methods for predicting multipath power or fading depth have been recently developed.

In Fig. 5.19b is presented relation between multipath power and elevation angle for different antenna gains. Although fading depth depends slightly on sea surface conditions, even if the incoherent is dominant, the simple model is useful for a rough estimate of fading depth. Fading depth, which is a scale of intensity of fading, is usually defined by the difference in decibels between the direct wave signal level and the signal level for 99 % of the time. It can be approximated by a 50 % to 99 % value for fading where the incoherent component is fully developed. Large fading depths usually appear in rough sea conditions, where the incoherent component is dominant. In Fig. 5.20a is shown the fading depth estimated by the simple method for antenna not exceeding for 99 % of the time and the corresponding C/M ratio for circular polarization at 1.5 GHz band under the condition of significant wave heights from 1.5 to 3 m. The antenna gains of 24, 20, 15, and 8 dB are functions of elevation angle with a fully developed incoherent component. The calculation is based on the theoretical method, where the shaded area covers the practical range of the sea wave slope, which depends on fading depth in rough sea conditions. Values estimated by this simple method give the mean values of those given in Fig. 5.20b.

On the other hand, as the theoretical model is not suitable for engineering computations using a small calculator, these simple prediction models are really useful for the approximate calculation of fading or interference. Such simple methods for predicting multipath power or fading depth have been developed by Sandrin and Fang (1986) and by Karasawa and Shiokawa (1988) for MMSS and Karasawa (1990) for AMSS.

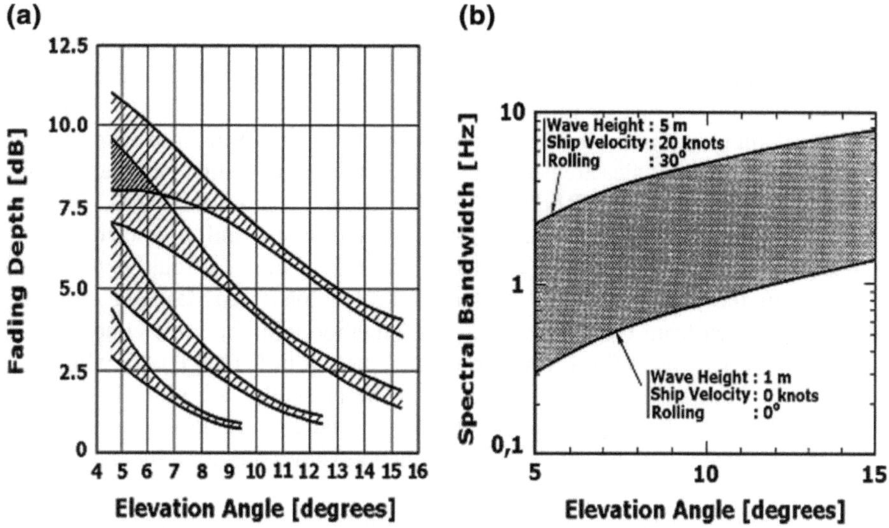

Fig. 5.20 Estimates of fading depth and spectral bandwidth—Courtesy of Book: by Fujimoto

Furthermore, the frequency spectral bandwidth of temporal amplitude variations enlarges with increasing wave height and elevation angle. Figure 5.20b shows the probable range of –10 dB spectral bandwidth (which is defined by the frequency corresponding to the spectral power density of −10 dB relative to the flat portion of power spectrum) of L-band multipath fading obtained by the theoretical fading model as a function of the elevation angle under the usual conditions of MMSC, namely significant wave height of 1 m–5 m, ship speed of 0–20 knots, and rolling conditions of 0°–30°.

5.8.4.4 Other Estimations of Fading for MMSC and AMSC Systems

The error pattern in digital transmission systems affected by multipath fading is usually of the burst type. Accordingly, a firm understanding of the fade duration statistics of burst-type fading is required. Mean values of fade duration (Φ_D) and fade occurrence interval (Φ_o) for a given threshold level as a function of time percentage can be estimated from the fading spectrum. A simple method for predicting the mean value from the −10 dB spectral bandwidth is available as a theoretical fading model. Predicted values of (Φ_D) and (Φ_o) for 99 % of the time at an elevation angle from 5 to 10° are 0.05 to 0.4 s for (Φ_D) and 5 to 40 s for (Φ_o). The probability density function of (Φ_D) and (Φ_o) at any percentages ranging from 50 to 99 % approximates an exponential distribution.

1. **Simple Prediction Method of Fading Depth**—According to theoretical analysis and experimental results made by the mentioned researchers in Japan, the

lowest elevation angle Earth-to-space path at 1.5 GHz RF-band satisfies the energy conservation law: [Power of coherent component] + [Average power of incoherent component] ~ Constant. If this expression is satisfied, the maximum incoherent power can be estimated easily by calculating the coherent power at u = 0. Otherwise, for a more accurate estimation, small modifications of some parameter dependencies are necessary. The modified procedure has been adopted in P.680 for MMSC and P.682 ITU-R Recommendations for AMSC.

In Fig. 5.21a is shown a scattergram of measured and predicted altitude dependence of fading depths (i.e., fade for 99 % of the time relative to that for 50 %) in the case of MMSC systems between measured data and predicted values derived from the simple calculation method with the same conditions. In this figure, Φ_{dp2} are values from the method set out in ITU-R P.680, while Φ_{dp1} are those from an alternative procedure of the prediction method for scattering angles. On the other hand, it is evident that the measured values given by these methods agree well with the same experimental values although the methods are rather approximate.

In Fig. 5.21b is shown the altitude dependence of signal fade depth not exceeded for 99 % of the time vs. antenna height onboard ships or aircraft. This experiment was obtained from measurements with a helicopter together with the calculated values from the simple estimation method of the solid line and the theoretical model of the shaded region in the figure. From the figure, it can be seen that the simple prediction method agrees well with both the theoretical model and measured data even in the case of the AMSC system.

2. **Fading Spectrum**—In system design, particularly for digital transmission systems, it is important not only to estimate the fading depth but also to know the properties of temporal variation, such as the frequency power spectrum. For MMSC systems, theoretical analyses were carried out in Japan and all

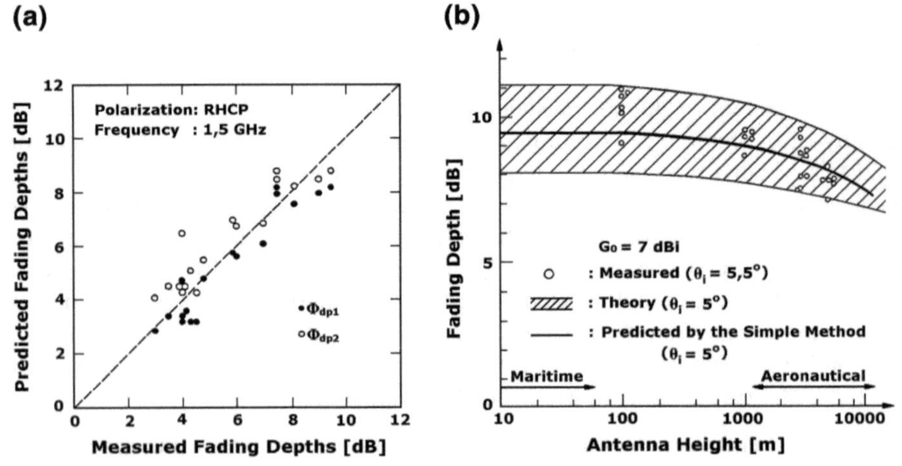

Fig. 5.21 Scattergram and altitude dependence of fading depth—Courtesy of Handbook: by ITU

parameters affecting the spectrum such as wave height, wave direction, ship's direction and velocity, path elevation angle, and antenna height variations due to ship's motion (rolling and pitching) were taken into account. In general, spectrum bandwidth is broader with increasing wave height, elevation angle, ship velocity, and the relative motion of the ship-borne antenna. The dependence of the spectral shape on antenna polarization and gain is usually very small. Moreover, since the speed of aircraft is significantly higher than that of the ships, the fluctuation speed of multipath fading in AMSC is much faster than that in MMSC, depending on the flight elevation angle measured from the horizontal plane. The calculated −10 dB spectral bandwidth is between 20 and 200 Hz for elevation angles of 5°–20°, for flight elevation angles 0°–5° at a speed of 1000 km/h.

5.8.4.5 Fading in LMSC System Due to Signal Blockage and Shadowing

Recently, in the USA, Canada, Australia, Japan, and Europe, domestic LMSC and GMPSC services have started. The main purpose of these systems is to extend MSC voice services to rural/remote areas where terrestrial/cellular services are not provided. In a typical urban environment, line of sight for cellular systems sometimes is not available due to blockage by buildings and other structures, when a MES can receive many waves reflected from these structures and conduct communication link using these signals. The LMSC system can be expected to use the direct line-of-sight signal from a satellite because of the high elevation angles. When the LoS is blocked by any obstacle, MSC is not available, but using path diversity the link is available because of overlapping of two or more signal from adjacent satellites. At present, path diversity from separate satellites is rarely used for GEO MSS, but Non-GEO MSS have an inherent capability to exploit diversity, because the number of satellites is large providing path diversity. Thus, Fig. 5.22 shows a typical propagation environment for scattering, multipath fading, shadowing, diffusion, and so on.

Therefore, to design LMSC system, one needs information about the propagation statistics of multipath fading and shadowing. A vehicle runs at a distance of 5–20 m from roadside obstacles using an omnidirectional antenna, which has azimuthally uniform gain but elevation directivity, or a medium or high-gain antenna with automatic tracking capability. Thus, signal blockage and shadowing effects occur when an obstacle, such as roadside trees, overpasses, bridges, tunnels, utility poles, high buildings, hills or mountains, impedes visibility to the focus of satellite. This results in the attenuation of the received signal to such an extent that transmissions meeting a certain quality of service may not be possible. At any rate, in the shadowing environments the presence of the trees will result in the random attenuation of the strength of the direct path signal. Hence, the depth of the fade is dependent on a number of parameters including tree type, height, as well as season due to the leaf density on the trees.

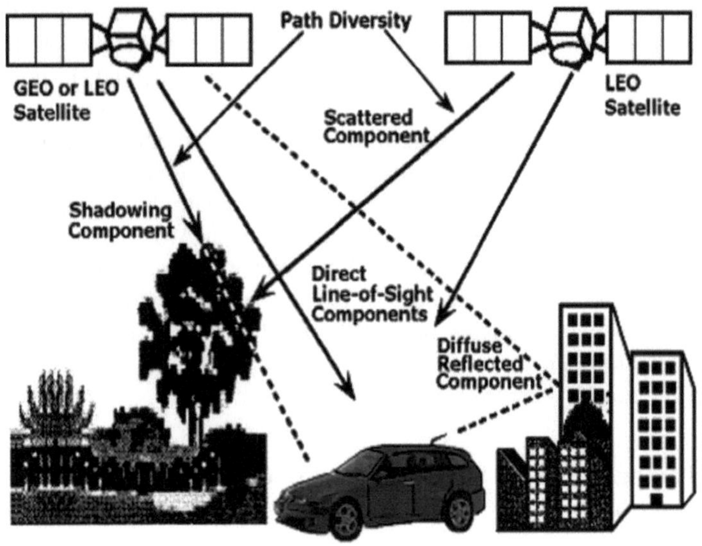

Fig. 5.22 Typical LMSC propagation environment—Courtesy of Book: by Ohmori

Whether a VES is transmitting on the left- or right-hand side of the road could also have a bearing on the depth of the fade, due to the line-of-sight path length variation through the tree canopy being different for each side of the road. In such a way, fades of up to 20 dB at the L-band may be presented due to shadowing caused by roadside trees. This shadowing by roadside trees cannot occur on modern highways because they are free of trees, only sometimes can shadowing appear by tunnels, very big constructions, bridges, and mountains or hills in narrow passages.

1. **Tree Shadowing**—Attenuation due to trees nearby the roads arises from absorption by leaves and blockage by trunks and branches. Absorption by leaves is a function of the type and size of leaves and the water content therein. Blockage due to trunks is primarily a function of their size. In addition to attenuation of the direct signal, trees also cause an incoherent component due to signals reflected and diffracted off the tree surfaces.

The overall attenuation from different types of fully foliated trees varies from 10.6 to 14.3 dB and the attenuation coefficient is from 1.3 to 1.8 dB/m. Measurements were conducted with MES Rx in a rural environment. Based on these average values, a frequency scaling law for the attenuation coefficient has been derived (Goldhirsh and Vogel) by

$$a_1 = a_0\sqrt{f_1/f_0}\ [\mathrm{dB/m}] \tag{5.54}$$

where a_1 and a_0 = attenuation coefficients at frequencies f_1 and f_0 [GHz], respectively. Hence, the range of variation for a_1 at 1.5 GHz is from 0.5 to 1.7 dB/m.

Moreover, trees without foliage attenuate less and the reduction in attenuation appears to be proportional to the total attenuation experienced when the tree is fully foliated. The received strength of the direct signal behind a tree will depend on the orientation of the signal path with respect to the tree. Therefore, the amount of absorbing matter lying along the path will determine the degree of attenuation and hence, on average, the length of the signal path through the tree shield can be considered a major factor in determining the signal level.

The path length is a function of the elevation angle and the distance between the receiver and the tree. At this point, the average attenuation behind an isolated tree can be estimated as the product of the attenuation coefficient and the path length through the tree. Thus, the path length through the tree canopy will depend on its shape and the orientation of the signal path within the canopy. Depending on the type being considered, the tree canopy may be modeled as any one of the shapes. Otherwise, for the intermediate elevation angle (20°–50°), attenuation is almost independent of elevation and dependence becomes important only at the higher and lower ends of the elevation angle range. By considering the path length variability as a statistical parameter, however, a tree can be modeled as giving an average attenuation and a distribution around it. Both the coherent and incoherent components will vary with the receiver position and complete decorrelation of the signal is expected over distances in the order of a few wavelengths.

2. **Building Shadowing**—Signal reception behind buildings takes place mainly through diffraction and reflection. A direct line-of-sight component does not usually exist and therefore shadowing cannot be defined unambiguously, as in the case of trees. However, shadowing may be loosely defined as the power ratio between the average signal levels to the unshadowed direct signal level. Otherwise, diffractions from buildings can be studied using knife-edge diffraction theory, which gives reasonable estimates. A concept view of knife-edge diffraction phenomena is shown in Fig. 5.23 for all losses caused by the presence of the obstacles as a function of a dimensionless parameter ν and in Fig. 5.24 which illustrates the geometry of the path for both the illuminated (a) and shadowed cases (b), in order to calculate the parameter ν, by using elevation and wavelength as follows:

$$\nu = \varepsilon\sqrt{2/[\lambda(1/d_1 + 1/d_2)]}\ \text{but}\ d_1 \gg d_2,\ \text{so}: \nu = \varepsilon\sqrt{2/\lambda(1/d_2)} = \varepsilon\sqrt{2d_2/\lambda} \tag{5.55}$$

The signal strength at the shadow boundary is 6 dB below the line-of-sight level. In the illuminated region, the signal fluctuations are experienced due to interference between the direct and the diffracted components. Hence, once inside the shadowed region, the shadow increases rapidly. An experimental investigation into building shadowing loss conducted by Yoshikawa and Kagohara in 1989 confirmed the applicability of the knife-edge diffraction theory. Measured signal strength behind a building at various distances was found to follow the prediction made, assuming a single diffraction edge.

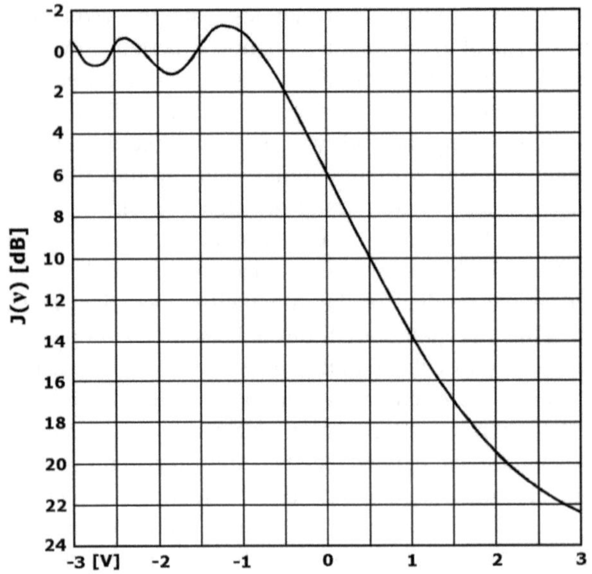

Fig. 5.23 Loss due to knife-edge diffraction—Courtesy of Handbook: by ITU

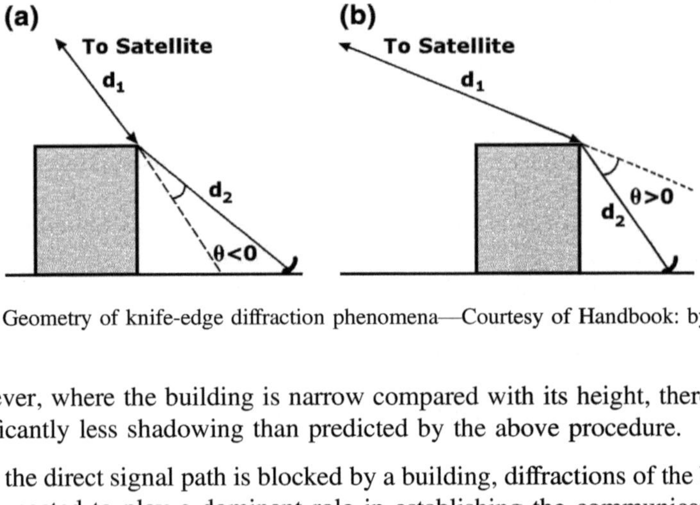

Fig. 5.24 Geometry of knife-edge diffraction phenomena—Courtesy of Handbook: by ITU

However, where the building is narrow compared with its height, there may be significantly less shadowing than predicted by the above procedure.

When the direct signal path is blocked by a building, diffractions of the buildings are not expected to play a dominant role in establishing the communication link, unless MES is close to the shadow boundary. Reflections may play a useful role in such situations, as happens in cellular systems. Building penetration depends on the type of exterior material of the building and the location inside the building. Thus, the loss through the outer structure, known as the penetration loss, is defined as the difference in median signal levels between that measured immediately outside the building at 1.5 m above the ground and that immediately inside the buildings at some reference level on the floor of interest. Measurements made at 940 MHz in a medium-size city in the USA indicate that on the ground floor of typical steel-concrete-stone office buildings, the average penetration is about 10 dB with a standard deviation of about 7 dB. While another set of measurements in a large city resulted in average ground floor penetration loss of 18 dB with a standard deviation of 7.7 dB. However, the overall decrease of penetration loss with height was about 1.9 dB per floor. The biggest average attenuations are 12 and 7 dB and standard deviations are 4 db and 1 dB for metal and concrete, respectively.

Attenuation through glass ranges from about 2 to 6 dB depending on the type of glass, i.e., plain glass produces less attenuation compared to tinted or coated glass, containing metallic components. Otherwise, the smallest average attenuation is through office furnishings, aluminum, and wood/brick of about 1, 2, and 3 dB, respectively. Losses within a building are both of distance from the exterior wall blocking the signal path, as well as the interior layout. Measurements have resulted in an inverse distance power law coefficient ranging from 2 to 4. The forthcoming ICO system has conducted experiments with satellite-borne signals whose final target is to improve and even to eliminate building shadowing.

5.8.4.6 Fading in AMSC System Due to Land Reflection

An experiment aboard a helicopter over land was carried out by receiving right-hand circularly polarized 1.5-GHz beacon signals from an IOR Marisat satellite at an elevation angle of 10°. Fading depths measured over plains such as paddy fields were fairly large (about 5 dB), nearly equal to that for sea reflection. However, fade depths measured over mountainous and urban areas were less than 2 dB. In the case of mountains, reflected waves are more likely to be shadowed or diffused by the mountains. As to urban areas, the shadowing and diffusing effects of reflected wave by buildings are also large. For this reason, the ground-reflected multipath fading in these cases is not generally significant.

Measurements of Sea-Reflection Multipath Effect—A study of multipath propagation at 1.5 GHz was performed with KC-135 aircraft and the NASA ATS-6 satellite. Otherwise, the signal characteristics were measured with a two-element waveguide array in the aircraft noise radome, with 1 dB beam width of 20° in azimuth and 50° in elevation. Namely, data were collected over the ocean and over land at a normal aircraft altitude of 9.1 km and with a nominal speed of 740 km/h. Coefficients for horizontal and vertical antenna polarization were measured in an ATS-6 experiment, where values for r.m.s. sea surface slopes of 3° and 12° were plotted versus elevation angle, in Fig. 5.25, along with predictions derived from a physical optics model. Sea slope was found to have a minor effect for elevation angles above about 10°. The agreement between measured coefficients and those predicted for a smooth flat Earth as modified by the spherical Earth divergence factor increased as sea slope decreased. The relationship between r.m.s. sea surface and wave height is complex but conversion can be performed. Namely, for most aeronautical systems, circular polarization will be of greater interest than linear. For the simplified case of reflection from a smooth Earth, which should be a good

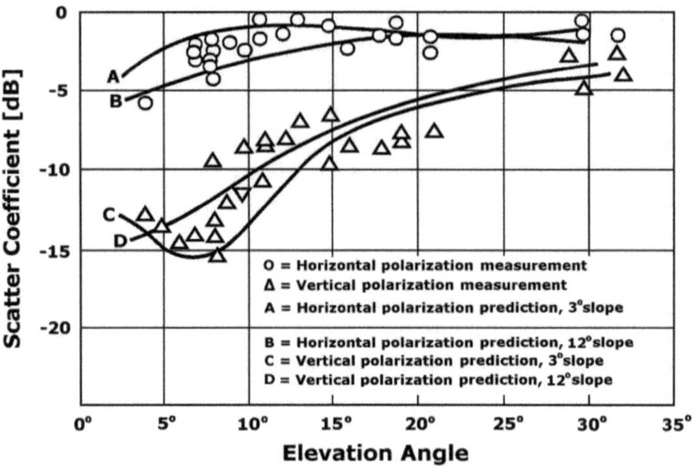

Fig. 5.25 Ocean mean-square scatter coefficients vs. elevation—Courtesy of Handbook: by ITU

assumption for elevation angles above 10°, circular copolar and cross-polar scatter coefficients (S_c and S_x), respectively, can be expressed in terms of the horizontal and vertical coefficients (S_h and S_v), respectively, by

$$S_c = (S_h + S_v)/2 \text{ and } S_x + (S_h - S_v)/2 \tag{5.56}$$

For either incident RHCP or LHCP is employed. Thus in general, the horizontal and vertical coefficients are complex values and phase information is required to apply the last equations to the curve, in Fig. 5.25.

5.8.5 Interference from Adjacent Satellite Systems

In GMSC systems for ships, vehicles, and aircraft, small mobile antennas are essential for operational and economic reasons. As a result, a number of low G/T value MES terminals with smaller antennas have been developed. However, such antenna systems are subject to the restriction of frequency utilization efficiency, or coexistence between two or more satellite systems in the same frequency band and/or an overlap area where both satellites are visible.

For coordination between two different systems in the same frequency band, a highly reliable interference evaluation model covering both interfering and interfered with conditions is required. Investigation into this area has been undertaken in particular by ITU-R Study Group 8. Advancement of such a model is an urgent matter for the ITU-R considering the number of MSC systems that are being developed in the meantime.

In GMSC systems, the desired signal from the satellite and the interfering signal from an adjacent satellite independently experience amplitude fluctuations due to multipath fading, necessitating a different treatment from that for fixed satellite systems. The main technical requirement is a formulation for the statistics of differential fading, which is the difference between the amplitude of the two satellite signals.

At this point, the method given in No 5 of ITU-R P.680 Recommendation therefore presents a practical prediction method for signal-to-interference ratio where the effect of thermal noise and noise-like interference is taken into account, assuming that the amplitudes of both the desired and interference signal affected by the sea-reflected multipath fading follow Nakagami–Rice distributions. In fact, this situation is quite probable in MMSC systems.

The basic assumptions of the intersatellite model are shown in Fig. 5.26, as an example of interference between adjacent satellite systems, where (a) is downlink interference on the MES side and (b) is uplink interference on the satellite side. This applies to multiple systems sharing the same frequency band. It is anticipated that the interference causes an especially severe problem when the interfering satellite is at a low elevation angle viewed from the ship presented in this figure because the maximum level of interference signal suffered from multipath fading increases with decreasing elevation angle. Another situation is interference between beams in multispot-beam operation, where the same frequency is repeatedly allocated.

5.8.6 *Specific Local Environmental Influence in GMSC Systems*

Local environmental influence is important for SES and AES equipped with beam width antenna. Many factors, with different kinds of noise sources, tend to make disturbances in mobile channels. Another factor that affects communication links is RF emission from different noise sources in the local environment. Thus, specific local mobile environmental factors can be noise contributions from various sources in the vicinity of the SES/AES and the influence of the ship's superstructure in the operation of maritime mobile terminals.

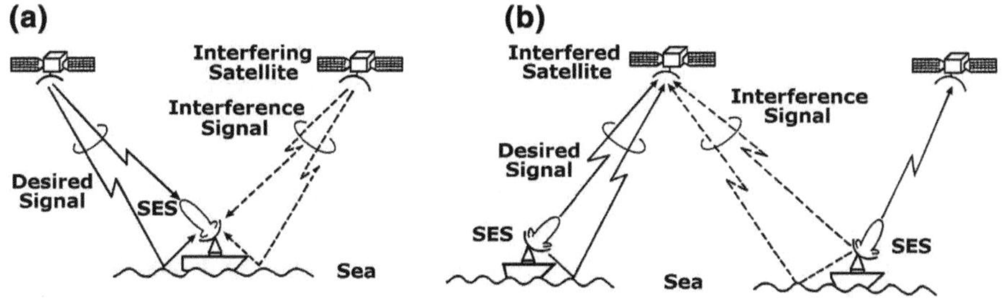

Fig. 5.26 Basic model for intersatellite interferences phenomena—Courtesy of Handbook: by ITU

However, some of these local environmental factors can affect SES/AES when a ship or aircraft is passing nearby the ground and, some of these are permanent noise sources. More exactly, these environmental sources include broadband noise sources, such as electrical equipment, motor vehicles, and out-of-band emission from powerful transmitters such as radars and ships HF transmitters.

5.8.6.1 Noise Contribution of Local Ships' and Aircraft' Environment

Some of the noise contributions from the local ships' environment are as follows:

1. **Atmospheric Noise from Absorption**—Absorbing atmospheric media, such as water vapor, precipitation particles, and oxygen emit thermal noise that can be described in terms of antenna noise temperature. These effects were discussed at the beginning of this chapter.
2. **Industrial Noise**—Heavy electrical equipment tends to generate broadband noise that can interfere with sensitive receivers. Therefore, a high percentage of this noise originates as broadband impulsive noise from ignition circuits. Namely, the noise varies in magnitude by as much as 20 dB, depending on whether it is measured on a normal working day or on weekends and holidays when it is lower in magnitude.
3. **Out of Band Emission from Radar**—Ship-borne and surveillance radars operating in pulse mode can generate out of band emission that can interfere with SES receivers. In general, such emissions can be suppressed by inserting waveguide or coaxial filters at the radar transmitter output.
4. **Interference from High Power Communication Transmitters**—High-power ships and terrestrial transmitters, for example, HF ship radio transceivers; HF radio diffusion and TV broadcasting can interfere with SES.
5. **Interference from Vehicles**—Under certain operational conditions, RF emissions from vehicles may impair Rx sensitivity. In one measurement, the noise emanating from heavy traffic has to be about −150 dB (mW/Hz) within the frequency band 1.535–1.660 MHz.
6. **Shipyard Noise**—Extremely high-peak amplitudes of noise of −141 dB (mW/Hz) was recorded from Boston Navy Yard, which was in full operation at that time. Thus, this noise is also a combination of city ambient noise and broadband electromagnetic noise from industrial equipment.

In the similar way, all five stated noise contributions from the local ship environment, except Shipyard Noise might affect instruments in the cockpit.

5.8.6.2 Blockages Caused by Ship and Aircraft Superstructures

Ship's superstructures can produce both reflection multipath and blockage in the direction of the satellite. For the most part, reflections from the ship's superstructure located on the deck can be considered coherent with the direct signal. The fading depth due to these reflections depends on a number of construction parameters including shape of the ship, location of the ship's antenna, antenna directivity and sidelobe level, axial ratio and orientation of the polarization ellipse, azimuth and elevation angles toward the satellite. Antenna gain has a significant influence on the fading depth.

In this case, low-gain antennas with broader beam widths will collect more of the reflected radio signals, producing deeper fades. Blockage is caused by ship superstructures, such as the mast and various types of antennas deployed on the ship. The geometry of blockage by a mast is presented in Fig. 5.27, where signal attenuation depends on several parameters including diameter of column, size of antenna, and distance between antenna and column. Accordingly, estimated attenuation due to blocking by a column-type structure is shown in Fig. 5.28 for antenna gains of 20 dB (a) and 14 dB (b), respectively.

In the case of blockage can be caused by aircraft superstructures, such as fuselage, wings, propellers, jet engines (engine nacelles), tails, and insignificant

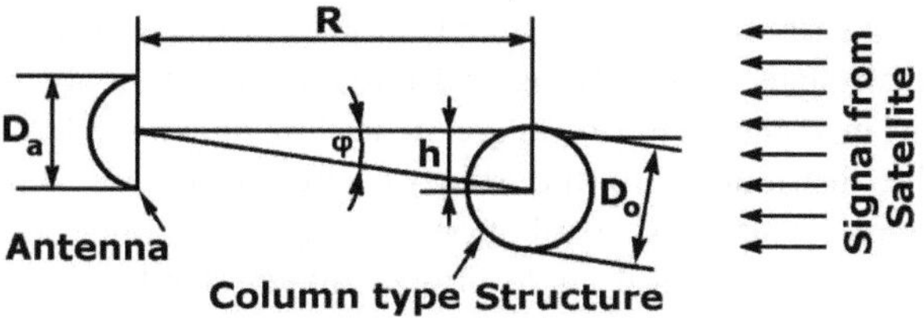

Fig. 5.27 Geometry of blocking structures—Courtesy of Handbook: by ITU

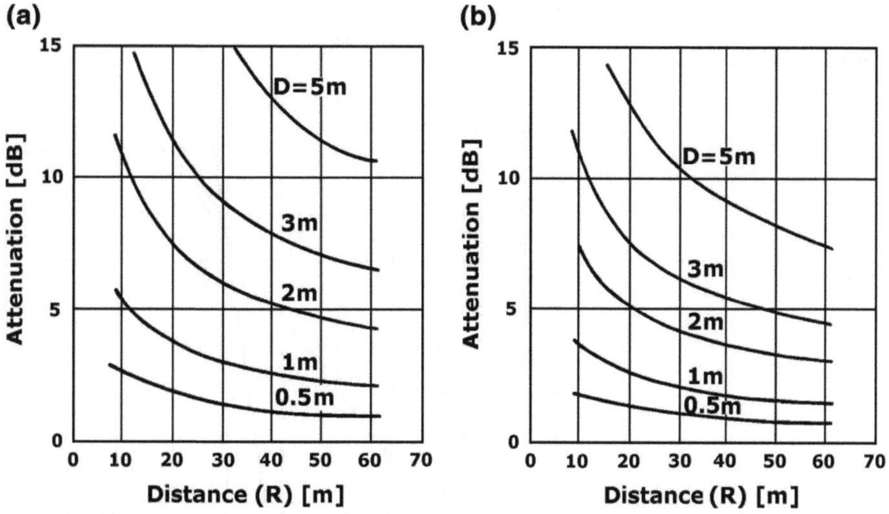

Fig. 5.28 Estimated attenuation due to blocking—Courtesy of Handbook: by ITU

various types of antennas deployed on the aircraft. Thus, signal attenuation depends on several parameters including diameter of wings, size of antenna, and distance between antenna and obstacles.

5.8.6.3 Motion of Ship's and Aircraft's Antenna

The motion of mobile satellite antennas is an important consideration in the design of ships and aircraft infrastructures. The received signal level is affected by the antenna off-beam gain because the antenna motion is influenced by the ship or aircraft motions in the space. The random mobile motion must be compensated by a suitable stabilizing mechanism to keep the antenna properly pointed toward the satellite. This is normally achieved either through a passive gravity stabilized platform or an active antenna tracking system. In either case, the residual antenna pointing error can be significant enough to warrant its inclusion in the overall link calculation.

Earlier experimental evidence suggests that the roll motion of a mobile follows a zero mean Gaussian distribution over the short term of the sea waves. The standard deviation of the distribution (σ_S) is a function of the vessel or aircraft characteristics and the sea or air state of the rolling and pitching. In Fig. 5.29 is presented distribution of the instantaneous roll angle of a ship under moderate to rough sea conditions. The distribution of the ship or aircraft motion approximates to a Gaussian standard deviation of distribution with σ_S = 5.42 value. Also shown in the figure is the distribution of roll angle of a passively stabilized antenna under the

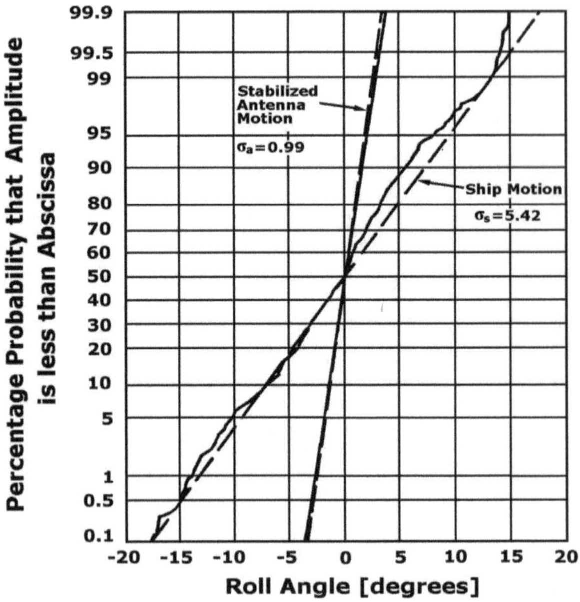

Fig. 5.29 Measured stabilized antenna motion—Courtesy of Handbook: by ITU

same conditions, which also follows a zero mean Gaussian distribution with a quantum of $\sigma_s = 0.99$. Solid curves in Fig. 5.29 represent measured values and dashed curves show calculated values for stabilized antenna motion over the sea conditions with wave heights of approximately 5 m.

The relation between the standard deviations of the two distributions depends on the design of the passive stabilizer. Although the ship's antenna motion is much reduced, depending on the antenna beam width, the residual pointing error may be large enough to produce appreciable signal fluctuations. Over a long period of sea waves time σ_s varies as a function of the sea surface conditions and its distribution can be approximately by either a log normal distribution or a Weibull distribution.

The same context can be used to explain the distribution of the instantaneous roll angle of an aircraft under moderate to rough drift conditions in the air.

5.9 Propagation for Space Mobile Broadcasting

One of the main objectives when designing a mobile broadcasting communication system over satellite or stratospheric platforms is to guarantee a specific QoS, which depends on given user requirements. Thus, a key point for this design is the quality of the channel that is present in the link. This channel varies in time and space, and in order to perform an estimation of the losses that the signal will experiment, it is necessary to consider all the factors influencing the signal attenuation.

A thorough understanding of the propagation mechanisms influencing radio waves propagating in the channel is therefore essential. The discussed space propagation mechanisms affecting the link produce various effects in the signal, which vary depending on the operating frequency.

It is important to emphasize considerable differences in the models operating at low radio frequency bands in relation to the ones operating in upper bands. Once these propagation effects have been analyzed and understood, an analysis of their nature and statistical behavior has to be performed, so that they can be represented and characterized through statistical models. Thus, by using these models, it is then possible to develop more realistic channel models, which are then applied when designing space communication systems.

For the case of the stratospheric communication platforms (SCP) systems, which have many similar effects and propagation conditions causing especially low orbiting satellite systems, it is necessary to consider environmental conditions defined by the scenario, as well as the phenomena affecting the link. In such a way, a model that clearly characterizes signal affectations will allow implementation of reliable and truthful evaluation schemes. This analysis sets the basis for a deeper channel characterization and the need to construct channel models, which are very useful tools when designing an SCP models.

Mobile broadcasting has a lot in common with two-way interactive mobile satellite systems for communication and navigation. The previous discussion reviewed propagation aspects for mobile broadcasting can be included for the general treatment on mobile broadcasting over satellite. In fact, there are two principal differences between mobile broadcasting and two-way voice or data services:

1. Mobile broadcasting must rely on one direction of information transfer, because there is no any possibility of requesting retransmission. As a result, when the signal finally fades out, the receiver has nothing to play except silence or what might have been prestored.
2. The programming must be delivered on a continuous basis, regardless of the location of the user. This is why both XM and Sirius include terrestrial relay transmitters to fill in dead spots in dense urban environments.

Accordingly, the new developments and discussion about a satellite digital broadcasting transmission system is important that consistently are helping to deliver video and audio bits to the mobile receivers. The satellite mobile receiver, in turn, must be able to reassemble the entire signal stream in order to decompress and convert from digital to analog system. Loss of bits causes dropouts at best and a complete loss of signal at worse. Thus, there is no graceful degradation aspect that can be accustomed to as FM listeners. The FM receiver only completely breaks lock if there is interference from a stronger signal on the same or adjacent channel frequency. In Satellite Digital Audio Radio Service, there is only one signal, but the fades can be deep and either rapid due to multipath or prolonged due to blockage. However, there are other reasons why FM is more resilient: Signal strength tends to be quite high, and the radio frequency range centered on 100 MHz is more able to bend and reflect in and around obstacles. Absorption by foliage is also much less than at L- and S-bands. In fact, with L- or S-band transmission, the signal into a fixed receiver with clear line of sight to the satellite will generally remain solid. In tropical regions around the geomagnetic equator, ionospheric scintillation can be a problem. Fades from this phenomenon can be 6 dB or more in seasons near the equinox. Fortunately, the satellite mobile broadcasting systems generally have margins of 8 dB or more, which should provide a commercially acceptable reception service. The Faraday effect is avoided through the use of circular polarization (CP), which thereby demands an acceptable axial ratio. Therefore, any chance of providing a continuous service under bed propagation conditions is small, thus the need for diverse paths containing the same broadcast.

Chapter 6
Ground Segment

In general, satellite ground segment is very important subsystem that is connecting Fixed Earth Stations (FES) with other FES via geostationary Earth orbit (GEO) or other types of satellites. In particular, mobile Earth stations (MES), fixed or mobile very small aperture terminals (VSAT), and TV (TVRO) broadcasting networks are connecting ground segment to the terrestrial telecommunication networks (TTN) via communication satellites. The ground segment in GMSC can be represented by conventional land Earth station (LES) for all mobile applications, coast Earth station (CES) for maritime and land applications, and ground Earth stations (GES) for aeronautical applications.

Thus, in the twenty-first century, satellite communication systems have become an essential and very important technology for modern commercial and government applications. Satellites are launched from Earth by the launchers in GEO, from high-flying airplanes or from ground based rockets. Once lunched, the payloads must reach proper elevation and escape velocity to be boosted into certain orbit. However, in order to maintain proper orbital path and communication operations, satellites are controlled by ground station on Earth that sends commands and receives status and telemetry from the satellite.

However, low Earth orbits (LEO) as a part of non-GEO satellites are visible for only a period of time from the point of view of an observer on Earth. They can dump data to the ground Earth when they pass by a ground station area. Therefore, the ground station plays a very important role in the communication with the satellites in the orbit. A mission control center (MCC) might use several ground stations to maintain the communication with many satellites. Although a ground station will have a lot of radio hardware to receive and send the microwave signal to and from satellites, the objective of this work was to focus on the software part of the ground station.

Global satellite coverage has proven to be extremely desirable in recent years in all sectors in both communication and exploration applications. Increased satellite coverage requires increased capacity of the ground segment, which is controlling the main operations in the life circle of satellites. The ground segment provides the

S.D. Ilčev, *Global Mobile Satellite Communications Theory*,
DOI 10.1007/978-3-319-39171-7_6

means and resources to manage and control the missions, to receive and process the data produced by the station instruments. The operational ground segment is in charge of the satellite housekeeping, the mission planning and telemetry reception from the satellites.

6.1 Definition of GMSC Services

This section shortly introduces the official definition of International Telecommunication Union (ITU) related to the various satellite services currently in operation.

6.1.1 Fixed Satellite Services (FSS)

According to the Radio Regulations (RR S1.21), the FSS is a radiocommunication service between given positions on the Earth's surface when one or more satellites are used.

These stations located at given positions on the Earth's surface are called Earth Stations or Radio Earth Stations (RES) of the FSS. The given position may be a specified fixed point or any fixed point within specified areas, such as rural and urban areas within one country or between two countries or continents. Stations located onboard the satellites mainly consist of the satellite transponders and associated antennas, which are providing satellite links between RES and FSS users. Figure 6.1 shows solution of Global or Regional FSS network with RES as are called gateways or space stations and interface to the TTN facilities of the FSS coverage, such as Internet, Integrated Services for Digital Network (ISDN), and Public Switched Telephone Network (PSTN).

At the recent time, with very few exceptions, all links between a transmitting Earth station and a receiving Earth station are affected through a single satellite. In fact, these links are comprised of two parts, an uplink between the transmitting station and the satellite and a downlink between the satellite and the receiving station. Thus, at the present time and in the future, it is envisaged that links between two Earth stations could use two or more satellites directly interconnected without an intermediate Earth station. Such a link between two Earth stations using satellite-to-satellite links would be called a multisatellite link (MSL). The satellite-to-satellite links will form a part of the intersatellite links (ISL).

The ISL network may be employed to provide connections between Earth stations in the service area of one satellite to Earth stations in the service area of another satellite, when neither of the satellites covers both sets of Earth stations. A set of space stations and Earth stations working together to provide radiocommunications is called a satellite system. For the sake of convenience, a distinction is made in the particular case of a satellite system, or a part of a satellite system,

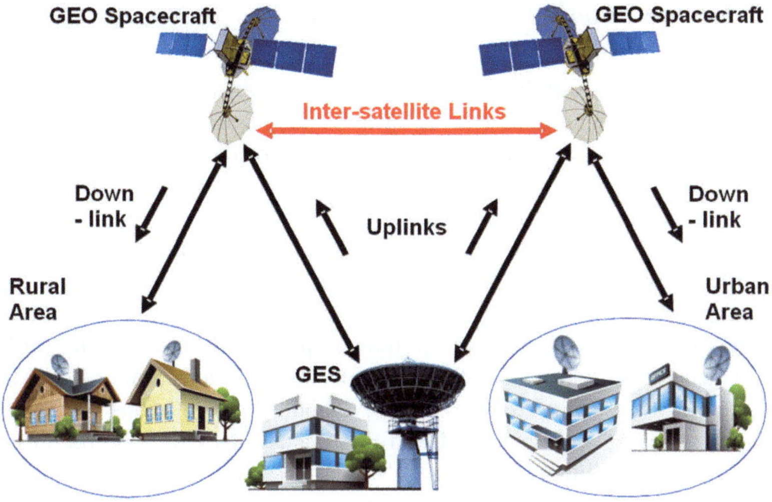

Fig. 6.1 Global or regional FSS network—Courtesy of Manual: by Ilcev

consisting of only one satellite and the associated Earth station that is called a satellite network.

The FSS also includes feeder links, i.e., transmission links from an Earth station located at a specific fixed point to a space station onboard satellite, or a feeder link is the portion of a broadcasting satellite system that provides the connection from the Earth to the satellite. In vice versa direction, satellite is conveying information for a space communication service other than for the FSS. This category includes, in particular, uplinks to the satellites of the Broadcasting Satellite Service (BSS) and up- and downlinks between fixed earth stations and satellites of the Mobile Satellite service (MSS).

Definitions of BSS and MSS are given in the same way, namely all types of any radio or telecommunications signals can be transmitted via FSS links: telephony (voice), facsimile (Fax), data, video or a mix of these signals in the framework of ISDN and PSTN, television and sound programmes, etc. In addition, the newest fixed and mobile satellite system is Digital Video Broadcasting-Return Channel via Satellite (DVB-RCS), which is providing service deploying VSAT in offices or onboard mobiles.

The latest communications satellite generations, operating in FSS frequency bands, are equipped with high-power transponders, which makes it possible to implement different broadcasting services direct to the general public for individual reception, such as Digital Audio Broadcasting (DAB) via satellite, direct-to-home (DTH) applications through very small receiving antennas known as television receiving only (TVRO), and for rural and community reception (professional and domestic applications).

6.1.2 Mobile Satellite Services (MSS)

The MSS applications use different frequency band allocations than the FSS. This service has similar techniques employed and many commonalities exist between these services and those of the FSS solutions. According to the Radio Regulations (RR No S1.25), this is a radiocommunication service between MES and one or more space stations, or between MES terminals by means of one or more space stations. In Fig. 6.2 is shown solution of Global or Regional MSS network with LES terminals as are called gateways or space stations and interface to the TTN facilities of the MSS coverage.

The MSS communication system may deploys GEO, medium Earth orbit (MEO), and LEO satellite constellations to establish connections of all MSS terminals, such as Ship Earth Stations (SES), Vehicle Earth Stations (VES), Aircraft Earth Stations (AES) and Personal Earth Stations (PES), with the ground Internet, ISDN, and PSTN facilities.

Therefore, MSS system includes maritime, land (road and rail), personal, and aeronautical applications providing global communication service, managed by the adequate traffic control centers (TCC). The MSS network may also serve to the disaster and search and rescue (SAR) facilities, rural communications, mobile multimedia, WAN, and so on. It is important to note that the modern mobile

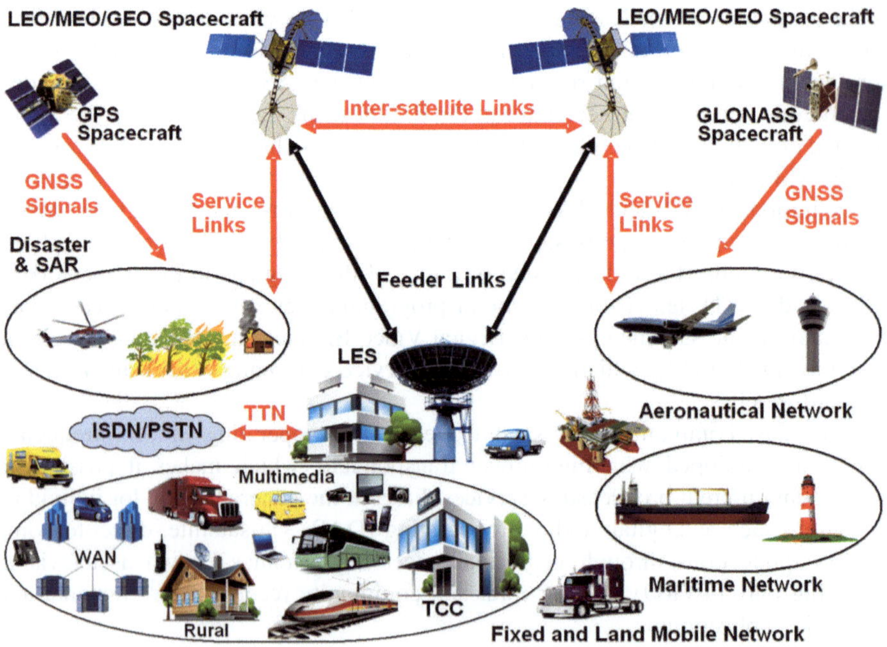

Fig. 6.2 Global or regional MSS network—Courtesy of Manual: by Ilcev

systems are implementing VSAT onboard ships, land vehicles, and aircraft, and that may even serve to the personal handheld phones and PC terminals as well.

The ISL network in MSS may provide connections from MES in the service area of one satellite to LES terminals in the service area of another satellite and vice versa, when neither of the satellites covers both sets of these stations. The MSS also includes feeder links, i.e., transmission service from LES terminals to the satellite, and service links, i.e., are providing transmission service from MES terminals to LEO/MEO or GEO satellites.

6.1.3 Fixed and Mobile Broadcasting Satellite Services (BSS)

This is a radiocommunication service in which signals transmitted or retransmitted by RES or space stations are intended for direct reception by the general public using VSAT for the TVRO reception onboard ships, land vehicles and aircraft, which scenario is illustrated in Fig. 6.3. The satellites implemented for the BSS are often called Direct Broadcast Satellites (DBS) for video and audio signals. The TVRO systems needed for BSS reception should be smaller than the ones needed for stated operation in the FSS system. The new TV system that can serve fixed and mobile TV broadcasting is high-definition TV (HDTV). Direct Audio Broadcasting (DAB) also needs very small satellite radio receivers.

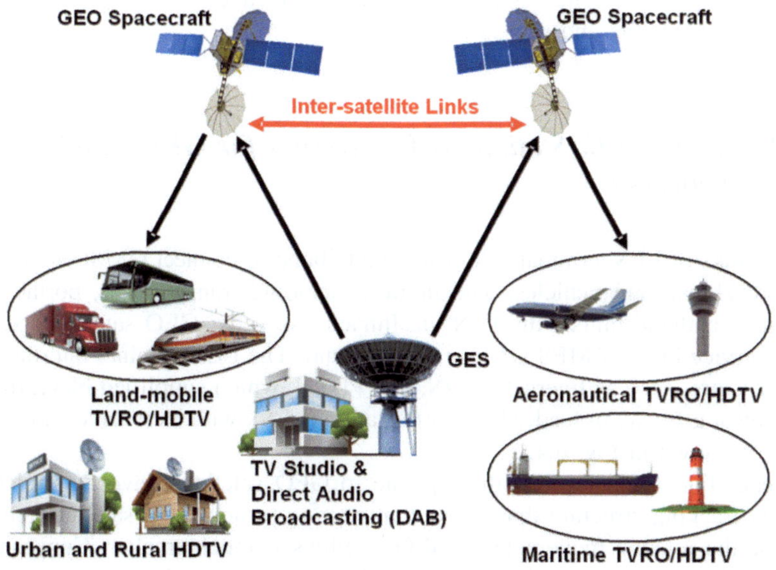

Fig. 6.3 Global or regional BSS network—Courtesy of Manual: by Ilcev

The direct reception shall encompass both individual reception known as direct to home (DTH) and community reception such as Community Antenna Television (CATV) and Satellite Master Antenna Television (SMATV). In addition, DVB-RCS has solution of Internet Protocol TV (IPTV) for both fixed and mobile applications. Some other specific satellite services may provide radiodetermination satellite service, radionavigation satellite not augmented and augmented service, meteorological-satellite service, and so on.

6.2 Divisions of Earth Stations in MSS

An Earth station in MSS sends and receives communication transmissions using a space objects. Earth stations can be in service of Inmarsat GMSC via GEO satellites and usually are called LES. While non-GEO GMSC Earth stations provide GMSC service via Iridium, Globalstar, or Orbcomm satellite constellations, they are referred to as Gateways.

They can transmit signals to space or when used for services such as pay TV, just receive. Standard Earth stations used for FSS and MSS generally transmit and receive. An Earth receive station is an Earth station that is only a receiver, so the majority of Earth stations for a BSS will only receive.

In addition to feeder link Gateway, satellite operations require Telecommand, Telemetry and Control (TT&C) centers in order to track and control the satellite constellation. There are at least two such sites in a given satellite mobile system serving as primary and backup facilities. They are used to provide command functions to satellites such as to perform orbit maneuvers, as well as monitor the status of the satellite and subsystems.

6.2.1 Land Earth Stations (LES) Design for GEO Mobile Networks

Feeder links of LES terminals are using RF C-band to connect GMSC transceivers onboard ships, land vehicles, and aircraft, including transportable, portable, and personal handheld units with TTN via Inmarsat or other GEO satellite operators, while service links of MES are using RF L-band. The LES terminal interface TTN infrastructure, such as Internet, ISDN, and PDTN, which simplified block diagram of a typical gateway or feeder link Earth station (LES), with its receive and transmit chains is shown in Fig. 6.4.

Feeder link Earth stations that operate to GEO satellites have a simpler LES antenna tracking structure than those that operate to non-GEO satellite networks, because the relative movement of GEO satellites is much smaller. Consequently, large Earth station antennas of about 10–15 m diameter can be economically used

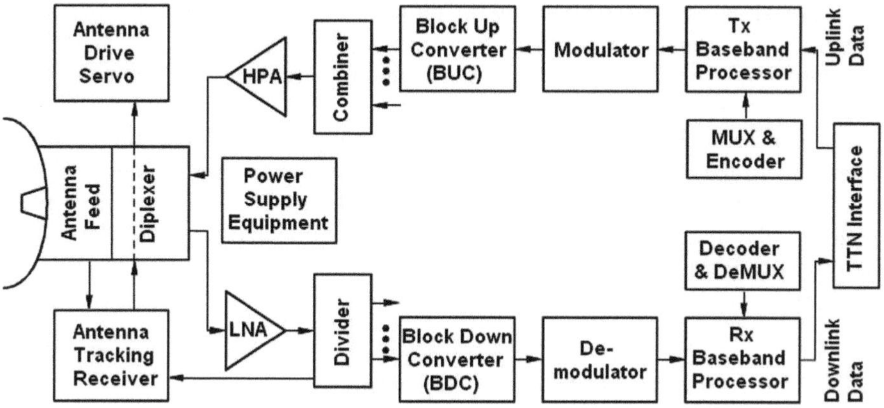

Fig. 6.4 Simplified feeder link LES block diagram—Courtesy of Handbook: by ITU

thereby improving link margins. Furthermore, only one feeder link Earth station antenna (and one spare for backup purposes) is required to handle all mobile calls from the coverage area of GEO satellite constellations.

Having said this, other feeder link Earth stations are normally added within the system to provide for resilience and to allow for national presence of operations within different regions. The GMSC service is managed by satellite operators, while these LES terminals are typically operated by service providers who interconnect traffic to the local terrestrial networks and who also market the mobile services within the region.

6.2.2 Gateways Design for non-GEO Mobile Networks

Feeder link of Gateway terminals or Earth station antennas for non-GEO satellite systems are generally smaller than those used for GEO operation, because of the lower path losses to the satellites and to allow for easier tracking ability. Feeder link frequencies normally used by Big LEO systems are in the 5/7 GHz bands or 20/30 GHz bands. Frequencies in the 15 GHz band are also available.

The design and location of feeder link Earth stations are related to the satellite constellation characteristics. For example, LEO satellite system such as Iridium having ISL network to route traffic between satellites would need fewer Earth stations than otherwise required. On the other hand, systems using Earth orbits without ISL, such as Globalstar and Orbcomm, will need more Earth stations than those using Iridium and MEO satellite constellations.

However, in the Iridium satellite system there are 1 or 2 Gateways station sites connected together via a global TTN and fiber optic network. Traffic is internally routed to the closest Gateway to the call destination before being routed through the public terrestrial networks. In the Globalstar system, there are between 30–40

Gateways sites planned globally with traffic routed directly through the public terrestrial networks at the gateways. Furthermore, the location of feeder link sites must take into consideration the existing national or regional telecommunications infrastructure available.

The number of tracking Gateway antenna installed at individual sites will vary depending on the number of satellites being tracked, the number of antennas lining up for the next satellites to become visible and the number being used as spare. Typically, the number of antennas installed at an individual non-GEO feeder link site can vary from about 2 to 6.

6.2.3 Components of Ground Segment

The ground segment, shown in Fig. 6.4, comprises the uplink and the downlink radio Earth station (RES) components, which include:

1. Multiple beam reflector ground antennas with antenna feed, diplexer and drive motors, as an outdoor equipment (ODE), for simultaneous connections with other satellites inside of their overlapping coverages, which consist of precision systems for satellite tracking. This precision system is managed by antenna controller and TT&C system for monitoring the performance of the satellites and receiving telemetry data from the satellites;
2. Radio frequency equipment (RFE) is containing low-noise amplifiers (LNA) with power dividers and high-power amplifiers (HPA) with power combiners;
3. Ground telecommunication equipment (GTE), receivers (Rx), and transmitters (Tx) as an In Door Unit (IDU), which contains downconverter with demodulators on receive side, upconverters with modulators on transmit side and baseband processing signals (BPS). There are two BPS units, first is on receive side containing demultiplexer and decoding, and second is on transmit side containing multiplexer and coding; and
4. Equipment for connection with the terrestrial network (TTN) and power supply units.

6.3 Ground Antenna, Components, and Control Systems

An antenna (plural antennae or antennas), or aerial, is an electrical device which converts electrical power of EM waves into radio waves and vice versa. It serves in fixed and mobile RES with a radio Rx and Tx or both in one as radio transceiver. An Tx supplies the electric current oscillating at RF or HF alternating current (AC) to the antenna, and the antenna radiates the energy from the current as EM waves. An Rx via antenna intercepts some of the power of an EM wave in order to produce and amplify a tiny voltage at its terminals.

6.3.1 Center Feed Antennas

Simplest form of axisymmetrical configuration is a paraboloidal reflector with a primary feedhorn located at the focus, which is shown in Fig. 6.5 (Left). However, this leads to a long waveguide run between the feed and the electronics box for antennas whose diameter is greater than about 3 meters. This is undesirable because it leads to reduction in signal power and increase in noise. A more compact configuration, especially for larger antenna diameters, can be realized by the introduction of a subreflector. The feedhorn is located at the rear of the main reflector, eliminating the need for long, potentially lossy, waveguide runs. This is known as Cassegrain antenna and is shown in Fig. 6.5 (Right). Thus, the subreflector is a section of hyperboloid situated within the focus of the main reflector.

However, the subreflector and associated support legs cast an effective shadow that affects the antenna efficiency and its sidelobe envelope. Typically, for a Cassegrain configuration, a decrease in peak gain of 0.1–0.5 dB can be expected. Also, the subreflector may not intercept all that is radiated from the primary feed and this could lead to degradation of the antenna sidelobe performance. To increase the percentage of energy intercepted by the subreflector, the antenna feed is configured to give a tapered illumination distribution at the subreflector. Moreover, decreasing or tapering the subreflector edge illumination reduces the overall antenna efficiency. This would require some compromise between the sidelobe performance and the antenna efficiency. It is feasible to achieve a reasonable wide-angle sidelobe performance and high antenna efficiency. An improvement in

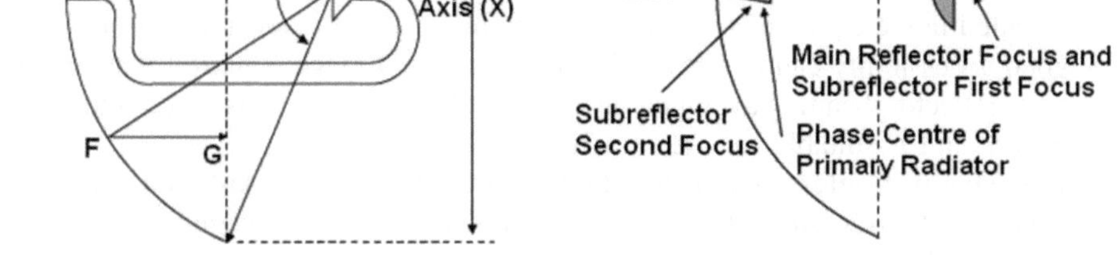

Fig. 6.5 Center feed and Cassegrain antennas—Courtesy of Handbook: by Intelsat

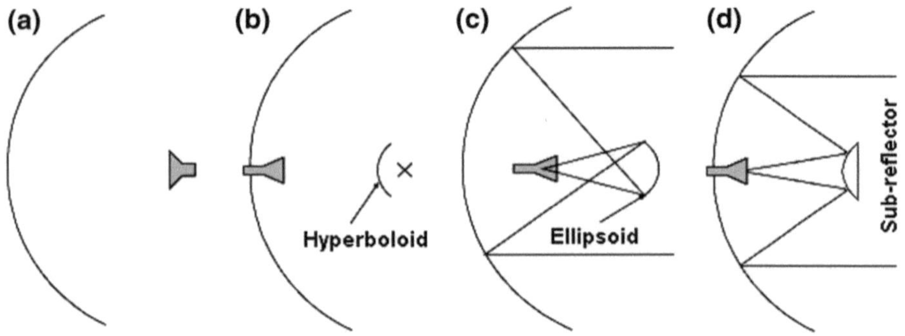

Fig. 6.6 Different types of ground center feed antennas—Courtesy of Handbook: by Intelsat

antenna efficiency can be achieved by shaping the reflector profiles and controlling the aperture amplitude distribution while still maintaining a uniform phase distribution. The near-uniform aperture distribution achieved will lead to higher sidelobes close to the boresight direction.

Here will be discussed the following types center or axisymmetric feed antennas:

1. **Center Focal Feed Paraboloid Antenna**—This is single reflector-type ground antenna with a simple configuration, low aperture efficiency and so reflector shaping cannot be applied, which is illustrated in Fig. 6.6a. It is generating high noise temperature due to large spillover power from main reflector and bad accessibility for large diameter antenna, because feed and LNA should be adjoined to primary radiator (horn). This type of antenna is fitted for installations at small size Earth stations.
2. **Center Feed Cassegrain Antenna**—This is subreflector convex (Cassegrain) antenna that provides high efficiency and low noise temperature because reflector shaping can be applied, which is shown in Fig. 6.6b. Its feed and LNA can be installed in the equipment room behind the main reflector, so for that reason rather good accessibility can be obtained. Frequency range of this antenna is narrower than that of 4-reflector beam waveguide feed type and is producing fairly good radiation pattern. This antenna is fitted for installations at medium size Earth stations.
3. **Center Feed Gregorian Antenna**—This is subreflector concave (Gregory) antenna with high efficiency and low noise temperature because reflector shaping can be applied, which is shown in Fig. 6.6c. Feed and LNA can be installed in the equipment room behind the main reflector, so at this point rather good accessibility can be obtained. Frequency band range is narrower than that of 4-reflector beam waveguide feed type. This antenna has good radiation pattern and is suitable for medium Earth stations.
4. **Cassegrain Center 4-reflector Fed Beam Waveguide Antenna**—This is advanced Cassegrain antenna fed by 4-reflector beam waveguide, which is shown in Fig. 6.6d. It provides high efficiency and low noise temperature over

super-broadband and good accessibility, because feed and LNA can be installed in the room free from El- and Az-rotations. No waveguide run and rotary joint is required, so transmit power can be about 2 dB higher at 6 GHz band for 30-m-diameter antenna. This antenna shows good radiation pattern and can accommodate large size Earth station antennas (D/λ about 500).

6.3.2 *Offset Feed Antennas*

The offset feed ground antennas, such as the offset Cassegrain and Gregorian, achieve a better radiation pattern because of lower aperture blockage. They are often known as non-symmetrical ground antennas and are generally used in small Earth stations because of construction problems and higher cost.

Here will be introduced two types of offset feed antennas such as:

1. **Offset Feet Paraboloid Antenna**—This ground antenna has excellent radiation pattern and shows low noise temperature because of no blocking features, which is shown in Fig. 6.7 (Left). It has excellent voltage standing wave ratio (VSWR) and is small size Earth station antenna (e.g., TVRO).
2. **Offset Feet Cassegrain Antenna**—This Earth station antenna is experiencing excellent radiation pattern because of no blocking issues, which is illustrated in Fig. 6.7 (Right). It also has high efficiency and low noise temperature because of no blocking possibilities and reflector shaping. This antenna provides excellent VSWR, small wind load if limited steerable amount is selected environments, and good accessibility because feed and LNA can be installed in the room free from El- and Az-rotations. It is fitted for installations in small and medium size Earth stations.

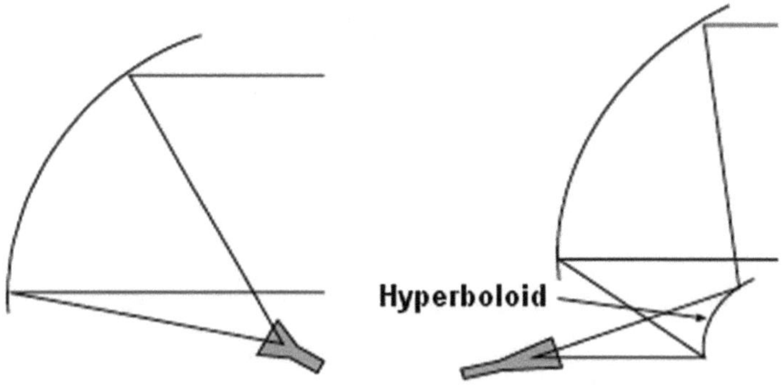

Fig. 6.7 Types of ground offset feed antennas—Courtesy of Handbook: by Intelsat

6.3.3 Models of GEO Ground RES Antenna Systems

An antenna with a feed in the center of the paraboloid (axisymmetric) reflector represents the simplest antenna configuration that is potentially capable of meeting the important RF specifications for RES applications. Thus, a major advantage of such a configuration is that mechanically, it is relatively simple, reasonably compact and in general, fairly inexpensive. The circular symmetry of the main antenna reflector leads to considerable cost savings in the manufacturing of the reflecting surface, backup structure and antenna mount. There are many types of ground antenna that can be deployed in RES infrastructures.

6.3.3.1 ViaSat 13.5 M Communication Earth Station Antenna

The ViaSat 13.5 m 8013 model antenna provides high-performance in RF of C-, Ku-, or DBS-band GEO satellite applications, shown in Fig. 6.8 (Left). The shaped Cassegrain reflector provides superior gain and sidelobe performance, meeting stringent of satellite operators' requirements. Precision stretch-formed and interchangeable aluminum panels are attached to a central structural-steel HUB with rigid, interchangeable structural trusses. The antenna panels are coated with a solar-diffusive white coating system that provides years of environmental protection while minimizing thermal expansion effects. The reflector back structure and subreflector spars are designed to exacting rigidity requirements under wind and gravity loads. The HUB provides a protective enclosure for sensitive electronics.

The ViaSat so-called Turning Head mount provides an efficient structure for supporting and positioning the feed and reflector system over a full 360° in 70°

Fig. 6.8 Models of GEO communication antennas—Courtesy of Manual: by ViaSat

continuous sectors. Elevation travel provides 0°–90° of continuous travel. Extended azimuth travel is available as an option. Antenna control system options range from adaptive step-track and program-track systems to full DC servo and monopulse tracking systems.

Main features of this RES antenna are as follows: Transmit operational RF at C-band is 5.850–6.425 GHz, at Ku-band is 13.75–14.5 GHz, and at DVB-band is 17.3–18.4 GHz; receive operational RF at C-band is 3.625–4.2 GHz, at Ku-band is 10.70–12.75 GHz, and at DVB-band is 10.95–12.75; transmit antenna gain is 57.1 dBi at 6,138 GHz, 64.2 dBi at 14.25 GHz, and 65.5 dBi at 17.85 GHz; receive antenna gain is 53.7 dBi at 3.912 GHz, 63.1 dBi at 11,85, and 63.3 dBi at 11.85 GHz; antenna travel for elevation is 0°–90° continuous; and antenna travel for azimuth 360° in overlapping 70° sectors >180° continuous (optional).

6.3.3.2 ViaSat 16–18.3 M Communication Earth Station Antenna

The ViaSat model 8118 is a 16.0 or 18.3 m GEO satellite antenna that provides superior performance through the use of precision stretch-formed reflector panels and a dual-shaped Cassegrain feed, which is shown in Fig. 6.8 (Right). Corrugated conical feed horns ensure excellent antenna gain and sidelobe performance. High-strength aluminum panels are durable enough to withstand rough handling and a range of environmental conditions.

The HUB also provides a protective enclosure for sensitive electronics. The high-strength structural-steel mount employs an elevation-over-azimuth geometry for easy pointing to any satellite within the visible orbital arc. The mount's stiff, rugged construction provides pointing accuracy for continuous operation, even under adverse wind conditions.

Main features of this RES antenna are as follows: Transmit operational RF is 5.850–6.425 GHz; receive operational RF 3.625–4.2 GHz; gain (Midband, Ref. Feed Horn) for 18.3 m transmit antenna is 60.3 dBi2 and for receive antenna is 56.6 dBi; and for 16.0 m transmit antenna is 58.9 dBi dBi2 and for receive antenna is 55.4 dBi.

6.3.3.3 ViaSat 9.1 M Broadband Earth Station Antenna

This RES is designed for the latest high-capacity GEO Ka-band satellites, and the 9.1 m antenna system offers exceptional broadband support to deliver high-speed Wi-Fi fixed and mobile connections for residential, commercial, and government satellite services, which is shown in Fig. 6.9 (Left). It is ideally suited for high-performance gateway applications.

The shaped Cassegrain antenna with precision-machined subreflector provides superior gain and sidelobe performance at Ka-band operational RF. Its improved high-precision, stretch-formed panels are supported by stiff radial trusses attached to a large central HUB. The reflector back structure and spars are designed for

Fig. 6.9 Models of GEO broadband antennas—Courtesy of Manual: by ViaSat

rigidity requirements to provide optimum performance under a wide range of environmental conditions.

All aluminum reflector panels, coated with a solar-diffusive white paint, minimize thermal effects. The oversized HUB is specifically designed to support multiple redundant suites of HPA, BUC and receive chain configurations. The extra-large doorway allows easy access from the large work platform for convenient maintenance of all the electronics. Besides, the redundant heating, ventilation, and air-conditioning (HVAC) system maximizes the electronics mean time before failures (MTBF) and service availability.

The rugged steel mount delivers Ka-band pointing accuracy in adverse wind conditions. The pedestal design features a precision azimuth bearing with dual drivers for very low backlash. Its control system offers full DC servo performance with adaptive step-tracking or monopulse autotracking for unparalleled tracking performance. For quick access and service, control system is conveniently located on the pedestal. Features of this antenna are: Receive RF is 17.7–20.2 GHz; transmit RF is 27.5–30.0 GHz; receive gain is 63.8 dBi; transmit gain is +66.7 dBi; and figure of merit (G/T) at 30° elevation is 39.7 dBi/K.

6.3.3.4 ViaSat 13.5 M Broadband Earth Station Antenna

ViaSat 13.5 m Ka-band antenna is suited for new GEO satellites. It offers exceptional broadband support to deliver high-speed Wi-fi connections for fixed and mobile residential, commercial, and government services, which is shown in Fig. 6.9 (Right).

6.3.4 *Models of non-GEO Ground RES Antenna Systems*

The LEO and MEO ground RES satellite antennas provide the accurate tracking function of sending and receiving voice and data by aiming it at Big LEO (Iridium and Globalstar) and MEO (O3B) mobile system for automatical communications with satellites. The ground antenna for Small GEO Orbcomm satellite system can provide transfer of data only. The major requirements are to track, control, and communicate accurately with the rapidly moving communication satellite around the Earth requires deployment of very high-quality ground RES antennas.

The Korean-based High-gain Antenna Company produces several types of LEO and MEO RES antenna terminals for mobiles reliable and quality service. The major functions of LEO and MEO satellite tracking antennas are to perform the task of controlling a satellite and communicating including TT&C facilities and so on. This type of RES antenna has different program, from 3.7 to 13 m in size, at intervals of one meter. In addition, each of these antennas covers RF at L-, S-, and X-band. Its wheel range is the elevation $0° \sim 90°$ or $-90° \sim 90°$ and the azimuth $\pm 180°$, the operation speed is from 0.02°/s to 16°/s, and its mount of Yoke & Tower type is provided. Here, will be introduced two antenna types:

1. **High-gain 11-m Antenna non-GEO Communication Earth Station Antennas**—This is RES antenna for mobile communication via LEO and MEO satellite constellations at L- and S-bands, which is shown in Fig. 6.10 (Left). In Fig. 6.11 (Left) is shown the drawing side view of this antenna, and in Fig. 6.11 (Right) is shown rear view.

The electrical specifications of this 11 m LEO and MEO ground antenna are as follows: two-port or four-port Tx/Rx, linear or circular polarized feed; full reflector and feed-deicing system with manual or automatic controls; motorized azimuth, elevation, and polarization drive system with controls and readouts; Servo system is

Fig. 6.10 Models of LEO and MEO communications antennas—Courtesy of Manual: by High Gain Antenna

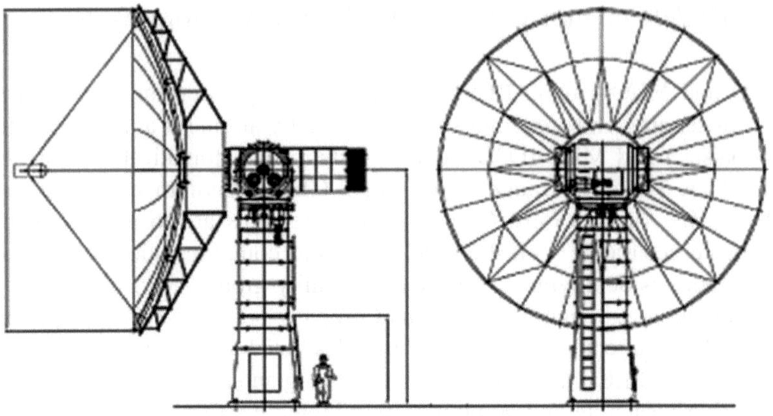

Fig. 6.11 Side/rear views of 11 m non-GEO antennas—Courtesy of Manual: by High Gain Antenna

included; Available for the sizes from 5.5 to 13 m, covering a full frequency of X-, L-, S-band and for other frequency ranges as well, RF range is 1.435 ~ 1.85 GHz (L-band) and 2.2 ~ 2.4 GHz (S-band); gain is 39.34 dBi for L-band and 43.90 dBi for S-band; G/T is 18.74 dB/°K for L-band and 23.3 dB/°K for S-band; Beamwidth is 1.26° for L-band and 0.82° for S-band; and power requirement is 55 kVA.

The mechanical specifications of this antenna are as follows: Continuous azimuth travel is ±360°; azimuth travel rate is 15°/s (Max.); elevation travel is −5°–185°; elevation travel rate is 15°/s (Max.); tilt travel is 10°–350°; tilt travel rate is 1°/s (Max.); reflector is aluminum built; surface accuracy is 0.5 mm (RMS); pedestal structure is steel;

Foundation size is 11 M × 11 M × 0.7 M; concrete volume 84.7 m^3; and soil bearing pressure is 10.000 kg/m^2.

2. **High-gain 7.0-m Antenna non-GEO Communication Earth Station Antennas**—This is also ground RES antenna for mobile communication service at L-, S-, and X-band via LEO and MEO satellite constellations, which is illustrated in Fig. 6.10 (Right). In Fig. 6.12 (Left) is shown the drawing side view of this antenna, and in Fig. 6.12 (Right) is shown front view. The most important functions of this and previous non-GEO antennas are to provide secure accurate tracking with the fast circulation of communication satellites and to realize not interrupted roaming.

Almost all electrical specifications of this ground antenna are the same of previous antenna, except the following features: RF at X-band is band 7.25 ~ 7.75 GHz for Rx and 7.9 ~ 8.4G Hz for Tx; gain is 35.78 dBi for L-band, 40.34 dBi for S-band, 53.6 dBi (Rx) and 53.78 dBi (Tx) for X-band; G/T is 15.28 dB/°K for L-band, 19.74 dB/°K for S-band and 33.46 dB/°K for X-band;

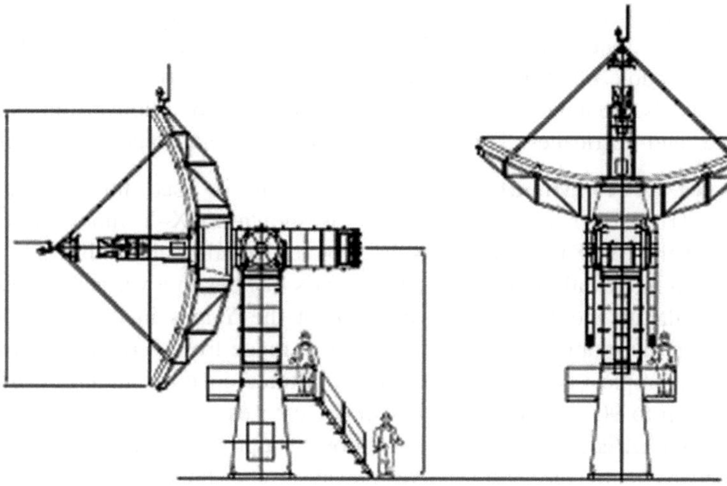

Fig. 6.12 Side/front views of 11 m non-GEO antennas—Courtesy of Manual: by High Gain Antenna

beamwidth is 1.90° for L-band, 1.30° for S-band, and 0.36° (Rx) and 0.34° (Tx) for X-band; and power requirement is 30 kVA.

Some of the mechanical specifications of this ground antenna are the same as previous antenna, except the following features: Continuous azimuth travel is ±270°; azimuth travel rate is 3°/s (Max.); elevation travel is −5°–95°; elevation travel rate is 3°/s (Max.); surface accuracy is 0.5 mm (RMS); foundation size is 7.0 M × 7 M × 0.6 M; and concrete volume 84.7 m^3.

6.3.5 Ground RES Antenna Mount Systems

There are variety sizes of ground RES antennas, from 1.2-m ultra-small antennas to 32-m ultra-large antennas, at intervals of one meter. There are various types of antenna mounts such as XY, AZ/EL, and Polar, including King Post, Yoke & Tower, and Wheel on Track type. The ground antennas provide services for GEO and non-GEO constellations, such as Small LEO, Big LEO, MEO, and HEO satellite services. The RES antennas perform satellite communications, navigation, observation, remote sensing, and monitoring according to their RF bands, such as L-, S-, C-, X-, Ku-, and Ka-band or antennas for multibands as well.

When selecting the location for an Earth station ground antenna of any size, the primary consideration is to ensure a clear view of the orbital arc, which allows the antenna to "see" the maximum number of satellites. Placing an antenna on a rooftop is often the optimal solution. However, extremely high winds can damage or destroy a parabolic dish antenna.

An Earth station antenna typically requires a rigid steel backup structure combined with an accurate dish surface. They are fitted with necessary bearings, gears, and drives to enable pointing accuracy within a few tenths of a degree. Thus, the structure must also be able to withstand extreme weather conditions, from excessive heat to cold, and hurricanes. Three common antenna mount types are as follows:

1. **Two-Axes (X–Y) Mount**—This mount system is used for medium-sized antennas (10–13 m). In Fig. 6.13 (Left) is shown an X–Y mount system, in which lower axis (X) is parallel to the ground. Rotation about this axis moves the antenna in elevation plane. The upper axis (Y) lies in a vertical plane and is perpendicular to the X-axis. The position of the Y-axis in the vertical plane depends on the rotation of the X-axis and can range from vertical to horizontal. This is the simplest type of mount, but is a limited coverage mount. It provides coverage up to 90° in the X-axis but in some cases, only ±5° in the Y-axis.
2. **Two-Axes (AZ/EL) Mount**—The location of a point on Earth surface can be described by using the azimuth-over-elevation coordinate system. Azimuth is defined as an angle produced by rotation about an axis, which is perpendicular to the local horizontal plane. The elevation axis rotates in the local horizontal plane as the azimuth angle rotates. A change in the elevation angle will cause a rotation of the antenna in the vertical plane. Installation of an AZ/EL antenna mount is relatively easy. The azimuth axis should be very nearly vertical to the local ground plane to minimize the change in elevation angle when azimuth is swept. In Fig. 6.13 (Middle) is shown a simplest form of AZ/EL mount. This type of mount has full elevation coverage and ±180° in azimuth.

There are two types of two-axes (AZ/EL) mount systems:

a) Fixed azimuth over elevation is shown in Fig. 6.14 (Left). This mount is designed for an antenna that is deployed for fixed service on a single satellite. The satellite antenna is pointed manually in elevation, azimuth, and polarization with the use of tools and requires an engineer on site to repoint.

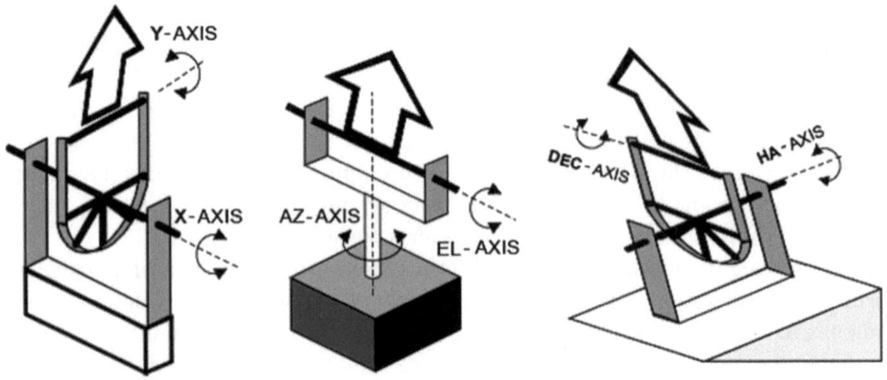

Fig. 6.13 Antenna mount systems—Courtesy of Handbook: by Intelsat

b) Two-axes motorized mount is shown in Fig. 6.14 (Middle). This is also an azimuth over elevation mount in which both axes of movement are motorized independently which allows a wider range of movement and pointing accuracy. These antenna mounts can be controlled remotely and when combined with a motorized polarization adjustment can offer full 3-axes pointing. There is also 3-axes mount with inclined orbit tracking, which uses a 3-axes mount combined with electronics and tracking algorithms to track a satellite in an inclined orbit. Typically this system provides track up to 10° Inclination.

Three main types of AZ-EL mounts are used for Earth station antennas, such as King Post, shown in Fig. 6.15 (Left), Yoke & Tower, shown in Fig. 6.15 (Middle), and Wheel on Track type, shown in Fig. 6.15 (Right).

3. **Polar Mount**—A polar mount has two axes of rotation as illustrated in Polar Mount—A polar mount has two axes of rotation as illustrated in Fig. 6.13 and 6.14 (Right). The first one is the hour angle axis, which is parallel to the Earth's axis. It is inclined in the north–south direction from the local horizontal through

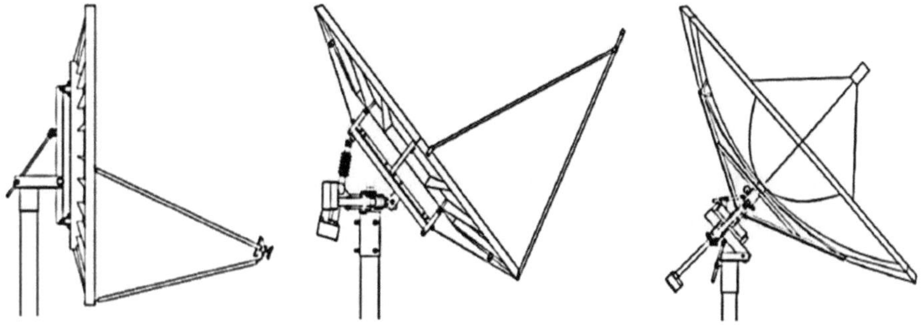

Fig. 6.14 Antenna mount systems—Courtesy of Handbook: by Sematron

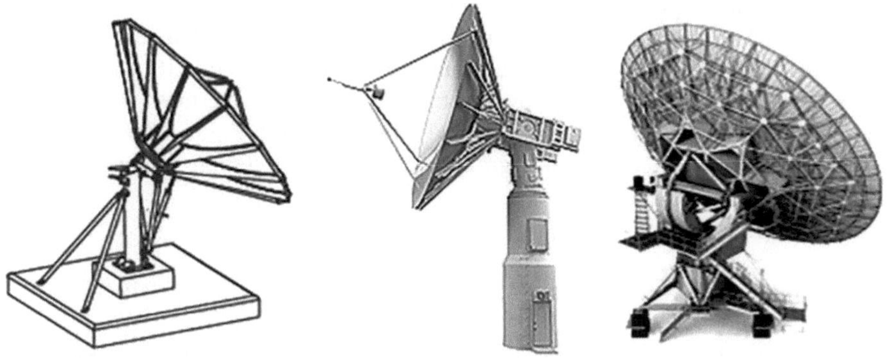

Fig. 6.15 King post, yoke and tower, and wheel on track antenna mounts—Courtesy of Manual: by Ilcev

an angle equal to the latitude of the site. Therefore, the hour angle axis is parallel to the ground at the equator and perpendicular to the ground at either the North or the South Pole, while the second angle is the declination.

In such a way, satellite longitude, Earth station longitude and latitude determine the amount of declination required. This mount uses a single motorized axis to slew the antenna so that the range of movement will follow an arc of sky that corresponds to satellite locations in GEO constellation. Able to be repointed remotely, this is the most cost-effective form of steerable satellite antenna and typically allows coverage on a 90° segment of the GEO arc.

4. **Ground Mount**—There are additional antenna mount systems, which mostly may suit the following applications:

a) Non-penetrating mount (NPM) is a freestanding mount designed to be placed onto hard standing or a flat roof and ballasted according.
b) Penetrating king post mount is a ground mount designed to be anchored into a suitable structural foundation for permanent installations.
c) Wall mount is a mount that attaches the satellite antenna to a vertical wall surface rather than the ground. This type of mount is convenient for smaller antenna sizes where suitable load bearing walls are available.

6.3.6 Main Ground RES Antenna Geometry and Parameters

Satellite ground Earth stations deploy dish antennas of 0.5–30 m in diameter. The dish surface contour is based on the equation for a parabola:

$$y2 = 4fx \tag{6.1}$$

where f = focal length, and x = coordinate along the axis of the paraboloid.

A paraboloidal surface contour satisfies the requirement that all the energy radiated from a launcher at the focal point toward the surface will be reflected to form a phase coherent plane wavefront across the dish aperture. Expressed in another way, path lengths ABC, ADE, and AFG, shown in Fig. 6.16, are all equal.

The important parameters of a ground satellite Earth station antenna are gain, beamwidth, and sidelobes.

6.3.6.1 Ground Antenna Gain

Satellite ground antenna gain is defined in the way, when a radio wave arriving from a distant source impinges on the antenna that collects the power contained in its antenna aperture effective (Ae). If the antenna were perfect and lossless, then the

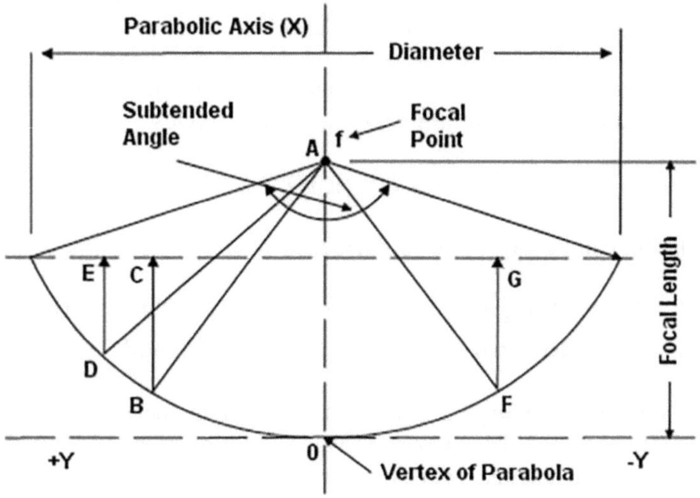

Fig. 6.16 Geometry of paraboloid—Courtesy of Manual: by Intelsat

effective aperture area Ae would be equal to the actual projected area A. For a circular aperture, the projected aperture is as follows:

$$A = \pi d^2/4 \tag{6.2}$$

where d = antenna diameter and the effective aperture area Ae = A (for an ideal antenna).

Taking into account losses and the nonuniformity of the illumination law of the aperture, the effective area is in practice expressed by the following equation:

$$Ae = \eta\, A \text{ and } Ae = \eta\, \pi\, (d/2)^2 \tag{6.3}$$

where η = antenna efficiency and $\eta < 1$.

Efficiency is an important factor in antenna design, so in such a way, special techniques are used to optimize the efficiency of Earth station antennas. Antenna efficiency is affected by:

a) The subreflector and supporting structure blockage;
b) The main reflector RMS surface deviation;
c) Illumination efficiency, which accounts for the non-uniformity of the illumination, phase distribution across the antenna surface, and power radiated in the sidelobes; and
d) The power that is radiated in the sidelobes.

Antenna aperture efficiencies between 55 and 75 % percent are typically attainable. Then, the on-axis antenna power gain (relative to an isotropic radiator) is given by relation:

$$G = 4\pi\, Ae/\lambda^2 \tag{6.4}$$

where λ = is the free space wavelength and π = 3.14159…., and Ae = effective aperture of the antenna. Substituting for Ae in (6.4) yields:

$$G = \eta\,(\pi\, d/\lambda)^2 \tag{6.5}$$

This relation can be expressed in decibels:

$$\begin{aligned} G_{dBi} &= 10\ \log\ \eta + 20\ \log\ \pi\ + 20\ \log\ d\ -\ (20\ \log\ \lambda)\ \text{or} \\ G_{dBi} &= 10\ \log\ \eta + 20\ \log\ f + 20\ \log\ d + 20.4\text{dB} \end{aligned} \tag{6.6}$$

where η = antenna efficiency, d = antenna diameter in meters, f = operating frequency in GHz and 20.4 dB = Constant value resulting from 10 log ($1*10^9*\ \pi/c$).

The definition of antenna gain is more straightforward if the antenna is considered during transmissions. However, according to the reciprocity theorem, the properties of an antenna are the same for transmission and for reception of radio signals. For example, an isotropic receiving antenna is able to receive uniformly from the same transmitter positioned in any direction and power.

6.3.6.2 Ground Antenna Beamwidth and Bandwidth

Beamwidth is a measure of the angle over which most of the gain occurs. It is typically defined with respect to the half-power beamwidth (HPBW) or −3 dB points of the main lobe in the antenna radiation pattern, and the scenario is shown in Fig. 6.17 and is given by the following relation:

$$\text{HPBW} = \lambda/d\sqrt{\eta}\ \cdot\ 57.29 \tag{6.7}$$

An Earth station with an antenna of 16 m (52 ft) and an efficiency of 70 % would have a beamwidth of 0.214° at 6 GHz.

With regard to the bandwidth, dish antennas are wideband devices. As seen from the gain equation, for a given diameter, the gain of a dish will increase as the frequency of operation increases. However, operation away from the design frequency will normally result in impaired performance due to the limitations of the feed and launch system.

6.3.6.3 Ground Antenna Sidelobes

While most of the power radiated by an antenna is contained in the main lobe, a certain amount of power can be transmitted or received in off-axis directions. Sidelobes are an intrinsic property of antenna radiation and cannot be completely

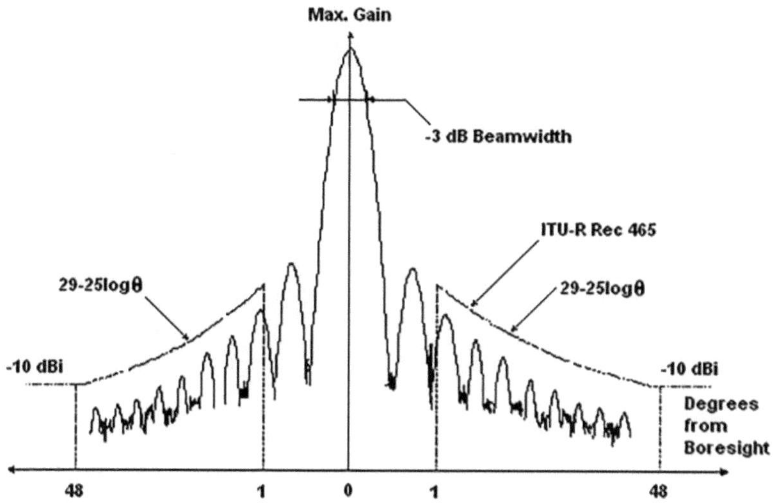

Fig. 6.17 Antenna Gain vs Diameter—Courtesy of Manual: by Intelsat

eliminated. However, sidelobes are also due to antenna defects that can be minimized with proper design. ITU-R Record 580-1, for instance module 1, defines the desired sidelobe envelope for different types of antennas. Thus, antennas installed after 1988 and with a ratio of $d/\lambda > 150$ must meet the following characteristics:

$$G = 29 - 25 \log \theta \, \text{dBi} \tag{6.8}$$

where θ = degrees from boresight and $1^\circ \leq \theta \leq 20^\circ$, d = antenna diameter (meters), and λ = wavelength for the operation frequency (meters).

For smaller antennas with d/λ between 35 and 100 (1.75–5 m for C-band and 75 cm–2.1 m for Ku-band) relation is as follows:

$$G = 52 - 10 \log d/\lambda - 25 \log \theta \, \text{dBi} \tag{6.9}$$

where for $(100\,\lambda/d)^\circ \leq \theta \leq d/5\lambda^\circ$.

Fore example, to calculate the transmit sidelobe envelope for a 1.5 m antenna operating at 14 GHz. (1 = 0.021 m) it is possible to get as follows: Ratio $D/\lambda = 70$; $G = 52 - 10 \log (70) - 25 \log \theta$; $G\text{dBi} = 33.5 - 25 \log \theta$; As $D/\lambda = 70$; then follows $1.4^\circ \leq \theta \leq d/5\lambda^\circ 14^\circ$.

Figure 6.17 shows a radiation pattern for a paraboloidal antenna and the recommended sidelobe characteristics. Diameter, operating frequency, and aperture efficiency affect antenna parameters, but sidelobe characteristic is one of the main factors in determining the minimum spacing between satellites and therefore, the orbit and spectrum efficiency. In Fig. 6.18 is shown a plot of antenna gain as a function of antenna diameters while using frequency and efficiency parameters.

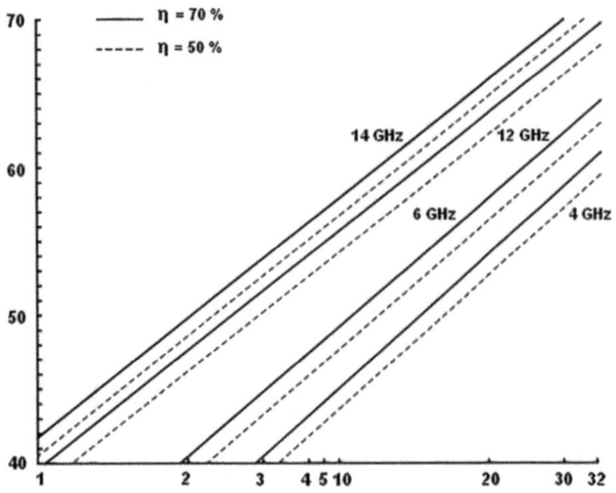

Fig. 6.18 Noise Temperature of an Antenna as a Function of Elevation Angle—Courtesy of Manual: by Intelsat

6.3.6.4 Ground Antenna Noise Temperature and Figure of Merit (G/T)

The system noise temperature of an Earth station consists of the receiver noise temperature, the noise temperature of the antenna, including the feed and waveguides, and the sky noise picked up by the antenna.

$$T_{Sys} = T_{Ant}/L + (1 - 1/L)T_o + T_e \tag{6.10}$$

where RES values are: T_{Sys} = system noise temperature, L = feed loss in numerical value, T_e = receiver equivalent noise temperature, T_o = standard temperature of 290°K, and T_{Ant} = antenna equivalent noise temperature as provided by the manufacturer.

The noise power into the receiver due to the antenna is equivalent to that produced by a matched resistor at the LNA input at a physical temperature of T_{Ant}. If a body is capable of absorbing radiation, then the body can generate noise. Thus, the atmosphere generates some noise. This also applies to the Earth surrounding a receiving ground station antenna. If the main lobe of an antenna can be brought down to illuminate the ground, then the system noise temperature would increase by approximately 290°K. Fortunately, however, synchronous satellites require vertical angles of elevation of 5° or more. If the directivity of the antenna is such that the ground absorbs 5 % of its radiated energy illuminates, then the same antenna used for reception would contribute 5/100 × 290°K, i.e., 14.5°K of noise.

Every satellite antenna has sidelobes, which are augmented by "spillover" from the dish edge and scattering of energy by the launcher or by the subreflector and its supporting structure. All of these factors influence the noise temperature of the antenna. The satellite antenna noise temperature is a complex function of antenna gain pattern, background noise, temperature of the sky, equivalent atmospheric

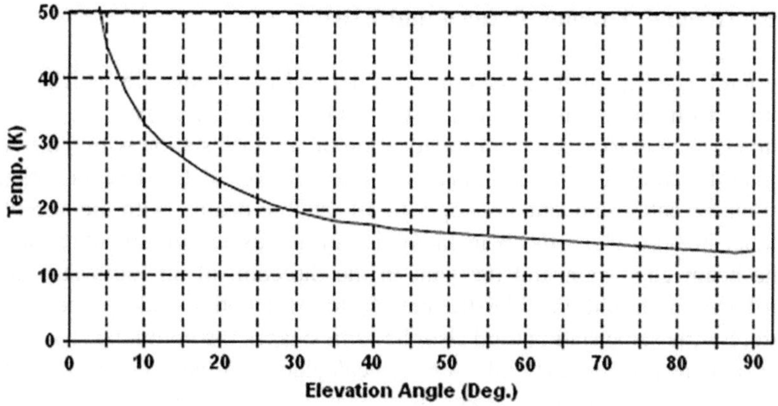

Fig. 6.19 Antenna gain versus diameter—Courtesy of Manual: by Intelsat

noise temperature, and noise temperature of the Sun. Thus, a typical curve variation of the antenna noise temperature with the antenna elevation angle is shown in Fig. 6.19. It is usually a minimum at zenith, typically 15–20°K for a low loss antenna with low sidelobes, which increases considerably as the elevation angle falls below 10°.

The next important parameter of ground antenna is Figure of Merit (G/T). In every radio transmission system, noise is a factor that greatly influences the whole link quality. The G/T_{dBK} is known as the "goodness" measurement of a receive system. Satellite systems require a specific G/T for all of the standard Earth stations for fixed and mobile service, such as presented in the Intelsat Earth Station Standards (IESS) documents. This means that providing the Earth station meets the required G/T specification, Intelsat or other satellite operators will provide enough power from the satellite to meet the characteristic of every service.

The ground antenna G/T values are expressed in dB relative to 1°K, so at this point the same system reference point, such as the RES receiver input, for both the gain and noise temperature must be used by the following equation:

$$G/T = G_{Rx} - 10 \log T_{Sys} \tag{6.11}$$

where G_{Rx} = receive gain in dB and T_{Sys} = system noise temperature in °K.

In such a way, as the antenna gain is frequency-dependent, the G/T must be normalized to a known frequency (normally either 4 or 11 GHz) by subtracting from Eq. (6.11) the factor 20 log f/f_o (f_o being 4 or 11) where "f" is the frequency in GHz.

6.3.6.5 Ground Antenna Pattern and EIRP

The ground satellite antenna pattern has a defined beam edge, in which values of equivalent isotropic radiated power (EIRP), G/T, and flux density are referenced. Adjustment factors to account for the location of an Earth station within the satellite beam may be applied to the link analysis. These factors, called beta-factors, aspect correction, or pattern advantage, apply to all satellite beams, with the exception of the global beams. Thus, the beta factor is defined as the difference between the satellite beam edge gain and the gain in the direction of an Earth station. Each station location should consider the beta-factors for uplink and for the downlink, because, for the same station (and also the same beam), the uplink beam coverage is not the same as for the downlink. A rough calculation of these factors can be made through the satellite beam coverage as shown in the correspondent satellite IESS where the lines or contours in the footprint represent 1 dB incremental from beam edge. A more accurate value can be obtained by requesting it from Intelsat, especially in those cases where fully steerable satellite beams are used.

The radiation pattern indicates the variations of gain with direction. For an antenna with a circular aperture or reflector, this pattern has rotational symmetry and is represented within a plane in polar coordinate form shown in Fig. 6.20 (Left) or Cartesian coordinate form shown in Fig. 6.20 (Right). The main lobe contains the direction of maximum radiation, while side lobes should be kept to a minimum. The angular beamwidth is the angle defined by the directions corresponding to a given gain fallout with respect to the maximum value. The 3 dB beamwidth, indicated in Fig. 6.20 (Left) by u3 dB, is often used. The 3 dB beamwidth corresponds to the angle between the directions in which the gain falls to half its maximum value, which is related to the ratio $\lambda = D$ by a coefficient whose value depends on the chosen illumination law. For uniform illumination, the coefficient has a value of 58.5°. Thus, with non-uniform illumination laws, which lead to attenuation at the reflector boundaries, the 3 dB beamwidth increases and the value of the coefficient depends on the particular characteristics of the law.

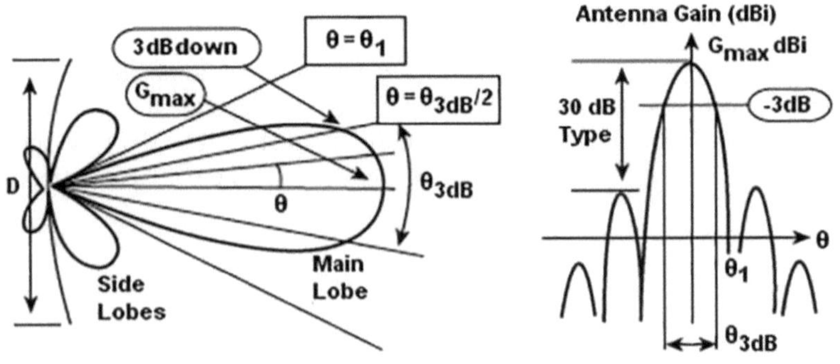

Fig. 6.20 Antenna radiation pattern—Courtesy of Manual: by Maral

The power radiated per-unit solid angle by an isotropic antenna fed from a radio frequency source of power P_T is given by $P_T = 4\pi$ [W/steradian]. In a direction where the value of transmission gain is G_T, any antenna radiates a power per-unit solid angle equal to $P_T G_T/4p$ [W/steradian]. The product $P_T G_T$ is called the EIRP value expressed in Watts (W).

The gain of the Earth station antenna, from Eq. (6.5), is as follows:

$$G_{max} = \eta(\pi\, df/c)^2 \tag{6.12}$$

where $\lambda = c/f$, c = velocity of light or 3×10^8 m/s and f = frequency of the electromagnetic wave (EW). Hence, expressed in dBi (the gain relative to an isotropic antenna), the actual maximum antenna gain is:

$$G_{max\,dBi} = 10 \log \eta\, (\pi\, d/\lambda)^2 = 10 \log \eta\, (\pi\, df/c)^2 \quad \text{[dBi]} \tag{6.13}$$

The power flux density in relation to distance (R) is given by:

$$\Phi_{max} = P_T\, G_{Tmax}/4\pi\, R^2 \tag{6.14}$$

The maximum EIRP of the Earth station (on the axis) is given by:

$$EIRP_{max} = P_T G_{T\,max} = 53.1\,\text{dBi} + 20\,\text{dBW} = 73.1\,\text{dBW} \tag{6.15}$$

Therefore, P_T at the input of the transmitting antenna multiplied by G_T of the same antenna is defined as the EIRP and is usually expressed in dBW. The latter also depends on the absolute power from the transmit amplifier and the loss in the combining and feed system. According to Fig. 6.21, the EIRP will be as follows: EIRP = PA*(power) − (Feed loss) − (Combining losses) + Antenna Gain, where PA*(power) = carrier power at power amplifier output flange.

In the link equation, by unfolding the kTB product under the logarithm, the carrier-to-noise ratios (C/N) becomes:

$$C/N = EIRP - L + G - 10 \log k - 10 \log T - 10 \log B \tag{6.16}$$

Fig. 6.21 EIRP calculation example—Courtesy of Manual: by Intelsat

The difference, G − 10logT is the figure of merit:

$$C/N = EIRP - L + G/T - 10 \log k - 10 \log B \quad (6.17)$$

where values L = transmission losses, G/T = figure of merit of the receiver, k = Boltzmann's constant, T = temperature in Kelvin, and B = carrier occupied bandwidth.

6.3.7 Ground Antenna Feed

An antenna with a feed in the center of the paraboloid dish (axisymmetric) represents the simplest antenna configuration that is potentially capable of meeting the RF specifications for RES applications. A major advantage of such a configuration is that mechanically, it has relatively simple design, reasonably compact and, in general, fairly inexpensive.

In telecommunications and electronics, an antenna feed refers to the components of an antenna which feed the radio waves to the rest of the antenna structure and in the air or in receiving antennas collect the incoming radio waves, convert them to electric currents, and transfer them to the receiver. Antennas typically consist of a feed and additional reflecting or directive structures (such as a parabolic dish or parasitic elements) whose function is to form the radio waves from the feed into a beam or other desired radiation pattern.

The antenna feeds of all Earth stations perform the same basic functions:

a) To shape the beam to provide the required uniform illumination of the main reflector.
b). To separate the transmit and receive signals with minimum loss and interference.
c). To convert from circular to linear polarization and vice versa for the downlink and uplink signals at C-band where circular polarization is employed.
d). To produce error signals that represent the degree and direction by which the main beam is off-track. This is not required for antennas that use a step-track system or are small enough not to need tracking.

In simple antennas, the feed usually consists of the feed antenna (driven element), the part of the antenna which actually converts the RF currents to radio waves or vice versa, and the feed or transmission line, which connects the feed antenna with the receiver or transmitter. Thus, the antenna feed is the combination of all of the components of the antenna that are used for the receiving and transmitting purposes of RF waves, in which sample for Intelsat RES transceiving antenna with associated components is shown in Fig. 6.22.

Antenna design should be made considering the maximum power transfer possibilities and the efficiency. The antenna feed impedance must be matched with the load resistance as the combination of the resistance, the capacitance, and the

Fig. 6.22 Antenna feed assembly—Courtesy of Manual: by Intelsat

inductance, which includes radiation and loss resistance. The feed impedance is the resistance offered at the input of the antenna to the signal and loss resistance is the resistance that is offered by the actual components of the antenna. The loss and the feed impedance must work together in order to achieve a proper working antennae feed. The efficiency, gain, and absolute gain are very important considerations in the design of the antenna feed. Thus, the total efficiency of the antenna describes the losses at the input terminal and through the antenna components. It accounts for the reflection, conduction, and the dielectric losses inside the antenna.

6.3.8 Ground Antenna Diplexer

The approach about antenna diplexers is the most efficient for combining receive and transmit signals. Each antenna diplexer uses two hybrids and two bands pass filters, which typical diagram of two-port diplexer/combiner is shown in Fig. 6.23. It consists of two diplexers (four are hybrids and four are band pass filters). The ports in each hybrid are labeled A to D. Each hybrid is the same type as previously described, so not only can it combine two signals into one, but it can also split a single signal into two components.

The signal input 2 (or 1) is split by hybrid 1 (or 3) and will appear in ports 1-B (or 3-B) and the same signal shifted 90° in port 1-C (or 3-C). Because the hybrid is a passive reciprocal device, it follows that if two versions of the same signal (one shifted 90° with respect to the other) are connected to two input ports (as for hybrids 2), then the signal will combine to one of the output ports (port 2-B). But, they will be canceled in the second port (port 2-C). The signal from diplexer 2 will enter hybrid 4 (diplexer 1), will split, and will be reflected by filters F2, and recombine to port 4-B reaching the combiner output.

In Table 6.1 is shown a typical insertion losses caused by the diplexer. Thus, the drawback with this approach is that every diplexer is designed for a specific frequency band, while it permits the use of smaller power amplifiers due to its low combining losses.

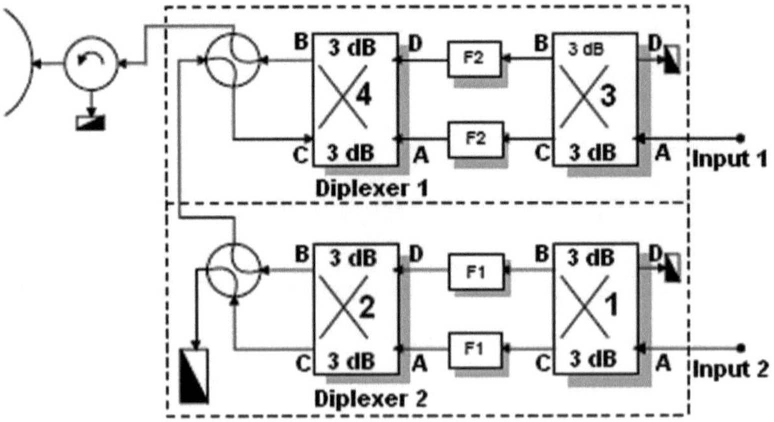

Fig. 6.23 Diplexer/combiner—Courtesy of Manual: by Intelsat

Table 6.1 Combining losses for Fig. 6.22

Input	Hybrid losses (to point B) (dB)	Diplexer loss (to point A) (dB)	Total losses (dB)
F1	3.0	0.8	3.8
F2	6.0	0.8	6.8
F3	7.8	0.8	8.6
F4 (SPARE)	10.8	0.8	11.6
F5 (TV)	–	1.8	1.8

6.3.9 Ground Antenna Tracking and Control Systems

During the early development of satellite communications, monopulse tracking of one form or another was used almost exclusively. However, since the mid-1970s through the present days, there has been a significant shift toward the use of step-track autotracking and modern programmed tracking, which will be shortly presented in the following sections.

The term monopulse was derived from radar terminology to describe a case in which angular error could be obtained from a single pulse. Thus, the type of antenna feed used is the higher mode detection feeds, which use the fact that unless the boresight axis of the feed is perpendicular to the incident EM field, higher order modes as well as the dominant mode will be produced within the feed waveguide. By detecting the presence of these higher modes, tracking information relative to angular displacement can be derived.

6.3.9.1 Antenna Step-track Autotracking

With the continuing improvement in the station-keeping accuracy of modern GEO fixed and mobile communication satellites, however, the opportunity arose to reduce significantly the complexity of the antenna autotracking system which, until recently, was predominantly based on the monopulse tracking technique described in the previous section. While this system is inherently very accurate, it is a complex, expensive, and difficult to maintain. To take advantage of the greater station-keeping accuracy of today's satellites, the monopulse tracking systems have been largely replaced in recent years by the newest step-track system, principally because of the latter's great simplicity and consequently lower cost.

The operational concept of the step-track system is quite simple. After acquisition of the satellite beacon (or pilot) signal, the antenna is commanded to make an initial angular move. By comparing the received signal level before and after the move, the direction of the next move can be decided. That is, if the signal level has increased, the antenna continues to be moved in the same direction. If the signal level has decreased, then the direction of movement is reversed. This process is continuous and alternates between the two orthogonal antenna axes.

System interface of antenna step-track autotracking with connection between the ACE, APE, and the servo drive system is shown in Fig. 6.24. For example, Standard B ground Earth stations use single channel per carrier (SCPC) only, so the signal often used by the autotracking system is the SCPC pilot, which is located in the middle of the SCPC band. On the downlink, this pilot is extracted by the SCPC terminal and passed onto the ACE. Thus, Intelsat spacecraft are equipped with different beacon frequencies.

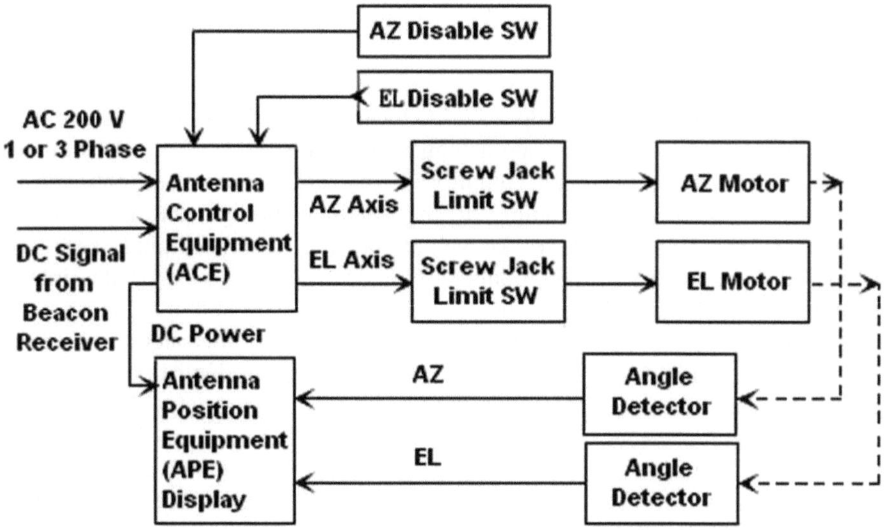

Fig. 6.24 Antenna control and drive system interfaces—Courtesy of Manual: by Intelsat

The main disadvantages of the step-track system are as follows:

1. Locating a beam maximum can never be as accurate as finding a sharp null; and
2. Tracking can be degraded by amplitude fluctuations in the received signal levels due to atmospheric perturbations. This would also have a detrimental effect on the receive signal at the satellite where the levels are required to be kept within ±0.5 dB of nominal.

These limitations can be overcome by choosing a step size that is sufficiently small, but not so small as to cause the antenna to continuously hunt the satellite as during moderate wind loading conditions. The biggest advantage of the step-track system is its simplicity, which means significantly lower costs because a special tracking communications feed to detect the higher modes is not required. The need for maintenance is also reduced.

6.3.9.2 Antenna-Programmed Tracking

With the advent of inclined orbit satellites, a programmed tracking system becomes more attractive and accurate. The antenna is under computer control and a permanent calculation of satellite orbital position is derived from pointing data (11 ephemeris parameters), thus eliminating the need for a satellite beacon. Also available are the so-called smooth stepback systems, which are step-track systems, using computer memories to memorize the satellite movement within the first 24 h of acquisition and then follow the memorized program. This feature is not recommended with inclined orbit.

6.3.9.3 Ground Antenna Telemetry, Tracking & Command (TT&C) System

To maintain the satellite at its assigned GEO location periodic adjustments are made by ground infrastructure station and antenna for Telemetry, Tracking & Command (TT&C), including control and monitoring, by using propellant onboard the spacecraft to enable repositioning of the satellite through the activation of propulsion jets. This procedure, known as satellite station-keeping, keeps the satellite drifting within certain bounds (like drawing a box). For the Intelsat satellites, station-keeping is performed both in the east–west and in the north–south directions. Thus, the amount of propellant placed on board, the satellite, therefore, determines the maneuver life of a satellite. If near-zero orbit inclination during the satellite launch phase (injection into GEO) is achieved, more fuel will be available for station-keeping maneuvers. For satellite, station-keeping is necessary a 13-m mono-pulse Cassegrain-type antenna with polarization diversity established for S-band TT&C, which scenario is shown in Fig. 6.25 antenna (Left) and RF/BB equipment for TTC (Right).

Fig. 6.25 TT&C ground antenna with RF equipment—Courtesy of Paper: by ETRI

To implement TTC&C facilities in any RES will be necessary to provide reception of satellite beacons, which are the only signals, apart from the satellite telemetry channel, which are produced by the satellite payloads. The majority of Earth stations use them for tracking and reference purposes. Each satellite transmits two satellite beacon signals in each of the frequency bands in the Pol Global Beam. However, the C-band beacons are modulated with telemetry data, whereas the Ku-band beacons are always unmodulated.

6.4 Ground Earth Station Radio Frequency Equipment

Radio frequency equipment in RES comprises low-noise amplifiers (LNA), power dividers, high-power amplifiers (HPA), and power combiners, shown in Fig. 6.4.

6.4.1 Low-Noise Amplifiers (LNA)

To receive the very weak signals from a satellite, the RES antenna must be connected to a highly sensitive receiver, i.e., one with very low inherent thermal noise. Before that, the receiver is connected to a low-noise amplifier (LNA) that amplifies a very low-power signal without significantly degrading its signal-to-noise ratio (S/N). An amplifier will increase the power of both the signal and the noise present at its input. The LAN units are designed to minimize the additional noise by considering tradeoffs that include impendence matching, choosing the amplifier technology, and selecting low-noise biasing conditions.

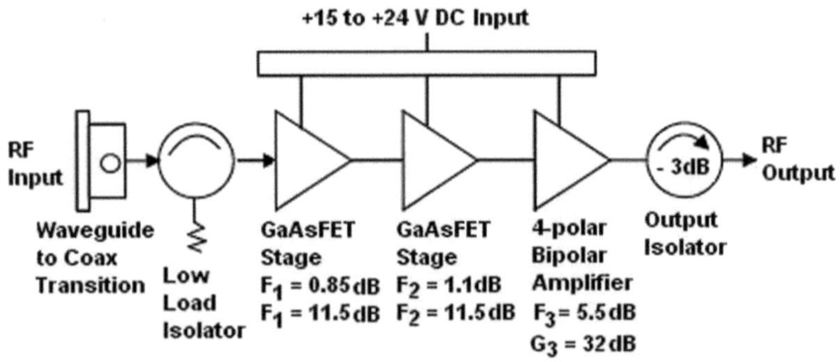

Fig. 6.26 GaAs FET block diagram—Courtesy of Manual: by Intelsat

Thus, the LNA devices are amplifiers that have a very good noise performance coupled with a wide bandwidth. This makes them essential for use as the first stage of a satellite RES receiving chain. Random movement of electrons causes thermal noise, which within semiconductors can be minimized by reducing the actual temperature of the LNA (by Peltier or cryogenic cooling method) or the current method of using uncooled field effect transistors (FET) amplifiers employing high electron mobility transistor (HEMT) technology. An FET amplifier uses one or more FET components, which main advantage used for amplification is that it has very high-input impedance and low-output impedance.

Bipolar transistors cause (shot) noise other than thermal noise and can provide mediocre performance at high frequencies. Besides, the noise due to field effect transistors is mainly of thermal origin and can be reduced by selecting the type of semiconductor used and the geometric characteristics of the transistor. In such a way, the performance in terms of noise factor is continuously improving due to the use of gallium arsenide (GaAs) and submicron lithography. The typical block diagram of GaAs FET is shown in Fig. 6.26.

The G/T system for an Earth station is virtually determined by the Noise Figure (F) and gain of the LNA, along with the antenna gain. The LNA is generally mounted as close to the antenna feed as possible so that the transmission line losses to the LNA will be at an absolute minimum. An LNA must also provide sufficient gain to overcome losses in the transmission line between the receiver and the LNA.

6.4.2 Power Dividers (Splitter)

Power dividers or splitters are electronic elements used in receiving radio technology and, when used in reverse transmitter technology, power combiners and directional couplers are passive devices used in the field of radio technology. They couple a defined amount of the electromagnetic power in a transmission line to a port enabling the signal to be used in another circuit. In such a way, when an Earth

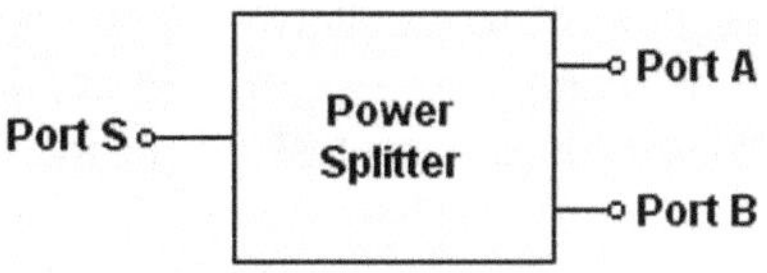

Fig. 6.27 Power splitter—Courtesy of Manual: by Ilcev

station must demodulate several carriers simultaneously, it is necessary to distribute the power at the LNA device output among the various converter channels. This is performed by a power splitter using passive devices (hybrid couplers or power dividers). The power splitter insertion loss L_{PS} adds to the feeder loss L_{FRX} from the LNA output to the power splitter input. The power split among the n-channels translates into attenuation by a factor n; hence, the total loss L from the LNA output to any converter input is

$$L = L_{FRX} L_{PS} n \tag{6.18}$$

Basically, a 0° splitter is a passive electronic device which accepts an input signal and delivers multiple output signals with specific phase and amplitude characteristics. In fact, the output signals theoretically possess the following characteristics: equal amplitude; 0° phase relationship between any two output signals; high isolation between each output signal; and Insertion loss, for instance, for 2 ports is 3 dB, for 8 ports is 8 dB, and for 48 ports is 16.8 dB, and so on. Thus, in the reverse mode, power combiner will also exhibit isolation between the input ports. The amount of isolation will depend upon the impedance termination at the combiner output or sum port.

The sample of splitter and combiner is illustrated in Fig. 6.27. Namely, when is used 0° power splitter, the input is applied to port S and equal outputs appear at ports A and B. When used as a power combiner, both inputs are applied to ports A and B and the sum taken from port S. The following signal-processing functions can be accomplished by power splitter/combiners:

1. Add or subtract signals vectorially;
2. Obtain multi-in-phase output signals proportional to the level of a common input signal;
3. Split an input signal into multioutputs;
4. Combine signals from different sources to obtain a single port output; and
5. Provide a capability to obtain RF logic arrangements.

6.4.3 High-power Amplifiers (HPA)

The order of magnitude of the power required at the transmitter output of RES terminal is 1 W or less per telephone channel and 1 kW per television carrier. The two main types of microwave tube used in Earth transmitting station power amplifiers are the Traveling Wave Tube Amplifiers (TWTA) and Klystron Power Amplifiers (KPA). Furthermore, in the case of small Earth stations, Solid-state Power Amplifiers (SSPA) are more and more used.

6.4.3.1 Traveling Wave Tube (TWT) Amplifiers

A TWT is an amplifier with a wide bandwidth and a power gain of typically 25–50 dB. The efficiency, in general, is a function of bandwidth and it ranges from 20 to 40 % with 20 % being a typical figure. It covers the entire usable band of the satellite (500 MHz or more) with the necessary uniformity of gain and group delay. Because of these features, the TWT appears to be the ideal power amplifier for Earth stations, since it allows several telephone carriers to be transmitted simultaneously with a single tube, irrespective of the repeaters and the frequencies allocated to these carriers.

It should be noted that the simultaneous transmission of several carriers in the same tube produces intermodulation noise components, which increase as the operating point of the tube approaches saturation. Since the maximum level of intermodulation components is subjected to specification, the degree of linearity of the tube parameter at the operating point is stipulated in each configuration. This leads to a back off of the operating point in relation to saturation and hence to a loss of available power. This loss during multicarrier operation can now be partly offset by using new equalizing devices known as linearizers.

In Fig. 6.28 is depicted a typical configuration of TWT device. The RF wave's electric field traveling in the slow-wave structure penetrates into the electron beam region and causes some electrons to accelerate and some others to decelerate producing a periodic velocity modulation in phase with the RF electric field. The electron bunches thus formed tend to concentrate ahead of the accelerating field and behind the decelerating ones. Because the average velocity of the electron beam is slightly greater than that of the RF wave, bunches will tend to move back into regions where the RF field will decelerate the electrons. As the electrons lose velocity, the energy lost by the electrons is transferred to the RF energy in the RF

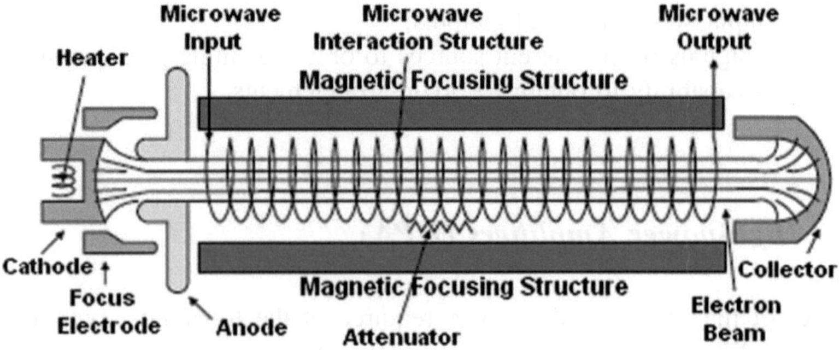

Fig. 6.28 Traveling wave tube (TWT)—Courtesy of Manual: by Intelsat

wave. This transfer of energy registers a constant gain in the amplitude of the RF wave per unit of length. At this point, a slow-wave structure will support RF energy traveling from output to input as well as from input to output.

The wave traveling from input to output will be amplified, and the wave traveling from output to input will not be amplified. However, in the presence of the inevitable reflections at the output and input couplers, some RF energy could be reflected back toward the input along the helix, and upon reflection from the input coupler, this signal will represent RF feedback. All practical TWT have sufficient gain for this feedback mechanism to result in self-oscillation. It is fairly simple to interrupt this feedback path by placing RF attenuation on one or more of the helix support rods. Thus, this attenuation is depicted in Fig. 6.28 by a resistive section (sever) and is formed by placing a carefully controlled pattern of a resistive material on the rods prior to their installation into the helix structure. Pyrolytic graphite (carbon) and titanium carbide are the most commonly used substances. The density of this attenuation pattern is selected to provide a very low reflection of RF energy so that any energy reflected from the output of the TWT is absorbed in the attenuation. The region of the helix structure containing the attenuation is called sever because the RF wave on the helix is terminated or severed at this point.

The following subsections describe the basic elements, construction, and functions of TWT are electron gun, cathode, heater, focus electrode, anode, slow-wave, and structure.

6.4.3.2 Klystron Power Amplifiers (KPA)

Klystrons are essentially narrow passband devices at about 40 MHz for 6 GHz klystrons, and 80 MHz for 14 GHz klystrons and sometimes more. These bandwidths suffice for traditional frequency modulated (FDMA mode) carriers, but may be inadequate for carriers with phase modulation and digital coding (TDMA mode). In any case, the choice of the klystron as the type of tube to be used generally entails the use of an amplifier for each of the carriers transmitted except for SCPC (and even possibly for SCPC and TV in certain domestic systems).

In Fig. 6.29 are depicted basic features of a typical multicavity klystron unit (5 cavities typical for a modern 3 kW KPA). Klystrons may be fitted with a mechanical (remote control) tuning device, whereby they can be adjusted to the operational center frequency (or to the center frequency of the repeater) so that this frequency can easily be changed.

In Fig. 6.30 is illustrated a typical block diagram of a KPA, which can also optionally connect to the control logic the following units: computer interface POB, remote control panel, and remote channel selector. Electrons emitted by the electron gun pass through the cavity gaps in each of the resonators and through cylindrical metal tubes (called drift tubes) located between the various gaps. Thus, in a Klystron amplifier, a low-level RF input signal is coupled into the first resonator, which is called the buncher cavity. The RF input signal excites oscillating currents in the cavity walls, which causes an electric field to appear across the buncher

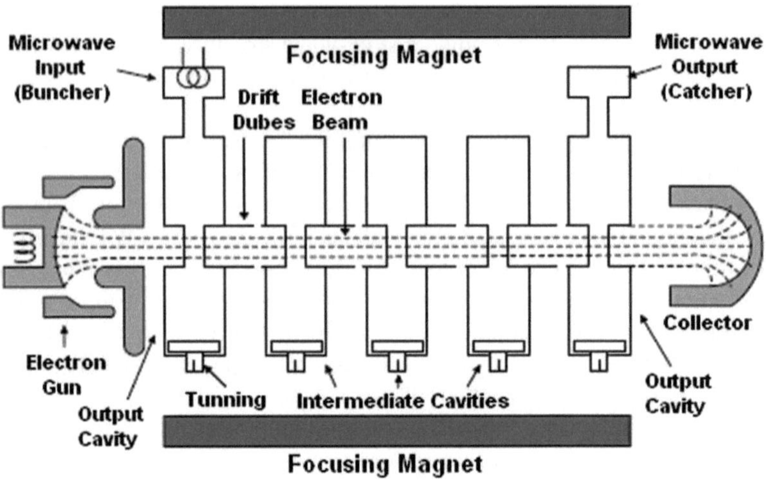

Fig. 6.29 Klystron—Courtesy of Manual: by Intelsat

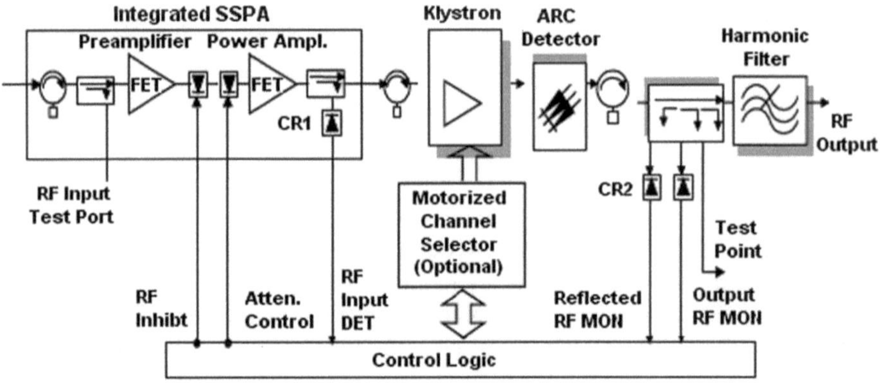

Fig. 6.30 KPA block diagram—Courtesy of Manual: by Intelsat

gap. Then, this excited electric field velocity modulates the electron beam creating bunches of electrons.

After leaving the buncher gap, the electrons proceed toward the collector in the drift tube region, passing through the intermediate resonators and the output (catcher) cavity. If the catcher cavity is of the correct size (i.e., tuned to the proper frequency), large oscillating currents will be generated in its wall that resulting in a RF output from the tube. To acquire the high gain and the required saturated power levels, intermediate cavities such as the second and third cavities are used. The electron beam is focused from the gun through the cavity gaps and drift tubes to the collector by means of a focusing structure, that is, either a permanent magnet for a short tube or a solenoid for a long tube.

6.4.3.3 Solid-State Power Amplifiers (SSPA)

In the case of small low-capacity stations, an SSPA normally with field effect transistors may suffice. Advances in FET technology, particularly Gallium Arsenide FET (GaAsFET), have significantly impacted satellite communications, Earth station as well as spacecraft applications as block diagram is shown in Fig. 6.26. The SSPA are available today to replace TWT in Earth stations and in the new-generation satellites (all solid state).

The SSPA offers the following advantages over TWTA: superior intermodulation distortion performance; higher reliability; lower maintenance and cost for spares; longer operating life; higher personnel safety—no dangerous high voltages; and lower power consumption.

Because of current advances in field effect transistors using GaAs and advanced circuit technology, recent power level of SSPA reached 100 W in 6 GHz band and 20 W in 14 GHz band. Solid-state power amplifiers are very reliable and economical and may provide an ideal solution for small Earth stations.

6.4.4 Power Combiners

In a typical Earth station, there could be more than one power amplifier connected to the feed port of the antenna serving a particular polarization. Under these circumstances, it is necessary to combine the output of the amplifiers into a single-signal path to the antenna feed. The output from the amplifier is usually in waveguide and is connected via one or more RF combiners. Thus, an RF combiner can be one of three types, namely hybrids, circulators, and diplexers.

The hybrid power combiner approach is the simplest and least expensive method for RF combining several signals. Hybrids used for this purpose are passive devices, and they are used to either split or combine the signal. A 3 dB hybrid combiner splits each input signal, so that 50 % of the input power (of input 1 and 2) is combined, both signals suffer the same loss, and the lost power is dissipated in the matched load. Similarly, a combiner shown in Fig. 6.31 splits each input signal so that 67 % of the input power in the cross-arm and 33 % in the through arm introduces a 1.8 dB loss to one signal, and 4.8 dB to the other.

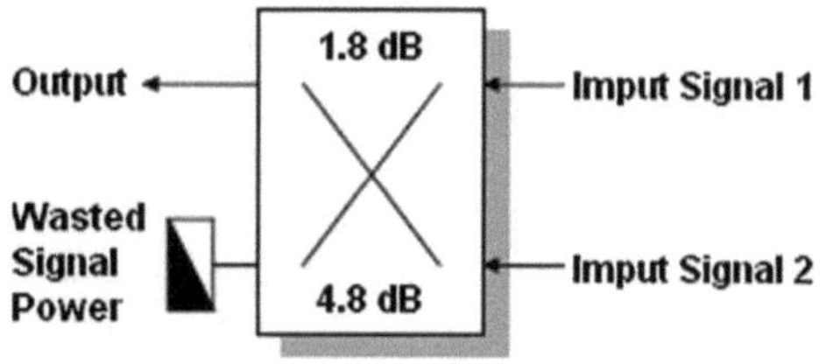

Fig. 6.31 Power combiner—Courtesy of Manual: by Intelsat

The signal wasted will be 33 % of input 1 and 67 % of input 2. This type of combiner is known as a Fixed Ratio. The power division in these types of combiners can vary from 0 to 100 %. A power combiner that allows the ratio to vary at a given time, according to the needs of the stations, is known as a variable ratio. The main drawback is the large amount of wasted transmitted power. The advantage is that the system is wideband.

6.5 Ground Earth Station Communication Equipment

The term telecommunication equipment usually refers to the equipment, which modulates the very high-frequency carrier with low-frequency signals (baseband) for emission and extracts (demodulates) these low-frequency signals on reception. Thus, the low-frequency signals may be analog telephone signals (usually multiplexed), digital signals, audio and video (television) signals, etc. The ground RES telecommunication equipment comprises the main equipment: receivers, transmitters, and uplink baseband processing signals. In particular, these equipment comprises frequency converter, modulating and demodulating equipment and signal-processing equipment.

6.5.1 Receivers (Rx) Subsystem

The receiver equipment is part of transceiver complex in any hypothetical ground Earth station. Every radio receiver as a part of RES transceivers contains RF equipment, such as Block Down Converters and Demodulators.

6.5.1.1 Block Down Converters (BDC)

Block downconverters change the radio frequency signals (e.g., 4 or 11 GHz) received by the antenna and preamplified by the LNA unit into intermediate frequency signals. These signals are then translated to the baseband in the demodulator. Ground Earth station receive electronics normally consist of an LNA at the antenna with a microwave cable connecting it to the indoor receivers. The receivers downlinks are equipped with downconverters to process either C-band or Ku-band, including new in use RF Ka-band.

Low-noise BDC consists of an LNA and a downconverter combined into one package. In Fig. 6.32 is illustrated an example of a low-noise BDC device where the first frequency conversion of a double frequency conversion occurs. The output IF frequencies are in a lower RF band and are fed to the receivers to final frequency tuning and further processing. Even though there can be problems associated with operating an BDC outdoors, where it would be exposed to varying climatic

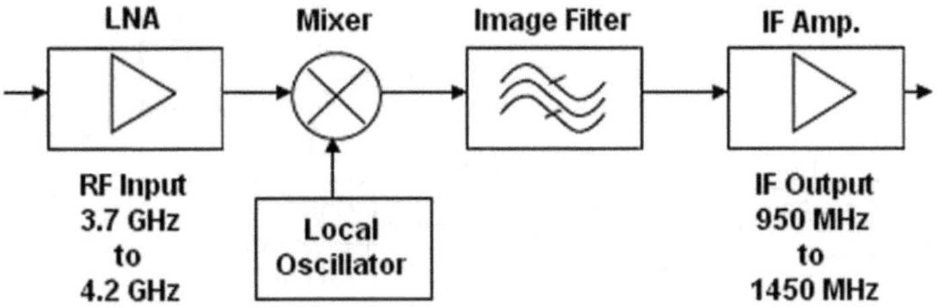

Fig. 6.32 Block diagram of BDC—Courtesy of Manual: by Intelsat

conditions such as temperature and humidity, there are significant advantages to moving the first frequency conversion outdoors, as follows:

a) The loss in coaxial cables increases as a function of the frequency as well as their length, by decreasing the frequency, cheaper cable can be used and the loss is reduced; and
b) One converter is shared by multiple receivers, making the receiver electronics cheaper.

6.5.1.2 Demodulators

The most essential portion of the baseband-to-IF chain in ground Earth station terminal consists of the demodulator and modulator or a combination of the two, called a modem. On the receive side, the demodulator detects the incoming carrier, synchronizes the data, performs error correction, and outputs a clean bitstream for the particular application. However, the RES transmit side works in the opposite direction. Threshold performance in terms of error rate and synchronization is determined at this level of the station.

Because there must be one modem for each frequency, multicarrier FDMA Earth stations can contain large quantities of these units, representing a sizable investment. In TDMA star networks, some frequency channels are shared by several Earth stations. There would need to be only a single modem per ground Earth station, since it can receive the bursts from the other Earth stations so long as they do not overlap in time. That is, of course, central to the operation of a TDMA satellite network as opposed to a CDMA network, where stations may transmit simultaneously on the same frequency.

There are other different modulation schemes that can be used to demodulate and modulate the data, such as QPSK, m-QAM, and FSK. In space communication, power is severely limited. FSK is not generally used, as it would require very high bandwidth to modulate our data resulting in very low bandwidth efficiency.

In Fig. 6.33 is presented the simplified block diagram of BPSK demodulator section. This demodulator determines the final performance BER, which is the ratio of incorrectly received bits divided by the total received bits during a given time

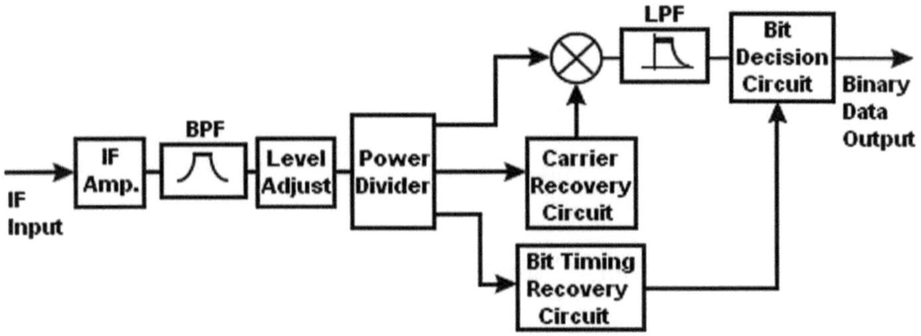

Fig. 6.33 Block diagram of BPSK demodulator—Courtesy of Book: by Elbert

interval. These errors are introduced when the received bandwidth contains instantaneous peaks of noise and interference that cancel the desired signal. The problem is further complicated by distortion of the bit pattern itself by bandwidth limiting, which is necessary to reduce the total noise power and channel impairments such as group delay and AM to PM distortion.

Therefore, the role of the RES terminal demodulator is to identify the phase or phase shift of the received carrier and to deduce from it the value of the bits of the transmitted binary stream. The demodulation unit can be as follows:

1. Coherent demodulation enables the binary stream to be reconstructed for both cases of transmission encoding, direct (BPSK, QPSK) and differential (DE-BPSK, DE-QPSK); and 2. Differential demodulation is always associated with differential encoding in receiver during transmission. At this point, the transmitted information can be recovered only if it is contained in phase changes.

6.5.2 RES Transmitters (Tx) Subsystem

The transmitter equipment is also part of transceiver complex in any hypothetical ground Earth station. Every radio transmitter as a part of RES transceivers contains RF equipment, such as Block Up Converters and Modulators.

6.5.2.1 Block up Converter (BUC)

A Block Up Converter (BUC) is used in the transmission RES uplink of satellite signals. It converts a band of RF from a lower to a higher frequency. Modern BUC units convert from the L-band to C-band, Ku-band, and Ka-band. Older generation of BUC units convert from a 70 MHz intermediate frequency (IF) to Ku-band or C-band.

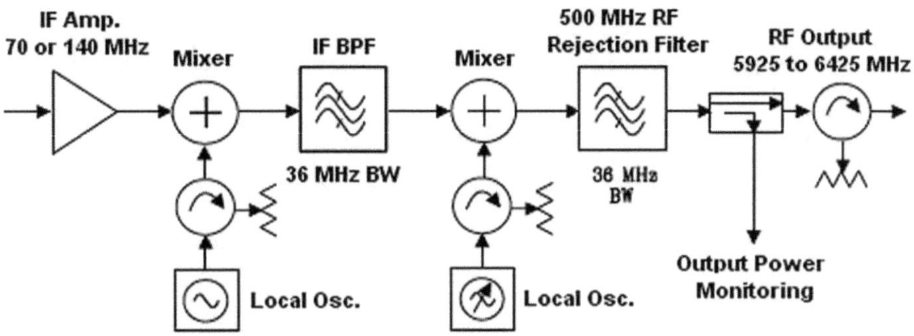

Fig. 6.34 Block diagram of BUC—Courtesy of Book: by Intelsat

Most BUC units use phase-locked loop local oscillator and require an external 10 MHz frequency reference to maintain the correct transmit frequency.

The BUC units are generally used in conjunction with LNB and in association with an orthogonal mode transducer (OMT). It being an upconverting device, makes up the transmit side of the system, while the LNB is the downconverting device and makes up the receive side. An example of a system utilizing both a BUC and an LNB is a satellite broadcasting system, used for DVB-RCS interactive telecommunications via VSAT, IPTV transmissions, and fast bidirectional Internet access via satellite.

In Fig. 6.34 is shown the block diagram of double upconverter, which change IF signals, such as 70 MHz, 140 MHz, 1 GHz, and so on, from the modulator into RF signals in the 6 GHz or 14 GHz band. These RF signals are then amplified by the power amplifier before being transmitted through the antenna.

Upconverters and downconverters are usually composed of RF filters, 2 cascaded mixers, 2 Local Oscillators (LO); one fixed frequency and the other variable frequency, IF amplifiers, possibly with automatic gain control, IF filters, group delay equalizers (GDE).

6.5.2.2 Modulator

In Fig. 6.35 is depicted a block diagram of a typical BPSK digital modulator, such as that used in a TDMA transmissions. Modulation is provided by mixing the data with a 70 MHz carrier and inverting the sinewave, when a change from 1 to 0 or vice versa is desired. The QPSK transmission can be obtained by adding a 90° phase shift to a second-modulated stream of data, thus doubling the use of the uplink frequency channel.

Reception works in the reverse direction, using the same mixing process but resulting in the original data being recovered. In such a way, additional circuitry is needed on the receive side to remove satellite link noise and resolve the bit pattern. The latter is accomplished by a decision circuit that, simply enough, decides if the

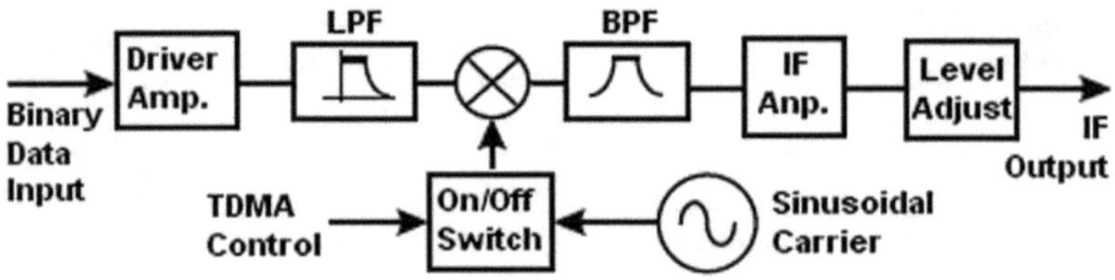

Fig. 6.35 Block diagram of BPSK modulator—Courtesy of Book: by Elbert

received bit is 1 or 0. The presence of noise and distortion in the demodulated bits causes the decision circuit to make a bad decision occasionally, a process that gives bit errors. It is the average rate of those errors, the BER, which determines received data quality.

The typical satellite modem (and terrestrial modem, as well) also contains the FEC coding and decoding circuitry that reduces the BER by factors of 100–10,000, depending on the strength of the coding technique. Other functions include burst synchronization for TDMA, baseband processing such as compression and decompression, including encryption and decryption, if appropriate.

6.5.3 Downlink Baseband Processing Signals (BPS) Equipment

The downlink BPS equipment is part of receiver complex of ground Earth station. Every radio receiver as a part of RES transceivers contains DeMUX and Decoder.

6.5.3.1 Demultiplexer (DeMUX)

Demultiplexer or DeMUX or file splitter is an electronic device taking an input analog or digital signals and selecting one of many data output lines, which is connected to the single input on the receiving end is often used with a complementary multiplexer. There are many types of demultiplexers, but here will be explained one of them.

The most common multistage frequency demultiplexer design is shown in Fig. 6.36. In this structure, the bandwidth of the input signal X(z) is divided into two sub-bands of equal bandwidths at the first stage. In the next stage, each sub-band is further divided into two new sub-bands of equal bandwidths. This process is continued until the last stage is completed. However, in order to divide the input signal in K channels, the demultiplexer must include M = log2 (K) stages. The basic signal-splitting element at each stage, designated $H_1,\ldots, H_M$ in Fig. 6.36, may simply be a two-channel ploy-phase Inverse Offset Discrete Fourier Transform (IODFT) demultiplexer.

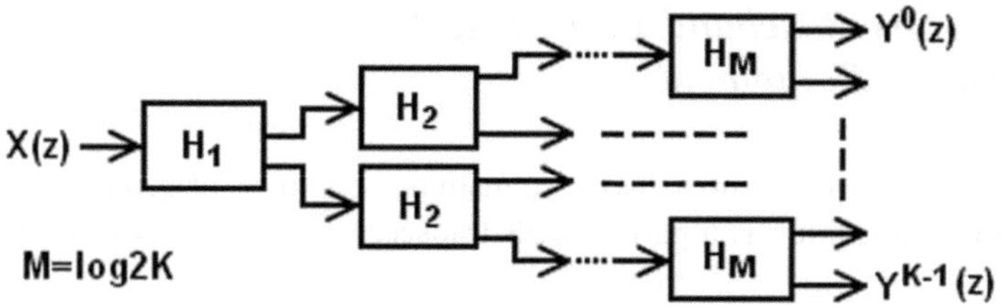

Fig. 6.36 Block diagram of demultiplexer—Courtesy of Manual: by Ilcev

6.5.3.2 Decoder

A decoder is an electronic circuit that changes a code into a set of signals. Thus, it is called a decoder because it does the reverse of encoding done by encoder, but we will begin our study of encoders and decoders with decoders because they are simpler to design.

There are many decoders such as: audio decoder that converts analog audio to digital form; binary decoder provides digital circuits such as 1-of-N and seven-segment decoders, decompressor (compression decoder) converts compressed data, such as audio, video, and images to an uncompressed form; instruction decoder is an electronic circuit that converts computer instructions into CPU control signals; and video decoder that converts base-band analog video to digital form.

The decoding operation in receiver is also performed on a block-by-block basis by the Low Delay Codebook Excited Linear Prediction (LD-CELP) decoder. Namely, upon receiving each 10-bit index, the decoder performs a table lookup to extract the corresponding codevector from the excitation codebook. In such a way, the extracted codevector is then passed through a gain-scaling unit and a synthesis filter to produce the current decoded signal vector. The synthesis filter coefficients and the gain are then updated in the same way as in the encoder.

The decoded signal vector is then passing through an adaptive postfilter, which has to enhance the perceptual quality. The postfilter coefficients are updated periodically using the information available at the decoder. The five samples of the postfilter signal vector are next converted to five A-law or μ-law PCM output samples, which is depicted in Fig. 6.37, where VQ Index is meaning index of Vector Quantization (VQ) and conversion is using Pulse Code Modulation (PCM)

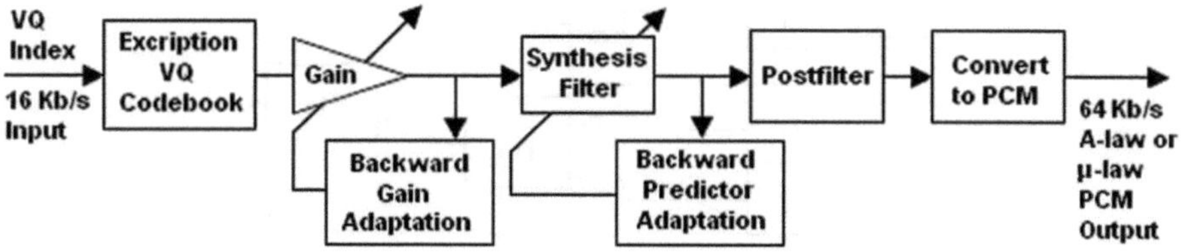

Fig. 6.37 Block diagram of decoder—Courtesy of Handbook: by ITU

6.5.4 *Uplink Baseband Processing Signals (BPS) Equipment*

The uplink BPS equipment is part of transmitter complex of ground Earth station. Every radio transmitter as a part of RES transceivers contains MUX and Encoder.

6.5.4.1 Multiplexer (MUX)

Multiplexing is the reversible operation of combining several information bearing signals to form a single, more complex signal. The signals that are combined in a multiplexer usually come from independent sources, such as subscribers in a telephone network. Prior to multiplexing, each signal travels over a separate electrical path, such as a pair of wires or a cable, whereas the multiplexed signal can be transmitted over a single communication medium of sufficient capacity. The reversibility permits recovery of the original signals, which often have different final destinations at the receiving end of the transmission link. This inverse operation whereby the original signals are recovered is demultiplexing.

In Fig. 6.38 is shown the most common multistage frequency multiplexer. It may be regarded as the dual structure of the multistage demultiplexer shown in Fig. 6.36. The filter bank takes a number of K narrow-band input signals and combines these to produce a broadband signal, using M = log2 (K) stages. At the first stage, the input signals are organized in K/2 pairs, and each pair is combined to one signal. At the second stage, the resulting signals are paired and combined. This process proceeds until the final output signal is provided at the output of the last stage. The basic signal-combining element at each stage may simply be a two-channel IODFT frequency multiplexer.

6.5.4.2 Encoder

An encoder is an electronic device, circuit, transducer, software program, algorithm or person that converts information from one format or code to another, for the purposes of standardization, speed, or compressions.

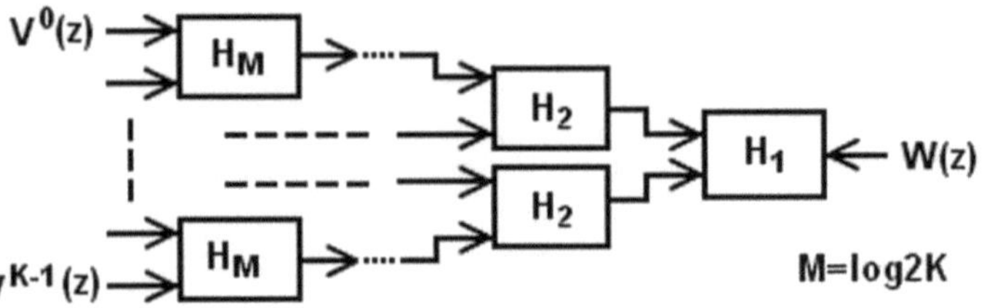

Fig. 6.38 Block diagram of multiplexer—Courtesy of Manual: by Ilcev

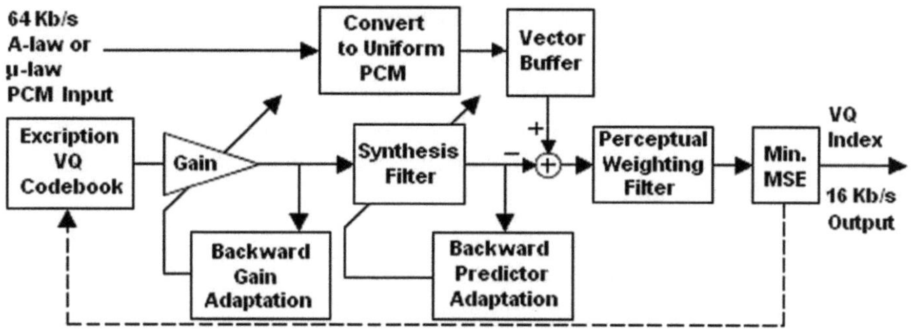

Fig. 6.39 Block diagram of encoder—Courtesy of Handbook: by ITU

There are several encoders such as: compressor encodes data (e.g., audio/video/images) into a smaller form; audio encoder converts analog audio to digital audio signals; video encoder converts analog video to digital video signals; and so on.

In Fig. 6.39 is depicted the block diagram of LD-CELP encoder, where Min. MSE is Minimum Magnetic Stripe Encoding. After the conversion from A-law or μ-law PCM to uniform PCM, the input signal is partitioned into blocks of five consecutive input signal samples. Thus, for each input block, the encoder passes each of 1024 candidate codebook vectors (stored in an excitation codebook) through a gain-scaling unit and a synthesis filter. From the resulting 1024 candidate quantized signal vectors, the encoder identifies the one that minimizes a frequency-weighted mean-squared error measure with respect to the input signal vector.

The 10-bit codebook index of the corresponding best codebook vector or codevector, which gives rise to that best candidate quantized signal vector, is transmitted to the decoder. The best codevector is then passed through the gain-scaling unit and the synthesis filter to establish the correct filter memory in preparation for the encoding of the next signal vector. The synthesis filter coefficients and the gain are updated periodically in a backward adaptive manner based on the previously quantized signal and gain-scaled excitation

6.6 General Infrastructure of Earth Station

The general infrastructure of an Earth station includes all premises, RES buildings and civil engineering works. In the premises space are located communication and TT&C antennas, while outside of the RES buildings are RF equipment, testing equipment, terrestrial interface points, and power supply units

6.6.1 Terrestrial Interface Equipment and Subsystems

Baseband data processing for converting the transmission formats of the terrestrial network into the satellite link formats and reciprocally at receive should be provided by the modem devices. The following terrestrial interface functionalities are included in a framing unit:

- Engineering Service Channel (ESC) equipment facilities;
- Redundancy switching for secured links;
- Data framing and synchronization; and
- Energy dispersion by scrambling.

The physical interfaces must comply with common standards, such as V.11, V.24, X.21 and HDB3, …) depending on the data rate. The ESC functions provide operation such as insertion/extraction of the dedicated channel into/from the normal traffic channel. Thus, synchronous multiplexing of overloads bits for the provision of ESC and alarms is also accomplished in the framing unit.

The clock of the data transmission along the satellite link may be provided by a network clock, a local Earth station clock, or from the incoming signal. A plesiochronous buffer copes with asynchronism, jitter, and Doppler effects.

The last function of the framing unit is to give some minimal stochastic characteristics to the data stream sent to the link. The two main purposes are scrambling for energy dispersion of the transmitted spectrum and of the receiver synchronization, and interleaving that permits spreading out bursts of errors generated by loss of synchronization in decoders. This latter technical approach is also used in the implementation of Reed–Solomon codes.

The terrestrial interface subsystem comprises one or several terrestrial interface modules (TIM), which connect to each user link and provide the appropriate buffering for data compression at the satellite transmit side (sub-bursts creation) and data expansion at the satellite receive side (sub-bursts extraction). Several types of TIM may be found depending on the type of terrestrial link to connect:

- 2 Mb/s or 1.544 Mb/s PCM links (telephony applications);
- Higher rate PCM links (telephony or digital TV applications);
- Various types of serial data links (EIA RS-232 or RS-422, ITU-T X.2 l, etc.); and
- 2 or 4-wire analog links.

In some cases, TIM units include additional processing for onboard voice compression or fax demodulation (telephony applications), or protocol spoofing (data applications). Some experimental systems include also ATM-oriented TIM units.

6.6.2 Power Supply Equipment

The satisfactory operation and service continuity of an Earth station depends on the correct design of its electric power supply. There are two main sources of power: 1. Main power supply, with stand-by capability; and 2. Uninterrupted Power Supply (UPS). Otherwise, can also be included batteries and solar power supply.

In addition, an auxiliary low voltage 24 V or 48 V DC source may be required to supply certain automatic equipment, while the main supply distribution network is via the station transformer unit. It is backed up by an independent generating set (or better still by two sets with 1 + 1 redundancy) driven by a fast-start 5–10 s diesel engine. This generator, which for large stations would have a power of usually about 250 kVA, supplies the whole station including the antenna motors, lighting and air-conditioning.

Maintaining the standby generator and keeping a stock of diesel fuel is one of the basic tasks in the management of the station. The purpose of the UPS, which receives its primary energy from the main power supply, is to provide a constant high-quality power supply (stable voltage and frequency with no significant transients), while the standby sets are starting up following a power cut in the distribution network.

6.7 Inmarsat GEO Land Earth Stations (LES)

The ground segment of Inmarsat GEO MSC operator comprises an integrated network of Land Earth Stations (LES), which are operated by Land Earth Station Operators (LESO), Network Coordination Stations (NCS), and a Network Operations Center (NOC). The LES sometimes referred to as Coast Earth stations (CES) in the maritime and land mobile environment and Ground Earth Stations (GES) in the aeronautical mobile environment, which links the Inmarsat's satellite constellations with the national and international telecommunication network. The LES are owned independently by telecommunication operators, which are often, but not always, the national signatory, the organization nominated by the government to invest in and work with Inmarsat of the country in which the LES is located. One LES can support multiple services providers. At present, there are about 40 LESO companies that are managing more than 40 LES terminals distributed around the globe, with at least one on every continent.

The LESO companies provide a link between the satellite constellations and the international telecommunication network. An LES is capable of handling many voice calls and data transfer to and from different MES terminals, such as Ship Earth Station (SES), Vehicle Earth Station (VES), Aircraft Earth Stations (AES), Transportable Earth Stations (TES), and Personal Earth Stations (PES), simultaneously over the different Inmarsat networks. Many LESO now offer service in all 4-ocean regions, through a global-sharing agreement with other LESO operators in ocean regions, which cannot be seen from their own location. The 4 ocean regions are as follows: 871 Atlantic Ocean Region—East (AOR-E), 872 Pacific Ocean Region (POR), 873 Indian Ocean Region (IOR), and 874 Atlantic Ocean Region—West (AOR-W).

For each Inmarsat satellite system and ocean region, there is an NCS that monitors and controls all communications. Each NCS communicates with the LES terminals in its ocean region, the other NCS and NOC terminals located at Inmarsat's London headquarters, making it possible to transfer operational information throughout the system. The NCS terminals are involved in setting up calls between an MES and an LESO facility. The LESO provider is responsible for adding an MES to, or removing an MES from, the MES authorization table that allows or disallows access to the space segment. In addition, LESO is providing and conducting service verification for different MES standards, such as IsatData Pro, IsatM2 M, Inmarsat-C and mini-C, Fleet 33/55/77, Fleet One, FleetBroadband, BGAN, and Maritime Global Xpress for SES terminals. For VES terminals, Inmarsat is providing transportable or installed mini-M and BGAN standards, while for AES terminals Inmarsat provides modern standards such as Swift64, SwiftBroadband, and Aero Global Xpress.

The LES terminal is a powerful land-based receiving and transmitting station serving in a GMSC system. Although LES is fixed and is not moving, it can serve but cannot be part of FSS. Some LES, such as Goonhilly, provides widely fixed links for FSS, and consequently the Inmarsat MSC network is a small part of overall LES providing service for Inmarsat-A, B/M, mini-M, D + and Aero MES. Each LES in the IGN is owned and operated by an Inmarsat Signatory in the mission to provide a range of service to all types of MES. There are more than 40 LES terminals located in 30 countries round the globe, but usually in the northern hemisphere, which is a small anomaly. The MES operator and shore subscribers can choose the most suitable LES, as long as they are within the same Ocean Region.

The fundamental requirement for each Inmarsat LES is that it be capable of communicating reliably with all MES terminals. There are two major types of LES: CES for maritime and land mobile applications providing service to all standards of SES and VES, and GES for aeronautical mobile applications providing service to all standards of AES terminals only.

The list of LES terminals, type of service and access codes, the country in which they are based and Ocean Region of operation are given in Inmarsat Operational Handbooks, in Admiralty List of Radio and Satellite Services and in ITU list of Coast Stations.

The technical side of typical Inmarsat LES (CES or GES) consists of three main features: the antenna system (left), the communication RF equipment (between Feed and ACSE), and ACSE unit (right), as illustrated in Fig. 6.40.

1. **Antenna System**—A typical LES antenna system for entire IGN would be a Cassegrain structure with a dish reflector of about 14 m diameter. At this point, each LES can have a minimum of one operational and one spare antenna in order to continue transmissions during maintenance. Some stations have more

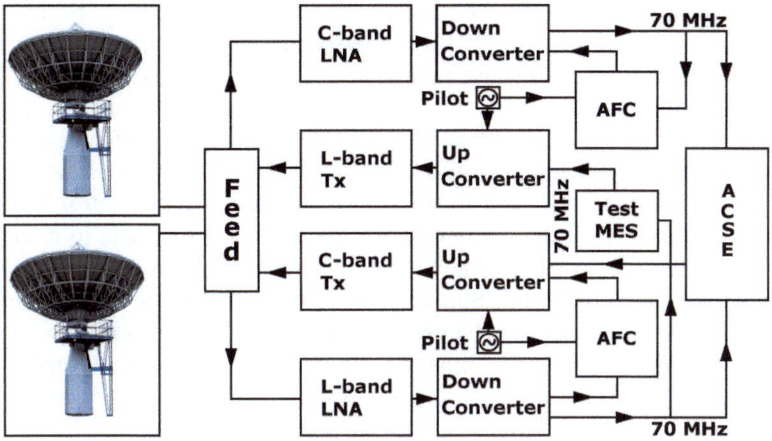

Fig. 6.40 Block diagram of Inmarsat LES—Courtesy of Book: by Ilcev

than two antennas, which depend on the Ocean Region covered and service provided. Thus, the antenna operates in both the L- and C-band to and from satellite, with gain requirements of 50.5 and 29.5 dBi, respectively. This antenna is designed to withstand high wind speed up to 60 m/h in its operational attitude and 120 m/h when stowed at 90°, and the parabolic dish would be steerable ± 135° in azimuth and 0°–90° in elevation angle. Tracking is ether by automatic program control or operator initiated. The antenna tracking accuracy of 0.01° RMS and repositioning velocity of 1°/s would be typical parameters for such a dish. The RF and baseband processing hardware design varies greatly with LES design and requirement. A single antenna may be used to transmit and receive L-band as well as the C-band signals or employing a separate L-band antenna avoids the need for a relatively complex feed system (to combine and separate the outgoing and incoming L-band and C-band signals) but this advantage must be weighed against the cost of procuring and installing a second antenna.

2. **Communication RF Equipment**—The equipment is situated inside of LES building and must be able to operate in Tx and Rx L-band link; to monitor the MSC L-band channel and respond to requests for frequency allocations by the NCS; to verify signal performance by loop testing between satellite and LES; and to receive the C-to-L Automatic Frequency Control (AFC). The AFC provides Tx and Rx direction control, which helps to keep MES as simple and as cheap as is practicable. A complete test of the LES equipment can be carried out without the cooperation of the MES, because a separate test terminal is provided at each LES for this purpose. The Inmarsat system requires AFC to correct for Doppler shift (caused by inclination of the GEO) and errors in frequency translation in the satellite and LES terminals.

The total frequency shift from the cause without AFC could be more than 50 kHz, thus is of the same order as the spacing between the NBFM channels and would be enough to cause failure of the system. The AFC reduces the RF shift to a few hundred hertz by comparing pilot carriers transmitted via the satellite with reference oscillators at the LES and using the difference signals to control the RF of the local oscillator associated with the up- and downconverter. A pilot transmitted at C and received at L-band is used to control the upconverter and thus offset the frequencies of the operational carriers to compensate for Doppler shift and satellite frequency translation errors in the ground-to-mobile direction. Similarly, a pilot transmitted at L and received at C-band controls the downconverter and corrects for Doppler shift and errors in translation in the mobile-to-ground direction. All RF errors are corrected except those arising from frequency instability of the MES up and downconverter, and Doppler shift resulting from the relative velocity of satellite and MES.

3. **ACSE**—Antenna control and signaling equipment (ACSE) is part of LES, whose principal purpose is to recognize request for calls, to set and release. This requires response to and initiation of in-band and out-bands signaling over satellite and terrestrial path. The next ACSE tasks are to recognize distress calls (from SES–AES) and preempt channels for them when necessary; to check that MES are on the list of authorized users and to bar calls (except all distress calls) from or to unauthorized MES; to switch voice circuits between TTN circuit and the LES FM channel modem; to switch Tlx circuits between TTN channels and the TDM/TDMA time slots; to determine Tx/Rx frequencies used by the FM channel unit in accordance with the channel allocations made by the NCS; to allocate TDM and TDMA time slots; and to collect statistics for billings, international accountings (for transit calls), traffic analyses, and management and maintenance purposes.

The MSC services offered by an LES vary depending upon the complexity of the station selected. For example, a typical LES could offer a wide range of services from and to the MES located in convenient ocean regions, such as two-way voice including fax/paging, Tlx, all data rate, Video, GAN/Internet, and Mobile Emergency service (Distress, Urgency, Safety and Medical assistance calls). The Inmarsat LES are operating in the 6/4 GHz band, and MES operate in the 1.6/1.5 GHz band. The LES terminals provide feeder link to the Inmarsat satellites, and MES terminals provide service link in reverse.

6.8 Iridium non-GEO Gateways

The Iridium network can provide voice and data services for all mobile applications, such as maritime, land, and aeronautical, including personal and determination service. Thus, in support of this service, a new type of mobile, referred to as a Satellite Data Unit (SDU) is deployed that will interoperate with the Iridium

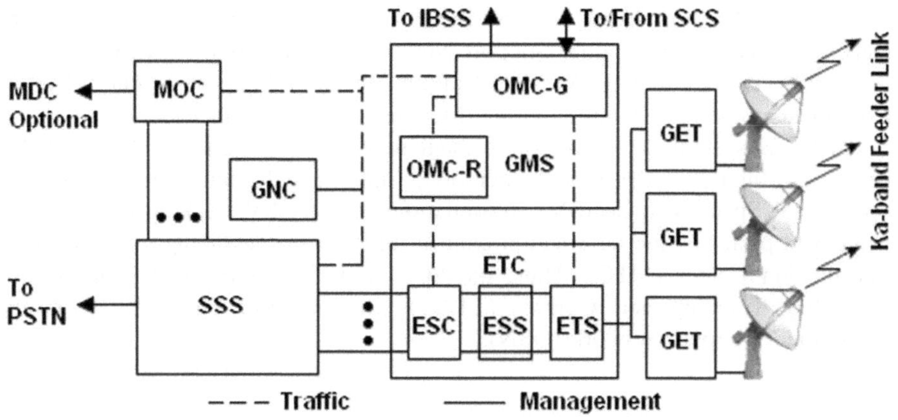

Fig. 6.41 Block diagram of Iridium gateway—Courtesy of Manual: by Ilcev

global satellite communications system and the existing mobile voice and data communication systems to users via satellite network and ground infrastructures. The three main components of the Iridium MSC service are as follows: Space, Ground, and Users segments.

The Iridium ground segment provides voice and data MSC service via 66 LEO satellites and two Gateways only, because utilizes Inter Satellite Links (ISL). The ISL network is providing connection of all Iridium satellites transferring MSC service and route network traffic to the Regional gateways and PSTN from any user station worldwide.

The Iridium gateway shown in Fig. 6.41 provides a number of the following roles:

1. Setup and maintain basic telephony services. The gateway provides the interface for all communications between Iridium subscribers and between Iridium subscribers and PSTN leased line users worldwide;
2. Interface to the PSTN or leased lines to the Iridium constellation via Gateways functions;
3. Provide messaging and data services, dial-up data, Short Burst Data (SBD), including Direct Internet access and Router-Based Unrestricted Digital Internetworking Connectivity Solutions (RUDICS); and
4. Messaging processing and services, management and billing functions.

The Gateway is split into two functional areas, such as the Gateway Management System (GMS) and the gateway operational elements. The Iridium gateway is comprised of the following network elements: Message Origination Center (MOC); GSM Switch Subsystem (SSS); Ground Notification Center (GNC); Gateway Management System (GMS); Ground Earth Terminals (GET); Earth Terminal Controller (ETC); ETC Transmission Subsystem (ETS); ETC Communications Subsystem (ECS); Web and Email Messaging Server (WEMS);

RUDICS; ETC Short Burst Data Subsystem (ESS); Public Switched Telephone Network (PSTN), Operational Support Network-Gateway (OMC-G); Operational Support Network-Radio (OMC-R), Iridium Business Support System; System Control Segment (SCS); and Optional Message Delivery Center (MDC).

The SCS system is the central management function in the Iridium satellite network, which provides global operation, support, and control services for the operational Space Vehicle (SV) constellation, delivers SV tracking data to gateways, and performs the termination control function of messaging use a Message Termination Center (MTC). Thus, the SCS comprises three main components: Telemetry Tracking and Control (TTAC) sites, Control Facilities (CF), and Operational Support Network (OSN).

6.9 DVB-RCS HUB Terminal

The DVB-RCS VSAT network has been designed to minimize the cost of scaling a satellite broadcast, broadband, and multimedia access via few HUB station solutions from terminal populations as small as several hundred VSAT units to tens or even hundreds of thousands of simultaneously logged-on VSAT units in urban and rural or remote areas, including for all mobile applications.

The HUB VSAT station is usually a relatively large, high-performance Earth station with an antenna diameter of anything between 6 and 9 m with tracking system, which consists of a control center to manage the network as well as microwave equipment, including an outdoor antenna for the transmission and reception of signals. Thus, a substantial amount of interfacing equipment necessary to support the wide range of terrestrial interfaces required at the HUB completes the installation. This equipment is usually mounted in several racks, which block diagram is shown in Fig. 6.42. The customer has choice of C, Ku, or Ka antennas and CIT RF equipment, what is depending on available satellite transponder RF band. The HUB can handles 1–10 forward link VSAT terminals and from one to several hundred return link receivers up to 200,000 terminals. Its architecture is also designed to accommodate upgrades and expansions, and to optimize stability in operation.

A Radio Frequency Terminal (RFT) contains the following components:

The transmit subsystem of HUB containing upconverters which change the 70 or 140 MHz IF to the required transmit frequency before feeding it to the HPA unit. In fact, if the HUB only uses a single carrier for data it is possible to use an SSPA otherwise a more powerful TWTA must generally be used. Uplink power control is often provided so that the power transmitted by the HUB can be increased to compensate for high link attenuation due to precipitation in bad weather. The receive subsystem consisting of an LNA with a noise temperature usually between 150 and 175°K (Ku-band) and a downconverter to change the received frequency to the IF frequency (70 or 140 MHz). A Network Control Center (NCC) controls and monitors the operation of the HUB and the IDU devices in the network.

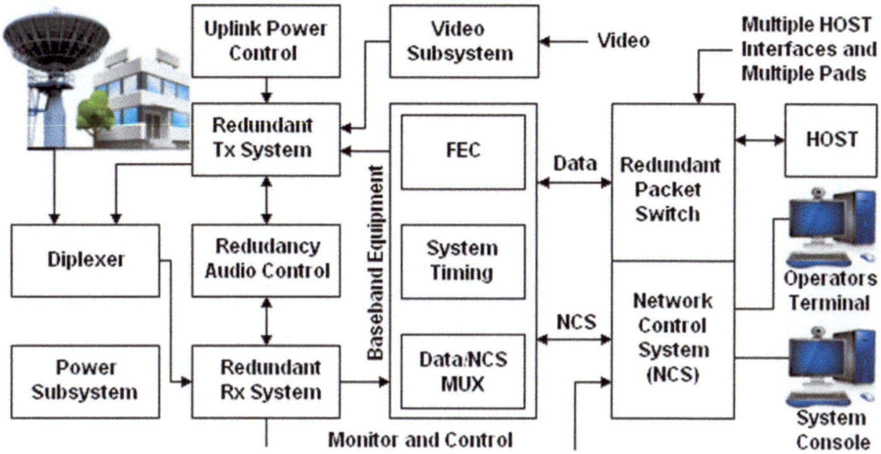

Fig. 6.42 Block diagram of DVB-RCS HUB terminal—Courtesy of Manual: by Ilcev

The HUB station consists of several main subsystems, except for the HUB antenna; these are usually fully redundant with automatic switchover in the event of failure:

A switch is generally a packet switch subsystem, which controls routing between host ports and the modulator and demodulator ports, as well as adding and reading header address information which controls routing to and from individual IDU devises. Moreover, HUB consists of one or more modulators that modulate the outbound carriers with the TDM stream, which is generated by the switch (each outbound carrier has a dedicated modulator), and then is included a bank of demodulators which receive the inbound carriers and extract the data packets and feed them to the switch.

Chapter 7
Users Segment

In general, Global Mobile Satellite Communications (GMSC) terminals can be located onboard ships, land vehicles, and aircraft and can be also used as transportable or handheld personal transceivers. Thus, as these mobiles can be in motion when the communications are in progress, there are some special considerations that need to be taken into account in the design and installation of the GMSC terminal onboard mobiles. The antenna should be mounted clear of any obstructions on the vehicle so as not to reduce the signal strength. However, if the antenna is directional, typically having a gain greater than 6 dBi, then it will also need to include a mechanism to automatically track the GMSC satellite. This may be accomplished by either mechanically rotating a fixed beam antenna, or by the use of electronic steerable or switchable beams in the case of phased-array antennas.

The GMSC terminal itself may be physically separated into two parts: an external sealed unit that houses antenna in the special plastic radome, radio frequency (RF) front end and possible antenna-tracking unit and an internal unit normally located near the operator that houses the other radio components. The LNA units are usually placed as close as possible to the diplexer so as to minimize the noise resulting from losses in the feed guides.

7.1 Overview of GMSC Applications

There are a wide variety of applications supported by the GMSC services for the general provision of voice, data, and video (VDV) communications to users of the maritime, land, aeronautical, and communities worldwide. In Fig. 7.1 are introduced the space, ground, and user segments for Inmarsat and DVB-RCS configurations of GMSC service.

The GMSC infrastructure can provide global coverage just with three geostationary Earth orbit (GEO) satellites to mobile earth station (MES) and fixed earth

S.D. Ilčev, *Global Mobile Satellite Communications Theory*,
DOI 10.1007/978-3-319-39171-7_7

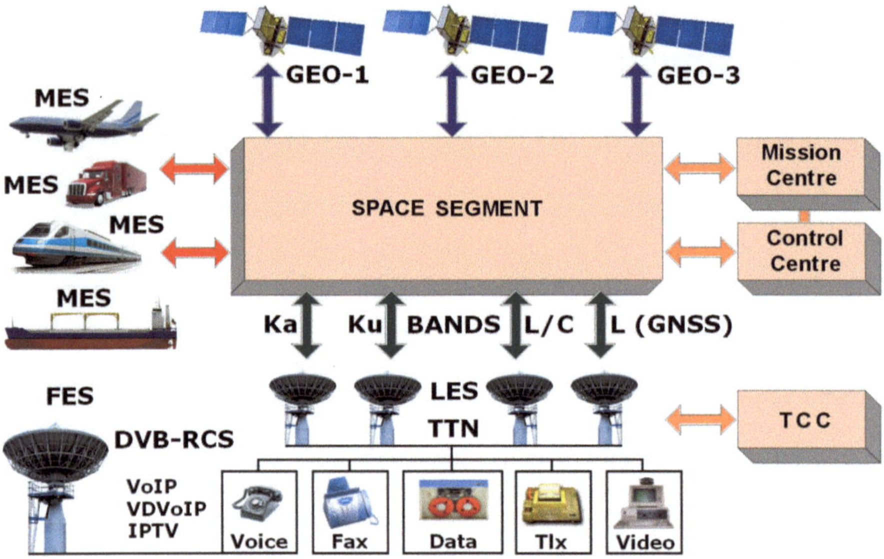

Fig. 7.1 Space, ground, and user segments of GMSC service—Courtesy of manual: by Ilcev

station (FES) via Inmarsat or DVB-RCS satellite constellations. The applications include emergency distress alerting and locating services, direct-dial telephone (voice), data transfer, facsimile, telex, electronic mails (E-mail), high-quality audio transmission, still pictures, videoconferencing, and compressed (digital) video. In general, the complete GMSC service can be included in a modern package known as voice, data, and video over IP (VDVoIP). Digital data rates currently available from different mobile satellite system (MSS) providers vary from tens of bits-per-second, which is suitable for short messaging services to 64 Kb/s, which is suitable for many high-speed data applications. Even higher bit rates are planned; for instance, the top Inmarsat GMSC onboard equipment is able to provide forward link transmission speed up to 50 Mb/s, while mobile DVB-RCS interactive VSAT several may provide forward link speed up to 775 Mb/s and even 1 Gb/s.

The GMSC equipment terminals come in a wide variety of shapes and sizes. They may be designed to be carried by the person, transported to a site, or fixed to a vehicle, ship, small craft, or an aircraft. They can vary from small paging terminals or handheld phones that fit on a trouser belt to very large terminals with directional dish antennas, requiring a stable platform for the operation to compensate for ship movement in rough seas. Typical users of GMSC applications are as follows:

1. **Maritime Users**—These users can be all floating objects at sea or rivers and lakes, such as operators of fishing boats, yachts, cargo ships, container vessels, ferryboats, drilling rigs, oil tankers, liquid natural gas carriers, and passenger cruise ships.

2. **Land Users**—In general, these users can be road and rail vehicles, while in particular, they can be journalists, traveling executives, disaster relief agencies, government officials, mineral exploration companies, and workers at remote sites, for example, in the building of hydropower projects, agriculture, mining, constructing companies, and tourism.
3. **Aeronautical Users**—These users can be longhoul airway operators, general aviation, corporate or private aircraft, and helicopters.
4. **Personal Users**—These users can be all persons carry handheld satellite terminals either onboard mobiles or ashore.
5. **Transportable Users**—These users can be transportable satellite terminals either inside land vehicles or as a temporary installation ashore.
6. **Portable Users**—The Broadband Global Area Network (BGAN) developed by Inmarsat is a portable GMSC transceiver that can provide Internet access via laptop or smart device.

7.1.1 General Architecture of GMSC System

The GMSC systems, whether they be designed to address maritime mobile, land mobile, or aeronautical mobile satellite services, all have certain common characteristics. Likewise, the satellites employed in these GMSC systems can be found in constellations using the GEO and non-GEO, such as medium Earth orbit (MEO), low Earth orbit (LEO), and high elliptical orbit (HEO), with widely differing perigees and apogees or even some combination of these orbital configurations known as hybrid satellite orbits (HSO).

It is more common, both globally and regionally, for the majority of GMSC allocations to be specified as generic MSS service, in the RR Table of Frequency Allocations, Article 5—rather than specifying the particular type of MSS applications, such as maritime, land, and aeronautical.

7.1.1.1 Functions of Feeder and Service Links in GEO GMSC Networks

Irrespective of which orbital configuration is employed, the GEO GMSC systems differ from fixed satellite system (FSS) service in one key aspect. Specifically, the FSS service uses one pair of RF bands, namely one band for uplink and one for downlink to connect fixed points via the satellite links.

However, the MSS service employs two pairs of links: one set known as MSS service links and the other set designated as MSS feeder links. The MSS service works in the way that MES transmits on the Earth-to-Space 1.6 GHz service link to

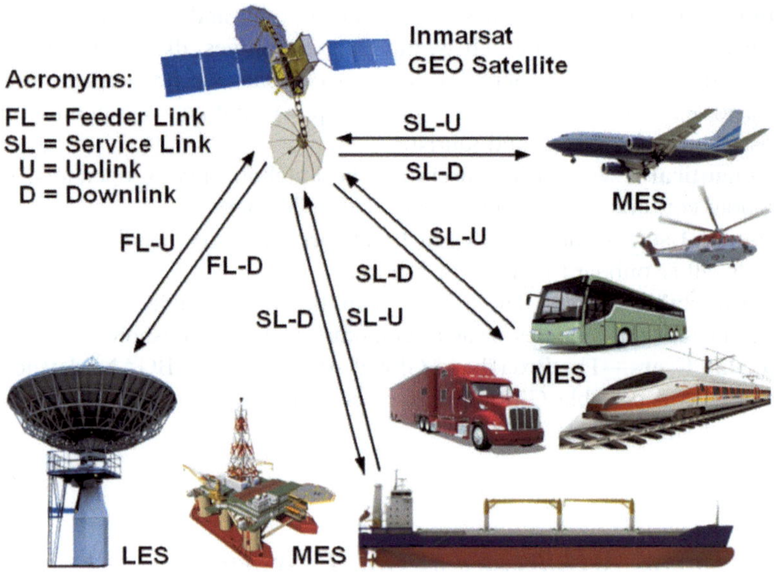

Fig. 7.2 GMSC feeder and service links—Courtesy of manual: by Ilcev

the satellite, and then, the satellite repeats that transmission on the Space-to-Earth 4 GHz feeder link to the Gateway fixed Earth station or LES. The LES terminal, in turn, is typically interconnecting the call or data transmission via the public switched telephone network (PSTN) to the party to whom the mobile wishes to connect. The landline party then sends back voice or data calls to the LES terminal, which transmits it via 6 GHz feeder link to the satellite transponder. Finally, the satellite transponder repeats that latter transmission backs again to the mobile terminal on 1.5 GHz, which originated the call via the service link.

In Fig. 7.2 is illustrated the tandem use of feeder links plus service links in an GMSC satellite link in relation to the Gateway (LES) station and the different MES terminals. In any MSS network, the service links connect the MES terminals to the satellite, whereas the feeder links are used to connect the land-based Gateway or feeder link LES to the satellite. It takes two pairs of links to complete the full MSS circuit, namely the forward link from LES out to the MES terminal, which comprises one feeder plus one service link as a pair of links and return from the MES terminal back to the LES, which comprises one service plus one feeder link as a pair of links. Quite often, the MSS networks utilize any of several FSS bands, generally higher in RF than the service link allocations themselves to operate their feeder links. For example, a particular MSS system could be designed to use segments of any of the FSS allocations at 5/7, 11/12/14/15, or 19/29 GHz for feeder link purposes. However, there are also particular FSS bands allocated by the ITU, which are specified as being designated exclusively for use by certain MSS feeder link operations.

7.1.1.2 Feeder Link Gateway Stations in Non-GEO GMSC Networks

Feeder link Gateway antennas for non-GEO satellites are generally smaller than those used for GEO operation, because of the lower path losses to the satellites and to allow for easier tracking ability. Feeder link frequencies normally used by Big LEO systems are in the 5/7 GHz bands or 20/30 GHz bands. Frequencies in the 15 GHz band are also available.

The design and location of feeder link Gateway station is related to the non-GEO satellite constellation characteristics. For example, LEO satellite constellation having InterSatellite Link (ISL) to route traffic between satellites would need fewer Earth stations (Gateways) than otherwise required. However, systems using LEO constellations will need more Earth stations than those using MEO constellations.

Furthermore, the location of any feeder link sites must take into consideration the existing national or regional telecommunications infrastructure available. Therefore, in the Iridium satellite system there are 2 Gateways serving 66 Big LEO operational satellites. Traffic is internally routed through ISL network to the only operational Gateway in USA, and the call destination is routed to the public terrestrial networks worldwide.

In the Globalstar, there are 48 Big LEO operational satellites and up to 60 Gateway cites worldwide with traffic routed directly through the public terrestrial networks. However, the Orbcomm satellite system is containing 36 LEO satellites and about 16–20 operational and few planned Gateway terminals worldwide with traffic routed directly through the public terrestrial networks. Thus, traffic of LEO constellations without ISL network is internally routed to the closest Gateway to the call destination before being routed through the public terrestrial networks.

The new developed MEO satellite constellation in 2014 known as O3b Networks for fixed and mobile satellite communications already completed about 10 gateway terminals and is planning to expand the total number of satellites from 12 to 20.

The number of tracking Gateway antenna installed at individual sites will vary depending on the number of satellites being tracked, the number of antennas lining up for the next satellites to become visible, and the number being used as spare. Typically, the number of antennas installed at an individual non-GEO feeder link site can vary from about 2–6. In addition to feeder link Earth stations, satellite operations require Telemetry, Tracking, and Command (TT&C) centers in order to track and control the satellite constellation. In fact, there are at least two such sites in a given satellite system serving as primary and backup facilities. They are used to provide command functions to satellites such as to perform orbit maneuvres, as well as monitor the status of the satellite and subsystems.

7.2 Inmarsat GEO Users Segment

Inmarsat GMSC user terminals can be located onboard ships, land vehicles (road and rail), and aircraft. As these mobiles can be in motion when the communications are in progress, there are some extra considerations that need to be taken into account in the design and installation of the terminals. The MES satellite antenna should be mounted clear of any obstructions on the mobile so as not to reduce the signal strength. Thus, if the antenna is directional, typically having a gain greater than 6 dBi, then it will also need to include a mechanism to automatically track the Inmarsat satellite. This may be accomplished by either mechanically rotating a fixed beam antenna, or the use of electronic steerable or switchable beams in the case of phased-array antennas.

The MES terminal itself may be physically separated into two parts: an external sealed unit comprising the radome which houses the antenna, RF front end, possible antenna-tracking unit, and an internal unit normally located near the operator which houses the other radio components. Usually, the low-noise amplifiers are usually placed as close as possible to the diplexer so as to minimize the noise resulting from losses in the feed guides.

7.2.1 Maritime Ship Earth Stations (SES)

The SES mobile terminals are shipborne satellite transceivers that have been designed for installation and operation onboard seagoing vessels. For instance, an example of such type of terminals are those forming a part of the Global Maritime Distress and Safety System (GMDSS), which are designed for high availability and other safety requirements. In this group of SES, terminals included Inmarsat mini-C or C as a part of GMDSS, but long-range identification and tracking (LRIT) system, Fleet One, FleetBroadband, and newest Global Xpress can be also its part. Thus, the LRIT installation is not a part of the GMDSS requirements either but its equipment, particularly Inmarsat mini-C or C terminals, can also be used to support the system.

The Ship Security Alert System (SSAS) is used by ships to transmit a security alert to a competent authority when the security of the ship is under threat or has been compromised.

The International Maritime Organization (IMO) legislation states that the ship's security alert system shall be initiated by pressing a special dedicated SSAS button on alarm panel and transmitting a ship-to-shore security alert to a competent authority designated by the administration, which may include the shipping company, identifying the ship and its location and indicating that the security of the ships is under threat or has been compromised. The SSAS is not part of the GMDSS communication requirements, but its equipment can be used to support the system. Ship security alerts can be sent using Inmarsat-C and mini-C, and relevant hardware and software can be added to the existing GMDSS terminals to support the service. Dedicated SSAS models are also available from MES manufacturers.

The GMSC system in a maritime environment has distinctive propagation characteristics and operating requirements. Maritime communications are particularly susceptible to multipath effects due to signal reflections from the structure of the ship or sea environment. Thus, methods employed to counter this effect include the use of modulation schemes that are tolerant to signal fading and intersymbol dispersion, and the incorporation of a significant fade margin, through the use of dynamic power control in the satellite link design. The required fade margin can be minimized by the use of data interleaving and forward error correction coding schemes.

Maritime user terminals can be classified into three main types:

1. Those having ship low-gain antenna (LGA) of approximately 6 dBi as typified by the Inmarsat Maritime-C terminals. In fact, these ship terminal types are characterized by very compact SES having an unstabilized and unsteered antenna subsystem, and they are very suitable for GMDSS solutions.
2. Those having ships with high-gain antenna (HGA) of approximately 15 dBi as typified by the Inmarsat Maritime-B, Fleet 33/55/77, Fleet One, FleetBroadband, and Global Xpress terminals. These terminal types would need to track Inmarsat satellites in order to counter the roll, pitch, and yaw motion of the vessels on which they are mounted. They must still be capable of correctly pointing to a satellite even when operating under heavy sea conditions by the deployment of special antenna-tracking devices.
3. Those having ship with medium- or intermediate-gain antenna (IGA) of approximately 6–15 dBi as typified by the Inmarsat Maritime mini-M and M terminals. These are scaled down versions of the high-gain antenna. A subset of these terminals is designed to operate only within the spot beam coverage areas of GEO satellites. These terminals typically use antennas that have wide beamwidths in the elevation (vertical) plane so that they need only track the satellite in the azimuth (horizontal) plane. Such terminals would be suitable for installation on pleasure craft because they are smaller, cheaper, and simpler than their fully stabilized counterparts.

In this section will be introduced only FleetBroadband and related solutions.

7.2.1.1 Maritime FleetBroadband (FB) SES

Above all is important to express that author of this book is inventor of this SES, because he already wrote about in his first edition of this book published by Springer in 2005. In 2000, he introduced Broadband Maritime Communication Service (BMCS) and submitted manuscript to Springer in 2003, while Inmarsat promoted its FleetBroadband in 2007.

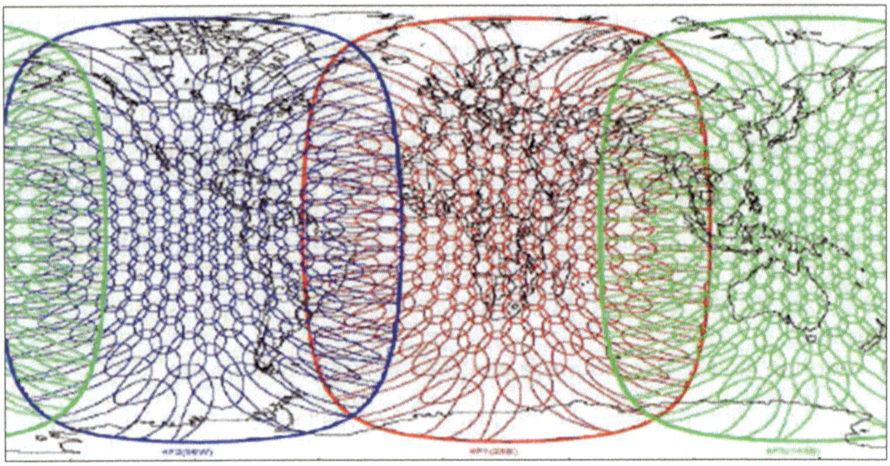

Fig. 7.3 Inmarsat-4 spot beam coverage—Courtesy of manual: by Inmarsat

The FleetBroadband SES terminal operates using the spot beam capabilities of the previous Inmarsat-4 and new Inmarsat-5 generation satellites. Thus, three satellites are deployed to provide global coverage located at positions 25° East, 143.5° East, and 98° West as shown in Fig. 7.3 and to connect ships to the terrestrial telecommunication network (TTN).

The use of spot beams enables Inmarsat to reuse spectrum across the coverage area which, combined with the dynamic network management, enables Inmarsat to optimize the use of satellite resources for all users connected to the network. FleetBroadband Standard IP data are a variable bit-rate service provided on a shared or contended channel and operating up to 432 Kb/s in each direction on a best-effort basis. Inmarsat system manages this service by dynamically allocating satellite resources to the Standard IP data service to ensure that all users experience the best possible performance.

Factors taken into account in managing this service include the following:

Total volume of traffic (and not just the activity in the channel); number of users in the spot beam and overall number of users; the level of activity in other spot beams; number of spare channels available and forecast use based on historical data; average bandwidth currently experienced; and how long has this pattern of heavy use been prevalent.

Thus, Inmarsat network is able to dynamically and transparently provide additional satellite resources to vessels demanding more capacity and restore and reallocate those resources when that demand has been satisfied.

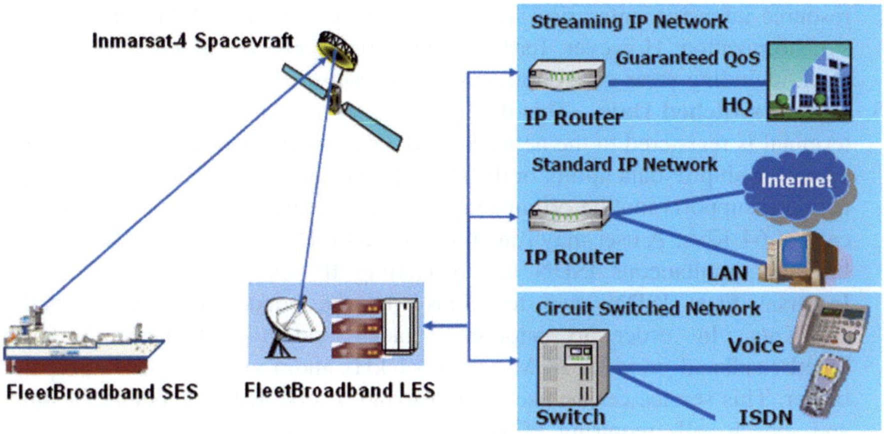

Fig. 7.4 Inmarsat-4 spot beam coverage—Courtesy of manual: by Inmarsat

FleetBroadband Onboard Services

FleetBroadband provides an on-demand range of services to suit all onboard applications. This SES system uses proven technology, and the terminal is quick and easy to install and operate. In such a way, unlike previous generations of Inmarsat terminals, services can be accessed and used simultaneously. FleetBroadband provides a total of three types of data connections, as shown in Fig. 7.4, which means that users can select the data connection that is most suited to your needs and carries your traffic in the most cost-effective manner. In general, this standard can provide the following services:

1. **Standard IT Data**—The Standard IP data service provides up to 432 Kb/s (maximum data rate will vary depending upon the type of terminal used) on a contended, best-effort basis. There are no guarantees associated with this service; thus, if the link is busy, with many active users, then the observed bit rate will be lower than if the link is quiet with little traffic. Standard IP is best suited to most typical onboard applications such as Internet browsing, e-mail, File Transfer Protocol (FTP), and also numerous maritime applications such as electronic charts, weather updates, engine monitoring, and many more. This is not Inmarsat Mobile Packet Data Service (MPDS), but new type of connection that will be used most of the time and for most applications.
2. **Streaming IP Data**—The Streaming IP provides an on-demand, guaranteed quality of service (QoS) connection at any one of 32, 64, 128, or 256 Kb/s. The capacity is not offered to other users and effectively delivers reserved capacity for a specific IP connection. It is available on demand and on a first come, first served basis. Thus, if the channel requested is unavailable, a user can request another channel with a lower bit rate. Streaming is very important for time-critical data transmissions such as live video and audio but unoptimized

resource-intensive enterprise applications such as Oracle, SAP, and database synchronization. However, further characteristics can be assigned to this standard, including error correction and application-specific routing instructions.

3. **Circuit-switched Data**—This data services are available from the moment the terminal is registered to the network, and data connections can be initiated from either the ship IP data uplink or the shore IP data downlink. The FleetBroadband network supports mobile originated and terminated ISDN circuit-switched data calls at 64 Kb/s. A user may run simultaneous ISDN and Standard IP sessions1 but not simultaneous ISDN and Streaming IP sessions. However, as with Inmarsat Fleet ISDN, two terminals may be bonded together to deliver 128 Kb/s. In order to support legacy modem and facsimile users, FleetBroadband provides PCM-coded 3.1 kHz audio via a 64 Kb/s transparent bearer. This service can be used to make and receive legacy modem, facsimile, and speech calls requiring PCM coding to and from the terrestrial PSTN or ISDN and also supports encrypted voice such as STU III. The 3.1 kHz audio services are provided at the terminal typically via either an RJ-11 analogue telephone connector or RJ-45 ISDN connector (where supported).
4. **Voice and Messaging Services**—FleetBroadband offers a direct-dial voice service using compression technology (the AMBE+2 codec3) delivering voice encoded at 4 Kb/s, which should not connect voice over IP (VoIP). This makes efficient use of satellite capacity while delivering good speech quality. It is possible to make a circuit-switched voice call while simultaneously using Standard and Streaming IP data services. It also provides a voicemail facility, which is comparable with most cellular networks. A subscriber's service profile can be provisioned so that call forwarding will divert calls to the voicemail server whenever the subscriber is unable to receive incoming calls. Each subscriber may receive a notification via SMS that they have messages waiting for them. In addition to the basic messaging service, subscribers can forward messages to another number, record a message and distribute it to one or more subscribers, and access their voice mail from any telephone, fixed or mobile. The FleetBroadband network incorporates an SMS (text) messaging application with a full range of messaging features. The SMS message format follows the standard 160-character structure, but it does not support concatenated SMS. Subscriber can send and receive SMS messages to and from other FleetBroadband terminals and terrestrial cellular networks 4 via your laptop or computer using the onboard launch pad utility.

FleetBroadband Onboard Equipment

There are two terminal types defined and type-approved for the FleetBroadband service. They are referred to as Class 8 terminal deploying HGA and Class 9 terminal using LGA. The key difference between Class 8 and Class 9 is the antenna.

For each type of FleetBroadband terminal, Inmarsat defines the air interface at the output of the terminal's antenna; the mandatory features and service types for each class of user terminal; and the performance requirements of the user terminal.

Manufacturers must meet all of these requirements in order to obtain type approval for FleetBroadband SES. The definition of other onboard equipment features such as physical connections, users interfaces, firewalling, and routing control is determined by each manufacturer according to specific market-driven needs. Thus, because of the possibility of different physical interfaces on terminals from different manufacturers, users should pay particular attention to the installation guidelines given in technical manuals.

The FleetBroadband SES provides most typical office applications such as e-email, Internet browsing, FTP, and numerous maritime applications. These onboard vessel applications are electronic charts, weather warnings (WX), navigation warnings (NX), and others, best suited to Standard IP combined with an Internet-based last-mile implementation.

FleetBroadband is a flexible and very versatile shipborne GMSC system capable of providing a cost-effective platform for the wide range of devices and applications that are to be found on a modern ship today. Its peripheral package is depicted in Fig. 7.5, where ADE is above deck equipment (Antenna Unit); BDE is below deck equipment (Satellite Transceiver); PoE HUB is Power over Ethernet, which integrates the built-in Wireless Fidelity (Wi-fi) router and private branch exchange (PBX); Digital Enhanced Cordless Telecommunications (DECT); cabin telephone unit (CTU); Secure Terminal Equipment (STE); Future Narrowband Digital Terminal (FNBDT); High Assurance Internet Protocol Encryptor (HAIPE); maritime safety information (MSI); communication, navigation, and surveillance (CNS); and ship traffic management (STM).

FleetBroadband terminal can connect the numbers of existing and optional communications equipment and peripherals, such as handsets, such as analogue, cordless, DECT, ISDN, and wireless Wi-fi network; VoIP peripherals, such as USB and Wi-fi phones; audio, such as conference systems; LAN and WAN devices, such as routers, HUBs, switches, and wireless access points; IP/Network cameras, such as remote surveillance; off-the-shelf video, such as Scotty-motion media, Polycomm and Sony IP; and legacy ISDN devices.

Besides, nowadays, many onboard systems are data network-enabled and can be connected to the onboard LAN. Such connectivity when used in conjunction with FleetBroadband will permit remote monitoring of onboard systems such as water filtration systems; weather sensors; preventative/predictive maintenance, engine telemetry sensors; refrigeration systems; and systems container loading/unloading monitoring sensors.

The distress alarm panel, as illustrated in Fig. 7.5, is supplied with power through the Ethernet interface using PoE (1), which is available in all the FleetBroadband systems, or alternatively (2) from a DC supply (10.8–32 V DC). The DC input is protected against reverse polarity. The alarm panel is used together with a Sailor and other FleetBroadband system and an IP handset designated for distress use. A push on the red FB distress button on the alarm panel initiates a

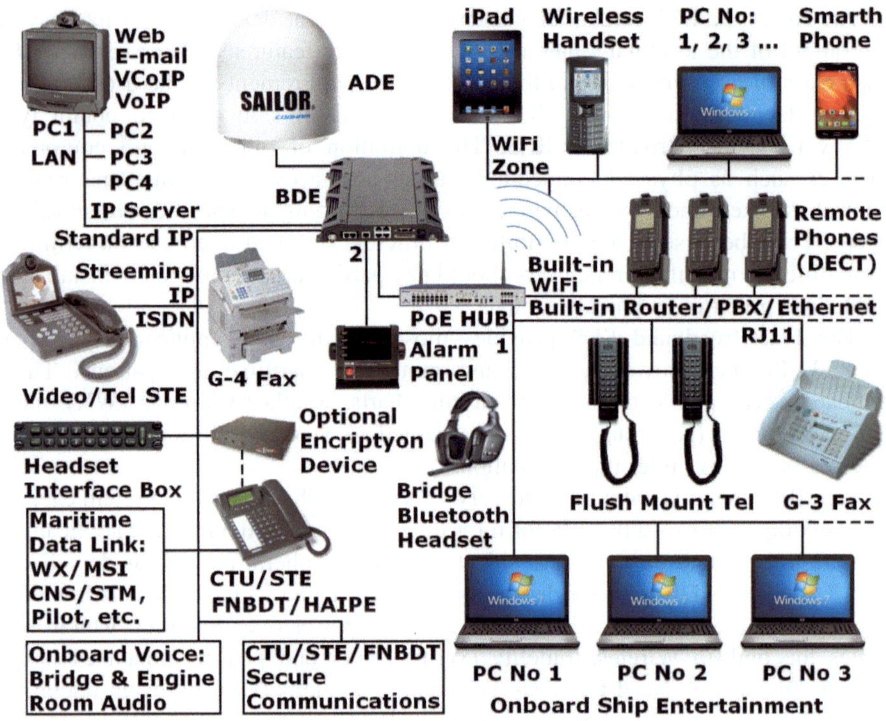

Fig. 7.5 Maritime FleetBroadband peripherals package—Courtesy of manual: by Ilcev

distress call from the distress IP handset. For further information on the use of the alarm panel and the voice distress (non-SOLAS) system, see the user manual for the voice distress (non-SOLAS) System.

In Fig. 7.6a are depicted main components that Furuno FleetBroadband Felcom 250 or 500 ADE units contains, such as (1) HPA unit; (2) interface assy; (3) antenna-tracking block (ATB) and LNA module; (4) GPS assy; and (5) antenna without radome, while in Fig. 7.6b are depicted main connectors of the Furuno FleetBroadband Felcom 250 or 500 ADE units, such as (1) RS-232C 9pin D-Sub female connector for ISDN UDI/RDI: unrestricted digital information (UDI) and restricted digital information (RDI); (2) National Marine Electronics Association (NMEA) in/out Phoenix 5pin (NMEA 0183 Ver. 2) for navigation equipment; (3) alarm output Phoenix 3pin for alarm equipment; (4) for power supply; (5) for antenna unit (antenna cable length selectable 30 m or 50 m); (6) LAN RJ45 ports for IP handsets, PC/network equipment/telephone HUB; (7) telephone RJ11 for telephones and facsimile devices; (8) telephone Phoenix 4pin for telephones and facsimile devices; and (9) Universal Subscriber Identity Module (USIM) card slot.

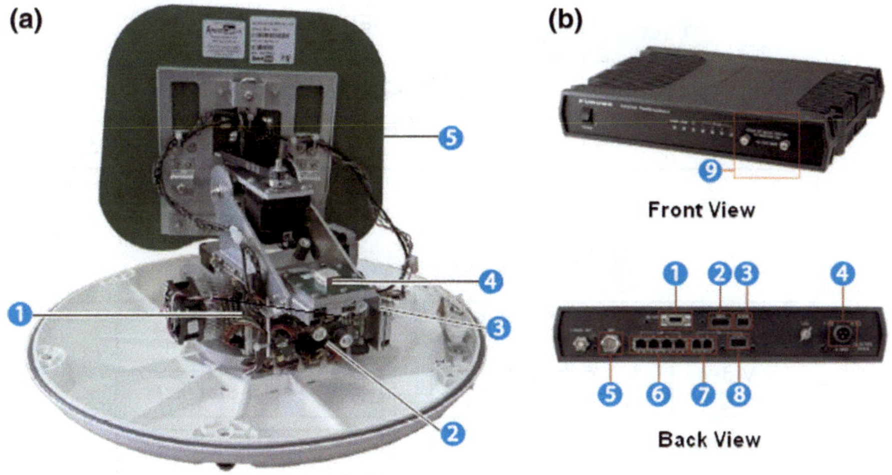

Fig. 7.6 ADE and BDE components—Courtesy of manual: by Furuno

7.2.2 Maritime SES Terminal Equipment and Installation

FleetBroadband is a robust GMSC system that will provide reliable communications across the globe in all weather conditions. However, in order to get the best performance out of the system, it is essential that the equipment, both ADE and BDE, is correctly installed. A well thought out and designed installation will ensure consistently high data throughputs and minimize, or even eliminate, outages due to shadowing. This section provides guidance on all aspects of the installation of the SES maritime broadband terminals of onboard ships together with some best practice recommendations.

FleetBroadband and other maritime broadband systems that provide simultaneous voice, high-speed data (HSD), and video communication via satellite through the mobile BGAN (Broadband Global Area Network) are very robust SES terminals. Mounting process of this SES type requires very serious preinstallation planning. The SES terminal equipment and configuration installed on the vessel is the responsibility of the customer, but it is expected that guidance will be provided by the Inmarsat distribution partner or service provider.

Usually, the ship antenna (ADE) structure can be mounted on the compass deck atop special antenna mast, while the transceiving equipment (BDE) with peripherals can be fixed in the navigation bridge or better in special radio station, whose configuration is shown in Fig. 7.7. The main elements of BDE are the following units:

1. **Cabin Interface Unit (CIU)**—With build in PC or system processor, it controls and monitors the whole system operations of the transceiver and direction of the antenna and also performs different task of baseband signal conditions between

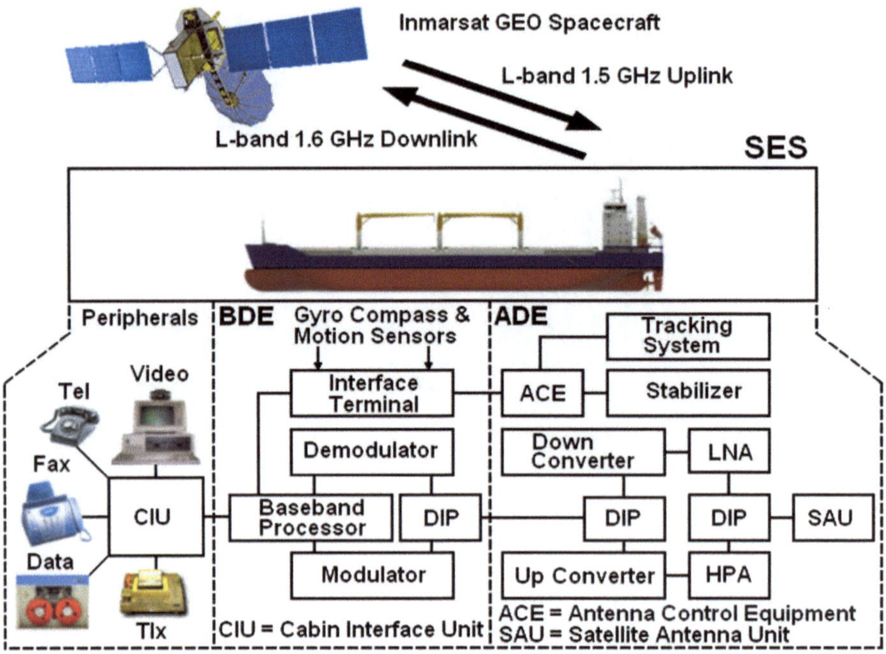

Fig. 7.7 Inmarsat SES, ADE, and BDE configurations—Courtesy of manual: by Ilcev

all obligation and optional BDE peripheral equipment (Tel, fax, Tlx, video, and data) and baseband processor.

2. **Baseband Processor**—Simply performs baseband processing of all transmitting and receiving audio, video, and PC/data signals. It comprises intermediate frequency (IF) amplifier, modems, and timing circuits for multiplexing up and down signals. In this case, the synthesizer produces the highly stable frequencies required for modulation and demodulation and for signal switching.
3. **Interface Terminal**—Connects navigation Gyro Compass and motion sensor units with CIU device for satellite tracking and electronic control of ADE.
4. **Modulator and Demodulator**—Both represent the nucleus of any transceiver. First modulates the baseband signal onto the IF carrier of transmitting signal, and second demodulates the baseband signal from the IF carrier of the receiving signal.
5. **BDE Diplexer (DIP)**—Enables direction of transmitting signals from modulator to the ADE and receiving signals from ADE to the demodulator.

The ADE is mounted below the waterproof container or radome on the stabilized platform. The following units compose the ADE assembly:

1. **ADE Diplexers (DIP)**—The first diplexer passes all transmitting signals to the up converter and from down converter to the demodulator. The second diplexer guides transmitting signals from HPA to the SAU and from SAU to the LNA.
2. **Up and Down Converters**—The up converter accepts the modulated IF carrier from modulator and translates it to the uplink transmitted RF via HPA, by mixing with local oscillator (LO) frequency, while the down converter receives the modulated RF carrier from the LNA and translates its downlink receiving RF to the IF.
3. **High-Power Amplifier (HPA)**—The HPA device provides amplification of transmitting signals by the traveling wave tube amplifier (TWTA) and klystron amplifier. The second HPA enables higher gain and better efficiency than TWTA but in smaller bandwidth of 2 %. The amplified uplink signal is going via DIP to the SAU.
4. **Low-Noise Amplifier (LNA)**—The LNA device provides initial amplifier stage of downlink signal coming from SAU via DIP without introducing much additional temperature noise. In this sense, the two most commonly used LNA products are new GaAs (gallium arsenide) FET (field-effect transistor) and old parametric amplifiers. Thus, the recently developed GaAs FET LNA enables very low-noise temperatures and takes advantages of its stability, reliability, and low cost.
5. **Antenna Control Equipment (ACE)**—This unit controls antenna stabilized platform (Stabilizer) and tracking system. It maintains the antenna direction toward the focus of satellite against any motion of the ship.

During SES installation, primary consideration should be given to onboard computers, hardware, and operating system, such as Windows, MAC, and Linux, including user applications and protocols, such as FTP, Internet Message Access Protocol (IMAP), regional HUBs, LES or points of presence (POP), Simple Mail Transfer Protocol (SMTP), Hypertext Transfer Protocol (HTTP), Transmission Control Protocol/IP (TCP/IP), or User Datagram Protocol (UDP), wiring, Ethernet, and other network connectivities, appropriate optimizations of applications and communications software, terminal installation, and in particular the location of the antenna.

7.2.2.1 Shipborne Antenna Mounting System

Ideally, the antenna should be installed on the highest point of the vessel with a clear view of the sky in all directions, and all possible steps should be taken to achieve this objective. However, if it is not possible to mount the antenna with an unrestricted view of the sky, then the antenna should be positioned so as to ensure minimum shadowing from the vessel superstructure such as funnels and radar. Such antenna positioning should take into account the typical shipping routes used by the vessel and the azimuth and elevation required for communication under way with the appropriate satellite.

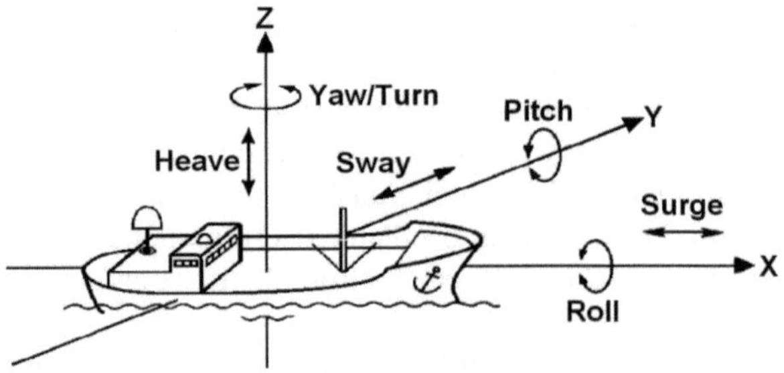

Fig. 7.8 Components of ship motion—Courtesy of manual: by Ilcev

In circumstances when shadowing might occur, it is useful and good practice to create a "shadow area" chart for use on the bridge showing at which azimuth shadowing may occur for each of the Inmarsat satellites to be used. Thus, if shadowing is a major problem, then consideration should be given to the installation of two antennas, one on either side of the ship superstructure, so operator can select the antenna with the clearest view of the satellite.

The SES antennas are relatively large and heavy directional structure, which comprises the mechanical assembly, control electronics and gyroscope, microwave electronic package, and the antenna assembly (dish, array, or similar) and has reduced considerably in both physical size and weight. The Maritime Mobile Satellite Communications (MMSC) SES antenna is generally fitted on the antenna pedestal pointing forward as indicated by the forward arrow on the base of the antenna. However, Land Mobile Satellite Communications (LMSC) and Aeronautical Mobile Satellite Communications (AMSC) antenna units more or less are similar. This pedestal or a platform has two horizontally, stabilized axes (X and Y), achieved by using a gyrostabilizer or sensors such as accelerometer or gyrocompass units.

The stabilized platform provides a horizontal plane independent of mobile motion such as roll or pitch. For example, all mobiles have some kind of motions, but ship motion has seven components during navigation such as roll, pitch, yaw, surge, sway, heave, and turn, which are shown in Fig. 7.8. Turn means a change in ship heading, which is intentional motion, not caused by wave direction, and the other six components are caused by wave motion. However, surge, sway, and heave are caused by acceleration.

Two-Axis Mount System (E/a and Y/X)

An antenna mount is mechanically moving system that can maintain the antenna beam in a fixed direction. In MMSC systems, the mount must have a function to point in any direction on the celestial hemisphere, because ships have to sail across the heavy seas. It is well known that the mount of the two-axis antenna configuration is the simplest mount providing such functions.

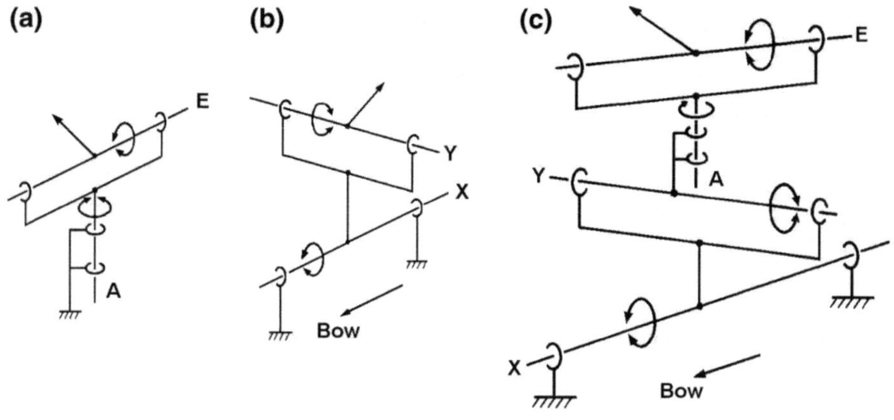

Fig. 7.9 Two-axis and four-axis mount systems—Courtesy of handbook: by Fujimoto

There are 2 typical mounts of the axis configuration: One is E/A (elevation/azimuth) mount and the other is the Y/X mount. Simplified stick diagrams of both mounts are given in Fig. 7.9a and b, respectively. Thus, in the E/A mount, a full steerable function can be obtained by choosing the rotation range of the azimuth axis (a-axis) from 0–90°. In the Y/X mount, a full steerable function is achieved by permitting the rotation angle from −90° to +90° to both the x- and y-axis.

In fact, this is the basic configuration for the ship utility, so a special function required for its antenna mount system is to compensate the ship motions due to sailing and ocean waves, and to keep the antenna beam in nearly a fixed direction in space. In the case of the pointing and tracking under ship motions, the required rotation angle range of each axis is from 0° to more than 360° for the a-axis, and from −25° to +120° for the e-axis with respect to the deck level, assuming that the operational elevation angle is de facto restricted above 5°. Otherwise, both mount types have several disadvantages.

Three-Axis Mount System (E/a/X, E′/E/a, and X′/Y/X)

The three-axis mount system is considered to be a modified two-axis mount, which has one additional axis. The three-axis mount of an E/A/X type, as shown in Fig. 7.10a, is the E/A mount with one additional x-axis. The function of the x-axis is to eliminate the rapid motion of the two-axis mount due to roll. However, in this system, the possibility of gimbal lock for pitch is still left near the zenith, when the e-axis is parallel to the x-axis. The three-axis mount of an E′/E/A type, as shown in Fig. 7.10b, is the E/A mount with an additional cross-elevation axis, E. In the mount system, the change of the azimuth angle is tracked by rotating the a-axis and the change of the azimuth angle is tracked by a combined action of the E and E′ axes. Hence, the E and E′ axes allow movements in two directions at a right angle.

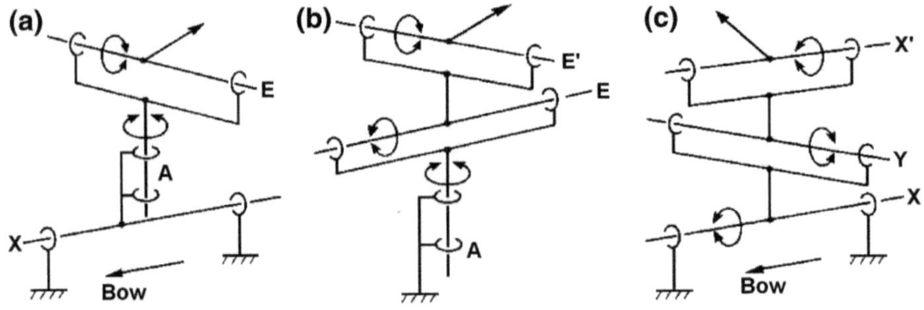

Fig. 7.10 Three-axis mount system—Courtesy of handbook: by Fujimoto

With an approximate axial control, this mount is free from the gimbal lock problem both near the zenith and near the horizon. The three-axis mount of an X′/Y/X type is the two-axis Y/X mount system with the x′-axis on it to obviate the gimbal lock at the horizon, as shown in Fig. 7.10c. When the satellite is near the horizon, the x-axis takes out the rapid motion due to yaw and turn. In this sense, the x′-axis rotates within $\pm 120^{o}$, so the x′-axis can only eliminate the rapid motion within the angular range. In general, this axis mount is rather more complex than that of the four-axis mount because steering and stabilization interact with each other.

Four-Axis Mount (E/a/Y/X)

The stabilized platform is made by the x-/y-axis to take out roll and pitch, and a two-axis mount of the E/A type is settled on the stabilized platform. This is the four-axis mount configuration, as shown in Fig. 7.9c. The tracking accuracy of this mount is the best solution because the stabilization function is separated from the steering function, and at any rate, four major components such as roll, pitch, azimuth, and elevation angle are controlled by its own axis, individually. The four-axis mount has been adopted in many SES of the current Inmarsat-A and B standards.

7.2.2.2 Omnidirectional Shipborne Antenna Mounting

When installing MSA, it is necessary to find a location on board ship that is as free from obstruction as possible. Thus, it is also important to maintain a certain distance from other communication antennas, especially radar installations. Normally, the best place for the MSA would be above the radar scanning antennas or far away from them. Otherwise, a minimum safe distance should be maintained to HF antennas of 5 m, to the VHF antennas of 4 m, and to the magnetic compass of 3 m.

The omnidirectional antenna is designed to provide satellite coverage even when the vessel has pitch and roll movement up to 15°. Thus, to maintain this coverage, the antenna should be free from obstructions in the area down to 15° below the

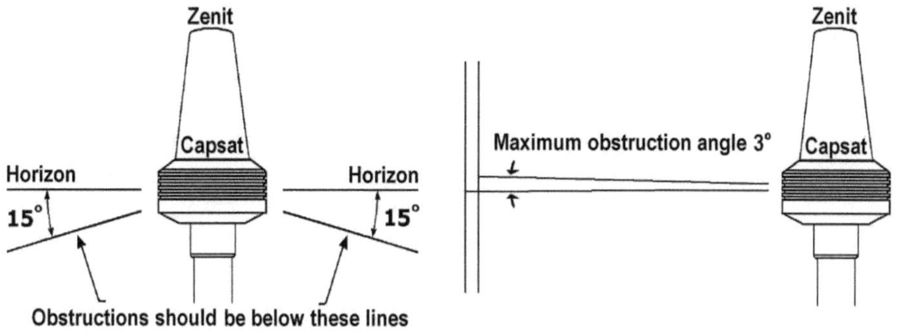

Fig. 7.11 Safe distance of Inmarsat-C antenna from obstructions—Courtesy of manual: by Sailor

horizon, as illustrated in Fig. 7.11 (Left). Since this may not be possible in the fore and aft directions of the vessel, the clear area can be reduced to 5° below the horizon in the fore and aft directions and 15° below the horizon in the port and starboard directions. Otherwise, any compromise in this recommendation will degrade performance. If an obstruction, such as a pole or a funnel, is unavoidable, the distance to these objects should be large enough, so that the obstruction only covers 3°. For instance, if the diameter of the obstructing object is 0.1 m, the safe distance should be about 2 m, as shown in Fig. 7.11 (Right).

The safety levels for the Thrane & Thrane Capsat-C Antenna Unit and similar Inmarsat-C aerials are based on the ANSI standard C95.1-1982. Namely, this standard recommends that the maximum power density at 1.6 GHz exposed to human beings should not exceed 5 mW/cm^2. Therefore, at maximum radiation output, the power from an Inmarsat-C antenna of 16 dBW EIRP corresponds to a minimum safety distance of about 30 cm. The future standard from the European Telecommunication Standard Institute (ETSI) concerning 1.5/1.6 GHz MES the recommendation will be maximum 8 W/m^2 (0.8 mW/cm^2) with a minimum safety distance of 62 cm at 16 dBW of EIRP.

7.2.2.3 Directional Shipborne Antenna Mounting and Steering System

The ship ADE unit consists antenna system mounted on a pedestal, RF, power, and control units, all covered by an antenna radome. Ideally, antenna should have free optical sight in all directions above an elevation angle of 5°. It must be placed as high as possible in the best position and to avoid blind spots with degradation or loss of the communications link.

The directional antenna has a beam width of 10° and ideally requires a free line of sight in all directions above an elevation angle of 5°. Possible obstructions will cause blind spots, with the result of degradation or even loss of the communications link with the spacecraft. So, complete freedom from degradation of the propagation is only accomplished by placing the antenna above the level of any possible obstructions. This is normally not feasible and a compromise must be made to reduce the amount of blind spots. The degree of degradation of the communication depends on the size of the obstructions as seen from the antenna; hence, the distances to them must also be considered. However, it should be remembered that the antenna RF beam of energy possesses a width of 12° angle cone and consequently, objects within 10 m of the radome, which cause a shadowing sector greater than 6°, are not likely to degrade the equipment significantly. Preferably, all obstructions within 3 m of the antenna system should be avoided. Obstructions less than 15 cm in diameter can be ignored beyond this distance. Knowing the route that the ship normally sails allows a preferable sector of free sight to be established, thus facilitating the location of the antenna unit.

The antenna beam must be capable of being steered in the direction of any Inmarsat GEO satellite, whose orbital inclination does not exceed 3° and whose longitudinal excursions do not exceed ±0.5°. This must be provided to point the antenna beam automatically toward the satellite with sufficient accuracy to ensure that the G/T and EIRP requirements, namely receive and transmit signal levels, are satisfied continuously under operational conditions. Careful and important consideration should be given to placing directional antenna in right place onboard seagoing ships. In particular, the focal point of the directional antenna must be pointing directly at the Inmarsat satellite being tracked without any interruption of the microwave beam, which may be caused by any obstruction on the ship.

Inmarsat specify that there should be no onboard obstacle that is likely to downgrade the performance of the equipment in any angle of azimuth down to an elevation of −5°, which is not easy to achieve. At this point, the SES design and Inmarsat installation guidelines give a theoretical antenna installation instruction mode satisfying this advice but with the disadvantage that the antenna is very high above the vessel's deck and would be impossible to install in such a way (see Fig. 7.12). This type of installation is not practical because a ship's antenna would certainly be adversely affected by strong, gusty wind and vibration and it would be difficult to gain access for maintenance purposes.

If ship's structures do interrupt the antenna beam, blind sectors will be caused, leading to degraded communications over some arc of azimuth travel. Otherwise, if

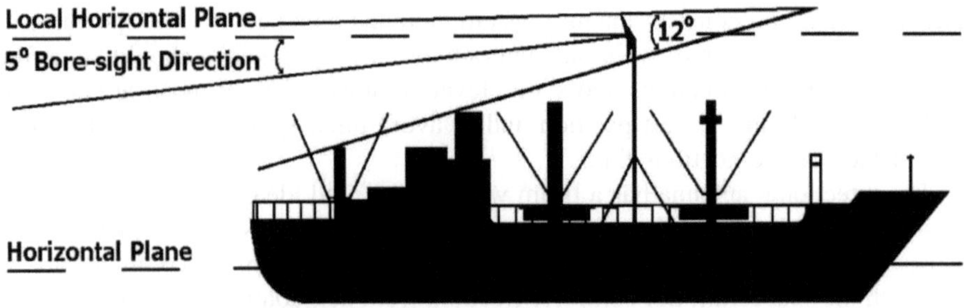

Fig. 7.12 Theoretical antenna installation—Courtesy of manual: by Inmarsat

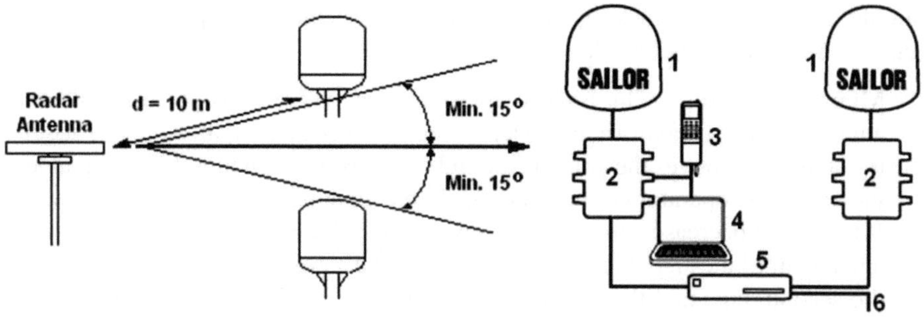

Fig. 7.13 Radar distance and dual antenna solution—Courtesy of manual: by Sailor

it is like that and as is often the case, it is impossible to find a mounting position free from all obstructions; the identified blind sectors should be recorded. It may be possible for the operator, when in an area served by two satellites, to select the satellite whose azimuth and elevation angles with respect to the ship's position are outside the blind sector. Obviously, this method is not enough practical because the satellite overlapping sectors inside of Inmarsat's four ocean regions cover relatively small areas. The best solution to avoid all blind sectors is to place the antenna unit on top of the radar mast or on a specially designed mast.

However, if antenna is placed in the same level of radar antennas, it is important to take into consideration interferences that radar antenna may cause. Since radar radiates a fan beam with a horizontal beam width of a few degrees and a vertical beam width of up to ±15°, the worst interference can be avoided by mounting the antenna at a different level, namely this meaning that the antenna is installed at minimum 15° above or below the radar antenna, which scenario is as shown in Fig. 7.13 (Left). Due to near-field effects, the benefit of this vertical separation could be reduced at short distances below 10 m between radar antenna and the Sailor FleetBroadband antenna. Thus, it is recommended to ensure as much vertical separation as possible when this antenna has to be placed close to a radar antenna.

When FleetBroadband SES antenna cannot obtain full line of sight (LOS) to the satellites because of blocking objects onboard ship, the new Sailor FleetBroadband Dual Antenna solution keeps connection. It maintains full LOS to the satellites with a combination of two Sailor 500 FleetBroadband SES antenna solutions installed in parallel at different locations together with Dual Antenna Control Equipment (DACE), as shown in Fig. 7.13 (Right), where 1 = ADE, 2 = master below deck equipment (BDE), 3 = IP handset, 4 = PC or laptop, 5 = Dual Antenna Control Equipment (ACE) and 6 = NMEA 0183 input.

7.2.2.4 Shipborne Antenna Tracking and Pointing System

The tracking system is another important function required of the antenna mount. Thus, the primary requirement for SES tracking systems is to be economical, simple, and reliable. Tracking performance is a secondary requirement when an antenna beam width is broad.

1. **Manual Tracking**—This is the simplest method, wherein an operator controls the antenna beam to maximize the received signal level. At first, the operator acquires the signal and moves the antenna around one axis of the mount. If the signal level increases, the operator continues to move the antenna in the same direction. If the signal decreases, the operator reverses the direction and continues to move antenna until the signal level is maximized. The same process is repeated around the second axis, and the antenna is held in both axes when the received signal level decreases. This method is suitable for LMSC units and especially for portable and flyaway communication terminals.
2. **Step Tracking**—Among various auto track systems, the step track system has recently been recognized as a suitable tracking mode for SES because of its simplicity for moderate tracking accuracy. The recent development of integrated circuits and microprocessors has brought a remarkable cost reduction to the step track system, whose principle is the same as that of the manual track. The only difference is that an electric controller plays the role of an operator in the manual track. The functional block diagram of the step tracking system is shown in Fig. 7.14 (Left). Sample-hold circuits are used to hold the signal levels, which are compared before and after the antenna has been moved by an angular step. If the level increases, the antenna is moved in the same direction and vice versa; if the level decreases, the direction will be reversed. In the other words, this process will be carried out alternately between two axes, whose good accuracy depends on the sensitivity of the comparators. As a result, the beam center is maintained in the vicinity of the satellite direction. Wrong decisions on the comparison of levels generally arise from the S/N ratio, and level changes due to multipath fading and stabilization error.
3. **Program Tracking**—This tracking concept is based on the open-loop control slaved to the automatic navigation equipment, such as a gyrocompass, GPS, the Omega, and Loran-C systems. In program tracking, the antenna is steered to the point of the calculated direction based on the navigation positional data. Since the satellite direction changes because of roll, pitch, and turn, a function to remove these rapid motions is required in the program track, whose block diagram is shown in Fig. 4.14 (Right). The error of the navigation equipment is

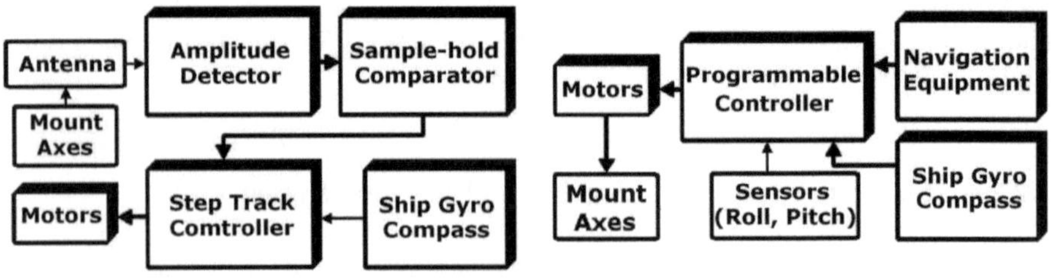

Fig. 7.14 Block diagrams of step and program tracking—Courtesy of handbook: by Fujimoto

negligibly small for the program track system, while its error mainly depends on the accuracy of sensors for roll, pitch, and turn, which is the stabilization error. At this point, an adequate sensor for the antenna program track system is a vertical gyro because it is hardly affected by the lateral acceleration. When the stabilization requirement is lenient, a conventional level sensor, such as inclinometer, a pendulum, and a level, may be used with careful choice of the sensor's location. The controller calculates the direction of the satellite to compensate for the ship's motions, which is affected by all components.

The simpler the axis configuration of the antenna mounts, the more complex the program calculation procedure becomes. Since the program controller has to execute calculations of many trigonometric functions, a microprocessor is a candidate for the controller. Thus, this program tracking is also applicable to the four-axis mount. Combination with the step track system is desirable because the error of the program track can be compensated by the step track and its rapid ship motion error can be compensated by the program tracking system.

7.2.2.5 Antenna Mast and Stabilization Platform

The mast has to be designed to carry the weight of the antenna unit up to 300 kg, depending on model design or manufacturer, which is depicted in Fig. 7.15 (Left). It must also be able to withstand the forces imposed by severe winds up to 120 knots on the radome and strong vibrations due to very rough seas on the whole ADE construction. The top end of the mast should be fitted with a flange with holes matching the bolts extending from the bottom of the radome. The flange must not be so large as to interfere with the hatch in the bottom of the antenna unit. The holes through the mast flange must be positioned symmetrically around the ship's longitudinal axis. If the height of the mast makes it necessary to climb up to the antenna unit, a ladder must be provided on the mast column. A guardrail must be attached to the upper section for safety purposes. If the height of the mast exceeds approximately 4.5 m, an access platform should be attached to the mast about 1.5 m below the radome bottom.

The radome completely encloses the periphery of the base plate assembly to protect the electronic and mechanical components from corrosion and weather. It is usually fabricated from high-gloss fiberglass and is electronically transparent to RF signals in the assigned frequency band. The radome is secured to the circular or square antenna stand (base plate) with several screws and can be removed easily without special tools.

The antenna-stabilized platform is housed inside the radome and consists in the electrical and mechanical elements, which is depicted in Fig. 7.15 (Right). There are two antenna control stepper motors. First, there is the azimuth step motor, which controls the position of the antenna reflector in the horizontal A plane (azimuth), and second, there is the elevation step motor that controls the vertical E plane (elevation). Each motor has four phase inputs coming from the drive circuit in the

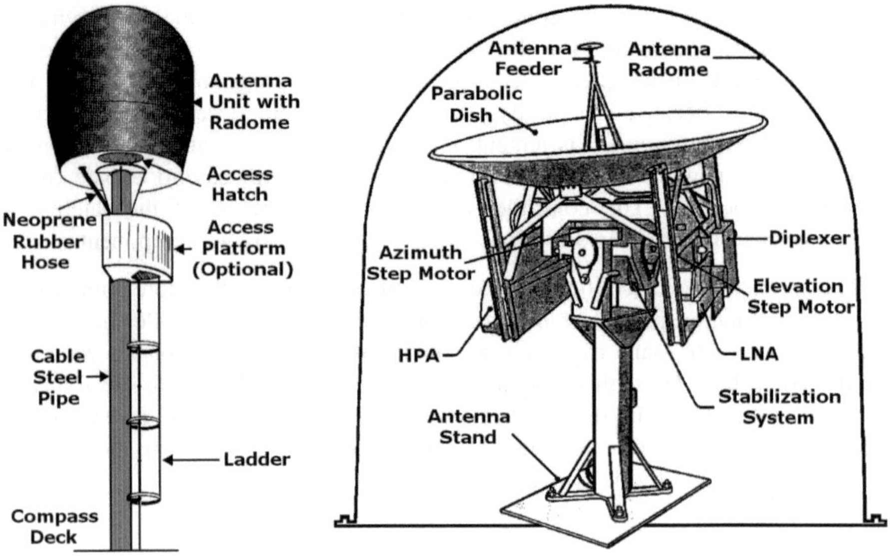

Fig. 7.15 ADE mast and stabilization platform—Courtesy of manual: by Ilcev

BDE Control Board and a supply voltage from the master power supply located on the antenna stand. As a stepper motor turns the antenna, it also adjusts the setting of the relevant sensor potentiometer. The sensor voltage supply is the reference voltage for the A/D converter on the Control Board.

Therefore, two stepper motors move the antenna in both azimuth and elevation angles and move the relevant sensor potentiometers, which provide feedback information on the position of the antenna. The stabilization system, or gyroscope with two gyro motors, stabilizes the platform for the antenna against the roll and pitch of the ship. A two-turn solenoid clamps the antenna platform to the gyroscope assembly. A diplexer passes the Rx signal from the antenna to the LNA and the Tx signal from the transceiver assembly to the antenna. The LNA amplifies the Rx signal, and the HPA amplifies the Tx signal. The antenna, a parabolic dish, radiates EM energy to and from the antenna feeder in Rx or Tx direction, respectively. Finally, the antenna assembly is mounted below the radome for protection purposes.

7.2.2.6 Antenna Location Aboard Ship

The ship's antenna unit should be located at a distance of at least 4.5 m from the magnetic steering compass. At this point, it is not recommended to locate the antenna close to any interference sources or in such a position that sources such as the radar antenna lie within the antenna's beam width of 10° when it points at the satellite. The ADE should also be separated as far as possible from the HF antenna

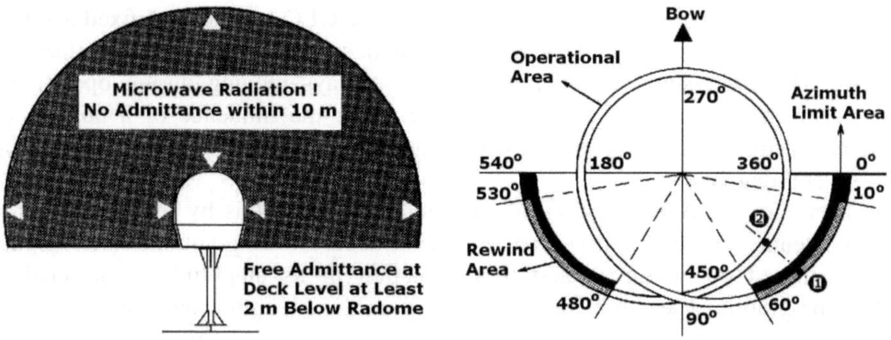

Fig. 7.16 Antenna radiation precautions and azimuth limit—Courtesy of manual: by Ilcev

and preferably by at least 5 m from the antennas of other communications or navigation equipment, such as the antenna of the satellite navigator or the VHF and NAVTEX antennas. In addition, it is not practical to place the antenna behind the funnel, as smoke deposits will eventually degrade antenna performance. Regardless of the location chosen for the antenna, it should be oriented to point forward in parallel with the ship's longitudinal axis when in the middle of its azimuth range, which will correspond to zero degrees on the azimuth indicator.

The EM RF signals are known to be hazardous to health at high radiation levels. Thus, it is inadvisable to permit human beings to stand very close to the radome of an SES when the system is communicating with a satellite at a low elevation angle. In this case, Inmarsat recommends that the radiation levels in the vicinity of the antenna should be measured. The crew members and passengers should not be admitted to areas closer than 10 m away from the antenna unit at desk level above 2 m, measured beneath the lowest point of the radome, as shown in Fig. 7.16 (Left). No restrictions, therefore, are required when the antenna radome is installed at least 2 m above the highest point accessible to crew and passengers. Hence, authorized personnel should not remain close to the antenna system for periods exceeding 1 h per day without switching off the RF transmitter.

However, radiation plan diagrams may be produced and located near the antenna as a warning for crew members, passengers, and ship's visitors, or distances from the antenna may be physically labeled at the relevant place.

7.2.2.7 Satellite Determination and Antenna Azimuth Limit

An Inmarsat directional antennas must be capable of locating and continuously tracking the GEO satellite available or selected for communication, namely if the ship has in view only one satellite or if the ship is in an overlapping position, respectively. Thus, Inmarsat-C has an omnidirectional antenna and does not need a tracking system. Locating and tracking may be done automatically, as in the case of an SES, or manually, as with a portable MES.

In fact, it is common practice to believe that the GEO satellites are fixed and that once the link has been established, it will remain so as long as the mobile does not move. However, ships or other mobiles are always moving during operational management of voyages, and satellites are under the influence of a number of variable astrophysical parameters, which cause it to move around its station by up to several degrees.

Therefore, an ADE tracking system must counteract this by repositioning the SES antenna at regular intervals and in case of need. At this point, the carrier signal is monitored continuously, and if a reduction in its amplitude is detected, a close-programmed search is initiated until the carrier strength is again at maximum. No loss of signal occurs during this process, which is automatically initiated. Obviously, the greatest antenna-tracking problem will arise when the SES is moving at speed with respect to the satellite. An Inmarsat antenna may be moved through any angle in azimuth and elevation as the vessel moves along its course. In this case, it is essential that electronic control of the ship antenna is provided. In practice, antenna control may be achieved by manual or simple electronic feedback methods.

1. **Manual Commands**—When the radio operator has selected manual control, elevation is commanded by up and down keys, whereas azimuth positioning is controlled by clockwise and counterclockwise keys. A command would be used when the relative positions of both the vessel and the satellite are known. Azimuth (A) and elevation (E) angles can be derived and input to the equipment, by using the two A and E charts of Inmarsat network coverage. Once the ship antenna starts to detect a satellite signal, the operator display indicates signal strength. Fine positioning can now be achieved by moving the antenna in A and E in 1/6th degree increments until maximum field strength is achieved.
2. **Automatic Control**—Once geostationary satellite lock has been achieved, the system will automatically monitor signal strength and apply A/E corrections as required in order to maintain this lock as the vessel changes course.
3. **Automatic Search**—An automatic antenna search routine commences 1.5 min after switching on the equipment, or it may be initiated by the operator. Therefore, the elevation motor is caused to search between 5° and 85° limits, whereas the azimuth motor is stepped through 10° segments. If the assigned common signaling channel signal is identified during this search, the step tracking system takes over to switch the antenna above/below and to each side of the signal location searching for maximum signal strength.
4. **Gyroscopic Control**—Lock is maintained irrespective of changes to the vessel's course by sensing signal changes in the ship's gyro repeater circuitry. Satellite signal strength is monitored, and if necessary, the A/E stepper motors are commanded to search for maximum signal strength.
5. **Antenna Rewind**—The antenna in the ADE is provided on a central mast and is coupled by various control and signal cables to a stationary stable platform. Thus, if the antenna was permitted to rotate continuously in the same direction, the feeder cables would eventually become so tightly wrapped around the

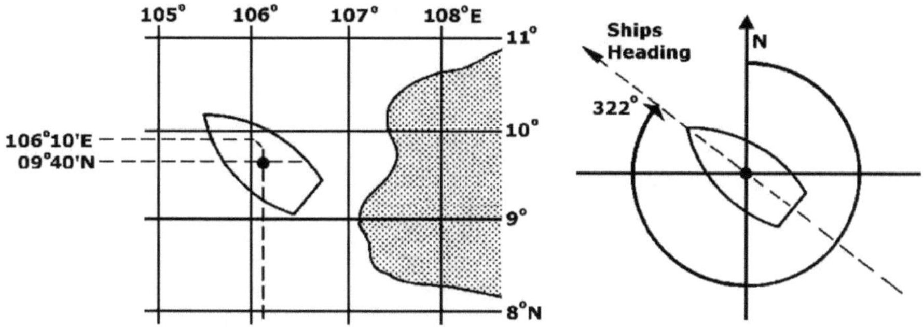

Fig. 7.17 Antenna pointing—Courtesy of manual: by Ilcev

central support that either they would prevent the antenna from moving or they would fracture. To prevent this happening, a sequence known as antenna rewind is necessary, as is shown in Fig. 7.16 (right). In fact, an antenna unit has three areas with rewind time of approximately 30s plus stabilizing time, giving a total of about 1.5 min:

(1) Operational area is the antenna-rotating limit in the azimuth plane. In fact, the antenna can rotate a total of 540°, which is shown as a white area in Fig. 7.17 (Left). Normally, the antenna will operate in the operational area, which is between 60° and 480°.
(2) Rewind area is necessary for the following reasons: If the antenna moves into one of the rewind areas, i.e., 10°–60° or 480°–530° (antenna azimuth lamp-lights) and if no traffic is in progress, the antenna will automatically rewind 360° to get into the operational area and still be pointed at the satellite, which is illustrated as a dotted area in Fig. 7.17 (Right). For example, the ship antenna moves from position 1 to position 2 and rewinds lamplights. If the SES terminal is occupied with a call and the ship turns so that the antenna enters the rewind area, no rewind will take place unless the antenna comes into the azimuth limit area. Definitely, if this happens, rewind will take place and the call will be lost. The azimuth-warning indicator on the operator display will light to indicate that antenna rewinding is in progress.
(3) Azimuth limit area is an important factor because when the antenna is in this area, the azimuth limits lamplights. If the antenna moves into the outer part of the azimuth limit area, i.e., 0°–10° or 530°–540°, rewind will start automatically, despite traffic in progress.

7.2.2.8 Antenna Pointing and Tracking

The directional reflector antenna is highly directive and must be pointed accurately at the satellite to achieve optimum receiving and transmitting conditions. In normal operation, the antenna is kept pointed at the satellite by the auto-tracking system of, for example, Saturn 3 SES. Before the auto-tracking can take over, the antenna must be brought within a certain angle in relation to the satellite. Thus, this can be obtained using the command "find" or by manually setting the antenna using the front push buttons on the terminal or via teleprinter command. For manual pointing, it is necessary to provide the ship's plotted position, ship's heading by gyro, azimuth angle map, and elevation angle map of the satellite.

1. **Ship's Plotted Position**—The plotted position is needed to decide which satellite can be used, namely which Inmarsat network area can be tuned: Indian Ocean Region (IOR), Pacific Ocean Region (POR), Atlantic Ocean Region-West (AORW), or Atlantic Ocean Region-East (AORE), depending on the ship's actual position, as shown in Fig. 7.17 (left). Sometimes, the ship can be in an overlapping satellite area covered by two or even three Inmarsat satellites. In this case, it will be important to choose convenient CES and to point the antenna toward one of the overlapping ocean regions.
2. **Ship's Heading by Gyrocompass**—The permanent heading of the ship determined by gyrocompass is needed for the antenna auto-tracking system (see Fig. 7.17) (right).
3. **Azimuth Angle**—The azimuth is the angle between north line and horizontal satellite direction as seen from the ship, as is shown by example of 259°, in Fig. 7.18 (Left). Thus, the actual azimuth angle for the various satellites due to the ship's plotted position can be found on the map, as shown in Fig. 2.5a.
4. **Elevation Angle**—The elevation angle is the satellite height above the horizon as seen from the ship, as is shown by the example of 38°, in Fig. 7.18 (Right). In this case, the actual elevation angle for the various satellites due to the ship's plotted position can be found on the map in Fig. 2.5b.

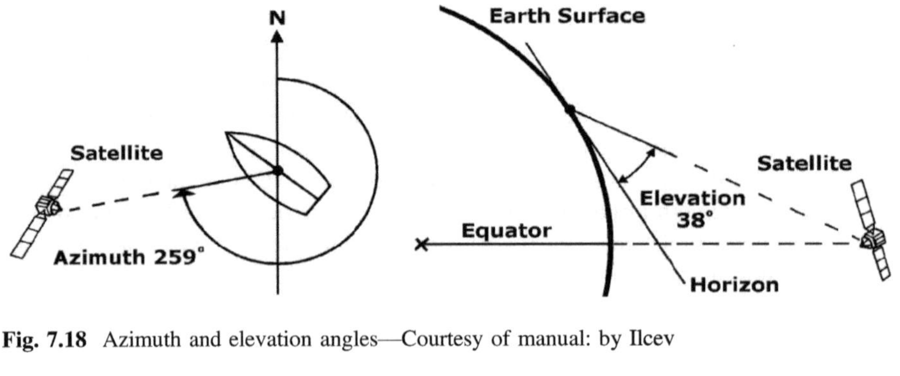

Fig. 7.18 Azimuth and elevation angles—Courtesy of manual: by Ilcev

7.2.3 Land Mobile Earth Stations (MES)

In a land mobile environment, the communications link between the satellite and mobile terminal is subject to impairments in the form of multipath effects arising from signal scattering and reflections and shadowing effects arising from signal blockage. This results in signal fading and phase variations in the carrier, the magnitude of which depends on whether the terminal is stationary or moving, the speed at which it is moving, and the environment between the terminal and the satellite. It is possible to provide sufficient link margin to compensate for fading or to allow for some in-building penetration of the signal. However, it is not practical in LMSC system to provide sufficient margin to still allow communications when the direct LOS of the link is blocked by buildings, heavy vegetation, and other obstacles.

In vehicular applications, such as road and rail vehicles, the antenna is normally placed on the roof and is typically of low or medium gain (less than 15 dBi). The latter is normally either a low profile motorized tracking antenna or a non-tracking rod antenna about a meter in height, which has an omnidirectional coverage in azimuth and fixed coverage in elevation (coverage suitable for use in temperate regions).

In this section will be described Inmarsat Broadband Global Area Network (BGAN) for vehicular and transportable applications. At the end of 2005, Inmarsat launched its BGAN service as the first high-speed wireless data solutions with voice and Internet available on a global basis. The BGAN service is accessed through onboard vehicle or portable broadband satellite transceivers with antenna easy to carry as a laptop computer. The BGAN network uses 3 GEO I-4 satellites and an optimized ground network, which interconnects with variety of TTN links at local BGAN users via LES terminal. The satellite coverage of BGAN network is the same as FleetBroadband, as shown in Fig. 7.3.

The BGAN satellite network and terminals employ bandwidth efficient modulation and coding techniques, capable of supporting variable bit-rate quality of service (QoS) and depending on the need of the applications. The BGAN provides a near-global coverage overlay for TTN, giving users service availability beyond the reach of IMT-2000 and Universal Mobile Telecommunications Service (UMTS) standards. A range of supported terminals, personal devices, and transportable and mobile units linked with entertainment, communications systems to remote base stations including local area networks for civilian and military applications including SCADA or M2M are discussed.

7.2.3.1 Land BGAN Transportable Earth Station (TES)

With the appropriate software and hardware, BGAN service supports all mobile data and ISDN solutions up to 64 Kb/s and Internet Protocol (IP) up to 492 Kb/s including secure encryption (STU III/STE) system important for corporate and

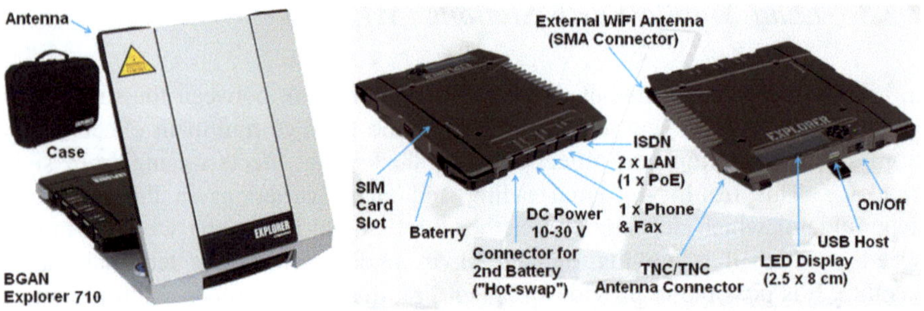

Fig. 7.19 Transportable Explorer 710 BGAN configurations—Courtesy of manual: by Cobham

military communications. It extends fixed or mobile office such as a LAN, Internet, and ISDN/PSTN in remote areas using multimedia semi-fixed transportable BGAN. The BGAN system provides a range of IP and voice, data, and video (VDV) transmissions covering almost all of the world land masses for civilian and military satellite communications, VoIP, Internet, videoconference, e-solutions, SCADA monitoring, and broadcasting in rural and remote environments.

The BGAN system offers few classes of solutions with large scale of service for portable, such as mobile and semi-fixed users via L-band I-4 GEO constellation and LES ground segment to other users, mobile or semi-fixed customers interfaced to the TTN infrastructure. The new transportable Explorer 710 BGAN announced in April 2013 provides on-demand streaming rates at 650 Kb/s out of the box, whose configuration with antenna is depicted in Fig. 7.19 (Left). This is the most powerful Class 1 terminal that provides high-speed new streaming service, whose components are shown in Fig. 7.19 (Right).

The BGAN Explorer 710 can be provided with BGAN network connected to different peripheral equipment as shown in Fig. 7.20. This unit also includes several advanced new features, including plug-and-play bonding, which can simply connect two Explorer 710 terminals with an Ethernet cable and double the throughput at speeds in excess of 1 Mb/s QoS. The new flagship in the Explorer series introduces smartphones to the world of BGAN units, enabling onboard mobile and fixed users to connect their own devices for voice calling and connectivity. Then, other features include a USB host interface, hot-swappable batteries, an easy-to-use LED display, and multiple interfaces to support a wide range of applications.

The white part of the Explorer 710, including the support bracket, is the detachable antenna module, which comprises a GPS and a BGAN antenna, while Bluetooth and wireless LAN antennas are integrated in the transceiver. It also provides a compass to help positioning the antenna, which Class 1 Cobham Explorer is shown in Fig. 7.21 (Left) and (Right).

The BGAN solution is a near-global Satellite Internet and Multimedia Network with telephony deployed to connect onboard mobile or fixed laptop or palmtop PC to broadband Internet in remote locations, although as long as LOS to the satellite exists, the terminal can be used anywhere. The value of BGAN terminals is that unlike other satellite Internet services which requires bulky and heavy satellite

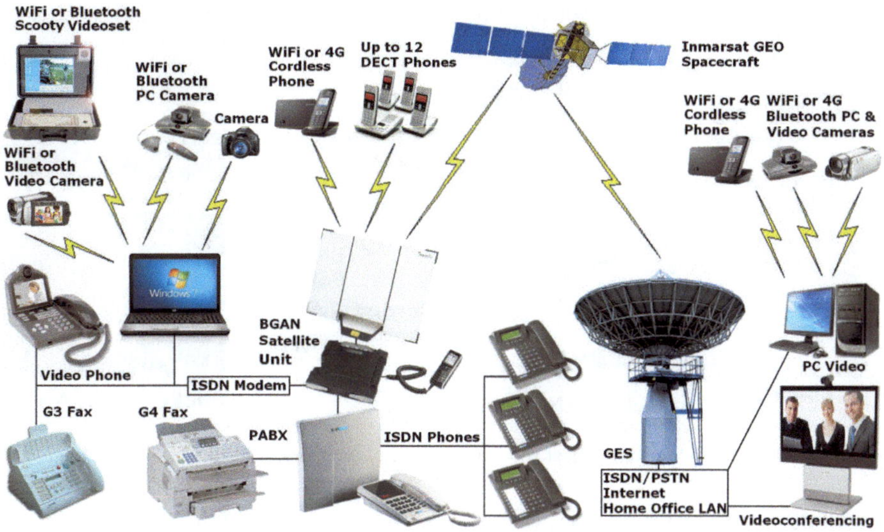

Fig. 7.20 Transportable Explorer 710 BGAN network—Courtesy of manual: by Ilcev

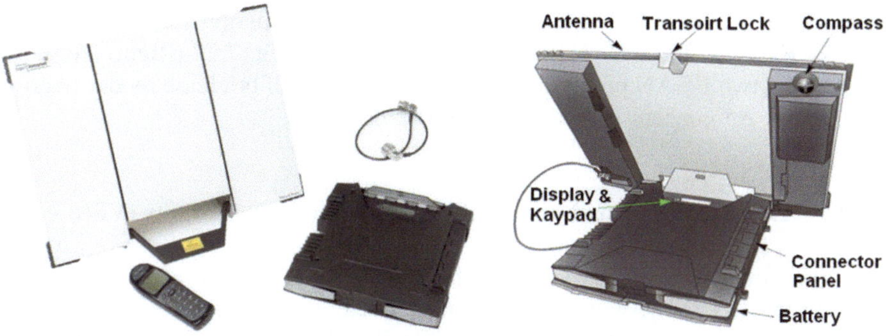

Fig. 7.21 Transportable Explorer 700 BGAN network—Courtesy of manual: by Cobham

dishes to connect, a BGAN terminal is about the size of a laptop and thus can be carried easily in the bag or vehicle.

Namely, BGAN is based on the Inmarsat Mobile Packet Data Service (MPDS) IP packet technology, so users only pay for the amounts of data they send and receive, and not for the amount of time spent online such as ISDN. So, this solution enables all users to stay always connected and standby on the Internet Webpages using a valid SIM card inserted in the BGAN Satellite IP Modem. Hence, this prescribed SIM card may be activated and used in other GPRS mobile devices for roaming between networks.

Other BGAN applications include File Transfer and File Transfer Protocol (FTP) for downloading and sharing files from the Internet, Web e-commerce for online ordering and procurement and providing secure end-to-end connectivity over

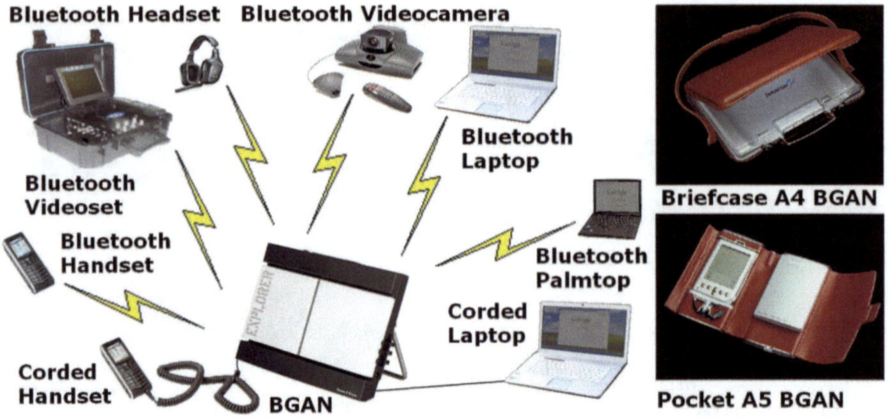

Fig. 7.22 BGAN LAN and smallest models—Courtesy of manual: by Ilcev

a corporate Virtual Private Network (VPN). Thus, if it is running as a secure and shared 144 Kb/s rate channel, BGAN operates at more than twice the speed of current ground GPRS (General Packet Radio Service) cellular phones.

There are other types of BGAN smaller terminals offering reduced service, such as classes 2, 3, and 4. In Fig. 7.22 (Left) is shown LAN peripherals and service of portable Explorer 300 BGAN designed by Cobham. In Fig. 7.22 (Right) are shown the smallest two BGAN prototypes: BGAN A4 concept of briefcase modes (Above) and NGAN A5 concept of pocket model (Below).

7.2.3.2 Land BGAN Road and Railway Vehicle Earth Stations (VES)

The VES for road and railway solutions is a series of L-band satellite terminals utilizing the BGAN network from Inmarsat, such as an Explorer 727 depicted in Fig. 7.23 (Left). Similar to the transportable BGAN, this terminal offers a broadband connection and voice service from anywhere in the world. Every terminal features easy setup and operation. Simply connect a PC or phone and point the terminal toward a satellite using the compass and sound indicator, and you are instantly connected from anywhere in the world.

In Fig. 7.23 (Right) is shown mobile BGAN RF-7800B-VU104 of Harris suited for commercial and military applications. This BGAN unit has similar technical characteristics as an Explorer 710, which streaming speed rates for both devices starting from 32 and reaching up to 492 Kb/s. In Fig. 7.24 is shown BGAN ground segment and network.

The basic Explorer 710 system consists of antenna and IP handset with cradle. The VES terminal, which contains the primary electronic parts, is designed for wall or desktop installation. It supplies 23.0–30.0 V DC to the antenna through a single coaxial cable. The DC input for the terminal uses both 24 and 12 V DC power supply. The Explorer 727 antenna is a mechanical tracking antenna, consisting of a

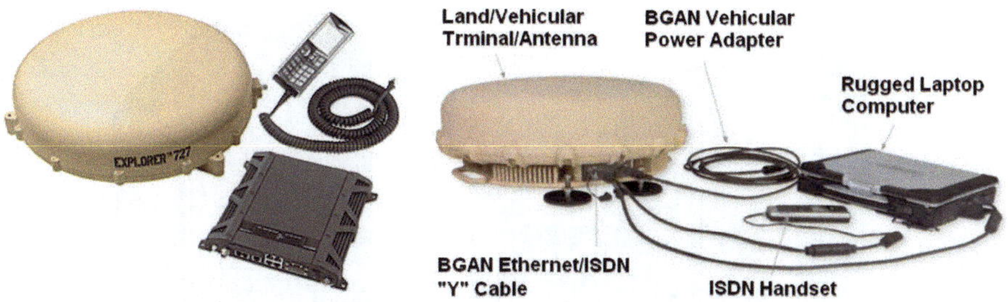

Fig. 7.23 Vehicular BGAN configurations—Courtesy of manual: by Cobham/Harris

Fig. 7.24 Mobile and transportable BGAN ground segment—Courtesy of manual: by Ilcev

two-axis stabilized antenna with RF unit, antenna control unit, and GPS antenna. The antenna is dedicated to the Inmarsat BGAN system and is designed for roof mounting on a vehicle. All communication between the antenna and terminal passes through a single coaxial cable. The IP handset connects to the LAN interface of the terminal and is power supplied with PoE through the LAN interface. The IP cradle serves as a holder for the IP handset, power supplied from PoE and connected to the handset with a coil cord and to the terminal with a standard LAN cable (Fig. 7.25).

7.2.3.3 Personal Earth Stations (PES)

Personal Earth stations can be divided into two categories, such as handheld Earth stations and docking Earth stations. Handheld MSS terminals are similar in size and appearance to normal terrestrial cellular phones. In fact, the terminals are commonly

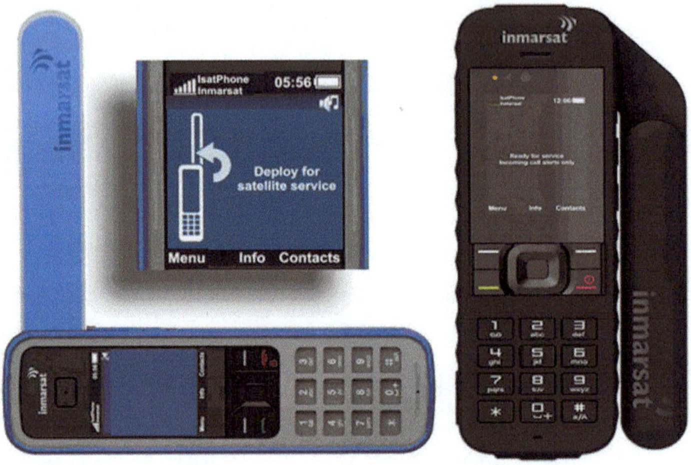

Fig. 7.25 IsatPhone Pro and 2 PES—Courtesy of manual: by Inmarsat

dual-mode phones that have the capability of operating to either the satellite or terrestrial mobile networks. When a call is placed, it is first connected through the local terrestrial mobile network. If, however, the terminal is outside the terrestrial coverage area, the call is then routed through the mobile satellite network. This method of operation allows MSS phones to complement the existing terrestrial mobile infrastructure by providing a regional or global extension to its coverage area. An important feature of these terminals is that they are assigned a single telephone number, which is used for both satellite and terrestrial cellular call reception. The terminals will generally require a direct LOS to the satellite. Other than this, their usage will be similar to terrestrial cellular operation.

1. **Handheld PES**—Inmarsat developed few handheld phone terminals, which can be used in urban, rural, and remote environments including onboard mobiles via Inmarsat-4 coverage, as shown in Fig. 7.3. Here will be introduced two PES units, such as IsatPhone Pro shown in Fig. 2.25 (Left) and IsatPhone 2 shown in Fig. 2.25 (Right).

IsatPhone Pro offers satellite telephony with Bluetooth for handsfree use, voice mail, text, and e-mail messaging. This phone is easy to use, which provides peace of mind when you are many kilometers from anywhere, it has the longest battery life with up to 100 h standby time, and GPS location data can be looked up or sent in a text message. It is an essential part of any traveler's kit; whether persons are traveling for business or pleasure, it is small enough to have on hand for a call back home to the family or for professional users in government, onboard mobiles, media, aid, oil and gas, mining, and construction sectors.

IsatPhone 2 is the latest addition to Inmarsat handheld satellite phone portfolio, offering a range of new features ensuring phone calls in any extreme and remote locations. The robust handset can be used in any weather conditions and any

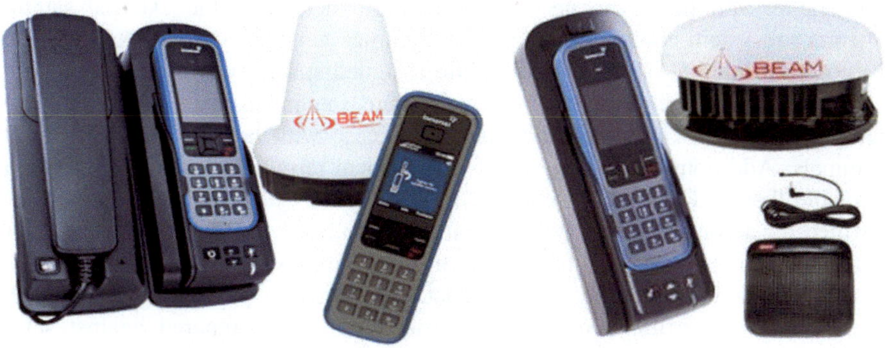

Fig. 7.26 IsatDock 2 Marine and drive PES—Courtesy of manual: by Beam

positions under Inmarsat-4 satellite coverage. The phone battery provides 8 h of continuing talktime and up to 160 h on standby. This phone gives users excellent voice clarity and call stability available on global basis over the three I-4 satellites, which have an operational lifetime into the 2020s.

– **Docking MES**—Inmarsat was developed several docking phone terminals, which can be installed onboard ships and road and rail vehicles via Inmarsat-4 satellite coverage. Here will be introduced two docking PES designed by beam, such as IsatDock 2 Marine shown in Fig. 7.26 (Left) and IsatDock 2 Drive shown in Fig. 7.26 (Right).

IsatDock 2 Marine is an intelligent docking station for the IsatPhone 2 that is specifically designed for maritime applications. The marine dock supports voice services via Bluetooth, RJ11/POTS for corded phones or PABX support, handsfree speakerphone, or the privacy handset. The phone handset is completely enclosed in the docking unit while still giving full access and functionality to the user. This phone supports the assistance alert, personal alert, and tracking functionality of the IsatPhone 2.

IsatDock 2 Drive is in-vehicle handsfree docking station, which provides a high-quality semipermanent installation good suited for land and railway vehicular applications for the Inmarsat IsatPhone 2. The docking voice station has in-built echo canceling and full duplex technologies to provide superior voice quality and professional handsfree or privacy calls.

7.2.4 Aeronautical Aircraft Earth Stations (AES)

Aircraft terminals, like the maritime terminals, also have high system requirements in the performance, availability, and integrity of communications. Aspects specific to aeronautical terminals include the need to cater for large Doppler shifts and temperature variations, the need for ergonomic design of the external satellite antenna subsystem components, so as to minimize atmospheric drag, and the need for compact internal RF units that fit into assigned spaces within the aircraft.

The terminals are likely to be much more integrated with other electronic equipment present on the aircraft than is the case with land or maritime mobile terminals. Additionally, a category of aeronautical terminals has recently become available that is designed for public correspondence calls only, and they are not certified for aeronautical safety applications or for use with air traffic management systems. Such terminals, typified by the Inmarsat mini-M service, are attractive for use in light aircraft as they are compact, the purchase price is lower, and the installation and certification process is much reduced compared against safety compliant terminals.

The AMSC terminals normally employ one of three types of antenna:

- Those having HGA with minimum gain of 12 dBi as typified by the Inmarsat Aero-H/H+, Swift64, and SwiftBroadband. The antenna, being directional, needs to track the satellite.
- Those having IGA as typified by the Inmarsat Aero-I service. Thus, these antennas are designed to exploit the higher power of Inmarsat satellite spot beams to receive Aero-H levels of service through smaller and cheaper terminals.
- Those having LGA as typified by the Inmarsat Aero-C and L service. In fact, these aero antennas are unsteered dipole or helix antenna having near omnidirectional gain.

Some of stated aero Inmarsat standards have been or will be abandoned, so in the following context will be explained new SwiftBroadband AES.

7.2.4.1 Aeronautical SwiftBroadband (SB) AES

In addition to the Maritime Broadband system, the author of this book in 2000 proposed aeronautical broadband solution as well. However, seven years later, Inmarsat introduced both FleetBroadband and SwiftBroadband terminals.

Prior to this event, Inmarsat aeronautical services introduced solutions on the second- and third-generation satellites, such as Aero-H, H+, I, L, Mini M Aero, and Aero-C, which are being referred to as “Classic Services.” The first aeronautical service, Aero-H, was introduced in 1991 and has become the de facto standard of cockpit data communication for many of the world’s airlines and private aircraft, giving Inmarsat a reputation of reliability and longevity. With the addition of Swift 64, in 2002, the first high-speed data service was added to the portfolio and has since proven to be an extremely popular solution for business and military users who need ISDN data connectivity.

Since its introduction, SwiftBroadband (SB) aero terminal has enhanced cockpit and cabin applications through one installation utilizing single AES satellite antennas. Inmarsat has established SB as the fastest growing aeronautical service ever introduced. This terminal provides global voice and high-speed data (HSD) service, simultaneously through a single SB installation onboard aircraft via Inmarsat-4 satellite spot beam shown in Fig. 7.3.

Here will be introduced the Cobham Aviator 700D system engineered to take advantage of the latest technologies of SwiftBroadband system for increased reliability and connectivity. The SB background data and standard voice channel allow users to access a shared data channel and a compressed high-quality voice channel at the same time. The Aviator 700D is an Inmarsat Class 6 system that supports circuit-switched standard voice, background IP up to 432 KB/s, Streaming IP at 8/16/32/64/128 Kb/s, SB X-Stream services, and ISDN service or 3.1 kHz audio.

The Aviator 700D brings onboard aircraft the following service: high-speed Internet with data rate up to 432 Kb/s, FANS 1A ready (compliancy in certification), and VoIP connectivity,

- Low-cost voice, built-in PBX supporting numerous handsets, ISDN connectivity, secure voice and data transmission, and Wi-fi capable.

This upgrade from AERO HSD+To Aviator 700 level D will allow in cockpit and cabin to get SB with cabin Wi-fi capability by installing a satellite communication system that is Level D qualified as a regulatory requirement for FANS 1A equipped aircraft. Installing communications equipment in a jet plane often means giving up other options. Not so with the Aviator 700D, which is no more than 8 MCU altogether. It is perfect to replace older and heavier onboard systems, freeing up space for additional payload or fuel savings.

The Aviator 700D installation also includes the following features: Interface with existing CNX-200 Router (pre-requisite for Dassault SB), 1 (or more) dedicated handsets wired directly to the SBU to use low-cost voice brought by SB (optional), SAT-AFIS functions capable and is DLNA Type F flutter installation by Dassault Aviation Design Office.

Configuration of SwiftBroadband AES

The modern transceiver of Cobham Aviator 700D FB configuration utilizes an HGA antenna, which package with peripherals, and LAN is shown in Fig. 7.27, where Above Cockpit Unit (ACU); Below Cockpit Unit (BCU) as transceiver with peripherals; Satellite Aerodrome Flight Information Service (SATAFIS); and Central Personnel Data File (CPDF). This terminal has the following features:

- Cockpit voice dialing from multifunction control and display unit (MCDU) is optional as a part of the flight management system (FMS) and.
- Cockpit data channel with data rates up to 432 Kb/s, low-cost and high-quality voice service, ISDN and VoIP as a connectivity Cabin Telecommunications Unit (CTU).

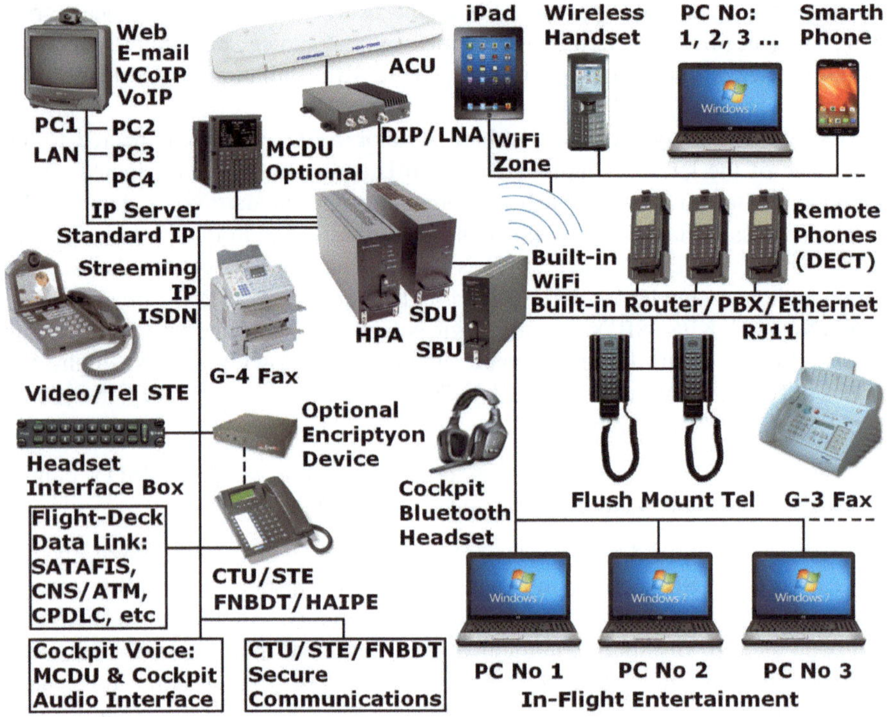

Fig. 7.27 Aeronautical SwiftBroadband peripherals package—Courtesy of manual: by Ilcev

- Interface to ACARS, Aerodrome Flight Information Service (AFIS), and communications management unit (CMU).
- Wired and wireless Web browsing, corporate VPN access and wireless use of personal digital assistant (PDA) or iPAD, and smartphones.
- Built-in router for intelligent connectivity support and multiple user support, 2 built-in PBX for supporting multiple handsets, built-in Ethernet switch for supporting multiple wired laptops, and built-in wireless LAN for supporting multiple wireless devices.
- Controller–pilot data link communications (CPDLC), Secure Telephone Unit (STU), Secure Terminal Equipment (STE), Future Narrowband Digital Terminal (FNBDT) for secure voice communication, and High Assurance Internet Protocol Encryptor (HAIPE).
- Blackberry on board (BOB) can be used to enable Wi-fi BlackBerry devices to stay well connected above 10,000 ft with aircraft satellite phone system. The BOB unit uses a simple carry-on laptop with Wi-fi local area coverage in the aircraft.

The Aviator 700ED SB system comprises the following main hardware:

1. **Satellite Data Unit (SDU)**—The SDU is the controlling unit of the satellite system, which is shown in Fig. 7.28 (Left). A configuration module (CM) and all the interfaces, except for the maintenance port, are located on the rear of this unit. This unit is seamless integration of Aero-H+ and Swift64 services into one terminal. As a compact 3 microcontroller unit (MCU) size, it has low weight and power consumption and is not requiring forced cooling. This terminal has built-in CTU that connects up to six (standard or wireless) handsets including two 2-wire (RJ-11/POTS) interfaces and no converter box required. It provides features such as intercom, conference calling, and call forwarding and also included ARINC 429 for IRS, built-in test equipment (BITE), attitude and heading reference system (AHRS), ACARS, MCDU, CMU, AFIS, CPDF, RS-232C port for portable data loader/configuration management terminal (PDL/CMT), detachable configuration module (CM), and Ethernet, ISDN, and RS422 connectivity. The dimension of this terminal is ARINC 404A 3/8 ATR short, mass is 7.2 lbs (3.5 kg), and consumption power is 28 V DC, 30 W typical, and 50 W maximum. The optimal environmental temperature is −25° to +55 °C and at altitude mean sea level (MSL) to 55,000 ft.
2. **High-Power Amplifier (HPA)**—The HPA device connects to the satellite antenna and amplifies the transmission signals of all channels simultaneously, which is shown in Fig. 7.28 (Middle). It has a very small size (3 MCU), low weight, and low power consumption; no external forced cooling is required and installation outside pressure area. The mass of this unit is 11.2 lbs (5.1 kg); the environmental temperature is −55° to +70 °C and can be used at altitude MSL to 55,000 ft. The HPA-7450 unit is a 2 MCU linear RF power amplifier, which amplifies the transmit RF signal generated by the SDU to the required power level to communicate with the satellite. Secondary functions of the HPA include providing regulated +28 VDC power to the antenna via the SDU and monitoring the diplexer and low noise amplifier/built-in test equipment (DLNA/BITE). It can support multichannel SB operation with HGA and IGA antennas. This unit meets Inmarsat (3rd Order) intermodulation specification of -38.7 dBc. Status LED displays provide visual output of HPA and DLNA status. This unit is qualified for installation in no pressurized and non-temperature-controlled environments. The unit interfaces ARINC 600 Connector, while for RF input and RF output is Threaded Neill–Concelman (TNC) Female.

The HPA unit provides Inmarsat type approvals of Class 6 B6CH01 and Class 7 B7CH01. It is important to say that airborne SB antenna systems have not built-in HPA and LNA inside radome, while shipborne FB antennas have them integrated within radome.

3. **SwiftBroadband Unit (SBU)**—The SBU device provides access to the SwiftBroadband services, which is shown in Fig. 7.28 (Right). A configuration module (CM) and all the interfaces, except for the maintenance port, are located on the rear of this unit. The features of this unit are IP packet or steaming SB

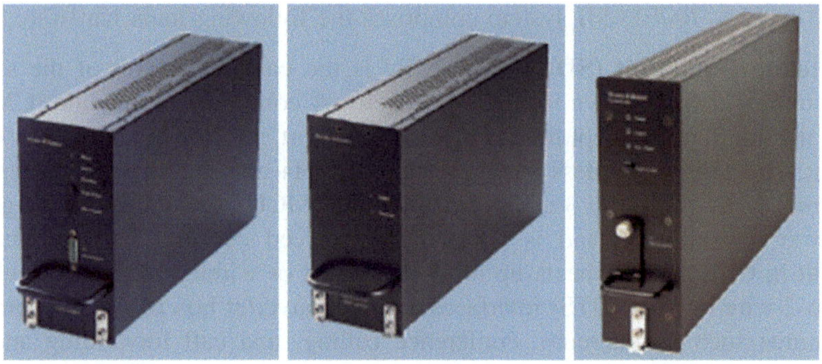

Fig. 7.28 SwiftBroadband hardware package—Courtesy of manual: by Cobham

data, Ethernet, and ISDN connectivity, built-in router with six Ethernet interfaces, and built-in wireless access point. It has low weight and power consumption, compact 2 MCU size, and no forced cooling required and can provide an easy integrations. The mass of this unit is 0.32 lbs (146 g); its environmental temperature is −25 to +55 °C and can be used at altitude MSL to 55,000 ft.

4. **Diplexer and Low-Noise Amplifier (DLNA) 7070**—The DAU-7070 unit is a Type F, ARINC-781, Inmarsat SB Diplexer, and DLNA, as shown in Fig. 7.29a. The DAU-7070 DLNA separates transmit and receive signals and amplifies the received signal. It ensures that specified signals in the transmit band from the HPA are transmitted to the antenna with minimal power leaks into the receive band LNA. However, this SB unit also suppresses unwanted harmonics generated in the HPA, providing protection against interference with other systems. It is ARINC-781 compliant and is compatible with Aero-H/H+, Swift64, and SwiftBroadband systems. It interfaces RF by the TNC Female, Tx by the N-Type Female, and Rx by the TNC Female.
4. **Satellite Data Unit (SDU) 7310**—The **SDU-7310** unit is **flange mount** packaged to do away with expensive and weighty ARINC 600 trays and connectors, as well as increase installation configuration options, which is shown in Figs. 7.29b and 1.31.

The unit simultaneously supports a telephone channel and data connections which speeds up to 200 Kb/s, offering affordable connectivity on any aircraft. The SDU-7310 includes a router, which enables simultaneous connectivity to multiple data users. This unit is flange mounted to reduce installation cost as well as increasing flexibility, and it is suitable for installation outside pressure vessel. This unit has built-in GPS Rx, internal calls between handsets, PPPoE Internet access, and configurable Power over Ethernet on all ports. It interfaces 2× Ethernet 10BaseT User Ports, 2× POTS by two-wire phone interfaces (two handsets per interface), and RS232 for remote monitoring and control.

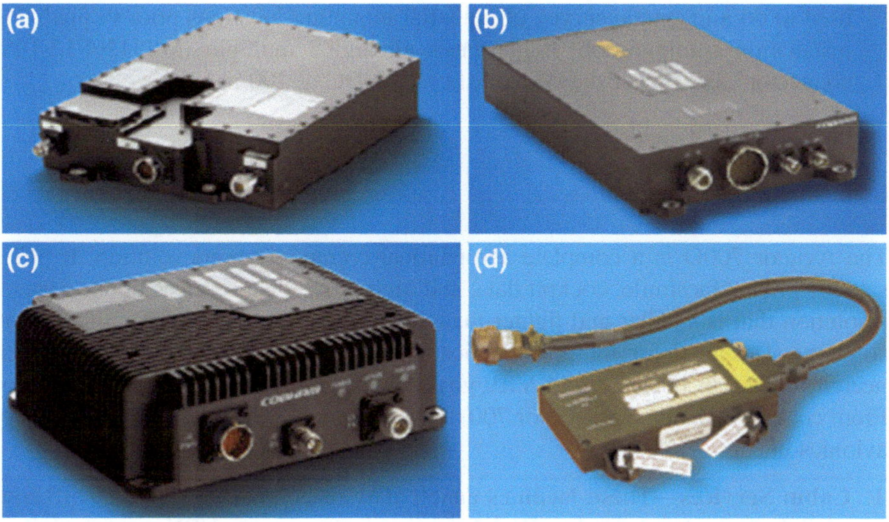

Fig. 7.29 SwiftBroadband hardware package—Courtesy of manual: by Cobham

5. **Combined HPA and DNLA 7260**—This unit is a combined high-power amplifier and diplexer–low noise amplifier (HPA/DLNA) packaged into a single, compact component, as shown in Fig. 7.29c. It forms part of a SwiftBroadband single channel system and reduces overall boxcount, system weight, and complexity. The system supports applications such as VPN, e-mail, phone calls, and Internet browsing. The SwiftBroadband transceiver is a packet-switched network that offers both streaming class (pay-perminute) services and background class (payper-Mbyte) services. A single box that performs 2 functions and reduces system size, used with a single channel SDU, can be used with a variety of HGA, IGA, and LGA (classes 6, 7, and 15). This unit can be mounted in non-pressurized and non-temperature-controlled environment. It uses full Inmarsat Tx and Rx band filtering for use with SwiftBroadband and interfaces antenna port by the TNC Female, Rx port by the TNC Female, and Tx port by the **N-Type Male**. It provides Inmarsat approval type Class 6 single channel B6CH01 and Class 7 single channel B7CH01. It can be used as optional solution together with SDU-7260, instead of using main solution combined with SDU-7330, HPA-7450, and DAU-7070.
6. **Magnetic Sensor Unit (MSU) 7250**—The MSU-7250 device is a stand-alone three-axis magnetometer and three-axis accelerometer used to provide magnetic heading information to a Satcom Reference Unit (SRU), (3-axis Magnetometer) shown in Fig. 7.29d. This unit improves the magnetic accuracy of an SRU device if the SRU cannot be mounted in a magnetically clean environment; however, it can be mounted in non-pressurized and non-temperature-controlled environment. The MSU is a sealed unit allowing mounting in extreme environments such as the aircraft's wingtip. However, this specific unit can also be

used to do a magnetic survey of potential mounting locations prior to finalizing installation configuration. It interfaces all digital multipin (MS3120F12-10S) connectors and power data diagnostics 28 VDC at 100 mA.

SwiftBroadband Service

The Aviator 700D is a complete aeronautical communications solution. It offers satellite phone, facsimile, cockpit data, and high-speed data services in one system, a solution that is smaller and lighter than any comparable satellite communication solutions. Designed for multiple application flexibility, the Aviator 700 is engineered for use in business and VIP, government, and military aircraft platforms. From cockpit to cabin, the Aviator 700 delivers the features and functionality every avionics need:

1. **Cabin Services**—These facilities enable data rates up to 432 Kb/s, VoIP connectivity, wired and wireless Web browsing and corporate VPN access, wireless use of PDA and smartphones for e-mail services, SwiftBroadband channel for simultaneous data and voice, built-in router for intelligent connectivity support and multiple user support, built-in Ethernet switch for supporting multiple wired laptops, built-in wireless access point option for supporting numerous wireless laptops and/or smartphones, built-in PBX for supporting multiple handsets and ISDN connectivity and, etc.
2. **Cockpit Services**—This service is providing MCDU, cockpit audio interface and data channel, AFIS/ACARS solutions, flight deck data link, automatic dependent surveillance (ADS) for safety services, CPDLC, voice communications, electronic flight bag (EFB), full flight plan, weather and chart updates, engine performance monitoring and fault reporting for major systems, general operational planning, crew reporting and general administration, etc.
3. **Passenger Services**—these features help passengers to get telephone connections just in seat, mobile, VoIP and text messaging, e-mail, intranet, Internet and instant messaging, secure VPN access, large file transfer, presentations, graphics, video, videoconferencing, in-flight news updates, etc.

In addition will be useful to introduce latest SwiftBroadband aero standard know as Cobham SB800, which delivers up to 864 Kb/s via dual channels and high gain antenna. The SB800 system includes an HGA, SDU, high-performance amplifier, and diplexer/LNA. It supports applications such as VPN, e-mail, telephone, and Internet browsing, as well as streaming data applications like video. Cobham antennas and components are lighter and smaller than most competitive SB solutions, saving your fuel while maximizing your stowage space and payload capacity.

Configuration of SwiftBroadband Peripherals

The Aviator 700 or any SwiftBroadband transceiver can provide to the cockpit pilot the following peripherals:

1. **2.4 GHz Cordless Handset**—The 2.4 GHz cordless handset system is used to receive and to make calls, as shown in Fig. 7.30a. In an instruction manual of Aviator 700, every pilot can see description how to use and to make a call with the 2.4 GHz cordless handset.
2. **Aviator Wireless Handset and Cradle**—The Aviator Wireless Handset can be used for making phone calls over an IP-based network, as shown in Fig. 7.30b. The handset connects to a wireless access point using wireless local area network (WLAN). The pilot operator can see in the user manual for a description of the keys and LED (display) of this handset and how to make calls.
3. **Sigma7 Handset and Cradle**—The Sigma7 handset is used to receive and to make calls by instruction written in description of user manual, as shown in Fig. 7.30c.
4. **Full Feature Handset and Cradle**—The full feature handset is used to make and receive calls and to control view status information and configuration data of parts of the Aviator 700 system, including the SDU, as shown in Fig. 7.30d. The cockpit operator can see operating instructions of the full feature handset for a description of keys and LED display of the full feature handset and how to make calls from this handset.

Fig. 7.30 SwiftBroadband peripheral package—Courtesy of manual: by Cobham

5. **Auxiliary Handset and Cradle**—The Auxiliary handset is used to receive and to make calls, as shown in Fig. 7.30e. It can be managed by using operating description of the auxiliary handset including a description of the keys and LED displays of the auxiliary handset and how to make calls.
6. **Multifunction Control and Display Unit (MCDU)**—An MCDU can be used optionally to make and receive calls, as shown in Fig. 7.30f. It also can access the Aviator 700 menu system for control and configuration of the satellite communication system from the cockpit. A subset of the MCDU menus and functions are identical to those of the full feature handsets. With the cockpit voice interface, which connects to the pilot headset via the cockpit audio management system, the pilot can make calls from the cockpit without using a handset. Operator can read instruction how to use the MCDU device.
7. **Integrated SwiftBroadband and Scotty Video Encoder**—This is ARINC 600-based video communications platform known as "Scotty Computer System (SCS)" from Scotty. This video encoder provides maritime, aeronautical, and other mobile service specialized in beyond line of site audio, video, ISDN mobile solutions, and data for communications and surveillance purposes.

7.2.5 Aeronautical AES Terminal Equipment and Installation

The AES interfaces onboard communication and antenna systems, such as duplex Tel, fax, data, and video equipment for aircraft, which standards meet the requirements of the ICAO and industry standards such as ARINC characteristics 741 as well as Inmarsat standards. In fact, ARINC 741 standard describes one physical implementation of the Inmarsat system, which characteristics 741 for AES comprises Above Cockpit Units (ACU) or antenna dish and Below Cockpit Units (BCU). The AES terminals are installed onboard aircraft as radio communication and antenna sets capable of communicating via satellite with ground Earth stations (GES) in the Inmarsat or other satellite operators via their satellite networks for accessing terrestrial telecommunication networks (TTN).

Same as maritime FleetBroadband system, aeronautical SwiftBroadband is also robust and important GMSC system that will provide reliable communications across the globe in all weather conditions. However, in order to get the best performance out of the system, it is essential that the equipment, both ACU and BCU, is correctly installed. In the case of AES terminal equipment installation onboard aircraft, there are not any obstacles atop fuselage or in tail to provide any sort of shadowing. However, it will be very important during installation of airborne antenna to be considered safe distance due to huge disturbances and outages from radar antennas.

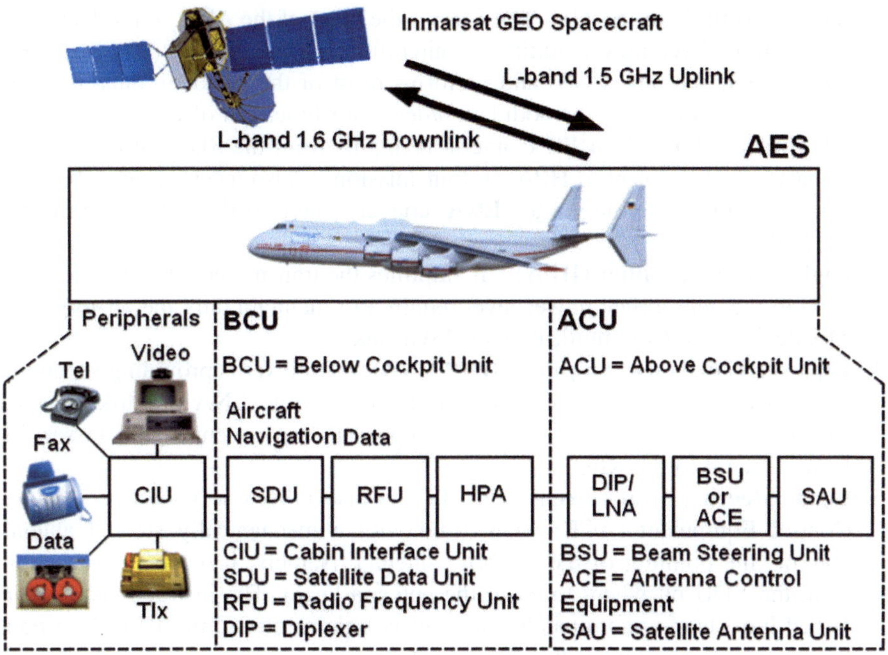

Fig. 7.31 Inmarsat AES, ACU, and BCU configurations—Courtesy of manual: by Ilcev

All stated aspects about shipborne antenna mounting system will be the same or similar for aircraft antenna mounting system, such as two-axis mount system (E/A and Y/X); three-axis mount system (E/A/X, E′/E/A, and X′/Y/X); four-axis mount (E/A/Y/X); shipborne antenna tracking and pointing system; satellite determination and antenna azimuth limit; and antenna pointing and tracking.

SwiftBroadband and other airborne broadband systems that provide simultaneous voice, high-speed data (HSD), and video communication via satellite through the mobile BGAN (Broadband Global Area Network) are very robust AES terminals. Thus, mounting process of this AES type requires very serious preinstallation planning. The AES equipment and configuration installed on the vessel is the responsibility of the customer, but it is expected that guidance will be provided by the Inmarsat distribution partner or service provider.

Usually, the aircraft antenna (ACU) can be mounted either on top or sides of the aircraft fuselage and inside the pilot cabin (cockpit). For instance, in some cases of wrong antenna installation can arise very serious problems caused by the blind spots phenomenon. In this section will be shortly introduced the occurrence of blind spots. However, the electronic elements of transceiver unit (BCU) with peripherals can be fixed inside the pilot cabin (cockpit), whose airborne configuration is shown in Fig. 7.31. The main elements of BDE are the following units:

1. **Satellite Data Unit (SDU)**—The SDU is the heart of the AES. It interfaces with other onboard avionics including the aircraft navigation system (GPS, GNSS or GLONASS, and new CNS) and performs most of the protocol, data-handling, modulation/coding, and demodulation/decoding functions of the AES.
2. **RF Unit (RFU)**—The RFU converts IF inputs from the SDU into L-band RF signals, which are sent to HPA for transmission. It also receives the L-band RF signal from the satellite via the LNA, converts them to IF, and passes them to the SDU.
3. **High-Power Amplifier (HPA)**—It amplifies the transmitted RF signal from the RFU to the appropriate power level required to maintain only the air-to-ground Mobile Satellite Communications (MSC) link.
4. **Diplexer/Low-Noise Amplifier (DIP/LNA)** —Diplexer is providing separation of Rx (receiver) and Tx (transmitter) signals, while LNA amplifies the RF signals received by the antenna to compensate for system signal losses and forward them to the RFU.
5. **Beam Steering Unit (BSU)**—The BSU is also otherwise known as Antenna Control Equipment (ACE) when used with a mechanically steered antenna controls the pointing of the airborne antenna. Namely, it receives instructions from the SDU on where to point the antenna beam. The instructions are converted into steering commands, either electronic or electromechanical, to point the antenna beam toward the desired satellite.
6. **Satellite Antenna Unit (SAU)**—The antenna as a part of ACU is the important component, which is mounted to the exterior of the aircraft fuselage enabling the system to transmit and receive RF signals to/from the satellite. In effect, there are the following three kinds of antennas specified by gain performance for use in the Inmarsat Aeronautical Mobile Satellite Communications (AMSC) systems: low-gain antenna (LGA), intermediate-gain antenna (IGA), and high-gain antenna (HGA).

7.2.5.1 Satellite Determination and Mounting System

For installation satellite antenna on an aeroplane follow the following general guidelines: always install antenna on top of fuselage as far away from obstructions (tail section) as possible; install HPA/LNA pack as close to the antenna as possible and AES transceiver in or near the cockpit (nose); and cross section of antenna has to meet maximum 20°, as shown in Fig. 7.32 (Left). Regardless of the location chosen for the antenna, it should be oriented to point forward in parallel with the aircraft's longitudinal axis when in the middle of its azimuth range, which will correspond to zero degrees on the azimuth indicator. The EM RF signals are known to be hazardous to health at high radiation levels, so it is recommended that technicians in airport have to be on safety distance from radiating satellite antenna.

Antenna mount atop fuselage can be top mount HGA or port and starboard side mount HGA system, as depicted in Fig. 7.32 (Middle).

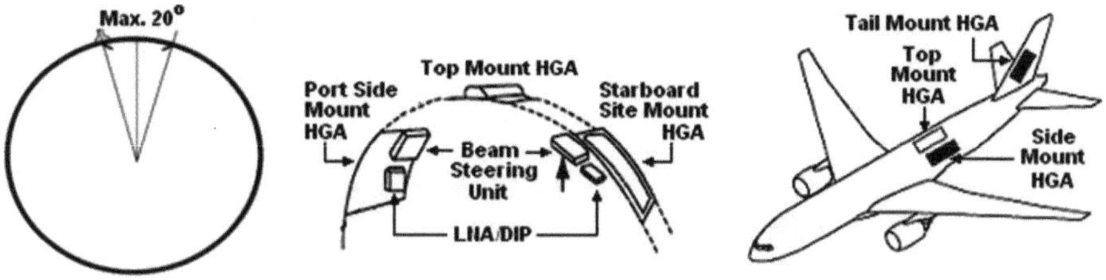

Fig. 7.32 Antenna cross section and mount solutions—Courtesy of manual: by Ilcev

Similar to shipborne antenna, when airborne HGA cannot obtain full LOS to the satellites because of same blockage of the satellite signals or more important due to low elevation angles as well, the new SF Dual Antenna solution can eliminate appearance of blind spots. In such a way, it maintains full LOS to the satellites with a combination of two Cobham Aviator HGA-7001 antenna solutions installed in parallel at different port and starboard side locations together with Dual Antenna Control Equipment (DACE), as is shown in Fig. 7.32 (Middle). In Fig. 7.32 (Right) are presented three main mounting of airborne HGA, such as top, side, and tail antenna mount.

7.2.5.2 Airborne Antenna Mounting and Motion System

The antenna gain is the fundamental parameter controlling antenna operation. The HGA CMA-2102 aircraft antenna system of the Canadian Marconi Company provides very good gain coverage, exceeding 12 dBi in 90 % and 9.5 dBi over 100 % of the Inmarsat system hemisphere, which extends from 5° upward over the horizon in all level of flight. Thus, this antenna provides very high-gain values of 14–17 dBi within 50 % of the hemisphere and has profound implications for the economics of the end user.

With superior receive gain, less satellite power is required for nominal service and more satellite channels can therefore be supported. High transmit gain improves multichannel operation capabilities by ensuring robust operation and the continuity of six-channel operation over a maximum coverage zone. The HGA CMA-2102 also exhibits good coverage below the Inmarsat hemisphere. Measured gain exceeds 9 dBi when looking down to −3° in the roll plane, while trials conducted in cooperation with Inmarsat verified that HGA has 10 dBi gains at 0° elevation and can maintain a voice call to at least −2°.

AES Antenna Mounting

As stated above, the airborne antenna assembly can be mounted on top of the aircraft or on the sides of the fuselage. Ideally, the best position for placing the

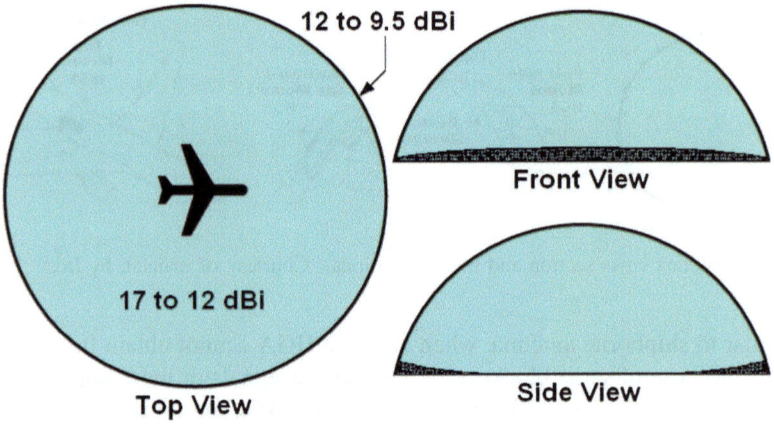

Fig. 7.33 Top-mounted airborne antenna—Courtesy of manual: by Canadian Marconi

antenna is a location on the centerline of the aircraft coordinate system, as is depicted in Fig. 7.33. In this case, the antenna assembly location should be moved forward or aft along the centerline from the ideal position to maintain proper separation from other antenna systems. All L-band antennas must be separated by at least 50.8 cm, with 1.5 m preferred. The airborne satellite antenna should be located at a distance of at least 5 m from GPS antenna and the magnetic steering compass.

At this point, it is not recommended to locate the antenna close to any interference sources or in such a position that sources such as the radar antenna lie within the antenna's beam width of 10° when it points at the satellite. The antenna should also be separated as far as possible from the HF antenna and preferably by at least 5 m from the antennas of other communications or navigation equipment, such as the antenna of the satellite navigator or the VHF and other related radio antennas. The top mount provides satellite coverage independently of the direction of aircraft travel.

Namely, the top view is 100 % clear, while front and side views have a very small shadowing on the lower levels of the hemisphere. This heading-independent antenna gain coverage is a very significant advantage, which may, in the future, become an important condition for all type of aircraft before being permitted to use the optimal flight paths made possible by satellite-based air traffic management (ATM).

On the other hand, typical side-mounted antennas have an inherent coverage deficiency due to their limited scan range. Thus, nose and tail bearings and shallow elevation angles to the satellite can reduce HGA gain values below 12 dBi, sometimes all the way to zero.

Thus, these triangular areas, called blind spots or "keyholes", occur in an aircraft's fore and aft directions, extending horizontally about 45° to each side at an elevation angle to the GEO satellite of 5° and rising to a point about 45° above the

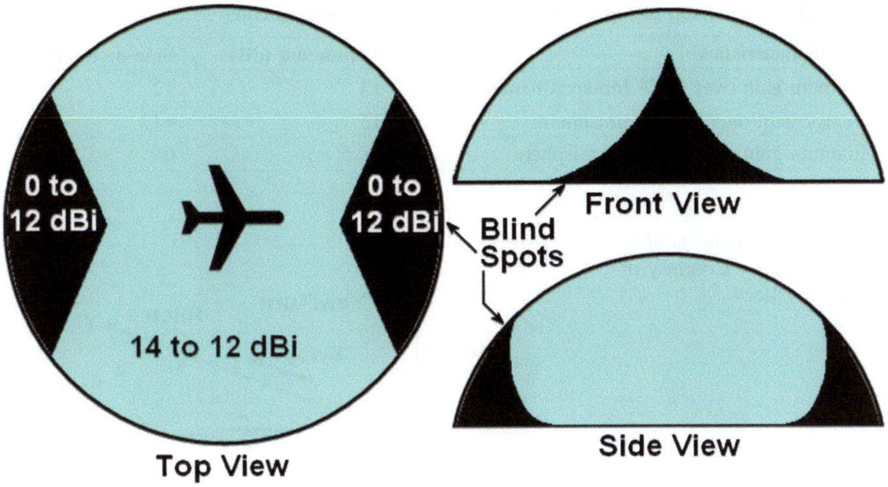

Fig. 7.34 Side-mounted airborne antenna—Courtesy of manual: by Canadian Marconi

nose and tail, as shown in Fig. 7.34. Recent operational experience by a number of airlines has confirmed that the blind spots cause problems with satellite signal reception, especially in certain areas of the world, such as the Pacific Rim.

Actually, these blind spots of the signal reception are the frequent cause of problems, including intermittent blackouts, dropped calls, and sometimes zero communications, even when the aircraft is well within the Inmarsat satellite coverage zone, particularly at high latitudes with low elevation angles. In any case, these operational problems have been important factors in the switch from side-mounted to top-mounted antenna system. Thus, side-mounted installations also require two feeds from the HPA, as well as a high-power relay, adding to system losses.

The top-mounted antenna can typically be installed with up to 1 dB less loss than permitted in ARINC-741, resulting in better communications and greater channel availability. Hence, Table 7.1 compares the top-mounted with a typical side-mounted antenna design.

Therefore, as shown in Fig. 7.34, the top view of a side-mounted airborne antenna is reduced in the nose and tail with shadowing of about 12 % in total, while both front and side views have shadowing spots of about 15 % each.

Components of Aircraft Motion

The GMSC antenna system is generally mounted on a stabilized platform, which has two horizontally stabilized axes (X and Y), achieved by using a gyrostabilizer or sensors, such as accelerometers or gyrocompasses. The stabilized platform provides a horizontal plane independently of mobile motion, such as roll or pitch.

Table 7.1 Comparison of top-mount and side-mount airborne antennas

Gain characteristics	Top-mount (dB)	Side-mount (dB)
Minimum gain over 75 % Inmarsat hemisphere	13	12
Average gain over 75 % hemisphere	15	<13
Minimum gain over 100 % hemisphere	>9.5	0

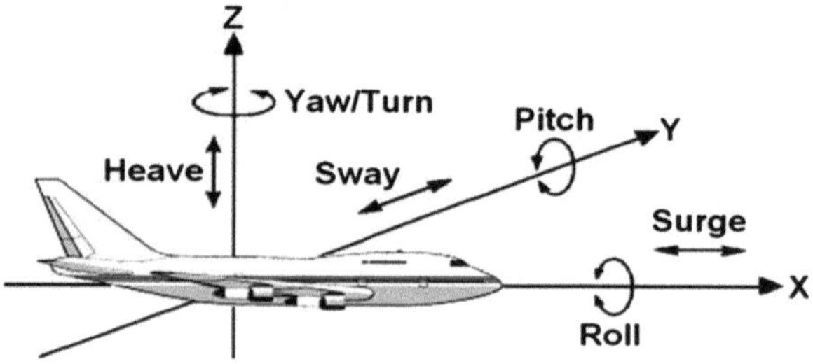

Fig. 7.35 Components of aircraft motion—Courtesy of manual: by Ilcev

For example, all mobiles have some kind of motion components during traveling, such as roll, pitch, yaw, surge, sway, heave, and turn, as shown in Fig. 7.35. Turn means a change in the aircraft's heading, which is intentional motion, not caused by wave motion, and the other six components are caused by wave motion. Surge, sway, and heave are also caused by acceleration.

7.2.5.3 Performance at High Latitudes and Low Elevation Angles

The MSC performance can be limited at high latitudes and low elevation angles, regardless of which HGA is used. For a GEO satellite orbiting at an altitude of about 35.786 km, the highest latitude that can be viewed is 81° 18′, corresponding to an elevation angle of 0°. For aircraft flying above this latitude, the satellite is below the horizon and unavailable for communications.

Hence, when the aircraft's longitude differs from that of the satellite, the maximum latitude is even less; at longitudes that are midway between two satellites, it can be as low as 71°. Even before the 0° elevation angle is reached, MSC signals suffer from several sources of increased loss at low elevation angles. These factors combine to reduce the number of available voice channels at 5° elevation compared to 20° elevation angle. High gain below the horizon can actually degrade an HGA's low-elevation performance because it increases the interference from signals reflected from the ground. Thus, side-mounted antennas can meet these limits only over uneven ground or rough seas, where multipath signals are diffused. At low elevation angles, the multipath rejection of HGA is further enhanced by its top-mounted design, since the antenna's predominantly vertical polarization ellipse helps to cancel out the effects of multipath reflections, mostly horizontally polarized.

The major source of antenna noise results from HGA "looking" at portions of the terrain, which are "hotter" than the sky. In any event, the top-mounted antennas controlled gain roll-off below the horizon, at all azimuth angles, giving much lower noise characteristics than side-mounted antennas, especially at low satellite elevation angles. In reality, the noise of this antenna is almost independent of elevation angle, whereas patterns of side-mounted antenna vary widely around the horizon.

At low elevation angles, it is critical that MSA can seamlessly transfer to the next satellite as soon as it has a higher elevation angle. The algorithm normally used to determine when handover should occur assumes that the satellite with the highest elevation angle is the optimal one but this is not necessarily true of side-mounted antennas, since the higher satellite may be in a blind spot fore or aft of the aircraft. This can cause delays in satellite selection and could lead to a temporary communications blackout. The HGA top-mounted unit has no blind spots and performs better than side-mounted during satellite handover.

7.2.5.4 Beam Steering Performances

Steering of the HGA is controlled by the Beam Steering Unit (BSU), which transmits steering commands to the HGA over an asynchronous EIA-RS-422A serial data link. The HGA interprets the steering command and steers the antenna beam to the desired position. Thus, the BSU translates antenna beam position data and beam change commands received from the SDU terminal in a standard digital format into signals needed to select phasing of the antenna elements that result in the antenna beam pointing at the desired satellite. The BSU is housed in an ARINC-600 rack-mounted 2-MCU package.

During subsystem operation, open-loop steering words and control words are received from the SDU. The steering words contain azimuth and elevation information for the direction in which the beam will be steered, using the coordinate system, which is shown in Fig. 7.36. In the proper manner, after receiving an open-loop steering word, the BSU calculates the beam number (one of more than 1000) and associated phase-shifter settings, which are then up linked to the HGA at a maximum rate of 20 Hz, via an RS-422A serial data links. However, should the rate of transmission of the open-loop steering words exceeds 20 Hz, the BSU buffers the incoming words unit it has uplinked and strobes in the phase-shifter settings for current azimuth/elevation value. If multiple steering words have been received during the processing time, only the most recent value is retained. Control words will be processed as they are received. Another function of the BSU includes reporting of the selected antenna beam and gain to the SDU.

Moreover, BSU's operational software meets the requirements of Level 2 Flight Essential category. In addition to operating software, the BSU terminal contains diagnostic software to assist ground maintenance personnel in fault isolation. The diagnostic software is capable of identifying faults to a more detailed level than required by ARINC 741.

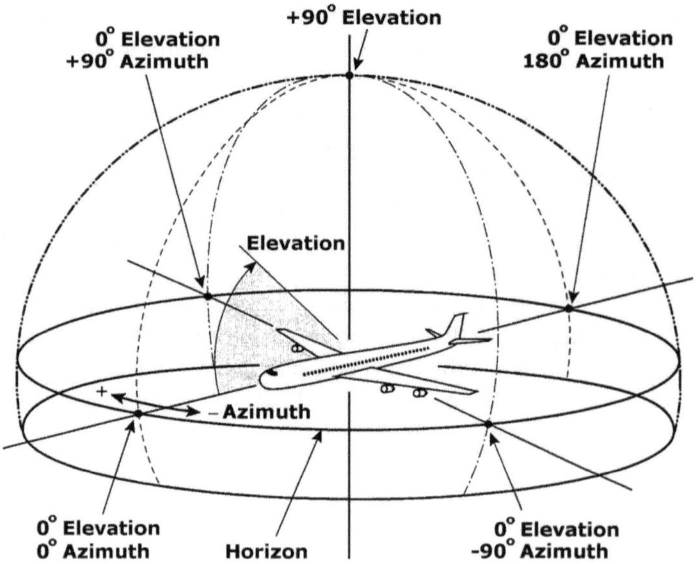

Fig. 7.36 HGA coordinate system—Courtesy of manual: by Canadian Marconi

7.2.6 Inmarsat Global Xpress (GX) Solutions

The concept of Global Xpress (GX) achieved in December 2015 as a new generation of Inmarsat-5 mobile broadband at Ka-band gives as follows: worldwide coverage via three satellites; much higher QoS and flexible bandwidth purchase models; major leaps forward as global system capacity is 20× Inmarsat-4 generation with throughput up to 50 Mb/s at forward link and 5 Mb/s in return link to a 60 cm antenna dish; independent from L-band constellation, but integrated as a service L-band to remain a major component of Inmarsat services; deployed for lower bandwidth and/or highly portable MSC (BGAN, GSPS, Inmarsat-C, SLDR, M2M, safety services); overlays Ka-band network to ensure high availability, continuity, and safety services for SwiftBroadband and FleetBroadband; and near-term launch of commercial service introduction Q4 2013 and global fleet Q4 2014.

The characteristics of three Inmarsat-5 global payloads are to provide 89 fixed beams each, which spot beam coverage is depicted in Fig. 7.37. The coverage is reconfigurable in orbit 2 × 72 × 40 MHz channels per satellite at >5 GHz aggregate bandwidth per satellite. Global traffic landed to three Inmarsat LES (Gateways), with site diversity high capacity payload (HCP), provides six steerable satellite spot beams. Orbital locations TBD based on frequency coordination is deploying the following RF bands: User service link uses Uplink at 29.0–30.0 GHz and Downlink at 19.2–20.2 GHz, while Feeder satellite link uses Uplink at 27.5–29.5 GHz and Downlink at 17.7–19.7 GHz.

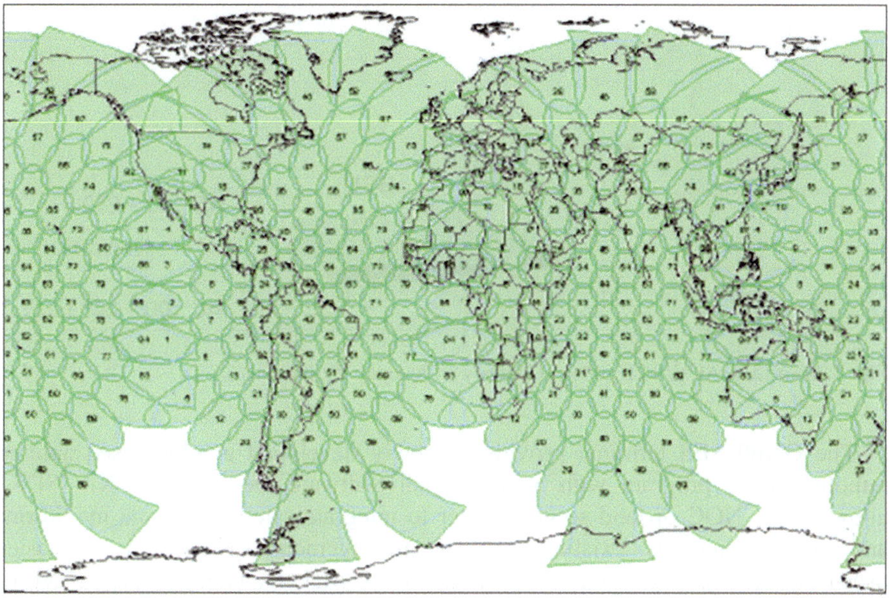

Fig. 7.37 Inmarsat-5 spot beam coverage—Courtesy of manual: by Inmarsat

7.2.6.1 Maritime XpressLink (XLink) System

Maritime XpressLink (XLink) system that connecting ship navigation bridges is the first project of GlobalXpress (GX) developed by Inmarsat; however, in the meantime, Inmarsat was designed GX for land and aeronautical applications. XpressL offers the maritime industry a compelling combination of high-speed broadband for a fixed monthly fee. It integrates the Inmarsat FleetBroadband with VSAT service to provide maritime users with increased performance and flexibility of unlimited data.

In Fig. 7.38 is shown network diagram of Intellian GX60 maritime XLink with antenna (ACU), integral core module (BCU), network and peripheral connection, Rx connection to BCU, Tx connection to ACU, and power supply lines for ACU and BCU from ship power generator. Initially, access of XLink was always available as IP data at up to 768 Kb/s and with a committed information rate (CIR) of 192 Kb/s since 2014 XLink became available for customers to receive double-bandwidth speeds with rates of up to 50 Mb/s.

Intellian GX60 contains a 65-cm Ka-band maritime stabilized antenna (ACU) and core module (BCU), which provides maritime broadband service via Inmarsat-5 satellite constellation. The GX60 HGA, efficient carbon fiber antenna reflector, and frequency-tuned radome ensure reliable and quality service for ocean going ships. In addition, the GX60 antenna supports a low elevation angle (−20°) capability to guarantee reliable connection at extremely high latitudes.

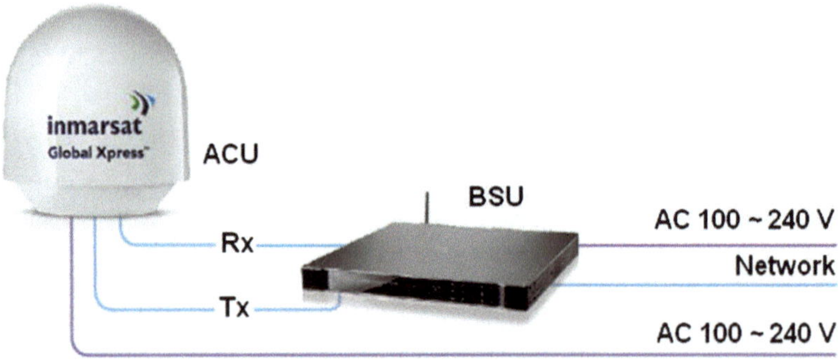

Fig. 7.38 XLink ACU/BCU units—Courtesy of manual: by Intellian

The GX60 SES terminal provides a newly developed graphic-based antenna remote control program with an additional Software Development Kit (SDK), allowing the NOC or land service center to integrate satellite antenna monitoring and control into its existing network management systems in an easier, user-friendly, and convenient manner. The GX60 ACU includes the integral XLink core module and a Wi-fi system for wireless monitoring and control of the system onboard the vessels. Besides, the GX60 comes with a 5 W BUC and a Ka-band LNB configured for circular polarized signals.

Frequency-tuned radome ensures efficient operations for Ka-band Inmarsat XLink and supports auto satellite and beam switching for seamless continuous coverage. The signal loss of the radome itself is minimized and the performance maximized with an optimized radome design that enhances the Ka-band system performance. Intellian's new generation gyro-free satellite search function enables the GX60 to acquire and lock onto the satellite without requiring a separate input from the ship's gyrocompass. The GX60 also supports auto satellite and beam switching for seamless continuous coverage.

7.2.6.2 Aeronautical Jet ConneX (JX) System

Jet ConneX (JX) is Inmarsat's latest Ka-band Global Xpress (GX) satellite communications service, delivering the highest data rates ever to business aviation industry. This airborne GX satellite high-speed broadband Internet enables crew and passengers onboard aircraft to stream videos, make phone calls, and watch live television via new generation of Inmarsat-5 satellite constellation. The JX service started with operations in Spring 2015 providing and supporting both voice and high-speed Internet including data streaming, videoconference, TV, and usage mobile phones via newest generation of Inmarsat-5 (I-5) over the Ka-band.

The Ka-band has more capacity than other L/C and Ku satellite networks and is more efficient in its use of bandwidth, meaning more affordable in-flight connectivity and simple billing options. The JX tail-mount antenna configuration with

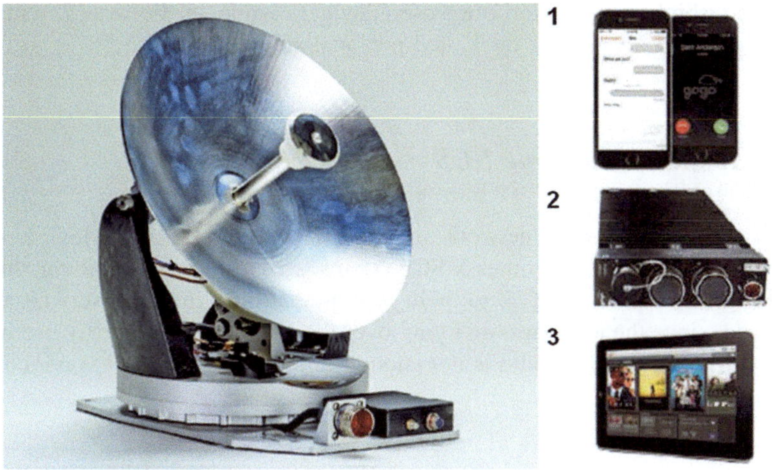

Fig. 7.39 Jet ConneX antenna and peripheral configuration—Courtesy of manual: by Honeywell

transceiver is illustrated in Fig. 7.39 (Left) and in Fig. 7.39 (Right) are illustrated Text and Talk solution (1) for usage personal phones onboard aircraft, Smart Router UCS 500 (2) as a media server and smart cabin system solution, and Cabin Display (3) for transmission movies, news, weather, moving maps, and so on as new in-flight entertainment (IFE) to passengers.

In general, JX service will provide data speeds similar to what is experienced in homes and offshore offices. It expands the possibilities for IFE to real-time TV and TV on demand as well as standard e-mail and Web browsing capabilities. This new standard offers extensive satellite coverage for customers fly across time zones, with continuous, consistent service as traffic is handed seamlessly across each Inmarsat satellite and from one to another.

General features supported by Jet ConneX include connections over the I-5 satellites; only 3 GEO satellites providing global, mobile Ka-band broadband with service speeds are 33 and up to 60 Mb/s, simultaneous high-speed data, and voice and continuous, consistent service across all time zones.

7.3 Iridium LEO Users Segment

The Big LEO Iridium system is a satellite-based network designed to provide truly global personal and mobile service of voice, facsimile, paging, and data solutions for mobile and personal applications, which also include the GPS capability. Iridium L-band service is a truly global satellite network with 66 cross-linked LEO satellites that form a worldwide grid. Thus, from the Arctic to Antarctica, a satellite is always in sight and ready to handle all calls. In this section will be shortly introduced shipborne and airborne user devices, while in Volume 2 will be in details described all mobile, deck, and personal solutions.

7.3.1 *Iridium Maritime SES*

Iridium maritime satellite network includes several types of GMSC terminals similar to the Inmarsat and other MSC systems. Because the oceangoing ships are not sailing in polar regions, so to them is sufficient only Inmarsat service. On the other hand, since the airplanes are flying over the North Pole, they may use Iridium service only, because HF radio is not enough reliable and Inmarsat is not covering beyond the 80° of Latitude.

7.3.1.1 Iridium Maritime SES Sailor SC4000

In Fig. 7.40 (Left) is shown the maritime Sailor Iridium single channel fixed terminal SC4000 designed by Cobham. This unit provides one-channel voice and data capability at a rate 2.4 Kb/s of O-QPSK modulation and at 10 Kb/s for basic Internet, e-mail, and PC-based SMS. The Iridium Sailor ST4120 transceiver features metal protective housing for professional maritime installations. The easy access to the SIM card readers allows crew members straightforward insertion of their own SIM cards. It is easy to install and is the central HUB of the entire Sailor SC4000 Iridium system.

The antenna as ADE has no moving parts and is designed to be highly weather-resistant. The helix omnidirectional construction is coated with glass-enforced epoxy, making the antenna able to withstand the extreme conditions at sea. The BDE subsystem consists of transceiver, Sailor SC4150 intelligent handset, and control unit, which features mobile phone style operation on a large LCD screen with a logical menu for ease of use. It can interfaced Tel handset, Tel/PBX, data RS232, and position information NMEA183 and can store up to 100 contacts with repeat call and call log functionality, and the high-quality loudspeaker ensures all calls are easily heard wherever operator is onboard. The same model has the possibility of multichannel service, providing 4 channels at a rate of 2.4 Kb/s with 4 separate helical omnidirectional antenna and the same interface solutions.

7.3.1.2 Iridium Maritime Pilot SES

Powered by the Iridium OpenPort broadband service, Iridium Pilot offers global voice and IP connectivity at speeds up to 128 Kb/s anywhere at sea through an optimized and durable antenna, whose configuration is shown in Fig. 7.40 (Right). With an always-on Ethernet connection and three voice lines that can be used simultaneously, Iridium Pilot delivers cost-effective and reliable data and voice

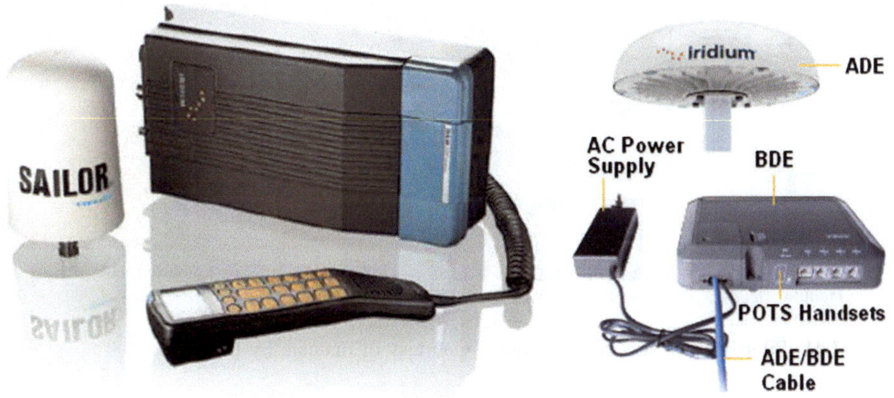

Fig. 7.40 Sailor SC4000 and pilot Iridium maritime transceivers—Courtesy of manuals: by Cobham/Iridium

communications. This maritime SES is ideal for oceangoing merchant ships, fishing, and yachting vessels.

Along with offering crew voice calls, ship captains can simplify and improve their onboard communications using Pilot SES for simultaneous data and three voice lines. Its Ethernet port provides a connection to a standard PC and a router or HUB for multiple PC units, allowing access to wide range of business applications, including FTP, VPN, and weather.

In Fig. 7.41 (Left) is illustrated the following components of Pilot SES configuration:

1. **Above Decks Equipment (ADE)**—This is SES antenna equipment that contains 6 dual element azimuthal arrays and 1 zenith element, electronically switched and phase steered. It provides the communication connection between the BDE and Iridium's satellites. The ADE will automatically locate and track Iridium satellites overhead.
2. **Below Decks Equipment**—This is SES transceiver unit with the appliance to which are connected telephone sets and own PC terminals. It is capable of handling three POTS/RJ11 telephone handsets and one Ethernet data connection.
3. **ADE/BDE Cable**—The cable connects the ADE to the BDE units.
4. **AC Power Supply Unit**—The AC power nodule provides power to the BDE.
5. **POTS Handsets**—Two handsets are provided with pilot equipment, a standard phone for installation on the bridge and one crew phone, designed for crew calling services.
6. **SIM Card**—The assigned SIM card validates the system at start-up, allows connection to the Iridium network, and enables the appropriate phone and data lines.

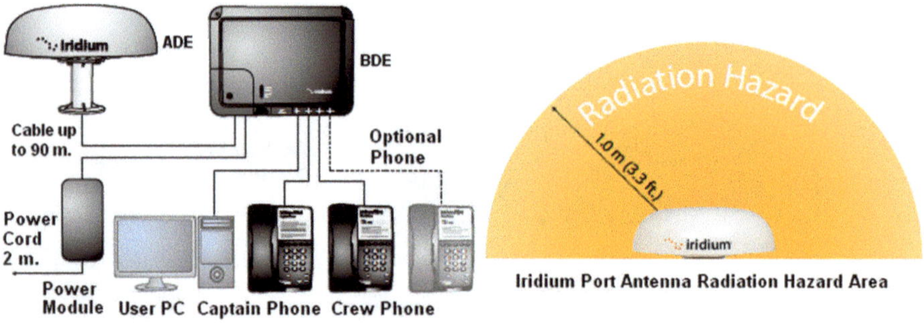

Fig. 7.41 Pilot Iridium SES configuration—Courtesy of manual: by Iridium

In Fig. 7.41 (Right) is shown the port antenna radiation hazard area. To avoid risk to personnel, keep safe distances between personnel and the antenna when the system is ON. Personnel must maintain a minimum separation distance of 1 m from the unit, and installers must place the ADE unit in a manner to maintain this minimum spacing requirement.

7.3.2 Iridium Aeronautical AES

Iridium aeronautical satellite network is very important service that is providing global coverage including both poles, which regions are not covered by Inmarsat. Here will be introduced two Iridium AES, such as ICG Sora and SkyConnect. The newest Iridium Aero OpenPort with other solutions will be introduced in Volume 2 of this book.

7.3.2.1 Iridium-Integrated Aeronautical ICG Sora IG+Terminal

The Iridium Aero solution provides communication and safety service via three-channel system integrated with peripherals and Inmarsat IGA SwiftBroadband antenna. It is packed with features that are important to conducting business on the fly. The ICG NxtLink ICS 220A is three-transceiver Iridium device that combines two channels of global voice and a dedicated cockpit datalink channel via a third short burst data (SBD) channel.

The ICG integrated airborne unit as shown in Fig. 7.42 is designed specifically to provide at flight crew with reliable communications facilities for both aeronautical operational control (AOC) and air traffic service (ATS) messaging. It combines the following units:

1. **ICG NxtLink ICS 220A**—This is a three-transceiver Iridium device that combines two channels of global voice and 2.4 Kb/s a dedicated cockpit datalink channel via a third SBD channel in a single 2MCU (Microcontroller

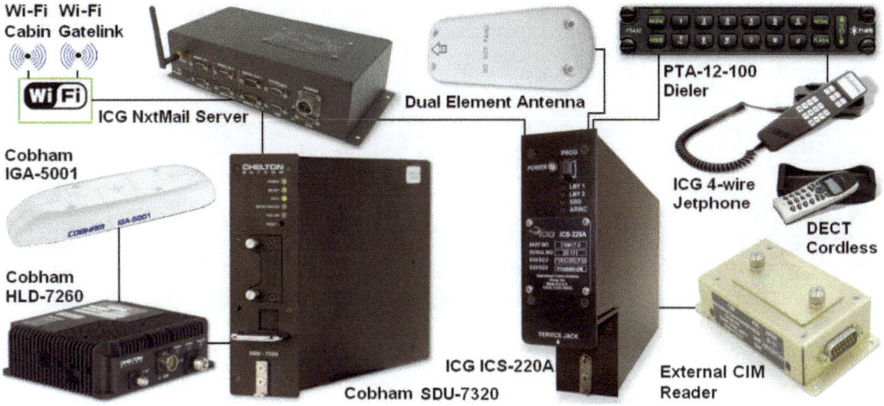

Fig. 7.42 ICG-integrated AES satellite terminal—Courtesy of manual: by ICG

Unit) and Line Replaceable Unit (LRU) to support ACARS, FANS messaging and Controller Pilot Data Link Communications (CPDLC) at the airport ATC cites.

2. **ICG NxtMail Server**—This Iridium unit provides a local wireless access point (WAP) permitting Wi-fi capable devices such as a smartphones and personal computers to access conventional Internet Protocol (IP) services.
3. **Cobham SDU-7320**—This is Inmarsat-2-MCU Satellite Data Unit integrated with ICG for providing a full-featured SwiftBroadband channel with data speeds at 332 Kb/s.
4. **Cobham HLD-7260**—This Inmarsat a highly compact unit that combines high-power amplifier (HPA) and diplexer/low noise amplifier (DLNA), delivering SwiftBroadband where other systems cannot.
5. **Cobham IGA-5001**—This is smallest intermediate-gain antenna (IGA) on the market today that delivers full SwiftBroadband capabilities. It is considered to be one of the optimum antenna solutions for Inmarsat's SwiftBroadband Class 7 service offering.

An additional voice/data transceiver is available for cabin crew and passenger services, credit card clearing, and providing communications in case of medical emergencies. This aircraft terminal permits connections of conventional telephony devices through either standard 2-wire "Tip and Ring" circuits or 4-wire audio connections and Digital European Cordless Communication (DECT) units. In such a way, telecommunication features include intercom calling, call transfer, conferencing, and follow-on dialing.

This device connects to standard aircraft systems via 4-wire audio for flight deck voice and ARINC 429 circuits for multipurpose control display unit (MCDU) dialing or datalink messaging exchanges. It has external SIM card readers making them a true LRU, and various types of handsets including cordless can be connected for cabin use (Fig. 7.43).

Fig. 7.43 SkyConnect AES with peripherals—Courtesy of manual: by SkyConnect

7.3.2.2 Iridium SkyConnect Terminal

The combine SkyConnect satellite telephones and tracking systems are becoming very mission-critical equipment for helicopter operations that require communications with disparate and ever-changing entities. Designed for the crews with a busy workload in helicopter cockpits, this device easily accommodates virtually any operational scenario, including fixed-wing missions supporting 2.4 Kb/s and voice and data at 10 Kb/s compressed. The SkyConnect satellite transceiver, L-band antenna, Mission Management Unit (MMU) Cockpit Dialer, tactical dialer, and flush mount cradle are shown in Fig. 3.27.

The MMU-II device is a computer-controlled dialer and is very flexible operation. It communicates with the transceiver over an RS-232 port and can be programmed in the aircraft with new message sets, or software. It has a four step-dimming feature to the display. The MMU makes it a breeze to send standard company-specific messages, such as number of passengers, current fuel load, patient status, estimated time of arrival (ETA), and more. There are two tactical dialers, a 6-position and an 11-position. The 11-position unit dialer has a backlit faceplate and the dimming feature to the panel lights, and the annunciators can be separate or together.

7.4 Maritime O3B MEO Users Segment

The medium Earth orbit (MEO) satellite constellation is developed in 2014 known as O3b Networks to provide GMSC service for fixed and mobile satellite communications users. The name O3B satellite system refers to the "Other 3 Billion" and is primarily intended as a backhaul service for terrestrial cellular phones in remote countries that do not have a wire line infrastructure. They will also be serving existing, high-end, and maritime markets such as the cruise and merchant shipping industry.

The constellation of 12–20 satellites will be in a circular MEO satellite constellation at a distance of about 8,063 km from the Earth surface. Each MEO satellite will have 12 fully steerable Ka-band spacecraft antennas of which 2 will target the terrestrial LES (Gateways) and 10 will target the coverage areas. Each beam will have a throughput capacity of 600 Mb/s in each direction. The beam footprints will have a diameter of about 600 km on the Earth's surface between 45°N and 45°S and will be dynamically steered as the satellite moves around the Earth, to cover the required service areas and skip over the unpopulated or non-contracted areas. Eight MEO satellites will be spaced 45° apart, orbiting around the equator, in a non-inclined orbit.

In Fig. 7.44 (Left) is depicted space segment of O3b containing minimum 8 spacecraft in MEO constellation at 8,062 km altitude, while in Fig. 7.44 (Right) is illustrated O3b ground segment with O3b LES (Gateway) antenna system. Building a next-generation O3b satellite constellation will be capable of offering to user better, faster, reliable, and more affordable connectivity.

Initial constellation of eight MEO satellites is revolving around the Earth, about 4 times per day providing continuous service coverage, namely when one satellite leaves, another satellite takes over without transmission interruption. Besides, its Ka-band beams of 700 km diameter are steerable around the globe, each delivering scalable options from 100 Mb/s up to 1.2 Gb/s.

Each beam is connected to a high-throughput teleport providing reliable and high-speed Internet at a competitive price, which high speed is 4× faster than GEO satellites. Low-latency roundtrip latency is between 120 and 150 ms (depending on latitude and environmental conditions): faster interactivity and superior voice transmission quality with lower cost up to 30 % cheaper than GEO service. Flexibility steerable beams can be placed anywhere 45° north/south of the equator.

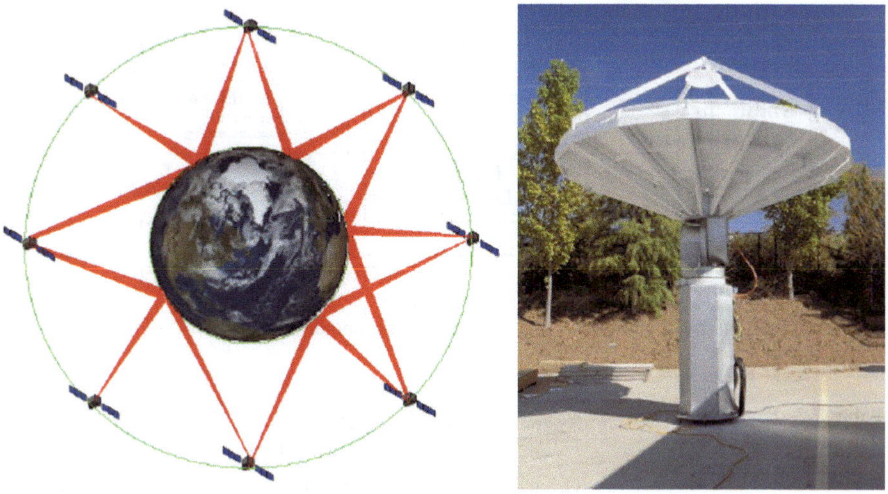

Fig. 7.44 O3b space and ground segment—Courtesy of manual: by O3b

Due to the much lower orbit, the latency of MEO satellite will be in the region of 130 ms, round trip, as opposed to about 720 ms for geostationary satellites. In fact, it will be a big improvement for phone calls and data flow rather than allowing GEO satellite constellation. The beams will target specific areas under contract, and in the case of maritime installations, the beams can actually follow the movement of the ship to optimize coverage. The ship will require two tracking antennas, each tracking a different satellite, and switching from one to the other as the satellites rise and set.

As a trial, O3B Networks have been awarded a contract by Royal Caribbean Cruise Lines to provide high-speed broadband service to their Oasis of the Seas, which is the world's largest cruise ship carrying 8000 passengers and crew. Two stabilized antennas, installed onboard the ship by Harris Caprock, will track each visible satellite and allow seamless handovers, as one satellite is setting and another rising. The satellite beams will also track the ship, keeping the ship centered in the beam, as it moves around the Caribbean. The system will use a ViaSat high-speed modem. There will be a third, hot standby antenna and modem onboard for redundancy. The low-latency and high-speed throughput will be vastly superior to the geostationary satellites used today.

In Fig. 7.45 is illustrated O3b Ka-band stabilized SES terminal with two 1.0 or 1.8 m shipborne stabilized tracking antennas and ViaSat high-speed modem, where term IFL is interfacility link and Xcvr is transceiver (Fig. 7.46). The O3b SES terminal deployed two types of shipborne satellite antennas as shown in Fig. 7.47, OrSat400 (Left) and OrSat300 (Right).

Both antennas are product of Orbit Company, which features are presented in Table 7.2.

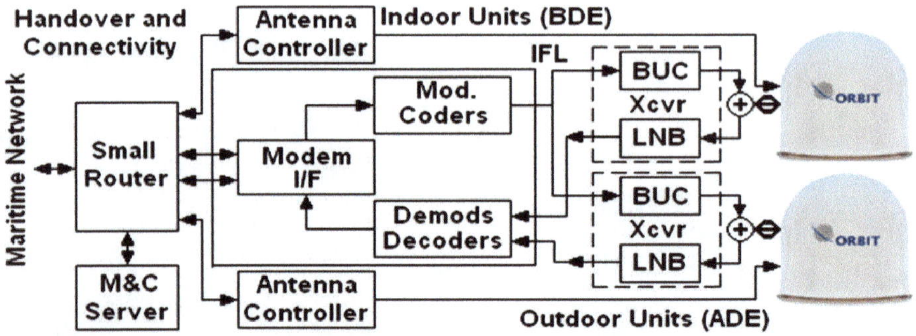

Fig. 7.45 O3b Ka-band stabilized SES—Courtesy of manual: by O3b

Table 7.2 Features of orbit O3b Marmite Antenna

System	Size (m)	Data rates (Mbps)	Optional BUC's
$\text{OrSat}^{\text{TM}}_{400}$	1.2	5–100	5 W/10 W/20 W
$\text{OrBand}^{\text{TM}}_{300}$	2.2	100–300	10 W/20 W/40 W

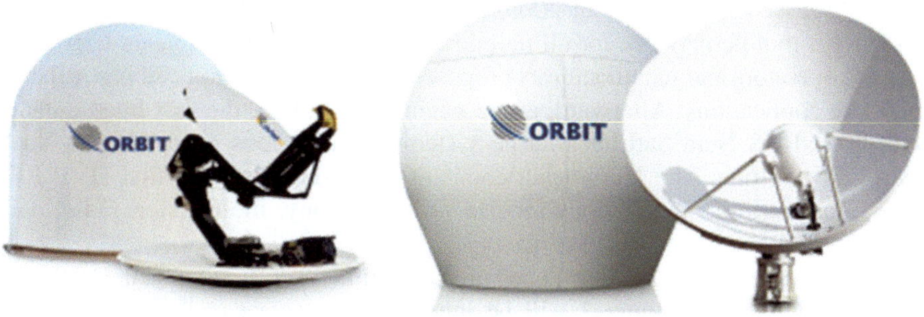

Fig. 7.46 Orbit O3b antenna system—Courtesy of manual: by O3b

The O3b satellite terminal is designed for seamless handover at end of pass and in case of blockage. Thus, 3rd hot standby satellite antenna for redundancy is included (for OrBand300), so a dedicated dual system selector provides instantaneous hot swap with a third redundant system.

The O3b MEO satellite system for maritime applications provides a real-time ship track monitoring via O3b Customer Portal. O3b satellite beams will follow ship tracks on ship's normal route, where cruise operator provides O3b with normal ship course. Then, beam tracking updates in real time if the ship has to change course, where O3b will maintain ship within beam center and ship provides latitude and longitude updates on 2 h intervals via in-band or out-of-band channel.

The satellite terminal O3b for maritime applications provides high-capacity throughput and delivery data rates up to 500 Mb/s, serves more users on ship at substantially higher data rates, and offloads business processes from ships to shore data centers and round trip latency between 120 and 150 ms, which is depending on latitude and environmental conditions, as shown in Fig. 7.46

7.5 Mobile DVB-RCS GEO Users Segment

Since 2005, many satellite operators worldwide initiated to provide mobile service via GEO satellite constellations at C-, Ku-, and Ka-band suitable for DVB-RCS S/S2 standards, whose configuration is shown in Fig. 3.30. The GEO satellite operators are Intelsat, Eutelsat, SES-NewSkies, PanAmSat, and other, with regional or global and spot beam coverage. The space segment configuration may consist of minimum three broadcast GEO satellites, while the ground segment infrastructure may have mission center to control the network, control center to manage spacecraft, and several LES terminals. Thus, the user segment is represented by FES and MES both interfaced to the ISP or TTN infrastructures and connecting users via digital voice, data, and video (VDV) facilities locally or worldwide.

As discussed in Chap. 3, in 2000, author of this book designed and proposed DVB-RCS mobile network for all mobile applications. Almost 5 years later, some satellite operators and manufacturers proposed their DVB solutions, in particular for maritime applications. Afterward, at the beginning of 2005, the first fitted out ship with DVB-RCS Nera SatLink Marine system was luxurious supper Motor Yacht M/Y Lady Marina. The Norwegian company Nera developed its first DVB-RCS VSAT for maritime applications for use with telephony, file transfer, radio, daily news delivery, and broadcast services.

In this section will be introduced just maritime and aeronautical DVB-RCS standards, while in Volume 2 will be introduced complete chapter about global broadcasting satellite system for all mobile applications.

7.5.1 DVB-RCS Architecture for Maritime Broadband

The solid maritime market demands robust, secure, and ubiquitous broadband connectivity delivered by terrestrial networks. Intelsat has developed global mobility solutions that rival the performance and reliability of land-based VSAT networks. The IntelsatONE Mobility Solutions via C- and Ku-band provide constant, modern, and reliable high-speed IP access for converged voice, data, and Internet applications on one anytime and anywhere platform.

1. **Global Intelsat C-band Coverage**—The Intelsat global C-band Coverage contains three GEO satellites for VSAT service: IS-701, IS-107, and IS-906 in tree ocean beam areas and four LES terminals, which network is shown in Fig. 7.47.

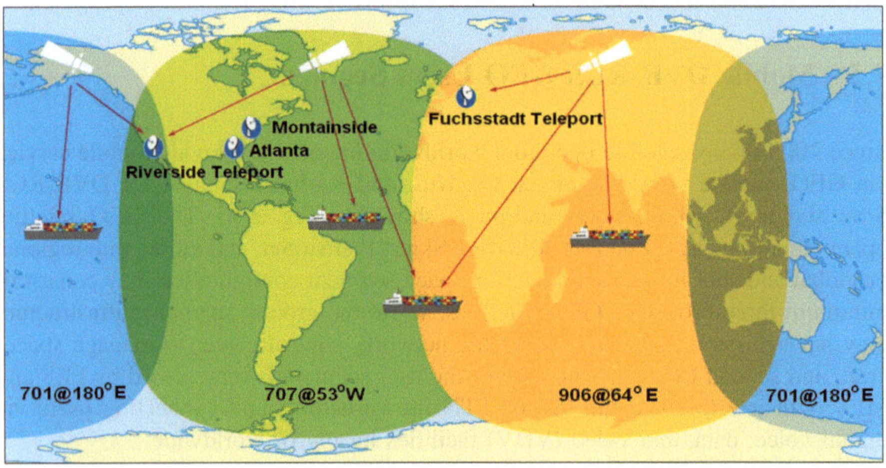

Fig. 7.47 Global Intelsat C-band coverage—Courtesy of manual: by Intelsat

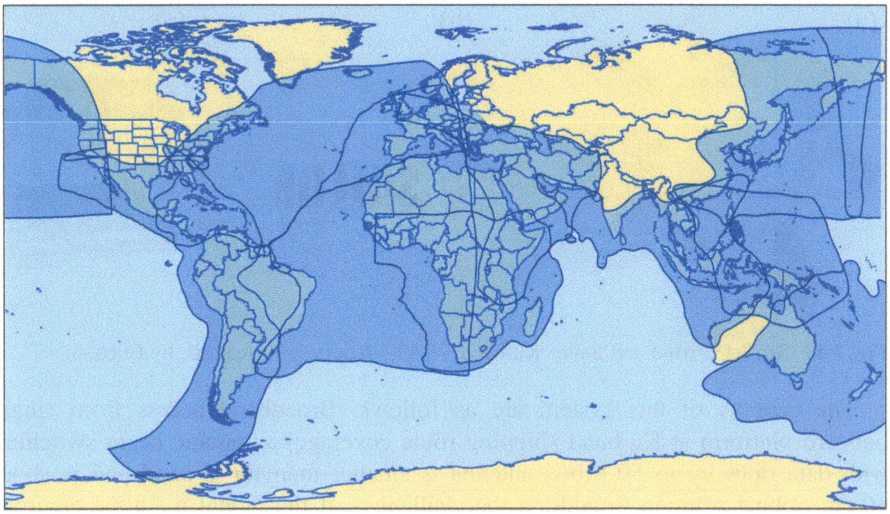

Fig. 7.48 Global Intelsat Ku-band coverage—Courtesy of manual: by Intelsat

The broadband maritime platform is IntelsatONE NBB (iDirect) C-band coverage, similar to the current Inmarsat-4 coverages, until 75° north and 75° south, which SeaTel 9707D VSAT is shown in Fig. 7.49a. Its benefits are as follows: Global C-band broadband access not affected by rain fade; automatic beam switching (ABS); data rates from 128 Kb/s to 4.096 Mb/s; cost-effective solution with predictable fixed monthly service charges per vessel; guaranteed committed information rate (CIR) values with variable max burst rates per vessel; bandwidth lease solutions with data rates over 60 Mb/s; and iDirect-based IntelsatONE platform. This system deploys iDirect VSAT Hub transceiver configurations with Intelsat terrestrial ABS infrastructure, which eliminates at all the need for human intervention as a vessel traverses between satellite beam coverage areas. The Intelsat single platform is always-on broadband access, which supports all converged VDV and Internet applications with global IP address with Multiprotocol Label Switching (MPLS).

2. **Global Intelsat Ku-band Coverage**—This solution is employing GEO six satellites and nine wide beams providing Ku-band coverage with IntelsatONE NBB (iDirect), which global coverage is shown in Fig. 7.48. The Above Deck Equipment (ADE) or antenna is shown in Fig. 7.49b. The Below Deck Equipment (BDE) Sailor 900 Maritime VSAT is shown in Fig. 7.49c, which Antenna Control Equipment (ACE) is shown (Above) and iDirect Evolution X5 Router (Below). Both, ACE and Router can be integrated with C-band SeaTel 9707D VSAT, while Sailor 900 VSAT can be tuned to both Ku and Ka-band frequencies already done by Cobham.

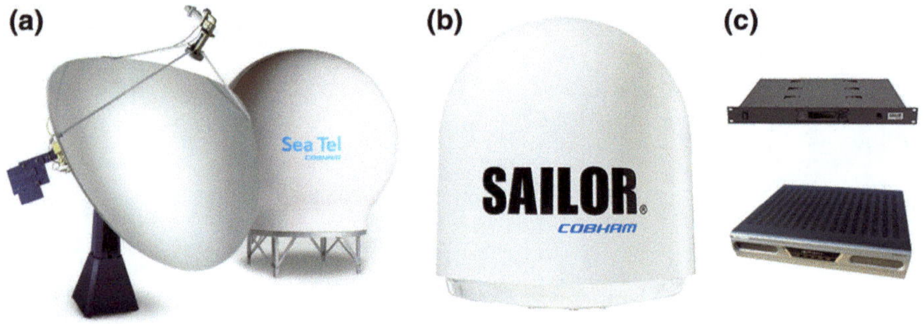

Fig. 7.49 SeaTel C-band and Sailor Ku-band VSAT—Courtesy of manual: by Cobham

The benefits of this system are as follows: Broadband access from single onboard platform at Ku-band shipping route coverage; automatic beam switching with data rates up to 50 Mb/s; antenna is smaller than for C-band and is about 60 cm; robust solution to address the challenges of the global maritime environment, covering VDV and Internet access with global IP/MPLS address; and manage backbone network which links all of Intelsat GES teleports. It is used to provide ocean vessels with a single IP address, which is maintained even when the vessel switches between beams that are uplinked from different teleports, when a special application session is maintained as the vessel crosses from one beam's coverage to another. The biggest problem of K-band is very high negative effects of fade rain on the propagation, so for that reason is possible to switch to Ka-band. The ocean vessel in navigation during very bad weather needs very reliable communication faculties. Thus, this problem can be solved with combination of usage different solutions onboard ships, such as Inmarsat, Iridium, or any C/L-band solutions.

In Fig. 7.50 is illustrated shipborne DVB-RCS VSAT configuration, which contains the following main components: antenna structure or ADE and BDE or DVB-RCS satellite router (modem) with peripherals:

1. **Above Deck Equipment (ADE)**—This structure contains three axis stabilized Ku-band antenna platform, automatic satellite acquisition, automatic polarization control, unlimited azimuth capability, 4 W Ku-band transmitter and Ku-band LNB receiver, 0.2° peak mispointing, connection between ADE and BDE via 2 coaxial cables, power consumption approximate 100 W (115/230 VAC), and the physical radome with antenna is approximately 114 kg and has a height of 149 cm and a diameter of 125 cm.
2. **Below Deck Equipment (BDE)**—This structure consists of DVB-RCS modem or router, Ethernet router and HUB, VoIP telephone adaptor, telephone set, single PC or multiple PC terminals in ship LAN, antenna cables, and power consumption approximate 200 W (115/230 VAC). To DVB-RCS antenna is associated to the ship Gyro Compass unit and Antenna Control Equipment (ACE) with Rx signal splitter for ACE and DVB-RCS, which manage tracking and controlling capabilities of satellite antenna.

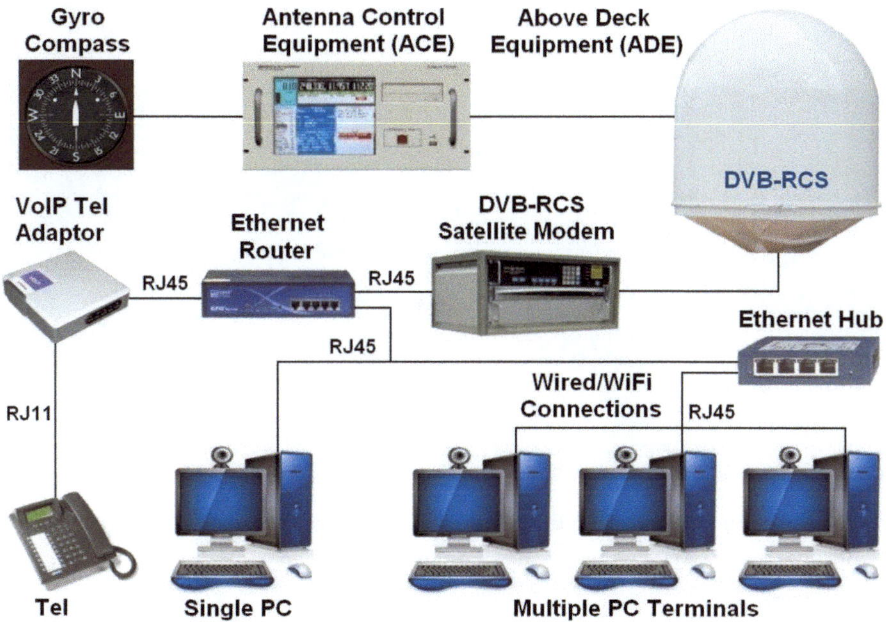

Fig. 7.50 Maritime DVB-RCS configuration—Courtesy of manual: by Ilcev

7.5.2 *ViaSat DVB-RCS Aeronautical Broadband*

In age of the BlackBerry and iPhones, captain, crew, and passengers onboard aircraft can expect limitless broadband connectivity, no matter where they are in the world. To achieve these solutions, ViaSat was developed the Yonder Mobile Broadband Service for all mobile applications, which in particular may help keep aircraft connected in the sky and on the ground, whose satellite coverage and flight lines are shown in Fig. 7.51.

This affordable, two-way broadband VDV communications service delivers new network performance on par with typical cable modem and DSL technology found in navy office or home, so the business jet literally becomes the office-in-the-sky offering: Place and receive cell phone calls just as you would on the ground; browse the Internet and Web or send and receive e-mail in real time; access certain company's network through a secure VPN; to cell phones, laptop, PDA, tablet computers, and other enabled devices via Wi-fi.

In Fig. 7.52 is shown onboard aircraft VSAT for the ARINC service on business jets, such as ViaSat modem, Rantec tail-mount antenna, and ACE. Thus, this VSAT includes an integrated transceiver, router, and a tracking antenna subsystem developed in partnership with Rantec Microwave Systems of Calabassas, CA. The antenna size for the ARINC service is as small as 30 cm, and it is usually accommodated in the tail of the aircraft.

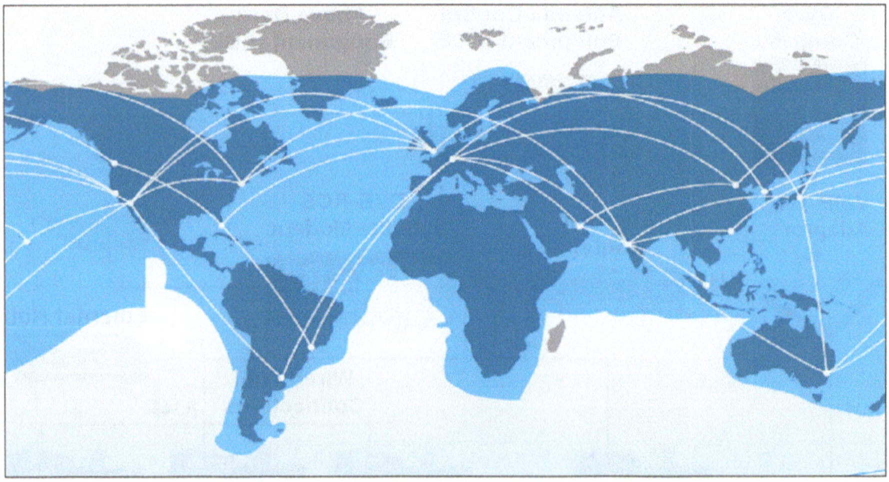

Fig. 7.51 Yonder aeronautical DVB-RCS satellite coverage and air routes—Courtesy of manual: by ViaSat

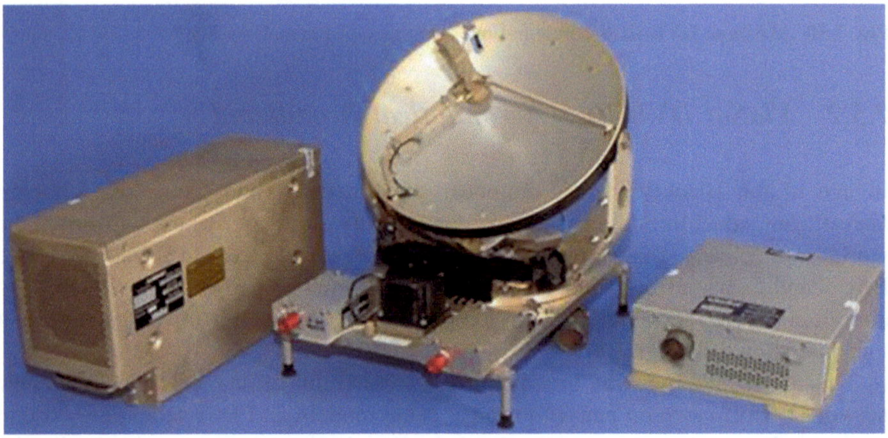

Fig. 7.52 ARINC DVB-RCS unit for Yonder airborne system—Courtesy of manual: by ViaSat

With Yonder, commercial airlines can offer true in-flight wireless broadband, enabled by ViaSat-1, the world's highest capacity satellite, and the ViaSat SurfBeam 2 ground system, such as higher speeds and more bandwidth for each customer, including lower installation and airtime pricing for service plan flexibility. The breakthrough bandwidth economics of ViaSat-1 create a world of possibilities for a personalized broadband experience, whether it is an "office-in-the-sky" or multimedia entertainment, such as LiveTV.

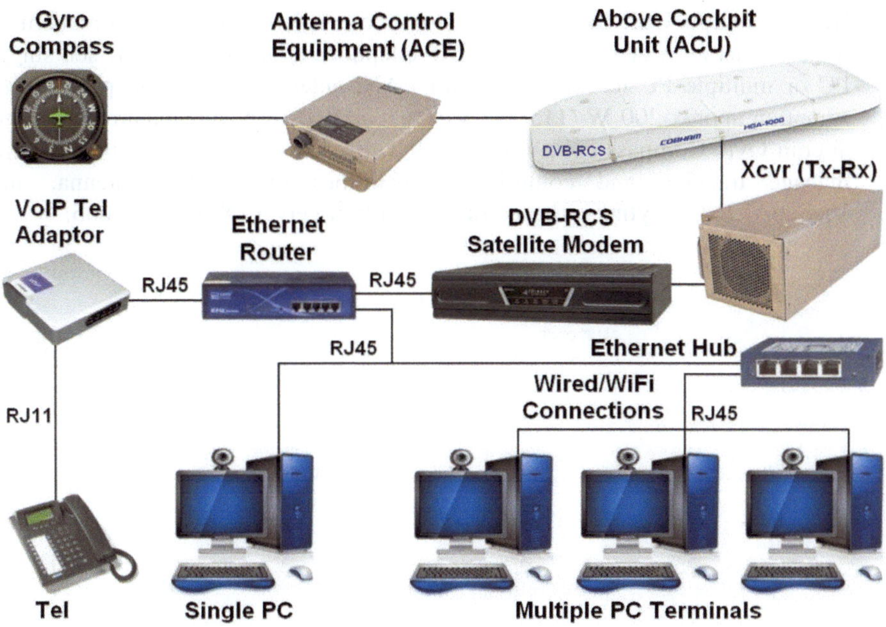

Fig. 7.53 Aeronautical DVB-RCS configuration—Courtesy of manual: by Ilcev

In addition, to the high-value passenger experience, the crew will be able to accomplish traditional manual tasks, such as cabin and flight deck operations, electronic flight bag (EFB) systems, and flight attendant reports in-flight using the satellite Internet connection.

The Yonder mobile network is providing high-speed Internet service to multiple customer classes operating on simultaneously around the world. Applications also include VoIP, and video, plus maritime security operations by the US Coast Guard and real-time video feeds from intelligence, surveillance, and reconnaissance (ISR) aircraft.

Similar to the shipborne DVB-RCS VSAT structure, it is designed the airborne DVB-RCS VSAT structure, as shown in Fig. 7.53. This VSAT configuration contains the following components: antenna unit or ACU and BCU or satellite router (modem) with peripherals:

1. **Above Cockpit Unit (ACU)**—This structure can contain tail or fuselage mount VSAT Ku- and Ka-band satellite airborne communication antenna. Here will be presented Gilat fuselage mount electronically steered array/phased array antenna (ESA/PAA), which is ultra-slim; low profile-phased array electronically steers the transmission and reception beams toward the satellite versus traditional mechanical antenna steering allowing operation even under the equator (no "key-hole" effect).

2. **Below Cockpit Unit (BCU)**—This structure consists of DVB-RCS modem or router, Ethernet router and HUB, VoIP telephone adaptor, telephone set, single PC or multiple PC terminals onboard LAN, antenna cables, and power consumption approx 200 W (115/230 VAC). To DVB-RCS antenna are associated aircraft Gyro Compass unit, ViaSat/ARINC Skylink airborne ACE 1007783 for manage tracking and controlling capabilities of satellite antenna, and ViaSat/ARINC Skylink airborne integrated transceiver (Xcvr) 1000005.

Erratum to: Global Mobile Satellite Communications Theory

Erratum to:
S.D. Ilčev, *Global Mobile Satellite Communications Theory*, DOI 10.1007/978-3-319-39171-7

The book was inadvertently published without removing the mention of Volume 1. The erratum book has been updated with the change.

The updated original online version for this book can be found at 10.1007/978-3-319-39171-7

S.D. Ilčev (✉)
Durban University of Technology (DUT), Durban, South Africa
e-mail: ilcev@dut.ac.za

S.D. Ilčev, *Global Mobile Satellite Communications Theory*,
DOI 10.1007/978-3-319-39171-7_8

References

1. Books

1. Acerov A.M. & Others, "Morskaya radiosvyaz i radionavigaciya", Transport, Moskva, 1987.
2. Acharya R., "Understanding Satellite Navigation", Elsevier, Oxford, UK 2014.
3. Ackroyd N. & Lorimer R., "Global Navigation – A GPS User's Guide", LLP, London, 1994. [10-11]
4. Ahson S.A. & Others, "VoIP Handbook", CRC Press, Boca Raton, FL, 2009.
5. Aragon-Zavala A. & Other, "High-Altitude Platforms for Wireless Communications", Wiley, Chichester, 2008.
6. Barnhart R.K. & Others, "Introduction to Unmanned Aircraft Systems", CRC Press, Boca Raton, FL, 2012.
7. Beckmann P., "The Scattering of Electromagnetic Waves from Rough Surfaces", Pergamon Press, London, 1963.
8. Berlin P., "Geostationary Applications Satellite", Cambridge University Press, Cambridge, UK, 1988.
9. Bertoni H.L., "Radio Propagation for Modern Wireless Systems", Prentice Hall, Englewood Cliffs, 2000.
10. Biezad D.J., "Integrated Navigation and Guidance Systems", AIAA, New York, 1999.
11. Blonstein L., "Communications Satellites, The Technology of Space Communications", Heinemann, London, 1987.
12. Brisibe T., "Aeronautical Public Correspondence by Satellite", Eleven, Utrecht, 2006.
13. Calcutt D. & Tetley L., "Satellite Communications, Principles and Applications", Edward Arnolds, London, 1994.
14. Calcutt D. & Tetley L., "Satellite Communications, Principles and Applications", Elsevier, Oxford, 2004.
15. Calcutt D. & Tetley L., "Understanding GMDSS", Edward Arnold, London, 1994.
16. Campbell J., "Understanding GMDSS", Waterline, Shrewsbury, 1998.
17. Chen X., "Antennas for Global Navigation Satellite Systems", Wiley, Chichester, 2012.
18. Chetty P.R.K., "Satellite Technology and its Applications", TAB, Blu Ridge Summit, 1993.
19. Curtis H.D., "Orbital Mechanics for Engineering Students", Elsevier, London, 2005.
20. Dalgleish D.I., "An Introduction to Satellite Communications", IEE, Peter Peregrinus, London, 1989.
21. Dardari D. & Others, "Satellite and Terrestrial Radio Positioning Techniques", Elsevier, Waltham, MA, 2012.
22. Davidoff M.R., "The Satellite Experimenter's Handbook", ARRL, Newington, 1984.
23. Del Re E. & Ruggieri M., "Satellite Communications and Navigation Systems", Springer, New York, 2008.

S.D. Ilčev, *Global Mobile Satellite Communications Theory*,
DOI 10.1007/978-3-319-39171-7

24. Diggelen V.F., “A-GPS, Assisted GPS, GNSS and SBAS”, Artech House, Boston-London, 2009.
25. Du Y., “A Satellite Ground Station Control System”, Technical University of Denmark, Kongens Lyngby, 2005.
26. Elbert B.R., “Ground Segment and Earth Station Handbook”, Artech House, Boston – London, 2001.
27. Elbert, B. R., International Telecommunication Management”, Artech House, London, 1990.
28. Elbert B.R., “Introduction to Satellite Communications”, Artech House, London, 1987.
29. Elbert, B. R., “Networking Strategies for Information Technology”, Artech House, London, 1992.
30. Elbert, B. R., “Private Telecommunication Networks”, Artech House, Norwood, 1989.
31. Elbert B.R., “The Satellite Communication Applications Handbook”, Artech House, London, 1997.
32. El-Rabani A., “Introduction to GPS”, Artech House, Boston-London, 2002.
33. Evans B.G., “Satellite Communication Systems”, IEE, Peter Peregrinus, London, 1991.
34. Everett J., “VSAT- Very Small Aperture Terminals”, IEE, Peter Peregrinus, London, 1992.
35. Feher K., “Digital Communications, Satellite Earth Station Engineering”, Prentice-Hall, Englewood Cliffs, 1983.
36. Flock, W. L., “Propagation Effects on Satellite Systems at Frequencies Below 10 GHz”, NASA, Washington, 1987.
37. Forssell B., “Radionavigation Systems”, Artech House, Boston-London, 2008.
38. Freeman R.L., “Fundamentals of Telecommunications”, Wiley, Chichester, UK, 1999.
39. Freeman R.L., “Radio Systems Design for Telecommunications (1-100 GHz)”, Wiley, Chichester, 1987.
40. Freeman, R.L., “Reference Manual for Telecommunications Engineering”, Wiley, New York, 1985.
41. Fujimoto K., “Mobile Antenna Systems Handbook”, Artech House, London, 2008.
42. Fujimoto K. & Other, “Mobile Antenna Systems Handbook”, Artech House, London, 1994.
43. Gagliardi R.M., “Satellite Communications”, Van Nostrand Reinhold, New York, 1984.
44. Galic R., “Telekomunikacije satelitima”, Skolska knjiga, Zagreb, 1983.
45. Gallagher B., “Never Beyond Reach”, Inmarsat, London, 1989.
46. Giambene G., “Resource Management in Networks - Optimization and Cross-Layer Design”, Springer, 2007.
47. Gordon G.D. & Other, “Principles of Communications Satellites”, Wiley, Chichester, 1993.
48. Grace D. & Mohorcic M., “Broadband Communications via High-Altitude Platforms”, Wily, Chichester, UK, 2011.
49. Grant A.E. & Others, “Communication Technology Update”, Focal Press, Boston, 2000.
50. Grewal M.S& Others, “Global Positioning Systems, Inertial Navigation and Integration”, Wiley, Chichester, 2007.
51. Griffin M.D. & French J.R., “Space Vehicle Design”, AIAA, Reston, 2004.
52. Group of Authors, “Current and Planned Global and Regional Navigation Satellite Systems and Satellite-based Augmentation Systems”, UN, New York, 2010.
53. Group of Authors, “ Earth Station Technology”, Intelsat, Washington, DC, 1999.
54. Group of Authors, “Earth Station Verification and Assistance (ESVA)”, Eutelsat, Paris, 2002.
55. Group of Authors, “Fifth International Conference on Satellite Systems for Mobile Communications and Navigation”, IEE, London, 1996.
56. Group of authors, “Fourth International Conference on Satellite Systems for Mobile Communications and Navigation”, IEE, London, 1988.
57. Group of Authors, “Global Maritime Distress and Safety System”, IMO, London, 1987.
58. Group of Authors, “GMPCS Reference Book”, ITU, Geneva, 2000.
59. Group of Authors, “Handbook - Mobile Satellite Service (MSS)”, ITU, Geneva, 2002.

60. Group of Authors, "Handbook on Satellite Communications", ITU, Geneva, 2002.
61. Group of Authors, "Manual for Use by the Maritime Mobile and Maritime Mobile-satellite Services", ITU, Geneva, 1999.
62. Group of Authors, "Manual on Mobile Communications Development, ITU, Geneva, 1997.
63. Group of Authors, "Radiowave Propagation Information for Predictions for Earth-to-Space Path Communications", ITU, Geneva, 1996.
64. Group of Authors, "Utilisation des satellites pour les recherches et le sauvetage", Cepadues, Toulouse, 1984.
65. Groves P.D., "Principles of GNSS, Inertial and Multisensor Integrated Navigation Systems", Artech House, Boston-London, 2008.
66. Ha J.C., "Mission Design and Implementation of Satellite Constellations", ESAI ESOC, Darmstadt, Germany, 1998.
67. Ha T.T., "Digital Satellite Communications", Macmillan Publications, New York, 1986.
68. Hadden A.D., "Personal Communications Networks, Practical Implementation", Artech House, London, 1995.
69. Haan S., "Meteorological Applications of a Surface Network of GPS Receivers", Universiteit Utrecht, Utrecht, Netherlands, 2008.
70. Heath S., "Multimedia and Communications Technology", Focal Press, Oxford, 1999.
71. Higgins J., "Satellite Newsgathering", Focal Press, Oxford, 2002.
72. Hofmann-Wellenhof B. & Other, "GPS: Theory and Practice", Springer-Verlag, Vienna, 1997.
73. Huurdeman A.A., "Guide to Telecommunications Transmission Systems", Artech House, Boston-London, 1997.
74. Ilcev D. S., "Global Aeronautical Communications, Navigation and Surveillance (CNS)", Volume 1 & 2, AIAA, Reston, 2013.
75. Ilcev D.S., "Global Mobile Satellite Communications for Maritime, Land and Aeronautical Applications", Springer, Boston, 2005.
76. Ilcev D.S., "Global Radio and Satellite CNS", Manual, DUT, Durban, 2014.
77. Ilcev D.S., "Mobile Antenna Handbook", Manual, CNS Systems, Durban, 2015.
78. Jagoda A. & Other, "Mobile Communications", John Wiley, Chichester, 1995.
79. Jamalipour A., "LEO Satellites for Personal Communication Networks", Artech House, London, 1998.
80. Jasik H., "Antenna Engineering Handbook", McGraw Hill, 1961
81. Johnson D.H., Dudgeon D.E., "Array Signal Processing: Concepts and Techniques", Prentice Hall, Englewood Cliffs, 1993.
82. Kadish J.E. & Others, "Satellite Communications Fundamentals", Artech House, Boston-London, 2000.
83. Kantor L.Y., "Sputnikovaya svyaz i veschanie", Radio i svyaz, Moskva, 1988.
84. Kantor L.Y. & Others, "Sputnikovaya svyaz i problema geostacionarnoy orbiti", Radio i svyaz, Moskva, 1988.
85. Kaplan D.E., "Understanding GPS Principles and Applications", Artech House, Boston-London, 1996.
86. Kaplan D.E., "Understanding GPS Principles and Applications", Artech House, Boston-London, 2006.
87. Kayton M. & Fried W.R., "Avionics Navigation Systems", John Wiley, Chichester, 1997.
88. Keattisak S., "Turbo Code Applications: A Journey from a Paper to Realization", Springer, Boston, 2005
89. Kelly M.C., "The Earth Ionosphere", Academic Press, San Diego, 1989.
90. Kolawole M.O., "Satellite Communication Engineering", Marcel Dekker, New York, 2002.
91. Kondoz, A. M., "Digital Speech - Coding for Low Bit Rate Communications Systems", Wiley, New York, 1994.
92. Kraus J.D. & Marhefka R., "Antennas", Mc Graw Hill, New York, 2001.
93. Law P.E., "Shipboard Antennas", Artech House, Washington, 1983.

94. Lee, J. S. & Others, "CDMA Systems Engineering Handbook", Artech House, Norwood, 1998.
95. Lee C.Y. W., "Mobile Communications Design Fundamentals", Wiley, New York, 1993.
96. Lee C.Y.W., "Mobile Communications Engineering", McGraw-Hill, London, 1982.
97. Lees G.D. & Others, "Handbook for Marine Radio Communications", LLP, London, 1999.
98. Lin S. & Others, "Error Control Coding: Fundamentals and Applications", Prentice-Hall, Englewood Cliffs, 2005.
99. Long M.E., "The Digital Satellite TV Handbook", Newnes, 1999.
100. Long M.E., "World Satellite Almanac", Howard W. Sams, Indianapolis, 1987.
101. Lawrence A., "Modern Inertial Technology: Navigation, Guidance and Control", Springer, New York, 1993.
102. Lukatela G. & Others, "Digitalne telekomunikacije", Gradjevinska knjiga, Belgrade, 1984.
103. Macario, R. C. V., "Personal and Mobile Radio Systems", Peter Peregrinus, London, 1991.
104. Maini A.K. & Agrawal V., "Satellite Technology - Principles and Applications", John Wiley, Chichester, 2007.
105. Malaguti, N., "Coordination of an Earth Station with Stations of Terrestrial Services as well as Earth Stations Operating in Opposite Direction," ITU-BR, Geneva, Switzerland, 2002.
106. Maral G. & Others, "Satellite Communications Systems", Wiley, Chichester, 1994.
107. Maral G. & Others, "Satellite Communications Systems", Wiley, Chichester, 2009.
108. Maral G., "VSAT Networks", John Wiley, Chichester, 2003.
109. Martin H.D. & Others, "Communication Satellite", AIAA, Reston, 2007.
110. Maslin N., "HF Communications", Pitman, London, 1987.
111. Miller M.J., & Others, "Satellite Communications; Mobile and Fixed Services" Kluwer A. P., Norwell, 1993.
112. Misra P. & Others, "GPS: Signals, Measurements, and Performance", Ganga-Jamuna Press, Lincoln, 2001.
113. Mladenov M.K. & Others, "Periferni ustroystva za personaliny kompyutri", Tehnika, Sofia, 1987.
114. Monroe J.W., "Marine Radionavigation and Communications", Cornell Maritime Press, Centreville, 1998.
115. Naugle, M., "Illustrated TCP/IP - A Graphic Guide to the Protocol Suite", Wiley, New York, 1999.
116. Nejat A., "Digital Satellite Communications Systems and Technologies – Military and Civil Applications" Kluwer Academic Publishers, 1992
117. Nenirovskiy A.S. & Others, "Radio-releinie i sputnikovie sistemi peredachi", Radio i svyaz, Moskva, 1986.
118. Noll E.M., "Landmobile and Marine Radio Technical Handbook", Howard W. Sams, Indianapolis, 1985.
119. Novik L.I. & Others, "Sputnikovaya svyaz na more", Sudostroenie, Leningrad, 1987.
120. Ohmori S. "Aeronautical and Maritime Antennas for Satellite Communications", CRC Press LLC, 2002.
121. Ohmori S. & Others, "Mobile Satellite Communications", Artech House, Boston–London, 1998.
122. Orehov A.A., "Radiopriemnie ustroystva morskogo sudna", Transport, Moskva, 1987.
123. Palter D. C., "Satellites and the Internet", SatNews Publishers, 2003.
124. Paluszek M. & Others, "Spacecraft Attitude and Orbit Control", Princeton Satellite Systems, Plainsboro, NJ, 2009
125. Parkinson B.W. & Others, "Global Positioning System: Theory and Applications I/II", AIAA, Washington, 1996.
126. Pascall S.P. & Others, "Commercial Satellite Communications", Focal Press, Oxford, 1997.
127. Pattan B., "Satellite-based Global Cellular Communications", McGraw Hill, New York, 1998.
128. Patzold M., "Mobile Fading Channels", Wiley, New York, 2000.

129. Pelton J.N., "Wireless and Satellite Telecommunications" Prentice Hall, Englewood Cliffs, 1995.
130. Perishkin I.M. & Others, "Sistemi podvizhnoy radiosvyazi", Radio I svyaz, Moskva, 1986.
131. Pirumov V.S. & Others, "Radio-elektronika v voyne na more", Voenoe izdatelstvo, Moskva, 1987.
132. Prasad R. & Ruggieri M., "Applied Satellite Navigation Using GPS, GALILEO, and Augmentation Systems", Artech House, Boston,
133. Pratt T. & Others, "Satellite Communications", TAB, Blue Ridge Summit, 1986.
134. Pratt T. & Others, "Satellite Communications", John Wiley, Hoboken, 2002.
135. Prentiss S., "Satellite Communications", TAB, Blue Ridge Summit, 1987.
136. Proakis J.G. & Others, "Communication System Engineering", Prentice Hall, London, 2002.
137. Proakis J.G., "Digital Communications", McGraw Hill, New York, 2000.
138. Radovanovic A., "PC modemske komunikacije", Tehnicka knjiga, Belgrade, 1991.
139. Rama B., "GPS/GNSS Antennas", Artech House, London, 2013.
140. Rappaport T.S., "Wireless Communications Principles and Practice", Prentice Hall, Englewood Cliffs, 1996.
141. Reed H.R. & Russell C.M., "Ultra High Frequency Propagation", Wiley, New York, 1953.
142. Reg A., "Unmanned Aircraft Systems", Wiley, Chichester, UK, 2010.
143. Remer D.D. & Others, "Global Navigation for Pilots", Aviation Supplies and Academics Inc., Washington, 1998.
144. Ricci F.J. & Others, "U.S. Military Communications", Computer Science Press, Rockville, 1986.
145. Richharia M., "Mobile Satellite Communications - Principles and Trends", Addison-Wesley, Harlow, 2001.
146. Richharia M., "Satellite Communications System - Design principles", Macmillan, Basingstoke, 1995.
147. Richharia M. & Westbrook L.D., "Satellite Systems for Personal Applications", John Wiley, Chichester, 2010.
148. Roddy D., "Satellite Communications", McGraw Hill, New York, 2006.
149. Rudge A.W. & Others, "The Handbook of Antenna Design", Volume 1 & 2, IEE, London, 1986.
150. Samama N., "Global Positioning – Technologies and Performance", John Wiley, Chichester, 2008.
151. Saunders S.R., "Antennas and Propagation for Wireless Communication Systems", Wiley, New York, 1999.
152. Sayegh C., "Mayor Earth Station Equipment", Intelsat, Washington, DC, 2014.
153. Schaphorst R., "Videoconferencing and Videotelephony", Artech House, London, 1997.
154. Schott J.R. "Remote Sensing", Oxford University Press, Oxford, 1997.
155. Schwartz, M., "Telecommunication Networks", MA: Addison-Wesley, Reading, 1987.
156. Sellers J. J., "Understanding Space - An Introduction to Astronautics", McGraw-Hill, New York, 2000.
157. Sheriff R.E. & Others, "Mobile Satellite Communication Networks", Wiley, Chichester, 2001.
158. Siouris G.M., "Aerospace Avionics Systems", Academic Press, London, 1993.
159. Sklar, B., "Digital Communications - Fundamentals and Applications", Prentice Hall, Upper Saddle River, 2001.
160. Smith P.C. & Others, GMDSS for Navigators", Butterworth Heinemann, Oxford, 1994.
161. Soleymani M.R. & Others, "Turbo Coding for Satellite and Wireless Communications", Kluwer, Boston, 2002.
162. Solovev V.I. & Others, "Svyaz na more", Sudostroenie, Leningrad, 1978.
163. Sonnenberg G.J., "Radar and Electronic Navigation", Butterworths, London, 1978.

164. Stacey D., "Aeronautical Radio Communication Systems and Networks", John Wiley, Chichester, 2008.
165. Stajic D. & Others, "Racunarske telekomunikacije i mreze", Tehnicka knjiga, Belgrade, 1991.
166. Stallings, W., "Business Data Communications", Prentice Hall, Upper Saddle River, 2001,
167. Stallings W., "Data and Computer Communications", Pearson, Upper Saddle River, NJ, 2007.
168. Sternfeld A., "Vestacki sateliti", Tehnicka knjiga, Belgrade, 1958.
169. Stremler F.G., "Introduction to Communication Systems", Addison-Wesley, Harlow, 1990.
170. Sun Z., "Satellite Networking - Principles and Protocols", John Wiley, Chichester, 2005.
171. Swan P.A. & Others, "Global Mobile Satellite Systems: A Systems Overview", Kluwer AP, Boston, 2003.
172. Tetley L. & Others, "Electronic aids to Navigation - Position Fixing", Edward Arnold, London, 1991.
173. Tetley L. & Others, "Understanding GMDSS", Edward Arnold, London, 1994.
174. Tomasi W., "Electronic Communications Systems", Prentice-Hall, New York, 2001.
175. Torrieri D.J., "Principles of Military Communication Systems", Artech House, Dedham, 1982.
176. Tsui J., "Fundamentals of Global Positioning System Receivers", Wiley, New York, 2005.
177. Van Trees H.L., "Detection, Estimation and Modulation Theory", Part 1, Wiley, New York, 1968.
178. Van Trees H.L., "Satellite Communications", IEEE, New York, 1979.
179. Venskauskas K.K. "Sistemi i sredstva radiosvyazi morskoy podvizhnoy sluzhbi", Sudostroenie, Leningrad, 1986.
180. Viola B., "L'operatore di radiocomunicazioni del servizio mobile marittimo", Trevisini, Milano, 1987.
181. Walke B. & Other, "UMTS - The Fundamentals", Wiley, New Jersey, 2003.
182. Walker J., "Advances in Mobile Information Systems", Artech House, Boston - London, 1990.
183. Wartz J.R., "Spacecraft Attitude Determination and Control", Kluwer AP, Dordrecht, 1984.
184. Waugh I., "The Maritime Radio and Satellite Communications Manual", Waterline, Shrewsburg, 1994.
185. Wellenhof-Hofmann B. & Other, "Global Navigation Satellite Systems (GNSS)", Springer, Wien, 2008.
186. Wertz J. R. & Other, "Space Mission Analysis and Design", Kluwer Academic Publishers, 1999.
187. Wilder, F., "A Guide to the TCP/IP Protocol Suite", Artech House, Norwood, 2002.
188. Yuen J.H. "Deep Space Telecommunications Systems Engineering", Plenum Press, New York, 1983.
189. Zovko-Cihlar B., "Sumovi u radiokomunikacijama", Skolska knjiga, Zagreb, 1987.
190. Zhilin V.A., "Mezhdunarodnaya sputnikova sistema morskoy svyazi - Inmarsat", Sudostroenie, Leningrad, 1988.

2. Papers

1. Agius A.A. & Others, "QHS – Characteristics of a Proposed Wire Antenna for an Satellite Personal Communication Networks (SPCN) Handheld Terminal", CCSR, Surrey, 1997.
2. Antonini M. & Other, "Stratospheric Relay: Potentialities of New Satellite-high Altitude Platforms Integrated Scenarios", IEEE Proceedings of the Aerospace Conference, IEEEAC, 2003.

3. Avagnina D., "Wireless Networks Based on High Platforms for the Provision of Integrated Navigation and Communication Services", IEEE Communications Magazine, 2002.
4. Bertran E. & Others, "Effects of Packet and SBAS Measurements Rate on the Emergency Control of an UAV", Data Systems in Aerospace Conference (DASIA.02), Dublin, 2002.
5. Blyenburgh Van P., "UAV Systems: Global Review", Avionics 2006 Conference, 1-52, Amsterdam, 2006.
6. Buracchini, "SDMA in Mobile Radio Systems", Internet, 2012.
7. Comparini M.C. & Others, "Microwave Equipment for Navigation Overlay Service", Alenia Spazio, Roma, 2002.
8. Copros E. & Others, "An Improved Space Segment for the End-State WAAS and EGNOS Final Operational Capability", Proceedings of the Institute of Navigation (ION), Alexandria, 1996.
9. Horwath, M. & Others, "Broadband Backhaul Communication for Stratospheric Platforms: the Stratospheric Optical Payload Experiment (STROPEX)", Proc. SPIE, Free-Space Laser Communications VI, San Diego, 2006.
10. Grewal M.S. & Others, "Test Results of Geostationary Satellite (GEO) Uplink Sub-System (GUS) Using GEO Navigation Payloads", Institute of Navigation (ION), Alexandria, 1999.
11. Grewal M.S., & Others, "Overview of the SBAS Integrity Design", ION Conference, Graz, 2003.
12. Grewal M.S., & Others, "Comparison of GEO and GPS Orbit Determination", ION/GPS 2002, Portland, 2002.
13. Horwath J. & Other, "Broadband Backhaul Communication for Stratospheric Platforms: the Stratospheric Optical Payload Experiment (STROPEX)", Free-Space Laser Communications VI, 6304, San Diego, 2006.
14. Ilcev D.S., "African Satellite Augmentation System (ASAS)", Collection of Papers, High Technologies, Researches and Industry Conference, St-Petersburg (Leningrad), 2010.
15. Ilcev D.S., "Airborne Satellite Antenna Mount and Tracking Systems", Microwave Journal, Norwood, 2009.
16. Ilcev D.S., "Antenna Systems for Mobile Satellite Applications", Journal of the Institute of Telecommunications Professionals (ITP), London, 2011.
17. Ilcev D.S., "Channel Coding used in Mobile Satellite Communications", Electronics World, London, 2011.
18. Ilcev D.S., "Channel Decoding and Error Corrections used in Mobile Satellite Communications", Electronics World, London, 2011.
19. Ilcev D.S., "Channel Processing used in Mobile Satellite Communications", Electronics World, London, 2011.
20. Ilcev D.S., "Characteristics and Development of Global Mobile Satellite Communications for Maritime, Land and Aeronautical Applications", Doctoral Dissertation, Belgrade, 2000.
21. Ilcev D.S., "Characteristics and Development of Maritime Satellite Communications Systems", Master Theses, Faculty of Electrical Engineering, Skopie, 1994.
22. Ilcev D.S., "Communication Stratospheric Platforms", Journal of the Institute of Telecommunications Professionals, Sunbury on Thames, 2010.
23. Ilcev D.S., "Development and Characteristics of African Satellite Augmentation System (ASAS) Network", Springer, Telecommunication Systems, Boston, USA, 2011.
24. Ilcev D.S., "Implementation of Local Satellite Augmentation System (LSAS) for Airport Infrastructures", Russian SIBCON, Siberian Federal University, Krasnoyarsk, Russia, 2011.
25. Ilcev D.S., "Implementation of System of Differential Correction and Monitoring (SDCM) for Russia and CIS Countries", IEEE 25th CriMiCo Conference, Sevastopol, Russia, 2015.
26. Ilcev D.S., "Integration in Communications and Navigation", Nautical Courier, Belgrade, Serbia, 1989.
27. Ilcev D.S., "Integration of Radio and Satellite Automatic Identification System for Maritime Applications", TransNav, Gdynia, Poland, 2015.

28. Ilcev D.S., "Introduction to Inmarsat GEO Space and Ground Segments", TransNav, Gdynia, Poland, 2015
29. Ilcev D.S., "New Aspects in Function of Aeronautical Communication, Navigation and Surveillance (CNS), 12th International Conference of Science and Technology AVIA-2015, Ukraine, Kiev 2015
30. Ilcev D.S., "Satellite CNS for Maritime Transportation Augmentation System (MTAS)", IEEE 19th Conference on Microwave Engineering and Telecommunication Technologies in Ukraine, Sevastopol, 2009.
31. Ilcev D.S., "Satellite DVB-RCS Standards for Fixed and Mobile Commercial and Military Applications", Microwave Journal, Norwood, USA, 2009.
32. Ilcev D.S., "Spacecraft Payload for MSC Networks", Microwave Journal, Norwood, 2009.
33. Ilcev D.S., "Stratospheric Communication Platforms (SCP) as an Alternative for Space Program", Aircraft Engineering and Aerospace Technology (AEAT) Journal, Emerald, Bingley, 2011.
34. Ilchenko M. E, "Application of High-altitude Platform Systems in Regions of Disaster and Emergency", 14th International Crimean Conference on Microwave and Telecommunications Technology, Sevastopol, 2004.
35. Ivancic D.W. & Others, "Application of Mobile-IP to Space and Aeronautical Networks", NASA, Cleveland, 2002.
36. Karapantazis S. & Others, "The Role of High Altitude Platforms in Beyond 3G Networks", IEEE Wireless Communications, 2005.
37. Pinkney F., "UAV Communications Payload Development", Proceedings MILCOM, 1997.
38. Razumovsky D. & Others, "Inmarsat's New Navigation Payload", ION, Portland, 2003.
39. Tuomas A., "Mobile IPv6 Security", Microsoft Research Ltd, Cambridge, 2002.
40. Zaharov V. & Others, "Smart Antenna Application for Satellite Communications with SDMA", Journal of Radio Electronics, Moscow, 2001.

3. Manuals

1. ACR, "Maritime EPIRB and PLB", Fort Lauderdale, FL, 2014.
2. ALRS, "GMDSS" - Volume 5, Admiralty List of Radio Signals, Taunton, UK, 1999.
3. ALRS, "Maritime Safety Information Broadcast", Taunton, UK, 2004.
4. ARINC, "ARINC System and Procedures", Set of Manuals, Annapolis, MD, 2012.
5. Beam, "IsatDock 2 Marine and Drive Handheld", Mulgrave, Australia, 2015.
6. Canadian Marconi, "CMA-2102 SATCOM High-Gain antenna", Ville St-Laurent, Canada, 1995.
7. Cobham, "Aircraft Antennas for LEO Satellite Systems", Lyngby, Denmark, 2009.
8. Cobham, "Alarm Panel FleetBroadband", Lyngby, Denmark, 2015.
9. Cobham, "EXPLORER 700 with BGAN LaunchPad", Lyngby, Denmark, 2014.
10. Cobham, "Inmarsat FleetBroadband, Fleet One and Global Xpress", Lyngby, Denmark, 2008.
11. Cobham, "Inmarsat LGA/IGA/HGA for AES", Lyngby, Denmark, 2009.
12. Cobham, "Inmarsat SwiftBroadband Diplexer and Low Noise Amplifier (DLNA)", Lyngby, Denmark, 2014.
13. Cobham, "Inmarsat Vehicular and Portable BGAN", Lyngby, Denmark, 2014.
14. Cobham, "Maritime VSAT for GEO Satellites", Lyngby, Denmark, 2014.
15. Cobham, "Radar Distance and Dual Antenna Solution for Sailor SES", Denmark, 2014.
16. Cobham, "SAILOR SC4000 Iridium", Lyngby, Denmark, 2009.
17. Cobham, "SwiftBroadband Hardware", Lyngby, Denmark, 2014.
18. Cobham, "SwiftBroadband Peripherals", Lyngby, Denmark, 2014.

19. Cobham, "Transportable Explorer 710 BGAN Configurations and Network", Denmark, 2014.
20. Cobham, "Vehicular BGAN Configurations", Lyngby, Denmark, 2014.
21. Cospas-Sarsat, "Cospas-Sarsat System Data", London, 2009.
22. Cospas-Sarsat, "Cospas-Sarsat LEOLUT, GEOLUT and MEOLUT Systems", London, 2009.
23. Cospas-Sarsat, "Description of the 406 MHz Payloads used in the GEOSAR System", London 2011.
24. Cospas-Sarsat, "Description of the 406 MHz Payloads used in the LEOSAR System", London 2011.
25. Cospas-Sarsat, "Description of the 406 MHz Payloads used in the MEOSAR System", London 2012.
26. Cospas-Sarsat, "Description of the Cospas-Sarsat Space and Ground Segment", London, 1996.
27. Cospas-Sarsat, "Instruction to the Cospas-Sarsat System", London, 2009.
28. CNS System, "Global Aeronautical Distress and Safety System (GADSS), Durban, 2002.
29. EB Communications, "Saturn 3 Installation - Elevation and Azimuth Angle Maps", Bergen, Norway. 2000.
30. EB Communications, "Saturn 3 Operators and System Description Manuals", Nesbru, Norway, 1986.
31. EMS Aviation, "Inmarsat LGA/IGA/HGA for AES", Ottawa, Canada, 2007.
32. Furuno, "Ship SES ADE and BDE Components", Nishinomiya-shi, Japan, 2014.
33. General Dynamic, "Multiband Aircraft Antennas", Washington, DC, 2015.
34. Gilat, "Aircraft Antennas for UAS Applications", Petah Tikva, Israel, 2015.
35. Gilat, "Land Vehicles VSAT for GEO Satellites", Petah Tikva, Israel, 2015.
36. Globalstar, "Description of Duplex Globalstar System", San Jose, CA, 2009.
37. Globalstar, "Description of the Globalstar SATPHONE System", San Jose, CA, 2009.
38. Globalstar, "Globalstar Simplex Data System", San Jose, CA, 2009.
39. Globalstar, "Globalstar Spot Satellite Trackers", San Jose, CA, 2009.
40. Globalstar, "Globalstar System - Payphone and Fixed Satellite Terminal", San Jose, CA, 2009.
41. Globalstar, "Maritime Terminal", San Jose, CA, 2003.
42. Globalstar, "Mobile and Handheld Systems", San Jose, CA, 2003.
43. Harris, "Vehicular BGAN Configurations", New York, 2014.
44. Honeywell, "Global Xpress Jet ConneX AES", Morristown, NJ, 2010.
45. Honeywell, "Inmarsat HGA Block Diagram", Morristown, NJ, 2010.
46. ICG, "Globalstar AEROCOM Terminals", Newport News, VI, 2010.
47. IMO, "Global Maritime Distress and Safety System (GMDSS)", London, 1987.
48. Inmarsat, "An Introduction to the Use of SATCOM for Air Traffic Services and Flight Operations", London, 1998.
49. Inmarsat, "Inmarsat-4 Satellite Constellation", London, 2005.
50. Inmarsat, "Inmarsat-5 Satellite Constellation", London, 2014.
51. Inmarsat, "Inmarsat Aeronautical Services Information", London, 2003.
52. Inmarsat, "Inmarsat Coverage Maps", London, 2015.
53. Inmarsat, "Inmarsat Data Services User's Guide", London, 1996.
54. Inmarsat, "Inmarsat FleetBroadband", London, 2008.
55. Inmarsat, "Inmarsat Global Xpress", London, 2015.
56. Inmarsat, "Inmarsat IsatPhone Pro and other Handheld Standards", London, 2015.
57. Inmarsat, "Inmarsat Maritime Handbook" London, 2002.
58. Inmarsat, "Inmarsat Maritime Services Information", London, 1997.
59. Inmarsat, "Inmarsat Mobile Standards", London, 2012.
60. Inmarsat, "SafetyNET User's Handbook", London, 1995.
61. Inmarsat, "Inmarsat Satellite Communications Services Users Handbook", London, 1987.

62. Inmarsat, "Inmarsat Standard-B High Speed Data Service", London, 1997.
63. Inmarsat, "Inmarsat Standard-C Maritime and Land Mobile Users Handbook", London, 1995.
64. Inmarsat, "Inmarsat SwiftBroadband", London, 2009.
65. Inmarsat, "TVRO Systems", London, 2015.
66. Intellian, "Global Xpress XLink SES", Irvine, CA, 2014.
67. Intellian, "Inmarsat Multiband VSAT Systems", Irvine, CA, 2014.
68. Iridium, "Airborne Transceivers", McLean, VA, 2006.
69. Iridium, "Implementation Manual for Iridium Satellite Communications Service", McLean, VA, 2006.
70. Iridium, "Mobile and Handheld Systems", McLean, VA, 2015.
71. Iridium, "Pilot Iridium SES and Maritime Transceivers", McLean, VA, 2006.
72. Jotron, "Aeronautical ELT Manual", Tjodalyng, Norway, 1999.
73. Jotron, "Maritime EPIRB and PLB", Tjodalyng, Norway, 2014.
74. JRC, "Inmarsat-C MES Operation Manual", Tokyo, Japan, 2001.
75. Kannad, "Aeronautical ELT", Guidel, France, 2014.
76. Kymeta, "Inmarsat HGA Block Diagram", Redmond, WA, 2015.
77. KVH, "KVH Tracphone 50 mini-M Manual", Middletown, Rhode Island, 1997.
78. KVH, " Maritime and Vehicular VSAT", Middletown, Rhode Island, 2015.
79. NASA, "Application of Mobile-IP to Space and Aeronautical Networks", Washington, DC, 2007.
80. NATO, "Rocket People - Creating a Rocket Industry", Washington, DC, 2008.
81. Nera, "F33/F55/F77 SES", Billingstad, Norway, 2003.
82. Nera, "Inmarsat-M SES", Billingstad, Norway, 2004.
83. Nera, "M2M Satellite Solutions" ", Billingstad, Norway, 2004.
84. Nera, "Saturn Bm C2 operator's Manual", Nera, Billingstad, Norway, 1996.
85. Nera, "SatLink Two-way Satellite Broadband Solutions", Billingstad, Norway, 2004.
86. Nera, "Saturn Bm Marine Class 2 SES", Billingstad, Norway, 2004.
87. Nera, "Saturn C Marine Version Operator's Manual", Billingstad, Norway, 1995.
88. Nera, "WorldCommunicator", Billingstad, Norway, 1999.
89. Nera, "Worldphone mini-M User's Manual", Billingstad, Norway, 1992.
90. NovAtel, "WAAS GUS Type 1 Receiver", Calgary, Canada, 2005.
91. NovAtel, "WAAS GUS Type 1 Signal Generator", Calgary, Canada, 2005.
92. NovAtel, "WAAS Reference Receiver G-II", Calgary, Canada2005.
93. O3b Limited, "O3b Space and Ground Segment", St. John, Jersey, Channel Islands, 2015.
94. Ocean Signal, "Maritime EPIRB and PLB", Margate, UK, 2014.
95. Orbcomm, " Maritime and Land Mobile Systems", Dulles, VI, 2004.
96. Orbcomm, "Orbcomm System Overview", Dulles, VI, 2008.
97. Orbit, " Airborne DVB-RCS VSAT Outdoor Units", Netanya, Israel, 2015.
98. Orbit, "Orbit O3b Antenna System", Netanya, Israel, 2015.
99. Orbit, " Maritime VSAT for MEO Satellites", Netanya, Israel, 2015.
100. Orbit, " Maritime VSAT and TVRO Antennas for GEO Satellites", Netanya, Israel, 2015.
101. Orbit, "Rail VSAT GEO Systems", Netanya, Israel, 2015.
102. P&L International, "Aircraft TVRO Solutions", Weddington, NC, 2015.
103. Qualcomm, "Globalstar Data User Guide", San Diego, CA, 2010.
104. Qualcomm, "Globalstar Duplex Video and Data Units", San Diego, CA, 2010.
105. Qualcomm, "Globalstar HARDCASE Terminal AEROCOM Terminals", San Diego, CA. 2010.
106. QEST, "Dual Band DVB-RCS Transceiving and Receiving Antennas", Holzgerlingen, Germany, 2015.
107. Raytheon, "GMDSS system operation manual", Kiel, Germany, 1998.
108. Rockwell Collins, "Tailwind Multiregion In-flight TV System", Cedar Rapids, IA, 2015.
109. Sea Launch, "User's Guide", Long Beach, CA, 1996.

110. Sea Tel, "Maritime TVRO for GEO Satellites for GEO Satellites", Hull, UK, 2015.
111. Sensor Systems, "Aeronautical ELT Solutions", Chatsworth, CA, 2014.
112. SkyWave, "Inmarsat IsatData Pro", Ottawa, Canada, 2010.
113. Specmat Technologies, "Aeronautical ELT", Rockford, IL, 2014.
114. Tecom, "Aircraft TVRO Solutions", Thousand Oaks, CA, 2015.
115. Tecom, "T-4000 High Gain Antenna System", Thousand Oaks, CA, 1995.
116. Telit, "Globalstar Telit Sat 550 and 551 Manuals", Sgonico, Italy, 2000.
117. Thrane & Thrane, "Inmarsat-C and Inmarsat mini-M SES", Soeborg, Denmark, 2000.
118. Thrane & Thrane, "Inmarsat LGA/IGA AES Antennas", User/Installation Manuals, Soeborg, Denmark, 2000.
119. Thrane & Thrane, "Capsat mini-M mobile telephone users manual TT3060A", Soeborg, Denmark. 1998.

4. Brochures

1. Advantech, "DVB-RCS Hubs and Mobile VSAT systems",
2. AeroVinronment, "SkyTower Space and Ground Segment", Monrovia, CA, 2006.
3. Artex, "Replacement Batteries for ELT and PLB", Prescott, 2011.
4. Beam, "Mobile Docking Satellite Terminals", Mulgrave, Australia, 2011.
5. Beam, "Mobile Satellite Terminals", Mulgrave, Australia, 2011.
6. Beam, "Mobile Tracking Satellite Terminals", Mulgrave, Australia, 2011.
7. Beam, "Transportable Satellite Terminals", Mulgrave, Australia, 2011.
8. Cobham, "Aeronautical SwiftBroadband", Fullerton, CA 2010.
9. Cobham, "Airborne HPA and HPA/DLNA Antenna", Fullerton, CA, 2011.
10. Cobham, "Aircraft Antennas for LEO and GPS Satellite Constellation", Fullerton, CA, 2011.
11. Cobham, "Block Diagram of Inmarsat Antenna HPA and HPA/DLNA Configuration", Fullerton, CA, 2011.
12. Cobham, "Broadband Transceiving Mobile Antennas", Fullerton, CA, 2011.
13. Cobham, "Inmarsat HGA Mobile Systems", Fullerton, CA, 2012.
14. Cobham, "Inmarsat LGA and IGA Mobile Systems", Fullerton, CA, 2012.
15. Cobham, "Maritime FleetBroadband", Fullerton, CA 2010.
16. Comtech, "Migration from DVB-S to DVB-S2 and Related Efficiencies", Tempe, AZ, 2006.
17. Cospas-Sarsat, "Cospas- Data Distribution Regions - Cospas-Sarsat MCC", London, 2011.
18. Cospas-Sarsat, "Cospas- Data Distribution Regions - Cospas-Sarsat OCC", London, 2011.
19. Cospas-Sarsat, "Cospas-Sarsat GEOLUT Performance", London, 2011.
20. Cospas-Sarsat, "Cospas-Sarsat LEOLUT Performance Specifications and Design Guidelines", London, 2011.
21. Cospas-Sarsat, "Cospas-Sarsat MEOSAR Implementation Plan - System Concept of Operations", London, 2011.
22. Cospas-Sarsat, "Cospas-Sarsat MEOSAR Implementation Plan - SAR/Galileo Payload Functions", London, 2011.
23. Cospas-Sarsat, "Cospas-Sarsat Mission Control Centre", London, 2011.
24. DGAC, "Integration of ETDMA within an ATN/IP Network", EC, Brussels, Belgium, 2011.
25. Ellipse TV, "Tracking Solutions for Aircraft Antenna Systems", Alpharetta, GA, 2011.
26. Emergency Beacon Corporation, "Replacement ELT/PLB Batteries", New York, 2011.
27. EMS, "Aircraft Antennas for LEO Satellite Systems", Ottawa, Canada, 2011.
28. EMS, "First Generation of LUT Station", Ottawa, Canada, 2011.
29. EMS, "Global Tracking Solutions", Ottawa, Canada, 2011.

30. EMS, "Inmarsat HGA for Helicopters and DVB Antenna", Ottawa, Canada, 2011.
31. EMS, "Integrated Satellite Transceivers and Peripherals", Ottawa, Canada, 2011.
32. EMS, "LEOLUT Functional Block Diagram", Ottawa, Canada, 2011.
33. EMS, "New Generation of GEOLUT Station", Ottawa, Canada, 2011.
34. EMS, "New Generation of LUT Station", Ottawa, Canada, 2011.
35. EMS, "Product Specifications of Aircraft Antenna", Ottawa, Canada, 2007.
36. ESA, "Satellite Technology - ISL of GEO Spacecraft Artemis", Darmstadt, Germany, 2009.
37. ETRI, "Mobile DVB-RCS Systems", Daejeon, South Korea, 2005.
38. ETSI, "DVB-S2 with ACM", Sophia-Antipolis Cedex, France, 2007.
39. ETSI, "DVB User Guidelines", Sophia-Antipolis Cedex, France, 2007.
40. ETSI, "Functional Block Diagram of the DVB-RCS System", Sophia-Antipolis Cedex, France, 2007.
41. ETSI, "Hierarchical Backward Compatible DVB-S2 System", Sophia-Antipolis Cedex, France, 2007.
42. FAA, "WAAS GEO Spacecraft and Ground Infrastructure", Washington, DC, 2004.
43. GT&T, "Satellite Teleport in the Hearth of Europe", Louvain-La-Neuve, France, 2000.
44. GT&T, "The Microsat Series", Louvain-La-Neuve, France, 1998.
45. GT&T, "VSAT Equipment", Louvain-La-Neuve, France, 2003.
46. Guardian Mobility, "Aircraft Tracking Satellite Terminals", Ottawa, Canada, 2011.
47. Honeywell, "SkyConnect Satellite Cockpit Terminal", Morristown, NJ, 2011.
48. Hughes, "Aeronautical Broadband Services Technology", MD, Germantown, 2008.
49. Hughes, "Broadband Mobility Solution", Germantown, MD, 2007.
50. Hughes, "Fixed and Mobile DVB-RCS Routers", Germantown, MD, 2006.
51. Hughes, "Hub Terminals", Germantown, MD, 2006.
52. ICAO, "Surface Wireless Communications", Montreal, Canada, 2011.
53. ICAO, "Technology Assessment for the Future Aero COM System", Montreal, Canada, 2011.
54. ICG, "Integrated Satellite Terminal", Newport News, VI, 2011.
55. ICS, "Globalstar ICS550 Satellite Telephone", Arundel, England, 1999.
56. ICS, "ICS550 Globalstar Satellite/GSM Telephone System", Arundel, 1999.
57. iDirect, "Fixed and Mobile DVB-RCS Routers", Herndon, Herndon, VA, 2006.
58. Inmarsat, "Inmarsat Coverage Map - The Position of four Inmarsat Satellite Ocean Regions", London, 2009.
59. Inmarsat, "Inmarsat I-4 Global Coverage Map for GMSC", London, 2010.
60. Inmarsat, "Inmarsat I-5 Global Spot-Beam Coverage Map for Digital Mobile Broadband", London, 2012.
61. Inmarsat, "Inmarsat Maritime Handbook - View of Inmarsat GEO Satellites", London, 2006.
62. Inmarsat, "Inmarsat Swift64 Aeronautical HSD Services - Facts", London 2001.
63. Inmarsat, "Introduction to FleetBroadband", London, 2008.
64. Inmarsat, "Introduction to SwiftBroadband", London, 2009.
65. Inmarsat, "IsatPhone", London, 2009.
66. Inmarsat, "Virtual Private Networking (VPN) over MPDS, London, 2009.
67. Intelsat, "Mobility Solutions", McLean, VI, 2006.
68. Iridium, "Handheld and Mobile Communication Products", McLean, VA, 2009.
69. Iridium, "Iridium Spacecraft", McLean, VA, 2009.
70. Iridium, "Personal Satellite Trackers", McLean, VA, 2009.
71. ITU, "Russian Radio Pioneer Popov Honoured by ITU", Geneva, Switzerland, 2009.
72. Jotron, "Tron 30S MkII and Tron 45S/SX", Tjodalyng, Norway, 1999.
73. Kannad, "Cospas-Sarsat ELT and PLB Terminals", Guidel, France, 2011.
74. Kannad, "Kannad 406 AS", Guidel, France, 2011.
75. Kannad, "Transmitting Antennas for ELT Distress Satellite Beacons (DSB)", Guidel, France, 2007

76. McMurdo, "Cospas-Sarsat PLB", Portsmouth, England, 2011.
77. McMurdo, "TSi Fixed and Mobile LUT Stations", Portsmouth, England, 2011.
78. Merl, "Replacement Batteries for Avionics ELT and PLB", Bath, England, 2011.
79. MOT, "MTSAT Multipurpose Spacecraft", Tokyo, Japan, 2004.
80. Musson Marine, "Distress Satellite Beacon Testers", Sevastopol, Russia, 2011.
81. Newtec, "Implementation of DVB-S2 into DVB-RCS Systems", Noordwijk, Netherlands, 2007.
82. Newtec, "IP over Satellite and DVB-RCS", Noordwijk, Netherlands, 2007.
83. Orbcomm, "Orbcomm Satellite", Dulles, VI, 2008.
84. Orbit, "Solutions of Tracking Aircraft Antenna Systems", Netanya, Israel, 2011.
85. Orbit, "System Layout of Orbit Technology Group Antenna", Netanya, Israel, 2011.
86. P&L International "Aircraft TVRO Solutions", Weddington, NC, 2011.
87. Quake, "Iridium Miniature Satellite Trackers", San Diego, CA, 2010.
88. Qualcomm, "Globalstar-GSP 288/2900 Fixed Satellite Phone", San Diego, CA, 2000.
89. Racal, "ASB for Aircraft TVRO Solutions", Holzgerlingen, Germany, 2010.
90. Rockwell Collins, "RC HSD-900D HFDL HF Airborne Data Radio", Cedar Rapids, IA, 2010.
91. Rockwell Collins, "Sat-906 System", Cedar Rapids, IA, 999.
92. Rockwell Collins, "Satcom-5000 System", Cedar Rapids, IA, 2000.
93. Rockwell Collins, "Satcom-6000 System", Cedar Rapids, IA, 2000.
94. Roscosmos, "SDCM - Current Status", Moscow, Russia, 2004.
95. RTCA, "Analysis of L-DACS1 and L-DACS2", Washington, DC, 2010.
96. SBIR, "PLB Terminals", Wright-Patterson, OH, 2011.
97. Scotty, "System SCS - Block Diagram of Scotty Video Encoder", Eisenstadt, Austria, 2010.
98. Scotty, "System SCS Package of FleetBroadband and Scotty Video Encoder", Eisenstadt, Austria, 2010.
99. Scotty, "System SCS Package of SwiftBroadband and Scotty Video Encoder", Eisenstadt, Austria, 2010.
100. Sea Launch, "Sea Launch - At-A-Glance", Long Beach, CA, 2000.
101. Sensor Systems, "Aircraft Antennas for LEO/GPS Satellite Systems", Chatsworth, CA, 2007.
102. Sensor Systems, "Transmitting Antennas for ELT Beacons", Chatsworth, CA, 2007.
103. SITA, "Aircraft Communications Systems", Set of Pamphlets, Vienna, Austria, 2010.
104. Sky Station, "Sky Station Stratospheric Platforms Project", Washington, DC, 2005.
105. DoCoMo 3G-Handset", Tokyo, Japan, 2006.
106. SkyWave, "Aero-D+ Equipment", Ottawa, Canada, 2010.
107. Specmat Technologies, "ELT Antennas", Rockford, IL, 2009.
108. Spidertracks, "Portable Tracking Satellite Terminals", Palmerston North, New Zeakand, 2011.
109. Taipei CNS/ATM, "Automatic Dependent Surveillance-Broadcast", Taipei, Taiwan (China), 2011.
110. TAO, "TAO Stratospheric Platforms Project", Tokyo, Japan, 2006.
111. Techtest Limited, "Cospas-Sarsat ELT Terminals", Hereford, England, 2011.
112. Tecom, "T-4000 High Gain Antenna System", Thousand Oaks, CA, 1995.
113. Tecom, "Block Diagram of KuStream 2000 Antenna Systems for DBS", Thousand Oaks, CA, 2005.
114. Thales Alenia Space, "Iridium Next - Secondary Payload ICD", Paris, 2010.
115. Thales Alenia, "Iridium Next - Secondary Payload ICD", Neuilly-sur-Seine Cedex, France, 2009.
116. ViaSat, "ARINC DVB-RCS Unit for Yonder Aeronautical System", Carlsbad, CA, 2006.
117. ViaSat, "DVB-RCS VSAT Hub Configuration", Carlsbad, CA, 2006.

118. ViaSat, "Fixed and Mobile DVB-RCS Routers", Carlsbad, CA, 2006.
119. ViaSat, "VSAT Configurations and Applications", Carlsbad, CA, 2006.
120. ViaSat, "Yonder Mobile-Aeronautical DVB-RCS System", Carlsbad, CA, 2006.

5. Periodicals

1. Horizon House Publications, "Microwave Journal", Norwood, 2000-2015.
2. IEEE, "Aerospace and Electronic Systems Magazine", Piscataway, 2005-2015.
3. IEEE, "Antenna and Propagation Magazine", Piscataway, 2005-2015.
4. IEEE, "Communication Magazine", Piscataway, 2005-2015
5. IEEE, "Intelligent Transport System Magazine", Piscataway, 2005-2015.
6. IEEE, "Transactions on Broadcasting", Piscataway, 2005-2015.
7. Phillips, "Via Satellite Magazine", Pittsfield, 1999-2015.
8. Springer, "Telecommunication Systems", Boston, 2006-2015.
9. Wiley, "International Journal of Communication Systems", Chichester, 2007-2015.
10. Wiley, "International Journal on Satellite Communications and Networking", Chichester, 2007-2015.

6. Web Sites

1. Advantech [www.advantechwireless.com]
2. ARINC [www.arincdirect.com]
3. ASAS [www.cnssystems.co.za]
4. Beam [www.beamcommunications.com]
5. China - GNSS Asia [www.gnss.asia/china]
6. Cobham [www.cobham.com]
7. Cospas-Sarsat [www.cospas-sarsat.int]
8. EGNOS [www.esa.int]
9. ESA [www.esa.int]
10. Eutelsat [www.eutelsat.com]
11. Flight Display [www.FlightDisplay.com]
12. Furuno [www.furuno.com]
13. GAGAN [www.gpsinindia.com]
14. Gilat [www.gilat.com]
15. Globalstar [www.globalstar.com]
16. GLONASS [www.glonass-iac.ru]
17. GPS [www.gps.gov]
18. Harris [www.harris.com]
19. Honeywel [www.honeywell.com]
20. Hughes [www.hughes.com]
21. ICAO [www.icao.int]
22. ITU [www.itu.int]
23. Inmarsat [www.inmarsat.com]
24. Intelsat [www.intelsat.com]
25. Intellian [www.intelliantech.com]
26. Iridium [www.iridium.com]
27. iDirect [www.idirect.net]
28. Jotron [www.jotron.com]
29. JRC [www.jrc.co.jp]

30. Kannad [www.kannad.com]
31. KVH [www.kvh.com]
32. McMurdo [www.mcmurdogroup.com]
33. MSAS [www.nec.com]
34. Newtec [www.newtec.eu]
35. NovArtel [www.novatel.com]
36. O3b MEO Networks [www.o3bnetworks.com
37. Orbcomm [www.orbcomm.com]
38. Orbit [www.orbit-tracking.com]
39. QEST [www.qest.de]
40. Quake [www.quakeglobal.com
41. Raytheon [www.raytheon.com]
42. Rockwell Collins [www.rockwellcollins.com]
43. Scotty SATCOM [www.scottygroup.com]
44. SDCM [www.sdcm.ru]
45. Sensor Antennas [www.sensorantennas.com]
46. SITA [www.sita.aero]
47. Telemeter Electronic [www.telemeter.info]
48. Thales Alenia [www.thalesgroup.com]
49. WAAS [www.gps.faa.gov]
50. ViaSat [www.viasat.com]

Index

S.D. Ilčev, *Global Mobile Satellite Communications Theory*,
DOI 10.1007/978-3-319-39171-7

MIX
Papier aus verantwortungsvollen Quellen
Paper from responsible sources
FSC® C105338

If you have any concerns about our products,
you can contact us on
ProductSafety@springernature.com

In case Publisher is established outside the EU,
the EU authorized representative is:
Springer Nature Customer Service Center GmbH
Europaplatz 3, 69115 Heidelberg, Germany

Printed by Libri Plureos GmbH
in Hamburg, Germany